7th Process Safety Management Mentoring (PSMM) Forum 2018

Topical Conference at the 2018 AIChE Spring Meeting and 14th Global Congress on Process Safety

Orlando, Florida, USA
22 - 25 April 2018

ISBN: 978-1-5108-6668-3

Printed from e-media with permission by:

Curran Associates, Inc.
57 Morehouse Lane
Red Hook, NY 12571

Some format issues inherent in the e-media version may also appear in this print version.

Printed by Curran Associates, Inc. (2018)

For permission requests, please contact AIChE
at the address below.

AIChE
120 Wall Street, FL 23
New York, NY 10005-4020

Phone: (800) 242-4363
Fax: (203) 775-5177

www.aiche.org

Additional copies of this publication are available from:

Curran Associates, Inc.
57 Morehouse Lane
Red Hook, NY 12571 USA
Phone: 845-758-0400
Fax: 845-758-2633
Email: curran@proceedings.com
Web: www.proceedings.com

TABLE OF CONTENTS

Some abstracts in this book refer to images that were not made available by the publisher

California's 2017 Process Safety Management Regulations for Petroleum Refineries: The Future of PSM in the United States?

Michael P. Wilson, Ph.D, MPH (1) William Hoyle (2); Donald Holmstrom, J.D. (3)

(1) National Director, Occupational and Environmental Health Program, BlueGreen Alliance
(2) Senior Investigator, U.S. Chemical Safety Board, retired
(3) Western Regional Director, U.S. Chemical Safety Board, retired

Photo courtesy U.S. Chemical Safety and Hazard Investigation Board. Richmond, Chevron. August 6, 2012

Abstract

On August 6, 2012 a pipe failure in the Chevron Richmond, California oil refinery crude unit led to a large fire that endangered the lives of 19 workers and caused over 15,000 area residents to seek medical attention for symptoms related to smoke and fire gas exposure. In the wake of the incident, the U.S. Chemical Safety and Hazard Investigation Board (CSB) and Governor Jerry Brown's Interagency Working Group on Refinery Safety called for substantial revisions to the Process Safety Management (PSM) regulations governing the state's 14 oil refineries. The California Department of Industrial

Relations (DIR) subsequently launched a three-year stakeholder process and promulgated a new, 24-part PSM regulation, *Process Safety Management for Petroleum Refineries, General Industry Safety Orders §5189.1.*[1] California adopted the new PSM regulation on October 1, 2017.[2]

The ~10,000-word new PSM regulation makes substantial changes to the 12 original (1992) PSM elements and adds nine new elements in the areas of damage mechanism reviews; hierarchy of controls analysis and inherent safety measures; safeguard protection analysis; process safety culture assessments; employee participation in PSM decision-making; root cause analysis for major incidents and near-misses; use of PSM indicators; mechanisms to improve accountability and transparency in PSM decision-making; integration of human factors; management of organizational change; and others. The new regulation draws heavily from the recommendations of the CSB, from the report of the Governor's Interagency Working Group on Refinery Safety, and from the Center for Chemical Process Safety (CCPS) publications, *Guidelines for Risk Based Process Safety* (2007) and *Inherently Safety Chemical Processes: A Life Cycle Approach (2009)*. The California Division of Occupational Safety and Health (Cal/OSHA) began enforcing the regulation on October 1, 2017 through increases in its PSM compliance staff that are funded by an annual refinery industry fee.

A RAND analysis of California's PSM regulation concluded that the combined compliance costs for the state's refineries would range between $20 and $184 million per year, with a point estimate of $58 million per year, spread across 14 refineries. When passed on to consumers, RAND reported that this figure would equate to a price increase of $0.004 per gallon of gasoline sold in the state. Based on the historical record in California, RAND found that each major refinery incident avoided would save a refinery about $220 million, not accounting for costs associated with worker fatalities and injuries or damage to surrounding communities. RAND concluded that the new regulations would substantially lower the risk of death among refinery workers and contractor workers, compared to the existing PSM standard.

California's new PSM regulation represents the latest and most sweeping regulatory reform initiated by a U.S. government entity in response to a major process incident. The regulation also offers an advanced model for industrial safety in the energy and chemical sectors more broadly, including in emerging industries in the clean energy sector that require the use of hazardous materials. In January 2018, the state of Washington (USA) proposed a revision to its own PSM regulation that closely reflects California's text. It is likely that California's 2017 PSM regulation will influence the national discussion on the role of regulatory policy in preventing major process incidents.

Background

Immediately following an August 6, 2012 pipe failure and fire at the Richmond, Chevron refinery, California Governor Jerry Brown convened an Interagency Refinery Safety Working Group, made up of representatives from 13 state, Federal and local agencies.[3] The final report of the Working Group,

---

[1] The California EPA developed a counterpart regulation under Program 4 of the state's Accidental Release Program (Cal/ARP), enforced by local county agencies known as Certified Unified Program Agencies (CUPAs). The Cal/ARP regulation was adopted along with the new PSM regulation on October 1, 2017.

[2] The new refinery PSM regulation does not apply to California's chemical facilities, which will continue to operate under the original PSM regulation that was adopted in California in 1992.

[3] Interagency Working Group on Refinery Safety members represented the Department of Industrial Relations (DIR), Cal/OSHA, Cal/EPA Secretary's Office, Air Resources Board (ARB), Governor's Office of Emergency Services (OES), Department of Toxic Substances Control (DTSC), State Water Resources Control Board (SWRCB), California Energy

issued in February 2014, concluded that "improving refinery safety is a goal strongly shared by government, industry, workers, and communities," and it called for changes in three areas to meet this objective:[(4)]

1) Emergency Response and Preparedness
2) Safety and Prevention of Hazardous Events
3) Community Education and Alerts

Changes to the state's Process Safety Management (PSM) regulation appear in Section Two of the report, *Safety and Prevention of Hazardous Events*. The report recommended that the following revisions to California's 1992 PSM regulation "be required as soon as possible" for the state's petroleum refineries:[(5)]

1) Implement inherently safer systems to the greatest extent feasible;
2) Perform periodic safety culture assessments;
3) Incorporate damage mechanism hazard reviews into process hazard analyses;
4) Conduct root cause analyses after significant accidents or releases;
5) Account for human factors and organizational changes; and,
6) Use structured methods, such as layer of protection analysis, to ensure adequate safeguards in process hazard analyses.

The U.S. Chemical Safety and Hazard Investigation Board (CSB) also called attention to weaknesses in the state's existing PSM regulation and recommended that both the regulation and the state's enforcement program be strengthened.[(6)] The CSB concluded, for example, that had the existing California PSM regulations required refiners to perform rigorous damage mechanism reviews (DMRs), the problem of sulfidation corrosion in the crude unit would likely have been identified and mitigated well before the catastrophic pipe failure in August 2012. Similarly, the CSB reported that there were no requirements of refiners to analyze the effectiveness of process hazard analysis (PHA) safeguards or to deploy inherently safer technologies, using a hierarchy of controls.

In developing its PSM revision for refineries, the California Department of Industrial Relations (DIR) conducted extensive outreach to industry and refinery workers. During 2014, for example, DIR convened or participated in 26 meetings or hearings pertaining to process safety at oil refineries. At each of these meetings, DIR presented the findings and recommendations of the Governor's report and CSB; described DIR's proposed revisions to the refinery PSM regulation; and listened to and recorded the views of meeting participants. DIR convened a technical PSM Advisory Committee, made up of

---

Commission, California Technology Agency (CTA), Department of Finance (DOF), Department of Public Health (DPH), Office of the State Fire Marshal (OSFM), U.S. EPA and Contra Costa County Health Services Agency.
Commission (CEC), California Technology Agency (CTA), Department of Finance (DOF), Department of Public Health (DPH), Office of the State Fire Marshal (OSFM), U.S. EPA and the Contra Costa County (CA) Health Services Agency.

[4] Governor Edmund G. Brown (February 2014). *Improving Public and Worker Safety at Oil Refineries: Report of the Interagency Working Group on Refinery Safety*. (Available: http://www.dir.ca.gov/dosh/interagency-refinery-task-force.html) Accessed March 22, 2017. pp. 24-33.

[5] Governor Edmund G. Brown (February 2014) *op cit.* p. 21.

[6] U.S. Chemical Safety and Hazard Investigation Board (October 2014). *Regulatory Report. Chevron Richmond Refinery Pipe Rupture and Fire.* Report No. 2012-03-I-CA. (Available: http://www.csb.gov/file.aspx?DocumentId=661) Accessed February 2018. pp. 96-98.

representatives of labor and industry. All Advisory Committee meetings were open to the public, who were invited to present their views before the Committee and DIR staff.

Regulatory Overview

DIR translated the recommendations of the Governor's report, the CSB, and Advisory Committee into a revised, 24-part PSM regulation for the state's 14 oil refineries, known as *Process Safety Management for Petroleum Refineries, General Industry Safety Orders (GISO) §5189.1*. The state's Occupational Safety and Health Standards Board held a public hearing on the proposed regulation on September 15, 2016; approved a revised version by unanimous vote on May 18, 2017; and adopted the regulation on October 1, 2017.[(7)]

The new PSM regulation represents the nation's first comprehensive rule pertaining to process safety. It requires refineries to take an integrated, structured approach to anticipating, analyzing and mitigating the hazards that underpin process incidents. In broad terms, it is intended to shift the practice of process safety management from a focus on controlling and responding to risks, to preventing risks from arising in the first place. In the same way, the regulation shifts the focus of the Cal/OSHA program from responding to emergencies and complaints toward working with—and requiring—refiners to build a comprehensive PSM program: one that continuously identifies and mitigates process safety hazards by deploying the most effective and enduring solutions, at the earliest possible point, well before a process incident occurs.

The Cal/OSHA program is meeting this objective with a larger and better-trained staff of PSM compliance officers, using funds generated from an oil industry fee. Most of Cal/OSHA's new PSM personnel are educated in engineering and other technical fields. All PSM staff receive several hundred of hours of PSM training before conducting an inspection. In previous years, a single Cal/OSHA compliance officer might assess a single element of a refiner's PSM program each year; under the new regulation, Cal/OSHA is coordinating its inspections with county compliance officers and U.S. EPA, such that an inspection is now able to address multiple elements of a plant's PSM program over a period of several weeks.

The new PSM regulation applies a mix of prescriptive and performance-based approaches. DIR Advisory Committee representatives pointed out that this approach is appropriate in a refinery setting, where thousands of potential risks must be identified, evaluated, prioritized and mitigated by applying engineering and management judgment. Because judgment is improved by involving the expertise of front-line workers (as noted within CCPS's four pillars of process safety, below), the regulation includes several provisions that require employee participation "throughout all phases" of a refiner's PSM program. As noted by the CCPS in 2007:

> *Workers are potentially the most knowledgeable people with respect to the day-to-day details of operating the process and maintaining the equipment and facilities.*[(8)]

---

[7] See the full text of the regulation at California Occupational Safety and Health Standards Board: http://www.dir.ca.gov/OSHSB/documents/Process-Safety-Management-for-Petroleum-Refineries-apprvdtxt.pdf. See the adopted regulation at California Code of Regulations (CCR) Title 8, General Industry Safety Orders(GISO) §5189.1, *Process Safety Management for Petroleum Refineries:* https://www.dir.ca.gov/title8/5189_1.html. A description of the purpose and necessity for each PSM element is provided in DIR's *Initial Statement of Reasons*: http://www.dir.ca.gov/OSHSB/documents/Process-Safety-Management-for-Petroleum-Refineriess-ISOR.pdf.

[8] Center for Chemical Process Safety (CCPS) (2007). *Guidelines for Risk Based Process Safety*. American Institute of Chemical Engineers (AIChE). Wiley: New Jersey. p. 124.

The Initial Statement of Reasons for the California PSM regulation makes a similar finding in presenting the rationale for including new employee participation authorities in the regulation:

> *Effectively integrating employee expertise into the refinery's PSM program is critical to ensure—and continually improve—process safety.*[(9)]

In addition to explicitly providing for the right of refinery employees to participate "throughout all phases" of PSM decision-making, the new PSM regulation gives the collective bargaining agent the authority to select its representative(s) who participate in the PSM program and in PSM teams. These teams are required by the regulation as part of conducting a process hazard analysis (PHA), damage mechanism review (DMR), hierarchy of hazard controls analysis (HCA), management of change (MOC), management of organizational change (MOOC), process safety culture assessment (PSCA), incident investigation, safeguard protection analysis (SPA) and pre-start-up safety review (PSSR). The regulation includes additional measures to improve transparency and accountability in process safety decision-making.

Altogether, California's approach is expected to lead to continuing improvement in process safety throughout the state's refinery sector, albeit with a learning curve that could prove to be steep for some of the state's refiners. While it is intended primarily to protect the safety of refinery workers and neighboring communities, the risk reductions achieved by the regulation also aim to help ensure the operational integrity and continuity of this important industrial sector in California, which relies on its domestic refining capacity to meet the great majority of the state's demand for processed fuels. Finally, beyond its application in the refining sector, the new regulation offers an advanced model for industrial safety and employee participation in the energy and chemical sectors more broadly, including in emerging industries in the clean energy sector that require the use of hazardous materials.

A more robust Process Hazard Analysis

The new PSM regulation contains nine new elements (subsections k, l, r-x) and retains 15 existing elements that have been substantially revised (Table 1).

Table 1. Twenty-four elements of California's PSM regulation. Subsections K, l and r-x are new in 2017.[(10)]

---

9 CCR Title 8, General Industry Safety Order §5189.1 *Initial Statement of Reasons*: http://www.dir.ca.gov/OSHSB/documents/Process-Safety-Management-for-Petroleum-Refineriess-ISOR.pdf

10 Table 1 applies to the full text of the regulation posted by the California Occupational Safety and Health Standards Board: http://www.dir.ca.gov/OSHSB/documents/Process-Safety-Management-for-Petroleum-Refineries-apprvdtxt.pdf.

Each of the PSM elements functions as part of an integrated engineering and management system, which is intended to drive continual improvement in process safety. The Process Hazard Analysis (PHA), which existed in the original PSM regulation, remains at the heart of the new regulation, but it now incorporates a more robust body of technical and managerial information, including the results of damage mechanism reviews (DMRs); findings from similar processes across the refinery sector; human factors; effects of organizational changes; and others (see Figure 1).

The PHA team is charged with generating recommendations for corrective actions, which then trigger implementation requirements and timelines in the regulation. An exception to this procedure occurs if the team identifies a process scenario or condition that could lead to a "major incident." This finding triggers a hierarchy of hazard controls analysis (HCA) and safeguard protection analysis (SPA), each of which link to the regulation's implementation requirements.

Importantly, the new regulation defines a "major incident" as "any event involving fire, explosion or release of a substance which has the potential to result in death or serious physical harm." Serious physical harm is defined, in turn, by California Labor Code Section 6432(e) as "*any injury or illness, specific or cumulative, occurring in the place of employment or in connection with any employment, that results in any of the following: inpatient hospitalization for purposes other than medical observation; the loss of any member of the body; any serious degree of permanent disfigurement; and impairment sufficient to cause a part of the body or the function of an organ to become permanently and significantly reduced in efficiency on or off the job, including, but not limited to, depending on the severity, second-degree or worse burns, crushing injuries including internal injuries even though skin surface may be intact, respiratory illnesses, or broken bones."*

It is readily apparent that preventing an incident that has the "potential" to result in "serious physical harm" introduces a more conservative, or protective, process safety performance standard compared to "preventing or minimizing the consequences of catastrophic releases," a phrase used in the 1992 version of the PSM regulation. The new California PSM regulation thereby lowers the threshold for corrective action in ensuring process safety.

Other PSM elements that might be incorporated into or otherwise affected by the PHA are mechanical integrity; process safety information; operating procedures; safety culture assessments; training; contractors; and others.

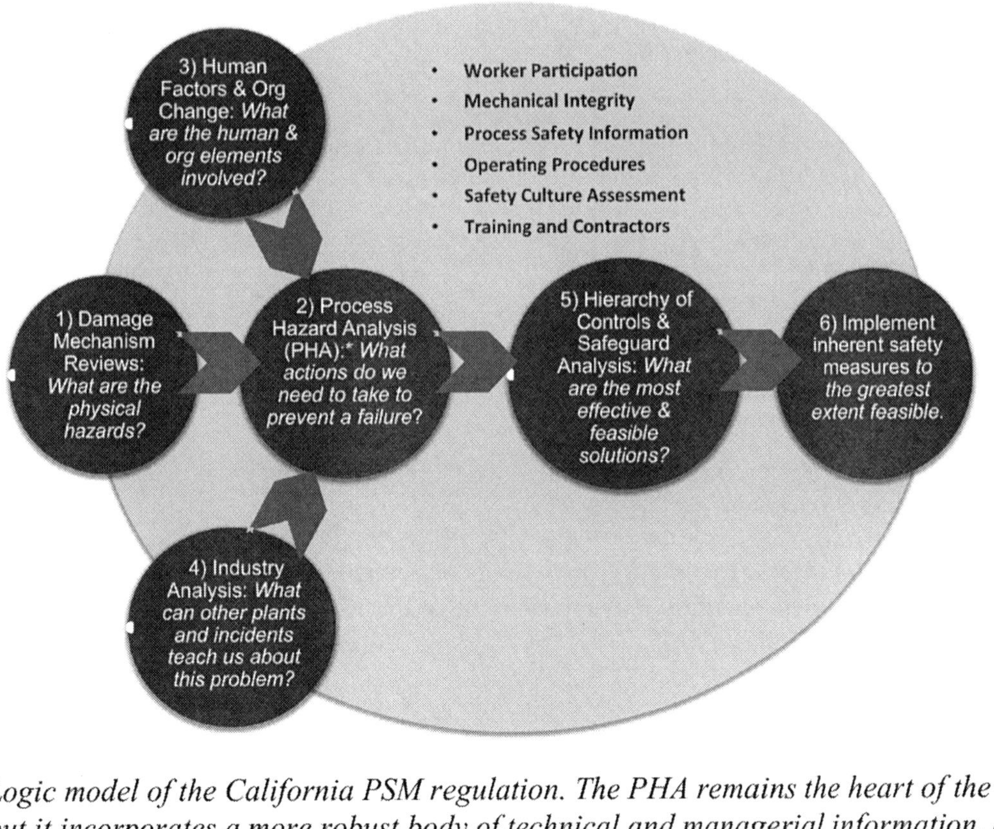

Figure 1. Logic model of the California PSM regulation. The PHA remains the heart of the PSM program, but it incorporates a more robust body of technical and managerial information, and it triggers explicit implementation requirements. If a PHA team identifies a process scenario or condition with the potential for a major incident, a hierarchy of hazard controls analysis (HCA) and safeguard protection analysis (SPA) are required, each of which trigger their own implementation requirements. Each of the bulleted elements may be integrated into or affected by the PHA. Several PSM elements are not included in the model.

Drawing on Best Practices

In addition to relying on the recommendations of the CSB and Governor's Interagency Working Group, California's PSM regulation draws heavily from the Center for Chemical Process Safety (CCPS)

Guidelines for Risk Based Process Safety (2007) which "reflects fifteen years of PSM implementation experience and well-established best practices from a variety of industries." [11] The CCPS *Guidelines* categorize the elements of an effective PSM program into the following four pillars of process safety: (1) commit to process safety; (2) understand hazards and risk; (3) manage risk; and (4) learn from experience (Table 2). The CCPS developed these areas of practice based on evidence that incident investigations in high hazard process industries "continue to identify inadequate management system performance as a key contributor to the incident," and that "audits reveal a history of repeat findings" that "indicate chronic problems whose symptoms are fixed again and again without effectively addressing the technical and cultural root causes."[12]

Table 2. Each element of the California PSM regulation supports one of the four pillars of risk-based process safety identified by the Center for Chemical Process Safety (2007).

Commit to Process Safety (a) Scope and Purpose (b) Application (q) Employee Participation (r) Process Safety Culture Assessment (v) Process Safety Management Program	**Understand Hazards and Risk** (c) Definitions (d) Process Safety Information (e) Process Hazard Analysis (k) Damage Mechanism Review (s) Human Factors
Manage Risk (f) Operating Procedures (g) Training (h) Contractors (i) Pre Start-Up Safety Review (j) Mechanical Integrity (l) Hierarchy of Hazard Controls Analysis (m) Hot Work (n) Management of Change (p) Emergency Planning and Response (t) Management of Organizational Change (x) Implementation	**Learn from Experience** (o) Incident Investigation – Root Cause Analysis (u) Compliance Audits (w) Division Access to Documents and Information

While Table 2 illustrates how each element of the California PSM regulation can be grouped within the CCPS's pillars of process safety, the regulation also updates this framework by including the "hierarchy of hazard controls" and "inherent safety," as recommended by the CCPS in *Inherently Safety Chemical Processes: A Life Cycle Approach (2009)*. This text compiles more than a decade of industry experience in the area of inherent safety. [13]

Consistent with this approach, the hierarchy of hazard controls analysis (HCA) is a foundation of California's PSM regulation, alongside the PHA. The HCA, combined with the regulation's implementation provisions, requires a refiner to identify, analyze and implement—to the greatest extent feasible—first and second-order inherent safety solutions to serious process hazards identified in the

[11] Center for Chemical Process Safety (CCPS) (2007). *Guidelines for Risk Based Process Safety*. American Institute of Chemical Engineers (AIChE). Wiley: New Jersey. p. I.

[12] Center for Chemical Process Safety (CCPS) (2007) *Risk-Based Process Safety*. The RBPS Subcommittee consists of members from Chevron Energy Technology Company, 3M Company, Celanese Chemical, The Lubrizol Corporation, Air Products and Chemicals, Inc., Rohm and Haas Company, DuPont, Eastman Chemical Company, Shell Chemical Company, Bayer Material Science BP, Eli Lilly and Company, BP, Monsanto Company, Olin Corporation, INEOS Olefins and Polymers USA, Rhodia, Inc. p. 1.

[13] Center for Chemical Process Safety (CCPS) (2007) *op cit.*

PHA. The HCA requires the refinery to assess potential solutions to these hazards by applying the hierarchy of controls, beginning with inherent safety measures, followed by passive safeguards, active safeguards, and procedural protections (Figure 2).

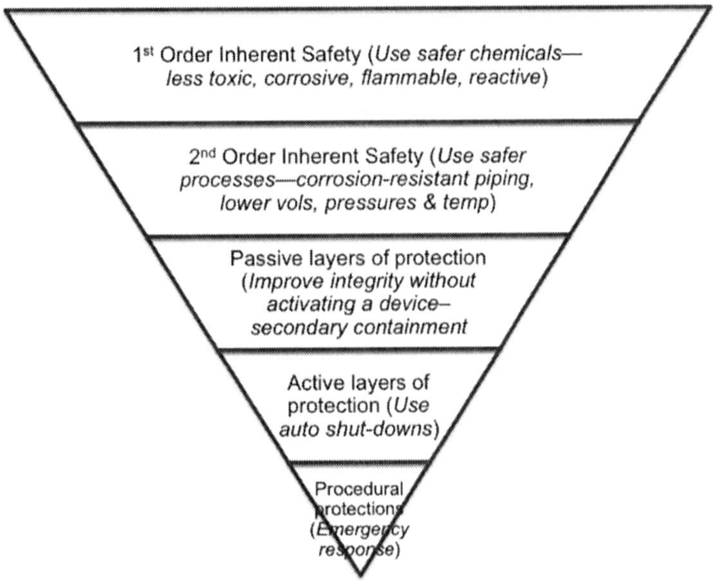

Figure 2. Framework of the California PSM Hierarchy of Hazard Controls Analysis (HCA).

For example, under the requirements of the HCA, an HCA team would need to address the risks posed by a hazardous chemical by assessing corrective actions in the following sequence and priority order:

1) First order inherent safety: Can the hazardous chemical be replaced with a safer alternative? Does the safer alternative introduce new risks at the plant or elsewhere up or down the supply chain? Can these risks be prevented or mitigated?

2) Second order inherent safety: Can the hazardous chemical be used in smaller quantities and/or under ambient temperatures and pressures?

3) Passive safeguards: Can the hazardous chemical be contained in piping and equipment that are more resistant to the corrosive effects of the chemical?

4) Active safeguards: Can devices be installed that automatically activate a device in the event of a unsafe condition?

5) Procedural actions: Is it feasible for employees to activate a device, such as a valve or fire monitor, to mitigate a release of the chemical?

Under the HCA requirements, it would likely not be permissible for a refiner to rely primarily on chemical release alarms, for example, or on procedures activated by employees as the primary solution to process safety problems identified in the PHA. While the regulation might allow these approaches to *augment* inherent safety measures or passive safeguards, they would not—in and of themselves—constitute an acceptable, baseline approach to corrective actions taken by a refiner.

Responding to a Persistent Problem

Incidents in the refinery sector demonstrate that improvements in process safety continue to be needed. In 2014, the CSB concluded that there had been "a considerable problem with significant and deadly incidents at petroleum refineries over the last decade." In 2012, the CSB tracked 125 significant process safety incidents at U.S. petroleum refineries, 17 (14%) of which took place in California.[(14)]An examination of reports submitted between 2007 and 2014 by petroleum refineries to the U.S. Department of Energy shows that the industry continues to experience process safety incidents on a regular basis.[(15)]

The federal regulations governing refineries have not been updated since the early 1990s, when the PSM regulations were first adopted in response to the 1984 industrial disaster in Bhopal, India, where a late-night leak of methyl isocyanate at the Union Carbide pesticide manufacturing plant killed thousands of residents—most of whom were sleeping at the time. In the intervening 25 years, PSM expertise by leading companies has advanced significantly, but the PSM regulation has remained static.

In the wake of the April 17, 2013 West, Texas disaster, the Obama Administration issued Executive Order 13650 in an effort to update the nation's process safety regulations, as administered by federal OSHA, EPA and Homeland Security.[(16)] In 2017, the Trump Administration delayed implementation of the newly updated EPA Risk Management Program (RMP) rules that resulted from this effort; there were no changes made to the federal OSHA PSM requirements.

RAND Economic Analysis

A RAND economic analysis of California's PSM revision at the time it was proposed concluded that implementing and maintaining compliance with the new regulation would cost the state's refiners between $20 and $184 million per year, with a point estimate of $58 million per year, spread across 14 refineries.[(17)] When passed on to consumers, RAND reported that this would equate to a price increase of about $0.004 per gallon in California.[(18)] RAND found that each major refinery incident avoided (as a result of improved PSM practices required by the new regulation) would be expected to save a refinery about $220 million, not including the potential costs associated with worker fatalities and injuries or damage to surrounding communities. RAND found that the improvements in process safety would also improve "system reliability, community relations, labor–management relations, and company reputation and public image."

In RAND's analysis, the largest potential economic benefit of the PSM revision would be the improved reliability of the state's fuel supply. This is particularly relevant for California, which refines nearly all

[14] U.S. Chemical Safety and Hazard Investigation Board (October 2014). *Regulatory Report: Chevron Richmond Refinery Pipe Rupture and Fire.* Report No. 2012-03-I-CA. p. 11.

[15] U.S. Department of Energy, Office of Electricity Delivery and Energy Relability. *Energy Assurance Daily*. [Available: http://www.oe.netl.doe.gov/ead.aspx] (Accessed March, 2017). (Note: For weekly summaries, go to *Download EADs* and scroll to *Petroleum*.)

[16] Executive Order 13650 (May 2014). Actions to Improve Chemical Facility Safety and Security—A Shared Commitment. Report for the President. DHS, USDA, DOJ, DOL, DOT, EPA. (Available: https://www.osha.gov/chemicalexecutiveorder/final_chemical_eo_status_report.pdf) Accessed Feb 2018.

[17] Gonzales D, Gulden T, Strong A, Hoyle W (2016). *Cost-Benefit Analysis of Proposed California Oil and Gas Regulations.* The RAND Corporation. Santa Monica, CA. (Available: http://www.rand.org/pubs/research_reports/RR1421.html) Accessed Feb 2018.

[18] Based on California's 2014 gasoline consumption rate of 14.5 billion gallons per year.

of its own fuels. RAND found, for example, that the California economy contracted by $6.9 billion in the first six months following the February 2015 explosion at the ExxonMobil refinery in Torrance, which destroyed the plant's electrostatic precipitator. During this period, Californians paid about $2.4 billion in the form of a $0.40 per gallon increase in gasoline prices.

Finally, RAND found that a refinery worker dies in many refinery incidents, and that in a few such incidents, multiple refinery workers die. RAND concluded that California's PSM regulation would result in a substantially lower death rate among refinery workers compared to the existing PSM standard.

Emerging Issues

In early 2018, the regulation is in its early stages of implementation. There is evidence that some refiners are successfully implementing the regulation and fully involving representatives of the United Steelworkers union (USW) in PSM decision-making. The USW represents refinery workers in most of California's refineries. In some cases, refiners are resisting key elements of the worker participation requirements. Cal/OSHA has opened PSM inspections in at least two refineries for complaints related to this element of the new regulation.

In addition to 14 refineries, there are about 1,800 chemical plants of various sizes in California. These chemical plants will continue to operate under the 1992 PSM regulation, California Code of Regulations (CCR) Title 8, *General Industry Safety Orders* (GISO) §5189. The Governor's Working Group on Refinery Safety identified several shortcomings with the existing PSM regulation, which has not been modified since its adoption in the early 1990s. The fact that the new PSM regulation applies to oil refineries but not chemical plants has by default resulted in a "two-tier" regulatory scheme in California, whereby refinery workers and communities now have greater regulatory protections compared to chemical plant workers and communities. This may need to be remedied in the future, at least for those chemical plants that present a substantial risk to worker and public safety.

Conclusion

California's new PSM regulation represents the latest and most sweeping regulatory reform initiated by a U.S. government entity in response to a major process incident. The regulation also offers an advanced model for industrial safety in the energy and chemical sectors more broadly, including in emerging industries in the clean energy sector that require the use of hazardous materials. In January 2018, the state of Washington (USA) proposed a revision to its own PSM regulation that closely reflects California's text. It is likely that California's 2017 PSM regulation will influence the national discussion on the role of regulatory policy in preventing major process incidents.

* * * * * * *

18 AIChE Spring Meeting
14TH GLOBAL CONGRESS ON PROCESS SAFETY
ORLANDO FLORIDA 2018

Jurisdictional Boundaries in Midstream – Where is the Line?

Lauren Mercer
Provenance Consulting, LLC
15995 N. Barkers Landing Rd.
Suite 145
Houston, TX 77079
lauren.mercer@provenanceconsulting.com

Prepared for Presentation at
American Institute of Chemical Engineers
2018 Spring Meeting and 14th Global Congress on Process Safety
Orlando, Florida
April 22 – 25, 2018

Jurisdictional Boundaries in Midstream – Where is the Line?

Lauren Mercer
Provenance Consulting, LLC
15995 N. Barkers Landing Rd.
Suite 145
Houston, TX 77079
lauren.mercer@provenanceconsulting.com

Keywords: Midstream, Terminal, Cavern, Jurisdiction, Boundaries, PHMSA, DOT, OSHA, PSM

Abstract

This paper will discuss the gray area between OSHA and DOT jurisdiction, including coverage of terminal operations, railcars/trucks, and underground storage applicability. The nuances of PSM applicability, including interconnectivity and colocation, hydrocarbon used as a fuel, atmospheric tanks, and safety systems will be presented. Furthermore, this paper will provide a history of jurisdictional boundary and PSM applicability cases/interpretations and will present examples of determining PSM jurisdiction and applicability. The intent is to pull all aspects of each regulations together in a single location to be used as a reference when evaluating the jurisdictional boundaries at a midstream facility.

1 Introduction

Midstream Operations are familiar with the applicability of the Pipeline and Hazardous Materials Safety Administration (PHMSA) regulation of the Department of Transportation (DOT) but may not be familiar with when, if ever, the Process Safety Management (PSM) regulation of the Occupational Safety and Health Administration (OSHA) applies to their operations. The intent of PHMSA is to cover the transportation of hazardous chemicals while the intent of PSM is to cover processes involving hazardous chemicals. But where is the line drawn between PHMSA and PSM? Since transportation is regulated by PHMSA, it is clear that pipelines are DOT covered. When the pipeline enters a midstream facility, the line between transportation and process becomes less clear.

2 Overview of Regulations

2.1 DOT PHMSA

2.1.1 History

The passage of the Natural Gas Pipeline Safety Act in 1968 marked the beginning of federal regulation of pipeline safety. The primary goal of the act was to develop comprehensive and enforceable safety procedures covering the pipeline transportation of flammable, toxic, and corrosive gases.[1] In September 1968, Congress created the Office of Pipeline Safety (OPS), under the Department of Transportation (DOT), to implement this new law. Then, in 1979, the OPS obtained jurisdiction over pipelines carrying liquids via the passage of the Hazardous Liquid Pipeline Safety Act. The purpose of the Act was to set minimum safety standards for pipeline transportation and facilities so as to mitigate risks to life and property.[2]

To regulate the two (2) pipeline safety Acts, as well as transportation of hazardous materials via other modes, Congress created the Pipeline and Hazardous Materials Safety Administration (PHMSA) within DOT in 2004. The agency's mission is to "protect people and the environment by advancing the safe transportation of energy and other hazardous materials that are essential to our daily lives,"[3] which only incidentally impacts occupational safety. In fact, in a 2008 consultation with the Occupational Safety and Health Administration (OSHA), PHMSA explained that the pipeline safety regulations were never intended to protect workers, unlike the standards promulgated under the OSH Act.[4]

PHMSA adopted the pipeline safety regulations (49 C.F.R. Parts 190-199) with the purpose of protecting public, property, and the environment. There are two (2) main PHMSA standards that are typically referenced, Part 192 *Transportation of Natural and Other Gas by Pipeline* and Part 195 *Transportation of Hazardous Liquids by Pipeline*.

2.1.2 Scope

The Pipeline Safety Act authorizes PHMSA to prescribe minimum safety standards for pipeline transportation and for pipeline facilities.[5] The Act defines "transporting gas" as:

> the gathering, transmission, or distribution of gas by pipeline or the storage of gas, in or affecting interstate or foreign commerce.[6]

The Act defines "transporting hazardous liquid" as:

> the movement of hazardous liquid by pipeline, or the storage of hazardous liquid incidental to the movement of hazardous liquid by pipeline, in or affecting interstate or foreign commerce.[7]

The Act defines a gas or hazardous liquid "pipeline facility" to include:

> a pipeline, a right of way, a facility, a building, or equipment used in transporting gas/treating gas during its transportation or used/intended to be used in transporting hazardous liquid.[8]

However, the scope of PHMSA Part 195 excludes the transportation of hazardous liquids through production, refining, or manufacturing facilities as well as the storage or in-plant piping systems associated with such facilities.[9] Part 195 also excludes the transportation through a materials transportation terminal if the facility exclusively transfers hazardous liquid between non-pipeline modes of transportation or between a non-pipeline mode and a pipeline.[10] Part 192 does not contain these exemptions or any similar exemptions. These exceptions will be discussed in further detail.

2.2 OSHA PSM

2.2.1 History

To address worker and workplace safety, Congress enacted the Occupational Safety and Health Act (OSH Act) in 1970. The Act created OSHA as an agency within the United States Department of Labor to assure safe working conditions for employees by setting and enforcing standards.[11]

In an attempt to reduce large scale industrial accidents, OSHA promulgated the Process Safety Management (PSM) regulation in 1992. The regulation provides "requirements for preventing or minimizing the consequences of catastrophic releases of toxic, reactive, flammable, or explosive chemicals."[12]

2.2.2 OSHA Preemption

The OSH Act applies to employment performed in a workplace in the United States and its Territories.[13] However, the Act includes a provision limiting OSHA's jurisdiction of working conditions that are regulated by other Federal agencies. Section 4(b)(1) states:

> Nothing in this Act shall apply to working conditions of employees with respect to which other Federal agencies... exercise statutory authority to prescribe or enforce standards or regulations affecting occupational safety or health.[14]

Both the Occupational Safety and Health Review Commission (OSHRC) and the courts have rejected industry-wide preemption under Section 4(b)(1).[15] In 2002, the United States Supreme Court in *Chao v. Mallard Bay Drilling* explained that coverage by another Federal agency does not preempt OSHA's jurisdiction, and the preemption only applies if the other Federal agency's regulations address occupational safety and health concerns.[16]

2.2.3 Scope

OSHA's PSM standard regulates processes that contain highly hazardous chemicals (HHCs) at or above the specified threshold quantities.[17] OSHA has established a threshold quantity (TQ) of 10,000 lbs. for flammable liquids or gases.[18]

OSHA defines a "process" as:

> any activity involving a highly hazardous chemical including any use, storage, manufacturing, handling, or the on-site movement of such chemicals, or combination of these activities. For purposes of this definition, any group of vessels which are interconnected and separate vessels which are located such that a highly hazardous chemical could be involved in a potential release shall be considered a single process.[19]

3 PHMSA or PSM

In midstream, there are three (3) systems where there may be gray area between OSHA and DOT: terminals, railcars/trucks, and underground storage. Each system will be discussed in detail in the sections below. Additionally, conflicting guidance presented by a PHMSA subcommittee for Midstream Safety will be discussed.

3.1 Memorandum of Understanding

In 1972, OSHA and DOT entered into a memorandum of understanding in which the agencies agreed to communicate when questions arise in the implementation of the OSH Act as those questions relate to the Section 4(b)(1) OSHA preemption.[20]

In 1992, OSHA, after consultation with the OPS, determined that pipeline safety regulation Part 192 addresses the hazards of fire and explosion and therefore the agency is precluded from enforcing the PSM rule over gas transmission and distribution.[21] In 1998, OSHA extended the preemption to LNG distribution and transmission facilities covered under Part 193 *Liquefied Natural Gas Facilities*.[22] However, it is important to note that the 1992 and 1998 Letters of Interpretation are specifically addressing the transmission and distribution processes and therefore natural gas and LNG processing facilities are subject to OSHA coverage.[23] Although commenters to the PSM final rule believed gas processing should be excluded from coverage, OSHA believed (and still does) that gas plants are appropriately covered by the process safety management standard.[24]

Generally speaking, OSHA starts where DOT ends. If PHMSA does not cover a process that contains a HHC, it needs to be evaluated for PSM coverage.

3.2 NGL Fractionation and Materials Transportation Terminals

3.2.1 Background – ONEOK

The ONEOK Bushton facility consists of fractionation operations, associated in-plant piping systems, and underground storage caverns (the ONEOK caverns will not be discussed here). The facility receives natural gas liquids (NGLs) from a PHMSA regulated pipeline, the NGL is then transported through the facility, and the product leaves via a PHMSA regulated pipeline. ONEOK had determined that, while the pipelines in/out of the facility and the piping up to the pressure control devices are PHMSA regulated, the in-plant piping leading up to any pressure control device and the fractionation system are not in the scope of PHMSA.[25]

In 2005, OPS attempted to conduct a pipeline safety inspection at the ONEOK Bushton facility, however, ONEOK communicated to OPS that the facility was not subject to PHMSA jurisdiction. Letters were exchanged between ONEOK and PHMSA and in 2006, ONEOK agreed to implement Part 195 to certain aspects of their facility. In 2008, after a release of NGLs and resulting fire, the OPS investigation alleged that ONEOK had committed safety violations in connection with the event.[26] In 2011 and 2012, PHMSA and ONEOK exchanged further letters regarding the jurisdictional status of the facility, resulting in PHMSA's determination that Part 195 regulates various parts of the facility, but there are portions that are exempt. More specifically, the in-plant piping used exclusively to transfer hazardous liquids between non-pipeline modes of transportation or between a pipeline and a non-pipeline mode is exempt, as is the fractionation-exclusive equipment and piping because PHMSA has not promulgated regulations regarding such operation.[27]

During the reviews of 2011 and 2012, ONEOK contended that their entire facility fell under the "production, refining, or manufacturing facility" exemption because it uses a refining process (fractionation). However, PHMSA disagreed with the notion that fractionating is refining because a change in chemical characteristics is not taking place. PHMSA concluded that the ONEOK facility is a materials transportation terminal that has a fractionation plant on its grounds and the presence of the fractionation plant (which is comprised of separation or processing equipment), does not exempt the entire facility.[28]

It is important to note that PHMSA defined a materials transportation terminal in one of the letters as "a facility where materials are received, stored and/or transferred, and in some cases processed and further transported" and indicated that such facilities can be regulated by PHMSA if hazardous liquids are received by pipeline and reinjected for continued transportation via pipeline (incident to transportation).[29]

In October 2016, the Acting Associate Administrator for Pipeline Safety issued the final order for the 2008 alleged safety violations. The order contained a conclusion that the interpretations issued by PHMSA are consistent with the Act's purpose and thus the facility is required to be in compliance with Part 195.[30]

Note that this conclusion addresses the materials in transportation and not any materials that are used to support the terminal (e.g. butane used in the regeneration of dehydrator desiccant). Such a system would remain under the jurisdiction of OSHA.

3.2.2 Where is the Line?

Although the letters and final order are clear that piping and equipment used exclusively for fractionation are exempt from PHMSA, PHMSA has yet to promulgate specific regulations covering fractionation systems. The official communications sent in this situation set a precedent and act as evidence of PHMSA's perspective on piping and equipment used exclusively for fractionation until such time that PHMSA releases a regulation. Furthermore, midstream processing facilities for hazardous liquids, previously thought to be outside the scope of PHMSA jurisdiction, now may be interpreted as regulated by PHMSA depending the modes of transportation in and out of the facility. Table 1 provides the governing regulation for different types of systems encountered in midstream, along with a schematic representation.

Table 1. NGL Fractionation and Materials Transportation Terminals

System	Governing Regulation	Representation
Piping and equipment used exclusively for fractionation	PSM	
Materials transportation terminal **(gas)**	PSM	
Materials transportation terminal **(hazardous liquid)** that receives liquid from a pipeline and re-injects for continued transportation via pipeline	PHMSA	

System	Governing Regulation	Representation
Materials transportation terminal (**hazardous liquid**) that is exclusively between non-pipeline modes of transportation or between pipeline and non-pipeline	PSM	
Systems that support the material transportation system (e.g. butane regeneration or butane blending)	PSM	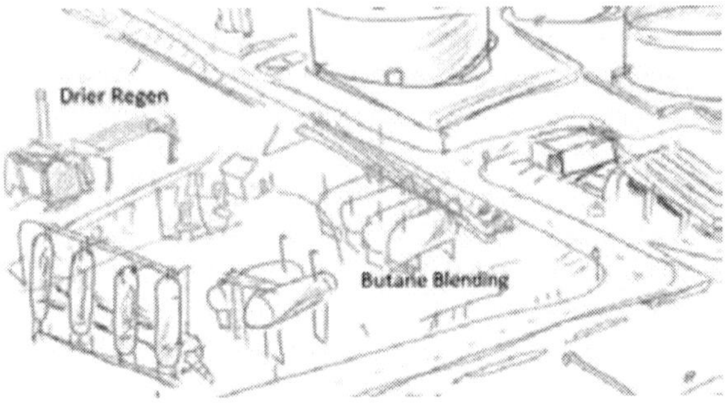
Lines that bypass a material transportation terminal (**hazardous liquid**)	PHMSA	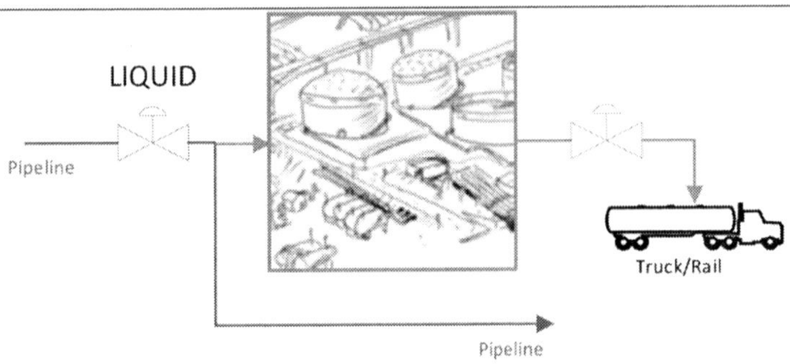
Materials transportation terminal (**hazardous liquid**) that is not exclusively between non-pipeline modes of transportation or between pipeline and non-pipeline (i.e. shared use lines are present)	Majority PHMSA	

While PSM may be the governing standard for the items above, the process will need to be evaluated to determine if it falls within the scope of the standard. Nuances of the PSM scope are discussed in Section 4.

3.3 Gas and Liquids Technical Advisory Committee

In 2014, the PHMSA Gas and Liquids Technical Advisory Committee presented on the need to create a workgroup of stakeholders to discuss the issues related to regulatory oversight of midstream processing facilities. PHMSA and OSHA are seeking understanding of the concerns of the midstream companies because the agencies are in agreement on jurisdictional lines: OSHA covers gas processing units and PHMSA does not have an interest in regulating processing units.[31] The subcommittee for Midstream Safety was formed with a goal to provide understanding of the boundaries between PHMSA and OSHA and ensure there are no gaps or overlaps in the regulation.[32]

The PHMSA subcommittee, consisting of representatives of federal agencies and industry, met on several occasions in 2014 and 2015. They presented their proposed guidance in August 2015, which was intended to provide clarity and is not considered to be a legally enforceable regulation. The proposed guidance was presented in the form of a set of frequently asked questions (FAQs) which are summarized below.[33]

3.3.1 Defining Processing and Processing Facility

In FAQ 1, the committee attempted to separate "processing" and "refining" by defining processing as:

> the treatment of products including, but not limited to, dehydration, removal of contaminants by separation or filtration, blending with other products, and heating or cooling units that separate or purify products and remove condensates by distillation.[34]

PHMSA considers the chemical conversion of crude oil into refined petroleum products as refining. Furthermore, the committee provided that a "processing facility" comprises units that have a processing function and meet the applicability of OSHA PSM.[35]

These definitions in isolation imply that midstream product terminals that treat hazardous liquids via dehydration would be subject to the PSM standard. However, an important distinction is missing – the mode of transportation in and out of the facility. As discussed in Section 3.2 previously, Part 195 exempts transportation of hazardous liquid through facilities located at a materials transportation terminal that are not pipeline-in/pipeline-out, leading to the belief that processing facilities that are pipeline-in/pipeline-out are covered under PHMSA and therefore preempted from OSHA PSM.[36] In fact, PHMSA discussed the inclusion of materials transportation terminals whose hazardous liquids are received by pipeline and reinjected for continued transportation via pipeline in the ONEOK letters presented in Section 3.2.1 previously.

The mode of transportation distinction only applies to hazardous liquids as Part 192 does not contain the same exemptions as Part 195. Therefore, the definitions provided by the joint committee, while contentious for hazardous liquids, can be assumed applicable to gas.

3.3.2 Predominant Use

The committee, in FAQ 2, delineates the boundary between pipeline transportation and a process facility with pressure control devices. PHMSA Part 192 and 195 will not regulate pipelines downstream of the first pressure control device entering the processing facility nor pipelines upstream of the last pressure control device leaving the processing facility (with an exception). In subsequent FAQs 3 and 4, the committee uses the term "predominant use" to provide guidance for shared in-plant piping. Piping that is downstream of the first pressure control device that is predominantly used to bypass processing is subject to regulatory oversight under PHMSA Part 192 or 195. Additionally, a pipeline that predominantly bypasses a pressure control device is subject to the regulatory oversight under PHMSA Part 192 or 195.[37]

In continuation with the discussion regarding the definition of process facility, PHMSA interpretations issued to ONEOK do not mention the boundary of a pressure control device and instead focus solely on the modes of transportation in and out of the facility. In contrast to the guidance provided here, in the ONEOK letters PHMSA used the term "exclusive use" instead of "predominant use" and stated that shared piping (piping that can bypass processing) is not exempt from PHMSA.

The ONEOK letters are focused on hazardous liquids (NGLs) and not on the regulation of gas operations. Therefore, predominant use, while contrary to interpretations for hazardous liquids, can be assumed applicable to gas.

3.3.3 Piping Related to Underground Storage

In the FAQ 6, the committee addresses piping associated with underground storage by defaulting to the purpose of the storage. Piping associated with underground storage used for the purpose of managing processing facility inventory is regulated by OSHA PSM while piping associated with underground storage used for transportation is regulated by PHMSA Part 192 or 195.[38]

Coverage of the underground storage itself will be discussed further in Section 3.5 below. Note that the FAQ is silent on possible preemption of the PSM standard for underground storage of natural gas incident to transportation discussed in Section 3.5.

3.4 Rail Cars, Trucks, and Associated Racks

3.4.1 Background

There are various regulations that may cover a truck or railcar, some of which fall under the Hazardous Materials Regulations in PHMSA. To avoid confusion with PHMSA Parts 192 and 195 referenced in this paper, these regulations will be referred to as "DOT" in this section.

In a 1994 Letter of Interpretation, OSHA explained that carriage of a covered chemical onto a worksite "would be covered by OSHA standards to the extent that it is not covered otherwise by [DOT]."[39] In a separate 1994 Letter of Interpretation, OSHA reiterated its stance that DOT would preclude OSHA so long as the cars are considered "in transit" by DOT. However, if a rail car or tank motor vehicle were to remain on a worksite and used to store threshold quantities of HHCs, the vehicles would be covered by the PSM standard."[40]

In general, DOT covers the transportation of a hazardous material, including movement by rail car or motor vehicle, loading incidental to movement, unloading incidental to movement, and storage incidental to movement.[41] DOT specifies that loading and unloading activities are covered when performed by carrier personnel or in the presence of carrier personnel.[42] Storage incidental to movement is defined as:

> storage of a transport vehicle...containing a hazardous material by any person between the time that a carrier takes physical possession of the hazardous material for the purpose of transporting it until the package containing the hazardous material has been delivered to the destination indicated on a shipping document...or, in the case of a private motor carrier, between the time that a motor vehicle driver takes physical possession of the hazardous material for the purpose of transporting it until the driver relinquishes possession of the package at its destination and is no longer responsible for performing functions...with respect to that particular package.[43]

DOT does not cover storage of a hazardous material at its final destination including storage of a rail car on private track.[44, 45] Additionally, DOT does not cover unloading when performed by a person employed by or working under contract to the consignee once the carrier has completed the delivery.[46]

Furthermore, as addressed in a 1993 Letter of Interpretation, the connection between the hose from a vehicle and the process to which it is being unloaded serves as the jurisdictional boundary between DOT and OSHA.[47]

3.4.2 Where is the Line?

Simply put, if DOT considers the rail car or truck in transit, OSHA is preempted. However, it is important that the facility have a basic understanding of the operations that DOT covers and does not cover, thus opening the door for PSM coverage. Table 2 provides the governing regulation for different types of systems encountered in midstream, along with a schematic representation.

Table 2. Rail Cars, Trucks, and Associated Racks

System	Governing Regulation	Representation
Loading for immediate shipment	DOT	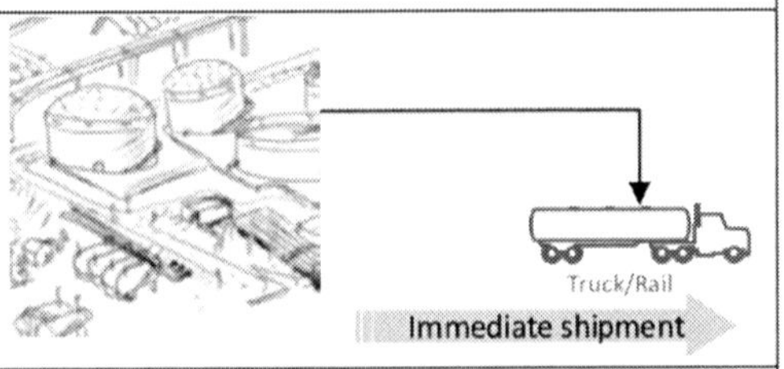
Loading but the transport vehicle is held on-site and the pre-transportation functions have been completed	DOT	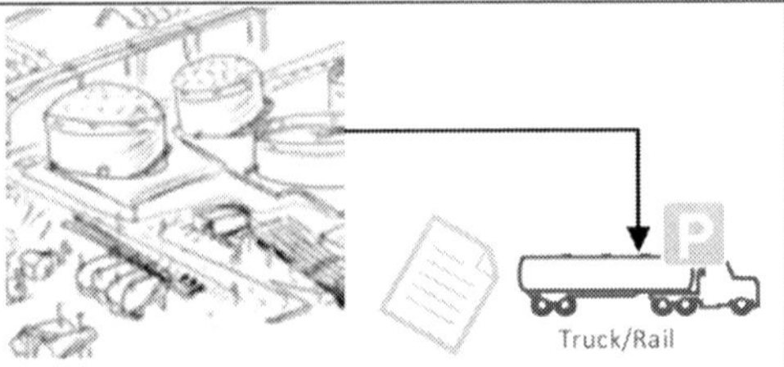
Unloading by carrier or under its direct supervision	DOT	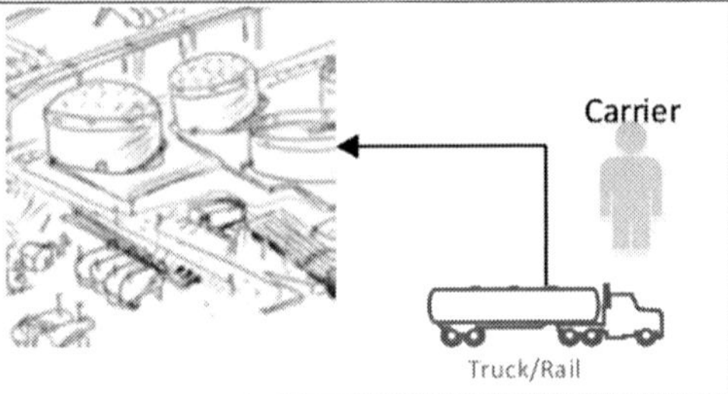
Unloading by consignee or direct contractor of consignee	PSM	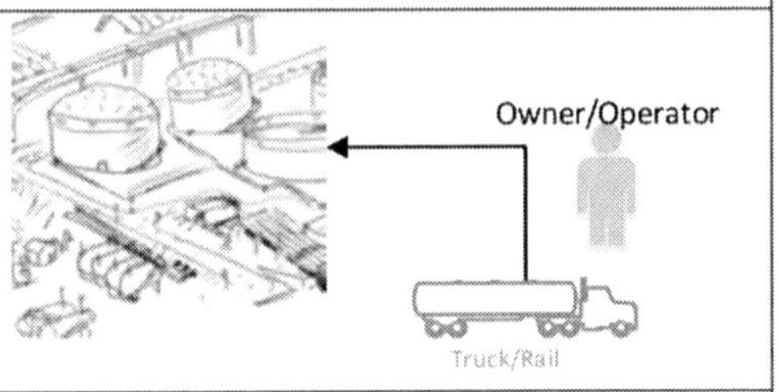
Storage of transport vehicle on premises or private track	PSM	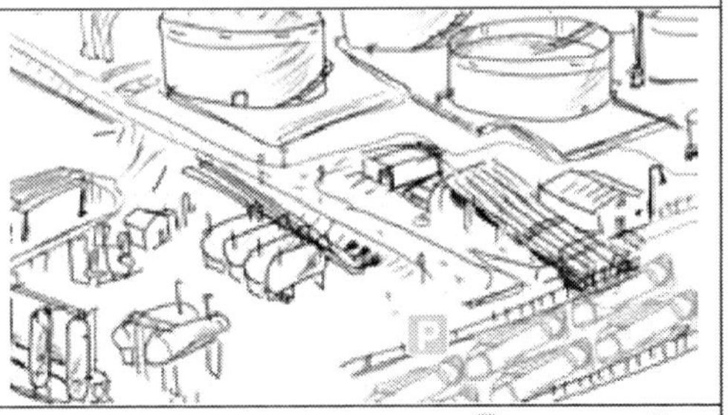
Storage of transport vehicle on non-private track	DOT	

System	Governing Regulation	Representation
Loading/Unloading racks	PSM	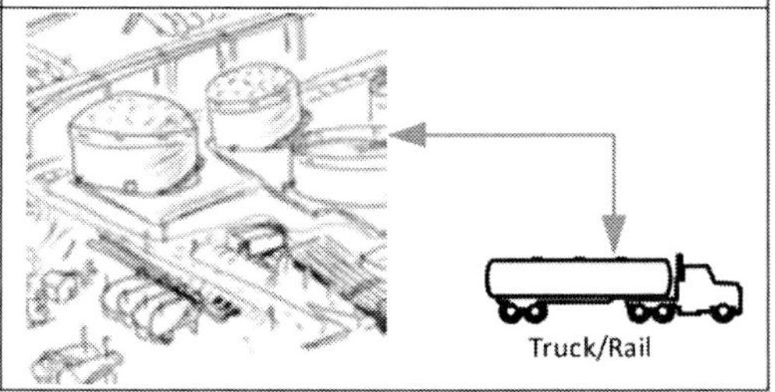

While PSM may be the governing standard for some of the items above, the process will need to be evaluated to determine if it falls within the scope of the standard. Nuances of the PSM scope are discussed in Section 4.

3.5 Underground Storage

3.5.1 Background

In 2008, OSHA received a request for interpretation regarding the extent of OSHA's jurisdiction over underground storage of gas and hazardous liquids. In preparing the response, OSHA consulted with PHMSA to determine if PHMSA has regulations covering working conditions at gas/hazardous liquid underground storage facilities.[48] PHMSA explained that "while it (or its State agent) may regulate the aboveground piping at an underground natural storage facility, its regulations would apply only until the last valve, before the downhole injection equipment and piping, and would not resume until the first valve aboveground for withdrawal facilities."[49]

In response to the Protecting our Infrastructure of Pipelines and Enhancing Safety (PIPES) Act of 2016, PHMSA issued an interim final rule (IFR), to revise the pipeline safety regulations in order to address critical safety issues related to underground natural gas storage facilities.[50] Under this IFR, "all intrastate transportation-related underground gas storage facilities will become subject to minimum federal safety standards [Part 192] and be inspected either by PHMSA or by a state entity..."[51] This IFR negates the statement in the 2008 Letter of Interpretation that, as of 2008, PHMSA had not exercised jurisdiction over working conditions at underground storage facilities.[52]

The new rule incorporates by reference American Petroleum Institute (API) Recommended Practices (RPs) regarding the underground storage of natural gas, and the non-mandatory provisions were adopted as mandatory, unless a deviation documents why compliance is not necessary for safety.[53] The adopted API RPs contain language regarding worker and process safety, satisfying the Section 4(b)(1) preemption.

The new verbiage can now be found in PHMSA Part 192.12.

3.5.2 Where is the Line?

Although OSHA has been silent on the PHMSA IFR, it is possible that OSHA will determine it is preempted from PSM coverage of underground storage of natural gas incident to transportation, especially given the inclusion of worker and process safety language. However, the IFR does not modify the coverage of non-transportation related underground storage or storage of hazardous liquid. Table 3 provides the governing regulation for different types of systems encountered in midstream, along with a schematic representation.

Table 3. Underground Storage

System	Governing Regulation	Representation
Underground storage of gas incident to transportation	PHMSA (assuming OSHA is preempted under the IFR)	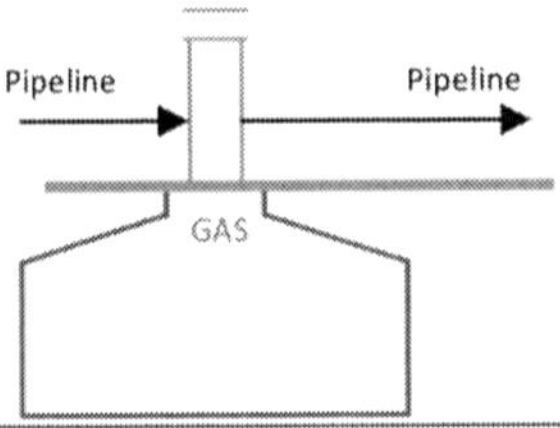
Underground storage of hazardous liquid incident to transportation	PSM	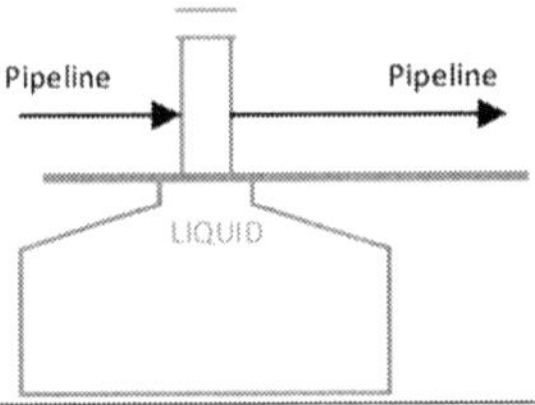
Underground storage of gas or hazardous liquid not incident to transportation	PSM	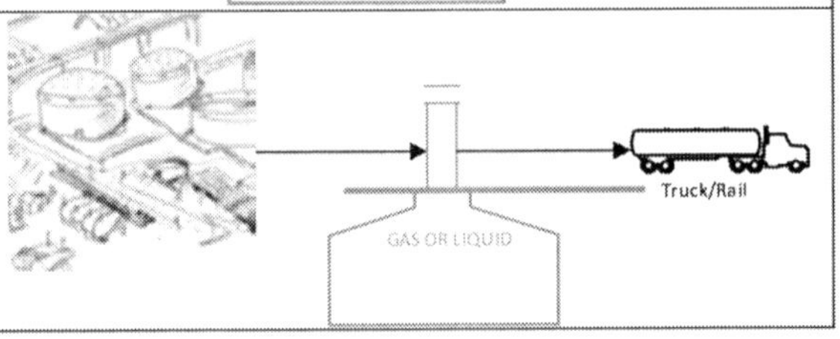

While PSM may be the governing standard for the items above, the process will need to be evaluated to determine if it falls within the scope of the standard. Nuances of the PSM scope are discussed in Section 4.

4 OSHA PSM Applicability Details

4.1 Defining a Process

4.1.1 Interconnected

Interconnectedness was integral to API 750 *Management of Process Hazards*, an industry consensus document which OSHA referenced when developing the PSM Standard.[54] During the rulemaking process, commenters believed that HHCs in a single, interconnected process should be covered by the PSM standard.[55] In writing the standard, OSHA presumed that all aspects of an interconnected process would participate in a catastrophic release and OSHA does not have the burden of proof to show the potential for a release with respect to interconnected (as opposed to separate) vessels.[56]

OSHA considers equipment occasionally connected to be included in the covered process, i.e. there are no minimum times for connections to be in place for equipment to be considered interconnected.[57] Furthermore, OSHA considers equipment connected by temporary flexible hoses as interconnected.[58]

In a 1997 Memorandum, OSHA explained that, although other elements of the PSM standard require engineering and administrative controls to prevent catastrophic release of a covered HHC, such controls cannot be used to determine the extent of a process (i.e. a point of separation).[59]

In midstream, the understanding of interconnectedness may come into play when evaluating a regeneration furnace connected to its storage vessel.

4.1.2 Co-location

For vessels that are not physically connected, owners and operators must make a reasonable determination about whether vessels are located such that they could be involved in a potential release. For example, when a nearby single fire or explosion could simultaneously compromise more than one vessel or container, they are considered co-located and part of the same process.[60] In contrast to interconnected equipment, "OSHA would be required to demonstrate as part of its *prima facie* case that unconnected but co-located processes are situated in a manner that a release from one process could contribute to the release of the other."[61]

In a 2014 Letter of Interpretation, OSHA stated that passive controls, depending on the hazard, can be taken into consideration when evaluating separation of vessels and provided an example of a physical barrier (a dike) that has been deemed acceptable. In the same Letter of Interpretation, OSHA reiterated that in a way similar to the evaluation of interconnectedness, engineering and administrative controls cannot be taken into consideration when evaluating co-location.[62]

In midstream, although most equipment falls into a process via interconnection, the understanding of co-location may come into play when evaluating bullet tanks (that are not incident to transportation) with different flammable liquids.

4.1.3 Aggregation

OSHA did not intend that facilities aggregate quantities of all covered chemicals but instead wanted the focus to be on the amount of a listed chemical in a plant that could be released at one point in time from an individual process.[63] Therefore, flammable liquids are treated separately from flammable gases for aggregation purposes. Threshold quantity (TQ) is determined by the aggregate amount of flammable liquids <u>or</u> flammable gases in a process.[64]

The TQ is determined by the amount of covered HHC(s) in the process at any particular time.[65] A process which has the capacity to contain a TQ or greater amount of a covered HHC but never contains more than the TQ amount, would not be covered by the PSM standard.[66] OSHA commented in a 1994 Letter of Interpretation that it is imperative for controls to be in place to assure the HHC remains below the TQ.[67] It is important to note that when determining a process (i.e. evaluating colocation and interconnectivity), controls cannot be considered, as discussed previously in Section 4.1.1 and 4.1.2.

In midstream, the understanding of aggregation may come into play when evaluating the threshold quantities at a terminal with different products.

4.1.4 Utilities and Safety Systems

Once it has been determined that a process is covered, the equipment, piping, hoses, or other interconnections that do not contain the regulated substance but are important for accidental release prevention must also be considered. Utility system failures (e.g. instrument air) have shown to cause upset conditions in processes. Therefore, facilities must evaluate utilities that serve a covered process to determine if the utility system can impact or affect a release of a HHC. If it is determined that said utility system can impact or affect a release, such systems would be covered under the PSM standard until a point where failure in the system is deemed to no longer affect a potential release.[68]

The same logic applies to safety systems that are installed outside of the covered process but provide mitigation measures. A process hazard analysis (PHA) should identify such safety systems.

4.2 Exemptions Relevant to Midstream

The PSM standard has five (5) exemptions; the three (3) that are frequently used in midstream operations are discussed here.

4.2.1 Atmospheric Storage Tanks

The scope of the PSM standard includes an exemption for "flammable liquids with a flashpoint below 100 °F stored in atmospheric tanks or transferred which are kept below their normal boiling point without benefit of chilling or refrigeration."[69]

In 1995, the Meer Corporation was cited for numerous safety violations including the claim that its storage of flammable liquids was excluded per the atmospheric storage tank exemption.[70] In the resultant hearing, the judge ruled that atmospheric storage tanks are exempt from the PSM standard, even if they are connected to a covered process. In 1997, OSHA issued a Memorandum addressing the *Meer Decision* and its contradiction to OSHA's consistent interpretations. However, OSHA stated that the decision would not be appealed because it is based on problems in the text of the standard.[71] Therefore, until the standard is revised, OSHA will not consider the quantity of flammables in atmospheric storage towards the threshold quantity of a process.[72]

In response to the *Improving Chemical Facility Safety and Security* Executive Order signed in August 2013, OSHA has published a request for information that includes policy change options that are under consideration. In the provided options listing, the first topic is "clarifying the PSM exemption for atmospheric storage tanks."[73] No further publications on this topic have been issued.

In midstream, understanding of the *Meer Decision* is relevant to documenting exemption of atmospheric storage tanks when they are not DOT covered or of unloading racks that connect exclusively to atmospheric storage tanks.

4.2.2 Hydrocarbon Used as a Fuel

The scope of the PSM standard excludes hydrocarbon fuels used solely for workplace consumption as a fuel so long as such fuels are not a part of a covered process.[74] In 1993 and 1995 Letters of Interpretation, OSHA provided guidance for furnace or heater systems fired by natural gas provided by pipeline from an outside source. OSHA explained that if the furnace contained other covered chemicals, then the furnace and pipeline may be part of the covered process to the extent they could be affected by a release of the covered material.[75, 76]

In midstream dehydration operations, a facility may use a furnace as part of their regeneration process. The natural gas firing the heater would need to be evaluated for coverage in such instances.

4.2.3 Normally Unoccupied Remote Facilities

The PSM standard does not apply to a "normally unoccupied remote facility" (NURF exemption), which is defined as:

> a facility which is operated, maintained or serviced by employees who visit the facility only periodically to check its operation and to perform necessary operating or maintenance tasks. No employees are permanently stationed at the facility. Facilities meeting this definition are not contiguous with, and must be geographically remote from all other buildings, processes or persons.[77]

In a 1998 Letter of Interpretation, OSHA agreed that a facility with employees at the site for an average of 1.5 man-hours per day and a total of 14.5 man-hours per week was normally unoccupied.[78] Then, in a 2005 Letter of Interpretation, OSHA referred employers to the 1998 Letter of Interpretation and stated that if employees do not exceed the averages listed in the 1998 letter, the facility would be considered NURF, setting these averages as an un-official limit.[79]

OSHA set another un-official limit by concluding that the facilities in question in the 1998 and 2005 letters satisfied the geographically remote requirement at four (4) and five (5) miles, respectively, from a potentially PSM-covered facility.[80, 81]

In midstream, the NURF exemption is typically seen used at cavern facilities. When a facility is evaluating the exemption, it is important to scrutinize the hours employees spend at the site closely.

5 Case Studies

5.1 Facility 1 – Focusing on Mode of Transportation

A product terminal receives various NGLs from a PHMSA covered pipeline system. Under normal operations, the NGLs are passed through dehydrators and the dried product is then transferred to storage vessels. The product is pumped to a truck dock where it is loaded to truck via hose connection. However, the product is occasionally sent out via a PHMSA covered pipeline system. The dehydrators contain beds of desiccant that need routine drying after significant moisture is collected. The regeneration process of the desiccant occurs in one dehydrator while the others remains in operation. Butane is used in the regeneration process; the butane is stored in a vessel, heated before being sent to the dehydrator, and then cooled before being sent back to the vessel. The facility is represented in Figure 1 below:

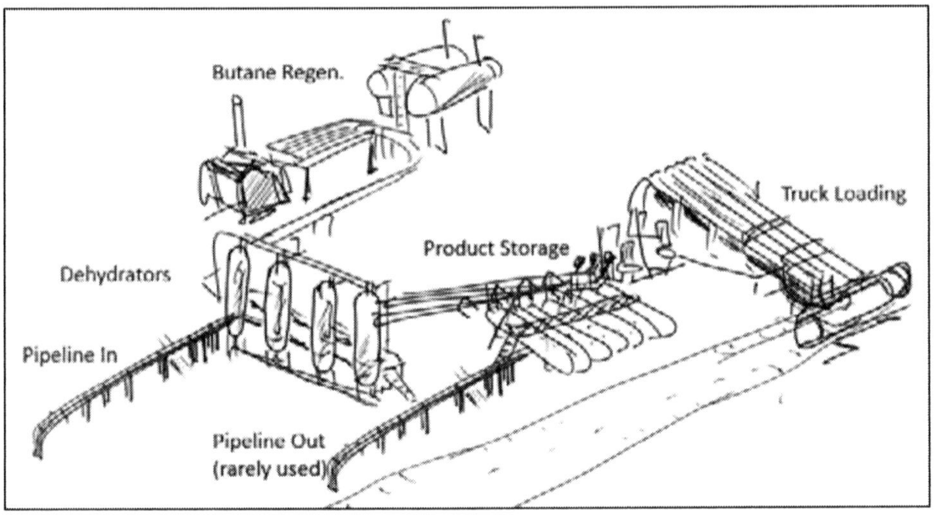

Figure 1. Terminal with Connection Back to Pipeline

If a facility were to follow the guidelines lined out by the PHMSA subcommittee for Midstream Safety discussed previously in Section 3.3.2, the dehydrators would be covered by PSM due to their predominant use of transporting hazardous liquids between a pipeline and a non-pipeline mode of transportation. However, the ONEOK case discussed previously in Section 3.2.1 set a precedence that if a terminal is not exclusively between non-pipeline modes of transportation or between pipeline and non-pipeline, the "shared use" equipment/lines would be covered by PHMSA. Because the butane used in the regeneration of the dehydrator desiccant is not in transportation, the butane regeneration system would not be covered by PHMSA and would therefore be considered under the jurisdiction of PSM. Additionally, the truck loading would be covered by PSM; assuming the truck is not stored at the facility, it is covered by DOT. The governing regulations are presented in Figure 2 below:

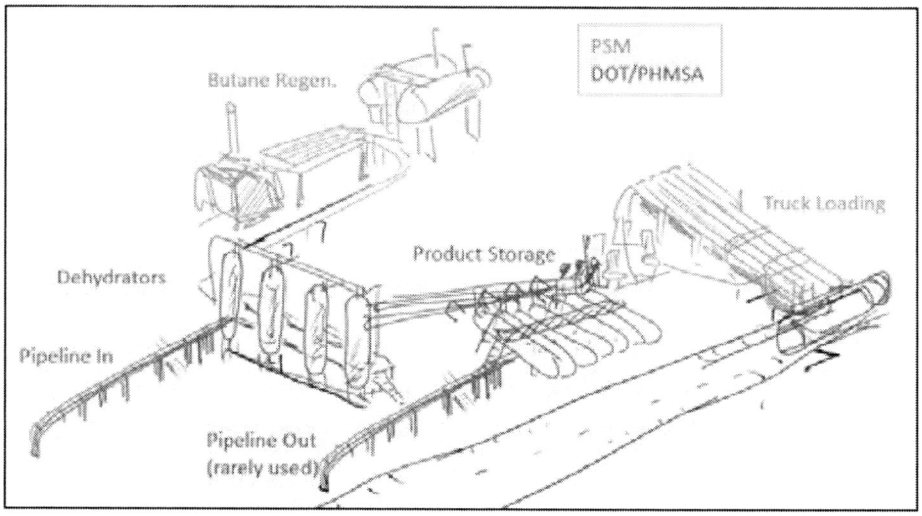

Figure 2. Governing Regulations of Equipment at a Terminal with Connection Back to Pipeline

5.2 Facility 2 – Focusing on Cavern Coverage and Fuel Exemption

A product terminal receives propane from a PHMSA covered pipeline system where it is stored in various storage vessels or a cavern before being reinjected back into a PHMSA covered pipeline system. Some of the propane is used as pilot-gas for the flare system. The facility is represented in Figure 3 below:

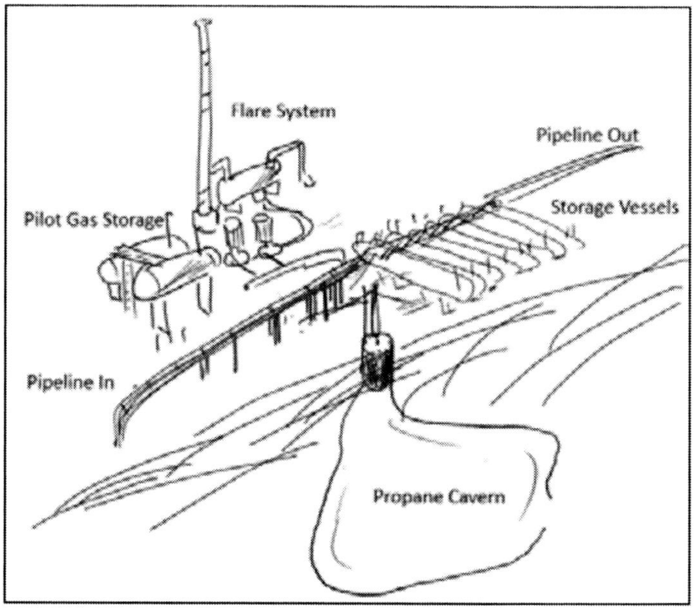

Figure 3. Terminal with a Cavern and Fuel Exemption

Because the facility is pipeline-in/pipeline-out, PHMSA is the governing regulation for the equipment/lines at the facility, except for the cavern. As discussed in Section 3 previously, PHMSA does not currently regulate underground storage of hazardous liquids. Therefore, the cavern is regulated by PSM; note that the well head is covered by PHMSA. Because the propane used as fuel gas is not in transportation, the pilot gas system would not be covered by PHMSA and would therefore be considered under the jurisdiction of PSM. However, as discussed in Section 4.2.2, PSM excludes hydrocarbon fuels used solely for workplace consumption as a fuel as long as the fuels are not a part of a covered process. In this case, the flare does not serve a PSM covered process and therefore would not be considered a PSM covered process itself. Therefore, the pilot gas system is exempt from PSM. The governing regulations are presented in Figure 4 below:

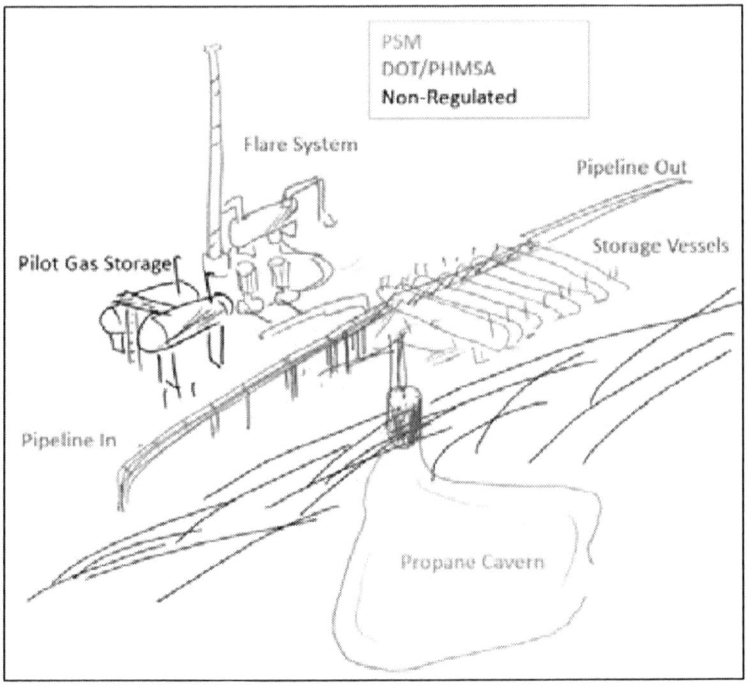

Figure 4. Governing Regulations of Equipment at a Terminal with a Cavern and Fuel Exemption

5.3 Facility 3 – Focusing on NURF Exemption

A storage facility receives propane via a PHMSA covered pipeline where it is stored in a cavern before being reinjected back into a PHMSA covered pipeline. Employees are not permanently stationed at the facility but operations personnel performs weekly rounds and monthly maintenance activities. In total, personnel spend an average of one (1) hour per day and a total of eight (8) hours per week at the site. The storage facility is located six (6) miles away from the regional offices, where employees are stationed. The facility is represented in Figure 5 below:

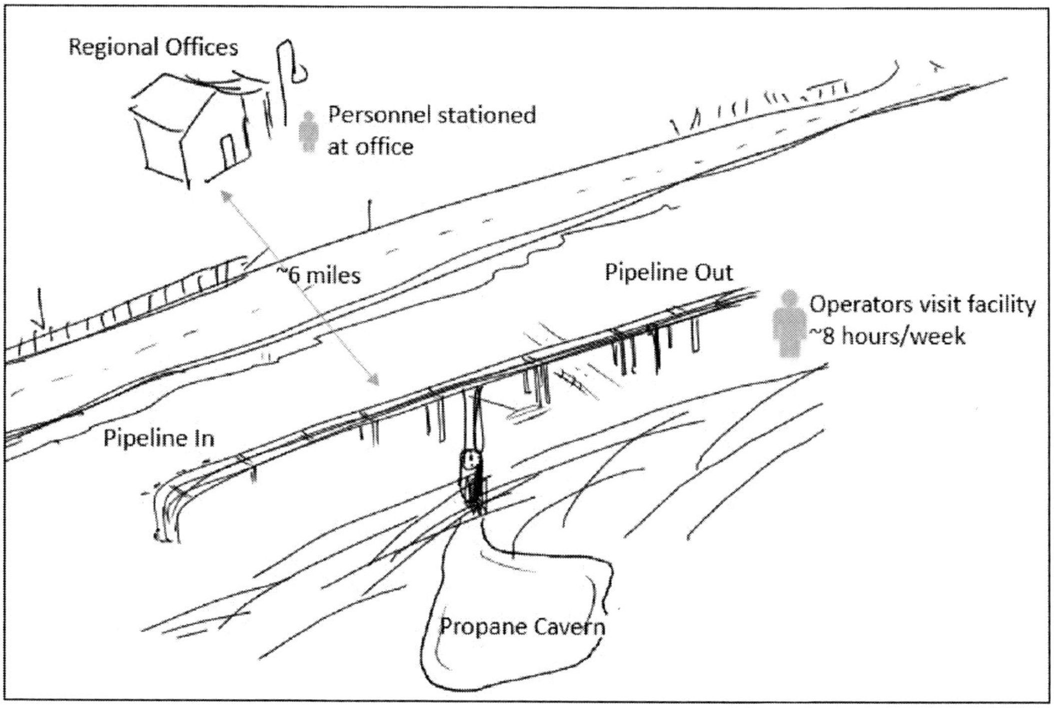

Figure 5. Cavern Facility

Because the facility is pipeline-in/pipeline-out, PHMSA is the governing regulation for the lines at the facility, including the cavern well head. As discussed in Section 3 previously, PHMSA does not currently regulate underground storage of hazardous liquids and jurisdiction would fall to PSM. However, as discussed in Section 4.2.3, PSM excludes normally unoccupied remote facilities. Because personnel spend less than the threshold of 1.5 hours per day and 14.5 hours per week, the facility would be considered normally unoccupied. Furthermore, because the facility is located further than the threshold of 4-5 miles from all other buildings, processes or persons, the facility would be considered remote. Therefore, the cavern is exempt from PSM. The governing regulations are presented in Figure 6 below:

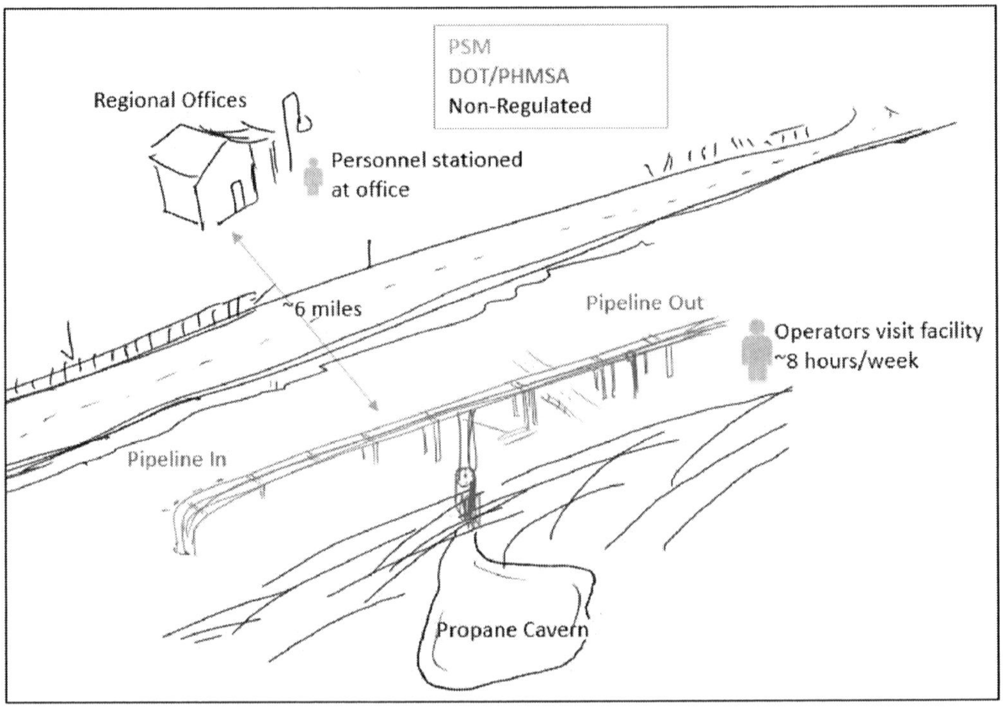

Figure 6. Governing Regulations of a NURF Cavern Facility

5.4 Facility 4 – Focusing on Defining a PSM Process and Aggregation

A product terminal receives propane from a PHMSA covered pipeline system. Once received, the propane is passed through dehydrators and the dried product is then transferred to storage vessels. The product is exclusively pumped to a truck dock where it is loaded to truck via hose connection. The dehydrators contain beds of desiccant that need routine drying after significant moisture is collected. The regeneration process of the desiccant occurs in one dehydrator while the others remains in operation. Butane is used in the regeneration process; the butane is stored in a vessel, heated before being sent to the dehydrator, and then cooled before being sent back to the vessel. Additionally, the terminal receives gasoline and diesel from a Refinery which is stored in vessels and exclusively loaded to truck via hose connection. The facility is represented in Figure 7 below:

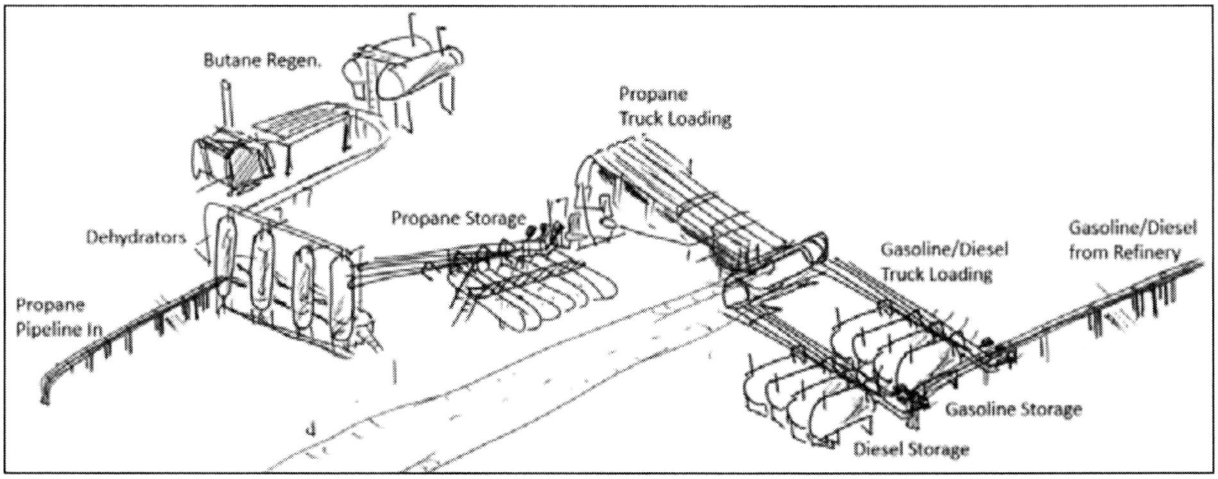

Figure 7. Terminal Exclusively Between Non-Pipeline Modes of Transportation

Because the propane system is exclusively between pipeline and non-pipeline and the gasoline/diesel system is exclusively between non-pipeline modes of transportation, the equipment/lines are covered by PSM, not PHMSA. The pipeline coming into the facility would be covered by PHMSA up until the facility's pressure regulator. Because the butane used in the regeneration of the dehydrator desiccant is not in transportation, the butane regeneration system would not be covered by PHMSA and would therefore be considered under the jurisdiction of PSM. Additionally, the truck loading would be covered by PSM; assuming the truck is not stored at the facility, it is covered by DOT. The governing regulations are presented in Figure 8 below:

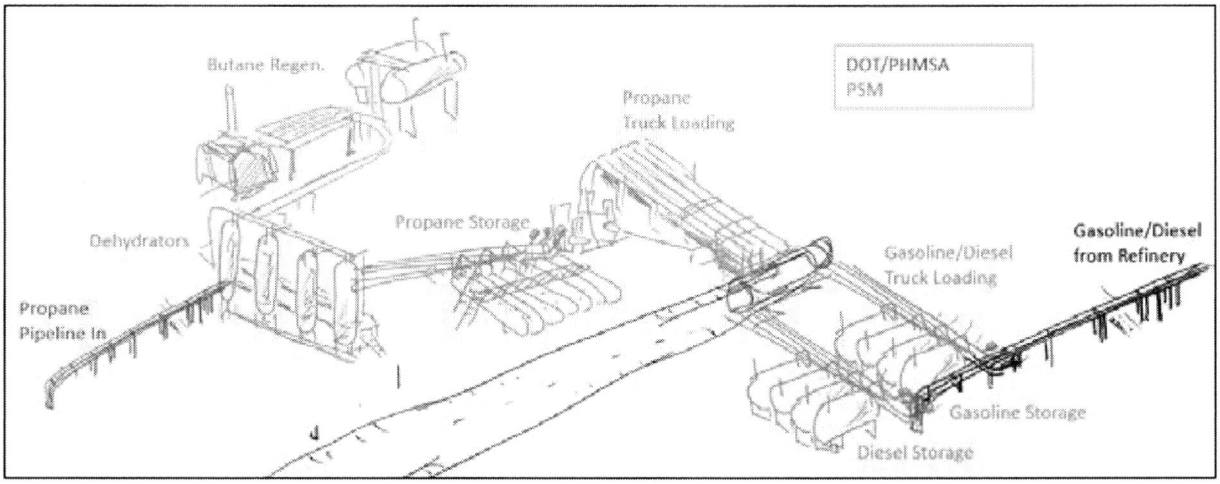

Figure 8. Governing Regulations of Equipment at a Terminal Exclusively Between Non-Pipeline Modes of Transportation

Now, that the governing regulations have been defined, the facility must define the processes at the facility by evaluating the interconnections of equipment and co-located equipment. The dehydrators, propane storage, and propane truck loading are all interconnected with each other. The butane regeneration system is interconnected with the propane system via the dehydrators.

This interconnected propane/butane system will be referred to as PSM Process 1. The gasoline storage and truck loading is interconnected, as is the diesel storage and truck loading. The gasoline system and the diesel system are co-located. This co-located gasoline/diesel system will be referred to as PSM Process 2. PSM Process 1 and PSM Process 2 are not co-located since they are separated by a road and a fire/blast from one is not expected to affect the other. The resulting processes are represented in Figure 9:

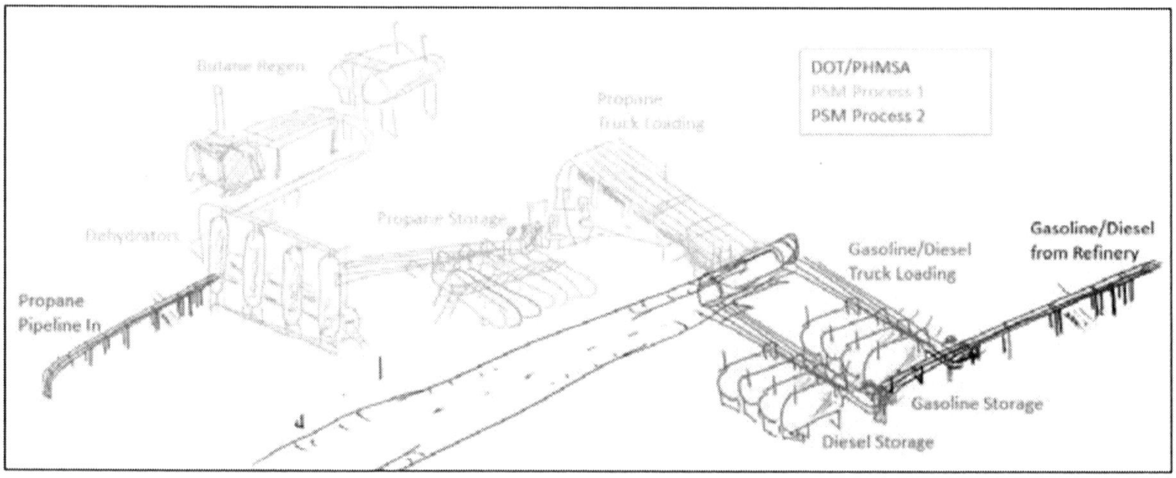

Figure 9. PSM Processes at a Terminal Exclusively Between Non-Pipeline Modes of Transportation

The final evaluation step is to calculate the inventory of the highly hazardous chemical in each piece of equipment and aggregate the quantities in each process. Then, compare the resulting quantity to the threshold quantity defined by PSM. The individual equipment inventories are shown in Figure 10 below and the aggregation is represented in Table 4. Note that diesel has flashpoint above 100 °F and is therefore not considered a flammable liquid under the PSM regulation.

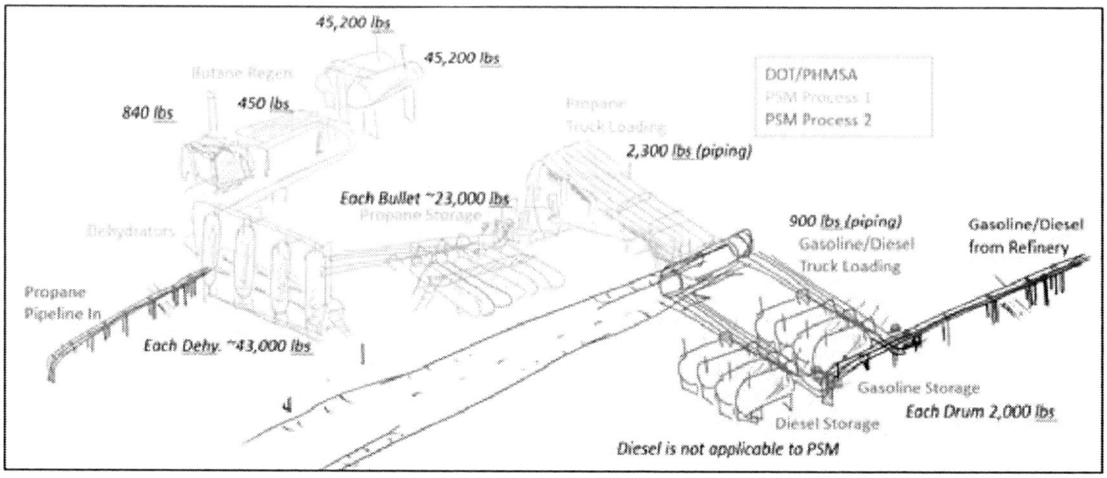

Figure 10. HHC Inventories in a Terminal Exclusively Between Non-Pipeline Modes of Transportation

Table 4. Aggregation of HHCs in a Terminal Exclusively Between Non-Pipeline Modes of Transportation

Flammable Liquid	Process 1 Inventory (lbs)	Process 2 Inventory (lbs)
Propane	269,000*	---
n-Butane	91,690	---
Gasoline	---	8,900
TOTAL	360,690	8,900

*Assumed that all six (6) storage bullets are in use and only three (3) dehydrators are in drying mode while one (1) is in regeneration mode.

Therefore, Process 1 is applicable to PSM while Process 2 is not, as reflected in Figure 11 below.

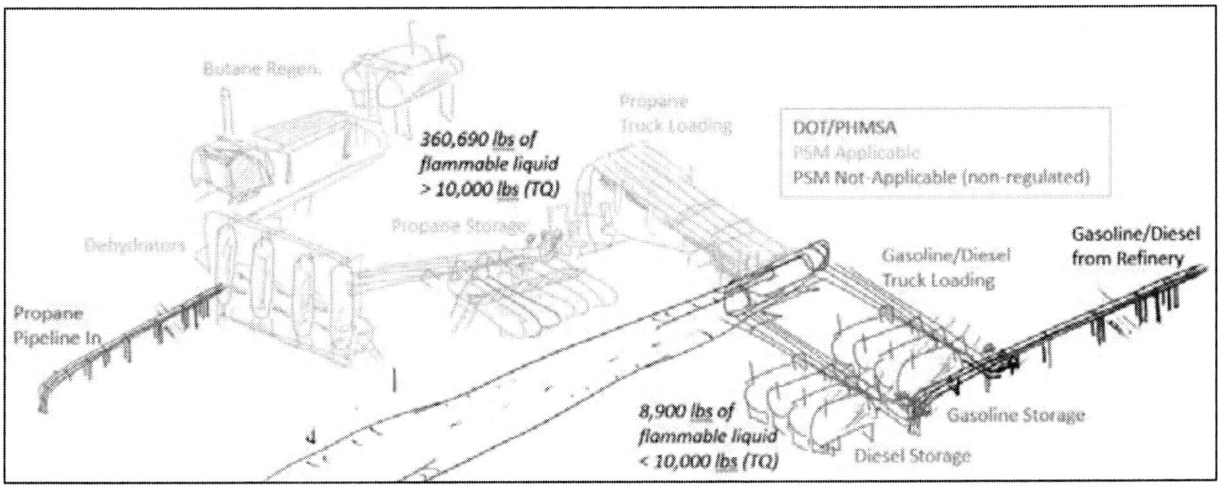

Figure 11. PSM Applicability at a Terminal Exclusively Between Non-Pipeline Modes of Transportation

6 Conclusion

When determining the jurisdictional boundaries for a midstream facility, a complete understanding of the history, purpose, scope, exemptions, and interpretations for <u>each</u> regulation is vital to setting the correct boundaries. By pulling all aspects of each regulation together in a single location, this paper can be referenced when evaluating the boundaries at a midstream facility.

7 Citations

[1] Statement by the President upon Signing the Natural Gas Pipeline Safety Act of 1968. August 13, 1968
[2] 49 U.S.C. § 60102(a)(1-2)
[3] PHMSA's Mission, Available at https://www.phmsa.dot.gov/about-phmsa/phmsas-mission Accessed on February 19, 2018.
[4] Letter from Richard E. Fairfax, Director of Enforcement Programs to Joanne B. Linhard, ORC Worldwide Consultant (October 7, 2008).
[5] 49 U.S.C. § 60102(a)(1)
[6] 49 U.S.C. § 60102(a)(21)
[7] 49 U.S.C. § 60102(a)(22)
[8] 49 U.S.C. § 60102(a)(3),(5)
[9] 49 C.F.R. § 195.1(b)(8)
[10] 49 C.F.R. § 195.1(b)(9)(ii)
[11] OSHA's Mission, Available at https://www.osha.gov/about.html Accessed on February 19, 2018.
[12] 29 C.F.R. § 1910.119 *Purpose*
[13] 29 U.S.C. § 653(a)
[14] 29 U.S.C. § 653(b)(1)
[15] Letter from Dorothy L. Struck, Acting Assistant Secretary to Michael Baly, III, American Gas Association President (October 30, 1992).
[16] Chao v. Mallard Bay Drilling, Inc., 534 U.S. 235 (2002)
[17] 29 C.F.R. § 1910.119(a)(1)
[18] 29 C.F.R. § 1910.119(a)(1)(ii)
[19] 29 C.F.R. § 1910.119(b)
[20] Memorandum of Understanding, OSHA and DOT (April 27, 1992 and May 2, 1992).
[21] Letter from Dorothy L. Struck, Acting Assistant Secretary to Michael Baly, III, American Gas Association President (October 30, 1992).
[22] Letter from Richard Fairfax, Director of Enforcement Programs to Richard Runyon, SCIENTECH, Inc. Senior Executive Consultant (December 9, 1998).
[23] Letter from Richard Fairfax, Director of Enforcement Programs to Richard Runyon, SCIENTECH, Inc. Senior Executive Consultant (December 9, 1998).
[24] Letter from Richard Fairfax, Director of Enforcement Programs to Ty J. Smith, Lesair Environmental, Inc. (February 16, 2005).
[25] Letter from Wesley Christensen, ONEOK NGL Pipeline, L.P. Senior Vice President, NGL Operations to John Gale, United States Department of Transportation, Director, Standards and Rulemakings (August 18, 2011).
[26] ONEOK NGL Pipeline, L.P. Final Order, CPF No. 3-2013-5014 (2016)
[27] Letter from Alan Mayberry, PHMSA Chief Counsel to Wesley Christensen, ONEOK NGL Pipeline, L.P. Senior Vice President, NGL Operations (November 28, 2012).
[28] Letter from Vanessa Allen Sutherland, PHMSA Chief Counsel to Vince Murchison, SNR Denton, US LLP (August 8, 2012).
[29] Letter from Vanessa Allen Sutherland, PHMSA Chief Counsel to Vince Murchison, SNR Denton, US LLP (August 8, 2012).
[30] ONEOK NGL Pipeline, L.P. Final Order, CPF No. 3-2013-5014 (2016)
[31] PHMSA Gas and Liquids Technical Advisory Committee, presented by Linda Daugherty, *Processing Facilities and PHMSA Jurisdiction – Midstream Regulatory Jurisdiction* (2014).
[32] PHMSA Subcommittee on Midstream Facility Safety, presented at the Gas Pipeline Advisory Committee and the Liquid Pipeline Safety Standards Committee, *Delineation and Regulatory Oversight of "Processing" – Oil and Gas Midstream Facilities* (August 25, 2015).
[33] PHMSA Subcommittee on Midstream Facility Safety, presented at the Gas Pipeline Advisory Committee and the Liquid Pipeline Safety Standards Committee, *Delineation and Regulatory Oversight of "Processing" – Oil and Gas Midstream Facilities* (August 25, 2015).

[34] PHMSA Subcommittee on Midstream Facility Safety, presented at the Gas Pipeline Advisory Committee and the Liquid Pipeline Safety Standards Committee, *Delineation and Regulatory Oversight of "Processing" – Oil and Gas Midstream Facilities* (August 25, 2015).
[35] PHMSA Subcommittee on Midstream Facility Safety, presented at the Gas Pipeline Advisory Committee and the Liquid Pipeline Safety Standards Committee, *Delineation and Regulatory Oversight of "Processing" – Oil and Gas Midstream Facilities* (August 25, 2015).
[36] 49 C.F.R. § 195.1(b)(9)(ii)
[37] PHMSA Subcommittee on Midstream Facility Safety, presented at the Gas Pipeline Advisory Committee and the Liquid Pipeline Safety Standards Committee, *Delineation and Regulatory Oversight of "Processing" – Oil and Gas Midstream Facilities* (August 25, 2015).
[38] PHMSA Subcommittee on Midstream Facility Safety, presented at the Gas Pipeline Advisory Committee and the Liquid Pipeline Safety Standards Committee, *Delineation and Regulatory Oversight of "Processing" – Oil and Gas Midstream Facilities* (August 25, 2015).
[39] Letter from H. Berrien Zettler, Deputy Director of Compliance Programs to John Anicello, Airco Gases Technical Manager, Regulatory Affairs (February 15, 1994).
[40] Letter from John B. Miles, Director of Compliance Programs to Joseph R. Schneider, American Institute of Chemical Engineers Professional Development Chairman (July 11, 1994).
[41] 49 C.F.R. § 171.1(c)
[42] 49 C.F.R. § 171.1(c)(2),(3)
[43] 49 C.F.R. § 171.1(c)(4)
[44] 49 C.F.R. § 171.1(c)(4)(ii)
[45] 49 C.F.R. § 171.1(d)(3)
[46] 49 C.F.R. § 171.1(d)(2)
[47] Letter from Roger A. Clark, Director of Compliance Programs to Gene F. Tappan, Reskitt & Coleman, Inc. Regulatory Affairs Manager (September 9, 1993).
[48] Letter from Richard E. Fairfax, Director of Enforcement Programs to Joanne B. Linhard, ORC Worldwide Consultant (October 7, 2008).
[49] Letter from Richard E. Fairfax, Director of Enforcement Programs to Joanne B. Linhard, ORC Worldwide Consultant (October 7, 2008).
[50] Interim Final Rule on Pipeline Safety: Safety of Underground Natural Gas Storage Facilities, 81 Fed. Reg. 243 (December 19, 2016).
[51] Interim Final Rule on Pipeline Safety: Safety of Underground Natural Gas Storage Facilities, 81 Fed. Reg. 243 (December 19, 2016).
[52] Letter from Richard E. Fairfax, Director of Enforcement Programs to Joanne B. Linhard, ORC Worldwide Consultant (October 7, 2008).
[53] Interim Final Rule on Pipeline Safety: Safety of Underground Natural Gas Storage Facilities, 81 Fed. Reg. 91861-91873 (December 19, 2016).
[54] Interpretation of OSHA's Standard for Process Safety Management of Highly Hazardous Chemicals, 72 Fed. Reg. 31453-31457 (June 7, 2007).
[55] Interpretation of OSHA's Standard for Process Safety Management of Highly Hazardous Chemicals, 72 Fed. Reg. 31453-31457 (June 7, 2007).
[56] Interpretation of OSHA's Standard for Process Safety Management of Highly Hazardous Chemicals, 72 Fed. Reg. 31453-31457 (June 7, 2007).
[57] Letter from Raymond E. Donnelly, Director of Office of General Industry Compliance Assistance to Grahame Barker, Baker Performance Chemical Incorporated (April 28, 1994).
[58] Letter from John B. Miles, Director of Compliance Programs to J.B. Evans, Union Carbide Corporation (September 27, 1995).
[59] Memorandum from John B. Miles, Director of Compliance Programs, Akzo-Nobel Chemicals – Limits of a Process (February 28, 1997).
[60] Interpretation of OSHA's Standard for Process Safety Management of Highly Hazardous Chemicals, 72 Fed. Reg. 31453-31457 (June 7, 2007).
[61] Interpretation of OSHA's Standard for Process Safety Management of Highly Hazardous Chemicals, 72 Fed. Reg. 31453-31457 (June 7, 2007).
[62] Letter from Thomas Galassi, Director of Enforcement Programs to Paul Crook, Fluro-Seal International, L.P. Safety and Health Manager (May 2, 2014).

[63] Interpretation of OSHA's Standard for Process Safety Management of Highly Hazardous Chemicals, 72 Fed. Reg. 31453-31457 (June 7, 2007).
[64] Letter from John B. Miles, Jr., Director of Compliance Programs to Susan Kunst, Boushell Law Offices of McKenna and Cuneo (July 18, 1994).
[65] Letter from John B. Miles, Jr., Director of Compliance Programs to Susan Kunst, Boushell Law Offices of McKenna and Cuneo (July 18, 1994).
[66] Letter from H. Berrien Zettler, Deputy Director of Compliance Programs to John Anicello, Airco Gases Technical Manager, Regulatory Affairs (February 15, 1994).
[67] Letter from H. Berrien Zettler, Deputy Director of Compliance Programs to John Anicello, Airco Gases Technical Manager, Regulatory Affairs (February 15, 1994).
[68] Letter from John B. Miles, Jr., Director of Compliance Programs to James B. Evans, Union Carbide Corporation (September 14, 1995).
[69] 29 C.F.R. § 1910.119(a)(1)(ii)(B)
[70] Secretary of Labor v. Meer Corporation., OSHRC 95-0341 (1995)
[71] Memorandum from John B. Miles, Jr., Director of Compliance Programs, Coverage of Stored Flammables under the Process Safety Management Standard (May 12, 1997).
[72] Memorandum from John B. Miles, Jr., Director of Compliance Programs, Coverage of Stored Flammables under the Process Safety Management Standard (May 12, 1997).
[73] Process Safety Management and Prevention of Major Chemical Accidents, 78 Fed. Reg. 73756-73768 (December 9, 2013).
[74] 29 C.F.R. § 1910.119(a)(1)(ii)(A)
[75] Letter from Roger A. Clark, Director of Compliance Programs to Garry Stubblefield, Petrocon Engineering, Inc. Senior Technician (January 8, 1993).
[76] Letter from John B. Miles, Jr., Director of Compliance Programs to Steve Neville, Analyst Associates Safety and Health Program Director (February 13, 1995).
[77] OSHA PSM
[78] Letter from John B. Miles, Jr., Director of Compliance Programs to Tim Bundy, State of Alaska DOL/OSHA (May 29, 1998).
[79] Letter from Richard E. Fairfax, Director of Enforcement Programs to Ty J. Smith, Lesair Environmental Inc. (February 16, 2005).
[80] Letter from John B. Miles, Jr., Director of Compliance Programs to Tim Bundy, State of Alaska DOL/OSHA (May 29, 1998).
[81] Letter from Richard E. Fairfax, Director of Enforcement Programs to Ty J. Smith, Lesair Environmental Inc. (February 16, 2005).

512337 Game Based Learning; Reinforcing Process Safety

Abstract:

The concept of process safety has been around a long time but a spate of major accidents in the last decade has further emphasized the need for better and wider training and education. This presentation will outline a novel approach to develop process safety knowledge and competence through Game Based Learning (GBL).

Making learning stick is a universal problem but, based on research that suggests people remember 90% of what they 'do', GBL takes your message out of PowerPoint and broadcasts it interactively to the trainee. A GBL session typically lasts 30-60 minutes, is hands-on and, by using every-day analogies for technical subjects and grounding them in the real world, it is accessible to all personnel across an organization. Indeed, GBL for major accident awareness has been run for school children as part of Science, Technology, Engineering and Mathematics (STEM) teaching. GBL comes into its own in high hazard industries where, generally speaking, improvement is about small margins. This presentation will touch upon a number of games developed by Risktec in recent years, covering common process safety themes such as layers of protection and safety-critical systems. These games can be used 'off-the-shelf' or tailored to suit a specific application, operation or asset.

539111 Tying-It All Together

Abstract:

Many facilities view process safety as a collection of elements and treat them independently. Process Hazard Analyses are conducted independent of safe upper and lower limits or operating limits. Management of Change procedures are written independent of the Mechanical Integrity program. Sometimes various process safety functions are performed by different departments within an organization without sufficient consultation. For example, alarm rationalization may be headed up by the control systems engineering department while operating limits are being set separately by Operations and Process Engineering. This leads to inefficiencies in implementing a process safety program. It can also lead to conflicting data and confusion within the organization.

The purpose of this presentation is to outline ways to tie various elements of a process safety program together into an integrated system. The relationship between operating limits, the various phases of operating procedures, and safe upper and lower limits will be described. Tying alarm rationalization into this mix will be discussed. The relationship between the safety systems description requirement under Process Safety Information and the safety systems description requirement under Operating Procedures will also be discussed. By using some of the techniques presented, a more integrated process safety program may be achieved.

The Uses and Users of PHA/HAZOP Results

William Bridges, Jeff Thomas, Paul Casarez
Process Improvement Institute, Inc. (PII)
1321 Waterside Lane
Knoxville, TN 37922
Phone: (865) 675-3458
Fax: (865) 622-6800
e-mail: wbridges@piii.com
e-mail: jthomas@piii.com

Prepared for Presentation at
American Institute of Chemical Engineers
2018 Spring Meeting and 14th Global Congress on Process Safety
Orlando, Florida
April 22 – 25, 2018

The Uses and Users of PHA/HAZOP Results

William Bridges, Jeff Thomas, Paul Casarez
Process Improvement Institute, Inc. (PII)
1321 Waterside Lane
Knoxville, TN 37922
Phone: (865) 675-3458
Fax: (865) 622-6800
e-mail: wbridges@piii.com
e-mail: jthomas@piii.com

Keywords: PHA, HAZOP, LOPA, SIS, Mechanical Integrity, Fire risk scenarios, safety critical equipment, operating procedures, safeguards, independent protection layers, IPLs, trouble shooting guides

Abstract

Process Hazard Analyses (PHAs) performed using methods such as HAZOP and What-if that are augmented by checklists, have become well establish as a core for understanding risk in a hazardous chemical process and other processes. Some see the PHA results as an end to them itself. But, the real benefit of performing a PHA lies in its usefulness within all aspects of controlling risk day-to-day.

This paper explains the different uses of the PHA results and who uses the PHA results. It charts the path for extracts of the PHA results (including the formal PHA report) to the rest of process safety implementation and process safety control. The many uses, intended from the start of hazard evaluations in the 1960s or discovered years later, will surprise many in the industry. Knowing the uses will help you implement process safety thoroughly and more efficiently, and this knowledge of the uses will change the amount of effort you put into the documentation of the PHA results.

Background

The increasingly complex and technical demands of process safety management (PSM) have placed a large demand on existing resources across the process industries. Success requires recognition of the synergies between process safety activities. Process hazard analyses (PHAs) are the "heart" of process safety because it identifies all process safety accident scenarios for the process and documents these scenarios. If the documentation in the PHA is clear enough and thorough enough, then it can provide critical input to the rest of the process safety elements.

1. The PHA report

Many consider the PHA report to be "The PHA". Actually, the brainstorming meeting with the team is much more important (100 times more important) to the company than the report itself because the brainstorming meeting is where the team uncovers accident scenarios and determines if the risk is tolerable. If the scenarios are not found, then organization will not know if it has sufficient safeguards against these scenarios.

The report is a record of the meeting and the resulting recommendations for lowering the risk. It is a short summary of the team discussions (10,000 words per scenario discussion that are condensed to 100-200 words per scenario).

Purpose and benefits of a thorough report

The report serves several key functions listed below:

- Provides a record of the meeting discussion, including a list of summarized accident scenarios discussed during the meeting
- Lists the recommendations for lowering the risk to tolerable levels (each of these recommendations must be resolved)
- Is one of the basis of judging the quality of the PHA meetings for internal and external compliance evaluations
- Is a valuable input to the other process safety activities and systems ← **Most of the paper will focus on this as the main purpose of the PHA report**

Figure 1 illustrates the relationship of PHAs to the rest of process safety

Writing Effective Recommendations

One key component of every PHA report, including executive summary reports, is the recommendation section. Recommendations help management decide how to lower the risk to tolerable levels. Recommendations many times feed directly into the engineering of the process, such as by specifying the limit scenario for sizing a relief valve or by judging when additional independent protection layers (IPLs) are needed.

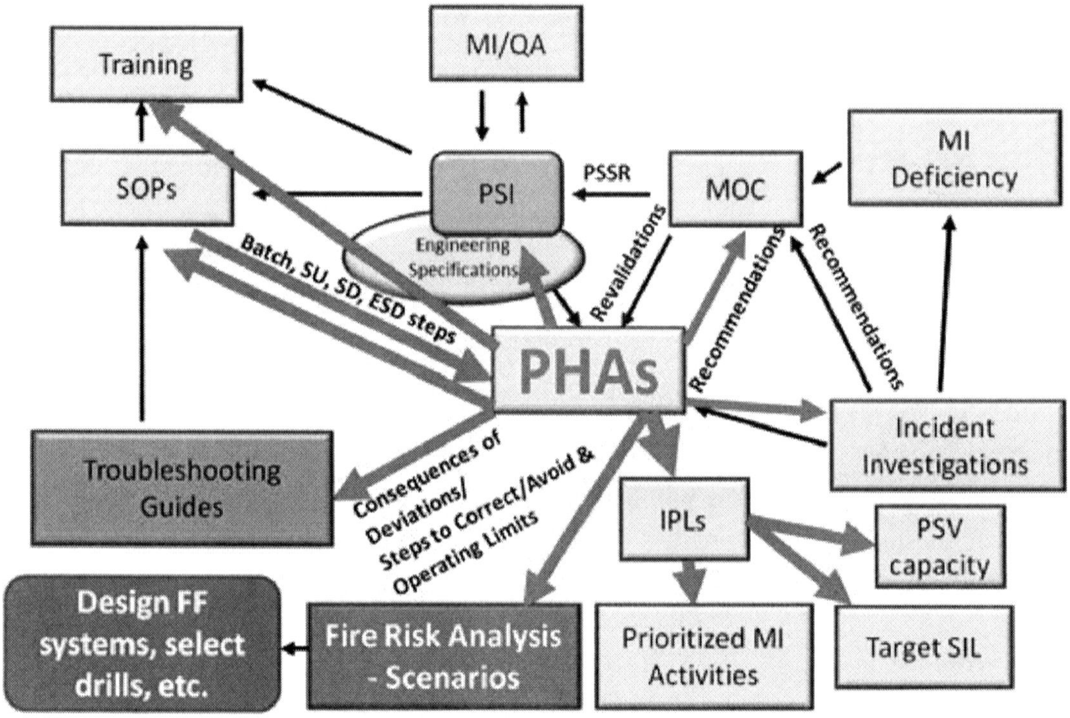

Figure 1: The Relationship of PHAs to the Rest of Process Safety Activities and Systems
© Process Improvement Institute, Inc. (2013-2018)

PII team leaders and scribes ONLY develop recommendations to bring the risk of a scenario to a tolerable level, so our PHA teams believe that all of their recommendations are necessary to achieve tolerable risk. However, not all recommendations will be accepted by management; they have the right and responsibility to reject recommendations that will not lower risk effectively, or to reject recommendations if the risk is already low enough without the recommended improvement. Therefore, it is best to write the recommendation in a style that allows management flexibility in deciding what will be implemented.

The best style for writing a recommendation is shown below. Always begin with a general statement of the concern so that management can address it in the best way possible. Then, provide management with a listing of specific suggestions (if possible)

Recommendation N°: Consider… (state general concern). Otherwise… (state the consequence that could occur). The team considered… (list the existing safeguards). Specific suggestions by the PHA team include ….

The *Otherwise* sentence is optional. If the existing safeguards are not listed, then management may think the team missed these safeguards and may judge that the exiting risk is tolerable. Specific suggestions from the team are listed without rating. (Example: *"Provide an independent high level switch to shut down the feed pump."*)

Report Content

In order to decide how to structure the report you must first decide what purpose the report will serve. If the report is strictly for compliance then a very thin report will typically suffice. If the report will be used to aid in the development of trouble-shooting guides or to support other process safety activities and management systems, then the clarity and quality of the report are very critical.

Introduction/Scope

This section should provide:

- The purpose of the PHA (e..g, to satisfy requirements of OSHA PSM, company standard XXX,)
- Whether the PHA is an initial PHA, a revalidation, or a re-do of a previous PHA
- What phase of the life cycle of the process (e.g., detailed design, pre-commission, post-commission, ongoing operations)
- The scope of the process(es) covered (including any procedures reviewed as part of PHA of non-routine modes of operation)
- Date(s) the PHA was conducted
- Name and position/title of team members present during the meeting(s)
- Any special circumstances affecting the scope, etc.

Method Used

This section should describe the method (e.g., HAZOP, What-if, FMEA) that was used to complete the PHA and a brief description of the method itself.

Executive Summary

This section should provide:

- Summary of recommendations perhaps with a table breaking down the type and number of recommendations if safety, environmental, and operability issues were considered.

PHA Tables

This section should include final copies of each of the completed PHA analysis tables (in the best format for each chosen method, HAZOP, What-if, 2 Guideword HAZOP of steps, etc.)

Appendices

Any documents that were used to support the PHA should be referenced. These may include:

- Previous incidents
- Checklist tables
- MOCs
- Drawings, P&IDs, etc. (ensure the same revisions used during the meeting are attached/referenced in the appendix)
- List of procedures (with revision number) used

To produce a report of good quality and excellent level of detail takes about the same amount of time after the PHA meeting as the meeting itself takes. But typically, only the leader and scribe write the report. The rest of the team members dedicate their time to providing input during the meetings and gathering data to close open items. The following table shows the typical responsibilities of the leader and scribe in preparing a high quality PHA report.

Table 1. Leader/Scribe report responsibilities

Scribe	Leader

Finish tables
• convert rough notes to words
• move info to the correct location in tables
• try to ensure consistency of words, links, etc.
	Finish recommendations
	• convert notes to words
	• add phrases to explain "why"
	• embellish if necessary
	• issue Preliminary Report of Recommendations
	
	Draft the rest of the report
	• introduction/scope
	• method used
	• executive summary
	• appendices
	  o previous incidents
	  o checklist tables
	  o list of documents
	  o etc.

Review leader's work	Review scribe's work
	Compile/issue Full PHA Report

Steps for completing the report

- Prepare summary of recommendations and have the team and management review
- Management can already assign responsibility and estimate completion schedule, or assign completion date for each recommendation (optional at this point)
- Draft report
    - Formalize logistical notes (i.e., attendance records, list of drawings, list of documents)
    - Develop detailed recommendations
    - Develop detailed hazard review notes/data (HAZOP or FMEA tables)
- Have draft report reviewed (distribute for employee review – optional at this time)
- Make corrections as necessary
- Finalize PHA report and obtain signoffs, distribute, and/or file official copies
- Distribute PHA report to affected employees for comment/review (if not done at draft stage). Obtain written proof of employee review.

2. Uses and Users of PHA Results (*with EXAMPLES*)

As stated earlier and as illustrated in Figure 1, there are many users of the results of the PHA. Table 2 is a partial listing of the users and uses, in relatively order of importance (the ordering is based on the combined experience of PII, who have helped more than 50 organizations implement process safety and who have led and documented thousands of PHAs of entire units or plants.

Table 2. The Primary Users and Uses of PHA/HAZOP Results

Rank	User	Use
1	Process engineering, Project Engineering, Designers, Instrumentation, EPC, etc.	▫ Identifying the Independent Protection Layers (IPLs) needed (including SIFs) ▫ Identifying the target SIL (PHA = SIL Study) ▫ Finding the limiting case for PSV sizing ▫ Vent, duct, flare, and dike sizing
2	Mechanical integrity - inspection, test, PM plans & schedules & procedures	▫ List of Initiating Events (IEs) and IPLs (all safety critical equipment) ▫ Information on component reliability or availability needed to reach the target risk reduction factor
3	Operations department, including operating procedures and focusing of training and drills	▫ List of Human IEs and procedure and human factors aspects to focus on ▫ List of Human IPLs that will need trouble-shooting guides and annual drills per operator ▫ Beginning portions of Trouble-shooting Guides (the procedures for responding to critical deviations)
4	Fire protection design and fire-fighting plans	Identifying fire scenarios for ▫ Fire-fighting system design; fire water capacity ▫ Fire-fighting scenarios for active drills ▫ Gas/fire detection requirements
5	Management of Change	▫ PHA describes current risk control envelope ▫ PHA is one basis to identify and document changes
6	PHA Revalidation	▫ Each PHA Revalidation builds upon the prior PHA
7	Incident Investigations	▫ PHAs identify possible accident scenarios to check (Did the initiating predicted occur? Did one or more IPLs fail?)
8	Auditors and quality assurance checks	▫ Did the PHA address the requirements ▫ Did the organization user results per listing above

A small fraction of PHA/HAZOP studies and related PHA/HAZOP reports meet the needs of all of the users listed above. Some people may say that is it possible for the PHA and report to meet all of the needs. But the authors and other experienced PHA

Leaders have been providing PHAs and reports that meet these needs for decades.

An organization that ensures their PHAs meet the needs of these users is one that understands how to control process safety. PII tests for this linkage during PSM audits. One way we do this is to extract about a dozen causes and a dozen safeguards (those that either are or appear to be IPLs) and then we check to see if these are in the list of Safety Critical Elements (SCE) and check to make sure that site is inspecting, testing, and maintaining each of these components. If not, then the site has a gap or perhaps the site fails to understand the relationship between the PHA and the rest of process safety.

The next few sections are examples of how to use the PHA to provide the information stated in Table 2 or how this information may be extracted from the PHA results and used in controlling risk at the site.

EXAMPLE 2-A: Determination of IPLs and SIFs during PHA/HAZOP as feed into Engineering/Design and into Mechanical Integrity[1]

One of most valuable outcomes of starting LOPA in the mid-1990s was the crystallization of the qualitative definition of an IPL. If the PHA/HAZOP leader was competent in the definition of an IPL, we found that IPLs could be just as easily identified in a PHA/HAZOP as in a LOPA. This was a significant outcome, since IPLs (along with the Initiating Events, IEs) are what we need to focus on to maintain tolerable risk for each scenario. In other words, we can focus our reliability/maintenance and operational efforts on IPLs to conserve resources while maximizing control of risk. SIFs are just one type of IPL and we found the same was true for identifying SIFs and setting their SILs. Also, as mentioned earlier, ANSI/ISA 84.00.01-2004 allows SIL to be set qualitatively.[2] *COMPETENCY*, as always, is Critical; the path to competency includes learning the rules for using this approach, so the rules are explained first in this paper.

Intentionally Achieving COMPETENCY in the Qualitative Definition of an IPL and SIF

First, the PHA Leader must become fully competent in the rules and descriptions of SIFs (of various SILs) and other IPLs. This is discussed in details in other papers by PII. Many PHA leaders are not competent in even how to conduct a PHA; in fact, about 90% of the PHAs reports that we have reviewed around the world are woefully deficient, especially with respect to finding scenarios during startup, shutdown, and online maintenance. Per the authors' experience, the path to the necessary competency is typically:

- Already be an experienced PHA leader, trained in all PHA methods, and capable of applying these methods to all modes of operation and capable to make sound qualitative judgments, along with the PHA team members, on when the number and type of IPLs is sufficent to control the risk. Achieving full competency as a PHA leader may require some remedial training on how to lead PHAs of startup, shutdown, and online modes of operation; or remedial training on how to uncover and discuss all plausible damage mechanisms.
- Attend a LOPA course to learn the basics of IPLs (including SIFs) as described in the previous sections. The key on qualitative risk judgment is to know when there are enough IPLs for the accident scenario under review.
- Get coaching (by someone already competent) during actual PHAs to learn how to help a team make judgments if safeguards meet the definition of an IPL (or SIF) or not and also on if there are enough IPls for the accident scenario (risk judgment). We know from thousands of PHAs over the past decades that a PHA team can make excellent risk judgments > 95% of the time, which also means that the IPLs and SIFs can be clearly identified > 95% of the time.
- Achieve competency, in the opinion of the competent coach on the skills above.
- In addition to the PHA team leader competency, the PHA team has enough understanding of either qualitatively risk judgment or LOPA risk judgment– just in time training by the PHA leader (we tend to accomplish this training across the first 5-10 accident scenarios we discuss)

Using the Qualitative Definition of an IPL and SIF

The competent PHA/HAZOP leader can now guide the PHA/HAZOP team through the following thought processes:

- IF (1) the safeguard meets the definition of an IPL and (2) if the team believes (qualitatively) this safeguard is critical to control risk to tolerable level (qualitatively),then add the designator " – IPL" to the right of the safeguard text. If the safeguard is not going to be labeled an IPL, then it can be run to failure; unless the safeguard supports an IPL, such as when a sight glass supports an LAH used in an Human Response IPL, in which case the sight glass will have some PM (such a periodic cleaning of the sight glass).

- IF an instrument is already in the ESD system or SIS and qualitatively meets the archetecture of an SIL 1, or SIL 2, or SIL 3, and also meets the definitions/rules for an IPL, then add the "- SIL-1" (or SIL-2, or SIL-3) to the right of the text.

Table 3: Excerpts from Petrochemical Process PHA at SS-TPC[3]

No.: 2 XXXX storage spheres xxx-T-XX A/B/C/D/E/F/G/H/I/J/K/L (1 of 12)					
#	**Dev**	**Causes**	**Consequences**	**Safeguards**	**Recommendations**
2.1	High level	Too much flow to one sphere from XX Plant (through their pump; about 40 bar MDH)	High pressure (see 2.5)	High level SIF with level sensors voted 2oo2, to close inlet valve **- SIL 1** Overflow thru pressure equalization line to other spheres (through normally open [NO] valve) **- IPL**	
		Misdirected flow - Liquid from xxx Plant(s) to spheres (see 1.4)	Overpressure of sphere not credible from high level, for normal operating pressure of the column (which is 1.75 MPa), unless all spheres are liquid filled and then thermal expansion of the liquid could overpressure the spheres	High level SIF with level sensors voted 2oo2, to close inlet valve **- SIL 1** Overflow thru pressure equalization line to other spheres (through normally open [NO] valve) **- IPL** Spheres rated for 1.95MPa (19.5 Bar, approx) and the highest pressure possible from the column feeding the spheres is 1.75 MPa Level indication and high level alarm in DCS, used by operators to manually select which tank to fill - **Human Response IPL**	
			Overflow into the equalization line will interfere with withdrawal from the column, but this is an operational upset only		
			Excessive pressure on inlet of high pressure liquid pumps, leading to excess load on pumps and trip of pumps on high pumps, causing trips of xxx, xxx, etc. - significant operability issue		
2.2	Low level	Failing to switch from the sphere with low level in time (based on level indication)	Low/no flow - Liquid from spheres through high pressure product pumps to the vaporizer (see 4.2)	Level indication and low level alarm, inspected each year, per government regulation **(not IPL; part of the cause)** Feeding from two spheres at all times, so unlikely for BOTH spheres to have low level at the same time **- IPL** Two level indication from SIS level transmitter, with low level alarm, with more than 60 min available to switch tanks (SIF driven alarm and response) - **possible IPL, if action of the operator is quick enough**	Rec 4. Make sure the Human IPL of response to low level in all spheres and tanks is described in a trouble-shooting guide (like an SOP) and practiced once per year per unit operator. This will make this response a valid IPL.
			Low/no flow - Unqualified liquid from spheres back to Plant (see 6.2)		

Table 3 is an excerpt from a PHA report that illustrates the identification of a safeguard as IPLs or SIFs; and this also defines the target SIL needed.

For the PHA illustrated in Table 3, in some cases the SIF that was installed for protection against scenarios during continuous mode of operation did not protect against even more catastrophic and much more likely consequences during startup or online maintenance. For such situations, additional IPLs, including SIFs specific to startup or online maintenance, were recommended by the PHA team.

At this site in China, the non-human IPLs identified were entered into the computerized maintenance management system (CMMS) and maintained as critical features in reliability/maintenance systems. The human IPLs put on a schedule to test (using live drills) once a year. All of these activities are to ensure the IPLs/SIFs deliver the PFD anticipated, while still ensuring reliable operation/control by not causing too many spurious trips.

EXAMPLE 2-B: Developing Trouble-Shooting Guides from PHA/HAZOP Results [4]

Trouble-shooting guides are a special form of operating procedure; they are written for the actions that we want the operators to take to recover from a process deviation, *before an emergency situation occurs.* They are called guides (rather than procedures) since rarely can we predict the process conditions at the time the action is required (so a rigid procedure with exact ordering of steps is not possible to write). Trouble-shooting guides (and training and drills of the action) are required for any action that is considered a Human Response IPL. The Action Limit is what we show as the Min or Max in a Trouble-Shooting Guide. The action limit triggers the demand to use the trouble-shooting guide.

- The response is typically still possible, but it is time dependent. The time available is called the process safety time (PST). The operator must complete the diagnosis, make the necessary change(s), and make sure they are out of harm's way by the end of the Maximum Allowable Response Time (MART).
- We usually set an alarm or a pre-alarm to trigger this action.
- This is usually before the shutdown triggers (ESD occur automatically) or release points (PSV set points) are reached
- The Min and Max shown in the Trouble-shooting Guide table is not the absolute safety limit for a system, but is instead some value that leaves us some time to take action to prevent from reaching the absolute limits.

- There is still time to prevent or avoid the final consequence that could occur if we reach the ultimate limits of the process. Usually, we want the MART to be ½ or less of the PST, and we want MART > 10 minutes for trouble shooting in the field/plant and MART > 5 min for trouble-shooting only from the control room.

If the unit has a good PHA/HAZOP, then it is best to extract information from the HAZOP (or What-If) analysis tables to start the development of each troubleshooting guide. The guide is then finished with input from the process experts.

Table 4. Examples of Creating a Trouble-Shooting Guide from a HAZOP Table

HAZOP Table Entry	Trouble-Shooting Guide Entry
Cause: *Bypass valve is open or passing*	Make sure the bypass is tightly closed
Safeguard: *Isolation valves for the vessel*	Isolate the vessel, if necessary
Safeguard: *Relief valve*	Make sure the relief valve block valves/relief path are open

The key categories of information needed in a trouble-shooting guide are:

- IMMEDIATE ACTION (by system or by operator)
- DECIDE IF ALARM is REAL
- FINDING and FIXING the CAUSE
- FIX or BYPASS PROBLEM

The example below is of a trouble shooting guide for one critical alarm/action.

Figure 2

Trouble-Shooting Guide

Alarm or Indicator:	**PAL 4446 – Low Pressure Alarm for Suction of Organic Feed Pump 40-PM-18.445**

Action Limit:	5 kPa		
Consequence:	Possible pump seal failure, releasing or spraying organic waste into the berm.		
Process Area:	FB&D Incinerator; Liquid Organic Liquid Feed	**Oper. Mode:**	Normal
Drawing #s:	D-400-PI-013		

IMMEDIATE ACTION (by system or by operator)

- DCS should shut down the organic feed pump (40-PM-18.445).
- From the DCS display, MAKE SURE the organic feed pump is shutdown.
- HAVE the field operator check for leaks near the organic feed pump.
- IF there is a large leak/release, THEN use the ESD switch to shutdown the unit and then follow/complete the shutdown and isolation procedure, OPS-ESD-117.
- IF there is a minor leak or no leak, THEN:
    - COMPLETE the rest of the trouble-shooting,
    - and DECIDE how to contain the leak for now,

DECIDE IF ALARM is REAL

- From the DCS, CHECK the pressure and feed tank level trends. IF the trends indicate the alarm if valid, THEN continue with finding the cause or fixing or bypassing the problem.

FINDING and FIXING the CAUSE

- CHECK valves upstream of the organic feed pump to see if any are closed too far, including checking ESD valves.
- CHECK, by feel with hand, if the heat tracing is on; IF Not, then TURN ON or open heat trace valves
- MAKE SURE nitrogen to the pump seal is at the normal operating pressure.
- CHECK if the line if plugged or frozen (skill)

ABC Chemical Company Prepared by: *Printed copy of this procedure is good for one job task duration.*	OPS-76-TSG-233	Incinerator Unit Revised 8/24/2015 Printed 2/21/16 Page 1 of 2

Trouble-Shooting Guide

Alarm or Indicator:	**PAL 4446 – Low Pressure Alarm for Suction of Organic Feed Pump 40-PM-18.445**

- CHECK if the line if plugged or frozen (skill)
- CHECK if the level is actually low, use the Organic Feed Tank.
- IF the cause is low level in the feed tank, THEN resolve the problem if necessary based on cause that is found (skill)

FIX or BYPASS PROBLEM

- IF necessary, SHUT DOWN the Unit to allow fixing of the problem.
- FOLLOW the proper procedure to resolve problems (repair procedure, line clearing procedure, etc.)
- IF the decision is made to continuing operation without all equipment in normal condition, THEN:
  - FOLLOW Temporary Operating mode, if there is a temporary procedure already written for this possible problem/condition
  - FOLLOW MOC procedures to obtain approval for any non-standard temporary operating procedure or mode

END

Optimal presentation of trouble-shooting information.

Ideally (or ultimately) such trouble-shooting information should be imbedded in the basic process control system (DCS) so that the operators can access the information on the DCS screen "on demand," with the click of mouse or key.

EXAMPLE 2-C: Documenting Changes in Baseline PHA/HAZOP to streamline Next PHA Revalidation (Evergreen PHA)[5]

An evergreen PHA is the result of using the baseline PHA report to the fullest for documenting MOC risk review results and for updating for learnings from incidents, as they occur (not waiting for the 5 year cycle to complete). See the table on the next page.

(A.3) (Reviewed 2005) Measuring Level on Bulk Storage Tanks

Drawing: SP 03.201.02

Item	Deviation	Causes	Consequences	Safeguards	Action Items
33.7	Step 2 (old Step 4) to remove measuring stick and record level performed incorrectly		(1998) No consequences of interest (2005) Operator could fall resulting in serious bodily injury (2005) Resin spill could damage valve	(2005) Railing around work area on tanks	
33.9	Skip step 4 (old step 6) to replace cap on quick connect	Typical causes of procedural errors (see Table A.1)	Release of process material if preparing to unload a tank wagon into the tank; fire and explosion hazard affecting a large area	Continuous ventilation The tanks are located in a containment area Checklist Generic and inherent safeguards protecting against or mitigating process material releases (see Table A.2) (2005) Air Monitors in Bulk room	'98Rec 26. See recommendation 19, item 29.12 (1998)
33.10	Step 4 (old Step 6) to replace cap on quick connect and pull ears up and secure Velcro straps performed incorrectly (poor seal when cap is replaced)		(2005) Quick connect cap can pop off during the filling stage and flammable material can spill out of tank; fire hazard affecting a medium area Incident Report 2/12/03 - Bulk Room - 878 Resin leak from valve on bulk resin tank during resin tanker unloading - (A.9) (Added Reval 2005) Incident Investigation Reports and Spill Reports 1998-2005 (linked to 82.11) MOC 92-2005-04 - 4/1/05 Replace portable tank valve lever stainless actuator to oval actuator brass construction - (A.8) (Added Reval 2005) Management of Changes 1998-2005 (linked to 81.23)	(2005) Velcro straps on valves	

This approach requires a high quality and thorough baseline PHA. It uses codes, in brackets, to note which MOC or incident the new information came from.

So, if the organization intends to keep their baseline PHA evergreen, then the PHA Leaders and Scribes of MOC risk reviews need to learn how to find and edit the baseline PHAs. This approach results in the baseline PHA being up-to-date with each single or grouping of revisions. The approach also saves considerable time as each 5 year revalidation cycle as the bulk of the updating is already complete well in advance, while the information is still fresh in the minds of the PHA leader and scribe.

3. IMPLEMENTATION PATH FOR USING PHA RESULTS

The path forward is not complicated, but it is hard work and requires organization discipline, primarily meaning that management acts upon a long term vision and not short-term vision. The basic steps to achieve the goals using PHAs are:

- Develop the organizational requirements for the linkages from the PHA/HAZOP results as listed in Table 2. Develop the standard to document these requirements and establish the roles and responsivities and accountabilities in this standard.
- Develop the guidelines and rules for documentation of PHA/HAZOP results, as illustrated in this paper and related papers, so that the basic data required in Table 2 is available in the PHA/HAZOP results.
- Have PHA/HAZOP Leaders attain the competencies necessary to deliver the quality and thoroughness of results needed, especially for identifying the IEs and IPLs (including SIFs and SILs) as illustrated in this paper and as described in detail in other papers by the authors.
- Develop one or more guidelines for extracting the data from the PHA/HAZOP reports to use the rest of process safety activities and management systems
    - o Guideline for Creating Trouble-Shooting Guides from PHA/HAZOP results
    - o Guideline for Extract critical IPLs and IEs from PHA/HAZOP Results to Identify Critical Engineered Features (or Safety Critical Equipment)
- Have staff develop the competencies needed for the extractions and use mentioned above.
- Develop the capabilities to keep the baseline PHA/HAZOP evergreen for each Change, one the MOC is actually implemented.
- Periodically audit and improve these systems

4. CONCLUSIONS

For many organizations, PHA/HAZOP results are an untapped gold mine of information. If the right protocols for performing and documenting PHA/HAZOPs are put in place and if the rules and guidelines for using the vast data in the PHA/HAZOP results are established, then HUGE gains in synergy from the PHA/HAZOP results are possible. These are gains that were intended from the beginning of process safety to help prevent incidents

5. ACRONYMS USED

AIChE – American Institute of Chemical Engineers
CCPS – Center for Chemical Process Safety (of AIChE)
GCPS -- Global Congress of Process safety
HAZOP – Hazard and Operability; as in HAZOP Analysis or HAZOP Study
IE – Initiating Event
IPL – Independent Protection Layer
LAH – High level alarm
LOPA – Layer of Protection Analysis
MOC – Management of Change
PHA – Process Hazard Analysis
PSM – Process Safety Management
OSHA – US Occupational and Health Administration
SIF – Safety Instrumented Function
SIL – Safety Integrity Level
SIS – Safety Instrumented System

6. REFERENCES

1. "Identify SIF and Specify Necessary SIL, and other IPLs, as part of PHA/HAZOP – or - Why it is not necessary to 'Boldly Go Beyond HAZOP and LOPA'", W. Bridges and A. Dowell III, *12th Global Congress on Process Safety,* Houston, TX, AIChE, April 2016.
2. *Functional Safety: Safety Instrumented Systems for the Process Industry Sector - Part 3: Guidance for the Determination of the Required Safety Integrity Levels*; ANSI/ISA 84.00.01 Part 3, 2004.
3. "Implementation of Process Hazard Analysis at SINOPEC-SABIC Tianjin Petrochemical Company Ltd, China", Homoud Al-Maymouni, Yunzhong Gao (both from SS-TPC), and W. Bridges, *11th Global Congress on Process Safety,* Austin, TX, AIChE, April 2015.

4. "Best Practices for Writing Operating Procedures and Trouble-Shooting Guides" W. Bridges and R. Tew, *13th Global Congress on Process Safety,* San Antonio, TX, AIChE, March 2017.

5. "Best practices for PHA Revalidations," W. Bridges, R. Tew, M Massello, *14th Global Congress on Process Safety.* Orlando, FL, AIChE, April 2018.

Practical Application of Bowtie Analysis; Enhancing Traditional PHA

James Sneddon

Principal Engineer

Risktec Solutions (Canada) Ltd.

Suite 1600, 144 4th Ave SW

Calgary, AB T2P 3N4

Abstract

The benefits of using bowtie diagrams for risk management have been realised by organisations world-wide across a variety of business sectors. They provide a readily understandable visualisation of the relationships between the causes of business upsets, the escalation of such events to a range of possible outcomes, the controls preventing the event from occurring and the preparedness measures in place to limit the consequences. More importantly, the preventive and mitigative measures are linked to tasks, procedures, responsible individuals and competencies, thereby demonstrating the crucial connection between risk controls and the management system for assuring their ongoing effectiveness.

This paper, and accompanying presentation, draws on Risktec's unparalleled experience in applying the bowtie methodology and is intended to be of interest to those who are new to the technique and experienced users alike. With the growing popularity of bowtie in North America, this paper focusses on application of the methodology and describes in detail its practical uses and benefits as well as potential pitfalls. Of particular note is resolution of the following, commonly asked, question: "What does bowtie do that PHA doesn't?"

Keywords

Bowtie Analysis, Barriers, Safeguards, Process Hazard Analysis (PHA), HAZOP

Prepared for Presentation at

American Institute of Chemical Engineers

2018 Spring Meeting and 14th Global Congress on Process Safety

Orlando, Florida

April 22 – 25, 2018

Practical Application of Bowtie Analysis; Enhancing Traditional PHA.

Author: James Sneddon; Principal Engineer, Risktec Solutions (Canada) Ltd.

1 Introduction

The exact origins of the bowtie methodology are a little hazy. The earliest mention appears to be an adaptation from the ICI plc Hazan Course Notes 1979, presented by The University of Queensland, Australia[1].

Undoubtedly, the Royal Dutch Shell Group was the first major company to integrate fully the total bowtie methodology into its business practices[2,3,4] and is credited with developing the technique which is widely in use today. The primary motivation was to seek assurance that fit-for-purpose risk controls were consistently in place throughout all operations world-wide.

Use of bowties has spread between companies, industries, countries and from industry to regulator, some examples of which are provided below:

- UK defence industry[5];
- French government[6];
- UK Health and Safety Executive[7,8,9a,9b];
- Australian State Regulator[10];
- Land Transport Safety Authority of New Zealand[11a];
- Petroleum industry international associations[11b,11c];
- European aviation industry[12a], and;
- Banking industry[12b].

This paper aims to demonstrate the practical uses and benefits of this versatile tool which can be used to qualitatively assess and demonstrate control of all types of risk in many industries and business sectors.

2 The Bowtie Method

The bowtie method provides a readily understood visualization of the relationships between the causes of business upsets, the escalation of such events, the controls preventing the event from occurring, and the preparedness measures in place to limit the business impact (see Figure 1).

Figure 1. Bowtie Diagram

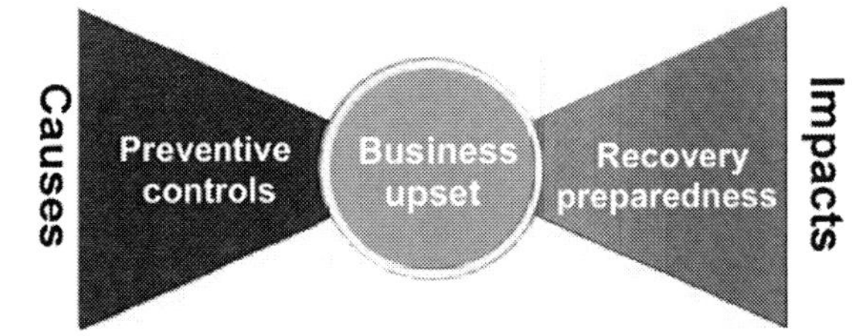

The method for building a bowtie diagram is well-documented[2,3,13,14,15,16,17], hence it is only covered briefly here.

In its most common use, the ultimate aim is to demonstrate control of health, safety and environmental (HSE) hazards; it is therefore necessary, firstly, to identify those hazards requiring bowtie analysis. Most companies involved in hazardous activities have an HSE management system[18,19,20] within which there will be formal procedures and/or guidance for identification of potential hazards and assessment of risks. Similarly, many companies have systems and standards for management of commercial, security, business continuity and corporate governance issues[21,22], to which the bowtie method is equally applicable.

Once hazards have been identified, the bowtie method can be applied to further assess risks and provide a framework for demonstrating their effective control. Typically bowties are developed by asking a structured set of questions which build up the diagram step-by-step (Figures 2 - 4):

Figure 2. Building the Bowtie – Steps 1 - 4

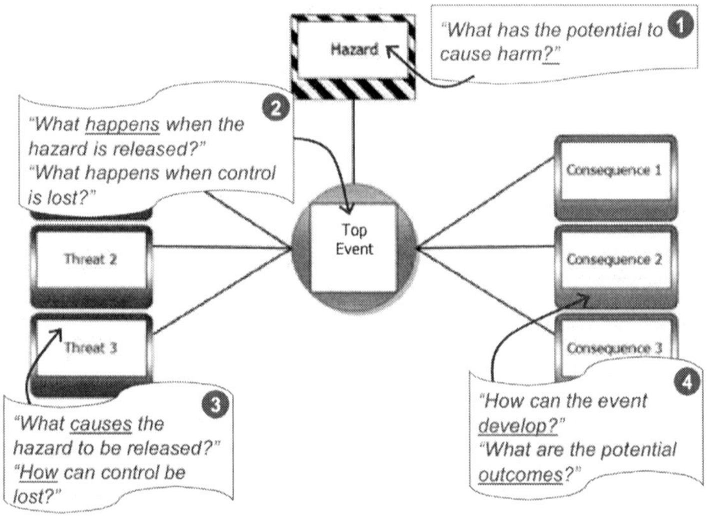

Figure 3. Building the Bowtie – Steps 5 – 8

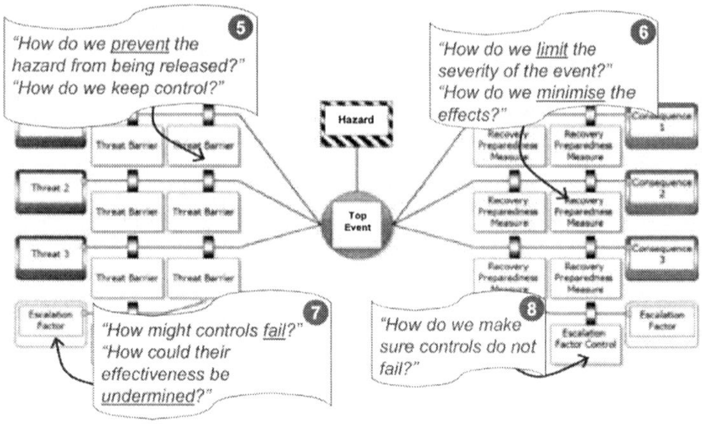

Figure 4. Building the Bowtie – Steps 9 – 12

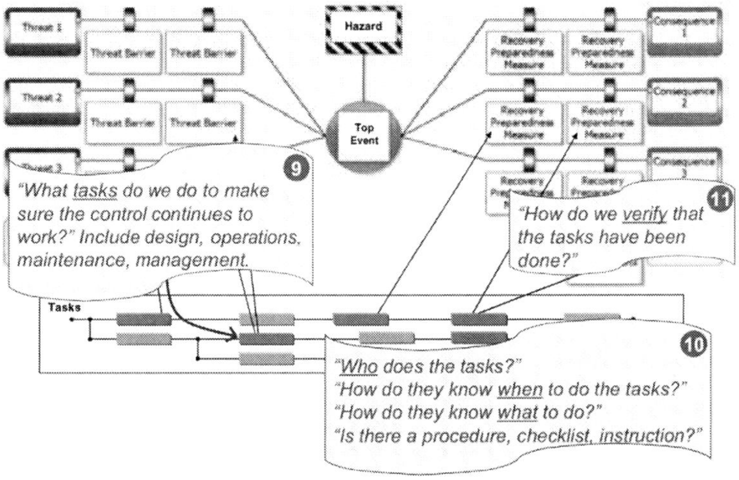

3 Application of the Bowtie Technique

Developing bowtie diagrams should be carried out in a structured manner in order to obtain quality information and best represent the actual risk control arrangements. Figure 5 summarizes an effective bowtie building process, which has been developed and refined through experience with a variety of companies, industries and work groups.

Facilitated workshops involving people who are regularly confronted with the risks have proven to be the most effective way of identifying real controls and capturing past incidents and current practice. Openness is an essential ingredient during these sessions if any weaknesses in controls are going to be uncovered. To encourage free discussion, the workshop needs to be run in an honest and engaging fashion, and, like HAZOP study for example, an independent facilitator can often help to create such an environment.

As is depicted in the first of the below steps, it is often possible to develop a series of 'framework' bowties to help facilitate discussion during the workshop. These framework bowties can be pre-populated using existing information generated through hazard identification / assessments exercises, and by mapping-out industry standard scenarios and their corresponding safeguarding.

Figure 5. Application of the Bowtie Technique

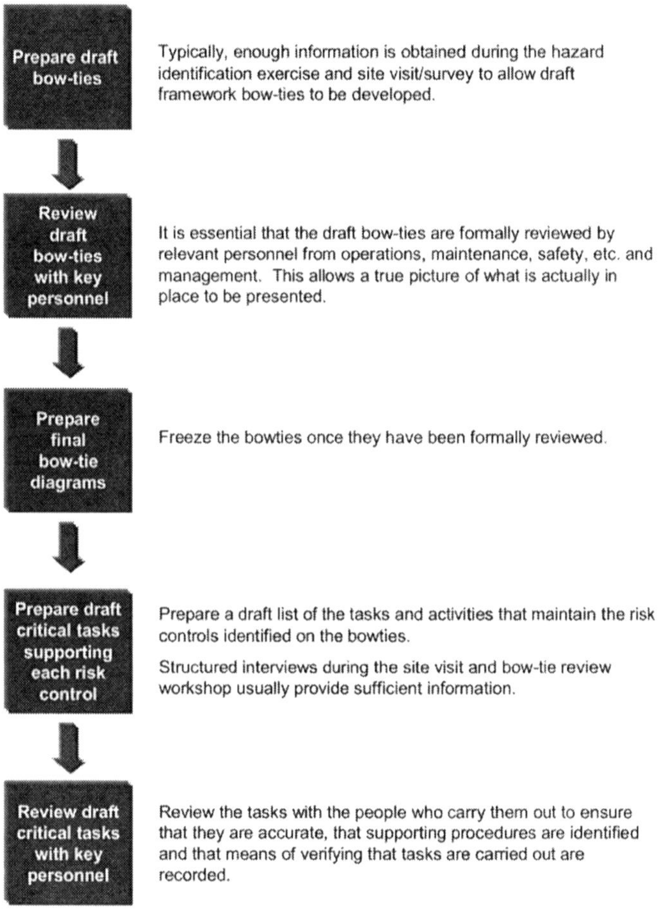

4 Practical Uses of Bowtie

The bowtie technique is incredibly versatile and has proven to be successful in many applications.

Logical Structured Approach

The structured approach of the bowtie forces an assessment of how effectively all initial causes are being controlled and how well-prepared the organization is to recover should things start to go wrong. This logical approach often identifies gaps and issues that are missed by other techniques.

Complete Risk Management

Risk assessments can have a tendency to concentrate on the level of risk only, rather than considering all aspects of the management of risk. The bowtie method however highlights the direct link between the controls and elements of the management system (see Figure 6). As such, it covers a much wider scope than HAZOP study, quantitative risk assessment (QRA), etc. alone.

Demonstration

Due to their origin, bowties are most commonly used where there is a requirement to demonstrate that hazards are being controlled, and particularly where there is a need to illustrate the direct link between the controls and elements of the management system (see Figure 6). For example, bowties have been used successfully in safety reports produced for compliance with the UK onshore chemical industry Control of Major Accident Hazard (COMAH) Regulations[23,24]. Indeed, the Competent Authority has stated that bowties are an acceptable means of demonstrating a systematic approach to hazard assessment[25].

The COMAH Regulations require that "*The demonstration should provide a clear link between the various accident scenarios identified and the measures which are in place to defend against them. The safety report should demonstrate how the necessary measures will prevent foreseeable failures which could lead to major accidents. There should also be a clear link to the SMSs which keep the necessary measures in place.*"[26].

There are other means by which such a link can be demonstrated (e.g. tables) but the bowtie provides the clearest graphical illustration which is easy to understand.

Figure 6. Bowtie Linkage to Hazard Assessment / HSE MS

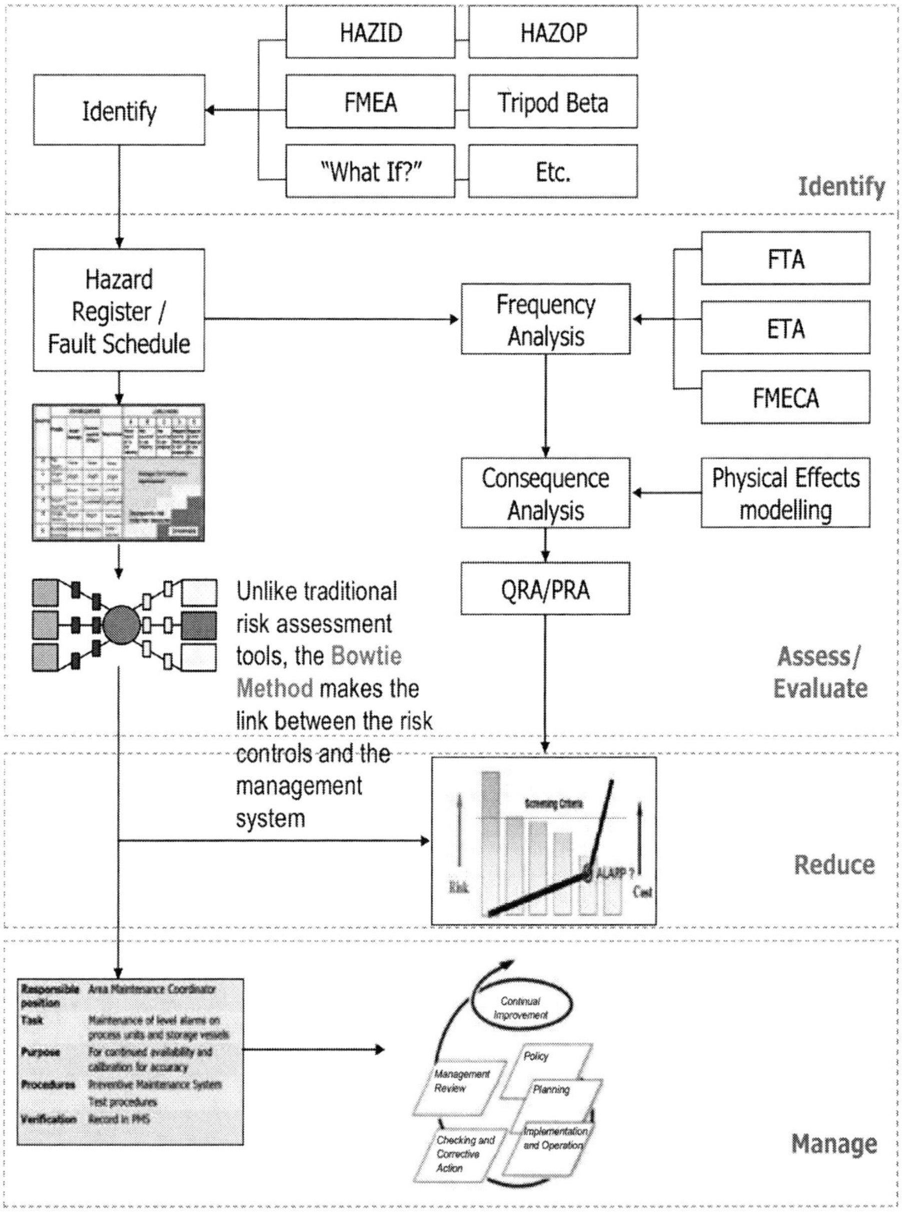

Communication

In its simplest, graphical form, the diagram can be understood by personnel at all levels of an organization, including those who are not connected with the day-to-day business operation being assessed. The bowtie lends itself to being displayed on posters highlighting key risk control issues. Pocket books and leaflets have also been produced for dissemination of the risk management message. There are examples of bowties being included in web-based HSE/Safety Cases as part of on-line training and information systems.

It is not always necessary to use sophisticated publishing techniques to get the most out of the bowtie method. Talking through the components of a particular scenario (causes, consequences, controls, tasks) whilst simultaneously sketching a bowtie layer by layer can clearly illustrate how the hazard is managed effectively, particularly to those who may not be familiar with the details of the operation. This approach has been used successfully during safety workshops, pre-job planning meetings, discussions with regulators and interested members of the public.

Bowties provide a robust, comprehensive yet simple means of 'rolling out' the main points from a risk assessment exercise or HSE/Safety Case. Workshops involving workforce teams can be taken through the bowties step by step, in order that they understand the significant risks associated with their place of work, the measures taken to manage these risks and the importance of their individual role in preventing and mitigating hazardous events.

Once personnel are comfortable with the bowtie technique, building a bowtie becomes a viable alternative to traditional hazard assessment tools such as checklists and prompts. The bowtie structure provides an excellent framework for 'brainstorming' sessions.

Critical Systems

Critical hardware systems can be defined as those which cause, prevent, detect, control or mitigate a hazardous event. The nature of a bowtie diagram means that these critical systems are clearly illustrated along the threat and consequence branches, providing a ready means of identifying systems which are critical to ensuring ongoing asset integrity. Critical systems can be linked to defined performance standards and means of verification.

Organizational Improvements

It is possible to use bowties in conjunction with organizational and cultural survey techniques to highlight the branches of the bowtie where organizational control is weak, i.e. controls are not effective. Recognizing that organizational failures are the main cause of accidents, bowtie allows for proactive, sustainable strategies for reducing and managing risk to be focused on weak spots, for example a higher level of inspection or auditing.

Similarly, bowties have also been used to ensure that no critical controls "fall through the cracks" after a company re-organization, merger or acquisition.

Also, the bowtie can be linked to incident investigation techniques which identify the organizational failures that enabled risk controls (defences) to fail (Figure 8).

Figure 8. Link to Incident Investigation

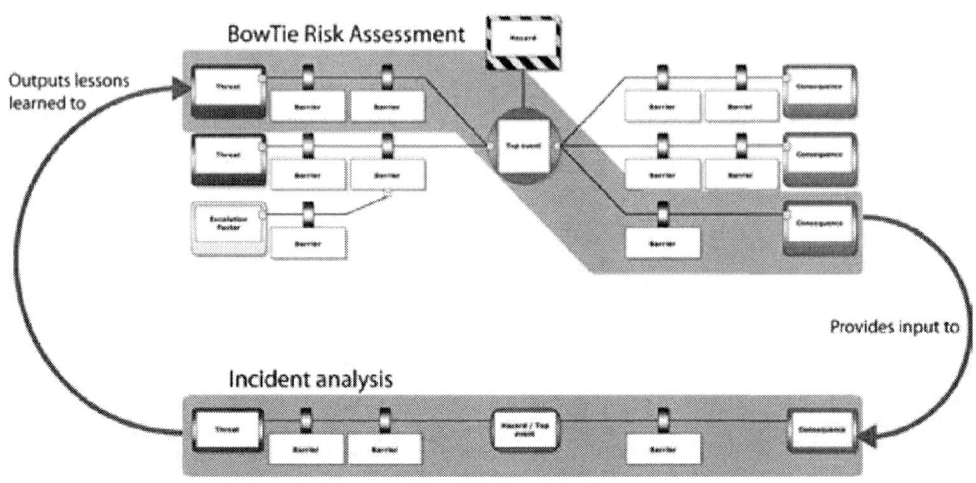

Specific Risks

Bowties are of value not only to meet formal regulatory or corporate standards but for undertaking thorough analysis of specific risks (e.g. non-routine activities or particularly problematic areas) to provide reassurance that everything that can reasonably be done to reduce and manage risk is actioned.

Procedures and Competence

A completed bowtie analysis includes a list of critical tasks undertaken to ensure ongoing integrity of the risk controls. As a minimum, the tasks need to be reviewed with the individuals responsible for carrying them out, but they also lend themselves to being incorporated into procedures, work instructions, individuals' job descriptions, etc. A thorough bowtie analysis can also detail, for each task, the input procedures required to undertake the task correctly and any records which verify completion of the task, as well as competence requirements for the role responsible for undertaking the task. The tasks can therefore be used to verify the adequacy of a company's competence assurance system; the competencies defined for an individual or role should align with the bowtie controls. In this fashion, bowties have also been used to manage handover and new-starter responsibilities.

Layer of Protection Analysis

Bowtie diagrams have also been used for conducting Layer of Protection Analysis (LOPA), a simplified risk assessment tool for determining if sufficient protection is in place and quantifying the residual risk where it is insufficient[27]. Well-constructed bowties, together with clearly defined rules for applying failure data for hardware and human intervention barriers, enable conservative estimates of residual risk to be compared against risk tolerability criteria. The approach tends to work well during design phases.

It is also possible to make direct links between the engineered safeguards on the bowties and specific work orders in a preventive maintenance scheduling and tracking system. In this way, critical hardware systems essential for ongoing risk management are prioritized.

5 Benefits of Bowtie Methodology

The bowtie method has three main benefits:

1. Clear communication and improved understanding - Visually illustrating the hazard, its causes and consequences, and the controls to minimize the risk, the bowtie can be readily understood at all levels, from senior managers and operations personnel, to regulators and members of the public. Bowties keep sight of the big picture and can capture the sequence of events as well as previous incidents.
2. Greater ownership - Recognizing that effective risk management is only possible if people are assigned responsibilities for controls. Bowtie workshops encourage participation and stimulate communication between key stakeholders, whether from the company, contractors or external parties, who all have a role to play in managing risk and yet may not be involved in more traditional techniques. When people feel involved they tend to 'buy-in' to the process. When action is taken based on what they say, people will take ownership. Bowties should especially be considered where lack of ownership of process safety by all levels of operations personnel may be an issue.
3. Efficiency gains – Realizing efficiency improvements through a number of different ways, for example: the method is less labor intensive than many other traditional techniques; it identifies where resources should be focused for risk reduction (i.e. prevention or mitigation); it can reduce the volume of safety analysis – it is true that a picture paints a thousand words; it can lead to a potential reduction in unnecessary/lower importance barriers (where fully justified); it helps to target maintenance, inspection and testing activities on critical hardware barriers; and it provides a 'corporate memory' to avoid reinventing the wheel every few years.

The approach also has a number of other benefits:

- 'Future proof' risk management - Illustrating not only what controls are currently in place today, but, through the use of critical tasks, why they will still be there tomorrow.
- Fit-for-purpose management system - Linking the elements of the organization's management system to specific controls to show how it ensures the ongoing management of risk. This avoids the development of over-burdensome management systems and unnecessary procedures, by focusing on those procedures required to support risk control.

- Practical approach - Focusing on risk management by people on a day-to-day basis, rather than analytical studies by technical risk specialists. All too often, risk analysis can become progressively more complex leading to 'analysis paralysis' which overwhelms the need to take positive action.
- Workforce involvement - Risk management is the responsibility of line managers and their people; all staff can see why what they do is critical for risk control.
- Logical structured approach - Considering all aspects of the management of risk, from initial cause to final consequence in a sequential manner. This logical approach often identifies gaps and issues that are missed by other techniques.
- Auditable trail - The diagrams and critical task lists provide a protocol around which auditing by internal departments or regulators can focus on what people are actually doing rather than the condition of physical systems.
- International application – The graphical-based approach is easy to implement with multi-national teams where language difficulties may otherwise hinder progress.
- All risks – The technique is not limited to assessment of HSE risks. Bowties have been developed for demonstrating management of security, information technology, business interruption and project risks. The possibilities are endless.
- 'Living case for safety' – Comprehensive bowtie assessments can be captured in a relational database which supports ready and wide access across an organization, and enables easier periodic updating. Links to current safety-critical procedures help to maintain any case for safety as a live tool rather than a document for risk practitioners that sits on the shelf.

6 Limitations of Bowtie Methodology

Of course, bowties are not the panacea for all risk management problems. If you want to quantify your level of risk in absolute terms then the bowtie method will not help directly. If you want to model complex inter-relationships between your risk controls, there are better ways than using bowties. If you want to identify individual safeguards for every line of every section of every unit of your process facility, then HAZOP study is the solution. But if you want to remove the mystique of risk management and obtain insights into your risk controls that are easy to understand and easy to communicate, and at the same time realize some efficiency gains, there is no better method than bowties.

7 HAZOP vs. Bowtie Analysis

As touched upon a number of times within this paper, a range of methods/techniques/tools exist whereby which risk assessment of a particular process, facility or task can be undertaken. One of the more common techniques applied throughout industry is HAZOP and, due to its status as a well understood proven technique, is often assumed to cover / encompass the range of information / analysis which would be contained within bowtie analysis; "If I'm already undertaking HAZOP/PHA, why do I need to perform bowtie analysis? What does bowtie do that PHA doesn't?"

- **HAZOP**: Excels at performing a detailed, structured review of the operation of a process, identifying the possible causes of a deviation from the design intent, the consequences and the engineering safeguards present.
- **Bowties**: Very flexible, graphical risk analysis method that allow for visualization of the relationship between Cause – Loss of Control – Consequence and the barriers in place to manage the sequence.

The following key differences between HAZOP and bowtie analysis exist:

- The graphical representation of bowtie allows for a much clearer, easier to understand representation of the risks and how they are managed. By making it easier to understand, it allows for more personnel to engage with this process which leads to better risk management; as the organisation understands what they are relying on and how well it is working, the better they can understand how it is performing and what the weaknesses are.
- Bowties are a very flexible method, and in addition to looking at process risks (which is generally where HAZOPs are employed), are also applied to a far wider range of risks including logistics,

construction, security etc. allowing for a common representation and understanding across an entire organisation.

- Bowties consider the full range of barriers that an organisation is relying on. HAZOPs tend to concentrate on the engineered safeguards in place, whereas Bowties will consider a wider range of safeguards e.g. training & competency, external protection, inspection & maintenance, etc.
- HAZOPs focus predominantly on what is happening inside the process, whilst providing a degree of assessment for external impacts and human factors (as directed by OSHA regulation). However, the flexibility of the bowtie approach allows for greater consideration of external events e.g. external impact, weather, human error etc.
- HAZOPS will tend to concentrate on the preventative safeguards employed to stop a sequence from happening i.e. to keep the process within the containment; Bowties allow for consideration also of the mitigative controls to model what happens once process containment is lost e.g. the emergency response measures such as detection, alarm, mitigation, firefighting, emergency plans etc.
- HAZOPs tend to stop with the static representation of risk i.e. what safeguards are present. Bowties allow for a deeper interrogation of the safeguards, to ask not only what is there, but why it will still work in the future. So Bowtie barriers can be linked to the persons responsible for performing maintenance, or to the performance requirements of critical equipment. This can then be further interrogated e.g. competency requirements for personnel to keep the barriers functioning, maintenance status of critical equipment.

None of the above is intended to say the Bowties should replace HAZOP. They are complementary tools. There is nothing better than HAZOP for ensuring that process systems are designed fit for purpose. Bowties follow on from this to allow for a through life picture, applicable to all levels of personnel, that can be applied to all risks.

8 Conclusions

This paper presents the lessons learned from application of the bowtie method across a number of business sectors. Our experience has shown that the bowtie is ideal for structured assessment and communication of risks, clearly demonstrates the link between control measures and management system arrangements and can be used to qualitatively assess and demonstrate control of all types of risk.

Bowtie should be used as a complimentary tool to the more established forms of PHA such as HAZOP. Bowties will take a robustly assessed design which has going through the appropriate level of assessment (e.g. via HAZOP, LOPA etc.) and ensure that the risk profile is being managed throughout the life of the facility.

9 References

[1] University of Queensland, Australia, Minerals Industry Safety and Health Centre, National Minerals Industry Safety and Health Risk Assessment Guide.

[2] Primrose, M.J., Bentley, P.D., van der Graaf, G.C., Sykes, R.M. Shell International Exploration and Production B.V., "The HSE Management System in Practice-Implementation," SPE 35826, 1996.

[3] Primrose, M.J., Bentley, P.D., van der Graaf, G.C. Shell International Exploration and Production B.V., "Thesis – Keeping the Management System "Live" and Reaching the Workforce," SPE 336034, 1996.

[4] Gower-Jones, A.D., van der Graaf, G.C., Milne, D.J. Shell International Exploration and Production B.V., "Application of Hazard and Effects Management Tools and Links to the HSE Case," SPE 36031, 1996.

[5] Gifford, M.J., Gilbert, S.M. Atkins Defence Systems, Barnes, I. Ministry of Defence, "The Use of Bow-tie Analysis in OME Safety Cases," ESAS 03, 2003.

[6] Couronneau, J.C., Tripathi, A, Transoft International, "Implementation of the New Approach of Risk Analysis in France," 41st International Petroleum Conference, Bratislava, 2003.

[7] Marine Risk Assessment, Prepared by Det Norske Veritas Ltd. for the Health and Safety Executive, Offshore Technology Report 2001/063, 2002.

[8] "Application of QRA in Operational Safety Issues," Prepared by Det Norske Veritas Ltd. for the Health and Safety Executive, Research Report 025, 2002.

[9a] "Lines of Defence/Layers of Protection Analysis in the COMAH Context," Prepared by AMEY Vectra Ltd. for the Health and Safety Executive. www.hse.gov.uk/research.

[9b] Guidance on Risk Assessment for Offshore Installations, UK HSE Offshore Information Sheet No. 3/2006.

[10] Victorian WorkCover Authority, Major Hazards Division, Major Hazards Facilities Regulations Guidance Note 14, Safety Assessment under the Occupational Health and Safety (Major Hazard Facilities) Regulations, 2002.

[11a] Review of the South Island Rail Coal Route, by Kellogg Brown and Root for the Land Transport Safety Authority of New Zealand, MET351-X-REP-001, Rev 1, June 2004.

[11b] Asset Integrity – the Key to Managing Major Incident Risks, The International Association of Oil and Gas Producers (OGP), Report No. 415, December 2008.

[11c] Petroleum and Natural Gas Industries – Offshore Production Installations – "Guidelines on Tools and Techniques for Hazard Identification and Risk Assessment," ISO 17776, 2002.

[12a] Safety Management System and Safety Culture Working Group (SMS WG) – Guidance on Hazards Identification, ECAST European Strategic Safety Initiative, March 2009.

[12b] Safety First – Scenario Analysis under Basel II, McConnell & Davies, April 2006.

[13] Shell International Exploration and Production, DHSE – Developing the HSE Case Training Course.

[14] "Demonstrating How Hazards Are Being Properly Managed," RISKworld, The Newsletter of Risktec Solutions Ltd, Issue 1, 2002.

[15] Lidstone, A., EQE International Ltd. THESIS: "The Health Environment and Safety Information System – Keeping the Management System 'Live' and Reaching the Workforce," IChemE Hazards XIV, Cost Effective Safety, 1998.

[16] Kandola, B., Sullivan, M, ABS Consulting, "The Use of Bow-tie Model and THESIS in the Control of Major Accident Hazards," FABIG Technical Meeting, October 2003.

[17] Trbojevic, Dr. V.M., EQE International Ltd., "The Use of Risk Assessment to Improve Safety Management Systems in Ports," Journal of The Dock and Harbour Authority, Volume 79, Nos. 889, 890, 891, 892, 1999.

[18] BS EN ISO 14001, Environmental Management Systems, 1996.

[19] BSI-OHSAS 18001, Occupational Health and Safety Management Systems – Specification, 1999.

[20] E&P Forum, "Guidelines for the Development and Application of Health, Safety and Environmental Management Systems," Report No. 6.36/210, July 1994.

[21] BSI PAS 56:2003, "Guide to Business Continuity Management," 2003.

[22] The UK Financial Reporting Council Combined Code on Corporate Governance, 2003.

[23] "The Control of Major Accident Hazards Regulations (COMAH)," Statutory Instrument No. 743, 1999.

[24] Considine, M. BP, "Demonstration in COMAH Safety Reports," 26th FABIG Technical Meeting, April 2002.

[25] Joint Guidance Prepared at a Workshop of Representatives of the COMAH Competent Authority and the Chemical Industries Association (CIA), Issue 1.1, March 2002.

[26] "The Control of Major Accident Hazard Regulations 2015," HSE Books, L111 2015.

[27] Dowell, A.M. III "Layer of Protection Analysis for Determining Safety Integrity Level," ISA Transactions 7 155-165, 1998

515832 A Quantitative Individual Risk Assessment Method in Process Facilities with Toxic Gas Release Hazards: A Scenario Library and CFD Combined Approach

Abstract:

People working in process facilities containing toxic gases may face a risk of gas release and poisoning. Many studies on Individual Risk (IR) have been carried out on a worst scenario basis. However, the worst scenarios-based approach cannot represent the real release risks and may overestimate the IR. In this study, an approach based on gas release scenario library and Computational Fluid Dynamics (CFD) is proposed to quantitatively assess IR in process facilities. By combining gas leakage probability and joint distribution probability of wind direction and speed, the release scenario library can be built. CFD method has been used to predict the concentration field of gas release and dispersion. And then the toxic gas concentration can be converted to poisoning fatality probabilities according to the dose-response model. A virtualized IR contour can finally be defined by the accumulative assessment of all release scenarios. A case study of the IR area classification in a natural gas process and carbon dioxide recycle terminal contains an ammonia refrigeration system has also been conducted. With the proposed methodology, the quantitatively classified IR level in process facilities can provide scientific references for safety decision-makers.

Key wordsï¼šscenario library, Computational Fluid Dynamics, toxic gas, individual risk, dose-response model, ammonia poisoning

Risk-Based Decision-Making in Offshore Production Facilities Design Using Enhanced Application of QRA

Daniel P Cheung
Technical Safety Engineer, Chevron USA, Inc. Houston, USA
dcheung@chevron.com

SreeRaj R Nair
Senior Technical Safety Engineer, Chevron USA, Inc. Houston, USA
snair@chevron.com

Jim Salter
Consulting Engineer, Chevron USA, Inc. San Ramon, USA
jsalter@chevron.com

Clive Robinson
Sr. Principal Consultant, Risk Advisory Services DNV GL, Loughborough, UK
clive.robinson@dnvgl.com

Zoë Wattis
Principal Consultant, Risk Advisory Services DNV GL, Loughborough, UK
zoe.wattis@dnvgl.com

Prepared for Presentation at
American Institute of Chemical Engineers
2018 Spring Meeting and 14th Global Congress on Process Safety
Orlando, Florida
April 22 – 25, 2018

Risk-Based Decision-Making in Offshore Production Facilities Design Using Enhanced Application of QRA

Daniel P Cheung
Technical Safety Engineer, Chevron USA, Inc.
Houston, USA
dcheung@chevron.com

SreeRaj R Nair
Senior Technical Safety Engineer, Chevron USA, Inc.
Houston, USA
snair@chevron.com

Jim Salter
Consulting Engineer, Chevron USA, Inc.
San Ramon, USA
jsalter@chevron.com

Clive Robinson
Sr. Principal Consultant, Risk Advisory Services DNV GL, Oil and Gas
Loughborough, UK
clive.robinson@dnvgl.com

Zoë Wattis
Principal Consultant, Risk Advisory Services DNV GL, Oil and Gas
Loughborough, UK
zoe.wattis@dnvgl.com

Keywords: Quantitative Risk Studies, Offshore Production Platforms, Data Analysis, Consequence Modelling, Facility Siting, Risk-Based Decision Making.

Abstract

Quantitative Risk Assessment (QRA) and related studies have been used for more than fifty years to assess risks from major accident hazard facilities. Historically, QRA has primarily been applied as a tool for demonstrating risk to the public and workforce to meet regulatory compliance. Over the years, QRA and its application have evolved and now play a significant role earlier in project design to compare concepts, optimize design and enable cost-effective risk management.

During the early design and engineering phases of offshore production platform projects, a number of critical design decisions are made including concept selection, capacity/throughput, technology and equipment selection, layout finalization, subsurface design, and type of facility (fixed or floating). All these have significant impact on the inherent risk of a facility. The use of QRA in early design decision-making enables the benefits of an inherently safer facility to be realized and can support effective risk management to achieve an optimal risk-based solution.

Developing the QRA early in the project design phase poses challenges due to the limited details of engineering design data available. The challenges have been overcome by a screening-based version of the QRA, which utilizes data and information gathered from previous production design projects to supplement the design data so that the QRA can be applied early in the project to support project decisions.

Traditional methods for gathering and collating input data for the QRA are labor intensive and time consuming. Recent developments have enabled electronic data extraction directly from native engineering documentation, for example from piping and instrumentation diagrams (P&IDs) and 3D Models. These enhancements in information processing allow greater flexibility of the application of QRA and enables the study to respond to design changes throughout the project design phase and give feedback to the design teams more rapidly than before.

This paper identifies the lessons learned and best practices in designing safer production facilities that can be realized from the key advancements in the process and application of QRA. This paper discusses the following advancements that have been developed and tested to improve the efficiency and quality of the risk studies:

- Enhancements in QRA tools through automated data extraction and processing of native engineering documentation.
- QRA screening techniques that utilize data from previous project designs to supplement sparse engineering data early in project lifecycles.
- The application of QRA results in early project phases to provide project teams with sufficient data to support decisions that are key to providing an optimal risk-based design that recognizes the inherent risk implications of concept selection and early layout decisions.

1 Introduction

Application of quantitative risk assessment (QRA) and related studies on offshore oil and gas major capital projects have traditionally been used to demonstrate facility compliance with regulatory requirements and corporate risk tolerances. In more recent years, Oil and gas companies have recently improved the effective utilization of QRA to aid in engineering design and proactively address key risk drivers in project decisions [1].

Realization in capturing risk reduction design opportunities earlier in the project has demonstrated substantial cost benefits though minimized late-project design changes and less dependence of add-on barriers during later phases of the lifecycle. Some key examples of early concept selection benefitting from QRA studies include facility processing capacity/throughput, technology and equipment selection, layout optimization, and subsurface design.

Another growing challenge for the industry is in the rapid improvements of production technology in the past twenty years, which have unlocked deeper and more complex oil and gas reservoirs. Those reservoirs contain oil and gas with higher operating pressures and temperatures along with complex enhanced oil recovery opportunities [2]. Along with the advancements comes added complexity of the facilities. There is recognition that we must proactively employ the best risk management technologies to be able to assure ourselves that we do not expose operators, contractors and the public to excessive levels of risk and that sufficient protection is in place to prevent major loss of life or containment [3]. Consequently, the focus for these complex offshore facilities has been placed on the QRA and improving our abilities to quantify and manage risk.

The following sections details the recent improvements in the QRA process to better integrate and apply risk conceptualization and management within early project design decisions and front-end engineering and design (FEED). Challenges in early QRA application and further discussion on how to overcome those challenges are identified through the structured approach of the QRA-lite. It is followed by discussion and examples of different offshore major capital projects on the additional benefits in early QRA-lite application such as leveraging inherently safer design concepts, improved execution strategies and economic benefits.

2 QRA Method Concepts

QRA techniques were first given wide application in offshore oil and gas in the early 1980's, prompted by a series of serious accidents including the Alexander Kielland capsizing in 1980 resulting in 123 fatalities and later, the Piper Alpha incident in 1988 resulting in 167 fatalities. Pursuant to regional regulations, these risk assessments were primarily carried out as regulatory assurance for a quantified cut-off criteria related to the risk impairment frequencies. Detailed risk analysis requirements were first defined in the UK HSE Case and eventually applied in other offshore regional locations. Since then, applicability grew for the QRA and provisions were laid down for a more integrated and dynamic use of risk analyses, with suitable quality controls on the process, covering the whole life cycle of the oil and gas production activities [4].

QRA is a means of estimating the total risk presented by a facility or operation. By identifying the potential major hazard scenarios and applying a frequency of occurrence, the QRA can provide a risk summation and estimated exposure to the workforce personnel and surrounding environment [5].

The intent of employing risk mitigation is to ensure that there are sufficient controls in place that realization of the scenario is appropriately unlikely. When projects are at the early stages of concept selection, the project team has the highest level of control of addressing risk and positively influencing the facility's societal and individual estimates.

Oil and gas operating companies strive to achieve the lowest risk exposure reasonably possible, and establish acceptable societal and individual risk criteria as thresholds for the project to follow to properly ensure protection of the people and surrounding communities. Should risk levels from QRA studies exceed the risk threshold, the project exercises hazard elimination or risk mitigation to reduce the risk to an acceptable level. Through adjustments in the installation of safeguards or mitigations, the operating company can determine the sensitivity of that change to the overall risk profile, and in a relative sense, quantify and justify the installation of that safeguard or mitigation to maximize risk reduction.

For offshore production platforms, high risk contributors differ from design to design, but generally are identified by the QRA to include well blowouts, riser/pipeline failures, export compressor or other process system leaks, collisions (helicopter and ship), and structural or marine related failures [6]. Focused spending to reduce these risks and efforts to develop technology for mitigation have been the focus for offshore process safety management.

3 QRA Application in Concept Selection and Early Design

Most of the facility lifecycle decisions required for design and engineering is taken during early project phases. Several factors are evaluated for decision making during design and engineering; for manned offshore installation projects, risk to personnel is a key factor. Early project decisions and design engineering are at times made without a right level of assessment of risk to personnel. This could lead to discovery of significant risk related concerns during detailed engineering or construction. Typically, the elevated cost from late project decisions are on the magnitude of millions of dollars for substantial changes in offshore scope; some examples for cost escalation include delays in schedule and addition of barriers (e.g. blast rated walls, high-integrity protection system, fire & gas detection system, and automated fire extinguishing systems).

Determining facility processing capacity, structure type (e.g. fixed, semi-submersible, spar, and tension leg), equipment selection, layout optimization, and subsea infrastructure and wellhead design are some of the key concept selections where QRA can help with risk based decision making early in the project design. However, the nature of early project design did not align well with the traditional QRA process, which requires detailed project documentation and is generally time consuming and unable to respond to the rapid evolution of alternative concepts. A number of improvements to the traditional QRA technique has allowed for further application in the design of offshore production facilities.

This section considers the advantages and challenges in conducting QRA in early phases of design, how to overcome the challenges and some examples of success and some missed opportunities signifying the relevance.

3.1 Key Advantages:

Early QRA application contribute to an overall improvement in a safer facility design to mitigate high consequence events, and substantial cost savings in design and project implementation. Early risk evaluation explores the ability to evaluate options and support risk based decision making, where use of consequence assessment as an input to engineering allows for inherently safer design. By identifying inherently safer options, the project can inadvertently eliminate the risk and reduces the need for relying on safeguards, which typically relates to realized overall cost savings.

Proper application of risk assessment in concept design and FEED avoids or minimizes late phase design changes to address safety concerns, which is a common driver of cost additions to large capital projects. By identifying concerns early, the project minimizes the need for add-on safety barriers (e.g. reduce likelihood and impact from explosion through ventilation, less confinement, less congestion whereby not to rely on a blast wall)

3.2 Challenges:

Although the advantages of early QRA utilization are well recognized, successful implementation is difficult to achieve due to a variety of obstacles. The ability to quantify and compare the relative risk contributions between alternatives continues to be a challenge in these early design decisions due to a lack of project data availability. Details are needed to reduce uncertainty of the QRA study results due to rigorous parts counts, hydrocarbon inventory and layout.

Additionally, the standard QRA approach could be intensive and time-consuming whereby failing to meet project milestones. A method is needed to provide more rapid, real-time risk quantification to be able to keep pace with the rapid changes in the early part of front-end engineering and design (FEED). Major process changes may take several weeks to analyze in a full QRA and these changes can occur several times in the first months of FEED. Many times, the design will have been updated with several critical changes before the QRA assessments have been completed and reported.

With a vast set of options to be analyzed that may not be fully defined, a need for specific boundaries and scope is essential to early QRA application success. Earlier application of QRA entails an earlier request for funding for risk studies, whose cost will need to be reasonably justified with the potential benefits of early application strategy. A quantification of costs for some of the examples provided is included for support.

3.3 How to Overcome the Challenges:

To address the lack of project data availability for the need of a detailed risk assessment, a fit-for-purpose QRA-lite technique is developed. By exercising the QRA-lite technique of organizing the primary components of processing units and high-risk contributors into a simplified model of building blocks, the complex makeup of offshore platforms can be well estimated for the purposes of early quantified risk-based decision making [3]. Behind each of these building blocks in the QRA-lite lies a foundation of predetermined parts counts for major assemblies (e.g.: compressors, production pumps, riser columns, separator units, etc.) that have been derived from a database of similar offshore processing units.

In addition, the building blocks contain pre-populated consequence models. These cover a range of operating pressures, temperatures and inventories, allowing the analyst to quickly develop estimates of release consequences without having to resort to the intensive effort required to develop these consequence models from scratch for each release location of interest.

Logistics and project management are keys to success for early application of QRA. Many of the studies involved with QRA impact design criteria and critical decisions, such as determining blast load criteria and the impact of alternate layouts, or determining the need for relocation for high pressure critical equipment such as boarding valves, export compressors or gas processing equipment. A critical review of the early phase project schedule and key project milestones is necessary to align the accelerated QRA schedule to support crucial design decisions. Early identification of key design decisions where risk could be a major factor (e.g. weight estimates, platform orientation, equipment layout, personnel exposure) is critical to ensure proper focus and dedication of the QRA on those aspects.

Practices of integrating the QRA consultant or the technical safety engineer overseeing the QRA process is suggested and has positive lessons learned from recent projects. This practice alleviates communication gaps and allows for quick processing of minor decision requests from the design team.

The remainder of this section will discuss how the QRA-lite has been implemented on a variety of offshore major capital projects to influence the decision-making process, allowing us to ensure the safety of the public and our workforce. These examples will explore the cost benefits of QRA-lite application for justification of an early application strategy. In each example, we will also see how decisions to exploit inherently safer design opportunities have realized cost benefits.

3.4 Examples Early QRA Application Opportunities

Use of consequence modeling and QRA in optimizing design and enabling risk based decision making can be carried out for both greenfield development projects as well as brownfield projects. This section identifies a few examples where the tool was successfully utilized in identifying key risk factors to design, and cases used for evaluating different engineering design options.

3.4.1 Example 1 – Enhanced Recovery

Technology for enhanced oil recovery has been identified as one of five priority areas for technological development in the offshore oil and gas industry [5]. Enhanced oil recovery methods such as water flooding, gas injection or polymer processes has been recognized as a cornerstone of offshore reservoir engineering due to its degree of impact to crude recoverability. However, the concept selection for enhanced recovery can introduce high risk contribution of various degrees depending on personnel exposure, platform layout, limited disposal options and reservoir impacts. These HSE risks are challenging to quantify without conducting detailed consequence modelling and the rigorous techniques of QRA.

The QRA-lite technique was readily applied to an offshore enhanced recovery project to provide preliminary risk quantification to determine which enhanced recovery technique was the lowest risk option and their respective impact to the operating risk levels. The enhanced recovery project identified evaluated two options: gas injection and seawater injection. For gas injection, it was understood that introducing reservoir gas of pressures upward of 12,500 psi at the injection compressor discharge could significantly increase the risk exposure to the workforce personnel; but to what degree was an uncertainty. In addition, risks were identified in flare management, potential hydrate formation within flowlines, depressurization temperature concerns at the injection headers, upsizing various production equipment and potential hull modifications that was challenging to quantify.

The alternative option for secondary recovery, seawater injection, would inherently eliminate the realized fire and explosion risks, but could introduce a variety of other hazards including hazards to the reservoir and subsea equipment. Injection of off-spec water with high concentrations of sulfates can sour the reservoir through sulphate-reducing bacteria (SRB), inadvertently introducing trace levels of H_2S to the flowline and topsides processing equipment. Introduction of trace concentrations of oxygen in the injected seawater can also cause flowline corrosion and impact the integrity of the subsea manifolds and risers [7]. With the variation of risk parameters present, it was impossible for the project to qualitatively assess the risk contribution among the alternatives and would require turning to quantitative risk assessment means.

A quantitative model was run with the QRA-lite technique for each case with the respective layouts for the gas and seawater injection modules and a baseline operating arrangement with representative processing modules and personnel exposure. It was determined that the individual risk per annum for the operations personnel had increased the overall facility risk levels by approximately 33% for the gas injection case, whereas only by 8% in the water injection case with all failure scenarios incorporated.

Although the gains in reserve additions that could be realized using developed gas injection techniques were greater, it was out of the company's practices to operate in risk levels above the company established criteria for safe operations, where risk mitigation measures would be difficult to achieve. The enhanced recovery project moved forward with the seawater injection case, where risk reduction measures could be established to control the identified risks of reservoir souring and flowline corrosion. This is a successful display of how early quantitative risk management supports a management decision to eliminate a high pressure natural gas hazard and enables the inherently safer design of an offshore facility.

3.4.2 Example 2 – Blast Protection

In the above project, further risk reduction opportunities were explored to reduce the explosion exposure through various shielding safeguards. A common challenge of offshore production platform design is addressing blast protection, as congestion of equipment and buildings are inherent risks contribution to high potential gas cloud formation. A QRA assessment was leveraged during the early FEED process to determine what level of blast protection for egress routes was optimal. Variations were run with blast walls from high risk process equipment, additional stairways, and installed solid floors. Additional cases were run with installed smokehoods to reduce smoke hazards and improved egress. Results are shown below in Figure 1 for the original platform configuration (case 1), inclusion of stairways on the outside of the platform (case 2), inclusion of shielding effects of equipment from jet/spray fires (case 3), solid floors to withstand blast overpressures of 5000 mbar (case 4), combination of above (case 5) and additional benefits of smoke hoods (case 6).

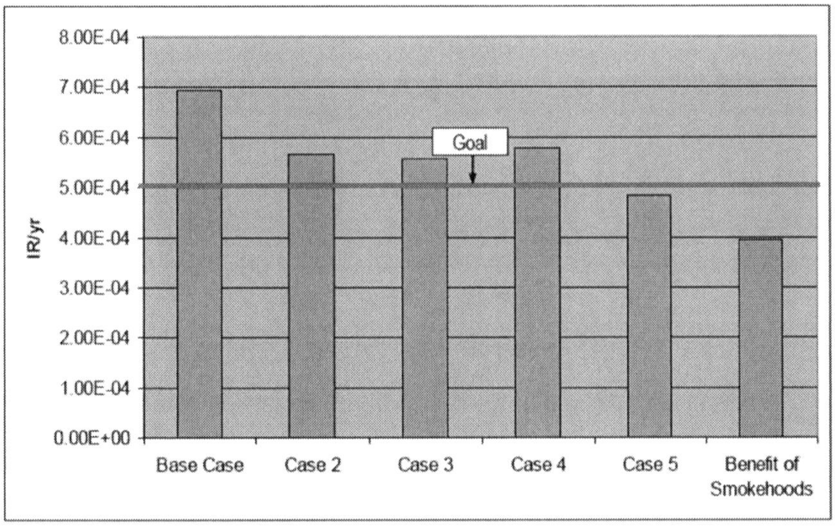

Figure 1: Individual QRA results for various sensitivity cases from QRA-Lite Application for Example 2 Project

The project proceeded to begin their engineering design with a base case of solid floors and proper shielding safeguards to ensure they remain under the tolerable risk threshold. Smoke hoods were considered on the frame on an as-needed basis. Application of early QRA has proven to drastically help address the inherent risks of blast and explosion protection, which have historically caused heavy project cost additions due to late installations of blast walls or relocation of buildings/equipment.

3.4.3 Example 3 – Gas Ventilation Dispersion with QRA-Lite Models

Identification of platform orientation and building siting with respect to the locations of hazardous sources is an early design criteria where the application of QRA-lite can greatly assist. Advertent effects of this design decision include ventilation and exhaust dispersions of operating equipment, where with the proper platform orientation, inherently safer options can be leveraged. Early application of ventilation modeling on a QRA-lite model can be used to determine the adequacy of the wind speed to natural dispersion of any potential gas releases. The goal for design is to minimize vapor accumulation through natural air movement through the facility by wind movement. This will inadvertently reduce dead zones where flammable gas clouds may result in a vapor cloud explosion. The impact of main obstructions including buildings, equipment cabinets, and structures were evaluated to determine suitable locations and orientation.

By using wind speed and wind direction specific to the location of the facility for modeling helps in evaluating and optimizing the layout. Specific considerations should be focused on benefits from locating Living Quarters and occupied buildings upwind of the facility to minimize the impact from potential gas clouds.

3.4.4 Example 4 – Equipment Layout with QRA-Lite Models

By using QRA-lite models that allow for detailed analysis of parts count, oil and gas inventory and release scenarios, one can provide detailed risk inputs for analyzing layout alternatives.

During the equipment layout studies in pre-FEED, an offshore project identified an elevated risk from high pressure gas release sources from export gas compressors that had been relocated to well ventilated areas with minimal confinement and congestion. Also, boarding shutdown valves (BSDV), which isolate oil and gas flow from high pressure reservoirs were located away from production areas to easily achieve adequate protection. The quantification was able to be represented through results from QRA-lite, where difficult layouts were analyzed and compared.

Various layout options will have an impact on the amount of pipe routing required by placement of vessels. This will inadvertently affect the oil and gas inventory on the topsides platform, effecting the overall release scenarios. Through the use of predetermined blocks in the QRA-lite tool, the amount of parts and inventory for each layout option were identified and applied as decision criteria in the form of risk quantification for the project team.

3.4.5 Example 5 – Equipment Selection with QRA-Lite Models

Equipment sizing and determining the number of trains and equipment sparing is a critical process decision in the early project phase. However, selections have a significant impact on the risk exposure through inventory and potential leak points. Through QRA-lite leak frequency assessments combined with consequence estimation, these various options can be modeled and the resulting risk contribution can be assessed. Evaluations between lesser number of equipment but higher inventory is pitched against higher number of equipment but with lesser inventory. Risk estimations along with the life cycle reliability requirements is a critical safety consideration in early design decisions.

For an example offshore platform production module, the project was analyzing between a 2x100% train or 3x50% trains to meet reliability criteria and risk criteria. Through leak frequency analysis and consequence estimation through preset processing units in QRA-lite, the team was able to identify the lower risk of 2x100% trains due to a lower potential leak source count although a slight increase in oil and gas inventory that was manageable through proper isolations, detection and emergency shutdown and blowdown systems.

3.5 Examples of Missed Opportunities Captured in Late Project Phases

Most of the changes or deviations for achieving additional risk reduction identified from risk assessments during later phases of the project could result in significant schedule delays and changes to the design whereby resulting in additional cost for construction and delay in getting returns. Below are examples of cases when projects were unable to capitalize on early QRA opportunities.

For offshore platform design, layout and orientation was mentioned in Section 3.4.4 as a critical early phase decision. A project case identified a scenario where the helideck location with respect to the turbine generator was inadequately placed. Due to close proximity, the generator exhaust limited the operational zone due to high turbidity and temperature changes in the helideck region. Additional layers of protection were added later for exhaust plume detection, and operational restrictions implemented for helicopter landing and takeoff as a procedural safeguard, limiting operations planning capability.

Late acknowledgement of hazard identifications generally leads to further layers of protection needed in late phase projects. In an example of a platform arrangement project where process equipment was present in areas with limited natural ventilation, additional gas detection equipment was required to meet regulatory compliance due to hazard recognition in a later project phase. Additional valves and equipment were included in the area during late phase design changes that were not accounted for in the initial QRA model, creating significant scope change and cost additions.

4 Enhancements to QRA

QRA was primarily used for design acceptance and carried out on a mature design after most of the design decisions and engineering were carried out. However, the shift to earlier usage of QRA requires the technique to process information more rapidly, such that real-time risk quantification is able to keep pace with the rapid changes in the early part of front-end engineering and design (FEED). Major process changes may take several weeks to analyze in a full QRA and these changes can occur several times in the first months of FEED. As a result, the design may have been updated with several critical changes before the QRA assessments have been completed and reported. Recent developments have enabled electronic data extraction directly from native engineering documentation, for example from piping and instrumentation diagrams (P&IDs) and 3D Models. These enhancements in information processing allow greater flexibility of the application of QRA with minor error margins giving feedback to the design teams more rapidly than before.

4.1 Automated Use of Engineering Data

In the development of a QRA a large amount of input data is required so that the details of the facility can be represented as accurately as possible. This comes from a number of sources including engineering design packages, environmental data, and onsite and offsite population distributions. The majority of the data comes from engineering information that is produced during the design and is then updated throughout the operation of the facility. Typically, there is a large amount of engineering information and this has to be interpreted by the QRA analysts in order to produce the necessary inputs for the QRA. If this part of the analysis could be automated then this would make the QRA process more efficient and reduce the time lags between design changes and producing QRA results, and allow the QRA analyst to focus on the value-added activities of analyzing the results and identifying potential mitigations. The reduction in time can avoid costs and delays associated with facility redesigns due to the lag time between designs and risk assessments and provide real-time perspectives of the risks associated with facility design parameters.

An automated methodology has been developed which makes use of design data from 2D drawings and 3D models. Several pilot studies have been carried out using different engineering design packages. The work described here makes use of Autodesk AutoCAD 2D drawings and Autodesk Navisworks 3D models. The methodology is based around a data processing tool and several supporting plugins written for AutoCAD and Navisworks (to assist with data entry, manipulation and extraction). Figure 2 shows the flow of data using the methodology.

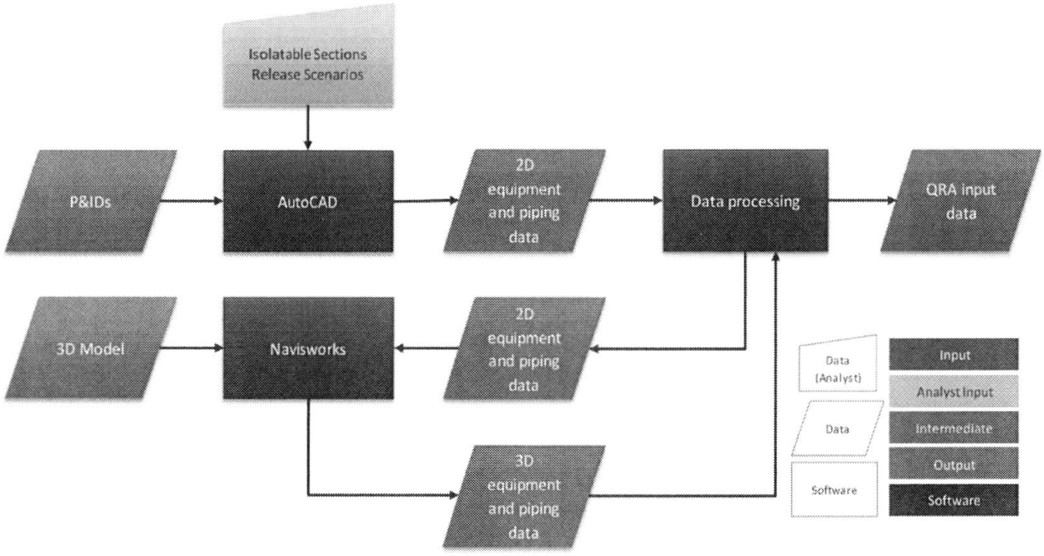

Figure 2 Automated QRA Methodology – Simplified data flow

The first stage in the process is to markup the P&IDs to show the isolatable scenarios and release scenarios. This is usually carried out either on paper copies or perhaps electronically on a pdf version of the P&ID. In the automated methodology this is done using the native file in AutoCAD. A typical marked up P&ID is shown in Figure 3 in which the different release scenarios are indicated by different colors.

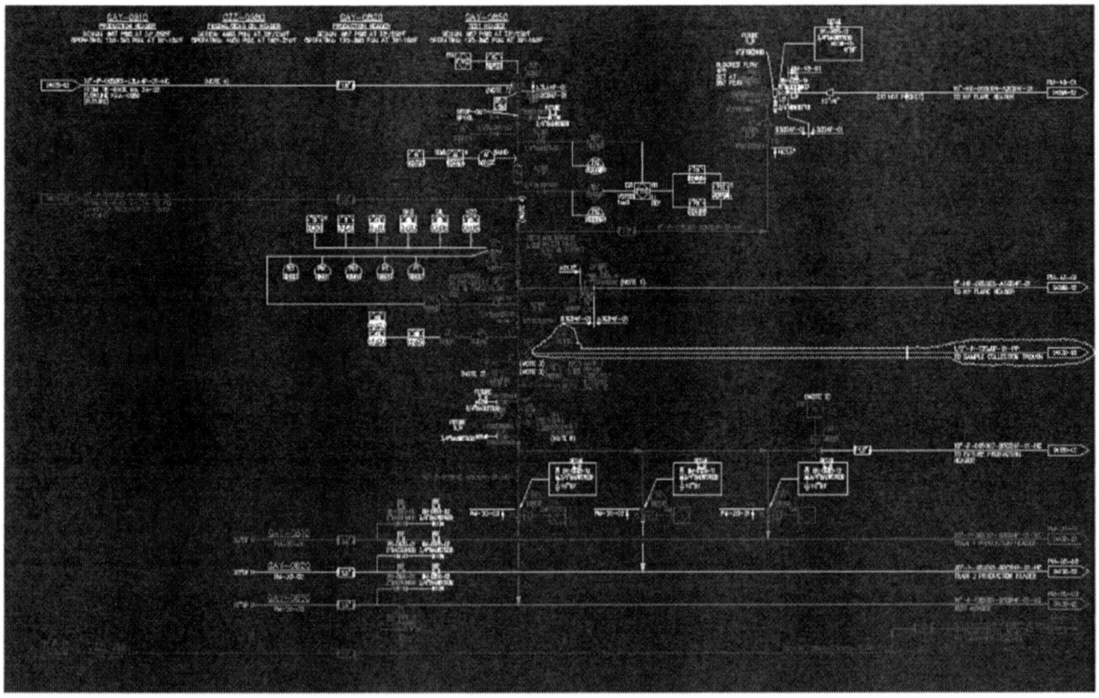

Figure 3 Marked up P&ID in AutoCAD

The data from the marked-up P&ID can be exported from AutoCAD to produce a data table in which the items shown on the drawing are grouped together by release scenario, thus enabling a parts count for each scenario to be carried out. This can then be combined in the usual way with historical leak frequency data to predict the release frequencies for each scenario.

If a 3D model is also available then the data extracted from the P&ID can be used in a number of ways to give further input data for the QRA. Any items that are identified with tags can be found in the 3D model and their locations identified. This information can then be used to produce release locations. Additionally, line numbers for pipes can also be found in the 3D model and pipe locations found. This can then be used to calculate pipe lengths and inventories. As the tags and line numbers have previously been associated with release scenarios this information can be calculated for each release scenario in the QRA. A final use of the data is to visualize the release scenarios within the 3D model. An example of this is shown in Figure 4. This visualization has been found to be extremely useful in verifying the marked-up P&ID data and in checking the release locations used in the QRA model are correct.

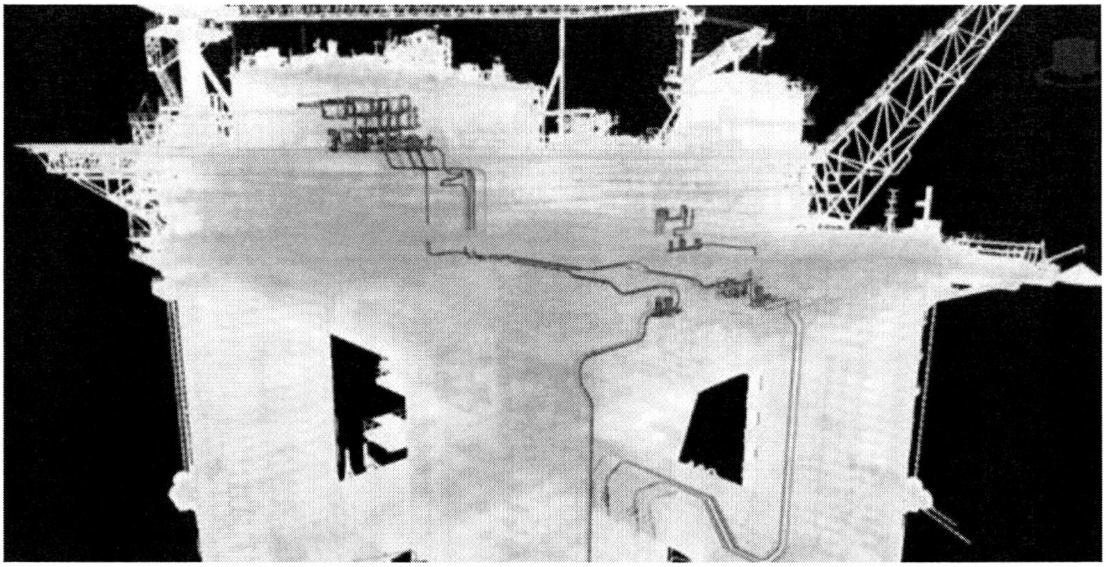

Figure 4 Highlighted release scenarios within 3D model

The above approach has now been used on actual projects and has proven to give time efficiencies and to assist in the verification of the input data for the QRA as a full audit trail exists. It has also been shown to be very efficient if changes are required either due to design modifications or changes to how the analysis is carried out, i.e. the representation of release scenarios and process isolation locations.

4.2 Analyzing Risk Contributors

Due to the complexities of QRAs large volumes of output data are often produced. There are often tens if not hundreds of release scenarios, each of which are analyzed for different release sizes, combinations of environmental conditions and operation of safety systems. Traditionally the results of the analysis are shown in a figures and tables within a report but due to the volume of data only a subset can sensibly be provided. Often when examining the results readers of the report will want to understand the contributions to the risk from different scenarios which requires access to the underlying data which is not in the report. They may also want to examine intermediate data such as consequence results.

To enable the efficient reporting and examination of the data, results from QRAs are now being made available via a web interface. An example of this is shown in Figure 5. This shows an example screen which gives the main QRA results.

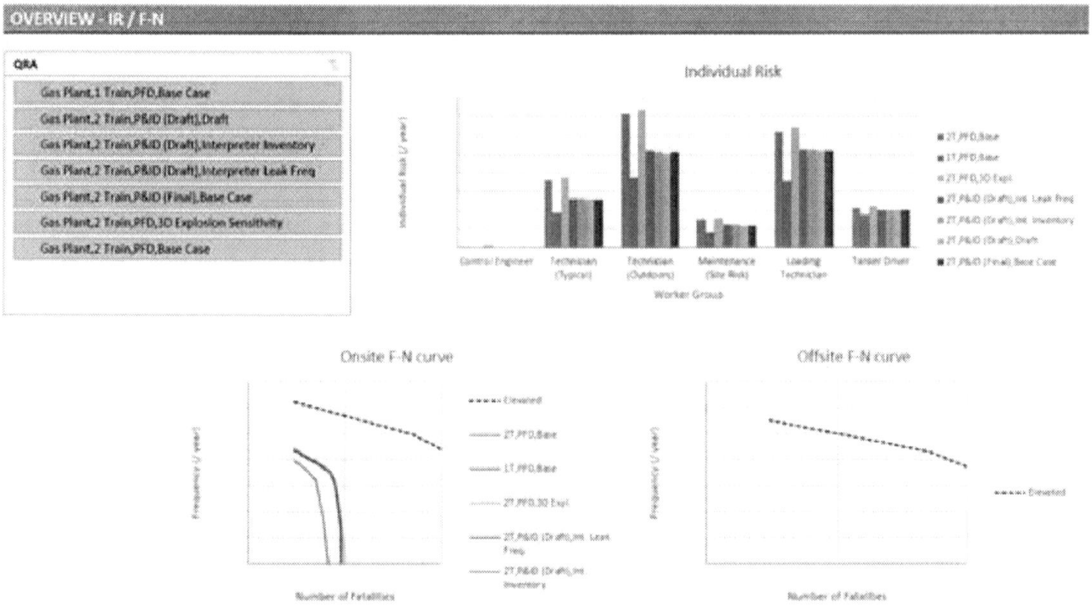

Figure 5 QRA main results

The data displayed is linked to a database so that the user can drill down to examine risk contribution in different ways. This can be done in many different ways as shown in Figure 6. The only limitation on how far the drill down of data can be carried out is the granularity within the database.

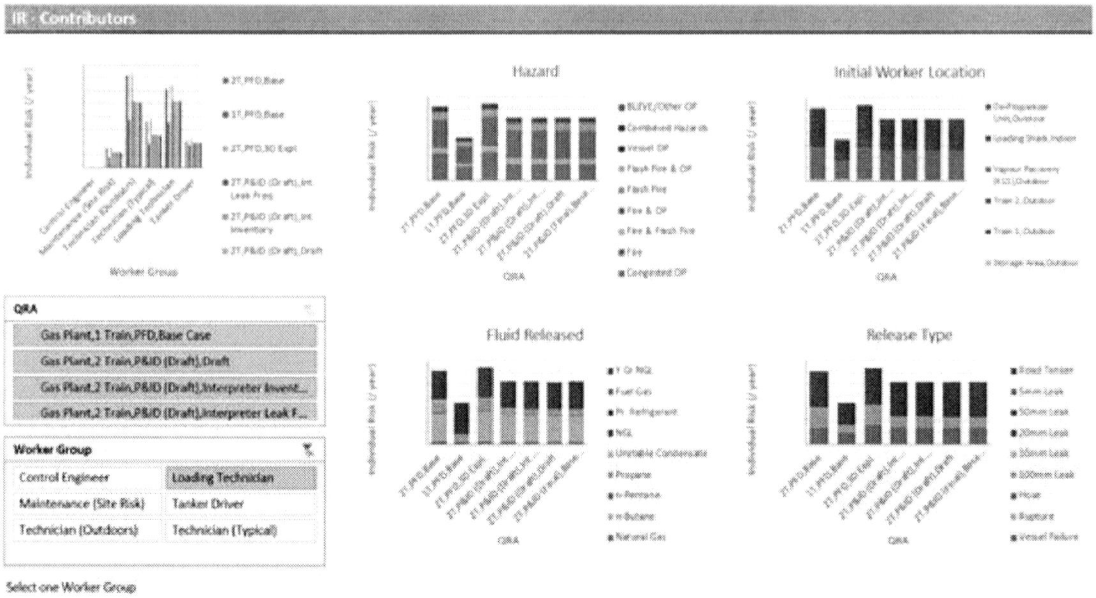

Figure 6 QRA detailed results

Other data such as consequence data can also easily be plotted on views of the facility as shown in Figure 7. As the location of release scenarios are all available within the database the views automatically show the correct location on the site. This makes assessments using this approach much more efficient than a traditional approach which can often involve reading distances from tables and comparing them against plot plans. As all the consequence data can be stored in the database it also means that the user can examine any of the data as shown in Figure 7 rather than just a subset of data provided in a report. Often this information is used separately from the QRA, for example by operational personnel and the easy access to the data assists in timely decision making.

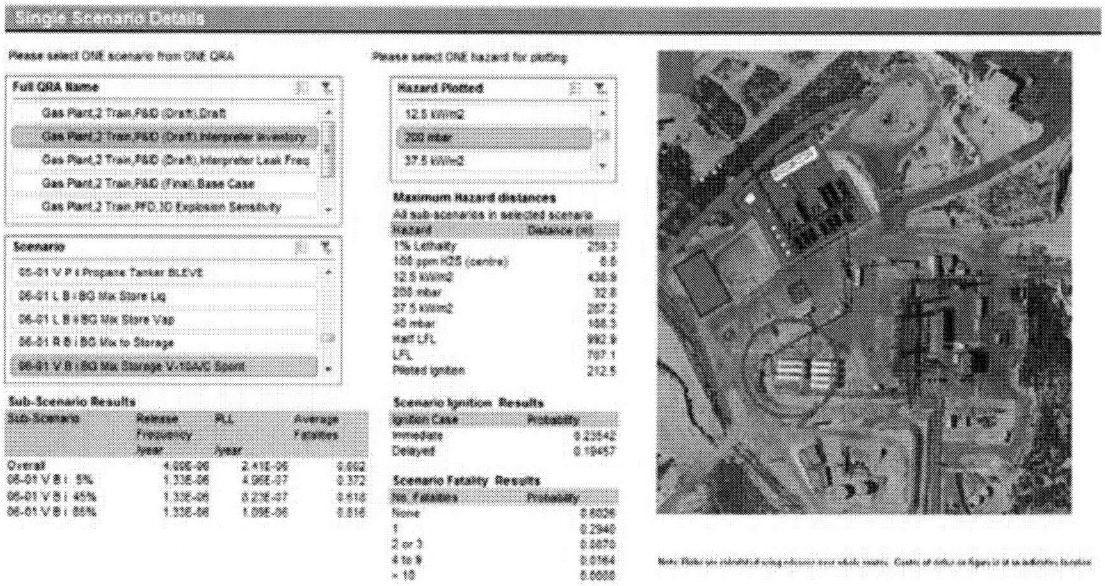

Figure 7 QRA consequence data

This approach has already shown itself to give added value to recipients of QRAs compared with a traditional report. Some other features have been implemented such as comparisons between revisions of the QRA due to design changes. The ability to drill down into the data enables the user to understand why risks have changed and thus make informed risk based decisions about facility modifications. Current work is also taking place to allow users the ability to run sensitivity studies on some parts of a QRA themselves.

5 Conclusions

Although QRA is a technique that has been used for many years, its ability to provide an objective measure of relative contributors to risk has not been fully recognized by oil and gas companies. Over the past decades, these companies have better recognized the inherent hazards in offshore production and have each shifted their focus on QRA utilization and improvements. It is now acknowledged that a company's overall goals of incorporating inherently safer concepts and designing safe facilities are supported by a more integrated QRA process. These goals align with the corporate safety integrity toward becoming the leading experts in fatality prevention and process safety management in the industry [8].

This paper has demonstrated how QRA can be used by projects to support decisions that will lead to inherently safer designs. However, for the method to be effective, it is critical that the studies be performed early in project design, ideally before a preferred option is selected. It is by early application that QRA can provide the greatest benefits, highlighting the potential pitfalls of some design options before significant effort is expended to develop a design during FEED. Early application will also avoid much of the cost associated with rework that may be required due to late application of the technique. Furthermore, it is critical that the QRA model continue to reflect the current design throughout its development such that decisions affecting safety can be fully understood.

In order to achieve this state, it is clear that the process of performing QRA needs to be further optimized. Approaches in automated parts count and improvements in transparency of risk curve results have significantly improved this process to provide data on the order of days to weeks rather than months; however, a real-time ideology is still not achieved. Further efforts are still needed to fully incorporate the QRA process behind every process flow diagram, process and instrumentation diagram and 3D model to provide a risk model that enables design teams to see the impact of design changes truly in real time.

Approaches to streamlining the QRA process are critical for developing risk-optimal designs which, in the long term, will protect the public, the workforce, and the reputations of the companies that continue to build ever larger and more complex plants, as well as the reputation of the oil and gas industry as a whole.

6 References

[1] Aven, Terje, "Risk Assessment and Risk Management: Review of Recent Advances on their Foundation", Elsevier European Journal of Operational Research, 2016.

[2] J. Erik Vinnem, "Risk Assessments for Offshore Installations in the Operations Phase," ESREL 2003.

[3] J. Salter, "Using Quantitative Risk Assessment to Make Billion Dollar Decisions," American Institute of Chemical Engineers, CCPS, 2014

[4] Risktec, "Quantitative Risk Assessment Across Major Hazard Industries," RISKworld Issue 13, 2008.

[5] P. Kang, "Screening Criteria and Considerations of Offshore Enhanced Oil Recovery," Energies, MDPI, 2016

[6] L.W. Lake, "Defining Enhanced Oil Recovery, Enhanced Oil Recovery," 1st ed. Prentice-Hall: Eagulewood Cliffs, NJ, 1989; pp 1-16.

[7] API, "Recommended Practices for Oil and Gas Producing and Gas Processing Plant Operations Involving Hydrogen Sulfide", American Petroleum Institute Recommended Practice 55, 2nd Edition, 1995

[8] M Sam. Mannan, "Stretch in Technology and Keeping the Focus on Process Safety for Exploration and Production in the 21st Century, Institute of Chemical Engineers Symposium Series No 156. 2011

[9] C. G. Ramsay, "Quantitative Risk Assessment Applied to Offshore Process Installations. Challenges after the Piper Alpha Disaster," J. Loss Prevention Process Ind., Volume 7, Number 4, 1994

[10] T. Kletz, "What You Don't Have, Can't Leak," Chemistry and Industry, pp. 287-292

[11] Trevor Kletz and Paul Amyotte, "Process Plants: A Handbook for Inherently Safer Design," Second Edition, 2010

[12] F. I. Khan, P.R. Amyotte, "How to Make Inherent Safety Practice a Reality," Canadian Journal of Chemical Engineering, 81, pp 2-16, 2003

[13] T. Aven, "Risk Assessment and Risk Management: Review of Recent Advances on their Foundation", Elsevier European Journal of Operational Research, 2016

513529 Holistic Planning Is Key for Good Safety in Projects and Construction Industry

Abstract:

Abstract

In todayâ€™s rapidly evolving and increasingly competitive global economy, construction companies are under tremendous pressure to function at the peak of performance, and do so cost-effectively. Many companies adopt cost reduction measures which often tend to provide short-term benefits, but over the long term, have a negative impact. "Schedule" is used as shield by every player during project management to hide all sins.

Owners while building new facilities put tremendous pressure on EPC companies to complete projects in record time. Delays in projects are not acceptable. In the process of meeting timelines many strategic and tactical errors are committed by various players. Errors by one player can have cascading effect on rest of the project and schedule. Such errors adversely impact profit margins of EPC companies, commissioning time lines of production facilities and profitability for owners.

In the changing environment EPC (LSTK/ EPCC, EPCM) companies are struggling. Challenges are many, like price pressure, schedule pressure, quality of resource, retention of technical talent, hiring of smart contractors. In order to be relevant in the changing environment they have to evolve. But most of them are trying to face challenges by doing few superficial things.

Leaderships often do not have realization that they themselves need to evolve to address challenges of changing environment. They need to pull the rest of the organization with focus on mindset and behavior. Secondly for evolving organization, strategies and processes need to be regularly updated and should be relevant. Irrelevant processes and slow response of organization impact moral of employees.

In this situation they must work on couple of fronts. They should facilitate evolution of best project management practices. Second mindset and behaviors. Leadership themselves need to take balcony view so they can adjust themselves and pull rest of the organization to the next level of word class performance

This paper highlights the challenges that were faced by EPC (construction) companies, short-term gains achieved through simple changes, long-term gains made through holistic planning and benefits realized in the process.

513015 Development of Chemical Accident Response Scenario for Public Officers in Korea

Abstract:

Nowadays, Korea is adopting the Incident Management System (ICS) in U.S. for managing complex disasters. In this study, a disaster response scenario which is focused on chemical accident is developed for acclimating public officers to the new response management system. 13 management units which participated in disaster response are selected for Korean response management system and their tasks are defined. Also, distinct characteristics of chemical accident compared with other accidents, such as diffusion and change in the extent of damage are applied to the response scenario.

Basically the scenario consists of timeline, accident situation, damage degree, main issues of specific time and the responses of management units. Wind direction, wind speed, location, accident equipment, accident materials and leakage are considered for making accident situation that rarely occurs. Because the ordinary or trivial accidents are treated well with no more special training, the rare accidents are worthy of training. The damage degree is determined by previous accidents that have similar scale of the scenario situation. The responses of management units are based on the assumption that the accident situation is solved smoothly by carrying out their work without any problems or failure.

In order to train the government employees effectively, 23 main issues are chosen by case study. Each of the problems is repetitive or significant problems when responding to chemical accidents. These issues are reflected in scenario that take into account the time that the problem occurs. By presenting the necessary countermeasures to solve the major issues that take place in specific time, it is possible to solve the problems occurring when responding to chemical accidents.

This research was supported by a grant [MOIS-DP-2015-10] through the Disaster and Safety Management Institute funded by Ministry of the Interior and Safety of Korean government.

506606 Improved Prediction of Alkyd Reactors Via Infrequent-Delayed Observations

Abstract:

Improved Prediction of Alkyd Reactors via Infrequent-Delayed Observations

Engr. Dr. Uzoh, Chigozie Francolins, Nnamdi Azikiwe University, Awka

Abstract

Due to lack of detailed kinetic model, the alkyd resin polymerization reactor problem has been a subject of extensive theoretical, simulation and experimental studies, given the demand of processes with better compromises between safety, productivity and quality, and the development of control methods. The problem of alkyd reactor polycondensation kinetics problems has been addressed. This was enhanced through a novel mechanistic approach and discrete-delayed measurements. The predictor provides adequate empirical-analytical basis for evaluating and operating alkyd batch reactor based on the present performance in order to monitor, control, correct the reactor and ensure prompt termination of the batch such that the final product quality index is maintained within specification. The current proposal considers modification of the existing models to include certain adjustable parameters whose values must be computed numerically such that the error measured between the actual and the predicted results are adequately minimized. The new approach was tested and validated in a reasonable number of reactions under realistic conditions. The results were found to be more reliable than the conventional approaches. Further detailed comparison of the results achieved through the error dynamics and RMSE revealed a significant overall improvement in the studied reactor calibration problems.

Recent Publications

1. **1. Uzoh, C. F.,** Onukwuli, O. D., 2016. Improved Prediction of Alkyd Reactors via Infrequent Delayed Observations. **Chemical Engineering Research and Design, 117 pg 168-176**. DOI: 10.1016/j.cherd.2016.10.016. **Elsevier.**
2. **2. Uzoh, C. F.,** Onukwuli, O. D., 2016. Optimal prediction of PKS: RSO modiï¬ed alkyd resin polycondensation process using discrete-delayed observations, ANN and RSM-GA techniques. **Journal of Coating Technology and Research**. DOI 10.1007/s11998-016-9881-6. **Springer**
3. **3. Uzoh, C. F.,** Onukwuli, O. D., 2016. Investigating the optimum unsaturated fatty acid content and oil length for auto-oxidative drying of palm-stearin-based alkyd resin. **Progress in Organic Coating, 101 pg 71-80. Elservier.**

1. **Uzoh C. F.**, Onukwuli O. D., 2016. Application of sinusoidal function and a 2^{5-1} fractional factorial array in the kinetics and optimization study of gmelina seed oil modified alkyd resin synthesis. **Chemical Product and Process Modeling**. DOI: 10.1515/cppm-2016-0003. **De Gruyter**.
2. **Uzoh, C. F.,** Onukwuli, O. D., Odera, R. S., Ofochebe, S., 2013. Optimization of polyesterification process for production of palm oil modified alkyd resin using response surface methodology. **Journal of Environmental Chemical Engineering**. 777-785. **Elservier**.
3. **6. Uzoh C. F.**, Onukwuli O. D., 2016. A surrogate model for acid functional group conversion in alkyd reactor. **South African Journal of Chemical Engineering. Elsevier.**
4. **Uzoh C.F.,** Onukwuli O.D., Madiebo Emeka, 2015. Kinetics of Castor Oil Alkyd Resin Polycondensation Reaction. Journal of **Chemical Engineering and Process Technology**

Biography

Engr. Dr. Uzoh, Chigozie Francolins is a Lecturer, Researcher and Chemical Engineer with expertise in Polymer Chemistry and Process System Engineering/Optimization. He has B.Eng, M.Eng and PhD in Chemical Engineering. He is a very dynamic and skilled researcher with excellent analytical and problem solving skills and strong aptitude for conducting research related to our current needs. He has used experimental process design and simulation to demonstrate sustainable production route powered by materials of regenerative resources. He has contributed tremendously to a broad understanding of novel polymeric materials for coatings applications. His impressive publication record includes numerous peer-reviewed articles in different ISI indexed journals, book chapter, and a number of presentations at important international conferences. He is extensively cited and has also formed the basis for a number of industrial and academic collaborations.

Email: cf.uzoh@unizik.edu.ng

510011 Managing Labs and Pilot Plant Safety Based on Risk Based Process Safety Management

Abstract:

Abstract

Pilot Plants are considered to be the important tool for tackling the scale up problems. Even though the quantity of the chemicals used is less compared to a commercial plant, they are capable of creating significant hazards and risks. Hence the hazards and risks posed by pilot plants are critical. The increasing number of accidents in research laboratories and pilot plants all over the world is a testimony to the above fact. Many organizations struggle to determine what guidelines they should apply to their own units. It is essential that the risks at pilot plants and laboratories are to be assessed and properly controlled to protect lives, property and environment.

If the elements of Process Safety Management (PSM) are implemented in a research and development facility accordingly safety can significantly improve. The key strategies suggested include properly conducted preconstruction safety reviews, leak tight design and construction, close attention to area classification, well designed ventilation, well planned storage, properly conducted maintenance, integrity management and strictly controlled procedures for the management of change.

Paper/Presentation Details

- Commit to Process Safety
- Understand Hazards and Risks in Pilot Plants and Research Labs
- Hazard Identification and Risk Analysis
- Manage Risk
- Asset Integrity and Reliability
- Training and Performance Assurance
- Management of Change
- Emergency Management
- Process Safety Performance Measurement Metrics
- Case Study
- PSM implementation in Pilot Plants and Laboratories in TRC

511345 Hybrid Mathematical Modeling and Deep Learning of Big Data for Optimal Sensor Placement and Tracking of Chemical Leaks

Abstract:

Leak incidents in chemical plants can easily escalate and spread to secondary and tertiary accidents, such as fire and explosions, and classified as one of major industrial accidents that may cause large personal injuries and material damage, if it not controlled reliably at the initial stage. A number of sensors are allocated near high-risk leak sources for earlier detection of leaks. Data reconciliation is performed to minimize nuisance alarms. In addition to data gathering, it is important to develop an active leak-source tracking system that promptly monitors the information and diagnoses about the leak source at the initial stage of the accident development â€" one of challenging inverse problems in safety. When the right source(s) of leak is diagnosed, it informs on-site response agents of the derived leak information, enabling systematic and efficient mitigation responses. Improving detectability and diagnostic performance of this kind of system contributes to the success rate of the whole response process and minimizes the consequence by the chemical leak and dispersion.

In this study, we propose an intelligent tracking system monitoring leaks spreading through a plant boundary and properly diagnosing the leak source in time. It learns the big data from dynamic simulations of representing probable leak scenarios, selected from risk analysis, and predicts the leak source by inputting a set of selected time-series sensor data to the deep neural networks. Training sets of data are obtained out of CFD simulations performed on the leak accident scenarios for actual chemical plant geometry and yearly distribution of weather conditions, and validated models are developed to predict leaking points by extracting concentration data from simulation results at virtual sensor locations.

Two types of popular network structure are studied to compare the best performance: deep feed-forward neural network (FNN) and recurrent neural network (RNN), reflecting on the time-series patterns of collected concentration data in real-time. The proposed tracking method based on deep neural networks not only shows high performance but also overcomes various problems of existing leak source tracking methods based on inverse vector tracking, such as local minimum, large demand time to track and essential use of sensitive mobile sensors, etc. Industrial case studies are discussed and it is expected that the proposed method can be widely used to replace the existing ad-hoc methods for leak source tracking.

511261 Stress Corrosion Cracking of Type 304 Stainless Steel Stem of Pressure Relief Valve in Crude Oil and Ethanol Environments

Abstract:

Stress corrosion cracking (SCC) occurs due to the combined influence of corrosive environment and tensile stress. SCC of stainless steel (type 304) has been widely studied in various aqueous solutions (such as acidic chloride solution, thiosulfate solution) at room temperature. However, it has not been studied in crude oil or ethanol environment, which is common in transportation of these chemicals in rail tank cars. There exist conditions where the stainless steel stem of the pressure relief valve under substantial tensile load is exposed to the corrosive environment. Pressure relief valve (PRV) is used a relief device from pressure build-up in the tank and the stem that retains a spring exerting a large static load (over a ton). Here, the chemical penetrates the crevice formed between the stem and spring retainer creating a latent corrosion site, where electro-chemical potential may initiate the corrosion cracking mechanism. In the present study, stainless steel corrosion containers with necessary support systems (*e.g.*, pressure gage, temperature gage, purging system, drainage system) have been fabricated to test the stress corrosion cracking of stainless steel stem exposed to crude oil and ethanol environments. The full size PRVs are mounted on the top of the container and the lower part of the PRV is immersed into the liquid to simulate actual conditions for several months. Actuators have been setup to create waves and splashes inside the tank. The potential crack generation and growth are have been monitored by combination of liquid penetrant method and eddy current testing apparatus. The experiments will enhance our understanding of why the stainless steel stems (type 304) are failing while tank cars are carrying crude oil and ethanol.

512740 Bowtie Risk Management Methodology in Chlor-Alkali Industry

Abstract:

This paper presents examples of practical implementation of the Bowtie barrier-based methodology for the management of major hazard risks in a Chlor-Alkali production plant with membrane technology. It describes how Bowties were built and used during the Major Accident Hazard Risk Assessment (MAHRA) conducted as part of the Safety Case preparation. Bowtie diagrams cover technical, procedural and human aspects of facility operation. Also, the role of the Bowties in the determination of Safety Critical Elements (SCEs) and Safety Critical Tasks (SCT) is described along with the information related to the responsibilities/accountabilities for ensuring the integrity of barriers and safeguards. Bowtie diagrams were built following CCPS best practice recommendations.

510942 Safety Assessment of Secondary Alcohol Oxidation with Hydrogen Peroxide

Abstract:

Safety Assessment of Secondary Alcohol Oxidation with Hydrogen Peroxide

Yue Sun[1], Lei Ni[2], Maria Papadaki[3], Wen Zhu[1], Juncheng Jiang[2] and M. Sam Mannan[1]

1 Mary Kay Oâ€™Conner Process Safety Center, Artie McFerrin Department of Chemical Engineering, Texas A&M University, College Station, Texas, 77843-3122, USA

2 Jiangsu Key Laboratory of Hazardous Chemicals Safety and Control, College of Safety Science and Engineering, Nanjing Tech University, Mail Box 13, No. 200, North Zhongshan Road, 210009, Nanjing, China

3 Department of Environmental & Natural Resources Management, School of Engineering, University of Patras Seferi 2, Agrinio GR30100, Greece

+1(979)985-0919, yuesun@tamu.edu

Abstract

Ketones are produced on massive scale in industry as solvents, polymer precursors, and pharmaceuticals. Current ways of producing ketone through alcohol dehydrogenation are energy intensive and usually expensive, hazardous and toxic due to employing strong oxidizing agents. In 1997, Kazuhiko Sato, et al., found a way of producing ketones via 30 % hydrogen peroxide oxidation of alcohol, and tungstate as catalyst along with a phase transfer catalyst. This method is considered as â€˜green chemistryâ€™ since it is organic solvent- and halide-free. A great deal of laboratory work was subsequently done focusing on different catalysts and experimental conditions as well as their effects on yields of different ketones.

However, there are safety concerns for scale-up regarding this reaction system because of its runaway potential. Thus, the purpose of this work is to conduct a comprehensive study of thermal and kinetic behavior of this reaction system. Calorimeters, such as DSC, Phi-TEC and RC1e, are used to help study, experimentally and theoretically, these reactions under normal operating and runaway conditions. In this paper, we report thermal hazard assessment of these reactions and discuss their classification according to Stoessel criticality diagram. Moreover, the influence of stirring rate on reaction heat generation is discussed. The findings will be further used to propose measures for safer design and scale-up of this reaction process in order to avert a potential runaway.

Keywords

Thermal runaway, safety assessment, hydrogen peroxide, secondary alcohol oxidation, calorimeter

511241 Performing Quantitative Risk Calculations for a Natural Gas Pipeline

Abstract:

Over the past several years, Quest Consultants Inc. has conducted quantitative risk analysis (QRA) and risk assessment studies for a range of pipelines in the United States and abroad. In most instances, the risk acceptance or tolerability criteria is defined by the individual risk (IR) to a person; often this risk is presented as location specific individual risk (LSIR). The LSIR is a measure of the risk to a person who is continuously at a specific location.

In recent years, there has been increasing dependence on the use of societal risk acceptance or tolerability criteria, including the risk associated with pipelines. Pipelines are often described as linear sources of risk, like highways and rail lines. The risk analysis methodology used to calculate the risk associated with fixed facilities (e.g., refineries and chemical plants) cannot be directly applied to linear risk sources.

This paper presents a risk calculation methodology that can be applied to linear risk sources and describes how the societal risk (in the form of F-N curves) can be calculated.

An Approach to QRA Using Modeling to Develop Event Frequencies

Bijan Najafi
JENSEN HUGHES, Inc.
2105 South Bascom Avenue, Suite 350, Campbell, CA 95008
bnajafi@jensenhughes.com

Adan Kiyama
JENSEN HUGHES, Inc.
8827 W. Sam Houston Pkwy. N., Suite 150, Houston, TX 77040-5399
akiyama@jensenhughes.com

Richard Anoba
JENSEN HUGHES, Inc.
309 E. Morehead St., Suite 210, Charlotte, NC 28202-2301
ranoba@jensenhughes.com

Paul J. Amico
JENSEN HUGHES, Inc.
111 Rockville Pike, Suite 550, Rockville, MD 20850-5109
pamico@jensenhughes.com

Prepared for Poster Presentation at
American Institute of Chemical Engineers
2018 Spring Meeting and 14th Global Congress on Process Safety
Orlando, FL
April 22 – 26, 2018

An Approach to QRA Using Modeling to Develop Event Frequencies

Bijan Najafi, JENSEN HUGHES, Inc., bnajafi@jensenhughes.com

Adan Kiyama, JENSEN HUGHES, Inc., akiyama@jensenhughes.com

Richard Anoba, JENSEN HUGHES, Inc., ranoba@jensenhughes.com

Paul J. Amico, JENSEN HUGHES, Inc., pamico@jensenhughes.com

Keywords: Risk Assessment, Risk Management, Human Performance, Accident Sequence Analysis, Accident Prevention, Systems Analysis, Risk-Informed Decision Making, Performance-Based Design, Internal and External Hazards.

Abstract

The traditional QRAs in the Oil & Gas industry rely on historical data to account for the frequency of the loss of containment. Such approaches tend to emphasize mitigation over prevention and do not take into account process/cultural/procedural contributors specific to each operation, which have important effects into the overall risk profile.

For over four (4) decades a variety of industries including, but not limited to: nuclear, aerospace, and DOE/DoD chemical facilities, have used an approach to QRA that allows for modeling of systems and human actions that may contribute to loss of containment and therefore, emphasize "prevention", as well as "mitigation". These approaches rely on historical data to develop and quantify the model for loss of containment. This is done for a variety of reasons that include:

1. Allows for modeling of complex, as well as simple, systems with significant dependencies.
2. Allows for modeling of human interface for actions that are intended to prevent or mitigate the accident, as well as those that may initiate and/or accelerate accident progression.
3. Provides insights into "prevention", as well as "mitigation" of risk.
4. Helps develop a quantitative risk management tool.

This paper discusses the application of such methods in the oil and gas, chemical processing, and manufacturing industries.

1 Introduction & Background

Risk assessment is typically done to assess and manage risks to human health and safety, property or business loss, and/or environmental damages. Other objectives derive from the need for risk assessments that vary based on the nature of the facility, its location, regulatory & insurance requirements, as well as other specific reasons.

Risk assessment has been used in various industries to accomplish various objectives, including:

1. Identify and implement improvements into the design, operation, and maintenance of the facility,
2. Assess resilience of the facility to internal (such as fire) and external (such as hurricane and earthquake) hazards,
3. Manage the risk through the life cycle of a facility to ensure that the design and construction objectives of the facility are maintained throughout its life cycle,
4. Use in operation & maintenance (O&M) to minimize risk and maximize operability, including improved reliability, availability, and maintainability (RAM).

To meet some of these objectives, a risk assessment will need to have the pedigree to support risk-informed decision making (RIDM). The pedigree is highly dependent on the complexity of the process, system, and man-machine interface.

This paper will offer case studies and insights where such risk assessments served these objectives.

1.1 Classical QRA in the Process Industry

Quantitative Risk Assessment (QRA) is widely used in the Process Industry as a decision tool to provide guidance to designers and operators on how to minimize risks, as well as for companies to achieve a balance between project costs and safety.

The use of QRA in the oil, gas, and petrochemical industry was initially based on the Probabilistic Risk Assessment (PRA) methodology used in the nuclear industry where complex processes and systems dictate detailed models for accident initiation and progression, system response, dependencies, and human performance.

One of the key distinctive differences between the QRA and PRA methodologies is the level of detail in which the accident initiation and preventive/mitigative systems, and human performance are modeled. In oil, gas, and chemical facility QRAs, the incident frequency is typically calculated based on failure rates gathered from historical data. This "lagging" approach introduces a series of uncertainties and errors to the QRA study in terms of a high-risk profile for the project.

On the other hand, the main approach currently followed by the oil, gas, and chemical industry to improve the accuracy of QRA has been focused on using complex Consequence Modeling tools. This has been achieved by developing highly sophisticated computational models to quantify the severity of the loss of containment events, such as the use of computational fluid dynamics

(CFD) modeling software. However, it must be acknowledged that the overall risk profile of a facility is a product of both likelihood and consequence. Therefore, considering the value of an increased investment on System Modeling may yield benefits not currently recognized by the oil and chemical industry.

1.2 An Alternative Approach to QRA

The Alternative Approach to QRA presented in this article is based on a thorough understanding of the sequence of events leading to an accident. All risk assessments model and analyze two distinct phases of an accident:

- Pre-loss of containment – This phase involves definition and characterization of the initiating events leading to the loss of containment, including "accident escalation" (a.k.a. "accident progression") scenarios. The initiating event could be a disturbance to the system which could include any deviation from the normal safe operating envelope. The disturbance may be caused by internal (pump failure, mal-operation of equipment, internal fire, etc.) or external factors (extreme weather, seismic, etc.). Depending on the complexity of the system, the ramifications of the pre-loss of containment scenarios can be modeled using fault trees or simpler tools such as FMEAs. The complexity of the systems is usually assessed based on the number of components involved in the containment system, the number of functional dependencies linking these components, and the human interface.
- Post-loss of containment - The final consequence following the progression of each event derived from the initial loss of containment is modeled using an event tree. The frequency of each consequence is calculated by multiplying the frequency of the initial loss of containment by each outcome frequency.

This Alternative Approach not only provides estimates and insights into risk, it is also a powerful tool in improving operability and protecting assets through better design, operation, and preventive maintenance. This has been demonstrated in many industries instituting formal risk management practices that have been accepted and encouraged by authorities having jurisdiction. In order to illustrate this Alternative Approach, this article is presented as follows:

Section 2.0 provides an overview of the Alternative Approach to Classical QRA.

Section 3.0 provides a summary of selected case studies where this approach was applied in the chemical, oil & gas, and process industries.

Section 4.0 discusses the potential challenges and the rewards of the alternative approach to highlight where and how it may be applied to satisfy varying objectives.

2 Overview of the Alternative Approach

The steps required for a typical QRA are shown in Figure 1. The alternative approach to QRA that has been used in several industries for over four (4) decades has similar steps; however, the differences lie in the nuances that, for the most part, are the result of complex dependencies, human interface, and internal/external influences altering the accident progressing.

It should be noted that the discussion in the Section is, for the most part, focused on the pre-loss of containment phase. However, some considerations, such as human performance, multiple objectives, and multiple hazards concepts apply to both.

Objective (s) – Similar to classical QRA, the first step requires defining key objectives of the QRA. It should be noted that such assessments may need to serve multiple objectives that at times may point to contradicting decisions. For example, risk vs. availability.

Modeling – The modeling is typically done in the two phases of the pre- and post- loss of containment because of the need for different tools and methods. This is also driven by the objectives of the assessment.

Initiation - The next step is identification and characterization of disturbances in the system and/or process that may negatively impact the stated objective(s). These disturbances may be the result of random equipment failures, human interference (errors of omission, for example), or external influences such fires, floods, and earthquakes. This step will require understanding and treatment of dependencies that are critical to the assessment, such as the impact of the initiator on the protective features of the system and/or process. For example, when a fire affects the instrumentation & control cables preventing operation of the safety functions by design or operator actions.

Progression – Progression of a disturbance to the system and/or process will involve a combination of random equipment failure, operator actions (failure to take an action or incorrect action due to faulty signal(s) or poor training), and events (e.g., size and location of a leak and physical/operational configuration of the plant at the time of the accident). Complexity of the dependencies among these factors is an important consideration in the approach selected for the assessment and could be very critical to the estimate of risk and the resulting decisions relative to risk management.

Quantification and Insights – This approach covers the same steps as the classical approach, with the following caveats.

- Identifies and implements findings and recommendations to reduce the risk profile of the project + *insights and recommendations for pre-loss of containment.*
- Demonstrates that the project meets the statutory risk requirement set by the authorities and/or the client's corporate risk tolerability criteria + *margin that is reflective of full list of contributors, such as human interface.*
- Confirms that more layers of protection/defense may not reduce the risk enough to justify the cost of the added protection/defense layers, and the risk is tolerable or follows the ALARP (As Low As Reasonably Practicable) principle + *more representative safety margin.*
- *And representation of uncertainty, qualitative, or quantitative.*

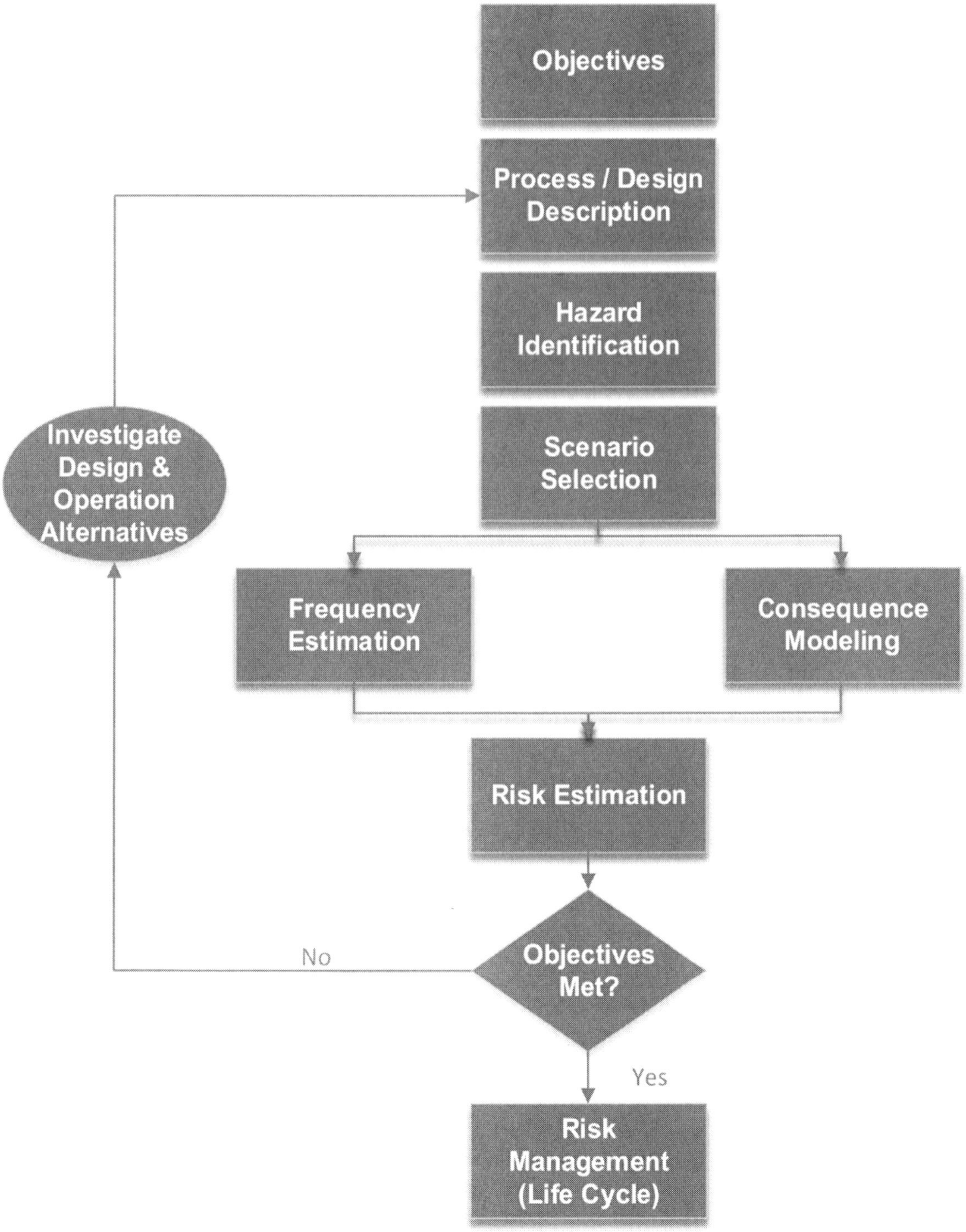

Figure 1 - Overview of the Alternative QRA Approach

3 Individual Case Studies

This section describes some the case studies that utilize the augmented alternative approach to QRA outlined above for various objectives and scopes. The discussion is tailored to protect confidential information of the client, while providing insights into the benefits and challenges of the proposed alternative approach to QRA.

3.1 I - Chemical Agent Disposal (Full Scope of Risk Assessment Against Internal and External Hazards) [1]

In 1985, Congress enacted a Law that directed the Department of the Army (DA) to establish a program to dispose of the United States stockpile of unitary chemical weapons and agents. In 1997, the United States ratified the Chemical Weapons Convention. This treaty committed the signatories to destroy all their chemical warfare material in an environmentally safe manner. To accomplish this objective, the U.S. Army used risk assessments to understand and control risks associated with storage and disposal.

Specific objectives of the program included:

- Assessment of the risk to public and site workers from various stages of disposal,
- Identify potential improvements in systems or operations that could further reduce the risks.
- Development of an integrated risk management program to gain additional insights as plant design and operation change, or additional insights into accident behavior become known.

Due to the complexities of the system and processes, the Alternative QRA approach was employed. This QRA involved all the elements discussed in Section 2 of this paper, and consideration of internal (fires) and external hazards such as: earthquake, aircraft impact, tornados & high winds, and hydrogen explosions.

The following insights were obtained for the assessment:

- Public risk from the process was dominated by the potential for a facility fire that affected agent inventories within the facility. This assessment included potential for the fire to spread across the facility, thus acting as one of the means of accident progression. *This is an example where an initiator can impair safety features of a facility design to reduce risk.*
- Seismically-induced fires contributed to the risk due to fire affecting large portions of the facility. In part, due to impaired state of the facility, including its fire protection

[1] Umatilla Chemical Agent Disposal Facility Quantitative Risk Assessment (https://www.nrc.gov/docs/ML0928/ML092890526.pdf) – participation from staff at JENSEN HUGHES.

capabilities under post-earthquake conditions. *This is an example of integrated impact of multiple hazards.*

- A non-negligible part of the risk was attributed to a temporary state of the facility operation even though the frequency and duration of those phases are not a significant part of the plant operation. *This is an example that consideration of various modes of facilities operation is critical to accurate estimates of risk. Infrequent and short-duration activities may be significant contributors to risk.*

3.2 II - Off-Shore Oil Platform (Integrated Risk Assessment and Online Risk Management) [2]

The oil and gas industry operates facilities on both land and sea to extract the hydrocarbons for storage and transport to refining and the market. The operations involve exploration, drilling, pumping, storing, and transporting the hydrocarbons. Each of these types of industrial operations entails risks associated with processing hydrocarbons.

Hazards and Accidents

Oil and gas facilities are subject to hazards that impact the safety of the platform and personnel, environmental containment of the hydrocarbons, and production operations. Each of these hazards shares some underlying characteristic that can be assessed and managed beneficially for each hazard, although unique aspects of each type of hazard also exist. For example, the risk of loss of piping integrity for production piping is hazardous to both personnel and platform safety, environmental containment, and production; but the risk of loss of an operating pump may only impact production risk, when no loss of containment ensues.

The design of an oil platform includes safeguards to prevent, control, and mitigate the effects of hazards to personnel and the environment. Each safeguard is required to provide a minimum level of performance, in terms of availability and functionality, and it is necessary that each safeguard maintains its respective performance criteria throughout its operational life, where ongoing operational monitoring is required, and the testing frequency and functional acceptance criteria to be applied.

Risk Assessment

Current risk assessment methods employed by the oil and gas industry include qualitative assessments and quantitative risk assessments provided for safety accident risk (those that involve potential loss of life and loss of the platform facility). Some examples of existing tools that are applied to the industry are Layer-of-Protection Analysis (LOPA) and Quantitative Risk Assessment (QRA). These tools are used and provide reasonable assessments of the risks associated with the safety aspects of platform operations.

These types of assessments are generally effective at assessing the safety and accident risk. However, given the common cause events that result in safety, environmental, and production risks, as seen by the common impacts from the hazards, benefit can be gained from expanding

[2] Based on a project completed for a confidential client.

the analysis to include a comprehensive, total integrated assessment. Such an integrated assessment of the risk to personnel and platform safety, risk of environmental release, and risk of production losses during platform operation would focus on the assessment of the common hazards and accident types.

Optimization

Given that the industry currently treats these risks using established methods, the natural question that occurs is "how does the application of these augmented methods and tools result in significant benefit beyond what is currently performed?" The answer lies in the ability to account for these hazards and goals with a combined assessment and with a risk assessment tool that can be used to understand the assessment. Because safety risk is typically the top priority for industrial facilities, improvements in safety can have a negative impact on production risk. Although this is the prudent way to approach industrial operations involving injury or fatality risk, the two do not necessarily have to be inversely related; the benefit of a combined assessment and tool would allow the oil and gas company to meet their safety goals and environmental impact goals while providing the ability to assess improvements that also decrease loss of production risks. At the very least, the integrated assessment and tool can be used to evaluate multiple options for decreasing the risk associated with safety and environmental impact while aiming to limit the risk of loss of production. This combined approach offers a best-of-all worlds approach to platform operation.

Scalability and Fleet Benefit

The oil and gas industry operates hundreds, if not thousands, of offshore production facilities. Although there is some unique design and operating characteristics of each of these facilities, the generic approach, using PRA methods and using an integrated model tool, can be effectively scaled and applied to a fleet of facilities with much less effort than that needed to develop the first risk model. A comparison of results between platform models may yield additional platform-specific insights that general safety studies may not identify. This would benefit the operating crew by providing specific risk assessment information rather than providing general rules or strategies for reducing or mitigating risks associated with platform safety, environmental containment, and production loss.

Online Risk Management

One key objective of the Integrated Risk Solution adopted by this project was to develop and install an online software tool to allow for monitoring system risk during the facility operation to direct online testing and maintenance. Such a tool was intended to improve risk through more frequent testing while improving reliability and thus better operability (reduced downtime) was achieved.

Figure 2 below illustrates the workflow of information employed by the online risk management software tool.

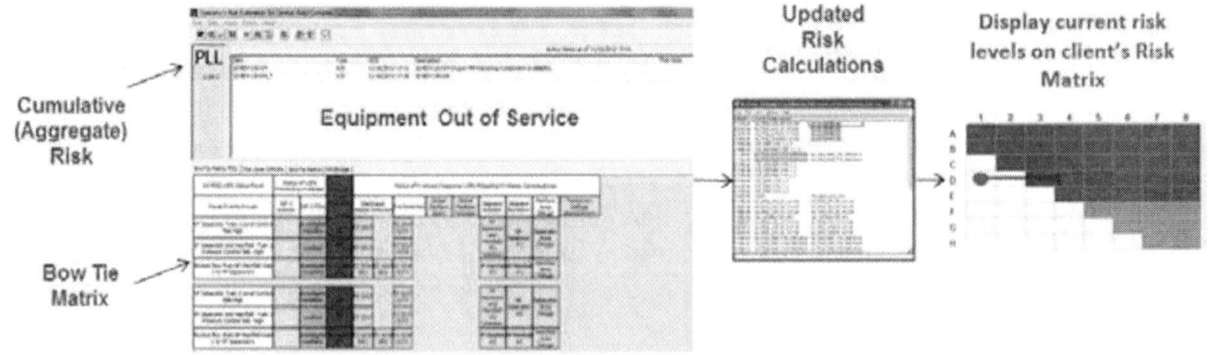

Figure 2 – Sample Workflow of the Online Risk Management Software

The online risk management project included the following outputs:

- Risk model that incorporates inputs from QRA and HAZOP/LOPAs studies
- Risk Model Building Notebook
- Risk Model Validation Cases
- Risk model on a central hosted environment

The Risk Model Building Notebook included:

- General model description and key assumptions
- Development of accident sequence model for hydrocarbon release and fatalities based on QRA and major accident risk (MAR) reports
- Development of initiating event fault tree logic derived from HAZOP and LOPA studies
- Development of fault tree integration logic from design data sources, e.g., well alignments, intersystem dependencies, etc.
- Development of equipment failure rates, release frequencies, and phenomenological split fractions based on facility data sources

The Model Building Notebook documents all risk model assumptions derived from risk assessment and technical reference documents, and was used as the basis for risk model verification and peer review.

3.3 III - Industrial - Refrigeration (Performance-Based Design) [3]

International attention on climate change has increased efforts in the development of low global warming potential (GWP) refrigerants to meet aggressive regulations focused on reducing greenhouse gases. There are limited acceptable options for low GWP refrigerants that have low toxicity, are non-flammable, and can achieve required energy efficiency goals when compared to

[3] Based on a project completed for a confidential client.

current refrigerants. For these reasons, the industry is evaluating the potential use of mildly flammable low GWP refrigerants. Currently, refrigerants used for Heating, Ventilation, and Air Conditioning (HVAC) applications are non-flammable; thus, the application of mildly flammable refrigerants introduces new risks to the industry. In the event of a system leak, the resulting refrigerant concentration of a mildly flammable refrigerant can reach flammable limits in small or confined spaces. Ignition of a leaked refrigerant mass can produce multiple hazard scenarios, such as ignition of surrounding flammable materials, damage to equipment or structures from over pressurization in a confined space, and potentially hazardous combustion products.

Due to the potential hazards previously mentioned, a systematic, augmented approach to risk assessment was employed that covered the entire life cycle from manufacturing, warehousing & transportation, installation, normal operation, service & maintenance, to decommissioning of these refrigerants. Such assessment involved:

1. Identification and characterization of the accident initiators, involving component or human interface.
2. Development and characterization of accident progression leading to ignition.
3. Evaluation of the available data, beyond equipment reliability or hazard frequency, to include influence factors altering accident progression.
4. Assessment of human interface to identify actions that may initiate, facilitate, or mitigate one or more accident scenarios, and quantification of human performance.
5. Sensitivity analysis to gain insights into design, maintenance & operations, as well as governing industry standards.

The assessment produced insights that included:

1. Improvements to the design; for example: through additional instrumentation and control devices,
2. Improvements to the training and certification for human interface, including service & maintenance, and
3. Understanding the limitations and uncertainty of data, models, and results to guide risk-informed decisions. At times, additional confidence in data is needed to guide decisions.

4 Challenges & Rewards

The Alternative approach to QRA discussed in the previous sections yields more benefits as the complexities and interdependencies of the system and process increase. The challenge thus being to choose the level of rigor in the analysis that provides the most cost-effective path to insights that satisfy the objectives. It is imperative to understand key contributors and address them, even in the simplest of the systems and processes.

In order to select the right tool for each project, the analysis must start with a clear understanding of the challenges and rewards in selecting levels of complexity and detail.

4.1 Challenges

The modeling of pre- or post-containment phase can pose a number of challenges that include: data, dependencies, and uncertainties.

Data Collection and Analysis

While obtaining equipment reliability data may have its own challenges, there are other types of data that may offer challenges for more robust analysis of accident progression in a complex system & process.

- Common cause failures, mainly on redundant equipment
- Ability to estimate non-statistical events (not equipment reliability) that are reflective of system/process operational stage or other non-random events
- Human performance, from actions before an accident (faulty maintenance) to post accident actions that either fail to follow procedure or take an erroneous action that accelerate and/or increase the risk. [4]
- Human action dependencies

Dependency Analysis

As the system and intricacy of process redundancies and dependencies increase, it becomes necessary to understand these dependencies. At times this can be a challenge that requires reliance on operational history, testing, and/or deterministic models. Examples include functional and support system dependencies.

Uncertainty Analysis

Understanding the source and magnitude of the uncertainty is a critical piece of any risk-informed decision making. From a decision on acceptable risk (ALARP) to its use to gain insights into system and/or process improvements. The practice of uncertainty covers a wide range from identification and characterization (i.e., graded) to full quantitative evaluation. This again will and can be tailored to the complexity of the system and process.

4.2 Rewards

The Alternative Approach to QRA has a number of rewards. Below are a few examples.

Risk Informed Decision Making - RIDM (Case Study I, II, and III)

In general, such approach helps provide more informative estimate of risk that can help with determination of ALARP.

[4] The International Organization of Oil and Gas Producers has published guidance materials to incorporate Human Reliability in risk assessments

- The data related to frequency of initiating events obtained from the Augmented QRA may serve to better define likelihood of low probability - high consequence events.
- Similar to the above, using historical accident frequencies introduces uncertainties since "two facilities are never the same"; therefore, there is uncertainty introduced when using generic data.
- Systems modeling may be a good option, particularly for novel processes and designs where generic data is not available or would misrepresent the reliability of the operation.
- In some cases when a higher degree of accuracy is needed, the use of systems modeling may reduce conservatism in QRA typically used in industry; therefore, this novel approach may yield cost savings in the design of safety related improvements.

Another example of an advantage of Risk Informed Decision Making (RIDM) is providing improvements in the design, operation and maintenance of a facility or process. While classical QRA provides plenty of opportunities for such practice for mitigation, the Alternative Approach brings the same balance to the pre-loss of containment phase for decisions to prevent accidental release of hazardous materials in the first place.

In another case study (not described in this paper) the same approach is being used as part of the design to serve improved reliability and operability.

Determination of Influence of Human Performance (Case Study I, II, and III)

Most systems and processes have significant human interface, even those that do not require actions to prevent or protect against accidents. Failure to consider these interfaces can potentially provide inaccurate estimate of risk and its contributors, and lead to inaccurate decisions on acceptability of the risk or system modifications.

At least in one of these case studies, human performance was one of the major contributors to the risk and the insights, requiring improved training.

Integration of Hazards Initiated by Multiple Impacts (Case Study I) / Objectives (Case Study II)

In some risk assessments, combined impact of the hazards (e.g., seismically induced fires) are a contributor to risk that cannot be ignored, as was demonstrated in Case Study I. It is critical to review and ensure that such scenarios are not ignored in systems and/or processes; that they may be significant risk contributors.

Where multiple, and sometimes competing, objectives are involved, an integrated approach that starts at the prevention level is critical in finding the right risk insights or design alternatives.

Risk Management (Case Study II)

Risk assessment and risk management are used across a wide range of industries with various complexities and objectives. The approach taken to understand and manage risk can vary from very simple qualitative approach, such as a "what-if" analysis, to more complex methods, such as Quantitative Risk Assessment or Probabilistic Risk Assessment. While a risk assessment is intended to ensure that a facility meets the stakeholder's objectives related to human safety, economic, and environmental protection, risk management aims to ensure that those objectives continue to be met throughout the facility's life cycle.

Quantitative or Probabilistic Risk Assessment is a powerful and necessary tool for Risk Management and Decision-Making where risk aversion is a desired objective. It allows a Company to:

a. Benchmark different designs against an objective set of criteria,
b. Help develop and implement Risk Management Programs, and
c. Define risk control strategies.

Quantitative and Probabilistic risk assessment techniques are designed to provide stakeholders with the ability to assess risk, as well as to prevent and/or minimize its consequences. These methods provide tools to evaluate risk to the public and workers at a facility, and to identify areas for cost-effective risk reduction. While in many cases they are aimed at reducing loss of human life in the event of a major incident, in other cases they are used to assess and prevent economic loss and environmental damage.

Risk management is an integration of the risk assessment and changes that the facility may go through in its life cycle, such as testing, maintenance, planned & unplanned outages, decommissioning, and others.

5 Conclusions and Recommendations

This article provides a series of case studies for risk assessments with differing objectives and systems/processes to demonstrate that the key to a successful risk assessment is to select the method that is best to achieve the stated objectives for the specific system/process. The solution does not always lie on complexity, rather than one that accounts for what is important in meeting the stated objective(s).

Consideration in selecting the right method include the following caveats:

1. Are multiple objectives involved and if they have interdependencies, how do they impact the insights and potential decisions?
2. Is the system & process vulnerable to external influences, such as fire and flood (some based on their location, as well as the system and operation), that may have significant contribution to risks and insights?
3. It may be more cost effective to prevent an accidental release of hazardous materials than to mitigate its consequences. This includes the public and regulatory impact of an accidental release, even with moderate consequences.
4. As the complexity of the system and process increase, consideration of dependencies and redundancies, including instrumentation & control, become a key part of risk assessment. Failure to understand and characterize them may lead to inaccurate risks and insights.
5. Impact of human performance factors to initiate, alter, or mitigate, an accident is likely a critical part of risk. This may be valid from simple to complex systems and processes requiring, at a minimum, identification and qualitative assessment of such actions.

6. The risk of a facility may vary significantly throughout its life cycle and operating phases. Such considerations may alter the estimated risks and insights along the trajectory of a facility's life cycle.

This article discusses tested methods to be included in a QRA to achieve one or more of these objectives.

6 References

1. ASME/ANS RA-S-2008 and Addenda, *Standard for Probabilistic Risk Assessment for Nuclear Power Plant Applications*, ASME/ANS 2008, 2009, 2013.
2. EPRI 3002003116, *An Approach to Risk Aggregation for Risk-Informed Decision-Making*, April 2015.
3. EPRI 1008051, *Outage Configuration Risk Management Consistency Study,* 2003.

18 AIChE Spring Meeting
14TH GLOBAL CONGRESS ON PROCESS SAFETY
ORLANDO FLORIDA 2018

LOPA: Performed When and by Whom

A.M. (Art) Dowell, III
Process Improvement Institute Inc. (PII)
adowell@piii.com

William G. Bridges
Process Improvement Institute Inc. (PII)
wbridges@piii.com

PROCESS IMPROVEMENT INSTITUTE

Prepared for Presentation at
American Institute of Chemical Engineers
2018 Spring Meeting and 14th Global Congress on Process Safety
Orlando, Florida
April 22 – 25, 2018

LOPA: Performed When and by Whom

A.M. (Art) Dowell, III
Process Improvement Institute Inc. (PII)
adowell@piii.com

William G. Bridges
Process Improvement Institute Inc. (PII)
wbridges@piii.com

Keywords: LOPA, IPLs, PHA, leader

Abstract

Layer of protection analysis (LOPA) was introduced in the mid-1990s by Art Dowell at Rohm and Haas Chemical Company (became Dow Chemical, now DowDuPont, Inc.) and by William Bridges at ARCO Chemical (now Lyondell-Basel) and JBF Associates. The first book was published in 2001 by CCPS. Since then, the method has swiftly grown in popularity for use in making risk judgments and in deciding on the SIL rating for an SIF. But, many users of LOPA do not know when to use LOPA and so they overuse this tool; and they do not know who should be doing LOPA, so they many times use a team, similar to or the same as a PHA/HAZOP team. This paper explains what the originators of LOPA intended and why, and also brings the industry up-to-date on the lessons learned from different approaches to using LOPA, related to when to do LOPA and who should do LOPA.

1. Introduction

The initial development of layer of protection analysis (LOPA) was done internally within several individual companies. However, once this method was developed and refined, several companies published papers describing the driving forces behind their efforts to develop the method, their experience with LOPA, and examples of its use (Bridges, 1997[1]; Dowell, 1997[2]; Ewbank and York, 1997[3]). In particular, the papers and discussion among the attendees at the October 1997 CCPS (Center for Chemical Process Safety, part of AIChE), International Conference and Workshop on Risk Analysis in Process Safety, brought agreement that a book describing the LOPA method should be developed.

In parallel with these efforts, discussions took place on the requirements for the design of safety instrumented systems (SIS) to provide the required levels of availability. United States and international standards (ISA S84.01 [1996], IEC [1998, 2000]) [4, 5, 6] described the architecture and design features of SISs. Informative sections suggested methods to determine the required safety integrity level (SIL) of a safety instrumented function (SIF), but LOPA was not mentioned until the draft of International Electrotechnical Commission (IEC) 61511, Part 3, which appeared in late 1999. These issues were summarized in the CCPS workshop on the application of ISA S84, held in 2000.

The first LOPA book was developed by a CCPS committee from 1997 through 2000 and was published in 2001[7] (Art Dowell and William Bridges were the co-originators and were principal authors of the book). In 2007, CCPS commissioned a new guideline book (1) to expand the list of independent protection layers (IPLs) and initiating events (IEs) and (2) to try to remedy some of the major issues noted in the use of LOPA. The new book has been discussed in other papers at past conferences; this book is *Guidelines for Initiating Events and Independent Protection Layers*, CCPS/AIChE, 2015[8]. William Bridges was the primary contractor/author of this book from 2007 to April 2012. Another companion book on related topics, *Guidelines for Conditional Modifiers and Enabling Events* [9], CCPS/AIChE was published in 2013; Mr. Bridges was a committee member and contributed to this book as well.

Since the first book was published, the method has swiftly grown in popularity for use in making risk judgments and in deciding on the SIL rating for an SIF. But, many users of LOPA do not know when to use LOPA and so they overuse this tool; and they do not know who should be doing LOPA, so they many times use a team, similar to or the same as a PHA/HAZOP team. This paper explains what the originators of LOPA intended and why, and also brings the industry up-to-date on the lessons learned from different approaches to using LOPA, related to when to do LOPA and who should do LOPA.

2 When do to LOPA? – Timing

The original LOPA book authors considered LOPA a single analyst job; after a PHA/HAZOP, for just a few scenarios (maybe after 100 HAZOP nodes, 1-10 LOPA scenarios would be done). Instead, the trend appears to be that companies (or perhaps their consultants) make LOPA part of the PHA (in-situ). If the PHA/HAZOP team is properly disciplined on what safeguards qualifies as independent protection layers (IPLs; using a qualitative definition of an IPL), then performing LOPA in situ is usually overkill. In some situations, an experienced qualitative team (HAZOP team) can make just as good or better judgment than provided by LOPA. LOPA is just another way to make a decision, has some pitfalls, and doesn't work for many types of scenarios.

In summary, the three major approaches seen in the industry on WHEN to do LOPA are:

- Do it during the PHA, scenario by scenario.
- Do it during PHA, at the end of a major HAZOP section (or group of sections), when scenarios are complete.
- Do it outside of PHA.

Every approach has different human resource requirements and also has its advantages and disadvantages. Each approach is described below.

2.1 During the PHA, scenario by scenario (NOT RECOMMENDED)

Some organizations do LOPA during the PHA, scenario by scenario. The authors have participated in this timing of LOPA.

For this option (which we DO NOT recommend), the personnel required is the full PHA team.

Practitioners of this approach suggests that the advantages are

- The scenarios are fresh in mind of personnel.
- If the Process Safety Information is in the minds of personnel (that is, the Process Safety Information is not well-documented), they may be able to answer questions more efficiently.

The Process Safety Information should be well-documented to ensure that the PHA is based on the correct information. And the PHA itself should be well documented so that all of the actions and future activities based on the PHA use the correct information.

The disadvantages of this approach are

- It is counterproductive to switch back and forth from the PHA brainstorming (inductive) thought process to the LOPA (deductive) thought process. An entirely different mindset and way of thinking are required for these two thought processes. It is important to keep the brainstorming mindset to identify hazards. A company should ensure NOTHING detracts from brainstorming, because if a scenario is missed, how will the company know if they have enough IPLs for that missed scenario?
- All the hazards and candidate IPLs have not been identified yet. The PHA team may or may not have pursued the cause to the ultimate consequence; sometimes the ultimate consequence is in a different node from the initiating cause and may not be fully understood until the team undertakes the LOPA on the initiating cause. Further, the LOPA done during the PHA of a node or step may generate recommendations for IPLs whose function may be provided by other candidate IPLs. The authors' experience is that it is better to have a complete picture of a process (or at least a group of sections of a process) before undertaking LOPA.

- Wastes time of PHA team. Some members of the PHA team are not necessary for the LOPA process. It is more efficient to do LOPA with a smaller group and bring in the needed expertise only for the scenarios where their knowledge is required.
- May lack needed expertise. Many PHA teams do not include full-time participation by instrumentation/control engineers, or maintenance engineering, supervision, or mechanics. That expertise may be brought in periodically during the PHA to answer questions that have accumulated. Consequently, the LOPA scenarios done during the PHA node may lack critical knowledge and skills, particularly, for understanding independence between the basic process control system and safety instrumented functions. Understanding independence frequently requires the control engineer to drill down into the specific hardware configuration (including wiring) and into the specific software configuration.

Consequently, this approach is **NOT Recommended**.

2.2 During PHA, at the end of a major HAZOP section or group of sections when scenarios are complete (NOT RECOMMENDED)

Some organizations have found it effective to evaluate the required LOPA scenarios with the PHA team at the end of the brainstorming for a major HAZOP section or group of sections. The authors have participated in LOPA done in this fashion.

The personnel required are typically the full PHA Team plus instrumentation and control engineer. The control engineer is needed to understand independence.

The advantages of this approach are

- The scenarios are fresh in mind of personnel.
- All the hazards and candidate IPLs should have been identified.
- If the Process Safety Information is in the minds of personnel (that is, the Process Safety Information is not well-documented), they may be able to answer questions.

As discussed in the previous section, the Process Safety Information should be well-documented to ensure that the PHA is based on the correct information. And the PHA itself should be well documented so that all of the actions and future activities based on the PHA use the correct information.

The disadvantage of this approach is

- It is counterproductive to switch back and forth from the PHA brainstorming (inductive) thought process to the LOPA (deductive) thought process. An entirely different mindset and way of thinking are required for these two thought processes. It is important to keep the brainstorming mindset to identify hazards. A company should ensure NOTHING detracts from brainstorming, because if a scenario is missed, how will the company know if they have enough IPLs for that missed scenario?
- Wastes time of PHA team. Some members of the PHA team are not necessary for the LOPA process. It is more efficient to do LOPA with a smaller group (on

average 1.5 staff) and bring in the needed expertise only for the scenarios where their knowledge is required.

- May lack needed expertise. Many PHA teams do not include full-time participation by instrumentation/control engineers, or maintenance engineering, supervision, or mechanics. That expertise may be brought in periodically during the PHA to answer questions that have accumulated. Consequently, the LOPA scenarios done during the PHA node may lack critical knowledge and skills, particularly, for understanding independence between the basic process control system and safety instrumented functions. Understanding independence frequently requires the control engineer to drill down into the specific hardware configuration (including wiring) and into the specific software configuration.

This approach is also **NOT RECOMMENDED**, though the issues with harm to brainstorming and lack of expertise are easier to manage with this approach.

2.3 Outside of PHA (RECOMMENDED)

The original LOPAs were done outside of the PHA, in many cases, a long time after the PHAs had been completed, and this is still the best approach (with the one improvement of shortening the time between PHA and LOPA).

The personnel required are typically a LOPA analyst, a process engineer (by phone or email, as needed), and other expertise – such as control engineering – by phone or email, as needed. In the authors' experience, the average is about 1.5 people per LOPA scenario.

The advantage of this approach is:

- Provides efficient use of personnel time. There are no large meetings with some personnel not needed and not participating.

The disadvantages of this approach are

- Requires well documented Process Safety Information. Remember that we need good process safety information to do a good PHA and to do all the continuing activities to maintain good process safety for the facility.
- Requires well documented PHA.

This is the **recommended approach** today and in the original LOPA textbook as well. This approach can be practiced immediately after the PHA completed while the PHA is still fresh in the memories of those who would support LOPA on a part-time basis.

3. When do to LOPA? – Triggering: What scenarios go to LOPA?

Similar to the decision on WHEN to do LOPA, there are three main approaches for how to decide WHAT SCENARIOS are selected for LOPA:

- Every scenario.

- All scenarios that have fatality as a consequence.
- All scenarios that have or require a SIF (to determine the target SIL)
- Scenarios recommended by the PHA team.
  - When the PHA team is not sure.
  - When the scenario is complex.

This section describes those approaches.

3.1 Every scenario

Some organizations require that every scenario be assigned a consequence severity and an initiating cause frequency. And they require that every scenario be evaluated by LOPA.

The apparent advantage of this approach is

- Appears to be thorough. In reality, it is overkill.

The disadvantages of this approach are

- Overkill, waste of PHA time.
- Review can easily become superficial because of the high workload for the PHA team.
- When coupled with doing the LOPA in situ with the PHA, with the full PHA Team, this approach has all of the disadvantages listed in section 2.1 as well.

This approach is **NOT Recommended**.

3.2 All scenarios that have fatality

Many organizations require that all scenarios that have multiple fatalities, or a single fatality be evaluated by LOPA. The authors have worked extensively in this system with mixed results:

The advantage of this approach is

- Conservative. It is a good starting point for risk reduction.
- Good practice for PHA teams that are not familiar with LOPA.
- Good practice for organizations are just starting to use LOPA.

The disadvantage of this approach is

- May be more work than required (waste valuable engineering/analyst resources)
- LOPA does not work for all fatality scenarios; it only works for scenarios with multiple types of IPLs available.

A PHA Team with the following characteristics:

- A PHA leader who is a well-trained and vetted LOPA analyst,
- An experienced, knowledgeable PHA team,
- Working on a well understood process technology,

Has proven capable of determining whether risk reduction is needed and how much is required without doing a formal LOPA. However, the authors have observed that many PHAs have not identified the obvious hazards, and consequently cannot correctly evaluate the required risk reduction.

All scenarios that lead to fatalities (for scenarios for which LOPA works) is the **Recommended** approach **when the PHA Leader is NOT a trained and vetted LOPA analyst,** because use of LOPA will continue to train the PHA Leader on independence rules and on process and human reliability.

3.3 All scenarios that have or require a SIF (to determine the target SIL)

Many organizations require that all scenarios that have a SIF or that the PHA think may require a SIF be evaluated by LOPA. The authors have worked extensively in this system with mixed results:

The advantage of this approach is

- Conservative. It is a good starting point for risk reduction.
- Good practice for PHA teams that are not familiar with LOPA.
- Good practice for organizations are just starting to use LOPA.

The disadvantage of this approach is

- Is definitely more work than required.
- LOPA does not work for all fatality scenarios; it only works for scenarios with multiple types of IPLs available.

A PHA Team with the following characteristics:

- A PHA leader who is a well-trained and vetted LOPA analyst, and with moderate competency in SIS fundamentals
- An experienced, knowledgeable PHA team,
- Working on a well understood process technology,

Has proven capable of determining whether risk reduction is needed and how much is required without doing a formal LOPA, including determining when a SIF is required and what the target SIL should be [10]. However, the authors have observed that many PHAs do not have the required expertise in their PHA team leader [11].

All scenarios that lead to SIFs is the **Recommended** approach **when the PHA Leader is NOT a trained and vetted LOPA analyst and does not have a fundamental understanding of SIS,** because use of LOPA will continue to train the PHA Leader on independence rules and will compensate for lack of understanding of SIS.

3.4 Scenarios recommended by the PHA team

As described in the discussion on the previous approach, a PHA with a leader who is a well-trained and vetted LOPA analyst (and who understands the fundamentals of SIS), and an experienced knowledgeable PHA team, working on a well understood process technology, may be able to determine the requirements for risk reduction and SIF/SIL for most scenarios without doing LOPA. Thus, the only scenarios that go to LOPA are those recommended by the PHA team. This recommendation is usually only made when the PHA team is confused on the scenario itself.

The advantage of this approach is:

- LOPA is done only on the scenarios that are uncertain or that are complex.
- Efficient.

The disadvantage of this approach is

- The PHA team may feel comfortable with serious consequence scenarios that do not have sufficient IPLs
- It requires a PHA leader who thoroughly understands LOPA principles, especially independence.
- The PHA leader should understand the basics of SIS.
- It requires knowledge and expertise on the PHA team (or in situ coaching by the team leader) to understand independence of initiating causes and IPLs. The PHA team must be aware when they do not know for certain that the LOPA requirements for independence are met, or are otherwise confused on the scenario. Then they must recommend that scenario for the formal LOPA evaluation.

This is the **Recommended Approach when the PHA Leader is a trained and vetted LOPA analyst**.

3.5 Observed Phases of LOPA Application

PII has observed that most companies tend to go through phases of the use of LOPA.

Figure 1Figure 1 illustrates these phases. First, a company that has not used LOPA in the past decides to use LOPA. Soon afterwards, they convince themselves (or consultants or regulators convince them) that if using LOPA for some scenarios is good, then using LOPA for many scenarios is better, and some companies eventually require use of LOPA for ALL scenarios. This use of LOPA is overkill, of course. On the other hand, the overuse of LOPA is good at training companies on the importance of (1) good PHA teams, (2) valid IPLs, and (3) solid programs for maintaining the PFD of stated IPLs. Eventually, the companies realize that the extra effort of doing LOPA, beyond the PHA team decision may not be justified for about 95% of the scenarios identified by the PHA teams. This transition may be partially due to improvements in the competencies of PHA team leaders (especially) and/or team members as they learn and use LOPA more.

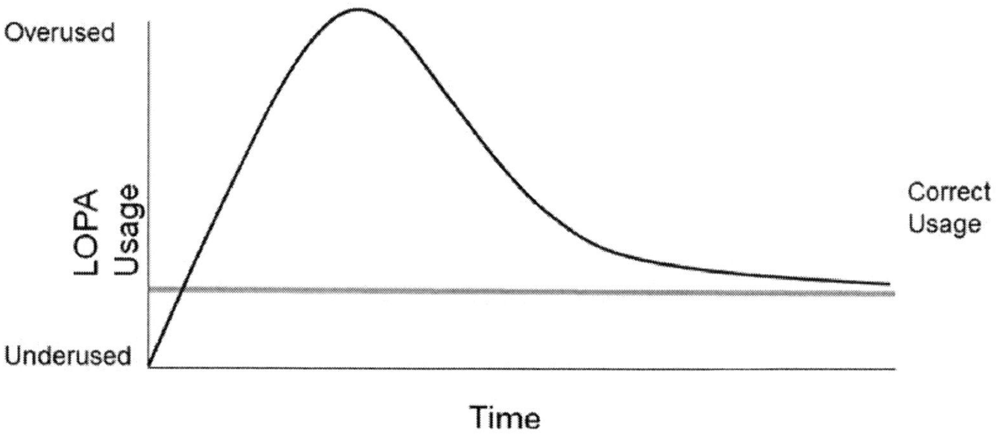

Figure 1: Typical Usage Rate for LOPA as application of LOPA matures within a company *(Courtesy of Process Improvement Institute, Inc. all rights reserved)*

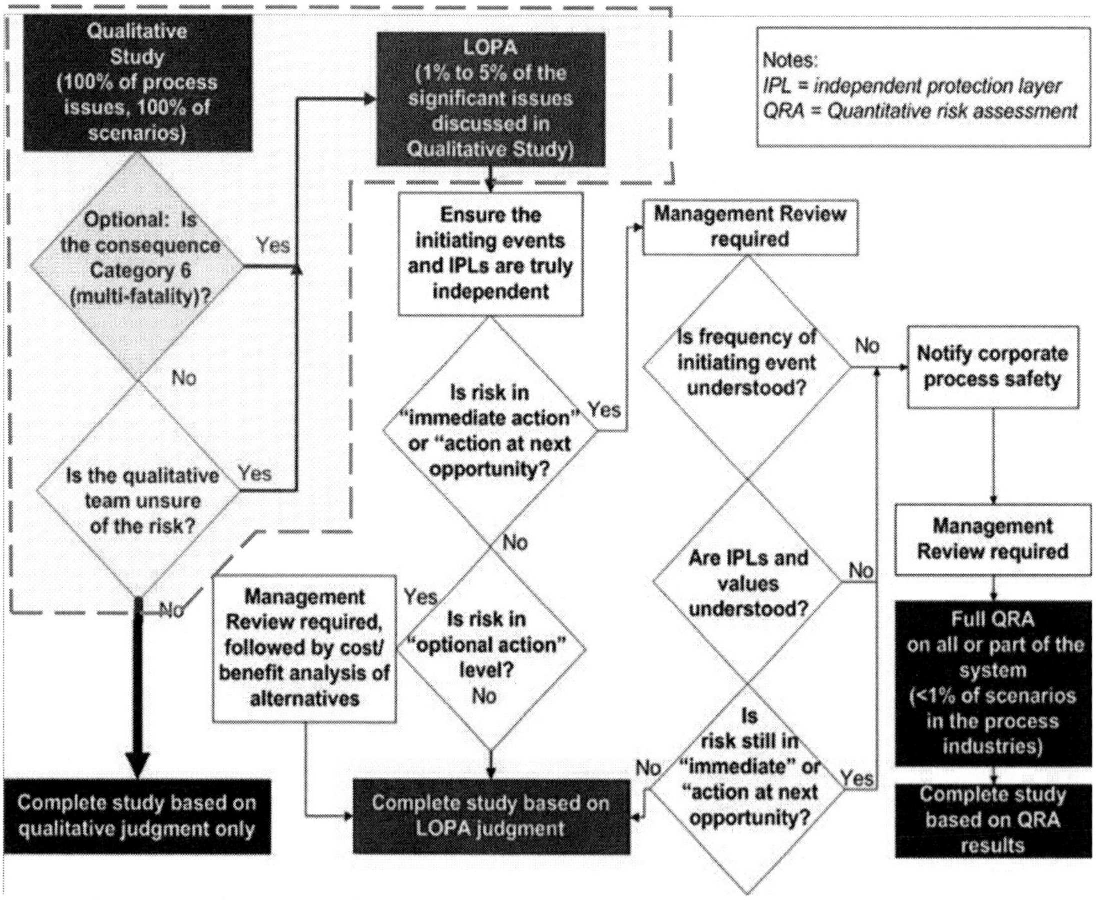

Figure 2: Criteria of when to use semi-quantitative/quantitative risk assessment

Figure 2 shows the decision-making process to go from Qualitative assessment to Quantitative. In particular, the criteria for going from Qualitative (PHA/HAZOP) to Semi-quantitative (LOPA) are highlighted. The approach recommended by PII (and the inventors of LOPA) is that shown in Figure 2.

4. Who performs LOPA?

The mention of a LOPA team in the first LOPA book was isolated and anecdotal; in fact the first LOPA book stressed the use of a LOPA analyst (singular). But many organizations now require a LOPA team (instead of a single analyst). Some companies used a LOPA team early because (1) the analyst trained in LOPA was not in the PHA session, so translation from the PHA team to the analyst was necessary in many cases and (2) LOPA was new, so more heads were needed to decide "Is this the right way to apply LOPA?" A more likely situation is Your company learned wrongly from other companies or from misleading consultants that a team is always required with LOPA.

However, if the LOPA analyst was on the PHA team or if the teams get used to communicating to the LOPA analyst(s), then one person can frequently perform the LOPA. Note that no brainstorming is necessary for LOPA, so the need for a team input (which may come from the LOPA analyst, if he or she was on the PHA team) is limited to confirmation of details of existing IPLs including configuration and independence, and to providing organizational preference for choosing IPLs and for detailed IPL configuration.

The three "LOPA team" configurations still seen today are:

- Full PHA Team
- Full PHA Team plus instrumentation and control engineer
- LOPA analyst with occasional into from process engineer, instrumentation engineers, other expertise by phone or text or e-mail mail, as needed, for an average 1.3 people per LOPA scenario, with the LOPA analyst being the only full-time staff doing LOPA

As mentioned before, the third option is the recommended approach.

5. Conclusions

The best approach for when to use LOPA, who performs it and on what scenarios is summarized in Table 1.

Table 1: Best approach for LOPA

WHEN: Timing	WHEN: Trigger	WHO
Outside the PHA	• Scenarios recommended by the PHA team *(If PHA Leader is a LOPA analyst)* • All scenarios that have fatality *(If PHA Leader is **NOT** a LOPA analyst and or not trained in fundamentals of SIS)*	• LOPA analyst • Process engineer (by phone or e-mail, as needed) • Other expertise (by phone or email, as needed)

6. References

1. William G. Bridges and Tom R. Williams, "Risk Acceptance Criteria and Risk Judgment Tools Applied Worldwide within a Chemical Company," *International Conference and Workshop on Risk Analysis in Process Safety*, October 21–24, 1997, Atlanta, GA, pp. 13–28. American Institute of Chemical Engineers. New York, NY. 1997

2. A. M. Dowell, III, "Layer of Protection Analysis: A New PHA Tool, After HAZOP, Before Fault Tree," *International Conference and Workshop on Risk Analysis in Process Safety*, October 21–24, 1997, Atlanta, GA, pp. 13–28. American Institute of Chemical Engineers. New York, NY. 1997

3. Rodger M. Ewbank and Gary S. York. "Rhône-Poulenc Inc. Process Hazard Analysis and Risk Assessment Methodology," *International Conference and Workshop on Risk Analysis in Process Safety*, October 21–24, 1997, Atlanta, GA, pp. 61–74. American Institute of Chemical Engineers. New York, NY. 1997

4. IEC 61508, Functional Safety of Electrical Electronic/Programmable Electronic Safety Related Systems, The International Electrotechnical Commission, 2010.

5. IEC 61511, Functional Safety: Safety Instrumented Systems for the Process Industry Sector - Part 1: Framework, Definitions, System, Hardware and Software Requirements, International Electrotechnical Commission, 2003.

6. ANSI/ISA 84.00.01-2004 (IEC61511-1 Mod), Functional Safety: Safety Instrumented Systems for the Process Industry Sector - Part 1: Framework, Definitions, System, Hardware and Software Requirements, 2004.

7. CCPS, *Layer of Protection Analysis: Simplified Process Risk Assessment*, American Institute of Chemical Engineers. New York, NY. 2001.

8. *Guidelines for Initiating Events and Independent Protection Layers*, CCPS/AIChE, New York, NY, 2015.

9. *Guidelines for Conditional Modifiers and Enabling Events*, CCPS/AIChE, New York, NY, 2013.

10. W. Bridges and A. Dowell III, "Identify SIF and Specify Necessary SIL, and other IPLs, as part of PHA/HAZOP – or - Why it is not necessary to 'Boldly Go Beyond HAZOP and LOPA'", *12th Global Congress on Process Safety*, Houston, TX, AIChE, April 2016.

11. A. Dowell III and W. Bridges, "More Issues with LOPA - from the Originators", *11th Global Congress on Process Safety,* Austin, TX, AIChE, April 2015.

511122 Calorimetric Study of Graphene Oxide Thermal Stability

Abstract:

Due to the immense potential of graphene for energy storage and composite filler applications, the large-scale production of graphene is of increasing commercial and academic interest. Synthesis route to produce graphene-like material involving graphite oxide (GO) is predominantly used because this method as shown potential for bulk production at high yield. This method involves the oxidation of graphite to GO and its subsequent reduction to reduced graphene oxide (rGO). However, prior studies have shown that GO can undergo explosive decomposition under certain conditions. There is no documented process safety incident specifically related to GO so far, but GO is an energetic material that can undergo explosive thermal reduction. A number of unanticipated process incidents have occurred due to inadequate study and understanding of energetic materials stored in large quantities. As industry is ramping up large scale manufacturing of GO, the motivation of this research is to investigate potential process safety issues with bulk GO storage and handling. Specifically, we have seen experimental data that increasing GO mass decreases the temperature at which material decomposes, also called T_{onset}. We have also been able to quantify the pressure release during decomposition. We will further examine the underlying causes of explosive behavior of bulk GO and propose safer storage and handling conditions. Studies are conducted in two calorimeters (Advanced Reactive System Screen Tool (ARSST) and Automatic Pressure Tracking Adiabatic Calorimeter (APTAC)) to understand the effect of storage temperature, impurities, pH, and process conditions on graphene oxide thermal stability. This research will be beneficial in assessing the hazards of GO and enhancing safety of rGO production processes over their life cycles.

Competence to treat Human Error by Distance Education EAD: Trip to the Sun!!!

Salvador Ávila Filho
Federal University of Bahia, Researcher and Consultant
Rua Graciliano de Freitas 43 Roma CEP 40444340 Salvador BA Brazil
avilasalva@gmail.com

Ivone Cerqueira
Federal University of Bahia, Posgraduate student
ivonecscerqueira@hotmail.com

Maria Lorena Sousa
Monsanto
PSM, partner from UFBA

Prepared for Presentation at
American Institute of Chemical Engineers
2018 Spring Meeting and 14th Global Congress on Process Safety
Orlando, Florida
April 22 – 25, 2018

Competence to treat Human Error by Distance Education EAD: a Trip to the Sun!!!

Salvador Ávila Filho
Federal University of Bahia, Researcher and Consultant
Rua Graciliano de Freitas 43 Roma CEP 40444340 Salvador BA Brazil
avilasalva@gmail.com

Ivone Cerqueira
Federal University of Bahia, Posgraduate student
ivonecscerqueira@hotmail.com

Maria Lorena Sousa
Monsanto
PSM, partner from UFBA

Keywords: Education – Human Factor – Culture – Process Safety

Abstract

One of the biggest challenges for the chemical industry, due the complex process characteristics, is the development of skills to perform critical tasks. The motivation for learning depends on the right information being transmitted effectively and through economically viable means. Distance Education (EAD) aims to building process safety teams and leaders and helps to avoid human errors during operations, decreasing incidents and near misses. The proposed EAD Program is represented a tour through the solar system leaving from Neptune and going to the Sun where the tripulants reach the maximum preparation level, that is, level 5 leadership. The EAD modules that address the human reliability and human factors have rules: after fulfilling each step, the crewmember win a reward, a certificate, seal or button, that means fuel to supply the spaceship and achieve the next planet.

This training will use manuals, books, articles, videos and studies directed to field learning. The time to each module will be 20 minutes. To obtain each award, should reach minimum 70% of the evaluations. The 20 minute modules have expected time for completion and, if there are longer delays, will bring the student to the previous level. The module will be lost, requiring repetition, if the time exceeds 4 days without classes or performance below 70%. The classes will be of minimum 20 and maximum 100 operators and should be coordinated, monitored and followed by the company leaders and engineers. A tutorial will be held on demand. There will be a work at the course final, and one presentation during a Workshop entitled Safe Behavior Cycle. The total module time will be 40 hours, namely, 120 training modules.

Learning levels will be classified as planets in the solar system and will represent that certain subjects have been learned and applied in the field. A media will talk about this trip and a folder will be printed. Registration is voluntary. The Company will train a group wich 10 staff members to assist and orientation the program, together with the 3 external educational specialists and

counselors who will work in the application of the 40 hours. The registration allows to acquire a password and to participate of the trip, of the course of 40 hours.
In Neptune, the course will be introduced and a test will be applied. Then going to Uranus will be treated the human performance factors (API770 and SPARH) where the documents will be discussed and the respective applications. A new test will be performed. The journey from Uranus to Saturn involves knowledge of principles and concepts of the danger areas, risk, culture and beliefs, habits, behavior, and stress. A manual will be delivered and to acquire the certificate of this phase, will be done test by module and at the trip end.
During this step, we will begin to discuss habits and decisions in the organizational environment. This is the phase of Cognition and Organization with the departure of Saturn and the destiny of Jupiter. The arrival to the largest planet in the solar system will be validated through the theme evaluation: human elements, organizational moments, cognitive functions, decision, executive function and examples of each discussed technique. Studies to achieve Jupiter involve learning through the second travel manual that presents practical exercises to be followed. A book presents the subjects already discussed in the manuals and tutorials and directed studies. This book just will be given to those who receive the fuel to follow the trip, that is, to leave Jupiter. The trip continuity, depends of more practical knowledge for the risk investigation, task, failure, human factors through the guidance of the third manual. The authorization to Mars access depends on the knowledge confirmation through testing.
Finally, we will leave Mars towards the planet Earth. Here ironically, the subjects required to enter Earth are: accidents, crisis and communication, how to investigate and avoid? Activities will be held in the field and will also be discussed scientific methodology for the CCPS work. During this stage, techniques applications for routine cases and accidents will be realized: GLP, Fireball, Texas City, and the manual 4 with the respective studies will be applied. The journey between Earth and Venus requires knowledge about case research that will turn into end-of-course work. This theme will be divided into: EAD presentations, orientation workshop, face-to-face meetings and final workshop to present the works. For this case, will be used manual 5 and test 7. Phew! Most of the group will be on Venus, and only the top 10% will travel to the Sun to be structured as level 5 leadership. To achieve the star-king have to acquire differentiated knowledge such as: dynamic culture; listening - empathy; to elaborate; group knowledge; informal leader; and self-management. This trip last stage, which is addressed 3 to 10 initial entrants, will be based on exercises and guidelines presented in manual 6. A test will be applied to approve this last phase and with the approval each team component, will receive an awards book.
This style of training is innovative due to the following characteristics: the challenge for goals achieving, within a journey through knowledge; the concepts for field application through directed studies and tasks to be carried out during the routine with the respective performances; the use of lessons learned from the accidents analysis, such as Texas City, Bhopal and Piper Alpha; the discussion about concepts and applications involving human factors in real cases; and the challenge to achieving level 5 leadership with limited vacancies and requirements. With proper guidance on operator teams and applications directed to routine and the equipment and processes of these same operators is intended to overcome risk aversion and to anticipate future events through warnings of failure and operational deviations. This training is based on the sociotechnical model discussion about the reliability of critical systems and follows part of the concepts indicated by the CCPS.

1 Introduction

One of the challenges for the highly complex chemical industry is the competencies development to carry out the critical tasks impacting occupational and process safety. The proposal of Distance Learning, stands out from the classroom because it is a training modality that promotes self-learning through systematically organized didactic resources, which is independent of the distance and time of the teacher and the student. However, student commitment and discipline are necessary, as well as tutoring.

In the context of training for the process industry, it focuses on an efficient method that reaches an important quantitative of worker ways to reinforce the barriers to avoid accidents and to lead to safe behaviors and assertive decisions in the demands out of the routine and emergencies evolution. The worker profile at the workplace occupies space for discussions, given the technology acceleration. In this context, [1] discusses the qualification and worker competence for industry job, highlighting the inclusion in the exercises training in the form of the situations experienced cases in the team internal tasks execution space, which pertains to one of the objectives of the EAD Program in Reliability and Analysis of Human Factors in the routine of the job. **(EAD – Distance Learning).**

In the workplace during the occurrence of failures, near accidents and accidents, as [2] points out in order to make the most appropriate decision in an emergency situation, the team acts with excellence, when trained in knowledge of safety concepts and principles, sociotechnical competence and human factors. To achieve the goals of a safe industry, it is necessary that these trainings are motivational and happen with a methodology that is easy to understand, accessible, attractive, playful and integrative. In order to do so, it is proposed the content of the modules of the EAD Program in a knowledge-seeking scenario, in a trip through the Solar System, which attends topics such as Reliability and Analysis of Human Factors in the routine of the work place.

[3] point to the importance of group learning, and discuss the factors that favor knowledge, when group training aggregates the motivation to participate in training, understanding that a group has common goals when they are part of a group same job. For this scenario, the EAD program in Human Reliability and Human Factors presents modules for the practice of case studies for the preparation of Conclusion work (FW).

[2] discusses new concepts, transporting to the routine of the researches, the routine of the factory floor and emblematic questions for an organization to live with the culture of safety and achieve resilience. Thus, discussion scenarios are presented with experts and other human factor engineering scholars who also adopt these new concepts as essential knowledge for process safety in highly complex industries.

2.1 New Concepts

New concepts will be addressed based on studies and applications in the processes in high complexity industrial plants. Among these concepts are: Human Factors (PSF) and Organizational Factors; Sociotechnical Systems (SST); Danger and Social Risk; Human Reliability Analysis (ACH); Reliability Analysis of Sociotechnical System (ACSST); Deviation, Human Error, Technical Failure (HAZOP Social & LODA & FTAH); Techniques of Analysis of Danger, Risk, Failure and Accident; Latent Fault and Hidden Fault: the digital SST fault; Environment Stressor and consequences (LODA & LESHA); Root Cause Analysis in SST; Failure of the Executive Function, cognitive failure (AFE); Operator's Speech and Risk Management (DO); Archetype &

Regionality & Culture Dominant technique; Organizational Communication, Culture of Guilt & Contingency. The risk assessment and reliability methods, depend on the human quantification and organizational factors. These should be analyzed according to the failure typology and risk activities. In order to investigate the failure, it is important to quantify human factors, according to hierarchical levels, cognitive behavior and human-machine interface, cognitive behavior when working in groups (interaction - PSF) and influence of organizational culture on human beings. In this context are inserted, the analysis of the techniques and the human performance in the performance of the tasks and the organizational environments.

The discussion of these complexities indicates the need to investigate the fault through operator discourse analysis (OD) and the process control and system production variables. Communication is considered an important tool for risk management. The mapping of events of abnormalities (MEA) is a method developed by [4,5] and aims to identify the root cause region based on routine information. It is understood that the information contained in the shift book is fragmented and needs to be reordered to indicate the actual causal nexus of the events. Gaps are covered by events that were addressed in later periods or even hypotheses based on process study, including service chains.

The Sociotechnical Reliability Model calculates the operations success probability, processes, man and equipment, says [6]. Communication influences the activities of the Leader and the team, and especially the organization. In contingency actions, it is expected beyond the competence of decision makers during an accident, an environment of trust and non-omission of failure reporting.

2.2. *Aims: Competence construction, a science*

This paper presents the characteristics of the distance education method in Human Reliability. In this context, it intends to promote the competencies development to investigate the human error root cause and to build teams and leaders focused on the process safety area. A methodology description is made through a solar system indicated in Figure 1 where the trip to the sun represents the graduations in competences acquisition. The goal is to train staff on accident and routine human factors, and prepare level 5 leadership [7] for crisis situations.

Figure 1 – Distance Learning in Human Reliability (EAD)

2 Literature Review

The literature review for discussion and the best practices understanding for a EAD model elaboration, applied to training in process safety with a human reliability focus and socio-technical systems, includes knowledge about the distance learning modality, concepts and Level 5 Leadership, Human Reliability, Competence, and Leadership Process assumptions.

2.1. Distance Learning - EAD

The learning activities developed systematically through Distance Education, using information and communication technologies, is increasingly accepted in training courses and specialization. According to [8] (teachers and students are separated physically in space and / or time, although this characteristic does not serve as indicative of face-to-face learning failure. EAD has the capacity to attend a larger number of participants.
It should be noted that the teachers (tutors) performance, as well as the didactic material used during the training, is directly associated with the performance and use of the content of the program that is adapted to be offered in the distance learning mode.

2.2. Process Safety – Human Reliability

[5] points out that Human Reliability is a challenge to try to predict emerges for scientists studying efficiency in chemical processes, difficult challenges such as the attempt to predict behaviors to avoid failure during the accomplishment of tasks, i.e. failure analysis and human error present in the plant process.
[9] also states that the productive systems performance is satisfactory when the observed results approach standards, indicating stability in the behavior of people, equipment, processes as well as community satisfaction.
The process control and safety systems development is based on historical events considered to be predictable, but reality shows that accidents can be caused by unforeseeable situations. [5]
The main factors that influence human reliability, says MUMMOLO (1999) are related to the environment, the technology type, the equipment HMI and the processes management. To analyze ways to increase this reliability in the industry it is essential to classify human errors and their relation to behavior in the work environment. The task planning is important for Human Reliability in equipment operation where there is a need for verification and monitoring during its execution. Human Reliability is also directly related to the production processes automation level.
The complexity to manage the human element in socio-technical environments is far greater than the complexity of equipment and processes industry control. The learning activities involve the concepts construction that are reaffirmed with the organizational motivation; in this field enter the abstraction and the imaginary, resulting in competence level to applied in the work environment. Thus, care must be taken to analyze human behavior in the workplace and especially, during high stress situation [5].

2.3. *Leadership and Level 5 Leader*

[3] define leadership as a behavior and [11] as a definition of leadership the way it is exercised through the procedure of a person who can influence others to achieve a certain goal. Beyond empathy and social ability, ability to relate to others, necessary components in a leader.
Selection and ongoing training are critical to the choice and professional ability development who is received in the company as a leader in a group. Knowledge of the organization's policy needs to be presented to the new contractor. As for the competence required for daily activities performance, it is linked to the process unit technology and the final product [5].
A Level 5 Leader has extreme humility to the point of assuming for himself certain responsibilities of not communicating a fault, a deviation, a near accident, attributing value the initiatives that have arisen from the group of operators, to the contingency of the event. This leadership has the determination and the fierce will to achieve goals, along with its work team, in addition to the technical competence that leads the group to trust, as an agent capable of developing quality safety activities. [12].

3 Characteristics of the Method and EAD modules

Each skill acquisition stage equates to a spaceship that encounters a particular planet in the solar system. Students who reach this planet will receive a certificate (confirming the acquired competence), a stamp and a "botton" of the equivalent module. These vouchers will have photos, or drawings representing the planet and its color.
After school study is very important. The materials for consultation will be manual by module, books that represent in more detail the techniques and methodologies related to human reliability, human factors, risk, process safety, specific risks and other information. Some critical procedures are basic materials for carrying out fieldwork, as well as articles and reference documents in the areas above. Confirmation of sufficient skills acquisition to achieve and maintain a secure behavior will be confirmed with a mandatory **FINAL WORK (FW)** with presentation (banner or oral) in the Safe Behavior Cycle (CCS).
The field activities development follows guided studies and confirmations to be answered in the following modules. The time for each module will be 20 minutes. And there are requirements for the acquisition of certificates, so you will need to: (1) pass 7 of the 10 questions in each module, ie 70% and (2) pass 70% of the answers in the exam for certification.
The certificates will be printed with an attached seal and each stage of the EAD has the color and the planet equivalent. The reward for success in approval!
The training time has slack for possibilities of delays due to the shift work operation (holidays, large play and small play) schedule. For example, the works that are modules part are realized in up to 3 months and have another 2 months to complete.
Between modules of 20 minutes there will be a minimum interval for reading of 4 days. This is sufficient time for review the material, directed study and field work which may involve tests and checklist. The rules to be followed symbolize the need for discipline in process safety. If the student of this course passes more than 4 days to carry out the module, he will lose the previous module. If this loss occurs in the first module of the series, it loses certification and will be retested. The classes will be 100 students to enable in case a refinery reaches at least 50% of the team in 3 years. These 100 students represent the number of people who want to reach the sun!

The FW will be accompanied by the operations leaders, refinery consultants and engineers allowing moments of discussion in the works elaboration. Part of the explanations about FW will be in EAD format, another part will be through lectures or workshop for the larger group, and finally session of specific doubts by team that will perform FW. It is important for engineers and leaders to be trainings to support this training. Table 1 presents the phases and topics covered, total hours, time and credits, and material to be used for the study. For each nominated Planet, this information was inserted for better understanding.

Table 1 – Description of Distance Leaning in Human Reliability & Human Factors

Planet	Subject	T (mx)	T acu	CR	CRa	Hs	Hs A	Detailed Subject	Instruct doc
Inside the Solar System	Registration	0	0	0	0	0	0	TRAINING MULTIPLIER of 5 Refinery and 5 from University	40 HS Foulder
Netune	Introduction	0,5	0,5	1	1	0,3	0,3	Presentation curse	Introduct, T1
Uranus	PSFs	2	2,5	5+5	11	1,7	2,0	API770 and SPARH	Documents adapted, T2
Saturn	Principles & Concepts	5	7,5	20+5	36	6,7	8,7	Hazard, Risk, Culture and Beliefs, Habits, Behavior & Stress	Guide 1 Test 3
Jupiter	Cognition and Organization	4	11,5	15+5	56	5	13,7	Human Elements, Organizational Movements, Conitive Functions, Decision and Executive Function, Exam.	Guide 2 Test 4 BOOK 1
Mars	Investigation	4	15,5	15+5	76	5	18,7	Risk, Task, Failure, Human Factors	Guide 3 Test 5
Earth	Accident, Crisis, comunication	4	19,5	15+5	96	5	23,7	Tools applied to cases of routine and accidents: PLG, Fireball, Texas City	Guide 4 Test 6
Venus	Conclusion Work - CCS, Tutorial	2	21,5	10+4	110	10	33,7	In 3 moments, EAD (2 hs), General presentation Methodology (3 hs) and Attendance 30 minut for group	Guide 5 Test 7
Sun	Leader 5 10%	5	26,5	20+5	135	6,7	40,4	Dynamic Culture; ear – empathy; elaborate; Groupal Knowledge; Informal Leader; Self mananagement	Guide 6 Bool 2 Test 8
Preparation	8 hs p 1 h					40 4	+40	= 444 HS	6 Mod, 2 Bks

4 Detailing Program

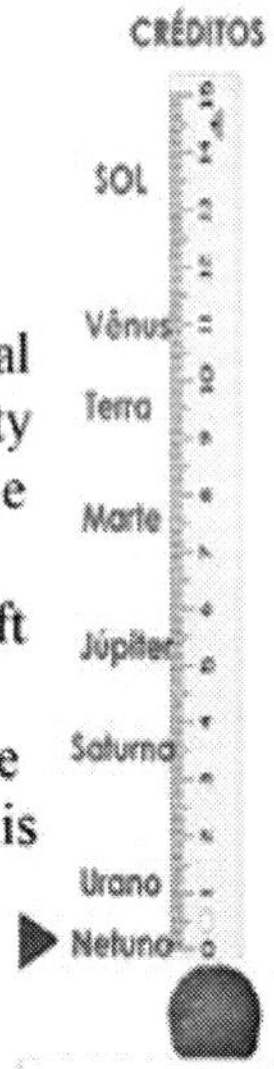

4.1. Introduction: Journey from Neptune to Uranus

We will be starting a trip towards the Sun ... This trip is real based on practical experiences in the industrial area, services and as a citizen dealing with related safety issues in the residential environment. The whole discussion aims to reduce the likelihood of human error, that is, increase human reliability.
Preparation for the trip requires a course for the STAFF that will accompany the aircraft maintenance with the basic knowledge necessary to travel.
The trip general description will be made to authorize the departure of Neptune. The traveler should attend and test 1 20-minute module of the program before starting this journey.
The ruler to measure which position on the trip indicates the planet, the planet figure and the position is number of credits in the course. The rule result is multiplied by 10 to indicate the number of credits.

4.2. Human Performance Factors PSF – from Uranus to Saturn

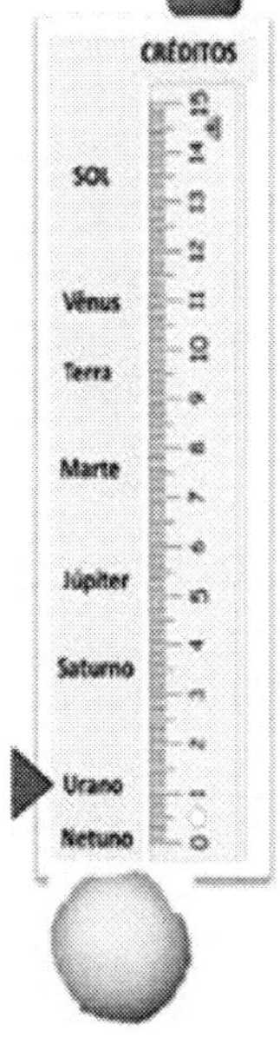

4.2.1. Travel Journey

After leaving Neptune with the known route of travel, it is important to monitor fuel and permits. For each stage, we will get more fuel (motivation) to move on! Will be issued certificate with button and celebrations!
The next trip is for Uranus where the API770 and SPAR-H will be presented for human reliability and performance factors in performing the task ...
A group perception about the PSF will be made at this moment and if it reaches with punctuation enough, a quota of fuel is fueled in the NAVE to go ahead ... The Certificate and button with the photo of Uranus will be given!
modules of 20 minutes will be completed.

4.2.2. Presentation: Human Performance Factors PSFs

API770
There are physical and psychological stressors that alter the level of stress and can lower the efficiency of the task ... some of them are: high risk, threats, extreme temperatures, vibration, repetition of movements, fatigue, stress, and others.
Situations that cause human error are: poor procedure; inadequate instrumentation; insufficient knowledge; conflicting priority; inappropriate feedback; conflict between practice and politics; availability of equipment; difficult communication; population stereotype violations; excessive mental burden; inadequate tools; excess functions; Excessive surveillance; and others.
SPAR-H
There are only 8 factors that are classified according to the situation found: procedures, task complexity, stress or stressors, experience and training, available time, ergonomics, and relationships at work.

The calculation of human reliability for execution is different from diagnosis. An accident or routine situation film is discussed. After learning is presented test for resolution where the PSFs will be classified, from least to more shocking ... a movie shows the situation for the analysis. A SPARH calculation is done for execution.

4.3. Principles and Concepts – from Saturn to Jupiter

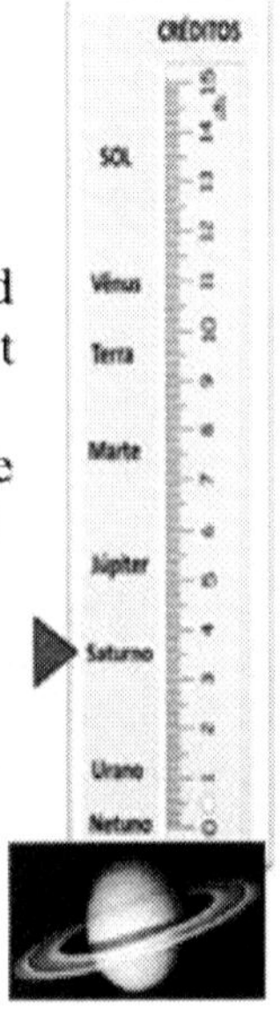

4.3.1. Travel Journey – Saturn

We will leave Uranus after understanding how to calculate human reliability and after knowing how to discuss the environments that promote human error. Do not forget that the rest between one stretch and another can only be 4 days!
From this moment, the foundations will be set up for the discussion about the safe or resilient behavior that reduces human error in the various economic, physical and social environments.
We intend through the discussion about principles and concepts to achieve Saturn. This trip will be long ...
We will be learning for 20 modules of 20 minutes each. Each set of 5 modules will be referring to a certain subject ..., danger-risk, culture-belief, habit, behavior and stress.
Recalling that certification is only achieved if 70% of activities are completed successfully and on time. If not, we return to the previous trip!

4.3.2. Presentation: Principles and Concepts

Danger and Risk
Imagine if we did not know about danger and risk being on an interplanetary journey like this, we would not be able to make choices for difficult decisions ... The size danger discernment and what happens in case of its release is fundamental to define the due actions. Here they are dubbed as safeguards in the prevention or problem correction. Sometimes, because events (unusual) visibility lack or decision lack, we end up not taking action at the right moment and provoking a chain reaction, starting then to contingency actions to reduce the impact of this chain of events.
Culture and Belief
It is important to understand the influence of cultures on the group to assess which biases in the decision or even which undue habits need to be corrected. The investigation of behavioral heuristics in the shift environment indicates which are the corrective-preventive educational measures to be carried out. The identification of archetypes leading to the accident allows action on them. Still, the understanding of what the shift group believes and does not believe facilitates the action and intensity of conviction.
Habits
Repetition of actions brings a habit. This repetition may be due to their usefulness to social coexistence, repetition may be due to unconscious secondary gains or due to family or local community learning, based on customs and beliefs. Certain habits are not appropriate for safety or for the team organization that performs this great trip that is the challenge for the company!
Behavior
What is the appropriate behavior for the organization? For the group? What behavior will allow to reach the objective of this trip despite environmental mishaps (meteor shower). Safe behavior is

determined by some factors that directly influence group life, competence, commitment, cooperation, and communication, as well as due control over (or for) stress levels.

Stress

The physical environment, social coexistence, work pressure, presence of danger and the human typology influence or are determinant factors on the level of stress and its consequence on the body and the task. We need to know how much stress impacts the activity to avoid loss of control!

4.4. Cognition and Organization – from Jupiter to Mars

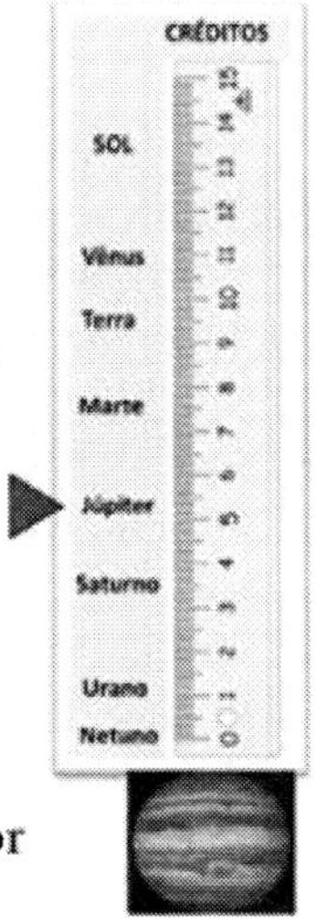

4.4.1. Travel Journey – Jupiter

We will leave Saturn after understanding aspects of hazard-risk, behavior, habit, culture and stress. This knowledge will directly aid in the decision-making process.

To proceed, all modules must be passed with 70% of the required score, without which you will repeat the previous module. This stage is important to know the characteristics and functions of human cognitive processing and its biases. This knowledge will allow you to reach Jupiter on this long journey!

This trip will consist of 15 modules of 20 minutes. Each set of 5 modules will refer to a particular subject ..., Human Elements, Organizational Moments, Cognitive Functions, Decision and Executive Function, with Examples for learning.

4.4.2. Presentation: Cognition and Organization

Human and Organizational Elements

It is important to introduce the main human and organizational elements that are normally involved in the Accident, as cause or as influence: Economic Bias, Risk Management, Task Planning, Cultural Traits, Conflicts between policies and practices and Process Design or Product-Process Technology.

For the task is proposed assessment technique based on standards review from the environment study, cognitive processing and risk activities control. This causes analysis relates the cognitive, emotional and intuitive functions with the elements that can cause the accident...

Organizational Moments and the Work Cycle

These moments are analyzed to verify the training type, education, procedure or ritual should be applied and to what intensity. The moments are: Free Thinking, Physical or Psychological Survival, Conceptual Construction and Experimentation - New Concepts, Decision on Task, Almost Impulse and Action (- rationalization), Intuition and Impulse Action - Emergence, Intuition and Rational Action, Repetition and Learning, organizational internalization values and social inclusion. It is important to discuss the life cycle of the operator in the company regarding skills: knowledge building, learning at work, work maturity and retirement.

Cognitive-Intuitive-Emotional Functions

Several cognitive and organizational psychologists discuss the best profile for the critical activities development in the industry or services. This interplanetary journey requires a combination of three main cognitive functions: perception, attention, and memory. The explanation about the memory functioning has complexities that involve the affective bond through the commitment level and the worker emotional balance. Thus, we discuss cognitive failures in the task including organizational, human and social movements. It becomes possible to find where the culture affects the decision.

Decision and Executive Function with Examples

The executive function analysis technique tries to bring measurements that allows to classify the team decision quality and the human error possibility. The decision functions are classified as cold (rational-cognitive) components and warm (emotional-intuitive) components. The processes for more assertive decisions in the dynamic environment are: socio-economic-affective utility in the activity, competence building, task planning and review, attention calibration, memory training and operations and the application of the company's intellectual capital and attitude to motor achievement and task control.

4.5. Failure Investigation - from Mars to Earth

4.5.1. Travel Journey – Mars

We will leave Jupiter after understanding how cognitive processing occurs in organizational relationships. After 15 20-minute modules and an assessment, passengers heading to the sun will be structured for the trip as a standard organization. They will have a certificate that represents this authorization and also the motivation to continue, there is fuel!!!

At this trip stage we will be discussing techniques of how to identify the fault considering the man existence and the factors that influence the human error appearance. Thus, these 15 modules will deal with the following subjects: Risk-Failure, Task and Human Factors. The techniques discussed have been applied in the oil industry and now we are preparing the travel team to reach the sun ... So, it is important that you reach the goals of each step in this learning!

We intend to reach Mars after learning how to investigate events including human factors.

This training will be illustrated through actual industry cases and aims to show that man can be the accident main safeguard!

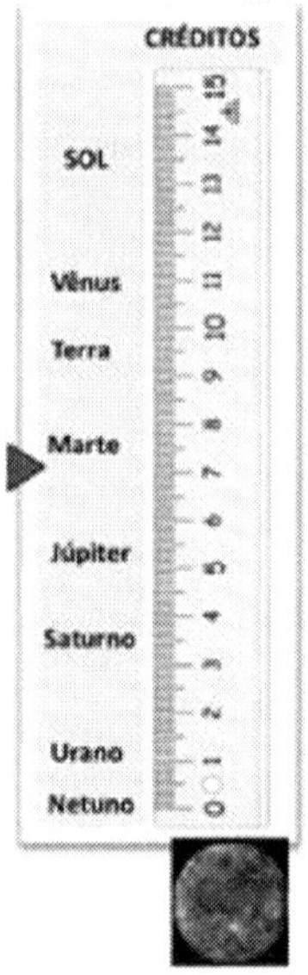

4.5.2. Presentation: Failure Investigation

Risk-Failure to Accident

The technology exists with its equipment, processes, controls and operations for the services accomplishment or for the products transformation. This technology can go into failure mode due to material fatigue, poor management decisions or human errors. If we can check the tendency to fail from the routine, it will be possible to correct the tendency for the danger release. The risk study try to avoid future failure. Reliability analysis allows building the current safeguard. And fault analysis indicates how best to avoid process loss. All of these techniques must include the subjectivity of human and organizational factors.

In the risk, you can analyze the business based on the failures that can occur in the team. In the risk one can verify what the tendency of the team in making human error.

Task to avoid losses

The writing of the pattern depends on knowledge and ability to handle the product or depends on the work with specific equipment. Analysis of the pattern to be followed with the respective steps, necessary training, controls and auxiliary memories can avoid human errors in critical tasks.

Several techniques that consider the communication complexity in different shift teams can be applied to analyze task failure. When discussing a trip to the sun, risks are discussed where any task variation may lead to the goal non-achievement.

Human Factors - Human Error

Several factors can lead to human error, these aspects are dynamic and variable, depend on the leaderships and cultures installed for the group. An important aspect is the stress level where all the knowledge learned may not be used for problem solving. Thus, it is important to review the trainings and choose the leaders for emergencies in extreme stress case.

Leadership can move the group toward increased competence, commitment, or even lowering the stress level, but it should be attuned to group movement and organizational resilience needs. The communications quality has high weight in the fault investigation and involves objective (record content) and subjective aspects (omissions analysis). Setting standards to achieve operational excellence requires experience and knowledge, i.e. competence. A cooperation level in the group is necessary to achieve organizational goals. During the journey to knowledge it is important to identify which factors influence performance and how to analyze them. Some techniques can be applied for analysis and for intervention. The main one is the operator's discourse analysis. The workstation observation can aid in the faults identification in the man interface with the process or the machine and also which tools can be improved.

Some human factors are related to the organizational culture, with managerial defects or still with the regional or global culture. Poorly established or understood priorities have undue consequences for safety. These factors that cross between the man and the organization affect the team commitment.

The productive system always involves technical and human-social factors (PSF-1) and is thus defined as a socio-technical system (SST-2). It is imperative to identify the social hazard / risk (APRSH - 3) for the possible failure or for the future accident. To maintain control of these systems, barriers must be protected from the human reliability calculation (ACH-4) and integration with the equipment characteristics, processes and operations (ACSST-5).

The accident investigation involves multiple dimensions' complex events of the fault involving progressive danger energy deviation, errors, failures, incidents until reaching the top event, the accident (LODA & FTA H - 6). With the classifying the risk level possibility and establishing a actions program in each enterprise phase (concept, detail, operation, crisis) (tPRFA - 7). These events or faults may often be non-visible (latent or hidden) indicating the digital configuration of the SST fault (FpA-8). A factor that makes dynamic failure is the physical-psychological stress level with a consequent unpredictable direction, therefore, it is difficult to calculate the human error probability (LODA & LESHA-9). The accident investigation and the failures requires methods that allow to increase the visibility on events networks possible to find the events root cause direction that are nodal and multiply the danger migration routes energy, plus a need to review the techniques to control the dynamic risk (FpA - 8).

The root cause study to avoid accidents in socio-technical high risk systems and complexity (CRsst-10) depends on methods based on cultural aspects, competence level, communication quality, attention calibration, technology type and planning task [13] The operator 's discourse (written, spoken, observed) (DO - 12) brings together these dimensions and allows together with the process and production variables to indicate unidentified directions in the top - down investigations carried out by the company. Process losses should be traced back to the executive function failures level [13] where commitment and cognitive quality are influenced by commitment, emotional balance, and intuitive dimension acceptance in decision making.

This investigation becomes more complex when a difference in the organization 's vision is established in relation to the operational practices resulting from the non - acceptance of regionalities or archetypes derived from inadequate rituals installed in the workplace (Heuristics A - 13). Thus, it is important to analyze the archetypes and regionalities to detect the executive function failure from a technical culture resulting from the cultural force field (Heuristics A - 13). At this moment, it is necessary to install a management based on organizational communication (PCO - 14) that moves in the different directions (BUp, TDown, horizontal and diagonal) and accessing the culture biases of guilt allows to define a standardization system closer to a tuning between the desired and the possible to be realized. In these cases, the top event probability is more remote and if it occurs, the contingency actions are more assertive.
In SST, it is difficult to identify the root cause to establish a barrier program!

Description of finger type model of social technical failure!
Production systems often analyze failure in a classical way where the factors are separated by departments or disciplines. That is, the fault is analyzed by a specialist in a single area. Thus, according to Figure 2, the process engineer analyzes the overload in the purification, the manager's assessor analyzes the incorrect decision, the Schedule and Control Production Engineer analyzes the failure to change the campaign, the maintenance engineer analyzes the lubrication failure and the Ergonomist analyzes the failure that causes fatigue. In fact, these failures are interconnected through common factors, have the digital fault format (fingertip model) as indicated in Figure 2 and should be analyzed at the same time to optimize the cost of research and improve the effectiveness in the barriers.

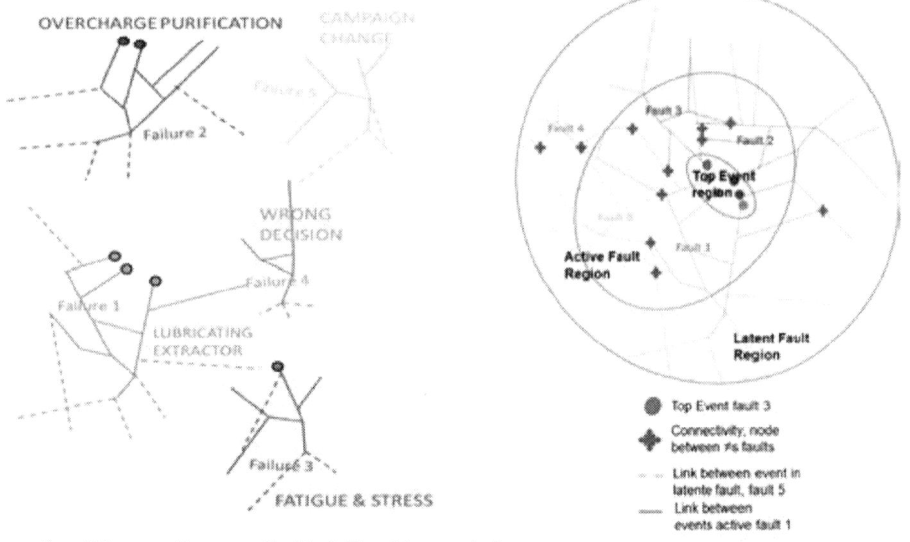

Figure 2 – Fingertip or digital Fault model

4.6. Accident, Crisis and Communication - Earth

4.6.1. Travel Journey – Earth

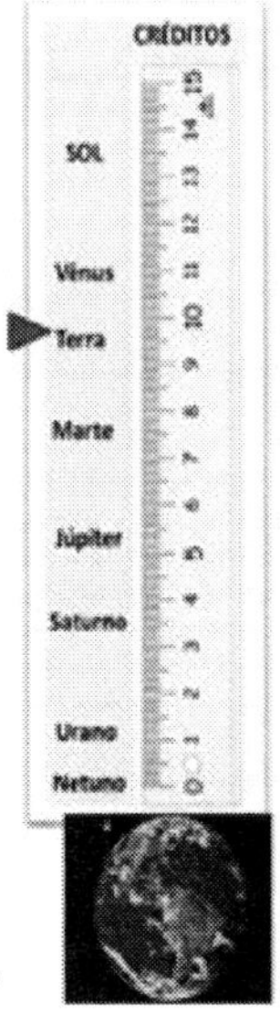

We will leave the Earth after understanding what are the human error probable causes and how we can find them since it involves objective aspects related to the task and subjective aspects related to culture and people. To proceed, all modules must be passed with 70% of the required score, without which you will repeat the previous module.

Just to recapitulate that for this long trip we leave Neptune already prepared from a previous knowledge of how difficult it would be to reach the sun. We had to gain fuel in the middle on the road through certificates and knowledge acquisition in: human factors, concepts and principles on hazard and risk, cognition and techniques for failure investigation and human factors. And now we are understanding what triggered the accidents occurrence and events based on lessons learned. This analysis intends to present all the stages from deviations, the process control lack, equipment failures, product leakage, accident and chain reaction provoking the crisis. This trip will consist of 15 modules of 20 minutes. Each set of 5 modules will refer to a certain subject: techniques application already learned for routine and accident cases: LPG, Fireball, Texas City.

4.6.2. Presentation: Accident, Crisis and Communication

Deviation, Process Uncontrol and Failure

The deviation should be avoided in order not to cause quality loss in the process and not to reduce the billing. The task should initially be planned, controlled and reviewed to avoid deviations in the process that initiate the failure process. The task environment is identified through its context analyzing the business objectives, the task type involved, the people who are part of the team and their behavior in the cities.

The partial failure study in equipment and latent failure in the team assists in the deviations identification that can lead to failure. Finally, if failure occurs, it is necessary to include human factors in the cause identification. For this case, will be studied the accident in the LPG industry until the top event, the fireball in the direction of the community.

Accident

In the accident approach, the events happen in a faster time and releasing much more danger energy. Thus, it is often important to establish procedures or install appropriate safeguards to take immediate action to correct the accidents consequences. The devices that indicate an accident onset need to have the human perception for immediate action, in that sense, communication is important.

Communication can be done in one alarm form or by sending messages directly to the group. The group must be structured to take action. When the released danger goes in the direction of causing harm to people, communication is considered important for evacuation.

The main techniques for investigating the failure were discussed on Mars. Now the techniques are directed towards the accident investigation. Among these we can mention: tree of failures, tree of events, BOWTIE, LOPA. Risk activities should have mastery over techniques and include human and organizational factors discussion.

Chain Reaction and Claim

Unfortunately, mainly due to the economic bias and the risk management lack, the principles of avoiding the hazard when released are not followed, thus bringing the accident uncontrolled to provoke a chain reaction. After triggering the chain reaction, it is important to look at the path of danger and seek to diminish its consequence.

The intention here is to reinforce the barriers to avoid the top event consequences and try to save as many people as possible. For the accident and the chain reaction will be used the Texas City film and the fireball case in the PLG industry.

4.7. Safe Behavior Cycle – FW, Venus

4.7.1. Travel Journey – Venus

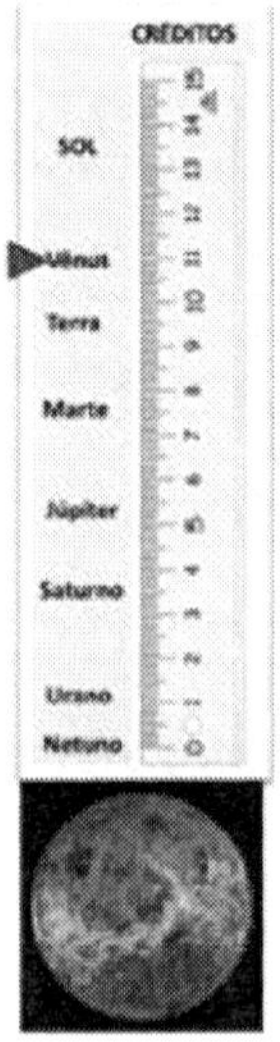

We will leave the Earth after experiencing the lessons learned and after analyzing these facts occurred in routine, risk, failure, incident, accident and crisis. At this point the tools discussed above will be tested for historical real cases. In this way, it is easier to apply in future cases that include human factors in the accident analysis. We had to gain fuel in the middle on the way with acquiring knowledge about: Applying the techniques for routine and accident cases: LPG, Fireball, Texas City. And now we are preparing scientific technical advice, applying previous knowledge, applying some scientific methods to elaborate the course completion work that will make up the CCS, Safe Behavior Cycle.

A practical case experienced by the operator or the group of 5 operators will be elaborated in the form of scenario, described and through the techniques learned and at the end recommendations will be suggested. This activity will last 10 hours be divided in three moments: EAD (6 modules of 20 minutes composing 2 hours); General presentation Methodology (3 hours) and attendance of 30 minutes per group of 5 composing a total of 5 hours.

4.7.2. Presentation: Safe Behavior Cycle – FW

EAD - 2 hours, 6 modules

Habits for safe behavior are followed by principles and basic concepts. A case is constructed to recall all the techniques discussed here, mainly, taking into account the human factors analysis in performing critical activities in the risk industry.

3 review modules! 3 modules on scientific methodology!

WORKSHOP & COURSE - 3 hours

For 3 hours, the group of 100 people divided by 3 groups, that is, 33 people will hold a workshop with awareness lecture (1 hour) and then with cases resolution with polls to be answered. These surveys will address: emotional balance, leadership-led profile, team aggregation level, commitment, the technical culture examination, and stress dynamics for critical situations.

Group Service - 30 minutes

Each group will have a time of 30 minutes in person to present their doubts and discuss how the the work format to be presented would be. As there are 10 groups of 5 per advisor composing a total of 50 operators. Therefore, there is a need for 2 advisors and 2 for the company to compose a total of 1 hour individualized service to prepare the FW and the banner for the CCS.

4.8. Leader Level 5 – Sun

4.8.1. Travel Journey – Sun

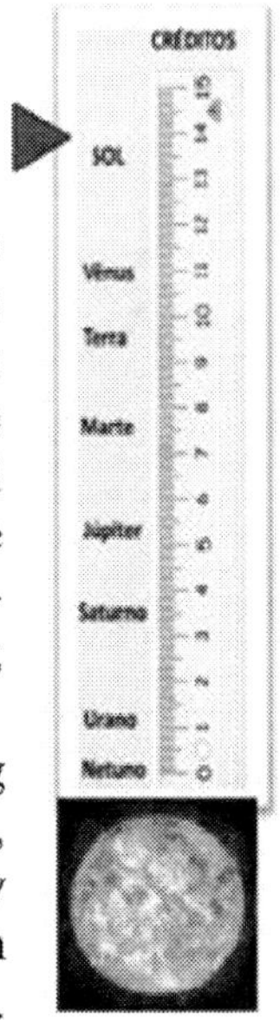

We will leave Venus with our experiment, testing our knowledge. I will be specialized in safe behavior with techniques and applications performed in the routine areas, risk, failure, incident, accident and crisis incorporating the human factors and errors analysis. 90% of the group concludes the work here, being aware of what is safe behavior, and knowing how to analyze when this behavior will tend to be insecure. Only 10% of the group will follow the star-king, the sun, will be the level 5 leaders. To reach this level, the subjects will be: dynamic culture; listening - empathy; to elaborate; group knowledge; informal leader preparation and activation; and models for self-management on shift.

This trip will be the last and as a level 5 leader, we will have the following characteristics: we will understand human factors and know how to manage them, we will have empathy and we will listen to others, we will have extreme humility and great care with the patrimony, with the people and the nature. We will be firm to review the group to meet the change demands, in organizational resilience search.

4.8.2. Presentation: Level 5 Leader

Dynamic Culture

The impact cultural changes recognition on behavior empowers the leader to adapt groups, tasks, and to review risks to anticipate failure situations, especially about communication and cooperation and commitment.

Another situation that is of great industrial concern today is the need to change behavior due to climate change and its impacts. This knowledge should promote in leadership the adaptation characteristic.

Listening - Empathy and Humility

The listening skills, building scenarios, observing, organizing information, projecting the future before defining action in the routine or in emergencies are interesting traits for leadership.

These characteristics are far from the induction of the current culture around the compulsion to buy, egocentrism, immediacy in the careers ascension, mixed feelings between the real and the virtual, intolerance and tribal exclusion.

Group knowledge

It is notorious that the leadership must know how to motivate specific knowledge and skills in the team to face the new cultural challenges. Thus, a skill-preparation scheme must address gaps for decision-making in complex situations.

The main items to be understood in the team are: behavior, culture, group relations, generation shock, automatism implications, multiculturalism, basic principles, habits and induction rituals, heuristic rules and organizational rules, technical culture, informal leadership, requirements and complexities.

The informal leader preparation and activation

Activities involving organizational resilience demand stability in formal leadership, leadership rotation on demand, informal leadership for rotating task forces, reviewing habits, and valuing the intellectual company's senior managers capital.

The informal leader is structured to pull activities and projects that are not routine, eventually or when necessary migrate to the organizational chart as formal leader and vice versa. Thus, for every 1 formal leader 2 informal leaders should be prepared in an attempt to keep the team active and resilient.

Models for self-management on shift

The crystallized structure meets the demand for activities repetition in the routine and the dynamic structure meets the demand for new external resolution and internal situations. Thus, a well-formed team for self-management must have both characteristics. Commitment depends on how much organizational values approximate individual and group values.

4.9. Complete Travel! (Figure 3)

In Neptune begins the journey with introduction on the course and the context involved, a test will be applied will start in Neptune. Follows Uranus with human performance factors discussions [14,15). A new test will be performed. The journey from Uranus to Saturn involves principles knowledge and concepts in the danger areas, risk, culture and beliefs, habits, behavior, and stress. A manual will be delivered and to acquire the certificate of this phase, will be done test by module at the trip end. At this stage the theme will be: habits and decisions in an organizational environment. This is the Cognition and Organization phase with exit from Saturn and destiny to Jupiter. The largest planet in the solar system arrival, will be validated through evaluation in the subjects: human elements, organizational moments, cognitive functions, decision, executive function and examples of each technique discussed. The direction for Jupiter involves learning through the journey second manual that presents practical exercises to be fulfilled.

A book presents the subjects already discussed in the manuals and tutorials and directed studies. This book will only be given to those who receive the fuel to follow the trip, that is, to leave Jupiter. The trip continuity depends on more practical knowledge in the risk investigation areas, task, failure, human factors through the third manual guidance. The authorization to enter Mars depends on the knowledge confirmation through testing. Finally, we will leave Mars for planet Earth, and the subjects required to enter Earth are: accidents, crisis and communication, how to investigate and avoid? Activities will be held in the field and will also be discussed scientific methodology for the CCS work. In this stage applications of the techniques for routine cases and accidents will be realized: GLP, Fireball, Texas City. In this case the manual 4 with the respective studies will be applied.

The journey between Earth and Venus requires knowledge about case research that will turn into final course work. This theme will be divided into: EAD presentations, orientation workshop, face-to-face meetings and final workshop to present the works. Most of the group will be on Venus, and only the top 10% will travel to the Sun and thus structured as a level 5 leadership. To achieve the astro-king it is necessary to acquire differentiated knowledge such as: dynamic culture; listening - empathy; to elaborate; group knowledge; informal leader; and self-management. This the trip last stage will be based on exercises and guidelines presented in manual 6. A test will be applied to approve this last phase and with approval each team component will receive an award book.

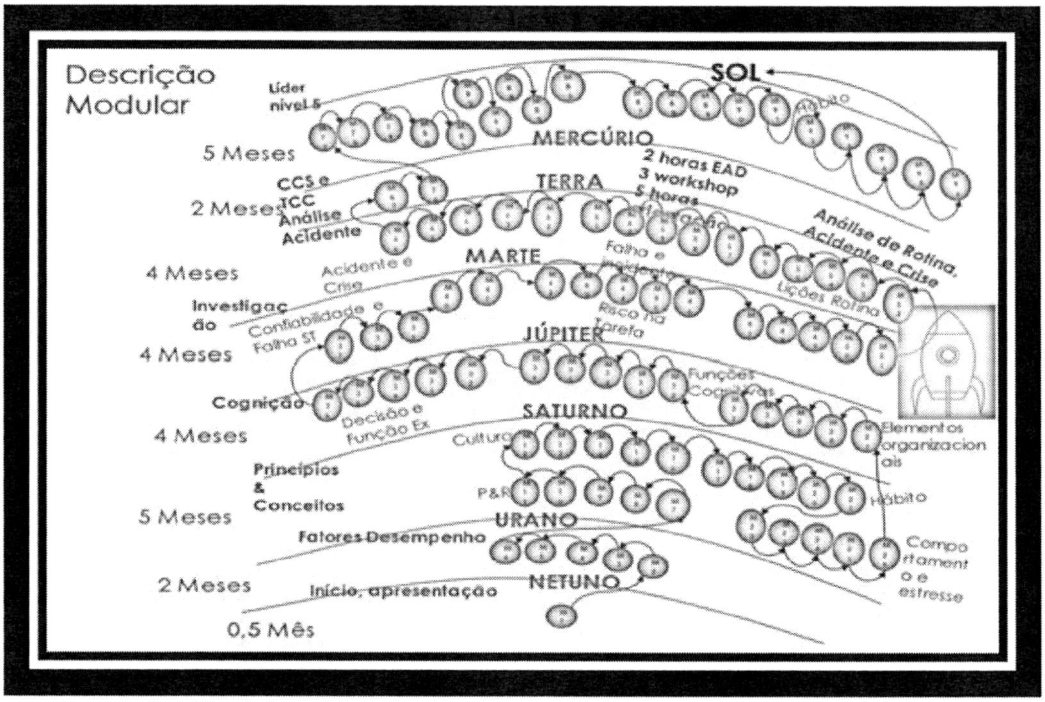

Figure 3 – Modules, Months, Subjects, Certifications and planets in EAD

5 Discussion about expected results, based in real case

The training accomplishment in an organization, has as one of the objectives, the activities execution for improvements that increase the organization performance. The greater the qualified people number within the company, the greater the flexibility in performing routine services and increasing productivity, as well as possibilities for improvements in their processes, since more people involved in operations tend to have a differentiated view on the services execution or decision making and that can aggregate information to risk analyzes, process failure possibilities and faster and safer ways to perform as activities.

In a medium to large company, it may be necessary to apply in training to attend 500 hours per person / year. This need for training requires a large investment to hire companies or qualified professionals to transfer the training content to a resources amount needed for the training to take place (location, transportation, overtime, food, materials, visual aids, etc.).

The online courses (e-learning) development at first may seem to be an organization with a higher cost, however, in the medium and long term, the dilution of these values is perceived, as well as providing a suitable content platform, adapted, standardized and customized to the needs and the business reality.

You can also add as an advantage the flexibility and better team time management, as it allows your employees and other employees, at their own workstation, to use time in a more strategic and intelligent way. Online training allows information standardization and content for all team members, regardless of the region in which team members are located.

5.1 "Ionizing Energy" A real case for Online X Training Online

The validation of the proposed distance education training in Human Reliability refers to a real case of training application in a medium-sized chemical industry in the subject Energy Ionizing. In the organization studied, this training would allow students to acquire knowledge about health, safety, environment, machines use and equipment, operational control and emergency actions, in addition to meeting a normative the region demand.

The classic way of acquiring this knowledge is through lectures with a qualified and qualified professional that would teach a face-to-face course with a minimum of 4 hours in duration for a team of 90 people (restricted to employees). Considering the people trained comfort, the availability of employees and the productive process, and the ability of the professional-coach to give attention and provoke discussions and leveling employee's knowledge, without compromising the quality of training and content, this training could only occur in 06 different classes with 15 people / class, in at least 3 distinct days considering 8h / work day and one hour interval (1h). The training frequency occurs every 2 years, and each new employee hiring. With the online training model application, maintaining the programmatic content, with audiovisual resources, playful interactive and planned and support material availability and question and answer tool, content retention evaluation and discussion forum through a module with 1.5h of training. The training model results from this Online Energy Ionizing X are presented in Table 2:

Table 2 – Comparison: face-to-face training and distance for ionizing energy.

Aspects	face-to-face training	ONLINE training
Training available for shift workers	Yes	Sim
ADM Staff Training Available	Yes. Restrict to Technical area	Yes. Technical area & Analysts
Training available for contractors	No	Yes
Number trained employees	90	188
Number of trained contractors	0	115
Man-hours / training	4h	1,5h
Extra resource / material available	Yes	Yes
Discussion forums during training	Yes	Yes
Content Fixer Rating	Yes	Yes
Content Utilization Index	75%	88%
% trained employees	30%	>90%
Training Frequencies	2 years	1 year
Flexibility training schedule	No	Yes
Access training content	Each 2years	In any moment

There is a significant increase in the number of employees trained, index of content utilization, and employees trained percentage (employees, contractors, trainees, etc.), the inclusion of contractors and analysts in the scope of training as well as an improvement in the index of content utilization.

6 CONCLUSION

This distance training on human reliability addresses the great demand for knowledge and practices to avoid accidents in process safety in the chemical, petrochemical, petroleum, nuclear, and air and rail transportation industries. The online training module will be available at any time for content access and support material.

The milestones elaboration or certificates to be fulfilled creates a motivation to conquer and through the travel configuration in the solar system brings a challenge to reach the sun and become level 5 leader in Process safety.

The online model will enable the training frequency to be increased in order to allow a knowledge revision, safety aspects and operational practices reinforcement at any time, and to increase the activities reliability involving training content, as well as discussions on the theme among those involved. All these items contribute to the unwanted events reduction, better the field tasks execution, safe practices uniformity and consequent unit safety culture increase.

7 REFERENCES

[1] ROTTA, Ivana Salvagni. Os Desafios da Organização do Trabalho: O Novo Perfil dos Trabalhadores e as Principais Tendências do Século XXI. UFSCAR, 2008. Disponível em: http://www.abepro.org.br/biblioteca/enegep2001_tr15_0483.pdf. Acesso 13 janeiro 2018.

[2] ÁVILA FILHO S., Sousa C. R., Carvalho A. C. Assessment of complexity in the task to define safeguards against dynamic risks. 6th International Conference on Applied Human Factors and Ergonomics (AHFE 2015a). Elsevier: Procedia Manufacturing. (2015a)

[3] DRAKE & SMITH . Ciência do comportamento na indústria. 1996

[4] ÁVILA FILHO, S. Metodologia para minimizar efluentes na fonte a partir da investigação de anormalidades operacionais: caso da indústria química. 2004. 115 f. Dissertação (Mestrado Profissional Gerenciamento e Tecnologias Ambientais no Processo Produtivo) –Teclim, Universidade Federal da Bahia, Salvador, 2004.

[5] ÁVILA FILHO, S. Etiology of operational abnornalities at industry: a model to learning.Tese (Doutorado em Tecnologia de Processos Químicos e Bioquímicos) – Federal University of Rio de Janeiro, Rio de Janeiro. 296 f.: il.Rotinas Limpas. 2010.

[6] S.ÁVILA FILHO & Drigo. E. Avaliação de Fatores Humanos Através do Comportamento, C4t. ABRISCO. 2017.

[7] COLLINS, Jim. Discussion Guide. JIM COLLINS.COM. 2002.

[8] MORAN, J. M. O que é Educação a Distância. Universidade de São Paulo. Disponível em < http://www.eca.usp.br/prof/moran/dist.htm Acesso em: 19 de janeiro de 2018.

[9] RASMUSSEN, J. Risk management in a dynamic society: A modeling problem. London: Elsevier Safety Science. England. v. 27. 1997.

[10] MUMMOLO, G., Affidabilitá dell'uomo, Affidabilitá degli impianti. Bari. Sicurezza degli Impianti industriali Politécnico di Bari, Itália. 1999.

[11] GOLEMAN, Daniel. LEADERSHIP THAT GETS RESULTS.2010. Disponível em https://www.uvm.edu/~facrsrcs. Acesso em 17 de janeiro de 2018.

[12] COLLINS, Jim. Global Leadership Summit. 2012

[13] Boring R, Blackman H. The origins of SPAR-H method's formance shaping factors multipliers. Idaho – EUA. August,

[12] ÁVILA FILHO, S. Learning environment to take operational decision in emergency situation.6th International Conference on Applied Human Factors and Ergonomics and the Affiliated Conferences, 2015. < Disponível em http://www.ahfe2015.org Elsevier Science Direct, 2015.> Acesso em 22 de novembro de 2016.

[13] ÁVILA FILHO, S. Assessment of Human elements to avoid accidents and failures in task perform, cognitive and intuitive schemes. 7o Global Congress on Process Safety 2011.

[14] LORENZO D.K. API770 – A manager's guide to reducing human errors, improving human in the process industries. API Publishing Services, Washington, 2001

[15] Boring R, Blackman H. The origins of SPAR-H method's performance shaping factors multipliers. Idaho – EUA. August, 2007. Francis, 1999.

511591 Common Bow Tie Errors and How CCPS Concept Book Rectifies

Abstract:

The bow tie risk analysis method is growing in its applications as it is a powerful tool for visually communicating major accident risks and the barriers deployed to prevent or mitigate these. The ease of communication can mislead users to think that bow tie creation is also easy, and this is not the case. A new CCPS Concept book, Bow Ties in Risk Management, thoroughly reviews how to create bow ties and provide detailed guidance on how to avoid errors.

This paper provides several examples of common errors seen in current bow ties. These cover structural errors where degradation factors and controls are misplaced onto main pathways,barriers that do not comply with guidance on required barrier core attributes, and it addresses incorrect hazards, top events, and consequences. The paper also covers poor treatment of human errors in bow ties by making these main pathway threats. These can lead to structural errors in the bow tie and a better means to treat human error is presented.

A final part of the paper presents a novel multi-level approach to bow ties that can be helpful got human error and mechanical integrity applications where deeper degradation controls are important to display.

512880 Joint Probability Density Estimation for Complex Variables and Its Application to Dynamic Risk Assessment Using Bayesian Method

Abstract:

Abstract

Risk assessment concept had been developed for nuclear industry in 1960s and then it got adopted in chemical process industry around 1980s. Since then it has been an integral part of making the chemical processes safer. To assess the risks associated with a process, the first step is identification and prioritization of process parameters that have crucial impact on the process operation. The second step is frequency and consequence modelling of rare events. An extensive literature is available for dynamic risk assessment (DRA) of rare events from accident precursor data, process history, and alarm databases using Bayesian method, Bow-Tie approach and multivariate techniques [1]. Bayesian method when applied with Copula analysis to model dependence structure of variables, is very efficient tool for DRA. However, the models developed still need to deal with highly non-linear, non-monotonic and complex nature of the variables [2].

The first part of the present work makes use of Shannon's information entropy [3] to identify the important process variables of a process based on maximization of mutual information between precursor and process variables. A cross-correlation analysis is performed to identify the process variables with similar impact on the process. After identifying the contribution of selected process variables to system, in the second part we model the frequency of rare events using Bayesian technique for identifying disturbances. A novel method is employed to get joint probability distribution function of failures of safety levels, which efficiently models and accounts for the complexity and non-monotonicity of variables. Then, Bayesian model is applied to get posterior estimate of safety level failure probability, and rare event frequency corresponding to upsets in process variables. Based on the weights of the process variables from the first part, and frequency of rare events from the second part, a risk indicator has been proposed to reflect the process health. This method is applied on Tennessee Eastman problem to identify faulty process variables, assign weights, and perform dynamic risk assessment.

Keywords: Dynamic risk assessment, joint probability distribution function, information entropy

References:

1. Pasman H. J., Rogers W. J., Mannan M. S. "Risk assessment: What is it worth? Shall we just do away with it, or can it do a better job?" Safety Science, Volume 99, Part B (November 2017), 140-155
2. Khan F., Hashemi S. J., Paltrinieri N., Amyotte P., Cozzani V. and Reniers G. "Dynamic risk management: a contemporary approach to process safety management" Curr Opin Chem Eng, 14 (2016), pp. 9-17
3. Shannon. C.E. "A Mathematical Theory of Communication" Bell Syst. Tech. J., 27 (1948), pp. 379-423

511134 Thermal Decomposition of Benzoyl Peroxide with Incompatible Materials

Abstract:

Organic Peroxides (OP) are widely used in petrochemical industry as initiators. Their unstable O-O bonds make them very useful, but also hazardous due to highly exothermic nature of decomposition, which can result in runaway reactions. Benzoyl peroxide (BPO) is one of the most commonly used OPs in industry and the hazards associated with its processing, storage and transport are revealed in many incidents. Benzoyl peroxide can decompose under favorable conditions and release large amounts of energy in short time. The decomposition reaction can ultimately lead to runaway reaction, posing great hazard of high temperature and high pressure. The most critical control factors of thermal decomposition and possible explosion of benzoyl peroxide are temperature, structure, and impurities or extraneous matter. This research is a comprehensive study of the runaway behavior of BPO hybrid system using isothermal and adiabatic calorimeters. The aim is the advancement of understanding the thermal decomposition of BPO under various conditions from both experimental and theoretical aspects. More specifically, this research will systematically study and develop thermodynamic and kinetic parameters related to decomposition of BPO with incompatible materials, such as HCl, H_2SO_4, HNO_3, H_3PO_4, NaOH, Na_2CO_3, NH_4HCO_3, and further the use of this particular knowledge to mitigate the risks in storage, transportation and manufacturing process.

510563 Maximum Explosion Overpressure of Methane/Air Mixture in Enclosures with Different Opening Degrees

Abstract:

TNO Multi-Energy method has a wide application in gas/vapor explosion. Many researchers use this method to calculate the overpressure vs. distance in many scenarios of gas/vapor mixture explosion. TNO Multi-Energy method gives a series of formulas to calculate gas explosion overpressure in different Confined Strengths. However, the Confined Strengths are only qualitative description without quantitative definition. Therefore, the non-quantitative relationship between Confined Strengths and opening degrees makes confuse when using the method in many cases. For many gas storage house or vapor generating workshop, it is hard to select confined strength according to the opening ratio of the enclosure. 56 kinds of methane-air mixture explosions in 4 different sizes of cubic enclosures with 14 kinds of opening degrees each were simulated by using CFD commercial software, FLACS. A correlation of the maximum overpressures in different opening degrees was proposed based on NFPA venting standard and FLACS simulations. To validate the correlation, series of tests were conducted in 100 cm ×100 cm ×160 cm enclosure with different opening degrees. 107 g methane gas which is about the 1.05 times of the flammable gas requirement of equivalent concentration was filled in the sealed box. The opening part of the enclosure was sealed with polyethylene film with thickness of 0.01 mm. The overpressure of four kinds of opening degrees was getting by NI data acquisition and computer system. The Confined Strengths parameters in TNO Multi-Energy method were quantified by using the proposed correlation. 10 Confined Strengths in TNO Multi-Energy method was quantified by the character of opening degree, *F* in this paper based on the correlation. The correlation and the quantified parameters can be used to calculate the maximum explosion overpressure of methane-air mixture in low blocking scenarios, such as storerooms, warehouses, or rooms with/ without opening areas, etc. It can be helpful as reference to the safety managers or relative researchers.

511314 Bridging Knowledge and Competency for Process and Occupational Safety

Abstract:

Bridging Hazard Recognition Knowledge and Competency

for Process and Occupational Safety

Understanding hazards and risk is essential for the success of process safety as well as occupational safety. Process safety involves managing energy transfers throughout a process life cycle in order to minimize the risk of releasing highly hazardous chemicals that can result in large-scale harm. The goal of the process industry is to manage safe and efficient transfers of energy that result in the safe manufacture of products. Occupational safety involves managing energy as workers perform jobs involving routine operations, maintenance, or repair in order to prevent personal injuries and illnesses, damage to facilities, and environmental harm. A consistent and systematic approach to hazard recognition can improve the efficiency and effectiveness of process safety and occupational safety. Establishing an effective safety culture requires solid hazard recognition and evaluation principles coupled with effective communication of process and occupational hazards. Successful process safety management requires that workers understand the technical documentation about a process (formally known as the process knowledge) and be able to conduct effective hazard identification and risk analysis.

The process knowledge is the resource that allows operators, maintenance workers, engineers, and other personnel to understand the various forms of energy involved so that the process can be monitored and controls adjusted to maintain the operation within design specifications. Hazard identification and risk analysis requires these personnel to recognize conditions, circumstances, and events that have the potential to cause harm. Process parameters occurring outside of operating limits, upset conditions, and emergency situations must be understood in terms of the states of energy within the process. It is necessary to understand how energy in a process is intended to behave, what catastrophic consequences can result from unintended conditions or interactions, and what interventions are required to prevent an incident.

An incident cannot take place without an energy transfer. Understanding all of the energy associated with a process or job and managing potentially harmful energy are necessary to ensure that incidents are prevented. A hazard can be defined as the potential for an uncontrolled release of, or an unwanted contact with, energy that can cause harm to people, the environment, and/or equipment and facilities. In some circumstances, a loss or lack of energy can initiate events that result in harm. Process and occupational risk both stem from the potential for energy transfer in an undesirable manner that can result in harm. The potential magnitude of harm is the general characteristic that differentiates process hazards from occupational hazards.

Process hazards are potential catastrophic events such as loss of containment (uncontrolled release) of highly hazardous chemicals resulting in spills, toxic vapor clouds, fires, explosions, and other large-scale events. These events can harm many people, the process unit, the processing facility, local communities, public areas, and the environment well beyond the facility boundaries. Process safety incidents often involve the shut-down of part or all of a process.

Occupational hazards usually include only one worker or just a few workers and can cause localized damage to the environment, equipment, or parts of the facility. The process often continues during and immediately after the incident. Energy contained within a process can present occupational hazards to operators, maintenance personnel, and others who are required to work with the process in order to perform adjustments, sampling, testing, repair, maintenance, and related process service work. When performing these tasks the workers have occupational risk stemming from their own physical actions, tools, equipment, and temporary use of external energy sources unassociated with the process such as motion, pressure, chemicals, or electricity that are used to accomplish the tasks.

The Management Oversight and Risk Tree (MORT) is a logic diagram that maps incident structure and includes energy as a prominent factor in all incidents. In MORT there are four conditions that are necessary for an incident to take place. An incident is not possible if any one of the four conditions is missing. Three of these conditions are: (1) the presence of potentially harmful energy, (2) the lack of adequate barriers and controls to manage the potentially harmful energy, and (3) the presence of vulnerable people or objects that can be exposed to the potentially harmful energy. The fourth condition is a sequence of events that leads to an uncontrolled release of energy, unwanted contact with energy, or incompatible interactions among the energies present. An incident occurs when people or objects are harmed because of this sequence of events.

There are five energy-based questions derived from MORT that can help workers recognize, evaluate, and manage hazards for a safe job outcome. These five energy-based questions are:

1. What is the job?

2. What energy is involved?

3. Where is the energy going?

4. How will we manage the energy for a safe job?

5. What are the stop-the-job triggers?

These questions are the key to accurately identifying hazards, determining the level of risk posed by the hazards, and formulating risk reduction options. All process operations and occupational tasks use energy that can be placed in one of the following categories— motion, chemical, pressure, heat and cold, gravity, biological agents, electrical, and radiation. These energy categories can be used as a guide for addressing the five energy-based questions. Answers to the questions will help determine what can go wrong, how bad it can be, and how often it might occur.

The first question ("What is the job?") focuses on identifying the goal and nature of the job and the steps required to complete the job. Important information for this question includes who will be involved, where and when will the job take place, what are the surrounding environmental conditions, and whether or not there are simultaneous operations. In addition, the answers to this question will determine whether or not special process procedures, permits (e.g., hot work, confined space entry), or additional resources are required. The second question ("What energy is involved?") focuses on the energy associated with the process or job. This question is intended to help workers understand all of the energy that is associated with the process or job. The third question ("Where is the energy going?") helps determine how the energy can be released or contacted and how interactions among different energy sources can occur. The third question also addresses several issues that are related to the potential transfer of energy and the resulting type and scale of harm that can occur. These issues include the state, magnitude, release direction, release duration, points of contact, interactions with other energy, and target resilience. The answers to this question help workers determine the hazards associated with the process, job, and environment where the job will be performed.

After the hazards have been identified, risk reduction efforts can focus on managing the energy that is identified as a hazard. The fourth question ("How will we manage the energy for a safe job?") addresses energy management techniques that will be used to manage all of the energy associated with the process or job. The fifth question ("What are the stop-the-job triggers?") helps predict unsafe situations that might arise during the running of a process or as a job is being done. These are situations that involve the possibility of an uncontrolled release, unwanted contact, or incompatible interactions among different types of energy. These situations can be called stop-the-job triggers and they require workers to stop the job or process or ensure that the potentially harmful energy involved is properly managed if the risk of shutting down the process is greater than continuing the operation.

Perceptions of risk vary depending upon a worker's knowledge of energy and whether the worker is focused on process or occupational hazards. This creates a need for common terminology about hazards so that the risk associated with a process or job can be effectively communicated, especially when process and occupational hazards are present simultaneously. Knowledge of the energy involved in a job or process is paramount to determining possible harmful outcomes. This knowledge also serves as the basis for managing energy for a safe job outcome. It must include an understanding of the type and magnitude of energies involved, release properties of the energies, interactions among energies, means of detecting energies, and other relevant characteristics of the energies. Effectively answering risk-based process safety questions in order to understand risk ("What can go wrong?", "How bad could it be?", and "How often might it occur?") is rooted in identifying and evaluating all of the energy involved. Teaching workers how to use an effective energy-based hazard recognition tool for both process and occupational safety provides an efficient and effective way to improve process and job knowledge. This will result in improved hazard identification competency, risk reduction plans, and communication among work groups with respect to hazards, risk evaluations, and subsequent risk reduction actions.

512741 Preparation of Safety Cases for Petroleum Refineries

Abstract:

This paper presents a strategy for the preparation of Safety Cases for Oil Refineries within a Safety Case regulatory framework. The strategy also ensures compliance with California oil refinery safety regulations "Process safety management for Petroleum Refineries", approved last year in U.S. The paper provides a 'road map' to the preparation of documented Safety Cases aimed to make this document the principal means by which the facility Operator controls and manages process safety risks during facility operation. Planning for safety case production work is also covered by providing recommendations to Operators on how this process must be conducted to ensure its robustness and to avoid 'compliance mentality' by challenging assumptions and established norms.

The paper recommends Operators to prepare a Safety Case production plan based on a 4-stage strategy which addresses technical, managerial and organizational aspects of process safety. Finally, examples are provided of the expected outcomes of the implementation of a Safety Case strategy for a Petroleum Refinery

512833 A CFD-Based Model for Investigating Thermal Runaway of MMA Suspension Polymerization in a Batch Reactor

Abstract:

Thermal runaway has been a major issue for the safety of polymer industry for decades. The gel effect that the dramatic viscosity increase hinders mixing and diffusion is recognized as a direct cause of thermal runaway. In this work, a three-dimensional computational fluid dynamics (CFD) model is developed to investigate thermal runaway process of suspension polymerization of methyl methacrylate (MMA) in a lab-scale batch reactor. Energy, momentum, and species transport equations are solved using a CFD software package. A mechanistic polymerization kinetic model is coupled with the CFD model, as a source term, using a user defined function (UDF) code. Gel effect is considered as a temperature-dependent factor of the kinetic model. Laminar flow regime is assumed as common practice for lab-scale suspension polymerization. Multiple reference frames (MRF) technique is employed to model the rotation behavior of the reactor impeller.

Considering the complexity of the proposed CFD model, the continuity, energy and momentum equations are first solved to guarantee numerical convergence. Subsequently the CFD model coupled with kinetic model and species transport equation is solved for conversion and temperature profiles. Experimental data from a Reaction Calorimeter (RC1e) with a similar reactor configuration in the literature are used to compare with simulation results. The temperature evolution and non-homogeneity inside the reactor during thermal runaway is unveiled by the CFD model. The model is then employed to predict the consequences of two common failure scenarios of batch reactors: (1) impeller failure, and (2) loss of cooling.

18 AIChE Spring Meeting
14TH GLOBAL CONGRESS ON PROCESS SAFETY
ORLANDO FLORIDA 2018

Best Practices for PHA Revalidations

William G. Bridges
Process Improvement Institute Inc. (PII)
wbridges@piii.com

Revonda Tew
Process Improvement Institute Inc. (PII)
rtew@piii.com

Matías A. Massello
Process Improvement Institute Inc. (PII)
La Plata, Argentina
mmassello@piii.com

PROCESS IMPROVEMENT INSTITUTE

Prepared for Presentation at
American Institute of Chemical Engineers
2018 Spring Meeting and 14th Global Congress on Process Safety
Orlando, Florida
April 22 – 25, 2018

BEST PRACTICES FOR PHA REVALIDATIONS

William G. Bridges
Process Improvement Institute Inc. (PII)
wbridges@piii.com

Revonda Tew
Process Improvement Institute Inc. (PII)
rtew@piii.com

Matías A. Massello
Process Improvement Institute Inc. (PII)
La Plata, Argentina
mmassello@piii.com

Keywords: PHA, Revalidation, process safety, PSM

Abstract

Process hazard analyses (PHAs) must be updated and revalidated every 5 years or sooner. PHA/HAZOP Revalidation is entirely different than a baseline or original PHA. The original textbook from CCPS on Revalidating PHAs was issued in 2001. This paper describes the current best practice approaches to revalidation and explains when to use each approach. The approaches are described in detailed flowcharts. Checklist for decisions making and quality control are provided, along with examples of completed documentation. This paper is based on completion of more than 300 Revalidations.

1. Introduction

A PHA Revalidation is a renewal at one point in time of an existing PHA to ensure the risk review for the entire unit/process is still valid. It is one stage during the lifecycle of a PHA of a process.

Prior PHA(s) + Update + Retrofit + Review (MOCs & Incidents) →
Revalidated PHA

Revalidations are done after the initial PHA and after the startup phase of a process. They are done throughout the life of a process until it is decommissioned. Reference Figure 1 below showing the types of hazard reviews during the life cycle of a process.

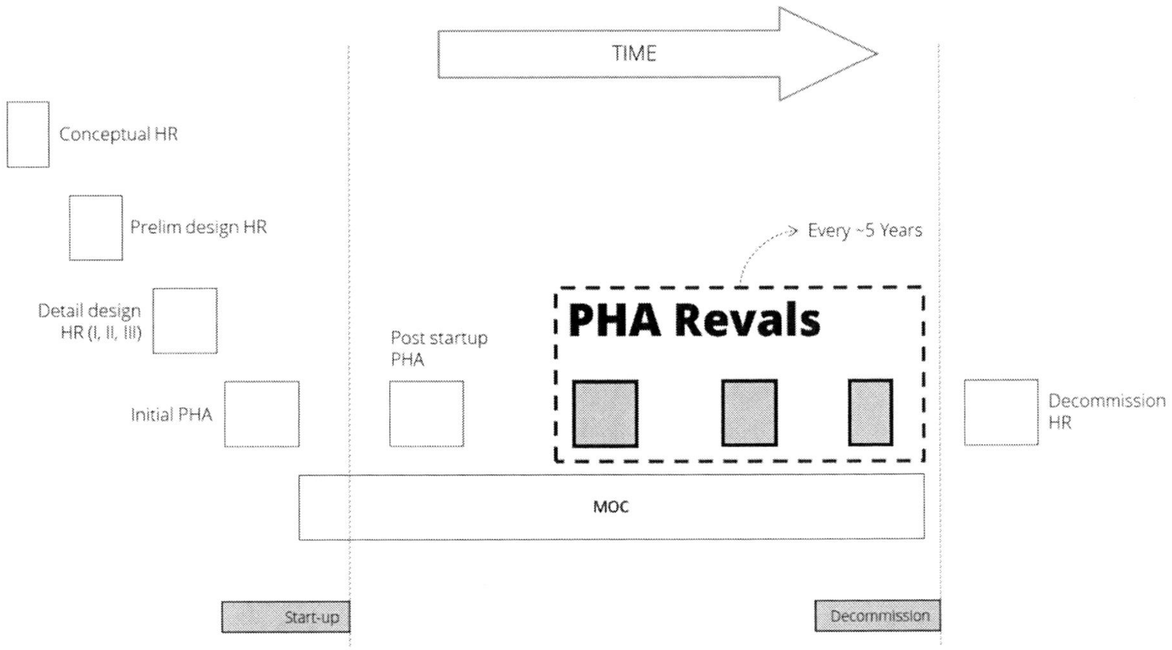

Figure 1. PHA Revalidations in a Unit's life cycle

Revalidating and updating the results of any previously conducted PHA(s) are both essential if risk is to be maintained. Therefore, a Revalidation PHA should be conducted on all "Initial" PHAs after five years, and then revalidated again each successive 5-year period. The primary objective of the Revalidation is to review the results of previously conducted PHAs and identify, evaluate and attempt to control any new hazards that have been introduced or discovered since the original PHA.
The original textbook, Revalidating Process Hazard Analyses, 2001, CCPS/AIChE [1], was written bases on the experience from a limmited number of Revalidations. Most PHAs between 1992 and 2003 were ReDo PHAs. This paper updates the state-of-the art on Revalidating PHAs and is based on thousands of revalidations by PII staff [2].

2. Why do we revalidate?

The underlying driver for revalidations is process changes. The PHA is updated to reflect the current process and to evaluate hazards of the current process and the controls for mitigating or eliminating the hazards. Operating history may also identify scenarios that the previous PHA did not address or that make documented safeguards and consequences invalid. These scenarios must be reviewed and added to the PHA.

Process changes

Changes are made to the process, materials, procedures, facilities, and equipment. MOC systems are implemented to review hazards associated with changes and to document the change to facilitate updating process safety information needed to maintain safeguards. Even with the MOC system, changes should be reviewed individually and then collectively to assess the quality of the hazard review completed at the time of the change.

Some changes are made to the process without using the MOC system. These uncontrolled and undocumented changes can invalidate the documented design basis of the process and can introduce uncontrolled hazards.

The accumulation of process changes since the last PHA can introduce hazards that may not have been identified or even possible (credible) during the review of a single process change.

Quality of previous PHA

Company guidelines and government regulations define requirements for PHAs. PHAs must address certain elements of any process such as human factors and process safeguards.

Furthermore, PHA methodologies can be incorrectly executed during the PHA meeting. Examples include not having a knowledgeable and experienced PHA team, or using only the What-if technique for a complex process unit.

Operating history & research

Since the previous PHA, new information emerges through additional years of operating the process and possibly through research and development prompted by incident investigations and near misses and through process improvement initiatives. Armed with new information, previous assumptions about causes, consequences and safeguards associated with process deviations may change. The revalidation process provides the opportunity to apply new findings to previous PHA results.

Previous PHA Recommendations

The revalidation team should discuss and study the recommendations generated from the previous PHA. The revalidation meeting is not the forum to resolve the recommendations. Furthermore, those with the authority to make resolution decisions may not be members of the PHA revalidation team. The team may evaluate why a recommendation is still Open. However, the team my find during the revalidation that the same recommendations resurface, further validating the need for the recommendation. Obviously, NO recommendations should still be open after 5 years, but sometimes they are.

Review of the recommendations also functions as an audit as the implemented recommendations are reviewed. Since all recommendations are considered process changes, the MOC system should be used to properly identify hazards and to update process information. If no MOC documentation is available, then the implemented recommendation must be considered an uncontrolled change and must be reviewed during the PHA revalidation.

Finally, review of the closure of the previous recommendations is necessary to help ensure that the previous PHA tables are updated for the new and revised safeguards that the recommendations typically result in. Resolution of recommendations may also result in elimination of accident scenarios.

Regulatory & company requirements
Government regulations and company requirements typically establish the need to revalidate, the frequency of revalidation, and the scope of the revalidation. Regulations and company requirements can also change, prompting PHA revalidations.

Decommissioning of process equipment
The processes involved in decommissioning process equipment (e.g., decontaminating, dismantling, disposal) present new risks that were likely not applicable during routine process operations. There may also be different/new deviations for emptying or cleaning the equipment. The scope and complexity of equipment being decommissioned will likely dictate the need for a new PHA in lieu of a revalidation.

3. How do we revalidate?

There are three major approaches to perform a PHA Revalidation:

Update (& Revalidate)
Updating the PHA to reflect process changes and incidents that have occurred since the previous PHA. If the plant has (1) a strong, effective MOC system and (2) a thorough, well-documented PHA, that was performed correctly with the right team, then updating and revalidating will be more streamlined. For uncontrolled changes and for documented changes where documentation is limited, and/or the hazard review is of poor quality, a more extensive review of the PHA section deviations is required.

Process changes are discussed while referencing the corresponding "node" or section, analyzing the effect of the change on the process deviation, and the causes, consequences and safeguards that are documented in the previous PHA report. The report is updated to reflect the process changes.

For incidents, discuss if the incident scenario was documented in the previous PHA. If so, determine if documented safeguards failed or if they were adequate. For incidents that uncover a new hazard, the team then applies the PHA methodology to the incident scenario and updates the PHA under the applicable section.

Evergreen: This is the simplest mode of revalidation, but only the companies with very mature PHA systems and good staff continuity can implement this approach. With this approach, a company updates the baseline PHA as the final step of each MOC (or every month); so the PHA is always Up To Date, or at worst about 1 month behind the latest MOC. This approach requires continuity of PHA team leaders between unit-sized PHAs and mini-PHAs for changes (MOCs). It also requires that the same software is used for MOC PHAs and unit-sized PHAs. This approach requires a high quality and thorough baseline PHA. It uses codes, in brackets, to note which MOC or incident the new information came from. See the excerpt on the next page.

(A.3) (Reviewed 2005) Measuring Level on Bulk Storage Tanks					
Drawing: SP 03.201.02					
Item	**Deviation**	**Causes**	**Consequences**	**Safeguards**	**Action Items**
33.7	Step 2 (old Step 4) to remove measuring stick and record level performed incorrectly		(1998) No consequences of interest (2005) Operator could fall resulting in serious bodily injury (2005) Resin spill could damage valve	(2005) Railing around work area on tanks	
33.9	Skip step 4 (old step 6) to replace cap on quick connect	Typical causes of procedural errors (see Table A.1)	Release of process material if preparing to unload a tank wagon into the tank; fire and explosion hazard affecting a large area	Continuous ventilation The tanks are located in a containment area Checklist Generic and inherent safeguards protecting against or mitigating process material releases (see Table A.2) (2005) Air Monitors in Bulk room	'98Rec 26. See recommendation 19, item 29.12 (1998)
33.10	Step 4 (old Step 6) to replace cap on quick connect and pull ears up and secure Velcro straps performed incorrectly (poor seal when cap is replaced)		(2005) Quick connect cap can pop off during the filling stage and flammable material can spill out of tank; fire hazard affecting a medium area Incident Report 2/12/03 - Bulk Room - 878 Resin leak from valve on bulk resin tank during resin tanker unloading - (A.9) (Added Reval 2005) Incident Investigation Reports and Spill Reports 1998-2005 (linked to 82.11) MOC 92-2005-04 - 4/1/05 Replace portable tank valve lever stainless actuator to oval actuator brass construction - (A.8) (Added Reval 2005) Management of Changes 1998-2005 (linked to 81.23)	(2005) Velcro straps on valves	

So, if the organization intends to keep their baseline PHA evergreen, then the PHA Leaders and Scribes of MOC risk reviews need to learn how to find and edit the baseline PHAs. At revalidation time (every 5 years), a PHA team of the proper composition is convened, but now their primary responsibility becomes making sure no changes were missed, making sure each mini-PHA had the right team structure, and making sure the Revalidation team consider the effect of the combination of MOCs and IIs since the prior PHA, as each MOC PHA team only considered the effect of that specific change on the baseline PHA

Retrofit (Update and Revalidate)

Retrofits are necessary when the PHA methodology was incorrectly applied or specific guidelines/requirements were not addressed, such as human factors considerations. In this scenario, the team would review each section of the previous PHA and correct any deficiencies. If human factors were not addressed, the team would go back and review each section, applying human factor concepts as causes. The team may also need to apply the global human factors checklist if it was not addressed before. In summary, if requirements were omitted, the previous PHA should be a candidate for a retrofit.

Retrofits are also necessary if there are holes in the previous PHA. In this case, the hole can be patched. Holes can include failure to evaluate one or more modes of operation or missing the hazard review of a selected equipment/sub-units. Retrofit could also include performing a facility siting review that was omitted in the previous PHAs.

Redo

A Redo is basically starting over and is more like doing an initial PHA. A Redo starts with up-to-date P&IDs and other PSI and operating procedures. Redo is needed where PHA methodology was incorrectly applied and cannot be corrected. For example, if the team composition was wrong or if a less extensive methodology was used for a very complex process, a redo is typically suggested. A Redo may also be necessary if there were substantial changes to company policy or government regulations regarding PHAs.

3.1. Deciding on the Revalidation approach to use

Figure 2 shows the decision making process to choose the Revalidation approach to use.

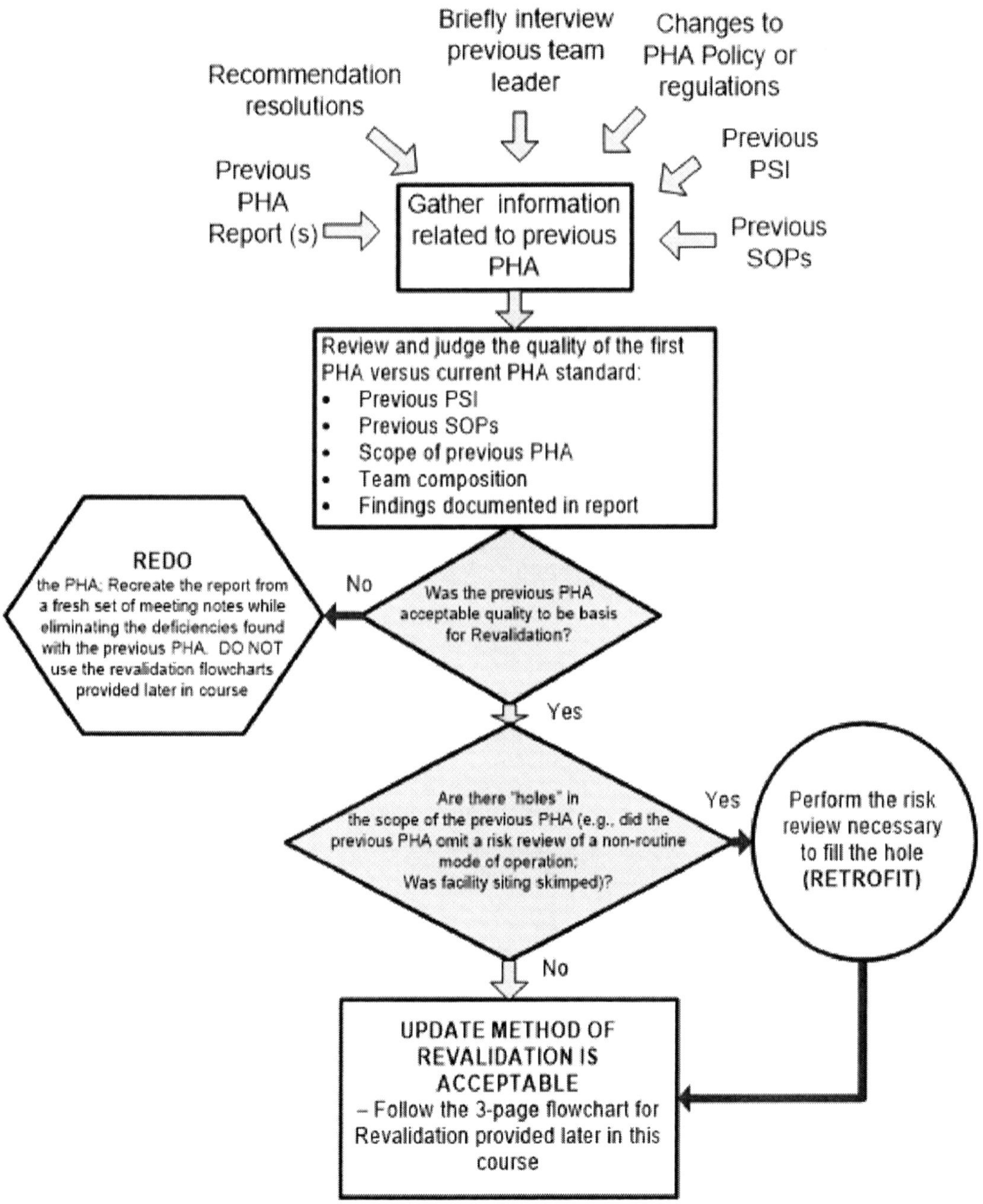

Figure 2. Decision making process for revalidation approach

To answer the question in Figure 2 "Was the previous PHA acceptable quality to be the basis for revalidation?" use the checklist shown in Table 1. A "No" answer to any of the questions indicates a Redo is likely necessary (sometimes two "No" answers make more sense).

Table 1. Determining the Quality of the Previous PHA

CRITERIA	Yes/No
1. Did the qualifications of the PHA leader/facilitator meet all company and regulatory requirements; i.e., was the leader/facilitator trained, experienced and competent in the PHA method used?	
2. Did the PHA Team make-up/qualifications meet all company and regulatory requirements; Did team include, as a minimum the disciplines listed below? • a stable team membership roster with minimal substitution of team members during the course of the review • operations team member(s) with adequate process and equipment knowledge including recent hands-on operating experience (preferably a senior hourly operator or at-risk operator) • an engineer with industry and specific process experience	
3. Was the prescribed PHA method appropriate for the complexity of the process studied; i.e., was the PHA method justly rigorous for addressing all potential hazards of the process?	
4. Was the PHA method used from the approved list below or was an appropriate equivalent methodology used? • What-If • Checklist • What-If/Checklist • Hazard and Operability Study (HAZOP) • Failure Mode and Effects Analysis (FMEA)	
5. Is the PHA documentation sufficient, or can sufficient documentation be reconstructed, to: • verify PHA leader/facilitator and team qualifications • indicate PHA team meeting dates • verify that previous incidents, including those with the potential for catastrophic consequences, were reviewed by the PHA team • verify the PSI utilized by the team was adequately available and up-to-date and accurate enough to ensure a thorough study • verify the PHA team's findings including daily worksheets, engineering and administrative controls cited (safeguards), the failure of controls (consequences), and their causes; do these findings appear to make sense and are they thorough, or does a significant number of critical scenarios appear to be missing? • verify that siting was addressed • verify that human factors were addressed • verify that a qualitative evaluation of a range of the possible safety and health effects was conducted (risk ranking or some other documented technique) *Note: if only a few of these bullets are an issue, they are probably "fixable," however, if several of them are an issue, then there is probably a sufficient lack of data to warrant a "No" answer.*	

4. Revalidation process

4.1. Preparation

The preparation (pre-meeting) stage of a PHA Revalidation has two main tasks (see Figure 3):

- Information gathering
- Information evaluation

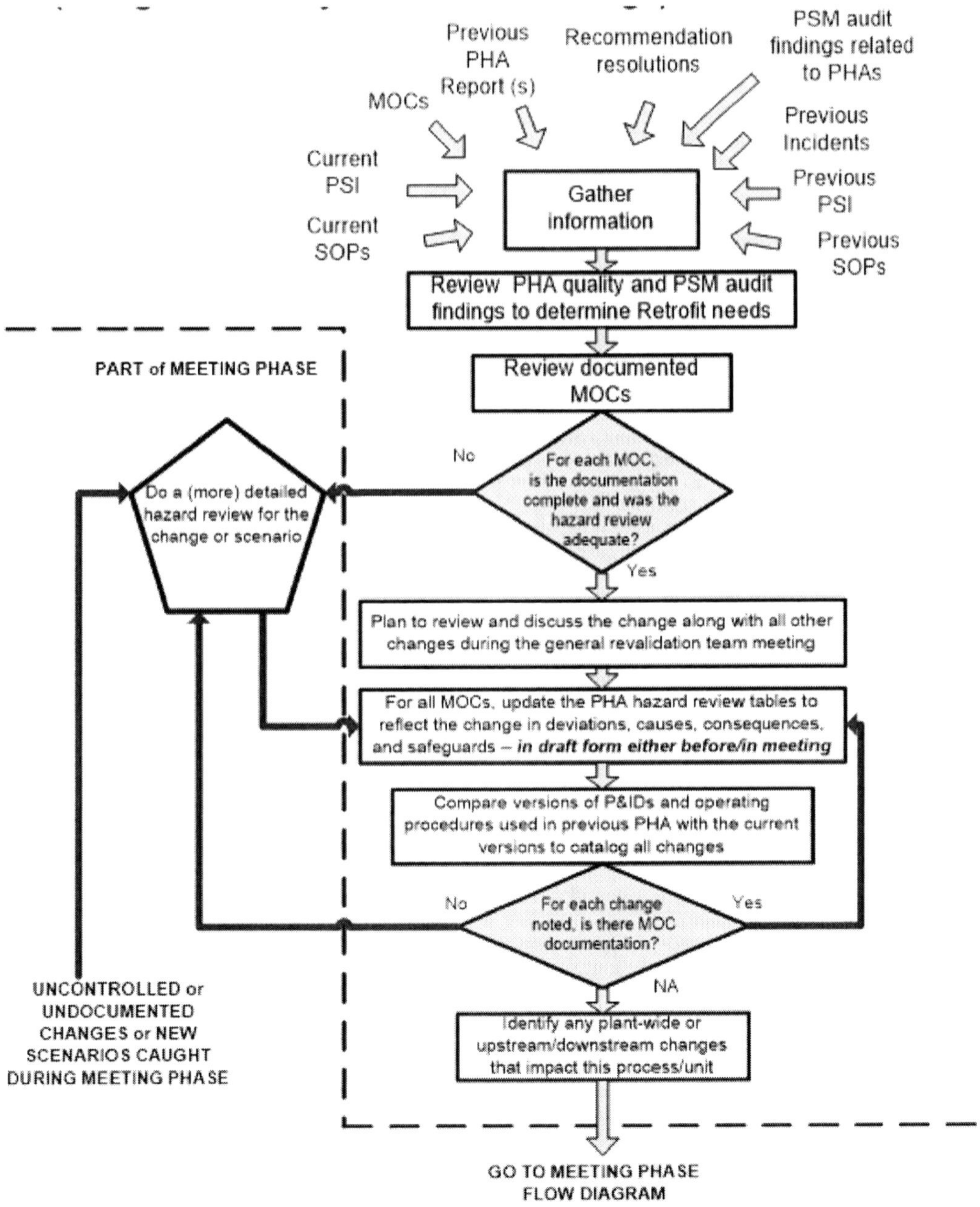

Figure 3. Revalidation: Pre-meeting activities

4.2. Meeting

Figure 4 shows a flowchart of the meeting phase activities.

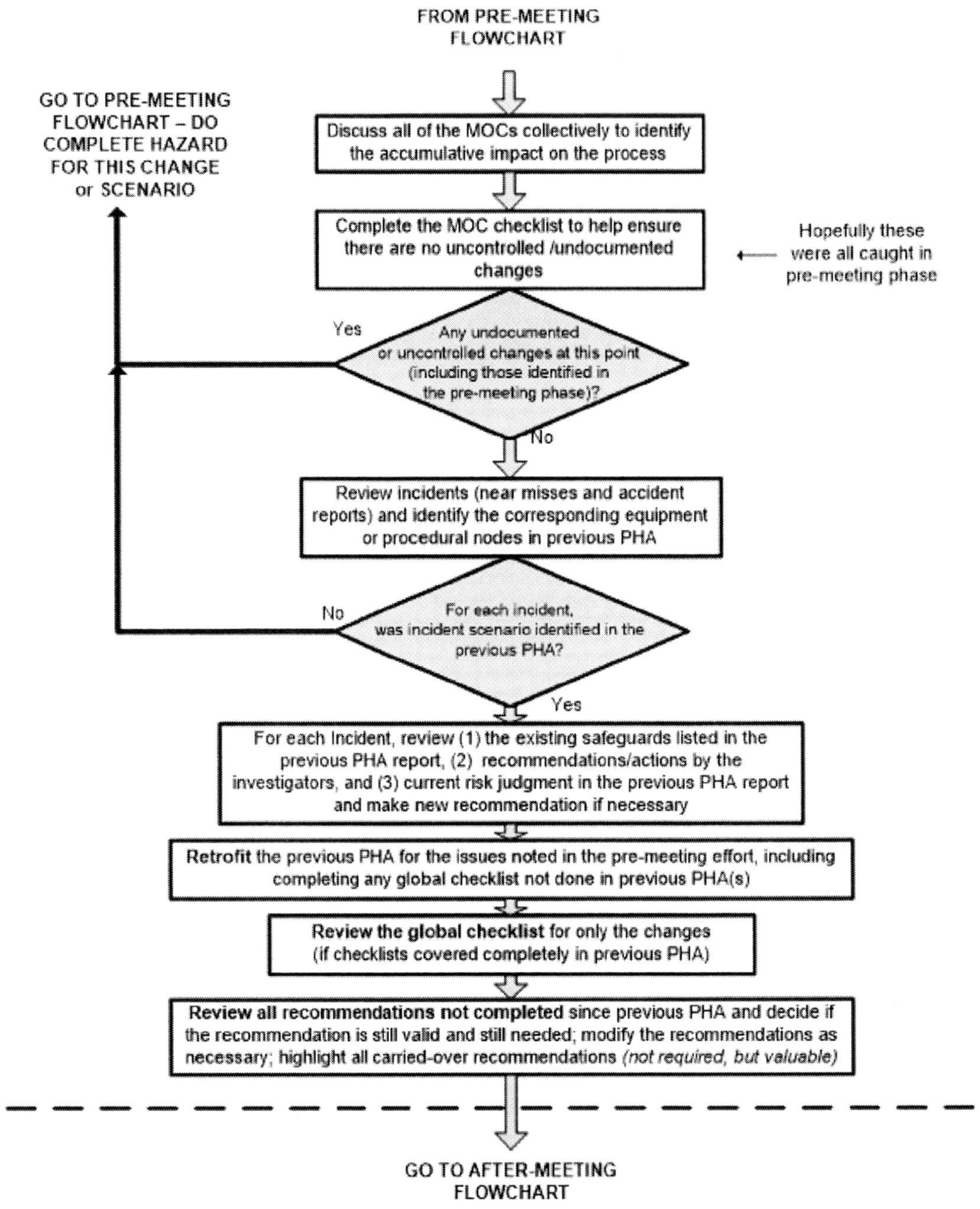

Figure 4. Revalidation: Meeting activities

4.3. After meeting

Figure 5 shows a flowchart of the after meeting phase activities.

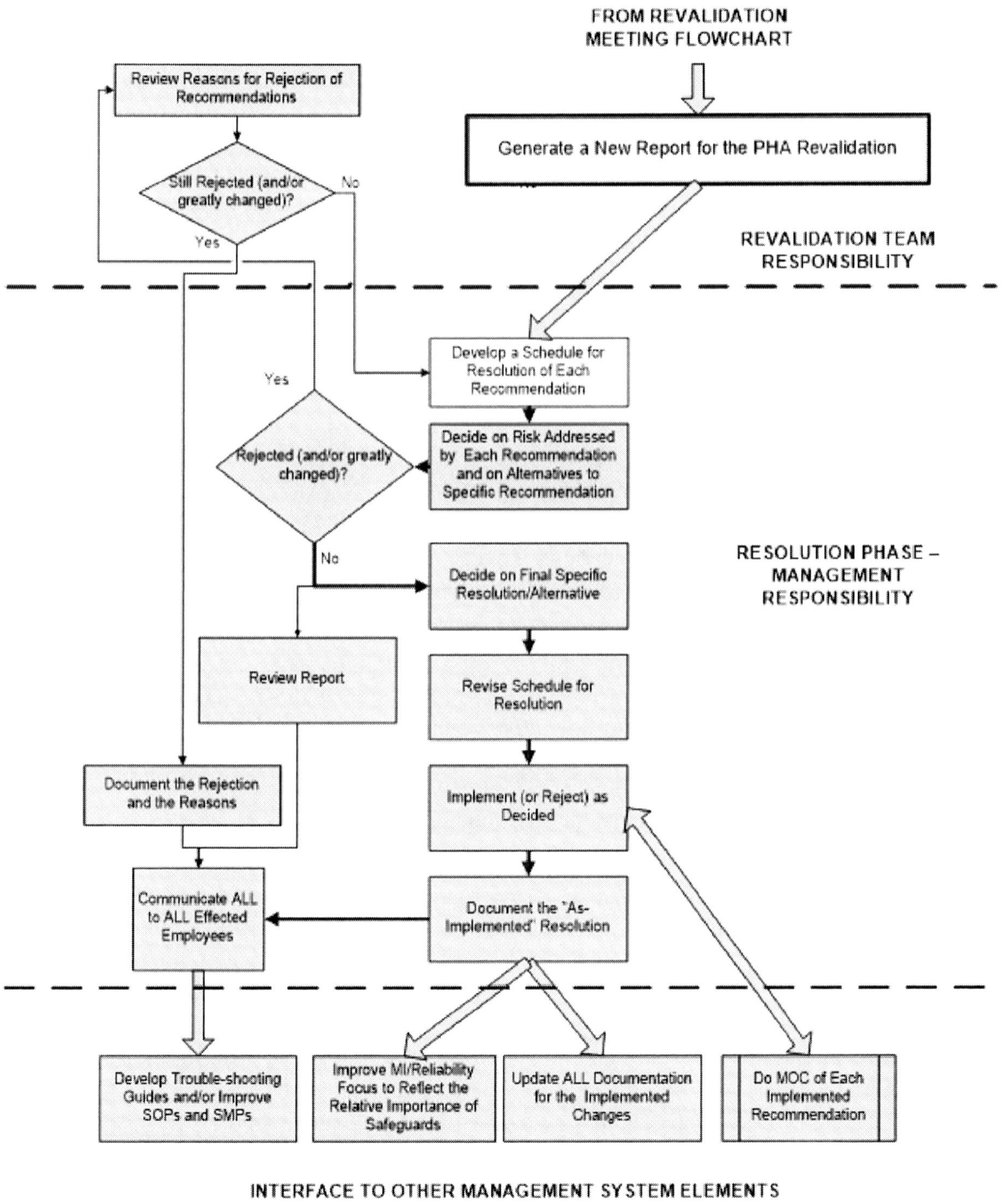

Figure 5. Revalidation: After meeting activities

5. Revalidation team

Nothing is more important to the success of the hazard evaluation revalidation than the Team. The team leader and scribe help ensure a concise revalidation while the team members do most of the brainstorming of accident scenarios based on their expert understanding of the process and their experience. The best techniques cannot compensate for a poorly staffed team.

5.1. Team Leader

The team leader is often viewed as the most important member of the hazard evaluation/revalidation team. But, the team leader is NOT more important than the others on the team; the leader's (facilitator's) role is difficult to learn because the principal job of the leader is to ensure the team has a good chance to uncover all accident scenarios while balancing the need of not wasting the team member's time. Typically, team leaders have the following traits:

- experience in process engineering, process safety, engineering standards
- trained and experienced in hazard evaluation technology
- trained in revalidation approaches
- unbiased with respect to the process design and operation
- skilled facilitator; does NOT let the team rely on him/her to make risk decisions
- well organized; saves time in the meetings at every opportunity
- good writing skills; able to summarize recommendations well

The best revalidation team Leaders are those that (1) make good judgments on revalidation approaches, (2) make good decisions on when to force a risk judgment by the team, (3) recognize when the team is wasting time on a discussion, and (4) make the brainstorming techniques "invisible" (the brainstorming seems to flow on it's own).

5.2. Scribe

The Scribe is optional for Revalidations, but it efficient to use one.

A good scribe takes the load off of the team by recording the notes of the hazard evaluation team discussion. A scribe needs to be able to listen to roughly 10,000 words of discussion on a complex accident scenario and be able to distill the discussion into roughly 100 words of written record. The scribe probably has the toughest job on the team and is critical to the team's success because the scribe allows the team to freely brainstorm. Key traits of a good scribe include:

- trained in analysis techniques
- trained in revalidation approaches and has the time to help the leader do the large amount of pre-meeting reviews
- attentive to details; able to decide quickly and without guidance where to show each step of the scenario in the hazard evaluation record
- bears complete documentation burden during the meeting (frees the leader to LEAD)
- well organized; saves time in the meetings at every opportunity
- excellent writing skills; able to distill the scenarios into concise descriptions

Sometimes the scribe role is filled by a Leader who is still in training. Other times we use junior engineers for this role since they will learn how "not" to operate the process. In a few cases, administrative assistants have been screened and taught how to scribe a hazard evaluation meeting (the jargon and the logic of an accident scenario can be difficult for non-technical persons to learn). In all cases, the scribe must have the time allocated for the large documentation load they will bear.

5.3. Team members

As with the initial PHA/hazard evaluation, the team members are the most important part of the revalidation team because they are the only ones with the expertise to brainstorm possible accident scenarios. Good team members can even compensate for poor leadership of the team. Certain team members (such as the unit operator or the process technology expert) are mandatory members of the team – if one of these mandatory members must be absent, they must be replaced "in kind" or the team meet must be halted until they return. Other team members (such as loss prevention or maintenance) may be part-time or temporary – their expertise is only needed for an hour of each 5-hour meeting-day to answer questions related to their expertise. The decision on who is necessary for any given discussion should rest with the team leader and can vary from session to session. The traits we want in our team members include:

- expertise in process design, operations, and maintenance
- listens attentively during discussions
- respects opinions of others
- actively contributes to discussion
- brainstorms — gets past status quo
- can't be distracted by other duties during meeting phase
- openness about previous problems in the process/operation

6. Time requirements

As expected, the time required for a PHA Revalidation is lower than the time required for the Initial PHA (unless the Revalidation is performed by a Redo).

For Revalidation there's a higher burden on the Leader (and Scribe) before the meetings since they have to gather and evaluate all the information available. However, meeting and documentation time are greatly reduced comparing with the Initial PHA. The better the quality of the Initial PHA (and the information required) the lower the time required for Meeting and Documentation stages.

Figure 6 compares the time requirements for Initial PHA and PHA Revalidation.

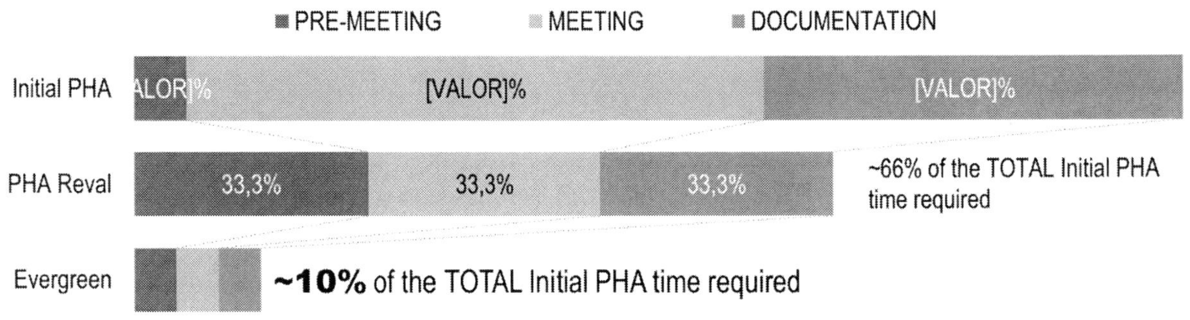

Figure 6. Time requirements comparison. Initial PHA vs. PHA Revalidation.

7. Conclusions

Revalidating and updating the results of any previously conducted PHA are both essential if up-to-date and accurate Process Safety Information is to be maintained. Therefore, a "Revalidation" PHA should be conducted on all "Initial" and "Capital Project" PHA's after five years. The primary objective of the "Revalidation" PHA is to review the results of previously conducted PHA's, and identify, evaluate and attempt to control any new hazards associated with the original PHA. Following the best practices described in this paper is a key step for achieving that goal in an efficient way.

8. References

1. *Revalidating Process Hazard Analyses*, 2001, CCPS/AIChE.
2. PII Course 9, *PHA Revalidation*, © 2003-current

511319 Active Safety Performance Dialogues at the Shop Floor

Abstract:

Process Safety KPIs have been researched and debated for long. A number of standards such as API 754, bibliography and surveys have been published to provide guidance on the effective adoption of KPIs.

Our experience with many leading firms suggests that, in most cases, companies have developed a set of KPIs but are still missing the most important point: KPIs are not only about tracking performance; they are about discussing and managing performance through active performance dialogues across all levels of the organization.

KPIs are often used to share information, for example during formal EHS Meeting and Risk Committees, on a periodic basis rather than effectively leveraged during the decision-making process.

This is even more true if we look at how performances are managed at the shop-floor level on a day-to-day basis. In the best cases; we have found that process safety performance is displayed through advanced LCD screens with limited use of active performance dialogues.

An effective performance management process through visual boards and active performance dialogues are essential components of the broader set of leader standard works which enable firms to instill simple and effective rituals and behaviors at line management level and allow easy integration of process safety in day-to-day work processes.

The goal is to provide the shop-floor with simple tools, capabilities and rituals that enable timely identification and resolution of safety challenges before they escalate.

Simple processes in turn raise the level of risk awareness and help shifting the organization from a "find & fix" to a "predict and prevent" culture.

18 AIChE Spring Meeting
14TH GLOBAL CONGRESS ON PROCESS SAFETY
ORLANDO FLORIDA 2018

Pitfalls of Using the Wrong Risk Matrix In PHA and LOPA

Kirk Busby
Kazarians & Associates, Inc.
100 West Broadway
Glendale, CA 91210
Kirk.Busby@kazarians.com

Mardy Kazarians
Kazarians & Associates, Inc.
100 West Broadway
Glendale, CA 91210
mkazarians@kazarians.com

Prepared for Presentation at
American Institute of Chemical Engineers
2018 Spring Meeting and 14th Global Congress on Process Safety
Orlando, Florida
April 22 – 25, 2018

Pitfalls of Using the Wrong Risk Matrix In PHA and LOPA

Kirk Busby
Kazarians & Associates, Inc.
100 West Broadway
Glendale, CA 91210
Kirk.Busby@kazarians.com

Mardy Kazarians
Kazarians & Associates, Inc.

Keywords: Hazard Identification and Risk Analysis, Risk Matrix, Process Hazard Analysis and Layer of Protection Analysis

Abstract

A Process Hazard Analysis (PHA) is an important tool for identifying and managing the hazards associated with a chemical process. Hazards may include safety issues, environmental impacts, business impacts, or reputational issues. It is now common to use a risk ranking scheme to discriminate among the potential hazard scenarios so that resources to reduce risk are allocated on a rational basis. Additionally, the PHA methodology has been extended to Layer or Protection Analysis (LOPA), which allows quantification of risk gaps based on a consistent methodology and industry data. The risk ranking process is driven by a company's individual Risk Matrix, which governs the decision-making process and ultimately determines the extent of mitigation measures necessary to reduce risk. If the risk matrix used during a PHA or LOPA session has not been carefully reviewed to ensure applicability, a number of issues may arise over the course of the study. On several occasions, the authors of this study had to request that company management revise their risk matrix to avoid unintended conclusions and decisions during the study. This paper aims to identify some of these traps and unforeseen consequences of using a risk matrix in PHA and LOPA.

Typically, hazard scenarios of the most interest in a PHA are those which have a high severity and a low likelihood of occurrence. Such scenarios are often subjected to the LOPA process to ensure that sufficient safeguards are in place. When looking at high severity, low likelihood scenarios, it is easy to unintentionally demand an overly conservative risk criterion. When this criterion is applied in a PHA or LOPA study, the results can lead to dramatic measures for risk mitigation, as a result of unacceptable risk levels or requirements for added safeguards which may not be warranted. Difficulties that the authors have faced in using corporate risk matrices have included (a) overly conservative risk ranking of high severity, low likelihood scenarios, (b) improper decision granularity (c) risk matrices unsuited for the specific operation under review (d) Severity and Likelihood Categories with Uniform Risk Decisions and (e) Increased Risk Decisions for Lower Severity, Higher Likelihood Scenarios. These and other related issues are addressed with specific examples from PHA and LOPA projects conducted by the authors.

1. Introduction and Background

Process Hazard Analysis (PHA) and Layer of Protection Analysis (LOPA) are critical tools for identifying and managing the hazards associated with a chemical process. Once a hazard scenario has been identified, a risk ranking process is used to categorize the risk of the scenario [Reference 1]. Typically, a pre-developed risk matrix is used to classify the severity and likelihood of each hazardous event, and to assign risk ranking. The risk matrix is often developed by company management as a decision-making tool to determine what represents a tolerable risk.

Risk matrices have been in use for many years, and have been adopted by a wide variety of industries. They provide a simple framework for performing a systematic review of risks, and are an approximation of the more rigorously developed risk contours using Quantitative Risk Analysis. Despite the length of time risk matrices have been in use, detailed analysis of their theoretical structure has only been undertaken in the last 15 years. Several authors, including Tony Cox [Reference 2] and Paul Baybutt [Reference 3] have explored the mathematical limitations and potential errors involved in using a risk matrix to quantitatively compare and rank risks in the process industry. In addition to the limitations and potential errors noted in those two articles, practical pitfalls have been noted by the authors of this article. Practical examples are given to illustrate some potential issues which may be encountered when using a risk matrix in PHA and LOPA.

When developing a risk matrix, there are a number of factors which must be considered in order to arrive at a matrix which is rational and appropriate for a business. In particular, the following conditions may lead to unintended consequences when performing a hazard analysis:

- Overly conservative risk ranking of high severity, low likelihood scenarios
- Risk matrices unsuited for the specific operation under review
- Improper decision granularity
- Risk decision not impacted by the severity and likelihood combination
- Potential to miss medium severity, but high risk scenarios

Using a risk matrix which has not been carefully selected can cause a PHA or LOPA study to severely over- or under-estimate the risks involved in a process, resulting in an unsafe or over-designed system. Careful consideration of the above points must be made when developing a risk matrix to ensure that the matrix is suitable for the particular operation under review.

2. Risk Matrix Construction

A risk matrix is typically presented as a table in which severity and likelihood rankings are assigned for a given scenario, and the combination is used to arrive at a risk level (sometimes called a risk decision). The first step in creating a risk matrix is defining individual distinct tiers for the severity and likelihood categories.

Severity of Consequences

When defining consequence severity categories, it is important to consider the full range of potential impacts on a system or operation. When dealing with hazardous chemicals or processes, it is important to consider impacts on safety and the environment, as well as business operations. Some companies choose to include a distinction between impacts to facility personnel and impacts to members of the public. Additionally, some companies also include severity categories for economic impacts (e.g., asset damage, loss of production, etc.). Table 1 provides an example set of severity categories for a large facility such as a refinery.

Table 1. Example Consequence Severity Category Definitions

Category	Safety	Environmental	Economic
A	Multiple fatalities	Major release requiring multiple years to remediate	>$500 million
B	Single Fatality	Major release requiring a year to remediate	$100-500 million
C	Permanent Partial Disability	Release requiring months to remediate	$30-100 million
D	Recordable Injury	Release requiring days to remediate, repeat permit violation	$10-30 million
E	First Aid Injury	Environmental permit violation	$2-10 million

This table represents one example of a severity definition matrix. Each company will have their own set of categories, which is relevant to the operation at hand. For example, some facilities do not differentiate between one fatality and multiple fatalities, while others may choose not to define separate categories for economic impacts – instead focusing their hazard analyses toward safety and environmental risks.

Likelihood Levels

Likelihood categories are defined in a similar manner. Typically, orders of magnitude are used to describe the potential yearly occurrence of an event. Some guidance may be provided to assist the PHA team in identifying the likelihood category most appropriate to a given event. Table 2 provides an example set of likelihood rankings.

Table 2. Example Likelihood Categories

Category	Occurrence	Description
1	10^{-1} per year	Likely - Expected to occur multiple times in the life of the facility
2	10^{-2} per year	Moderately Likely - May occur once in the life of the facility.
3	10^{-3} per year	Unlikely - May occur once when the life of 50 similar facilities is considered.
4	10^{-4} per year	Very Unlikely - Known to have occurred within the industry.
5	10^{-5} per year	Improbable - Has not occurred in the industry or has occurred, but current safeguards make the event less likely

Risk Levels

The selected severity and likelihood categories are then combined through the use of the risk matrix. The result will determine the risk level of a given scenario. Risk levels can be used by company management to determine which scenarios will require additional mitigation. Figure 1 presents an example of a risk matrix which may be used at a facility during a PHA or LOPA session.

	Likelihood				
Severity	5 (10^{-5})	4 (10^{-4})	3 (10^{-3})	2 (10^{-2})	1 (10^{-1})
A	Marginal	Undesirable	Undesirable	Critical	Critical
B	Marginal	Marginal	Undesirable	Undesirable	Critical
C	No Action	Marginal	Marginal	Undesirable	Undesirable
D	No Action	No Action	Marginal	Marginal	Undesirable
E	No Action	No Action	No Action	Marginal	Marginal

Figure 1. Example Risk Matrix

In Figure 1, four risk levels are defined in the risk matrix: Critical, Undesirable, Marginal, and No Action. It is important to clearly define the required actions associated with each risk level. Table 3 provides an example of required actions for each of the above risk levels.

Table 3. Example Risk Levels and Required Actions

	Required Action
Critical	For an operating system, management shall consider shutting the system down until proper measures are implemented that will reduce the risk to "Marginal" level or better.
Undesirable	Measures shall be implemented that will reduce the risk to "Marginal" level or better.
Marginal	Those risk reducing measures that do not severely impact the economics of the business may be considered for implementation.
No Action	This risk level is tolerable. No further action required.

Monotonic Nature of Risk Levels

Cells in the risk matrix should be assigned risk levels to follow "iso-risk contours" [Reference 2]. An iso-risk contour represents the line along which the risk presented by the combination of the severity and likelihood is the same. In the example presented above, the risk level increases from left to right and bottom to top, mirroring a set of monotonic iso-risk curves. In other words, in order to form a rational risk matrix, the risk level must continually increase when traveling along a single row or column in the matrix. This approximates what is commonly used in quantitative risk analysis (QRA) where the risk is estimated quantitatively in much greater level of detail.

Risk Tolerance Level

One of the most important distinctions on the matrix is the separation point between tolerable and intolerable risk. Similar to the risk level definitions and their assignments within the risk matrix, the separation point between tolerable and intolerable risk is selected by developers of the risk matrix. For example, in Table 3, the boundary between Undesirable and Marginal risk can be chosen for the separation point between "tolerable" and "intolerable" risks. This separation point is termed the "risk target." Scenarios which fall above this boundary (i.e., Critical or Undesirable risks in the case of the example) require additional mitigation to ensure that they fall into the Marginal or No Action risk levels, while scenarios which fall into these lower categories may be considered to be "tolerable risks."

Definition of the risk target is one of the critical decision points when developing a risk matrix. This target will determine the required protections to reduce risk to tolerable levels, and will have a large impact on the protection features implemented at the facility. This decision is a judgement made by company management on the level of risk the company or facility can tolerate.

Improper definition of any of the elements of a risk matrix will lead to poor hazard analysis results, and may lead to decisions which carry excessive risk, or which spend resources ineffectively in pursuit of reducing a risk that should be considered tolerable.

3. Common Risk Matrix Pitfalls

3.1. Overly Conservative Risk Ranking of High Severity, Low Likelihood Scenarios

The top left corner of the example risk matrix presented in Figure 1 contains the highest severity, lowest likelihood events. By their nature, these events can be the most difficult to analyze, as they are typically events which have never occurred, and are in a frequency range which is possible, but is often inconceivable to the analysts. These events represent a company's worst-case risk scenarios, such as a major release of toxic chemicals, resulting in multiple fatalities or wide-spread health impacts to the surrounding community. Such events are typically prevented by a number of protection layers installed specifically to ensure that the events are very unlikely to occur. The required protection level can be established using the risk target for a given consequence severity.

As an example, in the matrix defined in Figure 1, for a Severity A event, the corresponding risk target can be likelihood category 5 (i.e., $1x10^{-5}$ per year). A company using this matrix must provide sufficient protection layers to ensure that the scenario under consideration will not happen more frequently than once in 100,000 years. In this extremely low likelihood range, however, it is difficult to grasp the differences between various occurrence frequencies. This is experienced by the author and acknowledged in the literature [Reference 4]. That is, it may be difficult to conceptualize the difference between an event which presents itself once every 10,000 years, once every 100,000 years, or even once every 1,000,000 years.

This difficulty in grasping the extremely low likelihood regime can make it easy to develop an overly conservative risk target for the highest severity scenarios. Because these scenarios represent the most severe consequences, it is a natural tendency to want to require the likelihood of these severe events to be as low as possible (e.g., $1x10^{-6}$ per year.) While mitigating these events to be as infrequent as possible is a positive goal, care must be taken when setting these targets to avoid unanticipated consequences on facility design. It is possible to set a goal which will require an excessive number of protection layers (controls, trips, etc.), which may not significantly alter the risk decision.

In an example taken from industry, PHA and LOPA were to be conducted on the Front-End Engineering Design (FEED) of a proposed chemical production facility. The results of the LOPA would also be used to determine the necessary Safety Integrity Level (SIL) rating for instrument trip functions to be implemented at the facility. The PHA was to use an existing corporate risk matrix as the foundation of the analysis. In this case, the risk target for the most severe event (the equivalent of Consequence Severity A) was $1x10^{-6}$ per year. While performing the analysis, it was determined that the required risk target was leading to a large number of recommendations to add additional protection features, or to increase the SIL rating for a number of trip functions. When compared with current industry standards and best practices for facilities of a similar design, it was determined that the risk matrix was driving the facility to implement excessive instrumented features to meet the risk target of the owners.

In this example, the corporate risk matrix had been designed with overly conservative definitions for high severity, low likelihood events. The risk target for these events had to be carefully selected to ensure that it generates useful results. Selecting a highly conservative risk target, as it was shown in that case, will lead to addition of a large number of protection features in facility design. These features will help to prevent high severity scenarios, but may not be necessary according to common practices. In other words, these additional safety features may be protecting against concerns which are already sufficiently mitigated by existing safety features per accepted industry practices. Clearly, adding additional protection layers can greatly increase cost, and the analysts may need to examine the possibility of unanticipated safety concerns due to spurious trips or additional maintenance hazards.

Ultimately, placement of the risk target is a company-specific decision which requires careful consideration to avoid unintended consequences. The risk matrix must be reviewed to ensure that its effects on the operation under review are understood, and do not lead to unanticipated consequences on risk or facility design.

3.2. Risk Matrices Unsuited for the Specific Operation Under Review

When beginning a PHA for a specific facility or system, it is critical to ensure that the risk matrix is suitable for the size and scope of the review. In the above example, a company-wide risk matrix for a large corporation with multiple interests across various fields was used for the PHA and LOPA analysis. The matrix had not been reviewed to ensure that it was applicable to the facility under review.

In a similar vein, other portions of the risk matrix must also be reviewed to ensure that they are appropriate for the operation under review. For example, in addition to safety concerns, many companies choose to risk rank the impact of hazardous scenarios on facility assets or production time. Typically, the consequence severity categories for these events are assigned based on the dollar value of the impact. For example, in Table 1, a Severity A event is defined as an event which results in a loss of more than $500 million. While this may be appropriate for a large scale corporation, smaller operations may consider a significantly smaller loss to still result in a catastrophic impact on the business of that operation.

Conversely, if a matrix is designed such that the highest possible severity ranking is much too low, it may result in extreme protective measures to mitigate small operational concerns. For example, a theoretical company produces $10 million annually, and operates using a risk matrix in which the highest severity ranking correlates with a loss of $100 thousand. In this case, almost any event which results in plant downtime for a few days will be given the highest possible consequence severity rating, and will require the corresponding protection layers to prevent such events. The resulting analysis will be forced to recommend significant changes to system design and operation to avoid shutdowns, which will dramatically increase cost, while protecting against events which in reality have already adequate protection.

3.3. Improper Decision Granularity

Similar to the scenario described above, improper decision granularity can also lead to over- or under-estimations of the risk posed to a facility by a given scenario. If the severity or likelihood category scales are too broad, a hazard analysis will be unable to distinguish between high and low risk scenarios, which will result in a tight grouping of a majority of all scenarios within the same risk category (and therefore with the same risk target). In this event, high severity scenarios may not receive sufficient prevention or mitigation features, while low severity scenarios may require excess additional protection layers. Practical examples are provided below. Mathematical treatise of this concern is addressed in Reference 2.

It should be noted, however, that just because a majority of scenarios fall within the same risk category does not mean that the decision granularity is incorrect by default. Some facilities with simple processes, or with similar hazards throughout the process, may find that by the nature of the facility, a majority of risk will be categorized with the same risk ranking. Improper granularity is only a concern when risks of differing magnitude are grouped into the same risk category.

For example, take the same facility from Section 3.2 which generates $10 million annually. If this company uses the consequence severity categories presented in Table 1 of this report, an event which causes the facility to shut down for a year will be considered a category D event, while nearly all other less severe losses would constitute a category E event – the lowest possible scenario. However, in reality, a year-long plant shutdown would have serious impact on the company's finances, with the potential for the loss of ability to operate in the future, and should require sufficient protection layers to significantly reduce the likelihood of such an event.

3.4. Risk decision not impacted by the severity and likelihood combination

Some risk matrices may include severity or likelihood categories in which the risk decision does not change, no matter what ranking it receives. As an example, consider the following example risk matrix, with similar risk definitions as Figure 1:

Severity \ Likelihood	5 (10^{-5})	4 (10^{-4})	3 (10^{-3})	2 (10^{-2})	1 (10^{-1})
A	Undesirable	Undesirable	Critical	Critical	Critical
B	Marginal	Undesirable	Undesirable	Critical	Critical
C	No Action	Marginal	Undesirable	Undesirable	Critical
D	No Action	No Action	Marginal	Undesirable	Undesirable
E	No Action	No Action	No Action	Marginal	Undesirable

Figure 2. Conservative Example Risk Matrix

This matrix is skewed toward high risk scenarios. In particular, for Consequence Severity A, there is no possible risk ranking which falls into the tolerable range. This implies that if a Severity A hazard is identified, measures must be taken to totally eliminate the hazard from the design, rather than add protection layers against the event. In some cases, this decision may be intentional, if it has been determined that the facility cannot tolerate an event of a given magnitude, no matter how rare. However, it may also be the case that the risk matrix was designed to be highly conservative, without fully realizing the consequences implied for Severity A.

In his article "What's Wrong with Risk Matrices?", Tony Cox demonstrates that for an ideal risk matrix which minimizes the potential for risk ranking errors "all cells in the left-most column and in the bottom row are green (lowest-priority)" [Reference 2]. In Cox's example, he uses a risk matrix containing only three risk levels (i.e., only three color categories). When defining a risk matrix with more than three risk levels, this statement can be taken to mean that all cells which fall in the lowest likelihood or severity category should also fall within the tolerable risk range.

3.5. Potential to miss medium severity, but high risk scenarios

Risk matrices may also be arranged such that the risk decision for less severe consequences presents a higher risk than a more severe, less likely event. To illustrate, consider the risk matrix in Figure 4.

Severity \ Likelihood	5 (10^{-5})	4 (10^{-4})	3 (10^{-3})	2 (10^{-2})	1 (10^{-1})
A	Marginal	Undesirable	Undesirable	Critical	Critical
B	Marginal	Marginal	Undesirable	Undesirable	Critical
C	No Action	Marginal	Undesirable	Undesirable	Undesirable
D	No Action	No Action	Marginal	Marginal	Undesirable
E	No Action	No Action	No Action	Marginal	Marginal

Figure 4. Risk Decision Example Risk Matrix

Using this risk matrix, consider an example from industry in which a fire at a relief valve vent may lead to a fatality (Severity B) when personnel are working on an elevated platform near to the vent, or to a severe injury (Severity C) at grade. When evaluating such a scenario with multiple potential safety outcomes, typically the worst-case consequence (the fatality in this case) is evaluated to ensure that the facility is adequately protected against such an event. In this example case, the analysis determines that given the safeguards in place, the likelihood of personnel exposure in the remote area which results in a fatality is $1x10^{-4}$ per year, while the likelihood of personnel exposure at grade, leading to injury, is $1x10^{-3}$ per year. Using the risk matrix in Figure 4, we can see that this leads to a Marginal risk for the fatality case, but an Undesirable risk for the injury. If only the worst-case event was considered in this analysis, as is typically practiced, an undesirable risk may have gone undetected.

In this example, the risk matrix has been designed in such a way that a lower severity event may lead to a higher risk ranking than the worst-case event. Careful consideration must be taken when assigning risk definitions such as the one in Figure 4, to ensure that such decisions are noted and are adequately supported by the decision makers. In addition, it is critical to ensure that the hazard analysis methodology employed is capable of identifying events with multiple outcomes which fall into varying severity of likelihood categories, to ensure that the scenario which poses the highest risk is identified.

4. Conclusion

The concept of the risk matrix has proven to be a simple but useful tool in managing risk. It has been adopted by a wide range of diverse industries. It is an approximation of the more rigorously developed risk contours using Quantitative Risk Analysis and because of that it has some theoretical limitations that have been studies by various authors. However, given the number of facilities and processes within the process industry that must make risk decisions, the risk matrix approach has proved to be a useful approximation tool.

In applying risk matrices developed by various owner and operator companies, several discrepancies have been noted that have led to improper or overly conservative decisions. When developing a risk matrix, consequence severity and likelihood categories must be carefully chosen to ensure that they are applicable to the facility or system under review. These categories must accurately reflect the potential range of consequences possible in the system, and should be segregated based on the potential impact on continued operation of the facility.

The risk matrix is a critical tool in the hazard analysis process, and has a wide-ranging impact on system safety and design. When applied correctly, the risk matrix is a powerful tool in ensuring that a facility is operating at a tolerable risk level. However, using the incorrect risk matrix during a hazard analysis can result in systems which are significantly under-protected against process hazards, or may lead to a large number of recommendations for protection layers which do not alter the risk decision, and which may be costly and difficult to implement or maintain.

Before beginning any hazard analysis, whether in the design phase or for an existing process, the risk matrix to be used must be reviewed and evaluated to ensure that it remains applicable to the system to under consideration in order to obtain meaningful results.

5. References

[1] CCPS. *Guidelines for Hazard Evaluation Procedures.* Center for Chemical Process Safety. American Institute of Chemical Engineers. 3rd Edition. New York, NY. April 2008.

[2] L. A. Cox. "What's Wrong with Risk Matrices?" Risk Analysis 28, No. 2 (2008). 497-512.

[3] P. Baybutt. "Designing Risk Matrices to Avoid Risk Ranking Reversal Errors" Process Safety Progress 35, Issue 1 (2016). 41-46.

[4] CCPS. *Guidelines for Developing Quantitative Safety Risk Criteria.* Center for Chemical Process Safety. American Institute of Chemical Engineers. New York, NY. 2009.

510577 Application of Risk Based Process Safety (RBPS) to Drive Performance Improvement in the Semiconductor Industry

Abstract:

While most semiconductor manufacturing facilities do not contain quantities of highly hazardous chemicals in threshold quantities sufficient to be subject to the Occupational Health and Safety Administration's (OSHA) Process Safety Management (PSM) regulations, some organizations are applying Risk Based Process Safety (RBPS) concepts as a systematic means to not only ensure EHS risks are identified and quantified, but also to strengthen overall business performance and provide competitive advantages. Aging Fab facilities and infrastructure, ever-increasing production demands and rapid innovation and need for process modifications are also business drivers for RBPS. The core RBPS concepts not only provide a wholisitic approach to protect employees from catastrophic accidents and releases; but in most cases also drive production efficiencies, increase equipment/tool reliability and life span, promote quality improvements, and enhance business continuity measures. This is accomplished by application of a wide range of process safety management elements, as applicable to semiconductor operations, that fall under the following four basic pillars of RBPS: Process Safety Leadership/Commitment; Risk Assessment/Identification; Risk Management; and Learning and Continuous Improvement. This presentation will discuss the basic concepts of RBPS, including a brief review of the 20 elements as provided by the Center for Chemical Process Safety (CCPS)framework, and present the benefits of developing and implementing process safety management systems for semiconductor processes. Case studies will be presented, as applicable, to detail the advantages, as well as the challenges, of RBPS for the semiconductor industry.

512112 Benchmarking of Petroleum Tanker Discharge Operations at a Nigerian Coastal Terminal and Jetty Facilitates Optimisation of the Ship â€" Shore Interface

Abstract:

Benchmarking has progressively become entrenched as a requisite activity for process improvement and enhancing service delivery at petroleum jetties and terminals, most especially during tanker discharge operations at the ship – shore interface, as avoidable delays result in extra operating costs, non-productive time, high demurrage payments and ultimate product scarcity. The jetty and terminal in focus had been operational for 3 and 8 years respectively, with proper operational and logistic records maintained to evaluate their progress over time in order to plan and implement modifications and review of procedures for greater technical and economic efficiency. Regular and emergency staff meetings were held on a team, departmental and company-wide basis to progressively address major challenges that were encountered during each operation. The process and outcome of the resultant collectively planned changes carried out within the past two years forms the basis of this paper, which mirrors the initiatives effected to enhance operational and maintenance excellence at the affected facilities. Operational modifications included a second cargo recipt line designated for gasoline, product loss control at jetty and shore ends, enhanced product recovery and quality control, and revival of terminal–jetty backloading operations. Logistic improvements were incorporation of an internal logistics firm and shipping agency, fast tracking of discharge procedures for tankers, optimisation of tank vessel selection process, and third party product receipt and throughput. Maintenance excellence was achieved through construction of two new lay barges and refurbishment of the existing one; revamping of existing booster pump and purchasing of a modern one as reserve capacity; extension of Phase 1 of the jetty to accommodate two vessels and construction of Phase 2 for two more vessels; regular inspection, draining, drying and replacement of cargo hoses; corrosion management program for all process facilities; and an improved, properly planned and documented maintenance culture. Safety, environmental and security compliance were enhanced by installing state-of-the-art fire fighting facilities and equipment, seawater intake line construction as backup for borehole at the terminal, remediation of the shoreline and marine structures, modern spill containment equipment, improved housekeeping and accident prevention practices, and installation of hi-technology security enhancements, among others. The end result has been observed over the past 2 years to include improved tanker turnaround time, higher turnover on product sales, consistent product availability, greater indigenous human capacity utilisation by way of direct hires and contracts, as well as customer loyalty. The lessons learnt from this exercise would therefore serve as a model to be adapted by other operators of similar facilities, contractors, academics and consultants in a bid to deliver greater sustainability and profitability of operations at the ship – shore interface to this strategic industry.

505801 Failsafe-Ing Human Error

Abstract:

Errors fall into various categories, systematic, human and instrument error. A single error rarely leads to a catastrophic incident, however where errors form a chain of events then disastrous consequences can occur. A single error if analyzed correctly with telltale signals can ensure that a process safety management system is robust enough and has created sub levels within safety barriers. Research shows that more than 60% of major safety events have been a result of human error, this does not mean that humans should be removed from a process, rather systems should be in place that compliment human attributes. This study focuses on the importance of automated systems such as work permits, and takes a look at incidences that could have been prevented with the right safety features in place. It also provides an insight into the importance of doing this in an operationally efficient manner so that the process is simple and easy to follow. Time reduction of 40%, reduction in risk by 50% and reinforcement of safety barriers without making mass changes.

510901 Achieving Process Safety Management Best Practices - What Can You Do Differently

Abstract:

Since their promulgation â€“ the U.S. Occupational & Safety Health Act (OSHA) Process Safety Management (PSM) Standard, 29CFR1910.119, and the U.S. Environmental Protection Agency (EPA) Risk Management Program (RMP) Rule, 40CFR68 â€“ have resulted in improvements in plant's managing the release of highly hazardous flammable and toxic chemicals, fewer accidents and injuries, improved public health and safety, reduced environmental impacts, improved public image and employee morale, and other facility benefits. The OSHA PSM regulation and EPA RMP rule both require the development and implementation of a program to prevent catastrophic accidents for covered processes. These are "performance-based" regulations (meaning that the agencies state what to do, not how to do it) and are subject to ongoing interpretations and clarifications from the regulators. The prevention programs are wide-ranging and cover virtually every aspect of a company's operations. Companies are often unsure on what constitutes full compliance with the regulations.

In this presentation, the author will highlight common deficiencies found during compliance and risk based process safety audits and share industry best practices to address these deficiencies. These best practices are targeted not only to achieving regulatory compliance but also understanding how to take your PSM system one step further.

513027 Liquefied Natural Gas Vapour Dispersion Modelling Using Computational Fluid Dynamics Approach

Abstract:

Releasing at cryogenic temperature, Liquefied natural gas (LNG) dispersion is one of the most complicated problem in dense gas dispersion. Atmospheric boundary condition is an important factor affecting the dispersion process of LNG vapour by the effect of wind speed, surface roughness and atmospheric stability. Heat transfer from the surrounding and ground surface to the cold LNG vapour cloud will increase cloud temperature and reduce cloud density. The major effect of heat transfer to the LNG dispersion is the increasing of turbulent mixing process. Another relating heat transfer phenomenon is heat addition or heat removal due to the condensation or evaporation of water vapour. When the dispersion process occurs at sloping terrain with presence of obstructions, these factors will enhance gravity-driven flow and turbulent mixing. Computational Fluid Dynamics (CFD) is a powerful approach to take into account all of these factors for a sound modelling of LNG dispersion process.

Monin-Obukhov similarity theory assumes horizontally homogeneous atmospheric boundary layer and is widely used to model the profiles of velocity, turbulent kinetic energy and eddy dissipation rate in surface layer. These profiles are used as boundary conditions for the CFD simulation. Proposed model includes heat transfer from the ambient air to the vapour cloud by the diffusion of air to the boundary of the cloud. Change of material properties on temperature is also included. Parametric study of the proposed model is presented including (i) Effect of ground heat transfer models (ii) Effect of turbulent models. A study of the effect of impoundments in mitigating the distance to lower flammability limit (LFL) of LNG vapour cloud shows that impoundments can reduce the distance to LFL of the cloud to 40%.

510932 The Application of CFD Simulation in Evaluation of LPG Pool Fire Heat Flux

Abstract:

Liquefied Petroleum Gas is used widely in a great number of applications. The main components of LPG are propane and butane, which can be used safety if risks can be managed in acceptable levels. Vaporizing liquid LPG to gas is a critical step before using in a large number of applications, while safety is an essential factor that should be taken into consideration in this processing. The conflict in requirements of the minimum distance for LPG vaporizers to a container has been discussed for a long time, like NFPA 58 (â‰¥10 ft. or 3 m) and NFPA 59 (â‰¥50 ft. or 15 m). In order to have a better quantitative understanding of the heat flux from LPG pool fires and determine the appropriate distance between the vaporizer and containers, we have collected considerable experimental data about the main parameters of LPG pool fires. The results show that, for the pool fire sizes from 0.1 m to 10 m, the average emissive power is ranging from 35 kW/m^2 to 79 kW/m^2, and when the size is from 10 m to 20 m, the counterpart power is ranging from 47 kW/m^2 to 134 kW/m^2. The heat flux from pool fires to the surroundings decreases from 60 kW/m^2 to 2 kW/m^2 as the distance increases from 40 m to 80 m, and the heat flux is 9.7 kW/m^2 at the distance of 260 m with the largest pool size of 107 m in the literature. CFD simulation is an economic method to obtain a large number of data of parameters from LPG pool fire. To compensate the limited experimental data of the heat flux of LPG pool fires from different distances, CFD fluent has been applied to simulate LPG pool fires from the diameter between 1 m to 30 m, which gives the specific heat flux from different diameters. The data and modelings might not only help us understand heat flux from LPG pool fires and heat flux values which may lead LPG storage tanks to potentially fail, but also be used in revisions to the related standards.

511184 Process Hazards and Process Risk Investigation (as a prerequisite for operational discipline)

Abstract:

Process Hazards and Process Risk investigation as a prerequisite for operational discipline

In the field of safety and risk, it is widely known that Hazards and Risks are not the same.

Hazard and Risk studies provide the necessary information to adequately manage and control safety during the different phases within the lifecycle of a process plant from design to demolition. This information is required to control the risk during exploitation and as a knowledge database for a controlled management of change.

In 2014, because of a fatal accident, Umicore decided to change the approach of Hazard and Risk identification. Umicore divided the Hazard and Risk investigation into two different analyses.

The knowledge gathered during the process hazard assessment ~~analysis~~ (or PHA) is used in the further development of the chemical or physicochemical process in general, performing the process risk assessments (PRA), defining the required risk reduction measures (RRM) to achieve the acceptable risk level, developing the operational safety procedures, strengthen the management of change program in later lifecycle phases and to control knowledge management. In the PRA the translation of the identified hazards into risks in normal and abnormal operation conditions is further investigated.

In this paper, the authors will give a comprehensive overview of this unique combination of methods they use to investigate Hazards and link them to the Risk assessment, operational files, management of change and knowledge management to bring about operational excellence. They will also share the best practices including their in-house developed tool that assisted with consistency in adequate documentation of all identified hazards, risks analyses and risk reducing measures.

Marc Massant Umicore

Chris Israni ERM

511240 Advancing Asset Integrity in Upstream

Abstract:

Advancing Asset Integrity in Upstream â€" Providing the right focus

The presentation will outline the efforts Chevron Upstream has taken over the past decade to improve performance on Asset Integrity. The presentation will focus on two aspects of the Asset Integrity Program. The first will be the methodology to determine the scope of the Asset Integrity Program and the most important equipment to focus on. The second will be the governance process to cover variations. Chevron Upstream developed an Asset Integrity process to define expected minimum requirements and implementation is in progress. The Asset Integrity Process went beyond the classical mechanical integrity programs and covered all equipment types and equipment layers of protection. This included both the preventive and mitigative layers of protection using the classic bow tie techniques and leveraging existing process hazard analysis studies. The paper will outline the comprehensive Asset Integrity requirements including both the underlying principles and execution specifics; explain the status on the journey to full implementation and results to date; and, explore the learnings to date.

511234 Process Safety Emphasis for Integrity in Sulfur Recovery Units

Abstract:

PROCESS SAFETY EMPHASIS FOR INTEGRITY IN SULFUR RECOVERY UNITS

Zainab Kayat

Custodian Process Engineer

Group Technical Solutions, PETRONAS

Das Babua

Principal Process Technologist

Group Technical Solutions, PETRONAS

ABSTRACT

Increased severity of sulfur recovery units are required as refineries faces stringent specifications on motor transportation fuels together with dependency on heavy, sour crude oil feedstock. This has led to new demands on the processing capability of refinery sulfur recovery units (SRUâ€™s) where debottlenecking becomes a necessity. Emphasis on process safety whilst refinery pursued additional capacity seemed lagging behind affecting sustained business with lost opportunity and putting environment at a risk.

This paper provides the experience and perspectives of a refinery which went through numerous pitstop since it was revamped to a higher severity SRU using oxygen enrichment. Revamped had incorporated changes in the design of its thermal stage. Root cause failure analysis were conducted due to the continuous reliability issues to determine the process safety hazards that affects integrity and to mitigate leakage of highly toxic H2S gas arising from hot spots, runaway temperatures and leaks are addressed.

Method and Application of Cause-Consequence Analysis Based on System Reliability Theory

XIN Baoquan, WAN Lu
SINOPEC Research Institute of Safety Engineering
Qingdao Shandong 266101, China
xinbq123@163.com

Keywords: system reliability; cause and consequence analysis; failure rate; risk assessment; acceptable risk

Abstract

It is to solve the current practical problems such as the unclear selection of the target system, the chaotic logical relationship of the link events, and the unclear source of basic data in the "cause-consequential analysis method". Based on the reliability theory of people and machines in the system reliability analysis, the "motor overheating" event is taken as an example. From the point of view of the "safety control of the system for the occurrence and development of a motor overheating fire event", the "motor overheating safety control system" was determined as the target system. Through investigating and analyzing the functional composition of the target system, the reliability block diagram of the target system was obtained, and the logic relationship of the link events was determined. According to reliability theory, the probability of each basic event was investigated and calculated. Finally, the motor overheating risk assessment chart is drawn. Evaluation results show that the equal risk line in logarithmic coordinates is a straight line. This line is the critical line of acceptable risk. Research on the quantitative analysis and application details based on the reliability theory can improve the operability and application effect of the "cause-effect analysis method" and contribute to the popularization and application of this method.

1 Introduction

In the quantitative analysis of system security, the probability of occurrence of various events generally needs to be obtained based on system reliability theory and by analyzing the reliability of relevant equipment and people. These occurrence probabilities include the probability of an initiation event, the success or failure of a link event, and the occurrence probability of an event tree top event [1]. Therefore, the reliability theory is the basis of the system security quantitative analysis, and it occupies a very important position in the safety system engineering.

The Cause and Consequence Analysis (CCA for short) is based on the quantitative analysis of reliability theory. It can qualitatively analyze the cause of the accident and the possible consequences of the incident. It can also quantitatively calculate the probability of occurrence of each event and the risk rate of the consequences. Currently widely used in risk assessment and safety decision [2-3]. The CCA method is a combination of two system security analysis methods, the accident tree (FTA) and the event tree (ETA). Due to the complexity of the analysis and calculation process, there is less research in theory and application. In the process of actual use, the key issues such as unclear definition, logical confusion and unclear analysis are common in the selection of target systems, the determination of logical relationships in each link event, and the source of basic data in quantitative analysis. It leads to less attention to the details of the preparation, practicality, and low accuracy in the actual application, which brings a greater limitation to the application scope and effect of the CCA method [4]. Based on this, based on the systematic analysis of the CCA method, based on the system reliability theory, taking the motor overheating event as an example, the analysis process and application details of the CCA method are studied. At the same time, the risk of consequence events was calculated and evaluated based on the "Farmer Risk Assessment Chart". The analysis and calculation of the CCA method and the clearness of the evaluation process contribute to the promotion of the CCA method in practical applications and the resolution of practical problems.

2 Overview of the cause-consequence analysis

2.1 Method overview

The CCA method is an important analysis method that integrates the "forward push" characteristics of the incident tree because of the "reverse" characteristics of the incident tree and the "reverse reasoning" characteristics of the accident tree. It shows the relationship between the accident and many possible basic events. The advantage is the use of graphical methods developed from two perspectives. The forward is the outcome of the event and the backward is the basic cause of the incident. The accident tree and the event tree method are relatively cumbersome, and the CCA method can use a simple model to simultaneously represent the cause and consequence of the accident, and can perform quantitative calculation and risk evaluation.

The basic idea of the CCA method is:
(1) Determine evaluation events and draw event trees.
(2) Draw the accident tree with the start event of the event tree and the link event identified as failed as the top event.
(3) According to the qualitative and quantitative analysis results of the accident tree, the causes of the top events and the control schemes (i.e., minimum cut sets, minimum diameter sets) are obtained. And calculate its probability of occurrence.
(4) According to the results of the accident tree analysis, calculate the probability of occurrence of various consequences determined in the event tree.
(5) Evaluate the risk of the event by the product of the probability of the consequences and the loss of the consequences, and draw the evaluation result.

2.2 Analysis of existing problems

(1) The target system selection is unknown. In the process of causality-consequential analysis (CCA), the selection of target systems and the determination of evaluation events are the starting points and important links of CCA. However, due to the lack of thorough analysis of the definitions and basic issues, the analysis of the system under investigation was incorrect. Resulting target system and evaluation events are too large and not specific. It will directly result in deviation from the target when the system or event is investigated and analyzed, the analysis and development of the initial event is too large and loose, and the logic evolution process is incorrectly described.
(2) The logical relationship between the events is chaotic. Correctly establishing the logical relationship between the events is the key to successful analysis and problem solving by the CCA

method. The traditional method in the preparation of an event tree is "from the initial event, the event tree is drawn from left to right according to the event development process." The "event development process" in the method is the "logical relationship" of each link event. However, it did not specify how to determine the "event development process" in practical use. The ambiguity of the logical relationship makes it difficult to draw event trees.
(3) The source of basic data is unclear. It is the basis and key to quantitative analysis to correctly determine the basic data such as the occurrence probability or failure rate of each event. However, in practical applications, the probability data of each target event is generally given with reference to similar events, and then conventional calculations are performed. And less about the sources and methods of determining these basic data. This leads to poor accuracy and operability of the CCA method, and it is not able to achieve a good quantitative analysis.

3 Reliability theory and CCA steps

3.1 System reliability

To conduct systematic, accurate reason-consequence analysis, reliability theory is the foundation. Reliability refers to the ability of the system to accomplish specified tasks within the specified time and under specified conditions [5]. It is usually measured by reliability, including the reliability of human, a single machine, and system (subsystem).
Human reliability (r) including three steps of signal reception (r_1), signal processing (r_2), and response (r_3): $r= r_1 \times r_2 \times r_3$. Basic data of r_3 can be determined with reference to Table 1.

Table 1 Basic reliability value of r_1、r_2、r_3

Category	Content	r_1	r_2	r_3
Simple	3-5 variables, ergonomically considered comprehensive	0.9995~ 0.9999	0.9990	0.9995 ~ 0.9999
General	No more than 10 variables	0.9990~ 0.9995	0.995	0.9990~ 0.9995
Complex	More than 10 variables, ergonomically considered incomplete.	0.990~ 0.995	0.9990	0.990~ 0.9990

The reliability of a single machine refers to the probability that the equipment, components, components, etc., will perform the prescribed function within the specified conditions and time [6], expressed as R_M. Reliability of the system (subsystem) depends on the reliability of each subsystem or component [7]. According to the logical relationship of each subsystem or component in the system, the corresponding logic operation is performed on its reliability, and the reliability of the entire system can be obtained.

3.2 Key implementation steps of CCA method

(1) Selection of evaluation events
Any event that requires further analysis of its causes and possible consequences can be used as an evaluation of the CCA law. Usually, it is necessary to select a system fault, a device fault, a human error or a process anomaly that can easily cause the consequences of a larger or major accident. For example, the initial events of the event tree, the top events of the accident tree, and the intermediate events can all be used as the evaluation events of the cause consequences analysis.
(2) Draw the event-tree and the fault-tree

First of all, selected evaluation event is used as the initial event of the event tree. According to the ETA method, starting from the initial event, according to the temporal logic of the system components, from the left to the right, the two functional states are divided into successes and failures. Until the two values of all the units are combined, the event tree is finally constructed. Then the initial event of the event tree and the intermediate event of the failure of the security function are taken as the top event. The FTA method is used separately.

(3) Draw the cause and consequence map

Connect the plot tree with the event tree to form the cause - consequence graph.

(4) Probability of determining the basic data and calculating the consequences of the event

By means of investigation and system reliability analysis, the probability of the occurrence of the accident tree and the related data are obtained. According to the calculation method of occurrence probability of accident tree top event, the occurrence probability of event tree initiation event and failure link event is calculated respectively. Then the probability of occurrence of event tree is calculated.

(5) Rate of calculation of risk

According to the probability of occurrence of the events (*P*) and the loss (*U*), the risk of the accident is synthetically measured. The product of the two (*R*) is used as a risk rate: $R = P \times U$.

(6) Summary of evaluation results

Results of the evaluation include: ① Cause - consequence chart;②Calculation of the risk rate; ③ Risk assessment chart.

4 Applied research

A processing plant in China needs to introduce a motor system due to process improvement. CCA method is used to analyze the risk that the factory may be subject to. According to statistical data and literature [8], the failure probability of the heat dissipation system after motor overheating $P(B_0/A)=0.02$, and no-failure probability $P(B_1/A)=0.98$. In addition, the probability data of the initial events and other events need to be determined based on reliability theory.

4.1 Survey and analysis to determine evaluation events

The system safety analysis is carried out in combination with the operating conditions of the field. It is considered that the energy forms involved in the operation of the motor mainly include electric energy, mechanical energy and heat energy. The main events that may result in more serious consequences are the electrification of the shell, the exposure of the rotating parts and the overheating of the motor. The three can be used as the evaluation event of the cause consequences analysis (the starting event). Here, the most common "motor overheating" event is chosen as the evaluation event, and the cause consequence analysis is carried out.

4.2 Drawing ETA and FTA

Fire is the most common cause of motor overheating. Therefore, from the perspective of the "safety control of the system for the occurrence of a fire incident" in the process of "motor overheating and fire accident development," the "motor overheating safety control system" is determined as the "target system". By consulting the design and working principle of the target system and combining with the composition of the target system, the investigation and analysis of the working principle and safety measures of the target system were carried out around the initial event of "motor overheating". The target system includes three parts: the motor cooling system, the fire extinguishing system (artificial fire extinguishing and automatic fire extinguishing) and the fire alarm system after the failure. Finally, a reliability block diagram is drawn, as shown in Figure 1.

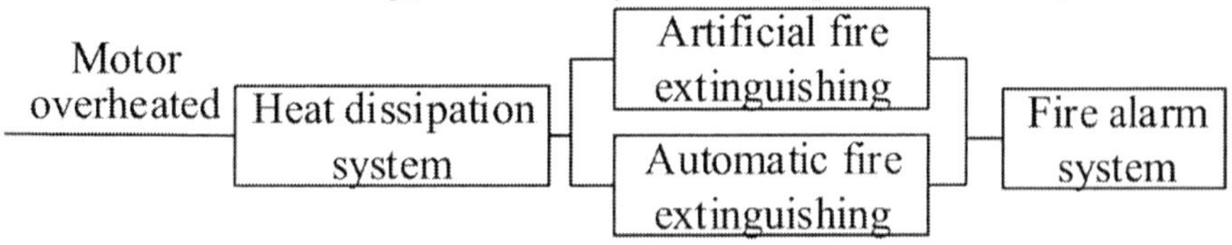

Fig.1 Motor overheating and fire system reliability block diagram

From the initial events of the system, relying on the drawn system control chart, according to the sequential logic of the system components, from left to right, the success and failure of the two states are listed one by one for subsequent functional units. And get the binary state combination of all units. The motor overheating event tree is finally constructed, as shown in Figure 2.
According to the event tree drawn, the start event "motor overheating" and the intermediate failure event are respectively taken as the top event. Through investigation and analysis, the cause events and their logical relations related to the top event are obtained. Then expand the intermediate events and basic events to compile the corresponding accident tree diagram.

4.3 Drawing cause - consequence diagram

Connect the event-tree and fault-tree to get a cause-consequence diagram, as shown in Figure 3.

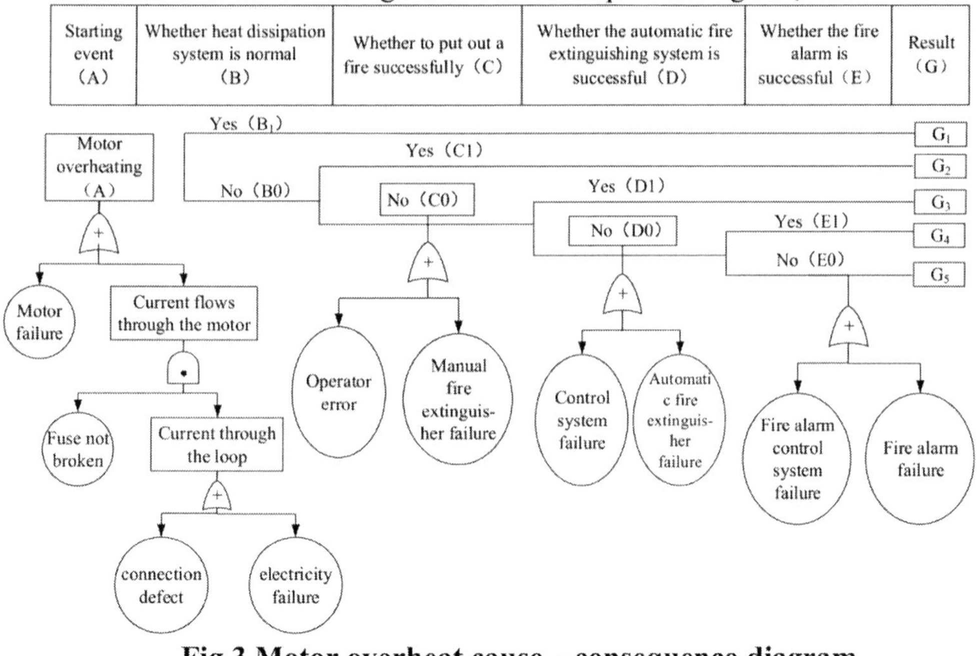

Fig.3 Motor overheat cause – consequence diagram

4.4 Calculate the probability of a consequence event

According to the cause-consequence diagram, based on the theory of human reliability and equipment reliability analysis, we investigate and calculate the occurrence probability and related data of fault tree [9-10].
Among the causes of motor overheating, the first is "motor failure (X_1)." The problem of motor failure rate should be attributed to the reliability of the equipment in the system reliability analysis. Therefore, it can be calculated based on the equipment reliability theory. The equipment reliability

refers to the probability like the equipment, components, components, etc., have completed the prescribed functions within the specified conditions and time. It is expressed by R_M [11]:

$$R_M(t) = e^{-\int_0^t \lambda(t)dt} = e^{-\lambda t}? \qquad (2)$$

Among them: $R_M(t)$—reliability of a single machine; (t)—the time function of the failure rate; t—running time of the machine. Calculation of λ: $\lambda=K*\lambda_0$. Among them, λ_0 can be determined according to Table 2 and Table 3. The coefficient K is determined according to Table 4.

Table2. American military manual MTL-HDBK-217A standard failure statistics, λ_0 (part) (unit: times / 10^6h)

Component name	Basic fault frequency	Average fault frequency	Upper limit of fault frequency
Motor	0.342	1.22	2.32
Circuit breaker	0.045	0.1375	0.949
Terminal	0.003	0.019	0.0436
Measuring instrument	0.135	1.3	15.0
Connection pole	0.01	0.063	1.02
Thermometer	0.05	1.0	3.3
Buzzer	0.05	0.60	1.3

Table3. Browning R.L. Browning fault rate recommended value (part) (unit: times / 10^6h)

Separation element	Measured value	Recommended value
Capacitance, resistance, line	0.001~1	1
Crystal semiconductor	0.001~1	1
Electrical connection, winding	0.001~1	0.001
Pipe fracture	0.0001~0.001	0.001
Weld disconnection	0.0001~0.01	0.001
Instrument indicator	10~1000	100
Motor and engine	1~1000	100

Table4. Value of the coefficient K

Use place	*K*
Laboratory	1
Normal indoor	1.1~10
Ships	10-18
Railway vehicles	13~30
Traction bus	13~30
Rocket test	60
Aircraft	80~150
Rocket	400~100

The initial reliability λ_0=1.43×10-6/h is initially determined. Combined with the environmental conditions of the motor, the severity coefficient K=10 is taken from Table 5. Finally, λ_1 = 1.43 x 10-5/h. A survey of the company's motor management system shows that its maintenance cycle T_l = 6 months = 4320 hours. The substitution probability calculation formula obtains the maximum failure probability $P(X_1)=1-R_M(t)=1-e^{-\lambda t}\approx\lambda_1 T_1=0.062$. According to the same calculation method, the probability of other basic events can be obtained. The summary results are shown in Table 5.

Table5. Basic event probability data calculation

Event tree link event	Basic events of fault tree	Probability of a basic event or equipment failure rate
Motor overheated (A)	Motor failure (X_1)	Motor fault frequency: $\lambda_1=1.43\times10^{-5}$/h; Set up maintenance cycle T_1=6 months =4320h. Maximum failure probability P (X_1) $=1-e^{-\lambda t}\approx\lambda_1 T_1=0.062$
	Wiring defects (X_2)	P (X_2) =0.19
	Power failure (X_3)	Power failure rate $\lambda_3=2.44\times10^{-5}$/h (Maintenance cycle T_3=6 months =4320h), Maximum failure probability P (X_3) $\approx\lambda_3T_3=0.105$
	Fuse unbroken (X_4)	Fuse failure frequency $\lambda_4=1.62\times10^{-4}$/h (Maintenance cycle T_4= 1 months =720h), Maximum failure probability P (X_4) $\approx\lambda_4T_4=0.117$
Operator didn't succeed in extinguishing the fire manually (C_0)	Manual fire extinguishing error of the operator (X_5)	P (X_5) =0.1
	Manual fire extinguisher failure (X_6)	Fault frequency of manual fire extinguisher $\lambda_6=1\times10^{-4}$/h (Maintenance cycle T_6=365h), Maximum failure probability P (X_6) $\approx\lambda_6T_6=0.037$
Automatic fire suppression system failed to extinguish (D_0)	Automatic fire extinguisher control system failure (X_7)	Automatic fire extinguisher control system failure frequency $\lambda_7=1\times10^{-5}$/h (Maintenance cycle T_7=2190h), Maximum failure probability P (X_7) $\approx\lambda_7T_7=0.022$
	Automatic fire extinguisher failure (X_8)	Automatic fire extinguisher failure frequency $\lambda_8=1\times10^{-5}$/h (Maintenance cycle T_8=2190h), Maximum failure probability P (X_8) $\approx\lambda_8T_8=0.022$
Fire alarm system alarm unsuccessful (E_0)	Fire alarm control system failure (X_9)	Automatic fire extinguisher control system failure frequency $\lambda_9=5\times10^{-5}$/h (Maintenance cycle T_9=1095h), Maximum failure probability P (X_9) $\approx\lambda_9T_9=0.055$
	Fire alarm malfunction (X_{10})	Fire alarm failure frequency $\lambda_{10}=1\times10^{-5}$/h (Maintenance cycle T_9=1095h), Maximum failure probability P (X_{10}) $\approx\lambda_{10}T_{10}=0.011$

Secondly, based on the data in the table, using the calculation method of the incident probability of the top of the accident tree, it can be calculated that the occurrence probability of the start event and the link event of the event tree is: P (A) =0.092/6 months; P (C_0) =0.133/365h; P (D_0) =0.044/2190h; P (E_0) =0.065/1095h。

The occurrence probability of five kinds of consequences events is calculated as:

(1) $P(G_1)=P(A)P(B_1/A)=0.092\times0.98=0.090$/6 months;

(2) $P(G_2)=P(A)P(B_0/A)P(C_1)=0.092\times0.02\times(1-0.133)=0.0016$/6 months;

(3) $P(G_3)=P(A)P(B_0/A)P(C_0)P(D_1)=0.092\times0.02\times0.133\times(1-0.044)=2.3\times10^{-4}$/6 months;

(4) $P(G_4)=P(A)P(B_0/A)P(C_0)P(D_0)P(E_1)=0.092\times0.02\times0.133\times0.044\times(1-0.065)=10^{-5}$/6 months;

(5) $P(G_4)=P(A)P(B_0/A)P(C_0)P(D_0)P(E_0)=0.092\times0.02\times0.133\times0.044\times0.065=7\times10^{-7}$/6 months。

4.5 Calculate risk value

According to the definition of risk value, the product of the probability of consequences and the loss of consequences is used directly as the risk value. Section 3.4 has the probability of occurrence of the consequences, and the consequences of loss are determined below.
Losses caused by accidents vary from company to company, and specific investigations should be conducted in conjunction with the company's own situation. The possible consequences of the overheating of the motor of the company and its losses are shown in Table 6.

Table6. Motor overheat consequences and loss table

Consequence	Explain	Direct loss	Shutdown loss	Total loss
G_1	Discontinued 2h	1000	2000	3000
G_2	Discontinued 24h	15000	24000	39000
G_3	Discontinued 1 month	1000000	744000	1744000
G_4	Indefinitely discontinued	10000000	10000000	20000000
G_5	Indefinitely discontinued, 10 casualties	40000000	10000000	50000000

Note: (1) Direct loss refers to the loss of property directly burned or caused. G_5 also includes $300 thousand per person for casualty pension. (2) Loss of work is the loss of $1000 per shutdown 1h. Loss of an unlimited period of production was about $1 million.

Add the "consequences" in Figure 3 to the first column of table 6. According to the survey of maintenance downtime or production time, casualties and property losses (including direct and indirect losses, etc.) are investigated for each corresponding consequences. And fill the result in the corresponding column of table 6. Finally, risk value is calculated according to the formula (1), and the result is shown in Table 7.

Table7 Risk rate of various consequences

Consequence	Risk value (US dollar /6 months)
G_1	270.0
G_2	62.4
G_3	401.0
G_4	200.0
G_5	35.0
Cumulative	$968 /6 months =1936 dollars / year

4.5 Risk Assessment

Appraised by Farmer's Risk Assessment Chart. The figure shows the accident probability as the ordinate and the loss value as the abscissa. Using a curve (i.e., equal risk line) as a safety criterion, the coordinate plane is divided into two parts. The upper right side of the equal risk line is the high risk area, and the lower left is the low risk area [12].
(1) Determine the safety standards: According to the actual situation of the company, assume that the safety standard is $300/6 months.
(2) Make equal risk line: When logarithmic coordinates are used, the equal risk line appears as a straight line. Based on this, we made the "$300/6 months" equal risk line, as shown in Figure 3.

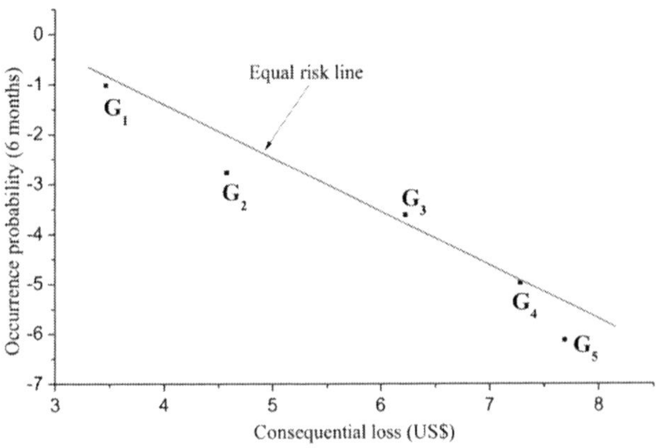

Fig.3 Motor overheat risk assessment diagram

(3) Draw the accident risk point: Take the logistic value of the accident consequence G1~G5 as the ordinate, and take the loss value as the abscissa, which is depicted in the risk analysis chart 3.
(4) Risk assessment: Select the consequences event where the risk point falls on safety risk standards, and propose corresponding countermeasures. As can be seen from the figure, if the $300/6 month is used as a safety standard, the risk of other consequences except G3 is acceptable. G3 should be further proposed to take measures to reduce risk.

Of course, in the actual evaluation process, the selected safety standards are not the same, the analysis results are not the same. For example: if the sum of the risk ratios for various consequences does not exceed $1000/6 months as a safety criterion, the above system may also be considered safe. Therefore, in the actual evaluation should fully consider the specific production conditions of different companies, and reasonably determine the risk standards.

5 Conclusions

(1) The CCA method lacks implementation details in actual use. There are widespread problems such as unclear selection of target systems, confusion in the logical relationship between link events, and unclear basic data sources such as failure rate and frequency of events. This limits the scope and effectiveness of the method.
(2) According to the system safety theory, the "motor overheating safety control system" was determined as the target system. Based on the reliability theory of people and equipment, a reliability block diagram is obtained according to the target system functions and control conditions. And draw the "motor overheating causes - consequences map." Clearly the logical relationship between the events. On this basis, based on the theory of human reliability and equipment reliability in system reliability analysis, the probability of incidents and the related data were investigated and calculated. The equipment failure rate and the probability of occurrence of each basic event, as well as the possible consequences and losses. Finally, the risk value is calculated and a risk assessment chart is drawn.
(3) The source of basic data in the quantitative calculation is studied in the reason-consequential analysis method. Based on the system reliability theory, the concrete feasible data determination method is proposed. However, since China has not yet established a complete human and aircraft reliability database, some of the data only originate from abroad. Adding technology and regional differences will, to some extent, affect the exact use of this method. Therefore, when using it, it should be corrected according to the actual situation.

6 References

[1] Bang Y H, Lee S D. A Study of System-Based Model for Evaluating EF (Exposure Factor) in Quantitative Security Risk Analysis[C]// International Conference on Convergence and Hybrid Information Technology. IEEE, 2008:661-666.

[2] Chen Wenying, Liang Ruohong, Wang Jia. Research on mechanical safety assessment model based on cause consequences analysis [J]. China Journal of safety science, 2010, 20 (9): 97-104.

[3] Qi Xin, Tang Chuan, Chen Feng Feng, et al. Study on risk assessment of geological hazards [J]. Journal of natural disasters, 2012, 21 (5): 33-40.

[4] Shayanfar M A, Barkhordari M A, Roudak M A. A new effective approach for computation of reliability index in nonlinear problems of reliability analysis [J]. Communications in Nonlinear Science & Numerical Simulation, 2018, 60:184-202.

[5] Tan Qinwen, Miao Dong Tao, Liu Jianping, et al. Study on [J]. China safety production science and technology based on the reliability theory of event tree analysis, 2015, 11 (6): 75-81.

[6] Xie Liyang. Mechanical reliability theory, and some issues of model [J]. Journal of mechanical engineering, 2014, 50 (14): 27-35.

[7] Gu Yingkui, Cheng Zi Xin, Li Jing. Reliability analysis of multi state system under performance level [J]. China Journal of security science, 2015, 25 (5): 68-74.

[8] Pham H. Handbook of Reliability Engineering [J]. Production Engineer, 2003, 90(430):804.

[9] Zio E. Computational Methods for Reliability and Risk Analysis [M]. World Scientific Publishing Company, 2009.

[10] Dusan, PROGOVAC, WANG, et al. Parameter estimation and reliable fault detection of electric motors [J]. Control Theory and Technology, 2014, 12(2):110-121.

[11] Chen Bao's family, Chen Xuefeng, Tian Hong Liang, et al. Real-time reliability prediction of equipment under small sample conditions [J]. Modern manufacturing engineering, 2015, 1: 1-8.

[12] Ávila S.F. Reliability analysis for socio-technical system, case propene pumping [J]. Engineering Failure Analysis, 2015, 56: 177-184.

511270 Study of the Effects of Flow Conditions on the Performance of Corrosion Inhibitors in CO_2 Environment

Abstract:

Major process safety incidents have been caused by corrosion all over the world. Despite the increasing knowledge in corrosion, efforts are still needed to understand different damage mechanisms and their control methods. One of the most common active corrosion mitigation techniques in refineries is the use of corrosion inhibitors. Film-forming corrosion inhibitors create a protective layer that can be influenced by different flow characteristics and flow regimes. These changes in hydrodynamic conditions have an effect on the performance of corrosion inhibitors; therefore, more study is required to understand the behavior of corrosion inhibitors under various flow conditions and corrosive environments. This work focuses on the study of different flow parameters such as flow velocity and its influence on corrosion inhibitors efficiency under CO_2 environment. The experimental work will be performed in a rotating cylinder electrode (RCE) that allows to simulate different flow effects. Electrochemical techniques and surface analysis will be used to understand the behavior of the inhibitor under different inhibitor concentrations and flow conditions. The objective is to analyze the adsorption mechanism of the selected inhibitor and its efficiency to help in the fundamental understanding of the performance of CO_2 corrosion inhibitors.

511306 Multi-Objective Stochastic Optimization for Preventive Maintenance Planning in Chemical Plants

Abstract:

Maintenance planning and process operations in chemical manufacturing facilities are subject to several sources of uncertainty ranging from volatile feedstock prices to uncertainty in the level of demand. In the context of assuring the integrity of assets in ageing chemical plants, the present research focuses on uncertainty in equipment availability and develops a novel multi-objective stochastic mixed-integer nonlinear optimization algorithm for preventive maintenance planning. The algorithm factors in uncertainty to arrive at robust optimal solutions in contrast with other approaches such as risk-based inspection and deterministic optimization. In addition, it simultaneously considers the two competing objectives of cost minimization and system reliability maximization to decide on the optimal maintenance frequency, set of online equipment and process flowrates at each time stage. The proposed approach consists of two main steps. In the first step, the system is represented as a dynamic Bayesian network to capture complex interactions between system components, identify critical failure pathways, and characterize the overall risk of system failure. In the second step, the multistage stochastic optimization subproblem is formulated over a shrinking time horizon to progressively incorporate decisions and information from prior time stages into the decision-making process. Following system representation and problem formulation, the overall multi-objective optimization problem is solved using the epsilon-constrained method to obtain the optimal maintenance policy. The overall optimization problem is non-convex and large-scale and computational difficulties are tackled within the algorithm using structure-based decomposition strategies such as Generalized Benders decomposition, or via metaheuristic optimization techniques such as simulated annealing. The results of the research include: (i) obtaining a robust optimal expected preventive maintenance plan and operating schedule, and (ii) provision of a Pareto front of optimal solutions from which the decision maker can select. This robust approach combining the techniques of Bayesian network-based risk assessment and nonlinear stochastic programming is illustrated with a case study and can be used to improve overall equipment availability and maximize plant productivity.

513110 Mechanical Integrity & Reliability - Lessons Learned from Industry Incidents and PSM Audits

Abstract:

Mechanical Integrity is a basic requirement of PSM/RMP and should be an integral part of any process safety program. This workshop will look at lessons learned from industry incidents and examine process safety management auditing, looking at the common deficiencies related to mechanical integrity and reliability. PSM key success factors will be discussed to highlight the core elements of successful process safety programs.

511062 What Is Your "ROA" or "Return on Audits"? A Different Approach to Understanding Toxic Risk

Abstract:

We measure and audit process safety performance in a number of different ways, yet our industry continues to have releases with actual and potential catastrophic consequences. Are our investments in these processes pointing us to actions that will drive loss of containment incidents to zero? Do you want to increase your "ROA" or "Return on Audit?" This presentation will share a novel approach for assessing toxic risks, including results from the initial application of this unique assessment process.

511214 Dynamic Risk Mapping Using Big Data Analytics

Abstract:

Dynamic risk mapping using Big Data Analytics

Prerna Jain*, M. Sam Mannan

Mary Kay O'Connor Process Safety Center

Texas A&M University, College Station, TX-77843-3122, USA

*pjain77@tamu.edu, +1 (979) 985 0773

Abstract

Process or energy systems such as onshore and offshore production platforms, chemical processing facilities, utility plants etc. are complex socio-technical systems. These systems have several sub-systems or multiple factors, complex interactions between system components, and their relationships. The dynamic operations and changing environment continuously impact the overall risk profile of these systems. Therefore, it is significant to analyze this emergent behavior and study these socio-technical systems considering both technical and social as well as internal and external influences. Not many researchers have explored the use of big data analytics to map the dynamic risk profiles of these systems. In this work, a systemic novel methodology is described and developed. For this purpose, the socio-technical system considered here is reproduced as a system of layers as shown in Figure 1. Based on this system of layers, dynamic risk profile is obtained by the identification of weak signals and utilization of the data generated in the facility from various sources such as historian, Central Maintenance Management System (CMMS), operational data, Process Safety Management (PSM) system.

Figure 1: Layered Dynamic Risk Analysis Framework (L-DRAF)

The evaluation of dynamic risk involves different steps similar to a Layer of Protection Analysis (LOPA) study. The methodology involves layer-wise analysis from plant layer to safeguards layer to calculate the final risk as low, medium or high. In the current paper, analysis for two case studies have been demonstrated to compare the differences between the conventional and the novel risk assessment methods. The advantages of the new method are discussed and summarized.

Keywords: Big Data Analytics, Process Safety; Dynamic Risk Map

510826 Application of Group Intelligent Optimization Algorithm in Gas Emission Source Identification

Abstract:

Dangerous gas emission may result in serious ecological, environmental, social and human life safety risks, thus gas emission source term identification is crucial for safety management. Based on the experiment data from static sensor distribution, the group intelligent optimization (GIO) algorithms including particle swarm optimization(PSO),ant colony optimization algorithm (ACO) and the firefly algorithm (FA), are compared to identify the parameters of the gas emission source, such as source strength and location parameters. The results showed that the FA has the best performance in terms of source parameter estimation, while the PSO and ACO algorithm are superior to the convergence rate, calculation efficiency, dependence on the initial value and boundary constraints. Finally, the GIO algorithm coupled with correlated matching of concentration distribution (CMCD) method is applied to the source term identification with mobile sensor based on a simulation scenario. The results indicated that it is a useful method to identify source term with high accuracy by GIO-CMCD method in the case of mobile sensor. The researches discussed in this paper can be applied to the management and assessment of the leak risk for the storage and transportation process of the dangerous gas or volatile materials.

18 AIChE Spring Meeting
14TH GLOBAL CONGRESS ON PROCESS SAFETY
ORLANDO FLORIDA 2018

Human Performance in Extreme Weather Events

Erin P. Collins
JENSEN HUGHES
111 Rockville Pike Suite 550, Rockville, MD 20850-5109
ecollins@jensenhughes.com

Elizabeth Cook
JENSEN HUGHES
ecook@jensenhughes.com

Peter McKnight
JENSEN HUGHES
pmcknight@jensenhughes.com

Prepared for Presentation at
American Institute of Chemical Engineers
2018 Spring Meeting and 14th Global Congress on Process Safety
Orlando, Florida
April 22 – 25, 2018

Human Performance in Extreme Weather Events

Erin P. Collins
JENSEN HUGHES
ecollins@jensenhughes.com

Elizabeth Cook
JENSEN HUGHES
ecook@jensenhughes.com

Peter McKnight
JENSEN HUGHES
pmcknight@jensenhughes.com

Keywords: Human Reliability Analysis (HRA), Quantitative Risk Analysis (QRA), Probabilistic Risk Analysis (PRA), Human Failure Events (HFE), High Winds (HW) Events

Abstract

Hurricane Harvey and its flooding aftereffects led to loss of both offsite and emergency back-up power at a chemical plant in Crosby, TX, which in combination with reduced site accessibility and availability of operations personnel, led to the plant's inability to keep their organic peroxides sufficiently cooled. The failure to maintain cooling caused the thermally unstable and heat sensitive compounds to undergo spontaneous decomposition, resulting in fires and explosions.

Recent efforts in the nuclear power industry to assess high wind (HW) effects through PRA/QRA and to develop coping strategies for loss of power scenarios may provide useful insights to the chemical process industry, particularly in the area of human reliability analysis (HRA). Based on operating experience, worker interviews, and HRA analyst judgment, an approach has been developed for applying adjustments to internal events human failure events (HFEs) to represent worker actions during high wind events.

High wind-induced initiating events considered in HW PRAs include tornadoes, hurricanes, and straight-line winds (i.e., thunderstorms and extratropical storms). Depending on the location of the site, some of these types of storms may be below the frequency threshold for consideration. The HW HRAs performed by the authors included procedure reviews and worker interviews to glean insights about worker response during a high wind event.

1 Introduction

Human reliability analysis (HRA) provides a method that allows the analyst to evaluate, prioritize, quantify, and improve the human interface with complex systems. HRA tools and techniques were pioneered by the commercial nuclear power industry but are based on fundamental precepts of human performance and data collection and analysis that are neither industry nor technology specific.

The HRA process begins with a task to identify actions performed by various workers and to define the actions in terms of human failure events or HFEs. These actions may include both failure to perform a required task or undesired actions with deleterious impact on system performance and safety. For the chemical process industry, this identification and definition step is expected to be very closely tied with the subsequent step of qualitative analysis because the same information gathering that helps define the events also feeds into the development of an "operational story" or HFE narrative that assists in further understanding the events. The information gathering involves reviewing the relevant incident reports, procedures or training manuals in order to become familiar with the process and expected actions, as well as to identify preliminary assessments of the part of the process that could go wrong and why.

HRA develops a baseline case for normal operations and then adjusts it to reflect the impact of so-called "external events" (meaning events that originate external to the facility site but impact facility operations), such as seismic activity or extreme weather events.

The topic of external events HRA gained increased relevance in light of the accident at Fukushima Daiichi, which prompted nuclear power plants in the U.S. and internationally to reexamine their plant's risk profile for external events. To provide guidance in this area, the Electric Power Research Institute (EPRI) published the report "An Approach to Human Reliability Analysis for External Events with a Focus on Seismic" in 2016 [Ref. 1], which provided a general framework to analyze HFEs for external events effects. Future work identified in that report was to further refine the method for various external events based on hazard-specific factors. Although the EPRI approach provides direction for external events as a whole, there is currently no guidance specific to high wind events for HRA best practices.

In light of recent experience with Hurricane Harvey impacts to process, site, worker and public safety, these concerns are believed to be relevant to the chemical process industries as well. Based on operating experience, worker interviews, and HRA analyst judgment, an approach has been developed for applying adjustments to baseline HFEs to represent worker actions during high wind events.

High wind events addressed in this paper are tornadoes, hurricanes, and straight line winds. Straight line winds or "straight winds" include thunderstorms and do not contain the rotational effects that are seen in tornadoes.

2 Hazard Bins

The EPRI approach provides guidance for development of various damage state bins with respect to seismic events. Earthquakes have different damage level bins from 1 to 4, with damage increasing by bin number depending on the intensity of the earthquake, as shown in Table 1. This is appropriate for earthquakes because seismic initiators may damage multiple systems, structures, and components (SSCs), even if they are located inside a Category I building, which has been designed to withstand the maximum potential earthquake stresses for the region where the plant is sited[Ref. 2]. On the contrary, high wind initiators do not cause damage to SSCs inside a Category I structure.

EPRI damage state Bins 1 and 2 do not address damage to safety or rugged non-safety (e.g. operability, environmental, quality and/or profitability factors) SSCs within a structure; that degree of SSC damage is addressed in the more severe damage state Bins 3 and 4. It was therefore determined that the high wind event related impacts to equipment, as well as availability of cues to workers, are best represented by EPRI plant damage state Bins 1 and 2. Further, for certain wind speeds, the high wind event is expected to cause a loss of offsite power (LOOP). This failure is defined in Bin 2 so the damage state identified by this bin seems to most appropriately reflect the damage caused by high winds.

Table 1. Summary of Damage State Bins [Ref. 1]

Damage State Bin #	External Event Damage State Description
1	No damage to the plant safety-related systems, structures, and components (SSCs) or non-safety SSCs required for operation. Limited damage to non-safety, non-seismic designed SSCs like residences and office buildings.
2	No expected damage to the plant safety-related SSCs or to rugged industrial type non-safety SSCs required for operation. Damage may be expected to non-safety SSCs not important to plant operations and to the switchyard (e.g., LOOP expected). Some falling of suspended ceiling panels.
3	Widespread damage to non-safety related SSCs and/or some damage expected to safety related SSCs.
4	Significant number of vibration trips and alarms requiring resetting.

3 High Wind Event Characteristics

Tornado events typically do not provide much warning to workers, with a lag-lead time of about 13 minutes [Ref. 3] from the time a tornado warning is issued to the time a tornado touches ground. Since some workers have never experienced a large tornado, they would probably not be truly prepared for the suddenness and violence of this storm. Most tornadoes last only a few minutes in any one location, although some storm systems have been reported to last considerably longer.

Unlike tornadoes, there is usually ample warning of a hurricane, perhaps days in advance. Since the site has hours or days to prepare, workers are able to review procedures ahead of the hurricane arriving onsite. In fact, procedures may even direct the plant to prepare for shut-down ahead of the hurricane arriving onsite.

Straight winds fall somewhere in between tornadoes and hurricanes in terms of timing. Like tornadoes, the duration of the straight wind event is much shorter than that of hurricanes However, unlike tornadoes, there is more advanced warning for straight winds events. As such, there may be time to prepare for the straight wind event (e.g., reviewing procedures, etc.).

4 Site-Specific HW Response

4.1 Operational Experience and Site Procedures

Operational experience can be extremely valuable in assessing actual plant impacts and operational responses during high wind events. In fact, operational experience has led to procedure changes in some cases. In one such case, a plant was operating at power during a hurricane. One of the units experienced water intrusion during the hurricane which eventually resulted in shutting down the unit. After the incident, procedures were reviewed and adjusted to initiate reduction to a lesser operating mode two hours prior to high winds arriving onsite. While not every plant has operational experience to draw from, when available these insights are beneficial to the high wind HRA.

Some site procedures may either instruct the that the plant be taken to shut down ahead of high winds arriving on site (in the case of hurricanes, when there is time to make these decisions) or may leave this decision to the discretion of the shift manager. Further, since there may be some warning before straight line winds and tornadoes (albeit, not much time), precautions can be taken ahead of these storms arriving onsite such as shutting down and prepping equipment that is out of service for post-incident restoration or repair, if possible. In the case of hurricanes, there is typically sufficient time to replace or restore critical equipment to service that may be necessary in the case of a LOOP (e.g., emergency diesel generators).

For the most part, high wind procedures typically involve actions to be taken prior to the high wind event (i.e., plant shut-down, securing outdoor equipment, etc.) and actions to be taken after the high wind event (i.e., damage assessment). Unlike an emergency operating procedure, for example, which provides a prescriptive response to the emergency/contingency event, the high wind procedure does not provide this type of detail. The worker response is more symptom-based and will utilize the existing off-nominal condition procedure set as necessary. The plant damage assessment may be performed concurrently with these procedures, but if a more urgent plant evolution occurs, the damage assessment could be placed on hold at the discretion of the shift manager.

4.2 Plant Damage Assessment

Plant damage assessments are required to be performed as soon as possible. For high wind events, the damage assessment may not be able to be completed until the event has concluded in order to be able to access unprotected areas. Upon completion of the damage assessment, it is assumed that workers will have a better understanding of the plant capabilities and available systems; thus, stress levels may be reduced at this point. It is generally assumed that the plant damage assessment is conducted within one hour of the onset of most tornado or straight wind events. Since hurricane events are sustained, it is more difficult to assume a specific time for which the plant damage assessment is completed. However, since hurricanes provide sufficient warning to workers, completion of the plant damage assessment is not deemed as critical when considering stress of workers as a result of the high wind event.

5 Key Issues for Worker Actions during HW Events

The key factors that have been identified for worker actions during high winds events are timing and location of the action. Timing adjustments depend on anticipated storm durations, which are estimated based on site-specific historical climate data and can vary greatly based on the type of high wind event. Since high wind events include tornadoes, straight winds, and hurricanes, specific sets of rules for each type of high wind initiator typically need to be developed. Actions in outdoor locations or structures that may not withstand high wind events may not be feasible during a high wind event. However, additional consideration is given such that some actions that occur in unprotected locations may be feasible after a storm has passed.

Consideration is also given to the diagnosis and decision-making (aka cognition) requirements to address the event and plant performance, stress, and execution of the actions as some storms may be sudden in nature (e.g., tornadoes), but other storms, hurricanes for example, will provide workers with advanced warning. Plant procedures may even direct workers to halt selected unit operations or shut down the plant prior to the arrival of a hurricane. Thus, in addition to the operational experience discussed earlier, interviews of workers can provide valuable insights into high wind response preparedness and issues related to HRA.

5.1 Feasibility

Like most HRAs, determining the feasibility of worker actions is one of the initial steps taken during analysis. Due to the nature of the high wind event, both timing of the worker action and location where the action takes place are integral to the context in which worker response is modeled during the high wind event.

Anticipated storm durations, which are estimated based on site-specific historical climate data, can vary greatly based on the type of high wind event. Further, the advanced warning for each type of high wind initiator varies, which can impact worker response as well. Therefore, specific sets of rules for each type of high wind initiator may need to be developed.

The location where the worker action takes place has an impact on not just the type of adjustment made to the HFE (e.g., control room action vs. field action), but also the feasibility of the action. Actions that occur outdoors or in an unprotected structure (i.e., a non-Category I structure) within the initial duration of the high wind event (e.g., typically 1 hour for tornadoes and straight winds and 2 hours or more for hurricanes – although these times are site specific) are not considered feasible and the associated human error probabilities are set to 1.0 (i.e., the action is failed in the risk model).

In some cases, longer term actions may be able to be completed after the high wind event concludes. This is especially true for tornadoes [Ref. 4] and straight winds, which have a much shorter duration than hurricanes. After the tornado is over, workers would typically be able to perform actions occurring outdoors or in unprotected areas. These considerations are based on the site-specific weather data which includes high wind event durations, as well as site procedures. Even if an action is feasible to be performed after the high wind event has passed, timing and performance shaping factor adjustments still need to be applied.

Other considerations for feasibility include cues for actions (such as alarms, displays of parameter limits, direction from management or procedures), manpower, and availability of equipment. Although these criteria should be reviewed as part of the feasibility assessment, they typically do not result in an action becoming infeasible.

Generally a minimum level of plant staff is required to be available to bring the plant to a safe and stable condition, if required. However, the ability of additional personnel to reach the plant after a storm is likely to be compromised by inaccessible roadways due to flooding and/or downed power lines and trees, so site preparations need to consider skeleton level staffing and onsite accommodations and provisions for emergency staff. Long term events in the HRA may need to consider fatigue related impacts on performance in extended conditions when skeleton staffing is unavoidable.

Finally, availability of equipment is addressed through the high wind fragility analysis and Layer of Protection Analysis (LOPA) or QRA modeling. For example, the QRA logic is structured such that the high wind fragilities are mapped to components in the QRA model. If the equipment required for manipulation is not available or failed due to the high wind event, then the failure logic propagates through the fault tree and no credit is taken for the associated worker action. For example, an action to align a backup air compressor may not be feasible if the air compressor is located outdoors and is damaged by the high wind event.

5.2 HW Event Timing

For actions that are feasible, timing penalties are applied to the baseline action to represent the delays that would occur as a result of the high wind event in workers recognizing and responding to the event or related plant conditions. Due to the unexpected nature of tornadoes, timing penalties have typically been applied to account for delays in working through procedures. For straight winds and hurricanes, however, workers would have time to prepare for the event and no such delay penalty is applied. One hour after the high wind initiator, it is assumed that workers will know the plant damage state and initial storm effects have a negligible delay on worker actions.

For actions in unprotected plant areas, timing adjustments are applied to account for increased worker travel time to the location where an action must be taken if pathways have been obstructed due to damage caused by the high wind event. For actions in a protected area or control room, however, such timing penalties are not usually applied since the environmental conditions within those areas would not be affected by the high wind event.

5.3 Diagnosis and Decision-Making

As noted previously, it is generally assumed that sufficient warning time exists for straight winds and hurricane events such that there is time for workers to review procedures ahead of the high wind event. This may include not only reviewing the appropriate severe weather or natural disaster type procedures, but also reviewing procedures for response to a Loss of Offsite Power, since LOOP is often a result of high wind initiators. Worker preparedness and their ability to diagnose and decide upon a response strategy is highly dependent upon procedures and training, however, so these need to be reviewed carefully from the perspective of environmental factors that could increase the time for cognition.

Tornadoes, however, may strike with little advanced warning. Therefore, cognitive adjustments are made for tornado-related HFEs but not for straight wind and hurricane HFEs.

Other HW event factors that may impact cognition include the following:

- High wind impacts on the plant are expected to cause an increase in worker workload.
- Multiple procedures may be required to address the event, such as extreme weather procedures as well as emergency operating procedures.

5.4 Stress

Due to the high wind event, it is expected that worker stress levels will be higher than under nominal (or even other more common or expected off-nominal) working conditions due to uncertainty in event duration or even considerations for personal safety. However, for longer term actions, this increase in stress may not be warranted, since once the plant damage assessment is complete and plant status and operability is better understood, stress levels may be reduced.

Taken as a whole, a summary of the adjustments that have been typically made to high wind HFEs by the authors (and applied using the EPRI HRA Calculator [Ref. 5]) are presented in Table 2. This table is merely a guide and these adjustments are not made in all cases, as warranted by analyst judgment.

Table 2. Summary of High Winds HFE Adjustments

Factor	Tornado	Hurricane and Straight Winds
	Timing	
Tdelay = time for arrival of cues to worker that action is required	Add 1-2 minutes to Tdelay.	Keep non-HW case selection
Tcog = time for diagnosis and decision-making	Depends upon level of worker preparedness via procedures and training	Depends upon level of worker preparedness via procedures and training
Texe = time for worker(s) to travel to location(s) of action and implement required action(s)	In-control room or protected area actions – Keep non-HW case selection	In-control room or protected area actions – Keep non-HW case selection
	Field or unprotected area actions – add 2-4 minutes to Texe	Field or unprotected area actions – add 2-4 minutes to Texe
	Cognitive Influencing Factors (e.g., adjustments to Cause Based Decision Tree Method [Ref. 6])	
Alarm/Indication Data Not Available (CBDTM Pc-a)	If alarm/indications are protected from storm damage, keep non-HW case selection	If alarm/indications are protected from storm damage, keep non-HW case selection
Alarm/Indication Data Not Attended To (CBDTM Pc-b)	Increase to high workload due to external event	
Alarm/Indication Data Misread or Miscommunicated (CBDTM Pc-c)	Keep non-HW case selection; more related to Human-Machine Interface issues than weather effects	
Alarm/Indication Information Misleading (CBDTM Pc-d)	Keep non-HW case selection; relates to cue matching procedures and/or training	
Relevant Step in Procedure Missed (CBDTM Pc-e)	Consider whether multiple procedures are being used	
Misinterpret Instruction (CBDTM Pc-f)	Keep non-HW case selection; relates to procedure providing sufficient information for task	

Table 2. Summary of High Winds HFE Adjustments

Factor	**Tornado**	**Hurricane and Straight Winds**
Error in Interpreting Procedure Logic (CBDTM Pc-g)	If procedure wording is clear and task is covered by hands on training, keep non-HW case selection	
Deliberate Procedure Violation (CBDTM Pc-h)	If crew believes that the instructions presented are appropriate to the situation (even in spite of any potential adverse consequences), keep non-HW case selection	
Peer Check Credit	Depends upon time available to take the action, manager and other worker concurrent assignments, and direction of procedures/training	
Stress	Assign at least moderate stress level	Assign at least moderate stress level

6 Uncertainty

After considering operational experience and worker interviews, a number of gaps still exist, leading to the following assumptions by the high wind HRA analysts:

- Timing information for delays in execution and cognition. Analyst judgment was exercised to increase cognitive and/or execution timing.
- Timing increases for different types of high-winds induced initiating events is a source of uncertainty. The timing adjustments have been treated similarly for the execution portion of the actions. However, it is not known if the execution timing adjustment for a tornado would be very different from a hurricane.
- It was conservatively assumed that any action which takes place outside of a protected structure is failed during a high-wind event. Given the focus of plants on worker safety, this assumption is generally appropriate.
- The HRA assumes the extent of damage should be known within 60 minutes of the onset of most tornadoes.

The final two assumptions may be re-assessed depending on the level of realism needed in the analysis. In general, these assumptions will not overstate risk for a well-designed plant that is inherently rugged and with actions that occur inside protected structures. However, these assumptions may overstate the risk for an older plant that has risk-significant actions performed in non-protected structures.

7 Additional Areas for Guidance

In addition to the assumptions made in existing high wind HRA analyses, there are some broader areas in which further consideration and guidance may still be warranted. For example, rainfall is not directly addressed in the high wind QRA but is addressed in external flooding assessments. Therefore, even though rainfall and high winds may co-occur during a high wind event, impacts from rainfall are not required to be considered. This reduces the level of realism in the case of hurricanes, for example, where there is a potential for both high winds and significant rainfall that could result in external flooding. Worker actions may be impeded by flooding and high winds during a hurricane event. However, only the effects of high winds are considered in the high wind HRA.

Further, as mentioned previously, long-term effects such as fatigue or worker availability may need to be considered to properly assess the reliability of worker actions that take place long after the start of the high wind event.

8 Conclusions

In the absence of guidance specific to the high wind external event, a methodology has been developed to conduct HRAs for high wind risk assessments. Site-specific information including historical weather data and worker preparedness using procedures and training helps shape the context in which the high winds HRA is conducted. There are some broader areas in which further consideration and guidance may still be warranted, such as external flooding.

9 References

[1] Electric Power Research Institute (EPRI) 3002008093, *An Approach to Human Reliability Analysis for External Events with a Focus on Seismic*, EPRI, Palo Alto, CA, December 2016.

[2] Regulatory Guide 1.29, *Seismic Design Classification for Nuclear Power Plants,* U.S. Nuclear Regulatory Commission, Office of Nuclear Regulatory Research, Washington, DC, Revision 5, July 2016.

[3] National Oceanic and Atmospheric Administration (NOAA), Available at http://www.noaa.gov/stories/tornadoes-101, accessed 8 November 2016.

[4] The National Severe Storm Laboratory, *Severe Weather 101: Frequently Asked Question About Tornadoes*, http://www.nssl.noaa.gov/education/svrwx101/tornadoes/faq/, accessed 30 March 2017.

[5] EPRI 3002004030, *The EPRI HRA Calculator® Software Manual*, EPRI, Palo Alto, CA, May 2014.

[6] EPRI TR-100259, *An Approach to the Analysis of Operator Actions in Probabilistic Risk Assessment*, EPRI, Palo Alto, CA, June 1992.

513101 Natural Gas Safety As in Response to Indutrial Events

Abstract:

In 2010 American Electric Power was faced with a critical business decision in the wake of the Con Agra and Kleen Energy Disasters. Bill Sigmon Jr. SVP EPFS made the decision that AEP will not use natural gas for cleaning natural gas piping, thus ending this practice in AEP and giving birth to the Natural Gas Safety Program. AEP then proceed to access its 40+ natural gas facilities and developed a list of corrective actions and best practice. As this work progressed, AEP was recognized by the Combined Cycle Journal for â€œThe Best of The Bestâ€ safety program in the industry. Work proceeded to correct areas in which natural gas venting could result in a flammable vapor cloud affecting personnel, purging, inerting, and gas reintroduction, over pressure protection, and hazardous electrical classification. As the program has evolved, tools have been made for dispersion modeling and leakage potential to bring more qualification to the analysis and increase support from all levels within the organization. AEP has taken this program to the Region Engineers and Plant Personnel along with comprehensive guidelines for Hazardous Area Electrical Classification along with tools and resources to implement and sustain the program. In 2015 AEP began using advance optical gas imaging and ultrasonic measurement to begin early leak detection within these complex piping networks. As AEP has retrofitted existing generating facilities, these guidelines and standards have been include improving ventilation, instrumentation selection and gas detection. In 2017, AEP has implemented a Leak Detection and Repair Program similar to the EPAâ€™s Methods 21 using optical gas imaging and quantification, PID gad detectors and ultrasonic leak detection.

513226 Learning from Experience: Preventive Potential of Near Miss Deep Understanding and Technical Analysis - Experiences and Concepts from Braskem

Abstract:

Process Safety Management (PSM) is a crucial sustainability feature for industrial companies who deal with dangerous products, especially considering the actual scenario of global market, high competitiveness, strong safety requirement and high world demand consume. Reinforced by the perception that the frequency of major process incidents in the last years in world is still higher than expected, it is necessary the Process Safety preventive initiatives that could make the PSM of a chemical and petrochemical organization a real efficient program. The focus on the management of effective main criticalities is one key aspect of a robust PSM program, reflected on the sustainable decrease of incidents occurrence.

Considering that major incidents normally occurs due to trivial failures in process safety management, the consolidation of a strong process safety preventive culture remains a challenge. For this culture to take hold, development systemic management practices underpinned by preventive measures are necessary. In this context, the development of practices based on the management pillar of 'Learn from Experience' is an important alternative, which the preventive approach is because of focus on leading metrics, where the deep understanding of minor process safety occurrences can be a strong contribution to a robust PSM program. The treatment of data from near misses occurrences, the critical analysis of their results and the comprehensive analysis internally in the organization are part of the strategy of preventive approach by using lessons from near miss events. In case of global companies as Braskem, that has different, business, site and facilities, this strategy has the potential to be protagonist in the way of a sustainable process safety preventive culture.

Braskem follows the metrics of API RP 754 (2nd edition), that classifies process safety events in four levels: Tier 1 (LOPC of greater consequence), Tier 2 (LOPC of lesser consequence), Tier 3 (challenges to safety systems) and Tier 4 (systemic or behavior deviation). The first two categories represent lagging indicators and the last two represent leading indicators, i.e. when the number of occurrences is higher, but the consequence impact is lower. Thereby, Tier 3 events metric, also called 'near miss', indicates the amount of occurrences that should be considered relevant process incidents, but not transformed in an incident, in most of cases, due to safety barriers performance. This Tier 3 metric is a precious source of information, which can indicate various types of failure tendencies and dominant causes, allowing preventive measures and making Tier 3 really "gifts from Gods".

Under a preventive perspective, this paper try to present concepts, initiatives and models developed by Braskem related to critical and statistics analysis of Tier 3 occurrences, with the main objective of showing how effective it is to generate preventive measures in order to avoid major process safety incidents (Tier 1 and 2).

510812 Analysis of Contributing Factors behind Injuries in Offshore Oil and Gas Platforms

Abstract:

Over 1900 injuries from offshore oil and gas facilities have been reported to the Bureau of Safety and Environmental Enforcement (BSEE) over the period of 2009 to 2016 [1]. That makes up 38% of all the reported incidents rendering injuries to be the most common type of incident that occurs in the Outer Continental Shelf of the US. These incidents have been accounted as lost time accident (LTA), injuries or restricted work/ job transfer (RW/JT) and has been investigated by BSEE. Current study looks at analyzing the investigation reports of more than 180 injury-related incidents in an attempt to identify the contributing factors behind them. The report provides a statistical representation of the various factors to show the degree of importance of several technical, operational, human and organizational contributors. Issues related to job safety analysis and procedures have been found to be the most concerning factors. An elaborate explanation behind each identified factor provides an overall view of what needs to be improved to reduce the number of such injuries.

References:

[1] Bureau of Safety and Environmental Enforcement (BSEE), https://www.bsee.gov, Accessed 10/10/2017.

512750 Alright, Who Has a Safety Moment?: Enabling Learning through a Systematic Offer

Abstract:

Alright, Who Has A Safety Moment?: Enabling Learning Through A Systematic Offer

It's a comment you hear daily or weekly at the start of your reoccurring team meetings... "Alright, who has a safety moment?". People will sit silently, waiting for someone else to speak up, or you will rack your brain to make something you read online that morning or heard on NPR on your way to work applicable to the conversation.

What if every day, every week, you were able to pull up a topic that was timely, directly applicable to your site, and enabled you to focus on embedding learnings within all level of your organization? Target Zero Today is your solution.

Target Zero Today is, in its most simple description, a series of four one-pagers that is published weekly by the BP Texas City Chemicals HSSE Department. It covers a variety of Process Safety and HSSE topics, with the focus on one topical area per week. The daily content builds on itself, linking to the previous days' information. The one-pagers will either become more technically detailed, or explore different facets of the topic. In the cases where multiple topics are aligned, an organized campaign is planned to build the content in the most effective way. Probing questions are included daily to promote conversation and retention of the topic.

Highlighted content can be categorized into three different areas, 1. Incident Reviews/Safety Flashes, 2. Process Safety and HSSE Topic Deep Dives and 3. Upcoming Site Activities. When a global, or site, learning needs to be disseminated, Day 1 of 4 will review the incident and Days 2 – 4 will discuss contributing factors, similar hazards at the site, and practices and policies in place to prevent the risk event from occurring

This paper/presentation will focus on the benefits of Target Zero Today, implementation of the high-level program, development of weekly topics, and learnings from over a year of delivery.

507322 Improving Performance of DCS Alarm System By Implementing Effective Alarm Management Philosophy

Abstract:

Improving Performance of DCS Alarm System by Implementing Effective Alarm Management Philosophy

Objective / Scope:

The DCS (distributed control system) alarm system at Gathering Centers of Kuwait Oil Company was not functioning as required. DCS alarm system was generating huge no. of alarms in which was out of operatorâ€™s alarm handling capacity. This was highlighted as one of the key contributing factors resulting in process upsets & incidents. A detailed investigation revealed lack of alarm management philosophy during design & operation stage as main cause of this poorly functioning system.

Methods, Procedures, Process:

This paper shares field experience of analysis of DCS alarm system and explains main root causes which resulted in such poorly functioning alarm system at oil and gas gathering centers. It provides guidelines to carry out alarm management exercise for improving DCS alarm system & for developing alarm management philosophy to implement and sustain efficient DCS alarm system. It also shares some of the actual alarm data and their analysis, which supports need of well-developed alarm management philosophy and continuous alarm management exercise for efficient DCS alarm system.

Results, Observations, Conclusions:

It was observed that majority of problems of alarm flood is contributed by selective alarm tags which generate huge number of alarms under category of frequent alarms, duplicate alarms and chattering alarms. The DCS alarms system improves significantly by resolving these alarms. It was concluded that a well-developed Alarm Management Philosophy and execution of well-defined alarm management approach is essential to ensure proper configuration of alarms starting from project stage and sustain it during operation stage of facility. The paper also shares quantified benefits due to reduction in no. of incidents, facility & equipment shutdowns and reduction in gas flaring after implementation of alarm management philosophy.

Novel/Additive Information:

The paper provides guideline to develop and implement Alarm Management Philosophy and justifies need for Alarm management philosophy to ensure implementation of efficient DCS alarm system and thereby achieving safe and smooth operation of facility.

506889 Communicating Major Incidents

Abstract:

Communicating Major Incidents

The way we communicate major incidents is broken.

Research into major incidents shows the lessons are almost completely forgotten three years after the incident. Unless you were personally involved—after three years no one remembers the details of the incident or the lessons learned. Incidents repeat themselves as if new.

To fix this problem, you need to do three things to your communication.

#1 Frighten Your Employees

Engineers/managers believe that scaring employees is unprofessional. This is wrong. What is unprofessional is repeating the same incidents over and over. A "high fear-appeal illustration" will result in at least 50% more behavior change. Behavior change is closely tied to an emotional reaction to the communication. You need to frighten your employees.

#2 Communication Should Contain Mostly Images Not Text

Human are not designed to read text, they are designed to look at images. Your lessons learned will be remembered up to eight times longer if your communication is dominated by hand-drawn illustrations, diagrams, and photos.

#3 Frontline Supervisors Should Present the Communication to Their Own Teams

The correlation between supervisors' informal communication of safety information and the number of major process safety incidents is huge ($r = -0.64$).

No other form of communication (print, video, electronic) delivers more behavior change than informal face-to-face communication between a supervisor and his/her team. That's why supervisors need to communicate the major incident. And that is why your communication should be designed so it is easy for frontline supervisors to use.

Using research and many samples, Dr TJ Larkin shows how to improve major incident communication.

511292 Evergreening - Managing Risk in the Face of Constant Change

Abstract:

Change is constant in processing facilities such as refineries and petrochemical plants. The drivers for change may vary from investment decisions, operational productivity initiatives, regulatory and compliance requirements, inspection and maintenance activities amongst other drivers. Every time a change is introduced to an operating facility, there is likelihood that risk is introduced to the facility if a strong management of change process is not in place or diligently adhered to. Over several changes, these risks compound which imperils the facility, its personnel and the community as a whole.

Evergreening is a structured program to manage a facilityâ€™s process risk exposure in the face of incessant change. The program involves distinct steps defining, assessing, aligning, developing and executing change in a structured approach, leveraging well defined processes, facilitated by proper tools and executed by appropriate resources. Evergreening looks to frontload expert engagement in the change management process to ensure better outcomes while also giving the ownerâ€™s opportunities for cost savings. This paper will review the practice of evergreening applied to pressure relief and flare systems with relevant examples and cases studies that validate the program.

506534 Emoc - Does theÂ e in Emoc Mean Easy or Erroneous?

Abstract:

Many companies are using electronic Management of Change (eMOC's) systems now. These systems make the Management of Change (MOC) process easier to manage and route eMOC's. However, do these systems make it easier or does it lead to errors in the MOC process? Does the ease of an eMOC system lead to a passive review of difficult changes

In this paper, a typical eMOC process will be presented as a model for review. The strengths and weaknesses of an eMOC system will be thoroughly detailed. Typical weaknesses of an eMOC are failure to thoroughly review the change, miss important signatures for the approval, not perform a risk assessment if needed, miss doing a required field inspection of the change in the plant, failure to attach appropriate documentation and failure to ensure that the MOC is closed out. Many audits findings include MOC's not properly closed out. Some major incidents came about due to temporary MOC not permanently fixed or the failure to accurately assess the change in the field.

The author of this paper will make suggestions to handle the weaknesses of an eMOC system and offer detailed solutions. The purpose of the paper is not to stop eMOC system use but to provide insight to the creators and uses of such systems to properly use them.

Experiencing operational process safety - while at university

Trish Kerin
Director, IChemE Safety Centre
Level 7, 455 Bourke St Melbourne, VIC, Australia
tkerin@icheme.org

Prepared for Presentation at
American Institute of Chemical Engineers
2018 Spring Meeting and 14th Global Congress on Process Safety
Orlando, Florida
April 22 – 25, 2018

Experiencing operational process safety - while at university

Trish Kerin
Director, IChemE Safety Centre
Level 7, 455 Bourke St Melbourne, VIC, Australia
tkerin@icheme.org

Keywords: process safety, education, academia

Abstract

In 2016 the IChemE Safety Centre led a project working with industry and universities to develop a list of desired learning outcomes for process safety in undergraduate qualifications. Over 2017 this list has been shared with universities all over the world and feedback sought on the practicality of delivering these learning outcomes. As part of this it was identified that several of the learning outcomes were based more in operational safety and it was thought this could not be taught at university as it is a non operational environment. This led to a rethink on how operational process safety could be taught. This paper will lay out a series of ideas that are being developed into resources by the IChemE Safety Centre. These resources will be available for universities to use in teaching operational process safety. They focus on how the laboratory can be seen as a small scale operational facility, requiring safety systems of work. This work should help to expose students to operational process safety as well as enhance the safety outcomes in laboratory work.

1 Introduction

The academic environment is quite different to an operational environment, and this often results is a sort of culture shock when a student graduates and moves into the operational workforce. This is apparent in all manner of aspects of work as an engineer, but particularly so in process safety. This culture shock in the workplace manifests itself in the perception that process safety is a bureaucratic activity that gets in the way of doing the 'real work'. If we are to improve process safety in the operational workplace, one step we need to take is to ensure it becomes a respected and valued aspect of the work.

2 Understanding more about process safety

As a first step, there needs to be a more fundamental understanding of what is involved in process safety during the undergraduate program. Some universities are doing excellent work incorporating process safety into the curricula, but others less so. A clear challenge is that the undergraduate program is in Chemical Engineering and not Process Safety, meaning that there needs to be a balance struck regarding the hours dedicated to learning the fundamentals on chemical engineering verses other aspects such as process safety. While it could be argued that process safety is a chemical engineering fundamental, from a pragmatic perspective, there are core subjects which must be taught and understood, such as reactive chemistry, heat transfer etc and only a certain number of hours in which to do so. The ISC has recently undertaken some work to develop a list of desired learning outcomes for undergraduate chemical engineers. This was done in consultation with universities and industry and will culminate in the release of a freely available document. This was released in March 2018. The covers the following areas as shown in Table 1:

Core element	Learning outcome
Process Safety Overview	Define the main concepts of process safety and describe the benefits of process safety to an organisation and to society
	Describe the similarities and differences between process and personal safety
	Describe the main elements of a Process Safety Management (PSM) system
Process Safety in Design	Define the concepts of inherent safety and list typical approaches to inherently safer process design
	Describe the benefits of multiple barriers and list typical barrier types for various process excursions
	Compare the advantages and disadvantages of risk-based design verses code/standard-based design in the overall PSM framework
Guidelines for Process Safety Risk Assessment	Describe the steps involved in the risk management process and define the main concepts related to risk management including hazards, failed protection layer, threats, consequences and effects
	Identify various process 'hazards' and develop 'risk' phrases in terms of loss of control of hazards, including identification of: • process hazards • the potential consequences of a failure to keep them contained/controlled • threats to containment of hazards • prevention and mitigation controls through the hierarchy of controls model

	Define the main concepts of common process hazard analysis tools. Describe qualitative and quantitative risk assessment techniques, their uses, benefits and limitations. Perform a hazard identification/ risk assessment for a process safety case study.
	Describe the concepts of As Low As Reasonably Practicable or / So Far As is Reasonably Practicable and describe the steps required to demonstrate that a risk has been reduced ALARP/SFARP.
	List typical factors that contribute to barrier effectiveness and explain the role of critical activities like monitoring, inspection and maintenance in managing barrier effectiveness.
Process Safety in Practice	Apply, adapt and/or create Process Safety Management System (SMS) elements as part of laboratory, practical or pilot plant activities.
	Research, investigate and summarise the application of Process Safety Management System elements as part of industrial training placements.
	Develop and apply regulations, standards, risk assessments, inherent safety techniques and risk-based decisions during Design Projects.
	Identify and evaluate causal factors in process safety incident case studies.

Table 1. Draft Learning Outcomes

In the development of these learning outcomes it was identified that additional resources may be needed to deliver some of the outcomes. This lead to a new way of thinking about how a university environment could become an operational environment. For this to have real impact, process safety needs to be viewed as an integral part of 'real work' for students.

3 Perceived value of process safety

For process safety to be recognized as an integral part of 'real work' its value needs to be established and taught at an early stage. While fundamentally this can be built from a study of previous incidents and their learnings, it goes beyond that. Like any workplace the academic setting requires active process safety leadership across the 6 elements of process safety, being, knowledge and competency, engineering and design, systems and procedures, assurance, human factors and culture. Undoubtedly there are academic institutions doing a very good job and managing these areas, but there is room for improvement. This is reflective of broader industry as well. Consistency in leadership across the elements helps drive a positive culture which starts to inculcate process safety as being a value, not a drain.

A key challenge in managing process safety in an academic setting is safety culture. Two factors impact this, being high turnover of student numbers each year and a perceived lack of risk. When considering the turnover of students, in an operational workplace, if one quarter to one fifth of the workforce turned over each year this would be a significant concern, but in academia this is the reality of each year level moving on and eventually graduating. Maintaining a positive culture in this level of turnover is very challenging. One way to manage this is to define and apply processes for the work, such as procedures and access rules. Defining and communicating the risks would also assist in addressing the perceived lack of risk.

4 University laboratories – an operational environment

A specific area in academic institutions that lends itself to being treated as an operational environment is the laboratory. This is because it contains pilot scale operational equipment, such as pumps, columns and heat exchangers to name a few. If the operation of these items of equipment is managed like a physical plant, it introduces a range of requirements for process safety in practice. This includes the following elements of a safety management system:

- Site induction
- Hazard identification and risk assessment
- Operating manuals and procedures
- Operational status, monitoring and handover
- Operational readiness and process start up
- Emergency preparedness
- Inspection and maintenance
- Safe systems of work – e.g. Permit to Work and Management of Change systems
- Incident reporting and investigation

With the exception of induction, hazard identification and risk assessment and operating manual and procedures, the remaining elements are usually not practically taught in an academic setting. Risk assessment is often taught as part of the design project and procedures are often used in laboratory activities. Site induction is utilised to allow access the laboratory. This needs to define mandatory safety requirements, such as personal protective equipment (PPE) standards, lone worker management and any special procedures specific to the site. This must be strictly enforced for all levels from undergraduate, post graduate and lecturers, regardless of their reason for entering a laboratory. For example in the operational workforce, a CEO would still be required to wear the correct PPE on a site tour, even if only walking through the plant and not actually operating any equipment.

By developing a simplified safety manage system what could be applied to cover these areas in university laboratories, it would teach undergraduates how to work with the types of safety systems that they will be exposed to in the workplace. This would provide context to the various aspect of safety that are often thought to be simply bureaucratic when the student first encounters them in the workplace. In the preparation for the experimental work, the context of the systems can be explained, which is often missing from the workplace, as the focus is to get the task completed. This simple setting the context may provide more value to the new graduate as they transition to the workplace.

In 2018 the ISC will be developing the materials for this simplified safety management system and make them available to universities for their use.

5 Added bonus – safer laboratories?

There have been numerous instances of major incidents occurring in university laboratories. Perhaps most notable are the fatal Sheri Sangji incident at University of California Los Angeles and the Texas Tech University laboratory explosion [1]. In both of these incidents there was a failure to identify the hazards clearly and take appropriate action. It should also be noted that both of these incidents occurred in graduate, not undergraduate laboratories, but the learnings should be transferrable. The application of a structured management system which addresses the hazards as listed above may deliver an additional outcome in creating safer laboratories for both undergraduate and graduate students.

6 Practical experience

Another area of difficulty for an undergraduate which impacts the development of process safety knowledge is the reduction in the number of industry placements available. This prevents the undergraduate from getting valuable experience during their studies. The current work by the ISC will likely not impact the number of overall placements offered, but there is an opportunity to better focus these experiences for both student and company. This can be achieved by providing some guidelines to the companies on what experiences they should be seeking to provide undergraduates and from the other side, guidance on what students should be looking for and asking about when on site. Items include understanding the background behind the induction material, the risk registers and major hazards on site and the basis of safety for the plant design and operation.

In 2018 the ISC will be developing guidance on what companies and students should seek to enhance the value of the experience and make them available to universities for their use.

7 Conclusion

There is a great deal on positive work going on in academic institutions internationally to improve process safety education, but there could still be more done. The ISC is interested in working with both industry and academic institutions to assist with this endeavor. By working together, we can achieve much more that working against each other. The ideas in this paper are not novel, nor are they particularly challenging, and they will not resolve ongoing issue with process safety internationally, but they are a step in the right direction to further enhance collaboration and understanding of process safety from a practical application.

8 References

[1] US CSB Case Study Texas Tech University Laboratory Explosion. Washington DC. 2011.

510757 Managing Risks Under Legal Thresholds

Abstract:

A key part of the initial assessment for determining a facility's applicability for the implementation the Occupational Safety and Health Administration's (OSHA's) Process Safety Management (PSM) standard and/or the United States Environmental Protection Agency's (USEPA's) Risk Management Plan (RMP) rules is to evaluate if the facility handles, stores, uses, or manufactures any of the listed chemicals exceeding an established threshold quantity. A common trend among facilities that have covered chemicals is to implement controls to ensure that they maintain these quantities below these established thresholds. Although this practice is common in industry, we often see companies stretching the limits of the threshold quantities by fractions of a percent, or using the OSHA atmospheric storage tank exemption for flammable liquids (Meer Decision) to avoid the legal obligation to implement PSM/RMP. However, OSHA and the USEPA can still hold the facility liable for not controlling an identified risk under the General Duty Clause. Facilities can struggle with managing these risks if they are not subject to the elements of the PSM/RMP rules. To mitigate these risks, Facilities can begin to implement key elements of the PSM/RMP rules to assist in controlling the hazards associated with highly hazardous chemicals. Some Recognized and Generally Accepted Good Engineering Practice standards such as the International Institute of Ammonia Refrigeration have developed codes on how to manage the risk associated with ammonia refrigeration if you do not meet the threshold quantities for PSM/RMP. This presentation looks at the key elements required to get facilities started in implementing a management system that caters to their specific needs based upon the complexity of their process.

-Brian Desrosiers, CSP, OHST

513589 Developing Probabilistic Barrier Failure Models to Predict Kicks While Drilling

Abstract:

Predicting a kick timely and efficiently is often a challenging task due to the complexities of drilling and other well intervention activities. Leading indicators based probabilistic barrier failure models for different stages of drilling are developed in this work to assess performance of primary well control barrier â€" hydrostatic head. These models would help to determine the probabilities of kick initiating events and identify key leading indicators for predicting kicks and preventing blowouts. The key causal factors for primary well control barrier (hydrostatic head) failure have been identified by conducting fault-tree analysis and analyzing historical incident data. Abnormal pore pressure and swabbing are found to be the major contributors for low hydrostatic head or mud column failure. To predict abnormal pore pressure and swabbing events effectively, leading indicators model are developed combining organizational, operational and real-time indicators. A Bayesian network tool is used to construct probabilistic well barrier failure models for these causal elements. The probability distribution for observing changes in real-time parameters when a kick is developing due to abnormal pore pressure or swabbing are also determined. These parameters are function of both kick initiating events and some influencing organizational factors. Proposed models have two major functions. First, with available leading indicators data probabilities of kick initiating events can be estimated and key parameters which need further improvement for successful kick prevention can be identified. And next, efficiencies of different kick detection parameters can be determined for a confirmed kick scenario and performance influencing factors for these parameters can be identified and improved for better kick detection efficiency. This study would allow both predictive (causes to effect) and diagnostic (effect to causes) reasoning of kicks and blowouts for better understanding of well control system while drilling. Developed risk models enable informed decision making with a relatively clear picture of the risk of barrier failure and provide useful information on actions required to prevent escalation of well control events.

513630 Challenges and Mitigation Measures of Safety Relief Revalidation

Abstract:

Relief system documentation is part of Occupational Safety and Health Administration (OSHA) Process Safety Management (PSM) 1910.119 (d) element Process Safety Information (PSI). Safety Relief Revalidation (SRR) is a complex process of updating the design of existing relief devices for code compliance. The SRR process involves variety of tasks such as review of unit process simulation, process chemistry, safe operating limits, process safety information (PSI), process hazard analysis (PHA), management of change (MOC), overpressure scenario identification, relief load calculation for all overpressure scenarios, evaluating pressure relief device capacities, and to identify the concerns. Safety relief design is a very specialized field and there are quite a few challenges associated with SRR tasks. This paper describes the SRR challenges and good industry practices to mitigate the challenges.

513175 Safety Culture Assessment in Ten OIL Companies in Trinidad and Tobago

Abstract:

TOPIC: SAFETY CULTURE ASSESSMENT IN TEN (10) OIL COMPANIES IN TRINIDAD AND TOBAGO

MOST SUITABLE SELECTION: 20th Process Plant Safety Symposium (PPSS) (T1A)

AUTHOR: MR. SHYAM DYAL – MANAGER HSE, PETROTRIN

YEAR: 2015

KEY WORDS: Safety Culture; Safety Culture in Trinidad Lease Operators

ABSTRACT TEXT:

An assessment of the safety culture of ten (10) petroleum companies in Trinidad and Tobago was conducted in 2015 with the following objectives:

Conduct a comparative study on safety culture

Benchmark safety culture results against an international industry benchmark database

Identify strengths and weaknesses in the safety culture and provide recommendations for improvement.

A Safety Culture Survey was developed using the Health and Safety Laboratory's (HSL) Safety Climate Tool which adopted eight (8) safety culture elements. The study was conducted for a six month to determine the status of Safety Culture within each of the ten (10) petroleum companies. Ninety percent (90%) of the companies surveyed met the 30% response rate standard.

The companies were benchmarked against the international industry benchmark database. Comparing the HSL 'All Industry Average' it was seen that only one company obtained benchmarks consistently above the HSL's average International Benchmark in all eight (8) elements and also obtained the highest benchmark of the ten (10) companies in five (5) of the eight (8) elements assessed. All ten (10) companies exhibited wide fluctuations of benchmarks between the eight (8) elements. These fluctuations between the eight (8) elements within each company and differences in trends between the ten (10) companies illustrate the uniqueness of the safety culture characteristics of each individual organization.

It was concluded that for several companies, the employees' perception of safety culture is average and there is room for improvement. Recommendations were provided to improve safety culture in all ten companies as well as the industry in general.

Key recommendations for improving safety culture in the companies were provided as follows:

- Introduce incentive programs, visible leadership by management through participation in inspections, tours, tool-box talks and committee meetings.
- Provide communication programs geared to all levels of the workforce; implementing hazard identification / risk management programs that are supported by incentives for motivation.
- Develop and implement a detailed Disciplinary Procedure; recognizing the role of human error during investigations.
- Make procedures more easily understood through the use of diagrams and pictures and focusing on training to ensure systems and expectations of procedural compliance are clearly understood.
- Foster a supportive work environment where workers feel empowered, valued and equal and able to challenge each other concerning unsafe behaviors.
- Ensure systems for incident reporting are clearly communicated, providing feedback to workers concerning the outcome of reports and devoting significant efforts to the collection of near miss data.

Word Count: 412

510675 Responsibilities of the Line and Support Organization in the Prevention of Catastrophic Process Safety Events

Abstract:

Many of the leading companies in the chemical, petrochemical, and petroleum process industries have experienced significant incidents in the past 10 years involving injuries, property loss, and even loss of life. These companies operate under OSHA and EPA Process Safety Management (PSM) regulations that were put in place over 25 years ago. Most have robust systems in place to meet regulatory requirements, however, in almost every case, it was found that the incident could have been prevented if PSM had been properly and thoroughly implemented. So the question is, why are important elements of PSM not being completed in a manner that would help prevent an incident? This paper will explore the role of leadership and how when they don't fulfill their responsibilities the PSM process often fails.

Premise:

1. PSM is a robust process that can prevent incidents when fully implemented. It is well understood, documented, and part of US Federal Regulations.
2. It takes people, cost, and capital to implement PSM. The number of people must be sufficient and they must have the necessary skills, including some with institutional knowledge. Cost and capital must be provided to sustain the process and equipment (both direct and ancillary) within design parameters.
3. Each person in the line organization and staff have specific responsibilities to assure that PSM process requirements are met.
4. Most critical to successful implementation is the line function, since there is control of people, costs, and capital.

Many companies operate with line accountability where leaders, managers, and supervisors are responsible for the actions of the line under them. But, this alone does not address the responsibilities for specific actions of individuals in the line.

A successful PSM program must be supplied with sufficient resources to get the work done. This starts with the budget for people, cost, and capital. Budget setting is a complex process to meet shareholder needs, as well as all regulatory requirements and production demands. In most companies the budget is set at a senior level, often well above the site level. Once the budget is in place, the tasks of monitoring performance and assuring that PSM elements are fully implemented fall upon a line leader. Changing conditions and unexpected demands create a challenge to fully implement the PSM process.

A successful PSM program also requires everyone to meet certain job responsibilities. But what are these responsibilities? What must each person in the line organization and support functions do in order to prevent catastrophic events? Below are examples of responsibilities for individuals in the line organization.

Manufacturing companies can be organized for a successful PSM program in a variety of ways. Below are some examples of the responsibilities for a wide range of levels. The responsibilities described are based on a wide range of businesses, both within and external to DuPont, that have solid PSM programs.

Examples of roles that will be expanded upon in the paper:

Business Leader

- Usually in charge of setting the budget. Conducts an annual (or semi-annual) review of PSM metrics including planned and unplanned personnel changes and contingency plans for key persons identified as being critical to the success of the PSM system. Other issues included are identification and replacement plans of major critical equipment, open PHA and incident recommendations, overdue inspections and tests, and other PSM metrics. Identifies areas that are resource deficient or could become resource deficient, and develops plans to correct.

Plant Manager/General Manager

- Meets with all employees annually in groups appropriate to the size of the organization. Covers a broad range of topics including a general business overview, status of major investments, significant initiatives such as a community outreach program, learnings from incident investigations and site audits, and status of the site employee and process safety management programs. Has the participants submit questions in order to answer some in the meeting and promises all questions will be responded to in an appropriate manner.
- Conducts a physical inspection of the facility. The purpose of this inspection is two-fold, it determines if the facility is "ship shape" and provides an opportunity for interaction with employees. Plant managers should emphasize the importance of following the operator's rules. They should pick a "cardinal" rule and discuss why the rule is in place and make sure the employee supervisor is part of the discussion.

Some examples of other positions in the line and staff organizations that are critical to the success of implementing a PSM system. These will be further developed in the full paper.

- Process Safety Management Leader
- PSM Element Leaders
- First Line Supervisor
- Capital Project Leaders
- HR Leader Specific Technology Leaders

Task & Communication in Just Culture, Oil Offshore exercise

Salvador Ávila Filho
Federal University of Bahia, Researcher and Consultant
Rua Graciliano de Freitas 43 Roma CEP 40444340 Salvador BA Brazil
avilasalva@gmail.com

Ivone Cerqueira
Federal University of Bahia, Posgraduate student
ivonecscerqueira@hotmail.com

José Michel Gagliano Ferreira
Federal University of Bahia, Posgraduate student
gagliano.fm@gmail.com

Edmara Drigo
Federal Institute of Alagoas, Posgraduate student
edmarasd@gmail.com

Prepared for Presentation at
American Institute of Chemical Engineers
2018 Spring Meeting and 14th Global Congress on Process Safety
Orlando, Florida
April 22 – 25, 2018

Task & Communication in Just Culture, Oil Offshore exercise

Salvador Ávila Filho
Federal University of Bahia, Researcher and Consultant
Rua Graciliano de Freitas 43 Roma CEP 40444340 Salvador BA Brazil
avilasalva@gmail.com

Ivone Cerqueira
Federal University of Bahia, Posgraduate student
ivonecscerqueira@hotmail.com

José Michel Gagliano Ferreira
Federal University of Bahia, Posgraduate student
gagliano.fm@gmail.com

Edmara Drigo
Federal Institute of Alagoas, Posgraduate student
edmarasd@gmail.com

Keywords: Just Culture, Blame Culture, Behavior, Oil Production offshore

Abstract

Safety in emergency depends on flexible, adaptable, empowered and proactive individuals who actively anticipate results situational of the needs to carry out work and protect the environment, assets and people. Effective management of routine and assets is a consequence of the tasks performed effectively. Task planning includes communication requirements that translate into superior management's willingness to perform operations on the shop floor. One way to demonstrate successful implementation of an efficient security standard is to develop a JUST culture (JC) environment within the organization. JC depends on how organizations deal with guilt and punishment. Developing and maintaining an environment of trust on the factory floor, with clear management actions, the definition of desirable standards to be achieved, and balanced and fair top-down decisions will create credibility in the eyes of employees. This work will analyze the activities developed in the offshore sector where the cognitive machine of the operator is shaken by the emotional balance due to the embarkation scheme having consequences in the circadian cycle and difficulties of relations in the work place. This study allows the initial mapping of regions of organizational efficiency and possible structuring of heuristics used for the future projection of this efficiency. The control of offshore activities requires preventive or corrective actions that are triggered by means of compliance key indicators, developed to control task and behavior, for the organization's resilience in the operational culture.

1. Introduction

Climate disturbances interact with vulnerability factors such as the economy, politics, and preexisting issues in each country or region. These issues are related to geographical location, dependence on agriculture and ecosystem services with their respective laws. It is important to discuss cultural adaptive capacity through the analysis of institutional, social, religious, economic and other aspects.
Safety is a dynamic and emerging property of our systems and is continually being recreated from the challenges of product, process and its controls and from the possible impacts to society and nature. The acceptance risk level is also considered dynamic. Safety is considered a process state and people where criteria and requirements are preserved to avoid danger release. In this case, situations are divided into routine or emergency scenarios. Control in the emergency situation is most critical due to high stress state and actions to be taken quickly. Emergency treatment depends on flexible, adaptable, skilled, and proactive individuals who actively anticipate situational needs to perform corrective and mitigation work. These characteristics depend on a culture resulting from various environmental influences.
The research has objective is the construction measure scheme to the perception of man acting in the routine control and emergency situations which suffers from guilt culture influence. This scheme cultural admits the omission as the worker defense and the loss of safety standard. It is intended to propose a transition form to a just culture avoiding failures, incidents, accidents in the oil production and alloys, which can cause energy losses or even fatalities

2. Behavior and Culture

This review intends to discuss the role of management in the promotion of standardization and the importance of feedback from the control of production systems. Present definitions of culture, safety culture, culture of guilt and fair culture to direct task-focused behavior. The main objective is to indicate how to promote cultural transformation.

2.1. Routine Control

Effective management of routine and assets promotes the performance of tasks effectively. The standardization process aims to prepare procedures according to best practices. Critical task analysis is important for energy savings (reduced rework and reprocessing) and to reduce the likelihood of an accident. But having a good procedure is not enough, you need to know how to apply it evenly in the shift group. Thus, the group must have been trained to carry out the activities and must possess sufficient knowledge and ability to treat the deviations in the routine.
Having knowledge and skill is also not enough, it is necessary to guarantee the motivation to act according to the demand even considering that the situations may be unusual. Having an employment relationship is also not enough, because over time the psychological contract will erode and decrease commitment to organizational goals.
The Organization must understand that there are cultural events that can cause behavior change and if they are not perceived by the leadership creates a climate of coexistence with the deviations. The lack of feedback from the operation makes it difficult to establish priorities to ensure the goals of the work. These omissions [1a] occurring in the routine may be due to the non-feeling of ownership by the task result. This creates a group failure with the break of the work link in the shift, important information is not communicated. The result of this failure is that the written

procedure is not valued, little time is applied in its elaboration and horizontal communication does not occur on the shop floor!
One way to recognize this problem that appears in communication failures is to know the heuristic rules that occur on the shop floor and compare with the formal intent of the organization. Informal rules can also be established on the staff involving the engineers or even at the managerial level. These informal rules can be identified through the operator's discourse that signals the cultural biases that alter the decision. It is important to define how the operators act in the informal rules from the operator's speech, manager and strategic. Hypotheses must be created about the informal rules of the operational routine. Thus, it is concluded that knowledge of the formal rules of work are necessary but not sufficient to avoid deviations. [2b]
Production planning involves the objective part of meeting agendas and organizational goals but also involves the subjective part of the recognition that the team will be committed and feeling possessive about the task. Thus, operational control involving the objective and subjective part can guarantee and improve productivity, reduce fixed and variable costs and increase the competitiveness of the organization. Task planning should include cognitive requirements to ensure communication [1a-3], thus translating the upper management's willingness to perform operations on the shop floor.

2.2. Culture and Power Cultural Field

Culture is a social system of shared and learned values and norms. In the Figure, the elements that build the organizational and security culture with the biases of regional and global culture are compared. Elements of organizational culture include assumed values, level of learning, group integration, level 5 leadership profile, type of communication, adopted archetypes, accepted and non-accepted rites, rites of entry and exit, and (religious and local customs and global adoptions), the established norms and the difficulties of complying with them, and the taboos accepted in part or in full.
According to Hofstede [4], culture is the collective programming of the mind that distinguishes the members of one human group from the other. Culture, in this sense, includes value systems; and values are among the building blocks of culture. These norms and values when negotiated authorize the acceptance of the individuals representing a form of action of the group. These norms and values, when imposed, create a resulting anomalous culture with no possibility of predicting future behavior.
A Culture resulting from a field of cultural forces (Figure 1) or we can call dominant culture result of: regional and global influence, personal characteristics, influence of the immediate and immediate leader, organizational influence and common sense as a citizen, family member settles on the factory floor. The commitments related to the psychological contract, the requirements to participate in the company, its individual intentions and level of satisfaction are discussed with a result in the Affective Socioeconomic [5] in Figure 2.
The organization needs to stay dynamic to understand the challenges and innovate in process, equipment and people technologies. Organizational paradigms are not static and therefore the high-risk industry must continually adjust its safety culture. The dynamics of behavior follow the cultural transformation of society and nature resulting from the adopted customs of globalization. Thus, behavior change and climate change are two major challenges to organizational sustainability.

Culture in the workplace, also called technical culture or operational culture [3] is the result of negotiations between organizational culture and security, regional behaviors and human typologies. Often patterns are created in the organization and from apparent responses it is believed that it has attained the desired culture, but unfortunately the cultural movement is a negotiation, it is not an obligation if the directors believe that what is written is being accomplished. We believe that in fair culture it is possible that the resultant negotiated is close to the operational culture intended by the organization.

[6] cites the negative influence of Brazilian culture on the organization, indicating that the result of paternalism, excessive flexibility, formalism and blind loyalty generates the anomaly of impunity in the face of human failures and errors. Already the biases of global culture induce egocentric behaviors, with memory failures, deviations of attention and sense of social exclusion of the group.

Different cultures lead to differentiated characteristics in the workplace, some that differentiate Safety. Society may differ along cultural dimensions, it is natural, the difficult is when the company does not accept the local cultural dimensions, believing that there are cultural stereotypes that need to be corrected. Social and religious structure and language influence by promoting cultural differences. These differences lead to specific individual values in the workplace and influence the informal rules at the workplace, thus changing their technical culture. This situation directly affects the outcome of the business.

Research on dynamic culture requires careful attention to the following current and global aspects: stereotyping, cultural context, education, work environment, workplace, distance of power, individuality, level of uncertainty, gender, temporal relation, and implications management. How to establish a strong security culture in the midst of so many inadequate characteristics and that induce behavior to the human error of Regional and Global culture?

Thus, it is proposed to analyze the culture to indicate the impact on management [4]: (1) strategic planning; (2) employee motivation; (3) security, image and profitability; (4) employee monitoring and control; (5) and decision making.

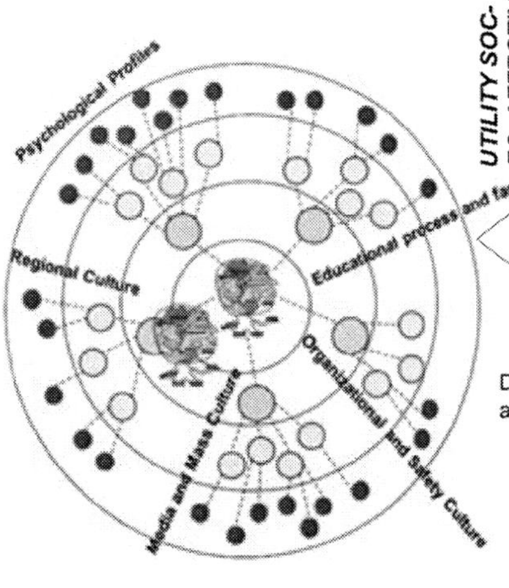

Figure 1 - Power Cultural Field - [7]

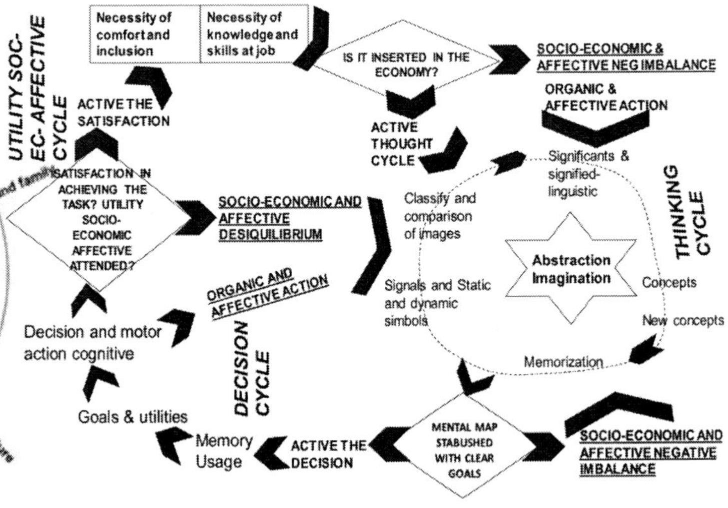

Figure 2 - Socio Economic Affective Utility- [7]

2.3. *Organizational Culture and Safety*

Organizational culture [5] involves a pattern of basic assumptions that the group has learned to solve its problems of external adaptation and internal integration, which has worked well enough to be considered valid and therefore to be taught to new members as the form to perceive, think, and feel about these problems. This relationship of characteristics, premises to build the organizational culture and security and the inherent difficulties in the regional and global culture are indicated in Figure.

The establishment of security as value depends on the organizational beliefs that are initiated from the induction of thought (policies) and validated from practices. The Employee starts to believe in the company when the responsibility of the actions is tested at all levels, from strategic to managerial to the floor of the factory, which is execution. As in industry risk is dynamic, processes have complexities, equipment ages and organizations change ownership, it is necessary to maintain an identity that does not change easily to the flavor of the winds but that when feeling the presence of storm prepare changes of criteria and values gradually. Thus, organizational integration prepares the changes to be implemented.

A certain organization X, in Brazil, is created to exemplify how the rules of behavior based on the cultural field of forces are constructed. The characteristics of the regional and global culture as well as the important elements for the safety and organizational culture are presented in Table 1 for discussion.

The establishment of security as value and the need to analyze the risks to establish preventive measures demands human behavior often unavailable in the organization. The team should be critical in the analyzes and discuss their positions in a clear and constructive way, often the team of workers in company X has the posture of spectator and avoids conflicts in the discussion about problem solving, without the necessary criticality in the elaboration of safeguards. The centralization of the manager and the blind loyalty of the employees generate a bidirectional function that balances in a position that is not appropriate for security.

Sharing responsibility in the team becomes difficult with centralization. Even more difficult is to believe that there is a sense of justice when managers want to be the owners of all outcomes without sharing with operational leaders and even with informal leaders. By the time an oil platform manager admits the difference between contractors and self-employed, it opens a chasm between operational routine and safety.

The accumulation of information for the work needs to be adjusted through the causal nexus established by concepts and new concepts according to what has been indicated about the cycle of thinking in the Figure. If the attention is dispersed or there is not the importance of establishing this research without the affective bond related to the task, the learning will be difficult.

Centralization hinders the promotion of integration and if the team does not believe it is difficult to establish a link due to the apparent difference between the organizational values represented by management and individual values.

The establishment of leadership requires: listening on emergent group events which indicates a certain humility [8]; the creation of a critical and questioning sense in the team, seeking a healthy environment for the discussion in search of solutions; determination in the delegation of activities, including transparent communication indicating means and ends; the planning of rituals and rites of entry-achieving goals to establish the group affective bond necessary for the work and that induces the cooperation with the commitment; care about the social inclusion of the new and old elements of the team.

The errors in the communication concerning the blame culture should be avoided: the feedback written and spoken with minimum quality and standard with technical and cognitive requirements well established. The strategic communication is made respecting the established operational culture and the managerial communication maintaining the sense of justice and the search for the actions from the study of root cause.
Some archetypes should be treated carefully to avoid the adoption of inappropriate rituals or even beliefs not appropriate for safety. Other archetypes should build security solutions to avoid future events. The archetypes that can cause damage are the hero, the princess and the negative, on the other hand, if encouraged at the right time, the archetype of teacher decreases the risk of the accident.
Knowledge about the operator's discourse and the analysis of the culture elements allow tracing the heuristic rules that normally create an environment unsuitable for safety following the negative characteristics of the regional or global culture. When identifying the heuristics that are inserted in the rituals performed by the team and that suffer the direct or indirect influence of the archetypes, it is important to establish safeguards that can be: to draw diversion to "deceive" the repetition of the action, to create a direct barrier to avoid realization, besides to reduce the danger energy. Some Heuristic rules adopted by the team but not appropriate for Security are indicated in Table 2.

2.4. Resilience

A culture that promotes adaptations of the task to the dynamic demands of society, technology and the environment can achieve Organizational Resilience. [9] discuss maritime resilience through organizational, individual, team and technical resources such as: culture, (organizational) training; roles, responsibilities (individual); education, skills, competencies and physical health (team); tools and machines (technical). Challenges to achieving resilience in maritime services include: team competencies, bureaucracy, language barriers, low coordination effort, training and job management.
According to [10], traditional security and risk management focus on improving reliability. But when systems adopted on ships become more complicated and interconnected, unexpected situations arise and are not resolved by the predetermined classical measures. Thus, it is suggested the development of infrastructure, human and organizational capacities to deal with these unforeseen circumstances [11]. The definition of criteria for the resilience of organizations requires the clear knowledge of the work considering that the systems are sociotechnical.

Table 1 – Analysis of Cultural Biases for Security and Organization (12a)

CULTURE	Regional Culture						Global Culture					
Elements of the Organizational culture safety / Features of Regional & Global Culture	Viewer posture	Avoid conflict	Centralization	Personalist	Blind Loyalty	Flexibility	Instant Behavior	Humor varies a lot at work	Scattered Attention	No affective bond in the task	Memory failure, porous	Social exclusion
Security as value												
Responsibility												
Learning												
Integration			C v									
Leader Level 5?	V	C	vnc	V	C					R		
Communication – c												
Archetypes: Hero, Prof., urgency, princess, negative												
Rites, Rituals, Myths – R												
Beliefs – C												
Norms – n												
Taboo												
Values – v												

Table 2 – Negative heuristics for security in task or critical activity

Heuristics	Approximate meaning...
send who can and obeys who has judgment	*It is better to carry out what is commanded when the position is superior. As a worker who is at a lower level in the hierarchy, one must obey in order not to be punished.*
do not mess with the winning team	*Do not seek to understand why, or how, just do not mess with the task, its criteria, parameters and standards, so the game does not turn against and agent miss!*
nobody moves my fish	*I protect all mine, no one stirs, otherwise adopt strategies against, all are saved, without exception!*
What do I get out of it?	*The minimum effort in the task activities or Gerson's law causes there to be a value for each activity ... Improvement activities have a high price ...*
It's just between us ...	*The strategic information, the leap of the cat, does not present itself to anyone, as long as it is advantage not to communicate, silence is made on the facts ...*

2.5 Migration from Blame Behavior to Just Culture

Proper accident investigation provides safeguards for the development of work tasks in a safer environment. The hypothesis considered in this study is that the Brazilian industry conducts the investigation of accidents anchored in the culture of guilt where those impacted are responsible for their occurrence. This is due to omission or commission behavior. This cultural framework is not interesting for competitiveness where the challenges are reached when there is motivation at work. According to [13], the maritime industry has not yet developed clear requirements for a just culture. A culture considered to be inadequate in the maritime industry is the culture that under-emphasizes the presence of danger, and therefore is an approximation to the culture of guilt. The

effective communication culture depends on how organizations treat guilt and punishment. A culture of non-guilt is also considered as non-viable and undesirable since it may indicate the non-correction of unsafe acts!
The construction indicated in Figure 3 presents a conceptual schema about the functioning of the culture of guilt and indicating the measures for the transformation towards a just culture. Government G and regulations in the world of work charge the entrepreneur AND an anticipated guilt that is transferred to the T worker through official and detailed procedures. The worker T, not to be penalized in the cases of accident do not communicate the failures individually or to the maximum point that is the omission group.

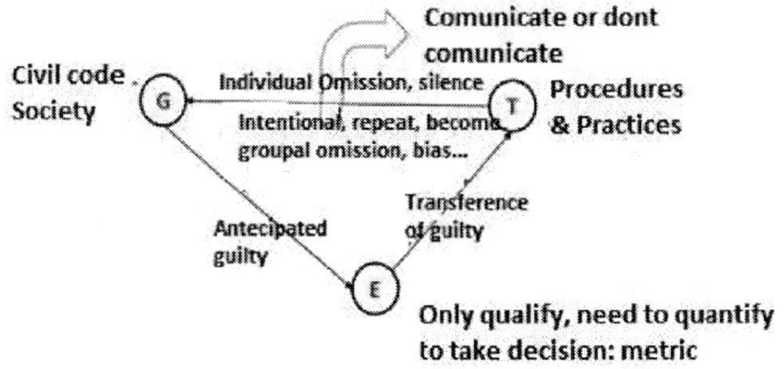

Figure 3 – Model of Blame Culture

Thus, in order to avoid the installation of a culture of guilt, it is important that the government is able to regulate risk gradually and demand more efficient standards and good practices with organizational and market learning. The employer, on the other hand, must dampen the bureaucratic machine in operations by valuing the construction of skills with field skills and the preparation of leaders with good listening skills. The worker should feel included in the team with challenges having, therefore, willingness to hit and communicate deviations before reaching the top event.
The organizational culture model in the case of company X should address the following undue behaviors: (1) minimizing the "way" for palliative solutions indicating limits for flexibility; (2) create opportunities for hazard identification training the team to avoid discipline and risk aversion; (3) avoid the tendency to transfer responsibility through awareness of operational risk; (4) adopt rules that can be reached in the routine for safe behavior with its fixation through daily practice; (5) emphasis on favorable cultural characteristics, such as creativity and team sense to learn from accidents and their experiences, generating strategies for intervention and correction of deviations [14].
One way to demonstrate successful implementation of an efficient security standard is to develop a "culture-fair-JC" environment within the organization. JC depends on how organizations deal with guilt and punishment. Developing and maintaining an environment of trust, with assertive management actions, setting achievable standards, and balanced and fair top-down decisions will create work credit.
According to [13], the key factors for the application of a just culture are: clear communication about what is not acceptable; the officialization of culture in the organization; registration systems (reports) linked to the system of fair culture; monitoring of records and actions; communication of culture in the organization. These definitions have not yet made clear how to achieve a just culture and how to relate to the task [13]. Procedures for fair culture include: fair culture code (accepted

and non-accepted behavior); data protection; selection and training of researchers; signaling positive behavior; addressing responsibilities of managers and supervisors at events; disciplinary proceedings after confirmation of the root cause study.
A Just Culture depends on learning phases, revising concepts and rituals to increase commitment. The technical control of the task is the result of Communication, Competence, Cooperation and Commitment (C4t), a technique to be elaborated from the research in the UFBA GRODin [15b]. In a Fair Culture the following aspects are installed or installed: (a) communication (standards and feedback) in the horizontal and vertical sense is clear and appropriate to local language schemes; (b) the skills installed and applied to the task are adequate to the demand, where there is resilience to the environmental challenges; (c) a climate of cooperation involves the team where fear is not in place; (d) the cognitive functions of task executors and controllers are aligned with the priorities directing the best decision alternatives. It occurs in the decision-making group (managers and executors) a comfort regarding the service of affective socio-economic utility [16a]

2.6. Culture of Safety and Human Factors

The golden rules or commandments proposed in consensus after this meeting with production and safety managers are outlined in Figure 4.
Parallel to a Culture of Energy, one must be able to identify the types of energy and perceive how losses occur. In the task one has to take into account where the energy is used, the event and moment of the beginning of the losses and what measures to prevent or avoid. The loss of energy does not have the same visibility as the material loss, requiring greater attention and calculations to map-estimating the amount lost and its impacts.
Lessons learned facilitate understanding about human error in past events to reduce the likelihood of occurrence in the future. People have different expectations and feelings about the work environment, so confinement can bring unfulfilled reactions if the emotional balance and comfort of the worker are not addressed.

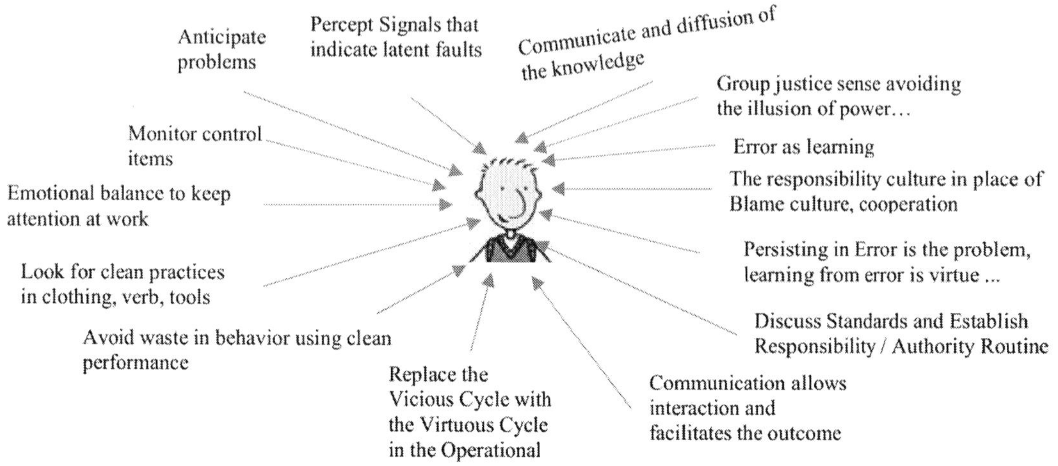

Figure 4 – Gold Safety Norms

One question, what are the environments and factors that make it possible to trust the adopted standards of the routine? How can these factors change the order of events and mechanisms leading to new events? We intend to discuss this dynamic to change the way of failure in the analysis of human factors. This variability may increase with increasing stress level. Thus, for a correct analysis of the risk of events and for the correct analysis of failure and reliability, it is important to include human and organizational factors, including the influence of stress on decision making or on impulse action. It is important to include general concepts that allow the understanding of performance on human factors. If it is difficult to analyze the variability of human factors, it is even more difficult to quantify this complexity, so care must be taken over methods for not believing that equipment is similar to people.
From the root cause analysis of 20 incidents it is concluded that human error is the result of conflicts, inconsistencies and decisions in the areas of management, knowledge, ability, complexity, perception and rules. Man, controls the task [1a] bringing through cognitive processing, decision and motor action, which are important results in the areas of asset control, risk management, avoiding energy losses and environmental impacts. The custom of moving around these goals indicates the trend of behavior in this region. But process losses can be much greater when we open the supply chain and analyze decisions at managerial and strategic levels leading to material, time, energy and image losses.
Failure analysis becomes less apparent the closer it gets to human and social relationships. Thus, communication, commitment, cooperation, and applied competence [12] are part of a discussion that includes the subjectivities of dominant cultures. Cultural biases are identified and need validation to aid in the identification of the variability of human behavior.

2.7. Culture transformation

As indicated in Figure 5, the transformation of culture from guilt to just culture requires a transition phase also called a learning culture [17] e [18]. Culture is something subjective that can bring as a result a comfort in the accomplishment of the task and the acceptance of group challenges to meet the organizational and environmental demands.
Initially the most important part for cultural transformation is the analysis of the communication of standards, rules and working methods. Often the operator does not know what the organization wants from him in terms of behavior or even in terms of the indicators for the success of the task. Some complex tasks must be analyzed to make clear the final state of the process and the environment after its implementation. Thus, critical tasks must also have indicators of meeting environmental, human and safety requirements.
The communication tools (operating instructions, checklist, occurrences, procedures, records, meetings with feedback, suggestions at the top management levels - top, middle management - medium, and factory floor - bottom) should be transparent. An investigation into the quality of technical and social communications should be open when there are many omissions with impacts of possible failure. A study to identify the gap between the competence demanded for the task and the competence performed will indicate difficulties in the decision that can cause failures and process losses. The lack of competence can be confirmed by the operator's speech.
Thus, differentiated levels of intervention are realized for cultural transformation. From concept training, creation of new concepts to unusual situations, perception and decision about failure processes, educational aspects that can act in the perception of addictions or undue behaviors, rituals created to unite the team and achieve organizational goals and other forms of action.

Recognition of the level of competence and communication authorizes the establishment of new team rituals around adjustments to the challenges. The formal and informal leadership should be updated as to the deficiencies, task measurements, cultural adjustments and review of educational models and rituals.

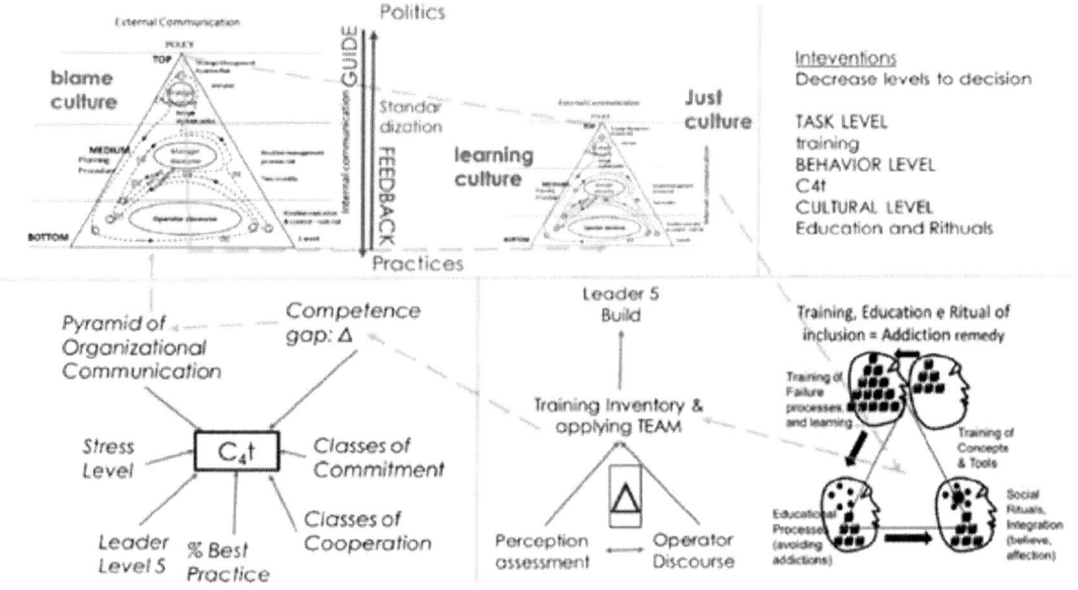

Figure 5 – Culture Transformation - [19]

3. Methodology

In work carried out in the offshore oil production industry for diagnosis and transformation of culture, a program of actions was indicated to achieve a Just Culture in Safety and Energy. The steps in this program are described in the block diagram of Figure 6.

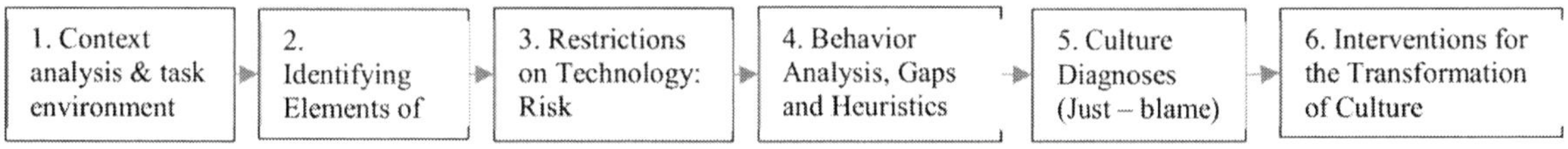

Figure 6 – Stages of implantation and transformation for a Just Culture

A Just Culture is the result of high average values for the calculation of the level or quality of cooperation, commitment, competence and communication [20], taking into account the control of stress level, and with management elements based on observation and empathy of the actual and emerging events in the task environment. Cultural biases are countered by C4t and mainly by the level of cooperation and transparency of communication. The leader who listens and observes and analyzes cannot be trapped in cultural biases or group vices.

The best quality of feedback and standards brings a reduction of the height of the organizational communication pyramid [20] by bringing the top closer to the bottom by increasing agility in revising the standards - for this case with reduced omissions. The reduction of the difference between offeror and claimant competencies can be made through more assertive competence diagnosis using operator discourse, fault mapping, and comparative examinations between knowledge inventory and human error over a period of time. Just culture induces commitment on the timeline and is grounded in efficient communication and a sufficient level of competence for the various tasks.

The activities developed in the offshore sector have particularities related to human factors and social relations [21]: the operator's cognitive machine is shaken by emotional balance, circadian cycle, harmonization and humanization in the workplace, as well as by the high physical distance and possibility of storms. To achieve excellence in the quality of services provided on board, new techniques that include human factors [22, 23] should be applied. Operations begin to be integrated when, through socio-technical analysis of activities [24d], greater human and operational reliability is achieved. Leadership needs to be adapted to analyze human factors through knowledge about executive function and impacts [16b-24d].

The understanding of what is just culture - JC [13] and culture of guilt - BC [25] allows us to investigate ways of migration between cultures with behavior change [26]. It is important to establish measurements and means for this transition, including decision making through routine observation, operator discourse and management strategies. Sociotechnical failure - ST [24] has a complex format where communication biases can modify the pattern. Thus, in a generic way, it is described in table 3 how to carry out the analysis of the culture for adjustment of behavior towards the right culture.

Table 3 – Method and Tools

Step	Description about generic applycation
Context Analysis and Task Environment	The reading of the shift book, the procedures and respective checklists allow to infer about probable mechanisms of failure in the workplace, in the application of procedures, and in the occurrence of incidents, and of accidents. The observation of the work station and the routine operation allows a critical analysis of the relations and aspects of decision in the task.
Identification of Cultural Elements for Cultural Dominance and Archetypes	The global and regional culture elements cited in figure 1 are identified in the operations routine. These elements affect rituals, beliefs, learning and team integration. It can make it difficult to establish values that are favorable to the organizational culture. Thus, the behavioral biases in the decision must be analyzed, such as: viewer posture, centralization, conflict avoidance, personalism, blind loyalty and excess flexibility. These biases promote situations with stories and archetypes related to impunity. On the other hand, instantaneous valuation, excessively variable or depressive moods, scattered attention, lack of commitment, lack of memory and a sense of social exclusion promote low cooperation and initiate communication and skills failures. The analysis of these cultural traits can appear from the identification of archetypes.
Restrictions on Technology, Risk Analysis, and Critical Task	Culture acquires traces of the applied technology in the production system and at the same time, the technology depends on the culture installed in the factory operation. It is important to identify what are the constraints of technology and what are the main risk scenarios for the elaboration of cultural transformation. The analyzes of the highest impact risks already performed are reviewed including human and organizational factors [16-21]. The task is classified as critical from the objective function analysis of social and economic risk. Thus, in the case of the metallurgical industry, the objective function is defined by the number of accidents with withdrawal (weight 50%), interruption of operations (30%) and quality of life (20%). Critical tasks and jobs are analyzed [1, 22] taking into account the risk and complexity in the environment. Constraints on cognitive perception, interpretation, and processing beyond constraints to task control are also analyzed through executive function investigation.
Behavior resulting from C4t: individual and group	Behavior analysis is based on the C4t tool, title of another article. The human type is identified from the investigation of human error. Commitment is defined from human resources data and complemented by shift supervision. Cooperation depends on the level of aggregation, disaggregation and resolution of small emergencies. Competence as discussed above depends on identifying gaps in knowledge and skills. These gaps are tracked from the operator's speech. Finally, the quality of the communication is analyzed in three hierarchical levels (bottom, medium and top) and in two situations: standard or feedback. The stress level directly influences the result of the 4 Cs. Leadership must identify ways to minimize these consequences.

Level rating Just Culture, perception	Polls prepared and applied to identify the level of fair culture or guilt culture in the organization, these are measurement measurement ends of the culture. It is important to define the dominance of antagonistic behaviors and also to establish the focus for accident and energy.
Interventions for the Transformation of Culture	A series of actions are applied to the organization and its analysis of results, through measurements that involve risk and task, indicates the success or need for review in interventions for cultural change.

4. Case: Offshore facility

This diagnostic research intends to collect production data (operator's speech) and process (variables) in critical periods of 3 to 6 months to enable the mapping of regions of organizational efficiency and makes possible the future projection of this efficiency with the transformation of culture. The control of offshore activities requires preventive or corrective actions, which are triggered by means of compliance indicators. Priority in investment decisions and levels depends on the risk of loss. Thus, the proposal is that after the research is done, key indicators (KPIs) should be developed to control the task and behavior, authorizing the organization's resilience within the operational culture.
The analysis of visits to production units in the industrial petroleum segment and the discussion about characteristics of the metallurgical industry, mining and human factors involved allowed the construction of the following exercises of application and analysis of projected results 23 24. The methodology is applied considering the specific characteristics of the Offshore Oil Exploration Industry. Most of the observations made for this work come from the [3]and from the Work published in the magazine Protection on social relations in oil platform [17].

(a) Context and Task Environment

The geographic distance from offshore oil wells to (the mainland, the capital and the city of origin of the workers) can be very large. This fact and embedded work time bring specific characteristics to the task environment. Stress is a real possibility and if the manager adopts a centralizing and autocratic stance, the possibilities of human error are diverse involving contractors and the audit team.
Oil drilling and exploration activities are complex because they involve flammable fluid, high pressures, low temperatures, sea activity with possible impacts to the environment and the possibility of serious events such as fire and pollution. The Fire Event History [27] with recurrence suggests that the investigation of the causes is not in the right direction. Some platforms operate remotely through automatic controls of control rooms on the continent thus increasing its complexity.

(b) Culture Elements

In the Organizational and Security culture care, must be taken regarding historical beliefs and migration to current beliefs. What people believed and believe? These beliefs are determined by events, discourses, and confirmations. Rituals should be studied to nullify movement in the direction of the accident, or in the direction of the danger route. Some archetypes deserve attention for the effect in favor of safety and energy saving or against. The Vision of the future must be confirmed by related actions, without which, disbelief occurs. The Mission compatible with the Vision and the existence of the Safety and Energy Procedures make it possible to conduct an audit.

The regional characteristics of Brazilian culture and the characteristics resulting from globalization in culture make up a multiplicity of characteristics that, if not worked, advance to the job. In the case of offshore oil production industry located at sea, regionalities may indicate the formation of closed socio-functional groups, hindering the interpersonal relationship. The behavior of excessive self-confidence in the work is marked bringing conflicts between contractors and contractors or still continent and sea, operation and safety and maintenance. Thus, there are problems of lack of social inclusion, that is, there is a lack of group membership.
In the Organizational and Security culture care must be taken regarding historical beliefs and migration to current beliefs. There have been major changes in the regulation of the oil market. Many people were motivated to leave the industry and especially the older ones. What are the consequences of these changes? What is asked in the current model is what people believed and believed now? These beliefs are determined by events, discourses, and confirmations. Rituals should be studied to nullify movement in the direction of the accident, or in the direction of the danger route. Some archetypes deserve attention for the effect in favor of safety and energy or against. The Hero archetype should be avoided! The Mission compatible with the Vision and the existence of the Safety and Energy Procedures make it possible to conduct an audit for the transformation of behavior and culture.

(c) Technology Restriction

The petroleum production process involves low temperatures, high pressures, liquid and flammable gas, transportation of oil and gas to the continent including subsea operations. The 12-hour shift regime embarked for 14 days and in the middle of the sea requires care to avoid stress and cognitive loss. Working with pumping, compressors, tanks, probe and other equipment is constant. Water processing and power generation are part of the activities of a platform. Risks of occurrence of process accidents are high as indicated by history. The reprocessing risks are smaller. Some tasks are considered critical as storage of hydrocarbon condensate or drilling and start of oil production.

(d) Behavior Diagnoses

Confidence in the outcome of the work of the operators requires a commitment to the organization that may reflect on blind or critical loyalty behavior. Blind fidelity may be the result of centralized management and critical loyalty may result from delegation management. In the case of platforms this depends on the "luck" of working with a GPLAT that delegates the activities and has a strong sense of justice.
Behavior is marked by a culture of guilt with many omissions. In the case of the oil industry where the number of procedures is large, the omission is probably due to the defense of the fault of the accident. The low-quality feedback on abnormal occurrences of the routine to revise standards and management feedback to revise organizational goals are those that most impact on the possibility of occurrence of accidents with withdrawal and routine with high losses of energy and product to the sea. The improper safety behavior has to do with the perceived (inverted or failed) risks in the work environment involved with transfer of flammable fluids, high system pressure, high costs for production.
Competency design depends on the type of technology and task to devise experience requirements, specific skills and knowledge that are required in the selection service and the development of the professional in the industry.

When the industry involves the employee in a climate of cooperation is a sign that the leaderships have been well established, role rotation is a reality, the will to achieve goals is translated into increased perception about the danger. Leadership should adopt a listening and negotiation stance between resources and achievements. The resultant of a cohesive group around the organizational goals inserted in the society indicates that the aggregation movement is greater than the movement of disaggregation in the routine. In the case of the oil industry, co-operation is a necessity due to the routine being carried out in a stressful environment (prolonged shift) and risks of fire and explosion allocated in the ocean where access is hampered.

(e) Classification of Just Culture Level

The oil industry is situated at 70% of fair scale culture, based on visits and interviews.

(f) Intervention to Culture Transformation

Interventions involve front-line sensitization and staff preparation to investigate the task and human errors. Rituals should be changed to an intermediate configuration between the economic result and the "union of the team". Failure processes should be documented in the form of training modules. Measurements of the task, behavior, and culture should indicate movements for or against just culture. Knowledge gaps are detected through the perception of the operation, the operator's discourse - human errors in the occurrences, and the quality of the decision through the examination of the technical culture.

5. Expected Results

Risks of processes that include human factors are identified and related to routine procedures and specific operations instructions. The tasks are controlled from performance indicators in the various areas: production, quality, asset management, safety & environment, productivity, time and others. It is considered a good performance when it exceeds 80% in average of the indicators of the task, except for the cases of security and environment where the indicators must be close to 100%. The risks that impact the man through his occupation are included in safety. The risks that impact the population, neighboring facility and biota including the risks of process accidents are discussed in the environment.

The first measurement to be monitored and used to control cultural transformation is related to the task indicators. The analysis of non-frequent events, or anomalies, also indicates the resultant in the form of indicators mainly when it involves process losses, fatalities and occupational and environmental fines.

The second measure to be followed is in the comparison of competence and communication-related behavior gaps. Where the quality of communication is measured and the difference between supply and demand for competence is verified. These gaps indicate that the basis for safe and energy-saving behavior is fragile.

Communication creates the task environment and competence enables it to execute successfully. The C4t analysis lacks the commitment that guarantees the full application of competence and improves the clarity of the standard and feedback. Finally, it is important to analyze the aggregation level of the team which generates an appropriate cooperation climate for the good result in the production activities.

The third measurement relates to the culture where it is sought to assess beliefs and values regarding the level of acceptance through the analysis of archetypes, their stories, characters that help and hamper organizational performance. The final result are indicators such as number of accidents with remoteness and or product leaks to effluent, which certainly not interesting to the company. The culture metric is divided into: final result, intermediate consequences, behavioral signs...

6. Conclusion

The Industry and Services Segment adopt organizational resilience posture when capture changes or environmental requirements and adapt to new restrictions. These dynamic corrections only become possible in a middle of Just Culture where the flow of communications occurs without large distances between the factory floor and the top management.

In the offshore Petroleum Industry, there are differences and coincidences of behavior that demand corrections in the project of organizational culture, or still, in the resultant cultural in the floor of factory. The isolation regime of the team in offshore oil production is characteristic that has to be worked seriously to minimize bad practices and negative results in the company. This treatment is possible only with learning, transparent communication and with a Just Culture installed.

Cultural transformation to avoid energy losses and losses in accidents requires actions, measurements and goals to be achieved in the field of Task, Behavior (C4t) and Culture.

7. References

[1a] Ávila Filho, S.; Pessoa, F. L. P . Proposition of review in EEMUA 201 & ISO Standard 11064 based on cultural aspects in labor team, LNG case. In: 6th International Conference on Applied Human Factors and Ergonomics, Las Vegas. AHFE 2015 Conference Proceedings (International Conference on Applied Human Factors and Ergonomics), 2015a.

[2b] Ávila Filho S., Sousa C. R., Carvalho A. C. Assessment of complexity in the task to define safeguards against dynamic risks. 6th International Conference on Applied Human Factors and Ergonomics (AHFE 2015a). Elsevier: Procedia Manufacturing. 3, 1772 – 1779. 2015b.

[2] Embrey, D. Preventing human error: developing a consensus led safety culture based on best practice. London: Human Reliability Associates Ltd. 14p. 2000.

[3] Ávila Filho, S. Etiology of operational abnornalities at industry: a model to learning. Tese (Doutorado em Tecnologia de Processos Químicos e Bioquímicos) – Federal University of Rio de Janeiro, Rio de Janeiro. 296 f.: il.Rotinas Limpas. 2010.

[4] Hofstede G. Escores das dimensões culturais de Hofstede. Disponível em: <http://www.geert-hofstede.com>. Acesso em 2017.

[5] S. Ávila Filho, M.N.E. Fonseca, A.L.A. Santos, C.N. Santino Analysis of Cognitive Gaps: Training Program in the Sulfuric Acid Plant. Proceedings of European Safety and Reliability Conference – ESREL.2016a.

[6] Motta, P.R. Contemporary management: the science and art of being a leader. 15. ed. Rio de Janeiro: Record. 256 p. 2004.

[7] Ávila, S. F.; Souza, C. R. O.; Silva, C. C. 2013. Worker behavior by changing dynamic culture. In: 9th Global Congress on Process Safety. San Antonio.

[8] COLLINS, Jim. Discussion Guide. JIM COLLINS.COM. 2002.

[9] Schraagen, J.M.C.; Van der Beek F.A. & Rypkema J.A. A. Assessing maritime resilience. SEAHORSE. Glasgow. 2016.

[10] Fang S. I., Sherazi U., Muhammad and Earthy J. Resilience engineering: developments in other industries and lessons learned for maritime safety. SEAHORSE. Glasgow. 2016.
[11] OECD - Organisation for Economic Co-operation and Development 2015. Human Performance under Extreme Conditions with Respect to a Resilient Organisation. Brugg: Proceedings of a CSNI International Workshop, NEA/CSNI/R(2015)16.
[12] Àvila, S. F., Bittencourt, E. S.: Operator Risk Assessment in the Oil and Gas Industry: Dynamic Tools to Treat Human and Organization Factors. In: Olson, K. F. Petroleum refining and oil well drilling. New science Publishers, New York. 2017a.
[13] Demosthenous V., Demosthenous S. Just Culture. SEAHORSE. Glasgow. 2016.
[14] Ávila Filho S. et al. Influence of Brazilian Culture in the Workers' Behavior and Operational Discipline. Proceedings of Brazilian Association of Risk – ABRISCO/PSAM. Rio de Janeiro, 2015c.
[15] Ávila Filho S., Ferreira J.F.M.G., Kalid R.A. Sousa C. R. O. Dynamics Operational Risk Management in Organizational Design, the Challenge for Sustainability. American Institute of Chemical Engineers. 2016 Spring Meeting:12th Global Congress on Process Safety. Houston, 2016b.
[16] Ávila Filho S. and C. Costa. Analysis of Cognitive Deficit in Routine Task: as a Strategy to Reduce Accidents and Increase Industrial Production. Proceedings of European Safety and Reliability Conference – ESREL. Zurich, 2015c.
[17] Reason, J. Realizando uma cultura segura: Teoria e prática, Work & Stress, 12: 3, 293-306. 2007.
[18] Dekker, S. Just Culture: equilíbrio de segurança e responsabilidade, Ashgate. 2007
[19] S. Ávila Filho. Melhorando a Tarefa e a Comunicação para uma Cultura Justa em Energia e Segurança, Exercício de Intervenção na Industria Metalúrgica e de Produção Offshore de Petróleo. ABRISCO. 2017.
[20] Drigo, E.S., Ávila Filho, S. Organizational Communication: Discussion of Pyramid Model Application in Shift Records. In: Jussi Ilari Kantola; Tibor Barath; Salman Nazir; Terence Andre. (Org.). Advances in Intelligent Systems and Computing. 1ed.: Springer International Publishing, v. 498, p. 739-750. 2017.
[21] Ávila Filho S. Risks in the Platform: Social and Human Guidelines Avoid Accidents and Alter Standards in Oil Production. Revista Proteção. 2013.
[22] Ávila Filho S. and C. Costa. Analysis of Cognitive Deficit in Routine Task: as a Strategy to Reduce Accidents and Increase Industrial Production. Proceedings of European Safety and Reliability Conference – ESREL. Zurich, 2015c.
[23] Lorenzo D.K. API770 – A manager's guide to reducing human errors, improving human in the process industries. API Publishing Services, Washington, 2001.
[24] Ávila Filho S. Reliability analysis for socio-technical system, case propene pumping. Journal Engineering Failure Analysis (JEFA), 56 (2015d), 177-184
[25] Gorini A., Miglioretti M.and Pravettoni G.. A new perspective on blame culture: an experimental studyjep Journal of Evaluation in Clinical Practice ISSN 1365-2753. 2012.
[26] Drigo, C4t
[27] ANP - Agência Nacional de Petróleo. Relatório de investigação do incidente de explosão ocorrido em 11/02/2015 no FPSO Cidade de São Mateus. Acessado em 27 de agosto de 2017. www.anp.gov.br/wwwanp/?dw=78834. Rio de Janeiro. 2015.

511585 Creating Process Safety Culture, Leveraging on CCPS Vision 20/20

Abstract:

CCPS Vision 20/20 envisions to achieve excellent Process Safety performance by 2020. This should be collectively supported by industry, regulators, academia and the worldwide community leveraging on resources, knowledge and skills in order to fulfill Vision 20/20.
Covering five core themes â€“ Committed Culture, Vibrant Management System, Discipline Adherence to Standards, Intentional Competency Development and Enhanced Application & Sharing of Lessons Learned, the assessment was carried out in PETRONAS across Businesses and Operating Units, inclusive of both Upstream and Downstream as well as Holding Companies.
The assessment covered various levels of management with respondents encompassing operators, technicians, engineers, middle management and top management from businesses across PETRONAS to ensure fair representation of result.
The assessment result provides us with an overview on how we evaluate ourselves against the five core themes, of which the essence is to gauge Process Safety implementation in PETRONAS. The assessment result was then mapped against PETRONAS HSE Management System (HSE MS) elements to gauge against the overall HSE MS effectiveness status.
The result was also benchmarked against other industry players, allowing PETRONAS to position itself with regards to Process Safety implementation. Ultimately, the assessment result was used to evaluate the level of Process Safety culture maturity in PETRONAS.
Once analyzed, the result allows us to identify gaps and room for improvements. Furthermore, it also validates ongoing initiatives to close the identified gaps and improve how Process Safety is being managed and implemented in PETRONAS. This assessment indirectly helps us to gauge and improve the level of Process Safety Culture in PETRONAS.
This paper describes PETRONAS Vision 20/20 benchmarking against CCPS, particularly on the assessment result and analysis conducted in PETRONAS to improve and pave the pathway towards achieving Process Safety excellence.

CRITICAL EQUIPMENT: A MATTER OF PRIORITIES

David Guss
Principal Consultant – DuPont Sustainable Solutions
DuPont (UK) Ltd, Stevenage, UK
david.a.guss@dupont.com

Prepared for Presentation at
American Institute of Chemical Engineers
2018 Spring Meeting and 14th Global Congress on Process Safety
Orlando, Florida
April 22 – 25, 2018

CRITICAL EQUIPMENT: A MATTER OF PRIORITIES

David Guss
Principal Consultant – DuPont Sustainable Solutions
DuPont (UK) Ltd, Stevenage, UK
david.a.guss@dupont.com

Keywords: Risk, Critical, Consequence, Strategy, Priority

Abstract

Many companies in the process industries use equipment criticality assessments as a vital tool to assist in setting priorities and optimizing maintenance strategies. The goal is to deliver high performance and availability in a cost-effective way. But what does the 'critical equipment' designation actually mean? And in practical terms, how should it be used? In the manufacturing industry, criticality is usually understood in terms of the magnitude of the consequences that would result from an equipment failure. Two fundamental approaches inform such designations: the risk-based approach and the consequence-based approach. Each offers different strengths and weaknesses, which are relevant to determining their most appropriate application in varying contexts.

More nuanced than a simple reflection of a company's tolerance for risk, the basis of a criticality designation is important to document, as it will influence a series of future decisions regarding the design, maintenance, and modification of equipment. Informed by DuPont's own experiences as a plant operator, this paper first outlines the broad elements that distinguish the risk-based and consequence-based approaches to equipment criticality. It then proceeds to explore how the nature of the consequences of failure in different scenarios—in particular, whether they are linked to business or environment, health, and safety (HSE) outcomes—should play a role in the process of equipment assessment and the actions that follow.

Ultimately, the paper argues that the risk-based approach is best for prioritizing where to perform costly reliability studies, and when and how to invest in spare parts and redundancy. In contrast, the strength of the consequence-based approach lies in determining maintenance strategy, as well as setting priorities for maintenance tasks designed to prevent events with high consequences of failure. When such consequences are HSE-related, the paper suggests that it is essential that maintenance strategies are formulated to prevent failure, and that HSE-critical maintenance tasks are treated with high priority, avoiding deferrals without proper scrutiny and risk assessment.

1 Introduction

Many companies in the process industries use equipment criticality assessments as a tool to assist in setting priorities and optimizing maintenance strategies. The goal, as ever, is to deliver high performance and availability in a cost-effective way. But what does the 'critical equipment' designation actually mean? And in practical terms, how should it be used? By definition, 'critical' denotes 'having a decisive importance in the success or failure of something'. In the manufacturing context, criticality is usually understood in terms of the magnitude of the consequences that would result from an equipment failure. The consequences of failure (CoF) are most often evaluated in relation to the likely impact of an incident on one, or all, of the following:

- Safety
- Health
- The environment
- Business (repair cost and/or lost production)
- Reputation

The specific basis of a criticality designation is important to document, as it will affect future decisions regarding the design, maintenance, and modification of equipment. Ultimately, two fundamental approaches are usually applied to determine criticality. The first is the risk-based approach, and the second is the consequence-based approach. In the context of the former, risk is understood as a function of not only the consequences, but also the probability, of equipment failure.

2 Risk and Consequence Based Approaches Discussed

RISK = CONSEQUENCES x PROBABILITY

As a result, it is possible for companies to calculate risk scores for each piece of equipment they operate. In turn, this allows organizations to rank equipment in order of risk, plotting the results on a risk matrix to determine whether the calculated risk is acceptable, or if further mitigation measures (for example, design changes or maintenance interventions) are required.

	C1	C2	C3	C4
F4	IV	III	I	I
F3	IV	III	II	I
F2	IV	IV	III	II
F1	IV	IV	IV	III

C = consequence, F = frequency,
I, II, III, and IV = risk score

Figure 1: A typical risk matrix

Figure 1 represents a typical risk matrix. Individual organizations are responsible for determining the scale of the consequence and probability axes, where specific bands of acceptability lie, and what action is appropriate in the case of different risk scores. For instance, does a score in the red zone mean that the risk is unacceptable and necessitates immediate mitigation activities? Unless a specific regulatory constraint exists, decisions around risk tolerance and management are left to operators. Even in the case of a moderately sized facility, performing such analysis properly is a significant task that might require numerous man-months of effort. So what is the value of criticality assessments? And what action should follow?

When the CoF are business-related, then the risk-based approach can be used to determine priorities for several activities:

- Deeper analysis: as processes like reliability-centered maintenance (RCM) and failure modes and effects analysis (FMEA) are time consuming, criticality analysis can help to prioritize the equipment that should be studied in detail. Operators can focus on high-risk items first, and use the studies to develop maintenance strategies that reduce the probability of failure, or mitigate the consequences.

- Investment in critical spares: once the risk ranking is evaluated, the highest risk equipment can be prioritized for rapid availability of spare parts, either from the site store, or through arrangements with a vendor.

- Investment in new equipment: in some cases, the risk may be deemed so great that it is actually cost-effective to engineer greater reliability into the equipment, or alternatively, build in some redundancy to ensure uninterrupted production.

Problems can arise, however, when assessing the role criticality should play in determining maintenance strategy. Historical failure rates are typically used to project the probability of future failure when evaluating risk levels for existing equipment. Equipment that has been reliable in the past will tend to be designated as low-to-medium risk, even if the CoF are high. But this raises the question of what influences past reliability. Is equipment inherently reliable due to its design or operational environment? Or is a well-developed and implemented maintenance strategy a more important factor? If the latter, and the CoF are serious, then predictive and preventive maintenance tasks should also be considered high priority. Any changes to these tasks (or their frequency) should involve high levels of scrutiny and approval.

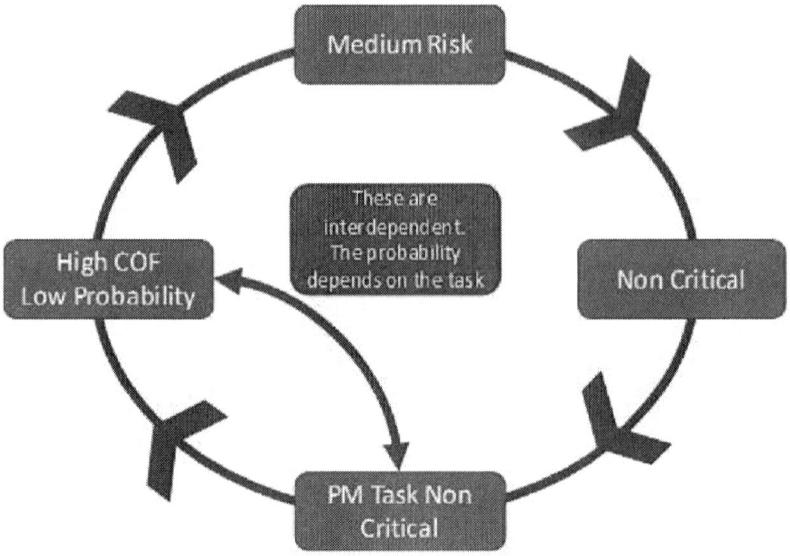

Figure 2: Regular maintenance often influences risk.

There is a real danger, however, that this becomes a circular argument. Designating equipment as non-critical based on risk may mean that the very maintenance tasks relied upon to keep that risk low become low-priority themselves. If delayed or deferred, this may lead, inadvertently, to an increased risk of failure. Ultimately, it is preferable to use a risk ranking to prioritize detailed analysis and investment, but at the same time base maintenance strategy and task priority on the CoF. If the CoF are related to environment, health, and safety (HSE) outcomes, then the inherent problem with the risk-based approach becomes more serious. Arguably, if the risk-based approach results in a high-risk designation linked to HSE, then equipment should not be operated at all unless mitigating action is taken.

The consequence-based approach makes more sense for HSE-related CoF, and is certainly a more cautious course of action. According to this method, all equipment with high HSE-related CoF (as defined by the organization) is deemed to be HSE critical, regardless of current or future probability of failure. Why does this matter? Using the consequence-based approach, an organization can mandate that maintenance strategies must be developed for HSE critical equipment, which minimize the probability of that equipment failing in a way that causes, or is unsuccessful in mitigating or preventing, significant HSE incidents.

At DuPont, we take this concept further by designating certain equipment types as 'default critical', including:

- Pressure vessels
- Pressure relief and vent systems and devices
- Safety interlocks
- Emergency devices (such as emergency shutdown systems and isolation systems)
- Fire safety systems and equipment
- High-energy rotating equipment

Equipment that is not placed on the default list can still be considered HSE critical if the relevant CoF meet criteria related to death or major irreversible health effects, significant property damage, or substantial environmental impact. Some may counter that this approach will inevitably generate a long list of HSE critical equipment, making it impossible to prioritize or focus on the vital few pieces. But while it is true that a facility could amass over 10,000 items of HSE critical equipment when adopting such a consequence-based approach, prioritization is not the point or purpose in this instance.

The rationale behind the HSE critical designation is to mandate an appropriate maintenance strategy. Maintenance tasks on HSE critical equipment are considered high priority. They cannot be deferred, or in the case of prevention systems such as interlocks, bypassed without high levels of scrutiny, risk assessment, and approval. If the risk-based approach is used to determine whether equipment is HSE critical, then maintenance deferrals on high CoF equipment may receive low or even minimal levels of scrutiny and approval.

High CoF equipment can—and indeed, should—be low-risk. The risk profile of this type of equipment increases when it deteriorates, grows unreliable, and demonstrates a greater likelihood of failure. This may be due to age, altered operational conditions (leading, for example, to higher corrosion rates), or even a previously unidentified failure mode. In such circumstances, it is essential to have a system of governance and management oversight in place that ensures thorough risk assessment is performed. This will determine what action should be taken—that is, whether to continue operating with risk-mitigation measures, or to cease operation and return the equipment to the as-designed condition.

3 Conclusion

Ultimately, criticality assessments can be a vital tool to determine priorities and strategy. The risk-based approach is best for prioritizing where to perform costly reliability studies, and when and how to invest in spare parts and redundancy. In contrast, the strength of the consequence-based approach is determining maintenance strategy, as well as setting priorities for maintenance tasks designed to prevent events with high CoF. When such consequences are HSE-related, it is essential that maintenance strategies are formulated to prevent failure, and that HSE critical maintenance tasks are treated with high priority, avoiding deferrals without proper scrutiny and risk assessment.

Four Major Gaps That Are Preventing Most Companies Worldwide from Achieving Excellent Process Safety Performance

William G. Bridges
Process Improvement Institute Inc. (PII)
wbridges@piii.com

Jeff Thomas
Process Improvement Institute Inc. (PII)
jthomas@piii.com

Prepared for Presentation at
American Institute of Chemical Engineers
2018 Spring Meeting and 14th Global Congress on Process Safety
Orlando, Florida
April 22 – 25, 2018

FOUR MAJOR GAPS THAT ARE PREVENTING MOST COMPANIES WORLDWIDE FROM ACHIEVING EXCELLENT PROCESS SAFETY PERFORMANCE

William G. Bridges
Process Improvement Institute Inc. (PII)
wbridges@piii.com

Jeff Thomas
Process Improvement Institute Inc. (PII)
jthomas@piii.com

Keywords: process safety, PSM, human factors, near misses, PHA, HAZOP, procedures

Abstract

Process safety requires implementing many management systems, specific engineered features, and operating and maintenance practices effectively. Most companies believe they have done just that, and yet major accidents continue to occur. Why is that? What is missing?

This paper looks at the statistics of major accidents, combined with results from audits and assessment from more than 50 chemical, petrochemical, oil/gas, and related processing companies world-wide. The paper illustrates the four major gaps that are common to the companies/sites that keep having major accidents, compared to those companies/sites that do not have such accidents:

- Accurate and clear operating and maintenance work instructions (procedures)
- PHA of all modes of operation, especially startup, shutdown, abnormal operations, and online maintenance modes
- Getting Near Misses reported and investigated
- Addressing human factors that are missing from most management systems

The paper describes some guidance for closing each of these gaps.

Introduction

"Process safety management (PSM) is the application of management principles to the identification, understanding, and control of process hazards to prevent process-related incidents"[1]. PSM entails development and implementation of programs or systems to ensure that the practices and equipment used in hazardous processes are adequate and are maintained appropriately. The primary categories of programs or systems have come to be called elements of PSM. However, the basic elements of PSM have been defined by many groups in a number of ways. Table 1 lists the elements of PSM systems from various industry and government groups. Many of the elements with different names have essentially the same meaning. For instance, "maintenance and inspection of facilities," together with some aspects of "personnel" practices, both under ACC's (formerly CMA's) Process Safety Code of Responsible Care™[2] are essentially the same as the single element, "mechanical integrity," under 29 CFR 1910.1190) [3].

Table 1. Comparison of PSM systems

OSHA 29 CFR 1910.119 EPA 40 CFR 68	AIChE/CCPS Risk-Based Process Safety (RBPS) Standard	Responsible Care© Process Safety Code
Management System	**Commitment to Process Safety**	**Management Leadership**
Employee Participation	Process Safety Culture*	Commitment*
Process Safety Information	Compliance with Standards	Accountability*
Process Hazard Analysis*	Process Safety Competency*	Performance Measurement
Operating Procedures*	Workforce Involvement*	Incident Investigation
Training*	Stakeholder Outreach	Information Sharing
Contractors	**Understand Hazards and Evaluate Risk**	CAER Integration
Pre-Startup Safety Review	Process Knowledge Management	**Technology**
Mechanical Integrity	Hazard Identification and Risk Analysis	Design Documentation
Hot Work Permit	**Manage Risk**	Process Hazard Information
Management of Change	Operating Procedures*	Process Hazard Analysis
Incident Investigation	Training and Performance*	Management of Change
Emergency Planning and Response	Safe Work Practices *	Facilities
Compliance Audits	Asset Integrity and Reliability	Siting
Trade Secrets	Contractor Management	**Codes and Standards**
	Management of Change	Safety Reviews
	Operational Readiness*	Maintenance and Inspection
	Conduct of Operations*	Multiple Safeguards*
	Emergency Management	Emergency Management
	Learn from Experience	**Personnel**
	Incident Investigation	Job Skills*
	Measures and Metrics	Safe Work Practices*
	Auditing	Initial Training*
	Management Review and Continuous Improvement	Employee Proficiency*
		Fitness for Duty*
		Contractors

* Contains some of the 6 elements of a Human Factors element, but all missing elements not covered

Note that the newest definition of process safety is CCPS's *Risk Based Process Safety (RBPS)*[4] replaces their earlier process safety definition. In the older definition from CCPS [1], there was an element on Human Factors, which brought strong focus to this necessary element; in RBPS, the human factors sub-elements are now spread across 6 different elements, though some human factors were inadvertently weakened in the transition to *RBPS.*

Although nearly the entire industry agrees that implementing PSM is the right thing to do, interpreting and converting the PSM requirements into practices is unique to each company, and even unique to each plant site. Not only can the requirements be interpreted differently for each site based on local needs, but each site also starts from a different point when they begin to implement a system that is consistent with this regulation.

This paper looks at the statistics of major accidents, combined with results from audits and assessment from more than 50 chemical, petrochemical, oil/gas, and related processing companies world-wide. The paper illustrates the four major gaps that are common to the companies/sites that keep having major accidents, compared to those companies/sites that do not have such accidents:

- Accurate and clear operating and maintenance work instructions (procedures)
- PHA of all modes of operation, especially startup, shutdown, abnormal operations, and online maintenance modes
- Getting Near Misses reported and investigated
- Addressing human factors that are missing from most management systems

The paper describes some guidance for closing each of these gaps.

1. Procedures Need to be More Accurate and Clear

Operating and maintenance procedures have always been crucial to the safety, quality, and productivity of process systems. With the advent of new safety and quality standards such as OSHA's process safety management (PSM) regulation, OSHA's personal protective equipment (PPE) regulation, EPA's risk management program (RMP) regulation, and the ISO 9000 quality standard, many companies are facing the daunting task of developing or upgrading their procedures to satisfy varied and sometimes complex and conflicting requirements.

Although there are overlapping characteristics that the various regulations and standards share, each approaches the procedure-writing process from a somewhat different perspective (e.g., quality or safety, protection of workers or protection of the public/environment), and the required level of detail for documenting the procedures differs greatly between these regulations and standards. For instance, the ISO 9000 quality standard advocates documenting procedures and work instructions that impact quality and suggests keeping the procedures as simple as possible, while the OSHA PSM regulation requires detailed procedures that address: all operating modes (startup, shutdown, etc.), operating limits, consequences of deviations, means to avoid hazards, safety and health considerations, and safety systems and their functions. How can companies develop and maintain procedures that ensure productivity and simultaneously satisfy the different regulations and standards? The key is to remember the ultimate goal of the regulations and standards: to reduce human errors that can impact quality, productivity, and/or safety.

Procedure-related errors are errors that occur because some characteristic of the procedure caused task performance to fail. **This is currently the most critical human factor at most sites since 90% of accidents have at least one root cause related to mistakes within procedures**. Reducing these procedure deficiencies can reduce human error rates by a factor of 2 to 20 [5].

1.1. Identified problems

Deficient Procedures are the most prevalent problem in process industries since procedures have not traditionally been developed from the perspective of optimizing human factors; instead procedures have been traditionally developed to meet a compliance requirement. Examples of procedure deficiencies (inaccuracies) include:

- Incorrect/incomplete/nonexistent (most procedures we have audited have been only 70-75% accurate – the inaccuracies include missing critical steps, steps as written are not what needs to be done, or the steps are out of sequence) [5]
- No/misplaced/incorrect information in warnings (for example, a warning should **never contain** the action to take; it should instead **emphasize** the action to take) and warnings should be located BEFORE the step applicable to the warning
- Poor format and presentation rules

1.2. Objective

The flowchart (**Error! Reference source not found.**) and the procedure writing checklist (Table1) describes the basic sequence of steps for writing effective procedures. Note that some basic rules are contained in the flowchart, such as "operators write operating procedures," "maintenance craft-persons write maintenance procedures," and "Lab technicians write lab procedures." Procedures are NOT written by engineers or department superintendents. This is a critical to ensure procedures are written in the user's common language and are written at the right level of detail.

Figure 1. Best Approach to developing written procedures, once the unit has decided which procedures are needed (© PII 2003-2017) [6]

The procedure content controls the human error related to "procedure-based" error – if content is wrong, the trainee is very likely to learn the step wrong (trainer uses the procedure as a basis for training or the trainee refers to the procedure step that is wrong when executing the procedure).

Even if the content is correct, not following page formatting best practices can still increase human error rates. If the content of the step is wrong it does not matter if the presentation is clear. If the content is perfect, you can further reduce error rates by 50% to 70% by presenting the information properly.

Rules for developing content[7]

- The procedures should be written at the level of someone who has just completed the basic training for that task. Do not write the procedure for someone just hired or for the 10+ veteran.
- Clearly identify ahead of time what activities need procedures to help ensure error rates are controlled low enough, and identify which activities are considered common "skills" of all of the staff.

  Example: Starting a pump is typically considered a "skill." This means a procedure step can simply be "Start the benzene recycle pump (P-119)." The procedure does not need to explain how to start the pump. But, starting a pump is a skill that nonetheless must be learned; it requires doing several sub-steps related to positioning of intake and discharge valves, how the pump is throttled at startup, checking local pressure gauge, etc. So, a training module is needed for starting pumps of various types, and the operator needs to learn this necessary skill. Once the skill is learned, it will be applied to so many various cases and so often, that a procedure is no longer needed as a "refresher" on how to "start the pump."

- The first draft of the procedure must be walked-down in the field by another user. Simply reading through the procedure does not catch enough of the mistakes.
- The revised draft should be reviewed in the field by a technical staff person, such as a process or operations engineer (for operating procedures).
- The procedure needs to be checked to ensure it follow the page format and step writing rules described in the next section. This ensures the accurate steps are clearly presented.

The target is to reach an accuracy of 95% or better[7] (so, no more than one wrong or missing step out of 20 steps). The following have been observed in the field by PII staff by direct data collection at more than 110 sites/plants around the world, based on walk-down review of several to dozens of procedures at each unit/plant:

- At 95% accuracy or better and when the same procedures follow 80% of the rules for procedure clarity (presented next), then most users will follow the written procedures and will try to keep the procedures up-to-date
- At about 85% accuracy or less, about half of the users stop using the procedures
- At about 75% accuracy or less, less than 10% of the users will refer to the procedure or will try to keep it up-to-date. **So, the written procedures are not very effective.**
- Unfortunately, the typical operating procedure walked down by PII staff (accompanied by senior operators, a process engineer, and a shift supervisor) is about 75% accurate (so one step in four is missing or wrong). Usually these inaccurate procedures also follow less than half of the best practices for procedure format, presented next [7].

Procedure Clarity

The importance of the accuracy of the procedure steps were discussed above. The clarity of the procedure steps (how they are written and how the page is formatted) is also important. Following best practices for step and page format reduces human errors by a multiplying factor of 3 to 5. [5], [7]

The best practice rules for writing and validating procedures have been published for many years *(see Bridges & Williams, 1997; Madden & Bridges, 2016, 2017)* [7],[8],[9]. These rules include best practices for formatting of the pages and steps. These have been gradually improved over the past decades and now are incorporated into PII's training materials.[6]

Table 2. Procedure quality checklist (courtesy PII, copyright 2003-2017)[6]

#	Issue	Response
	Procedure Content Checklist	
1	Is the procedure drafted by a future user of the written procedure? (Engineers should not author procedures to be used by operators or maintenance staff.)	
2	Is the procedure validated by a walk-down in the field by another future user of the procedures?	
3	Is the procedure reviewed and commented on by technical staff (engineers or vendors)?	
4	Is the procedure checked versus the Page and Step format rules below?	
5	Is a hazard review of step-by-step procedures performed to make sure there are sufficient safeguards (IPLs) against the errors that will occur eventually (when a step is skipped or performed wrong)?	
6	Is the content measured using "newly trained operators" to judge the % of steps that are missing, steps that are confusing or wrong, and steps that are out-of-sequence? (A score of 95% accuracy of content is good; 98% should be the targeted average.)	
	Page Format Checklist	
1	Is the title of the procedure the largest item on the page?	
2	Is the procedure title clear and consistent with other titles, and does it uniquely describes the topic?	
3	Are the document control features the smallest items on the page?	
4	Are temporary procedures clearly identified?	
5	Is white space used effectively? • Is there one blank line between each step? • Does indentation help the user keep their place? • Are the margins large enough to reduce page congestion?	
6	Is type size is 12 pt font or larger?	
7	Is mixed case used for words of steps, with ALL CAPS used only for special cases (such as IF, THEN, AUTO, and WARNING)?	
8	Is the step number very simple (single level of number used)? Only an integer?	
9	Have sections or information not necessary to performing the steps been moved to the back or to another part of the manual or training guide?	
10	Are section titles bold or larger than the text font? Do sections have clear endings?	
11	Is the decision on electronic presentation versus hard copy correct? Are electronic linkages to procedures clear and accurate and easy to use? If paper is chosen, is the color of the paper appropriate?	
12	Is the overall page format (such as Outline format or Two-Column [T-Bar] format) appropriate to the use of the procedure?	
13	Are play script features added for tasks that must be coordinated between two or more users? • Play script is normally used when there are two or more hand-offs of responsibility for steps.	
14	Are rules followed for formatting of Warnings, Cautions, and Notes? (See annotated rules, such as Warnings are for worker safety and Warnings must clearly standout from rest of page.)	

Step Writing Checklist		
1	Is each step written as a command?	
2	Is the proper level of detail used throughout? This is judged based on: • *Who* will use the procedures • Same level of detail used in similar procedure steps	
3	On average, is there only one implied action per instruction? Best practice is to average 1.2.	
4	Does the procedure indicate when sequence is important? • If sequence matters, each step should be numbered (with an integer or letter) • If sequence does not matter, bullet lists should be used	
5	Are only common words used? Apply "education" level test (5 grade reading level is best)	
6	Do all acronyms, abbreviations, and jargon aid understanding? • Develop a list of such terms for use in procedures *and* communication. • Use terms that users use (within reason)	
7	Is each step specific enough? No room left to guess/interpret: • The meaning of a word or phrase (Check vs. Make sure) • The intent of a step or series of steps • A desired quantity or value • To what equipment the step applies	
8	Is the procedure free of steps that require in-your-head calculations? • Values expressed as ranges rather than targets with error bands • Conversion tables, worksheets, or graphs provided where needed	
9	Are graphics to the user's advantage? • No explanatory paragraphs or lengthy instructions that could be replaced by a picture • No impressive graphics that provide no real advantage	
10	Are references to the user's advantage? • No lengthy explanations or instructions that could be replaced by branching to a reference • No references to a procedure that references still another • No gaps or overlaps between this procedure and a referenced document • If branching, must branch to a procedure, not to a specific step in a procedure	
11	Are rules followed for writing warnings, cautions, and conditional steps? • No more than 2 per page • No actions within a warning or caution (actions must always be numbered steps) • Warnings and Cautions contain descriptions of potential consequences	

For a complete description of best practices, refer to PII's paper *"Best practices for writing operating procedures and trouble-shooting guides"*. [7]

2. PHAs (Process Hazards Analyses) Need to Close Two Major Gaps

Process Hazards is a broad term. The relationship between many components define process hazards. For a PHA team, Process Safety Information (information on the chemicals, technology and equipment) and operating procedures are necessary to identify specific components of and analyze a hazard scenario. These components include: process deviations (parametric and procedural), their causes, the process safety consequence and various safeguards for detecting deviations and causes, preventing causes, detecting and mitigating consequences and intervening safeguards. A thorough review of incident investigations also offers an opportunity to further analyze process hazards.

2.1. Identified problems

Over the last twenty years the CSB has brought needed attention, through its investigations of industry incidences, to two significant gaps in many current PHAs; where the teams did not identify and analyze:

- Process hazards unique to non-routine modes of operations resulting in catastrophic consequences, and
- Process hazards related to Damage Mechanisms.

Though the intent of the US OSHA PSM regulation has always been for PHA teams to analyze the hazards for all modes of operation, OSHA did not give specific requirements for addressing damage mechanisms and only recently have they increased their enforcement attention on PHA of non-routine modes of operation.

Industry has responded by modifying the *Guidelines for Hazard Evaluation Procedures*, 3rd Edition, 2008, CCPS/AIChE[10] to improve the coverage of these two main issues. API in turn issues guidance on addressing damage mechanisms[11]. Industry best practices now exist for addressing both of these weaknesses. Now, with proper approaches, a team with process safety competencies and process knowledge can thoroughly address the issues during a PHA.

Another less significant gap noted by regulators and by the US CSB has been an incomplete consideration, within the PHA, of previous incidents.

This section of the paper lays out the case from the US CSB and US OSHA on these PHA gaps:

- Poor or no coverage of hazards during non-routine modes of operations
- Poor or no coverage of damage mechanisms

2.2. Objectives

2.2.1. Hazard Evaluation of Non-Routine Modes of Operation

Non Routine modes of operation include start-up, shut-down, online maintenance and other abnormal operations. The hazard evaluation of non-routine modes of operation involves reviewing procedures using a HAZOP, simplified HAZOP, or What-if analysis to uncover potential accident scenarios associated with non-routine operations, for continuous or batch operations. Human error is more likely and more critical during non-routine operations. By analyzing procedural steps where human error is more likely, and where human error or component failure could lead to a consequence of interest, risk can be reduced. The hazard evaluation team's objective is to evaluate the risk associated with skipping steps and performing steps wrong.

The purpose of a hazard evaluation of non-routine modes of operation (governed by written procedures) is to make sure an organization has enough safeguards for the inevitable instance when a step is either performed wrong or skipped (inadvertently or due to shortcutting or other reasons).

Industry has found that a HAZOP or what-if analysis, structured to address procedures, can be used effectively for finding the great majority of accident scenarios that can occur during non-routine modes of operation[10],[12],[13],[14]. Experience shows that reviews of non-routine procedures have revealed many more hazards than merely trying to address these modes of operation during the P&ID driven hazard evaluations.

General guidelines for analyzing non-routine modes of operation or batch processes

- Define the assumptions about the system's initial status. "What is assumed to be the starting conditions when the user of the procedure begins with Step 1?"
- Define the complete design intention for each step. "Is the step actually 3 or 5 actions instead of one action? If so, what are the individual actions to accomplish this task?"

- Don't analyze safeguard steps that start with ensure, check, verify, inspect, etc., or where the consequence of skip is "loss of one level of safeguard/protection against…" There is no reason to analyze these steps since they will show up as safeguards of deviations of other steps. This approach is similar to not analyzing a PSV during a HAZOP of continuous mode (i.e., during a parametric deviation analysis); instead the PSV is shown as a safeguard against loss of containment.
- Together with an operator before the review, identify the sections of the procedures that warrant use of:
  - 7-8 Guide Words (extremely large consequences can happen if deviations occur)
  - 2 Guide Words (the system is complex, mistakes are costly, or several consequences could occur)
  - On others, use What-If (no guide words or guide phrases; for use on simpler or lower hazard systems)
- Decompose each written step into a sequence of actions (verbs)
- Apply guide words directly to the intentions of each action

The following preparation steps may also be needed prior to the procedure PHA:

- Walk through procedure in the plant with one or more operators to see the work situation and verify the accuracy of the written procedure.
- Determine if the procedure follows the best practices for "presentation" of the content; the best practices will limit the probability of human error
- Discuss generic issues related to operating procedures, such as:
  - staffing (normal and temporary)
  - human-machine interface
  - worker training, certification, etc.
  - management of change
  - policy enforcement
- Review other related procedures such as lock out/tag out and hot work.

If the procedures are NOT >90% accurate, then redo procedures before the PHA.

2.2.2. Damage mechanisms

Recently, US CSB and more and more regulators are making analysis of Damage Mechanisms (DMs) a PHA requirement. This requires that potential process damage mechanisms and their potential consequences of interest be identified. Damage mechanisms can be broken down into three main categories. These categories and examples of each are shown in the following table.

Table 3. Types of Damage Mechanisms [15]

Category	Examples
Mechanical	• Mechanical loading failure • Mechanical fatigue • Buckling • Cracking • Embrittlement • Ductile fracture • Brittle fracture
Chemical	• Corrosion - Uniform - Localized - Pitting
Physical	• Thermal failures - Creep - Thermal fatigue - Transformation

The team may use aids such as piping specs, established literature and standards, any applicable MOC documents, etc. to help identify DMs. The PHA leader has several options of how to address damage mechanisms, but PII preferred approach is to discuss during HAZOP under "Loss of Containment" deviation in each node/section.

Damage mechanisms can (and sometimes ***should***) be reviewed prior to the PHA. For example, all MOCs should consider damage mechanisms and their potential consequences of interest before the change is approved. The MOC should also require updates to process safety information which might be relevant to protect against DMs.

Coverage of Damage Mechanism (DM) for each major section [14]

As mentioned above this is PII's preferred method for thoroughly covering all DM within a PHA/HAZOP of an entire unit. The benefits are that the team can more easily catch changes in DM from section to section and they can more easily identify when unique safeguards, such as Remote Isolation Valves, segregated containments (dike), and unique materials of construction are needed. To facilitate this, a list of generic causes of loss of containment and a generic list of safeguards against loss of containment are covered in the Loss of Containment (LOC) deviation of each section.

For each node (each line, each vessel, each column, etc.), the PHA team should discuss and document each damage mechanism listed in Table 3 (as a cause of loss of containment), consequence of the failure if the damage mechanism occurs, and the safeguards in place to prevent the damage mechanism, detect the mechanism before failure, prevent the release, detect and response to the release, and mitigate or contain the release.

The documentation style varies between PHAs. There are two different styles that have been acceptable in the past.

- One style (Example A, Table 4) uses a reference to a summary table of typical causes (this summary table is not shown but is similar to Table 5 of this paper) instead of listing each individual damage mechanism. In Example A, the same approach is used for safeguards. This approach saves redundant text and some time, but it requires the leader and scribe to be diligent to cover everything in the Typical tables that were referenced in the Cause and Safeguard columns. A modified approach is to relegate some generic causes and safeguards to Typical tables for reference and then to carefully list the specific cause and/or safeguards of interest for LOC for each specific node.

Table 4. Example A – Referencing Generic Tables for Typical Causes and Typical Safeguards [14]

Dev	Deviation	Causes	Consequences	Existing Safeguards	Recommendations
9.10	Loss of containment	Accelerate corrosion External fire High pressure (linked from 9.7) Typical causes of loss of containment (see Table A.1)	Release to atmosphere leading to potential injury of workers and/or community	Pressure Relief Valve's 231A, B, C on Reactor and no valves in line to reactor Most lines / connections are welded construction; only a few flanges Drills conducted each year on evacuation, rescue and isolation Emergency Response personnel are trained at SABIC FTC Generic safeguards protecting against or mitigating process material releases (see Table A.2)	Safety 7. Consider changing the ITPM schedule for managing most PSVs whose inspections are too infrequent based on industry standards and best practice. For instance, the current inspection frequency for PSV-8220 on the Ammonia Receiver (I-2005G) is 9 years, whereas, consensus codes typically recommend testing/inspection every 1-4 years for PSVs in highly toxic services.

- Example B (**Table 5**) does not use or reference a generic or Typical list of causes (damage mechanism) or Typical safeguards, but instead develops a specific listing for the LOC deviation of each node. This style has proven easier to justify to regulators and other outside reviewers but takes more documentation effort.

Table 5. Example B – Specific Listing of Causes and Safeguards [14]

Dev	Deviation	Causes	Consequences	Existing Safeguards	Recommendations
1.9	Loss of containment	Corrosion/erosion External fire and/or flame impingement Gasket, packing, or seal failure Improper maintenance Material defect Operator failing to close or inadvertently opening a valve to the atmosphere (e.g., a valve at a hose connection) Railcar inadvertently derailed Valve leaking to the atmosphere High pressure (linked from 1.5) Acid corrosion caused by high concentration of water (linked from 1.8) High ambient temperature External impact (such as from a mini-engine or another railcar)	Catastrophic release of chlorine from a ruptured railcar Steady release of chlorine from a ruptured connection Steady release of chlorine from a leaky connection High pressure caused by thermal expansion of liquid chlorine if railcar is also over-full	Chlorine repair kit Derailer and warning flag to prevent impact by a mini-engine or another railcar Limited vehicular access to area Maintenance/operator response as required, including isolation if needed Operator periodically monitoring the railcar valves while unloading Personal protective equipment in the area Plugs installed in all chlorine valves to the atmosphere when the valves are not in use Relief valve on each railcar for mitigating releases caused by overpressure Supplier maintenance of railcars (per strictly enforced US DOT requirements) Video monitoring of the unloading area Concrete crossties on rail spur Dike preventing any combustibles spilled nearby from reaching the unloading rack area Concrete railroad ties in chlorine unloading area to prevent fires near railcar	10. Consider installing a chlorine detection system in the unloading and vaporizing area to help detect chlorine releases (especially at likely release points) 11. Verify that periodic maintenance and inspections are being performed in accordance with Chlorine Institute recommendations 12. Review the drainage system for the unloading are, and identify the areas that may be affected by a large chlorine release 13. Consider prohibiting the use of heavy equipment (e.g., cranes) in the unloading and vaporizing area unless special precautions to prevent equipment damage are enacted 33. Consider providing a high pressure alarm for each vaporizer 40. Consider providing a water deluge system in the unloading area to help mitigate chlorine releases from the railcar

Regardless of method, the team leader and scribe must ensure the team rigorously discusses all of the damage mechanisms in Table 5 and adequately documents the results of the damage mechanism review in the PHA/HAZOP analysis tables.

For a complete description of PHA best practices, refer to PII's paper *"Recipe for a Complete Process Hazard Analysis"*. [14]

3. Near misses Need to be Reported and Enough of them Investigated

We must learn from accidents and near misses to prevent recurrence. The first step in the learning process is investigation to determine the causes and underlying reasons why accidents and near misses occur. A thorough investigation of root causes will identify the management system weaknesses. Learning which management system weaknesses are leading to near misses and accidents is one of the highest value activities in which a company can invest, and learning from near misses is much cheaper than learning from accidents. Many chemical companies have implemented process safety management systems, and now they are beginning to focus on getting near misses reported and on root cause analysis. This is a very exciting trend. Unfortunately, the chemical industry gets very few near misses reported (the chemical industry is certainly not the only industry with this problem).

To understand more about near misses and getting them reported, it is best to first review the basic definitions.

> *An incident is either an Accident or a Near Miss.*
>
> *An Accident is a sequence of unplanned events and conditions that result in harm to people, environment, process, product or image.*
>
> *A Near Miss is an unplanned sequence of events that could have caused harm if conditions were different or is allowed to progress, but did not in this instance.*

Given a consistent understanding of the definition of a near miss, it is possible to estimate how many near misses should be reported for every accident. Studies in several industries indicate that there are between 50 and 100 near misses for every accident. Also, data indicates that there are perhaps 100 erroneous acts or conditions for every near miss. This gives a total population of roughly 10,000 errors for every accident. **Figure 2** illustrates the relationships between accidents, near misses and non-incidents.

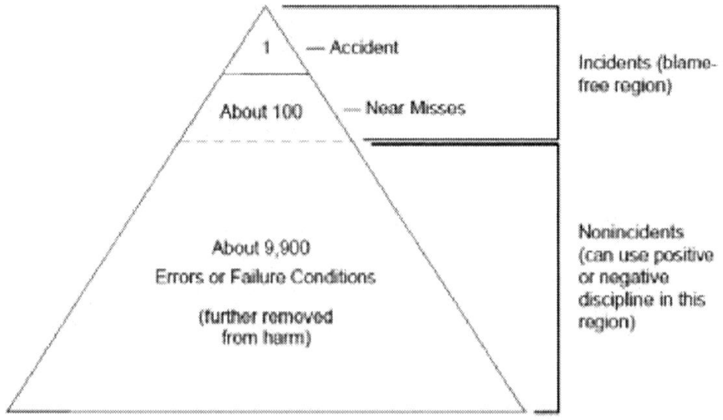

Figure 2. Relationship between Errors and Potential or Actual Impact [16]

Investigating near misses is critical to preventing accidents, because near misses share the causes and root causes of accidents; they are one or two barriers away from the loss/accident. We are very likely preventing many apparently unrelated accidents when we prevent the ones that are obviously related.

3.1. Identified problems

During the past 15 years, we have asked more than 5,000 students of our process safety management (PSM) courses and 3,500 students from our investigator leadership training courses how many near misses they get reported for every accident. The students represented about 400 companies, predominantly in chemical-related process industries (chemical, polymer/plastic, petrochemical, refining, oil and gas exploration, pharmaceutical, pulp and paper, etc.). The answers are quite disturbing. More than 95% said that their ratio of near misses reported to accidents reported falls in the range of 0 to 20. Less than 5% of the individuals said the ratio was greater than 5, and less than 2% said the ratio was higher than 10. Students noted that fear of disciplinary action, lack of management commitment, and lack of understanding of the difference between a near miss and a non-incident were the main reasons why near misses do not get reported. [16]

In conducting about 150 PSM audits and 7000 process hazard analyses (PHAs) during the past 22 years, we have found ratios from 0 to 105. For the first half of the 1990s, more than 90% of the facilities we talked to had ratios in the range of 0 to 0.5, and more than 95% had ratios in the range of 0 to 1. In the last half of the 1990s, more than 90% of the facilities had ratios in the 0 to1 range, and more than 95% had ratios in the 0 to 2 range. In addition, only a few (less than 2% of the facilities) had a ratio higher than 5. By 2005, many facilities had ratios in the 20 to 100 range and each of these had seen great gains (more on this in future articles), but the average across all companies we talked to increased only slightly. By 2011, the ratio of near misses to accidents (any loss event) was about 2 for companies we deal with. But by the end of 2011, many major companies had ratios well above 50! [16]

Our auditors and PHA leaders commented that the primary reasons for lack of reporting were about the same as those found in the survey. The barriers to getting near misses reported, discussed in detail later, are:

- Fear of disciplinary action.
- Fear of teasing by peers (embarrassment).
- Lack of understanding of what constitutes a near miss versus a non-incident.
- Lack of management commitment and lack of follow-through once a near miss is reported.

- An apparently high level of effort is required to report and to investigate near misses compared to low return on this investment.
- There is No Way to investigate the thousands of near misses per month or year!
- Disincentives for reporting near misses (e.g., reporting near misses hurts the department's safety performance).
- Not knowing which accident investigation system to use (or confusing reporting system).
- Company discourages near-miss reporting due to fear of legal liability if these are misused by outsiders.

The good news is that near-miss reporting appears to be improving, and the industry appears to recognize most of the barriers to near-miss reporting. The bad news is that the ratio is still very poor and improvement appears slow!

The number of companies that participated in the informal surveys suggests that the results are a statistically significant representation of the chemical industry.

A formal (written) survey was developed and e-mailed, faxed, and/or mailed to more than 100 companies. Of these, more than 12 replied originally in 1997. Since then, we have reviewed data in detail from more than a dozen major companies. These 25+ companies or affiliates provided data from more than 400 facilities, including more than 150,000 employees in manufacturing and operations. The data were from the chemical industry, polymer industry, refineries, drug/pharmaceutical companies, pulp and paper mills, petrochemical companies, and oil exploration/production.

Table 6 summarizes the results of both the informal and formal surveys.

Table 6. Summary of survey results [16]

Survey Items	**Informal surveys**		**Formal Survey** (or actual organization data)
	Classroom polls	**Found during Audits and PHAs**	
Current Near-miss Reporting Ratio Range (0-5 years prior)	1 to 105	2 to 20	0 to 105
Previous Near-miss Reporting Ratio Range (5-10 years prior)		1 to 5	NA
Current Near-miss Reporting Ratio Average (0-5 years prior)	2-3	3	5
Previous Near-miss Reporting Ratio Average (5-10 years prior)		1	NA
Goal for Near-miss Reporting Ratio (average)	20+	NA	20
Theoretical Upper Limit on Near-miss Reporting Ratio (average)	About 100	NA	100
Number of Participating Companies	400	300+	25+
Number of Participating Facilities	8500+	500+	400+

3.2. Objective

A company should strive to reach a ratio of 50-100 and investigate about 20 near misses per accident. This will provide a statistically significant sample of all incidents (and all important errors) and provide a company with sufficient feedback on which management system weaknesses are causing the errors and component failures. Various companies with different cultures have achieved high ratios with great return on investment.

Table 7 shows a brief description of possible solutions to overcome each one of the barriers identified in the survey.

Table 7. Summary of Survey Results[16]

Barriers	Solutions
Fear of disciplinary action	Implement a policy to NOT punish individuals when their errors lead to accidents and Near Misses.
Fear of teasing by peers (embarrassment)	Ensure that all employees understand the importance of near-miss reporting; demonstrate, through feedback of lessons learned.
Lack of understanding: Near miss vs Non-incident	Develop a list of "in-context" examples that illustrate what you consider to be Near Misses)and what you consider to be non-incident
Lack of management commitment and lack of follow-through on reported near misses	Hold management accountable for achieving a Near-Miss reporting ratio
Apparently high level of effort is required to report/investigate Near Misses	Ensure that the data are entered in a database and queried regularly. Share the results with employees so they can see the value of the reported near misses
There is no way to investigate the thousands of Near Misses per month	Let front-line foremen or supervisors decide if a Near Miss or accident needs to be investigated
Disincentives for reporting Near Misses	Ensure that goals and incentives are not tied to lower incident rates (since this discourages reporting), but instead provide incentives for high Near-Miss reporting ratios
Not knowing which accident investigation system to use	Have ONE incident reporting system with ONE approach
Company discourages Near Miss reporting due to fear of legal liability	Involve legal on major Near Misses and accidents to ensure the results are protected as much as possible under attorney/client privilege.

For a complete description of the possible solutions for each barrier, refer to PII's paper *"Gains from Getting Near Misses Reported"*.[16]

4. Key Human Factors are Not being addressed

All accidents (or nearly all, if you consider that there are some natural phenomena that we either cannot guard against or choose not to guard against) **result from human error**. This is because humans govern and accomplish all of the activities necessary to control the risk of accidents. Humans influence other humans in the process – not only do humans cause accidents (unintentionally) by making errors directly related to the process itself, but they also cause errors by creating deficiencies in the design and the implementation of management systems (i.e., we make errors in authorities, accountabilities, procedures, feedback, proof documents and continual improvement provisions). Ultimately, these management systems govern the human error rate by directly causing or indirectly influencing the process. The process-related activities where errors have the most influence include:

- Designing a process
- Engineering a process
- Specifying components
- Receiving and installing equipment
- Commissioning
- Operating a process
- Maintaining, inspecting and repairing a process
- Troubleshooting and shutting down the process
- Managing process, procedure, materials, facility and personnel changes

The 10 human factors categories to be controlled by companies are:

- Available Time (includes staffing Issues) – for responses only
- Stress/Stressors (includes staffing issues)
- Complexity & task design
- Experience/Training
- Procedures
- Human-Machine Interface (includes tools)
- Fitness for Duty
- Work processes & supervision
- Work Environment
- Communication

4.1. Identified problems

Recent major accidents have highlighted the need for increased focus on human factors. The U.S. Chemical Safety Board (CSB) cited (US CSB Video, 2006) human factor deficiencies as one of the main contributors of the catastrophic accident at the BP Texas City Refinery in March 2005[17]. The human factor deficiencies included lack of control of worker fatigue, poor human-system-interface design, poor communication by radio/phone, out-of-date and inaccurate operating procedures, and poor (no) communication between workers at shift handover. The CSB cited similar issues from many other accidents[18] and has urged industry and the U.S. OSHA (the regulator) to pay much more attention to human factors. As a result, the recent U.S. OSHA National Emphasis Program for Refineries included human factors as one of the 12 core elements.

Implementing human factor engineering and policies to prevent accidents is not a new concept. Nearly all (or all, from a more complete perspective) of the causes and root causes of major accidents in the past

30 years have been the result of poor control of human factors. This has been cited in many root causes analysis reports and papers concerning these major accidents.

Process Safety Management (PSM) systems based on OSHA's PSM standard are likely lacking the fundamental human factor elements and implementation guides that, if applied across the applicable PSM elements, would work together to reduce human error. The *Risk Based Process Safety* guideline from the CCPS/AIChE (2007) does contain these human factor requirements.

Many site and company staff do not know there are standards for the control of human factors. However, at last count there were more than 300 non-governmental standards (such as from ANSI, ISO, IEEE, etc.) for the control of human factors, and there are more than 100 government regulations and standards and guidelines. Some of the better government regulations cover aviation, marine operations (shipping) and aerospace. An organization should devote resources to finding and implementing such best practices.

The following is a description of each of the human factors mentioned above and specific ways to control each. Our data from review of more than 2000 incidents reveal that:

- 90% of accidents had at least one root cause related to deficiencies in written work instructions.
- 70% of accidents had at least one root cause related to miscommunication between workers or between workers and supervisors.
- 40% of accidents had at least one root cause related to excessive fatigue of the worker. [5]

4.2. Objective

Not all errors can be prevented. Since the beginning of time, humans have tried to control error rate with more or less success. Human errors have been measured for hundreds of years. Psychologists have studied why humans make mistakes and have gradually put a science around human error probability. Today, the best models for control of human error in the workplace are generally agreed to be related to control of human factors; in turn, these have grown out of what was previously denoted performance shaping factors. What can be done to control each Human factor is discussed in greater detail next.

For a complete description of each human factor, its importance and how to control them; refer to PII's paper *"Human factors and their optimization"* [5].

Time available for the task

Task design is directly linked with this category. Process design, procedure development and worker scheduling need to take into consideration not only the time it takes to complete each task but also the time it takes to manage usual as well as potentially unusual or unexpected situations. True; we do not create norms for exceptions, but exceptions sometimes are more usual than realized.

Stress/Stressors

A well-designed, organized and managed work process helps maintain and promote individual health and well-being, but stresses will arise through no fault of the organization. Supervisors should be trained to recognize stress in workers and, in severe cases, reassign them and help them get assistance to deal with the stress and in other cases, reduce their job pressures for a day by shifting assignments. A wellness program may also be necessary to help cope with stress. Simple approaches, such as training workers to recognize and deal with stress, can help most individuals. Approaches to deal with stress include positive self-talk and practicing relaxation techniques (i.e., breathing control).

Complexity & task design

If roles and responsibilities for accomplishing tasks are not clearly defined, then there will be a risk of serious errors. It is also important that employees and engineers/designers understand the characteristics of the work elements involved, how each element passes information to the other, how each person involved communicates, and that each person has guidelines to adhere to during task design and procedure development.

For optimal performance, task designers often must integrate human and automated equipment in their task analysis.

It is crucial for task designers also to have an understanding of the appropriate allocation of roles and responsibilities to the various participants in a task. Understanding the characteristics and limitations of all of the humans and automated systems involved in the task (particularly critical tasks) will allow for establishing additional controls where needed.

The task analysis process includes *Task definition* which is the initial operational description of the operator tasks required for execution of a given system function; *Performance analysis* of all tasks with respect to human performance requirements; *Error and workload analysis* to be included in the design process, particularly when the likelihood of unacceptably operator error is high; and *Evaluation* since evaluating the design of a new or complex task is important to ensure that the task can be effectively performed as designed. For very high consequence scenarios involved complex human interaction, it is likely prudent to invest in a human reliability analysis (HRA).

Experience/training

Training to master the Knowledge and Skills and Abilities (KSAs) required for a job may be obtained from a variety of sources. Companies provide the training or it is obtained from other sources (e.g., trade schools, universities, contractor organizations, etc.) to reach the skills and knowledge established in job descriptions. There are several factors that may result in personnel not mastering the required KSAs for a job. These include course design and delivery methods, course completion, practical skills demonstrations or simulation, and training frequency. Course design begins when the learning objectives have been identified. The design process consists of determining the delivery methods (simulator, on-the-job training, etc.), number of hours required to cover the materials, instructor qualifications, etc. Although some methods, materials and instructors may be more effective or efficient than others, the important issue is that the course content is complete and addresses all of the relevant KSAs, so that the learning objectives are met and the KSAs mastered.

Another determinant of KSA mastery is course completion. Although this factor appears obvious, there are often competing demands on personnel that may pull them out of training at times. As a result, they may miss the instruction related to specific KSAs. Testing may not identify the KSA deficiency because it is impossible to test mastery of all KSAs. Sampling techniques are used to generate examinations. If attendance and participation are not controlled, some personnel may miss training on specific KSAs and testing may not identify the deficiencies before an error is committed.

Another factor affecting KSA mastery is forgetting. An individual's ability to perform a task will degrade over time unless the relevant KSAs are refreshed. Proficiency training (i.e., refresher) will be required for some tasks to maintain the level of mastery that was demonstrated following initial training. One function of training programs is to identify those tasks that require proficiency training.

If certain tasks are performed frequently, proficiency training may be unnecessary. By performing a task, personnel practice the task and obtain feedback on where they have weaknesses. Task performance refreshes the KSAs and successful task performance verifies that proficiency has been maintained. Furthermore, since humans learn by comparison to similar activities, as a worker practices one task, they are learning about similar tasks. Care must be taken here to avoid the practice drifting away from what the procedure requires. Also, if new operators are trained by experienced operators, you must ensure that the new employee is trained according to the procedure, not whatever practice is actually being used. (Of course, ideally, the practice and procedure will be identical.)

Operating and maintenance procedures

Many procedures do not follow best practices for controlling human error, and so the written procedure actually "contributes" to increased error rates. Additionally, many organizations do not have guides on how to troubleshoot (what to do when process deviations occur). The best practice rules for writing and validating procedures have been published for many years. [6], [7], [8], [9]

Procedures have not traditionally been developed from the perspective of optimizing human factors; instead, procedures have been traditionally developed to meet a compliance requirement to have written procedures. Examples of procedure deficiencies (inaccuracies) include:

- Incorrect/incomplete/nonexistent (most procedures we have audited have been only 70-85% accurate – the inaccuracies include missing critical steps, steps as written are not what needs to be done, or the steps are out of sequence)
- No/misplaced warnings or misuse of warnings (for example, a warning should never **state** the action to take; it should instead **emphasize** the action to take an the statement of the action should be in a numbered step)
- Poor format and presentation rules

Human-machine interface

Design for operability refers to designing the Human-System Interface (HSI) to be consistent with the abilities and limitations of the personnel who will be operating it. Weaknesses in the design processes can result in an HSI that is not well suited to the tasks that personnel must perform to ensure plant safety, resulting in increased workload, decreased performance by personnel, and an increased likelihood of errors.

Design for maintainability refers to designing the HSI and associated plant equipment to ensure that personnel are able to perform necessary maintenance activities efficiently. Weaknesses in the design process can result in systems that impose excessive demands on personnel for maintenance and, therefore, are prone to maintenance errors or problems with reliability and availability. *If it is hard to reach, workers will make more errors, including the error of deciding it is not worth it.*

Design for flexibility refers to the way that changes, such as upgrades to the HSI, are planned and put into use. A new HSI component may require the user to perform functions and tasks in new ways. Skills that the user developed for managing workload when using the former design, such as ways for scanning information or executing control actions may no longer be compatible with the new design. The new HSIs must also be compatible with the remaining HSIs so that operators can use them together with limited possibilities for human error. Also, HSI modifications may not be installed or put into service all at one time, causing the user to adapt to temporary configurations that are different from both the original and final configurations. Weaknesses in HSI implementation can increase operator workload and the likelihood of errors.

Fitness for duty

Implement a company fitness-for-duty program with a primary responsibility for detecting and preventing impaired personnel from performing tasks that may affect productivity and safety. Medical evaluations of personnel, behavioral observation programs, employee assistance programs, and drug and alcohol testing are used to detect impairment. Weaknesses in this program may allow impaired personnel to have access to vital areas in a plant where they could commit errors.

Overtime Policies and Practices – Most companies establish limits for work hours to reduce on-the-job fatigue and the potential consequences for poor task performance. Routine authorization for work hours in excess of those recommended may result in fatigued workers. Furthermore, a practice of excluding training or meetings that occur outside of an individual's normal work schedule from work-hour limitations will also contribute to fatigue. But in many cases these have weak enforcement, especially during shutdowns when the company is under stress to re-start soonest.

Shift Scheduling – Shift scheduling may also affect the likelihood that personnel will show performance decrements due to fatigue. A change in the assigned shift or a rotating shift schedule will disrupt circadian rhythms and may increase the likelihood of errors. So, a company must choose the proper shift rotation to allow adjustments to sleep patterns.

Work Processes & Supervision

Written plant policies and procedures are meaningful only when they are enforced; otherwise, worker practices are the policy. Once discrepancies are tolerated, individual workers have to use their own judgment to decide what tasks are necessary and/or acceptable. Supervisors and managers should vigorously enforce plant policies/procedures and ensure that plant policies/procedures and practices are revised as necessary to be consistent. To enforce well, a supervisor must (1) lead by example, (2) watch workers often to ensure they are following company policies and procedures, and (3) determine the correct course of action (including recommending changes to policy and procedures) when workers are found deviating from procedures.

Work environment

Programmatic causes of task environment errors are typically found in the company's processes for designing human-system interfaces or in managing maintenance activities. Other programs may also be implicated. Common programmatic causes of task environment errors include:

Industrial Hygiene and Radiation Protection – These programs are responsible for ensuring that task environments have been evaluated to identify hazards and that needed controls are implemented to minimize exposures. Weaknesses in these programs may result in personnel working in task environments that are conducive to errors.

Work Planning and Control – Weaknesses in the work planning and control system may allow work to be planned without consideration of adverse environmental conditions and performed without the necessary compensatory measures. For example, communication devices may not be provided in noisy environments to support task performance. For tasks that involve unusual physical positions or cramped workspace, additional time to complete the task may not be scheduled. Rest breaks for hot and cold environments may not be planned into the work, or additional temporary lighting may not be provided if the work site is not adequately lighted.

Procedures – Weaknesses in the company's procedure development process may result in the design of procedures that are inappropriate for the conditions in which they will be used. For example, procedures that may be used at night, outside and in the rain should be laminated and the type size should be larger to ensure the procedure can be read. Procedures that will be used in vibration conditions may also require larger type size than procedures read in the stationary environment of the control room.

Human Factors Engineering – Weaknesses in the human factors engineering program may result in the installation of new equipment or systems without consideration of task environment characteristics. For example, the impact of control room lighting on the visibility of digital displays or effects of vibration on the legibility of dials or gauges at local control stations should be considered before installation.

Communication

Successful communication requires several steps. The sender first develops the intention to communicate either verbally or in writing. The sender then composes a message that presents the meaning as clearly as possible. The receiver must pay attention to the message and then interpret its meaning. If the communication is successful, the receiver interprets the message consistently with the sender's intended meaning. Table 8 summarizes the rules for successful verbal communication.

The similarity of the meanings given to the message by the sender and receiver can be verified through feedback. An example of feedback verification in verbal communication is when the receiver "repeats back" the message and the sender either agrees with the receiver's repeat back or corrects it. Verification feedback serves an important error-checking function in the communication process. It also allows supervisory oversight of communications to catch errors before they have consequences.

A sender and receiver must both be active for communication to be effective. The sender and receiver share responsibility for ensuring successful communication. However, when companies analyze the causes of events, errors in sending messages are more often identified than errors in receiving. The reasons for the difference are unclear. A company's investigation should consider sending and receiving errors and corrective actions should address both to be effective.

Shift handover is another critical aspect of communication. The status of equipment, repairs that occurred, problems faced, upcoming chores, etc., are all key information the oncoming worker will need to know. One good guide on shift turnover is U.S. DOE 1038-93. [19]

Table 8. Rules for effective verbal (face-to-face, radio, phone) Communication (© PII, 2012)

#	Communication rules
1	Senders MUST require repeat back by the receiver; ask questions to make sure of understanding.
2	Receivers of messages MUST repeat a message they receive.
3	Communicate one task at a time.
4	Sender & receiver must use the standard name and component number (and building name and number) and/or for related reference material (drawing, procedure name/number) in each message.
5	Sender must require receiver to give feedback of foreseeable conflicts. Sender (if message is to subordinate) should follow-up in the field if possible, to make sure message was understood.
6	Receiver must request clarification anytime they believe the message is confusing.
7	Do NOT give verbal instructions to workers who do not have the demonstrated skills to correctly understand and perform the task assignment (unless you plan to supervise/coach the task yourself).
8	Communicate to the most senior member on the work crew.
9	Know your audience & change message accordingly to compensate for lack of knowledge (when appropriate; see other related rules).
10	Use units (and use approved units, such as SI) for process parameters.
11	Always count from left to right when giving instructions related to multiple choices of equipment.
12	Use approved jargon only.
13	Use the words **check**, **make sure**, and **actions** consistently (see procedure writing rules).
14	When possible, write out the task (use a special procedure or work order) rather than communicating verbally. Use formalized templates for consistency and make sure the receiver is provided all information (such as work orders and drawings) necessary for the task.
15	Perform pre-job meeting with work crew (including with contractors or construction company).
16	Perform walkthrough at the location where the work will be completed.
17	Have a backup communication method. Don't rely on one mode of communication. In emergency situations have hand signals as backup to loss of verbal communication.
18	Talk to the operator or technician who is doing the work (no delegation of work); but see Rule 8.
19	Receiver (worker or group) must report back when work is complete.
20	If confusion exists in implementing a task, the shift supervisor (and higher, if supervisor is unavailable) must be contacted to make sure he/she understands the problem.
21	Do NOT use a PA (public address system) for process instructions, since this method of communication does not allow repeat-back.
22	Use Open questions and non-confrontational questioning methods when requesting clarification.
23	When communicating remotely (by phone or radio), if the message is not understood on the second attempt at clarification the Sender must find the Receiver and communicate face-to-face.
24	For large or complicated jobs, in addition to repeat back, the sender must ask the receiver for an assessment of the pre-job briefing to ensure the workers are not confused.
25	Workers and supervisors must keep a shift log to aid in turnover between shifts. Workers and supervisors and support staff (if necessary) must have a minimum or 15 minutes overlap with a relieving shift. *See shift turnover standard for more details.*

Conclusion

Closing the gaps in these 4 areas has a greater than 100:1 payback!

Acronyms and Abbreviations

ACC - American Chemistry Council
AIChE – American Institute of Chemical Engineers
API – American Petroleum Institute
CCPS – Center for Chemical Process Safety (of AIChE)
CMA - Chemical Manufacturers Association
CSB – Chemical Safety Board
DM – Damage Mechanisms
FMEA – Failure Mode and Effects Analysis
GCPS -- Global Congress of Process safety
HAZOP – Hazard and Operability; as in HAZOP Analysis or HAZOP Study
HRA – Human Reliability Analysis
HSI – Human-System Interface
KSA – Knowledge, skills and abilities
LOC – Loss of containment
LOPA – Layer of Protection Analysis
MOC – Management of Change
OSHA - US Occupational and Health Administration
PHA – Process Hazard Analysis
PSM – Process Safety Management
RBPS – Risk Based Process Safety
SOP – Standard Operating Procedure

References

[1] *Guidelines for Technical Management of Chemical Process Safety,* CCPS/AIChE, 1985.

[2] "The Cost and Benefits of Process Safety Management: Industry Survey Results," W. Bridges, *Process Safety Progress*, AIChE, January 1994.

[3] "Necessity of Performing Hazard Evaluations (PHAs) of Non-normal Modes of Operation (Startup, Shutdown, & Online Maintenance)", W. Bridges and Mike Marshall (US OSHA), *18th Annual International Symposium, Mary Kay-O-Connor Process Safety Center, College Station, TX, October 2015.*

[4] *Guidelines for Risk Based Process Safety*, CCPS/AIChE, 2007.

[5] "Human factors and their optimization" W. Bridges and G. Collazo-Ramos, *8th Global Congress on Process Safety*, Houston, AIChE, April 2012.

[6] PII Course 5, *Writing effective operating and maintenance procedures*, © 2003-current

[7] "Best practices for writing operating procedures and trouble-shooting guides", W. Bridges and R. Tew, *13th Global Congress on Process Safety*, San Antonio, AIChE, March 2017.

[8] "Effective Procedure Writing – A Two-Part Series (Part 1: Reducing Risk of Worker Errors; Part 2: Addressing Process Safety Management Requirements for Operating Procedures)," *Chemical Engineering Progress*, 1997, W. Bridges with T. R. Williams (both of JBFA).

[9] "Best Practices for Writing Operating Procedures and Trouble-Shooting Guides", L. Madden and W. Bridges, *12th Global Congress on Process Safety*, Houston, TX, AIChE, March 2016.

[10] *Guidelines for Hazard Evaluation Procedures*, 3rd Edition, 2008, CCPS/AIChE.

[11] "API RP 571 - Damage Mechanisms Affecting Fixed Equipment in the Refining Industry", API, 2nd edition, 2011.

[12] Bridges, W.G., et. al., "Addressing Human Error During Process Hazard Analyses," *Chemical Engineering Progress*, May 1994.

[13] "Necessity of Performing Hazard Evaluations (PHAs) of Non-normal Modes of Operation (Startup, Shutdown, & Online Maintenance)", W. Bridges and Mike Marshall (US OSHA), *18th Annual International Symposium, Mary Kay-O-Connor Process Safety Center, College Station, TX, October 2015 and 12th Global Congress on Process Safety, April, 2016.*

[14] "Recipe for a Complete Process Hazard Analysis – Especially Addressing the Key Demands from US CSB", R. Tew and W. Bridges, *13th Global Congress on Process Safety*, San Antonio, AIChE, March 2017.

[15] PII Course 8, *PHA/HAZOP Leadership training*, © 2003-current.

[16] "Gains from Getting Near Misses Reported", W. Bridges, *8th Global Congress on Process Safety*, Houston, AIChE, April 2012.

[17] US Chemical Safety and Hazard Investigation Board (CSB), *Anatomy of a Disaster: Explosion at BP Texas City Refinery,* published 2006.

[18] US Chemical Safety and Hazard Investigation Board (CSB), *"Pesticide Chemical Runaway Reaction Pressure Vessel Explosion – Bayer CropScience, LP"*, published 2008.

[19] "DOE-STD-1038: Guide to Good Practices for Operations Turnover", US DOE, 1998.

510485 Functional Safety Lifecycle Software - Why Industry Needs Them and the Benefits They Bring

Abstract:

2018 Global Congress on Process Safety

Functional Safety Lifecycle Software â€“ Why industry needs them and the benefits they bring

David Hansen, CFSE

Global Functional Safety Expert

Mangan Software Solutions

Process industry functional safety concepts have been applied for decades now. Though these concepts are elegant and logical, process industries continue to struggle to apply them effectively and efficiently. Organizations have had disappointing results in attempting to provide those most responsible for risk access to understandable process risk information, breaking down organizational silos, assuring resources are allocated where the risks are greatest, ensuring functional safety protection layer design quality and consistency, managing changes, managing the large amounts of data involved in process safety management and feeding real world performance back into the risk analysis and design.

This presentation will discuss common functional safety and process safety barriers to success with real world examples and how functional safety lifecycle software can be used to overcome those barriers.

David Hansen biography

David Hansen is a Global Functional Safety Expert with Mangan Software Solutions. David has over 30 years of process industry experience in both operations and projects with a focus on functional safety. David's experience includes functional safety planning and leadership for some of the world's largest energy projects. He has also spent many years supporting automation and functional safety related operations and maintenance activities.

510852 State of the Art Development of Emergency Depressurization System for High CO2 Cryogenic System

Abstract:

Over the last few years, several studies have been conducted on cryogenic technologies such as cryogenic distillation and supersonic nozzle in CO2 separation for fields containing more than 40% of CO2. Based on the studies, these new cryogenic technologies have shown to have high potential in separating CO2 from natural gas offshore to be utilized under carbon, capture, storage and utilization (CCUS) project.

The new cryogenic technologies are currently being tested for the proof of concept. Hence, a pilot plant, which is a scaled down version of the technology was developed. One of the major challenges faced during the pilot plant testing is the emergency depressurization philosophy as the process involves CO2 solids handling which is uncommon to the industry standard. Depressurization of high CO2 fluid at cryogenic temperature would lead to possibility of CO2 solid formation, hence potential blockage of process equipment and venting/flaring line.

Due to the thermodynamic properties of CO2, there is a risk associated with CO2 solid formation that can lead to blockages in the system. This paper focuses on the experimental approach to understand the phenomena of CO2 solid formation under expansion, which includes the PVT measurement conducted for mixtures containing CO2 and hydrocarbons under low temperature to predict the phase envelope, interfacial tension measurement (IFT) and nucleation test to predict the solid CO2 formation delay as well as the agglomeration of the solid CO2 in the system. These experimental studies have given a dynamic insight in understanding the scientific phenomena further which can be used to optimize the design of the blowdown system. The experimental data is used to validate the equation of state in modelling the blowdown simulation for CO2 solid prediction. The optimized design of blowdown system will be used for the pilot plant testing of the CO2 separation technology and its future offshore installations.

511219 Safety Cases for Aerial Drone Work in Industrial Plants

Abstract:

New technology may also be disruptive technology; aerial drone technology fits this model. As such it is incumbent upon every person involved in the use of aerial drones in the industrial workplace to apply this technology in as safe a manner as possible, while developing and applying the standards of practice in the workplace. A Safety case is arguably the best means at hand to assess and identify risks and safeguards prior to use of this disruptive technology.

> "A **Safety Case** is a structured argument, supported by evidence, intended to justify that a system is acceptably safe for a specific application in a specific operating environment." 1

This presentation presents three safety case scenarios, and comments on unique aspects identified during the developments of the safety cases. Arguments for the applicability of the safety case and for the use of bowtie methodolgy for hazard assessment within the safety case are presented. The three scenarios include: equipment inspection at altitude, aerial documentation in incident investigation, and response monitoring during emergency response.

1 *Defence Standard 00-56 Issue 4 (Part 1): Safety Management Requirements for Defence Systems*. UK Ministry of Defence. p. 17.

511350 Faster and More Accurate Responses for Chemical Accidents Using Machine Learning Technique

Abstract:

Today, chemical accidents are great concern of worldwide owing to recent chemical disasters. In order to minimize loss of life, it is crucial to develop a rapid and correct evacuation order when chemical accidents such as toxic gas releases occur near populated areas.

In this paper, a method and results are presented to perform the prediction of emergency evacuation orders by utilizing reliable and good-quality data from chemical incidents. The Hazardous Substances Emergency Events Surveillance (HSEES) system has collected and analyzed information about acute releases of hazardous substances that result in a public health action, such as an evacuation in the U.S. Data were analyzed using the Naïve Bayes classifier(NBC) and Back propagation algorithm. This study was performed with the aid of Rapidminer 7.5 software, a big-data analysis tool. The analysis included over 61,563 incidents cases among which approximately 10% are the cases of actual evacuation orders. The study was also carried out to verify overall percentage agreement, sensitivity, specificity and Area Under Curve(AUC) values. As results, it is figured out that effectiveness using back propagation algorithm is better than that using NBC. Furthermore victimsâ€™ info have been used for evaluating algorithms. The results from this study would help future evacuation orders in case of chemical disasters for better accuracy.

511226 Post Graduation in Process Safety Management Based on Rbps PSM System

Abstract:

RSE Consulting has been implementing the Post-Graduation in Process Safety Management for the past two years in Brazil, and there are three classes in progress. The course is the only one in the country registered in the Brazilian Education Ministry, and it aims to train professionals to implement and maintain a strong safety culture, in a 524 class-hoursâ€™ workload. Process Safety is one of the mainly growing areas in the world, increasing the demand for trained professionals. This specialization could enable Brazilian Industry to prevent major accidents and save lives. The course has a professional team of Teachers and Coordination. There are more than 10 professors with outstanding performance and recognition in the themes related to the modules.

The course is available in two modalities: traditional and transmitted classes as webinars, providing a complete and aligned training with the state of the art in the area. The whole course is composed of 18 modules, and it has as technical reference the four pillars and twenty elements of Risk Based Process Safety management system proposed by the Center for Chemical Process Safety (CCPS). The modules are: . This program also outlines the importance of Human Reliability and Performance Influencing Factors for improving Process Safety. Students are encouraged to identify how project and process must be managed at different stages of installation life cycle. Additionally, students have classes about entrepreneurship on process safety, where they can learn Strategies for Business Sustainability; Dynamic Model for Assessing the Conditions of Process Safety Barriers; Development of Research and Study Centers; and others. Thereby, it is possible to say that the course provides a complete and aligned training with the most up-to-date techniques in the area.

Only two years after implementation, the post-graduation course has 64 enrolled students, covering 17 different graduation courses, mainly Engineers. Another relevant fact is that 43 companies are involved in the specialization of these professionals, a variety of 16 industrial sectors. The main sectors affected in improving process safety practices in Brazil through the post-graduation course are Chemistry and Petrochemical, which represents 42% of the studentâ€™s workplace. Of the total companies already mentioned, 17% are members of the CCPS, multinational companies of great importance in the global scenario of Chemical Process Safety, such as DOW, Monsanto and Cristal.

The First Post-Graduate Group is committed with Process Safety. They already contributed for improving and enhancing Process Safety Knowledge through the development of a Tridimensional Training Matrix on Process Safety Applied for Chemical and Petrochemical companies. This study was published on the 13th CCPS Global Congress. Therefore, this postgraduate course aims to train professionals for the implementation and continuous improvement of the Process Safety Culture and Risk Management.

513144 Application of Bayesian Networks in Emergency Response Analysis of Typical Leakage Accidents

Abstract:

With the agglomeration development of chemical industry, chemical industrial park has become the mainstream direction for the development of chemical industry in the future, and plays an irreplaceable role in the global economy development. Due to the relative centralization of enterprise layout in the chemical industry park, the occurrence of accidents can lead to catastrophic domino effect. In the control and management of accidents, accident prevention is important, but rapid and effective emergency response of the chemical industry park after accidents also plays an important role in reducing accident losses and controlling the expansion of accidents. The completeness of emergency resources in chemical industry park is one of the important influencing factors of emergency response speed and quality. In recent years, Bayesian network has been widely used in quantitative risk assessment and accident consequence analysis. Although the method has been applied in emergency response analysis, it has not been used to analyze the impact of emergency response on accident development in different accident stages.

The completeness degree of the equipment of the main emergency materials and emergency personnel, which are important components of emergency resources, are selected as the indicators of emergency response capacity. In order to analyze the impact of different emergency resources on the accident consequence in emergencies, development trend of emergencies under different situations of emergency response was studied based on Bayesian network. The Bayesian network model from the beginning to the consequences of the accident was established for a typical leak accident selected. The nodes of emergency response were added at different stages of the accident in the model on the basis of the analysis of influence factors of emergency response, such as the preparation of emergency materials, the consummation degree of emergency organization and etc. The effects of different emergency response processes on accident consequence were analyzed when emergency material and emergency personnel are variously equipped respectively or simultaneously at different stages.

The results can provide a quantitative analysis basis for the establishment of the index system of emergency resources in the chemical industry park, and realize the scientific equipment of emergency resources. It can also guide the emergency command to recombine the most necessary emergency materials and make reasonable personnel arrangements according to the characteristics of the accident, and make quick decisions and command in order to effectively respond to emergencies.

513601 October 1, 2017 â€“ Paradigm Shift in PSM/Calarp Requirements for California Refineries â€“ What Are the Requirements and What Does Management Need to Know?

Abstract:

Paper 479475, presented at the 2017 Global Congress on Process Safety, provided a snapshot of the significant changes being made to the PSM/RMP Regulatory Framework in California and the beginnings of a rescheduling and reprioritization of federal PSM/RMP initiatives. Post-publication, State-PSM and State-RMP (CalARP) requirements applicable to California refineries were finalized and made effective on October 1, 2017. Cal/OSHA's requirements are driven by a separate PSM regulation focused on petroleum refineries, and CalEPA/CalOES requirements include a new program level (Program 4), initially applying primarily to petroleum refineries. These two California requirements are very similar, and minor differences will be identified in the paper. Whereas the pendulum has swung towards increased PSM/RMP regulatory requirements for California refineries, the Federal OSHA and U.S. EPA updates to Process Safety Management (PSM) and Risk Management Plan (RMP) requirements as part of the PSM/RMP Modernization/Expansion Programs have been rescheduled and reprioritized.

The changes associated with the Cal/OSHA and CalEPA/CalOES programs are extensive and are not entirely synchronized between the agencies. The focus of this paper will be to provide relevant background information, a summary of the changes, and tips for addressing key technical and regulatory challenges, as well as providing the reader with a "roadmap" for effective implementation. The timing of GCPS-2018 and the update provided in this proposed paper is important for:

- Unveiling the Final Regulatory Requirements in California

- Identifying Strategies for Addressing these Requirements with Minor Differences

- Providing the Entire GCPS Audience with an Understanding of Why These Changes, Initially Focused on California Refineries Could Become Important for All U.S. Refineries

- Providing an Update of U.S. EPA and Federal OSHA Initiatives

The contents of this paper will be important to communicate to GCPS attendees in the April 2018 time frame, and key topics include:

- Background – Pre-2012 PSM/RMP Universe

- Brief Overview and Key Elements within the PSM/RMP Program Modernization Initiatives

- Initiative Status and Potential Impact on Plant PSM/RMP Programs

- Compliance with Different Scheduler Requirements

- Potential Impact on Smaller Refineries

- Strategies for Effective Implementation

As part of the paper and the April 2018 presentation, select tips to streamline/focus the application of some of the more challenging new elements, as well as a comprehensive Summary Matrix of the more complex scheduler requirements for the new California regulations, will also be provided.

511198 Unrevealed Failures of Safeguards Used in Process Hazard Analysis - Improving and Understanding Functionality

Abstract:

Safeguards credited in Process Hazard Analysis (PHA) can have enhanced performance, when their potential for unrevealed failures are reviewed revealed in the PHA sessions. The safeguard performance improvements can include additional monitoring of its "state of operability" prior to when called upon to function and by the predicted demand frequency for appropriate maintenance intervals.

Unrevealed failures of safeguards are a common "root cause" in process safety incident investigations. They are defined as systems or equipment that can remain in a failed state of operation without the knowledge of operations. The distinction between "double jeopardy" and unrevealed failures are presented with examples and definitions to improve understanding of these issues when conducting PHAs.

Safeguards that may have unrevealed failures include switches to trip a piece of equipment's operation or alarm, check valves, and basic process control systems. The severity of consequences (hazards) associated with credited safeguards are factors for determining the interval of maintenance and the appropriate use of tools (additional devices). Additionally, the demand to function and equipment age (run to failure) issues should be included in the analysis of unrevealed failures.

The identification of these types of deficiencies (safeguards with unrevealed failures) in PHAs and operation systems can improve overall system performance in reducing and/or controlling process safety risks.

508393 Controlling Static Discharge Using Antistatic Additives

Abstract:

Controlling Static Discharge using Antistatic Additives.

Abstract

The presentation will discuss controlling static discharge in non-fuel, non-conductive/low-flash point fluids. The scope of the presentation will concentrate on industrial producers, distributors, consumers of these fluids. Furthermore, the presentation will concentrate on the coatings industry where the largest proportions of these fluids are used. Common Static-Accumulating Flammable Liquids That May Form Ignitable Vapor-Air Mixtures:

- VM&P naphtha
- Cyclohexane
- n-Heptane
- Benzene
- Toluene
- n-Hexane
- Xylene
- Ethyl benzene
- Styrene

Static discharge has caused catastrophic explosions and fires (U.S. Chemical Safety Board, Static Spark Ignites Explosion Inside Flammable Liquid Storage Tank No. 2007-06-I-KS).

Issues

- Nonconductive flammable liquids can accumulate static electricity during transfer and storage.
- Static sparks can readily ignite flammable vapor-air mixtures inside storage tanks.
- Material Safety Data Sheets (MSDSs) often do not adequately communicate hazard data and precautions.

Key findings:

- Safety publications omit antistat options. (see Process Safety Beacon, August, 2016)
- US affiliates of multinational corporations have different guidelines for antistatic use than their parent companies.
    - NFPA®77 and the American Coatings Association's, *Generation and Control of Static Electricity in Coatings Operation* has guidelines that are different than EU regulations ATEX- Appareils destinés à être utilisés en ATmosphères EXplosibles and ESIG's *Flamability: Safe Working With Industrial Solvents.*
    - The presentation will highlight these differences and how they contribute to the disparity in antistatic additive applications.
- Normal Bonding and Grounding May Not Be Enough! Companies that handle, transfer, and store flammable liquids should contact manufacturers to determine if these liquids can accumulate dangerous levels of static electricity, and if they can form explosive vapor-air mixtures inside storage tanks. If so, *extra precautions*—beyond normal bonding and grounding—may be necessary. Antistatic Additive is an extra precaution.

Recommendations: Knowledge members can use

Who adds the Antistat?

- How much Antistat is added?

How is the Antistat added?

511247 Influence of the Particle Size Distribution on the Dispersion Process of Combustible Dusts in the 20 L Sphere

Abstract:

The 20 L sphere is one of the standard apparatuses that is utilized to characterize the explosibility parameters of combustible dusts, gases and vapors. One concern about the use of this experimental setup for the analysis of combustible dust clouds is related to the variation of the particle size distribution of the powder. This physical property changes due to the agglomeration and deagglomeration of the solid aggregates occurred during the dust dispersion. These phenomena determine that a high turbulence in the cloud represents both enhancing and quenching effects for the ignition and the initial propagation of the combustion flame. Therefore, this condition might lead to a high uncertainty level during the estimation of the severity and ignitability parameters of dust explosions and induce the bad design of process equipment.

This study evaluated the dispersion and the explosion severity of dust clouds formed with wheat starch agglomerates of different Particle Size Distributions (PSD). The standard dispersion test was performed in the 20 L sphere with the standard rebound nozzle with a triplicate test at the dust minimum explosive nominal concentration (30 g/m^3). The turbulence levels of the dust dispersion were analyzed with a Particle Image Velocimetry (PIV) technique. Furthermore, the PSD variations of three samples with different initial aggregation levels (d_{50}: 68.2, 270 and 420 μm) were measured before and after dispersion by using laser diffraction. Additionally, a qualitative analysis of the integrity of the particles after the dispersion was done using microscopy.

From the PIV analysis, it was found that the evolution of turbulence in dust clouds is similar for agglomerates with different PSD. On the one hand, the mean horizontal velocity remains close to 0 m/s with small variations due to the flow recirculation inside the sphere and the partial symmetry of the body. On the other hand, the mean vertical velocity is determined by the injection conditions of the standard test since it has a negative value at the beginning of the dispersion (0-10 ms) that becomes positive (10-100 ms). Lastly, at the end of the dispersion, the mean vertical velocity has a negative value for each agglomerate, but close to 0 m/s (sedimentation stage).

Subsequently, the granulometric and microscopic analyses established that an important deâ€'agglomeration in the three starch samples since their diameters d_{50} were reduced until 19 Âµm approximately. This fact implies that the external forces that are exerted on the solidâ€™s surface are strong enough to overcome the adhesion forces of the agglomerates. This phenomenon occurs due to the high turbulence induced by the pressure difference between the pressurized air of the dust reservoir and the air located within the explosion chamber. Moreover, this condition determines the evolution of the velocity field within the sphere due to a baroclinic effect that is an important source of turbulence and vorticity. These results confirm the relevance of the ignition delay (standard value: 60 ms) on the experimental test and allow adjusting this parameter according to the physical properties of the dust sample.

513733 Influence of Intertization on the Determination of the Minimum Ignition Energy of Combustible Dusts

Abstract:

This research work aims to contribute to the improvement of safety conditions against flammability risks of solids in processing industries.
Flammability of dusts from several materials, such as straw, wood chips, almond shell, olive pit, coal, thermally dried sewage sludge, vine, flour, starch, pineapple, coconut shell, etc. has been characterized. This characterization has been carried out through the determination of the Minimum Ignition Energy (MIE) of the samples, obtaining information about the ease of ignition of the dust cloud. Tests were firstly carried out under ambient conditions (air: 21 % v/v O_2 â€" 79 % v/v N_2), the atmosphere was then partially inerted reducing the oxygen content to 15% v/v and reaching 9% v/v for some sample. A specific gas mixer has been developed to obtain the desired oxygen and nitrogen ratios.
Only a few samples had a MIE value in air below 1000 mJ (samples with lower moisture content). However, these samples presented a MIE value at 9% v/v O_2 above 1000 mJ. The sulphur sample showed a high flammability, characterized by a MIE value less than 1 mJ when ignited in these tests for any oxygen concentration from pure air to 15% O_2, showing a sharp decrease in its flammability when the oxygen content dropped by 15%.
Partial inerting of the air by reducing the oxygen content causes an increase in the value of the MIE and results in the ignition being more difficult to achieve since more energy is required in the spark.
The results of this research contribute to show the importance of determining some flammability characteristics of combustible dusts in order to improve safety in industrial processes.

511246 Analysis of Combustible Dust Flash Fires on Personal Protective Equipment Fabrics

Abstract:

Combustible dust flash fires present unique hazards to fabrics utilized in personal protective equipment (PPE) garments. NFPA 484 (2015) *Standard for Combustible Metals* states that "Testing has shown that the 84 kW/m^2 heat flux used in the NFPA 2112 *Standard on Flame-Resistant Clothing for Protection of Industrial Personnel Against Short-Duration Thermal Exposures from Fire* garment-equipped mannequin flash fire test is only 25 percent to 32 percent of an iron dust flash fire heat flux" and suggests that garments designed to comply with NFPA 1971 *Standard on Protective Ensembles for Structural Fire Fighting and Proximity Fire Fighting* or ASTM F2621 *Standard Practice for Determining Response Characteristics and Design Integrity of Arc Rated Finished Products in an Electric Arc Exposure* may be more appropriate but that there are "no test data to substantiate their performance for metal dust fires."

In addition to radiative heat transfer from the flame, combustible dust flash fires also produce high-temperature solid particles that may adhere to fabrics and produce significant and prolonged heat transfer to the garment. While this additional heat transfer unique to combustible dust flash fires may be significant, the prescriptive performance requirements of PPE fabrics required by standards are based primarily on flammable gas-fueled flash fires, and have not been thoroughly evaluated for combustible dust flash fires. Therefore, these standards may not be directly applicable.

In this study, in an effort to analyze the additional hazards that combustible dusts flash fires pose on PPE fabrics, various PPE fabrics are exposed to combustible dust flash fires using a custom flash fire apparatus. Organic and metal dusts are dispersed using an injection system similar to a standard Siwek 20-L combustion chamber, and subsequently ignited into a flash fire directed at the PPE fabric. The performance of the PPE fabric is evaluated using heat flux measurements and compared to a flash fire from a flammable liquid mist.

Use and Comparison of Different Passive Fire Protection Assessment Methods for LNG Plants

Sanjay Ganjam
Manager of Process Safety, Kiewit
sanjay.ganjam@kiewit.com

Raymundo Rodriguez
Sr. Fire protection engineer, Kiewit
Raymundo.Rodriguez@Kiewit.com

Andrew Staszak
Senior Engineer, MMI Engineering
AStaszak@MMIEngineering.com

Prepared for Presentation at
American Institute of Chemical Engineers
2018 Spring Meeting and 14th Global Congress on Process Safety
Orlando, Florida
April 22 – 25, 2018

Use and Comparison of Different Passive Fire Protection Assessment Methods for LNG Plants

Sanjay Ganjam
Manager of Process Safety, Kiewit
3831 Technology Forest Blvd., The Woodlands. TX. 77381
sanjay.ganjam@kiewit.com

Raymundo Rodriguez
Sr. Fire protection engineer, Kiewit
raymundo.rodriguez@Kiewit.com

Andrew Staszak
Senior Engineer, MMI Engineering
astaszak@MMIEngineering.com

Keywords: Small scale, LNG, structural analysis, fireproofing, PFP, vaporizer

Abstract

LNG plants pose hazards due to pool fire, jet fire and cryogenic temperature brittle fracture and embrittlement of supporting steel. In a Fire Hazard Evaluation, the first step is to evaluate location and types and area of fire hazard. Fire hazard protection analysis considers mitigation methods and protective systems to protect against these hazards including instrumented shutdown systems which prevent process excursions and prevent loss of containment.

However, after loss of containment some of the passive methods are effective in limiting the extent of damage and prevent escalation of the incident. Passive fire proofing (PFP) is an excellent tool that can be used to protect equipment and pipe steel supports from effects of fire and cryogenic exposure. PFP is not directly mandated by code, but generally regarded as part of best practice design. Usually such applications are based on gross extents extending 20-30 feet in all directions from a hazard source and coating every structural member within. This can result in higher costs and its use questioned by cost conscious projects in a competitive market.

In this paper we demonstrate different methodical techniques of identifying the hazard and specifying the PFP requirements. These techniques usually include detecting pool, jet fire and cryogenic hazards of LNG and evaluating their impact on load bearing structural members. Then the amount and location of PFP application is optimized to prevent failure and escalation of the incident thus achieving cost reductions. With such optimization, PFP be considered for smaller LNG projects and applied in a cost effective way precisely where needed using advanced structural analysis. Using these techniques cost saving of 40% or more can be achieved compared to traditional tools and methods.

1 Introduction

LNG plants are built in different capacities to serve various needs. They are usually categorized into small scale, medium scale and large scale base load plants. In the categories of small to medium LNG plant are liquefaction plants for domestic transportation and heavy machinery use and peak shaver plants. Peak shaver plant typically liquefy LNG from pipeline during summer months, store in a LNG tank and vaporize in winter to meet peak gas demands. Large scale LNG plants are base load plants built for LNG storage and export purposes. Most hazards in these LNG plant are jet fires, cryogenic jets along with pool fires.

In USA LNG plants are designed to meet with various revisions of NPFA 59A which calls for extent of fire protection to be based on evaluation using fire protection principles and identified hazards within the facility. The fire protection analysis shall determine equipment, type and methods necessary for protection of equipment and structures from fire exposure. In later version of NFPA 59A in the risk assessment sections the code mentions the criteria for property damage due to radiant heat from fires specifies that for steel structures the temperature shall not exceed 1000 °F over the duration of fire. So to protect equipment and structural steel from exposure to radiant heat it is necessary to use fire proofing to limit the damage. Code does not clearly call for fire proofing but to really limit the temperature of steel to 1000 °F for which passive fire proofing is the most effective method of protection. Using passive fire protection is common industry practice to protect against fire and cryogenic exposure in a LNG plant.

In this paper we would discuss different ways LNG plants small, medium and large scale facilities can consider using passive fire protective as an effective tool to protect assets and personal from fire and cryogenic hazards. We have also demonstrate that there are different level of analysis that projects have at their disposal. Each of these levels involve different level of effort and result in different level of PFP protection and at the same time affecting the risk level of the project.

2 PFP Specification Methodologies

In the US, LNG plants are designed to meet the criteria of various revisions of NPFA 59A which calls for fire protection to be based on generic fire protection principles and hazards within the facility. The fire protection analysis shall determine equipment, type, and methods necessary for protection of equipment and structures from fire exposure. In recent versions of NFPA 59A, the risk assessment sections provide criteria for property damage due to radiant heat from fires and specifies that for steel structures the temperature shall not exceed 1000 °F over the duration of the fire. Thus, the code does not clearly call for fire proofing, but to really limit the temperature of steel to 1000 °F fire proofing is the most effective method of protection.

As discussed above, passive fire protection specification and applications within hydrocarbon containing facilities is not specified by regulation but rather governed by guidance and best practices (commonly addressed as 'RAGAGEP' or Recognized and Generally Accepted Good Engineering Practices). The challenge with this is that any number of best practice approaches could be technically sufficient and deemed acceptable, depending on the individual design team, project approach, risk acceptance for the facility, etc. This results in numerous best practice approaches, each having their own benefits and shortcomings. This section will outline and discuss the most common approaches which can and typically are used to identify passive fire protection requirements. The methods discussed here have been aligned into a tiered process (five elements) such that financial benefits of each can be quantified and compared.

The authors note that there are numerous alternative references (global) and individual corporate requirements which provide guidance on the identification and assessment of hazards, quantification of fire effects, selection of fire protection, etc. but the focus of this paper in on general best practice methods typically employed for US based onshore facilities.

2.1 Method-1 (M-1): Qualitative Selection (Minimum) Specification

While there are several sources of guidance for the selection, use, testing, and performance evaluation of PFP products, API 2218 is most widely applied guidance for the use of PFP within petroleum and petrochemical processing plants. API 2218 focuses on liquid pool fire exposures for defining structural fireproofing. Despite recent updates (2013) which included Annex C, now discussing jet (or spray) type fires, limited guidance is provided on how to address these events. As such many facilities are still specifying PFP based on simple qualification of pool fire potential, based on a high level reviews of equipment areas, operating conditions, etc. and focusing on liquid events. API provides a flow chart of the process.

With this, an evaluation and selection of PFP can be developed through qualitative evaluations and the use of spacing, placement, and fire envelope guidance provided in API 2218. Inputs to this type of PFP specification are typically biased by experienced-based design rules and outside influences from corporate or insurance guidelines. While this author's personal experiences have seen a full spectrum of results from these assessments, the focus of this method outline is solely on the minimum acceptable approach.

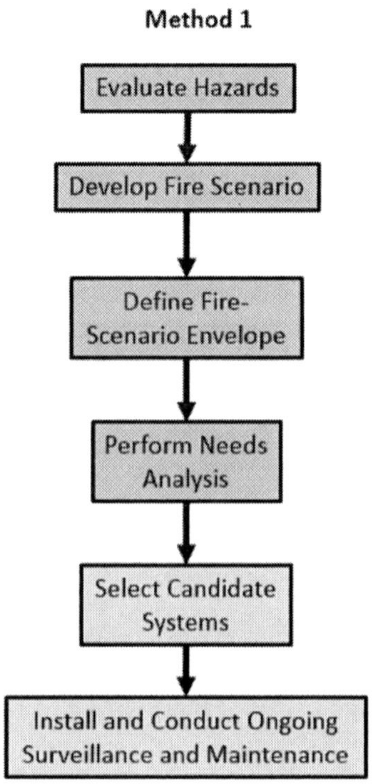

Figure 1: Qualitative PFP Specification Method Flow Chart

2.2 Method-2 (M-2): Fire Hazards Analysis Specification

Recognizing the need for improved evaluation of fire hazards, API provides guidance and references to conduct a more quantitative assessment of fires and their extents in the form of a formal 'Fire Hazards Analysis' or FHA. This FHA approach is generally regarded in the industry as the standard of information which should be utilized to develop PFP specifications. While an FHA can and does take many forms, the base assessment encompasses the following elements:

- hazard evaluation, quantifying available inventories;
- selection of fire scenarios including the calculation of release rate and fire extents; and
- selection of PFP based on impacts and desired duration of protection.

Most facilities conduct these assessments as part of (or utilize the data from) Facility Siting Assessments (API 752) conducted for Process Safety Management programs. These assessments usually choose an event size representative of a major or significant event (such as 2 inch release or 20% of pipe size) and evaluate fire extents as steady state. The process flow diagram can be updated to reflect the elements of this method, shown below.

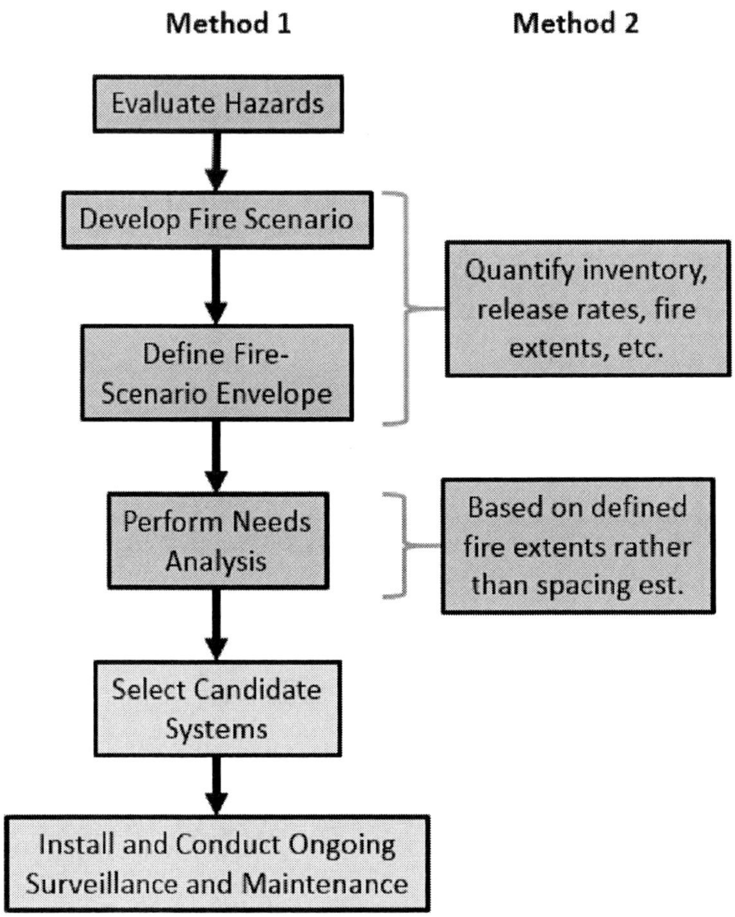

Figure 2: Fire Hazard Analysis PFP Specification Method Flow Chart

This provides a significant improvement in protection over a qualitative selection approach, as both pool and jet fires are addressed, but typically results in a conservative specification or overestimation of required passive protection. Although there is no explicit frequency analysis in this method, we know that due to reduction in fire consequence effects and built-in conservative assumptions and calculations forming the specification, the risk to the facility is greatly reduced below the tolerable limits. From a protection standpoint this may be beneficial, cost and maintenance issues could be overly restrictive.

2.3 Method-3 (M-3): Improved Fire Hazards Analysis Specification

Significant improvements can be made to an FHA (Method-2) based specification by increasing the understanding of each hazardous scenario and improving fire extent calculations. This method aims to put physical restraints on events by quantifying the release process (beyond steady state) and accounting for integrated safety systems which may be present in the design. The following additional elements beyond those discussed in Method-2 should be included/address:

- (optional) define variation of potential release sizes (e.g. small, medium, large or minor, moderate, significant) based on specific equipment and piping in the vicinity of the release;

- define available inventories based on automatic control systems, isolation points, and physical process barriers (e.g. accounting for phase changes/separation in large vessels and columns);
- account for pressure changes during hazardous releases due to limitation in pressure producing equipment (e.g. pump performance curves and/or automatic trips), active relief systems such as blowdown, phase changes, etc.;
- conduct transient release rate modeling and develop transient fire extents based on improved release information; and
- selection of PFP requirements based on time dependent fire extents.

The result is a (or set of) transient fire extents (described in flame length below but can alternatively be classified as a distance to flux) which can be used to specify the requirements for passive protection.

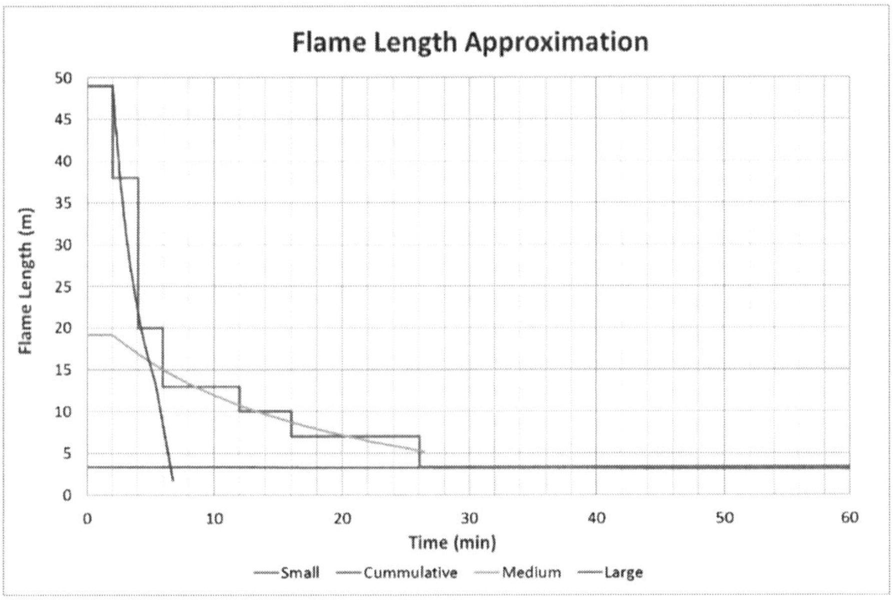

Figure 3: Example of Transient Release Modeling for Improved Fire Representation

When performing fire hazard analysis using improved consequence modeling, one must incorporate certain improvements and methods to accurately represent the leak scenarios. For example, leak flow through a hole is mostly transient in nature, where the release rate is high during the initial phases and as the pressure and inventory reduces, the leak flow rate decreases. Also the analysis should consider detection of abnormal situation, shutdown and isolation philosophy in the design to define the isolation sections and the inventory associated in each scenario. Usually certain time passes before a leak is detected either by gas detection, alarms of process, and initiation of shutdown. Shutdown could be automatic based on programmed logic, such as gas detection or manual by operator action. The analysis should consider such delays in development of transient scenarios.

This typically allows for improved protection over Method-1 by accounting for both pool and jet fires and improvements over Method-2 by including physical event limits and safety features which can significantly reduce the time duration of fire events. Which in turn, reduces the requirements of passive protection. Again, the process flow diagram for this method can be updated to include necessary elements, shown below.

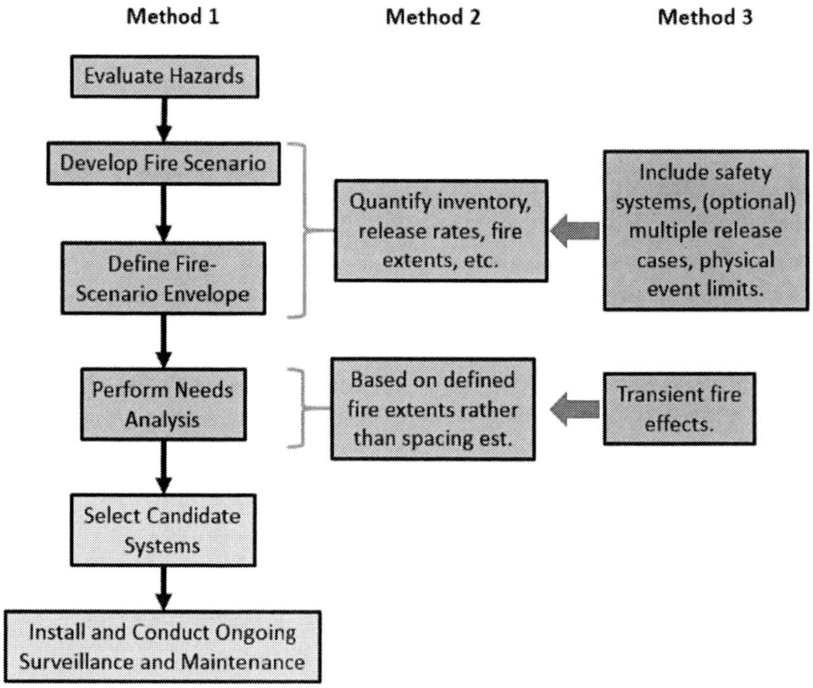

Figure 4: Improved Fire Hazard Analysis PFP Specification Method Flow Chart

It is important to note that the approach to quantifying fire effects, such as fire shape and radiation distances, have been intentionally neglected in this paper. The differences between approaches, such as simple phenomenological modeling and advanced fire modeling (computational fluid dynamics) are well documented and discussed. As with the methods described here, any improvements in the fire modeling and representation, would lead to improvements in the passive protection scheme regardless of the method selected.

2.4 Method-4 (M-4): Performance Based Specification

To this point, all methods have focused on the quantification of the hazards and not yet on the specific effects of hazards to structural steel. So far, protection needs were based on simple exposure and time limits.

Further improvement of a passive protection scheme can be achieved by identifying and selecting PFP requirements based both on the transient hazard extents (Method-3) and optimizing the performance of individual structures based on a performance criterion. This method requires the selection of performance criteria (critical to final scheme requirements) and advanced structural analysis to be performed. The general method includes the following elements:

- Method-3 conducted to identify appropriate transient fire cases;
- selection of structural performance criteria (e.g. deflection limits base on span, remaining structural capacity against global collapse (tension, compression, shear in individual elements);
- development of a structural model for each critical structure;
- analysis of fire loads on structures (including directional nature of jet fires); and
- selection of PFP requirements based on structural member by member optimization (requires iterative routine).

In this stage more advanced structural analysis models are required to verify performance and the conduct optimization. Optimization can be done through simple engineering driven iterations, use of numerical optimization searching routines, or more advanced directives such as machine learning. Regardless of the optimization approach, verification of the structural survival against selected criteria is critical.

Before optimization can be performed, preliminary analysis to identify endurance times of structures and critical failure locations (based on scenario transients, directions, origins) must be conducted for fire and cryogenic events. Figure 5, shows the results of this analysis for a particular fire scenario. The results of which identify failure of the system within 1.5 mins of the fire, and highlight areas of the structure (red) where initial PFP optimization efforts should be focused.

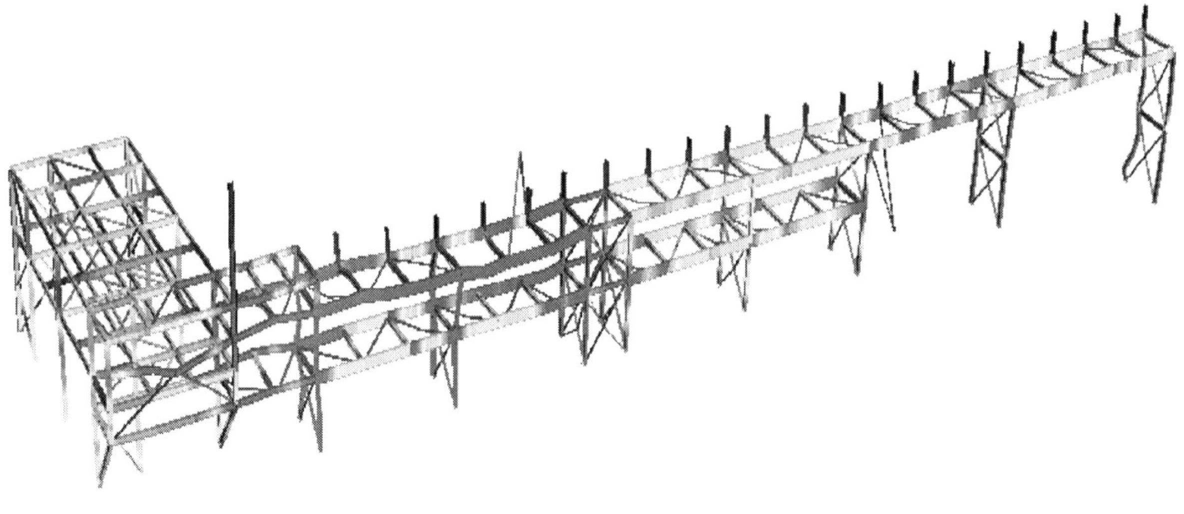

Figure 5: Preliminary failure analysis of unprotected structure due to fire load.

Similarly, failure due to cryogenic exposure from the same event, identifies system failure within 5 mins, and provides indication of optimization (PFP protection) precedence.

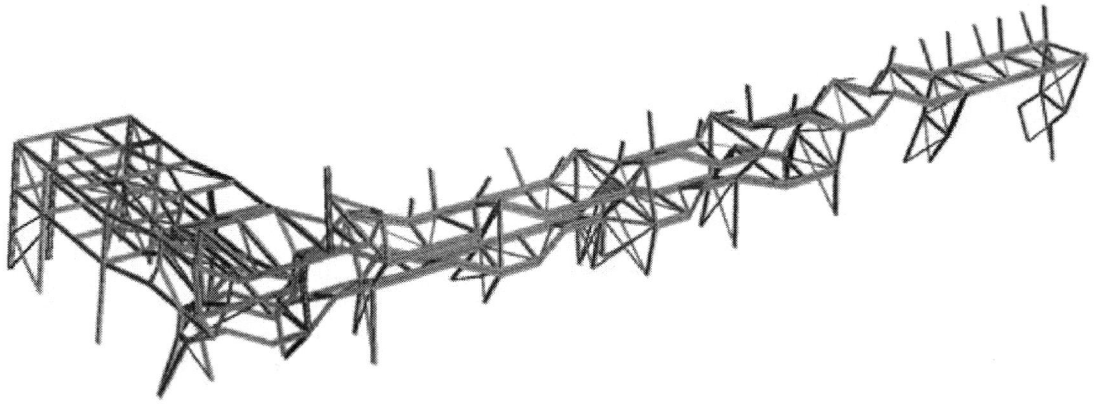

Figure 6: Preliminary failure analysis of unprotected structure due to cryogenic load.

The general optimization efforts should then review and account for variations in loads (live and dead), critical load bearing members versus secondary supports (i.e. redundant cross members required for lateral and buckling support), individual element size and thermal heat up durations, and construction restraints which may necessitate 3-sided versus 4-sided PFP coating. The figure below provides an example of verification of structural survival times with and without PFP applied, with the colors showing variation of individual element critical core temperature -CCT (top) and plastic utilization (bottom) during a consequence event. As the load bearing capacity of the steel is lost leading to failure, optimization considers this requirement and applies PFP only to those members which are necessary to meet capacity criteria. Variations in PFP thickness are considered and altered to meet endurance criteria.

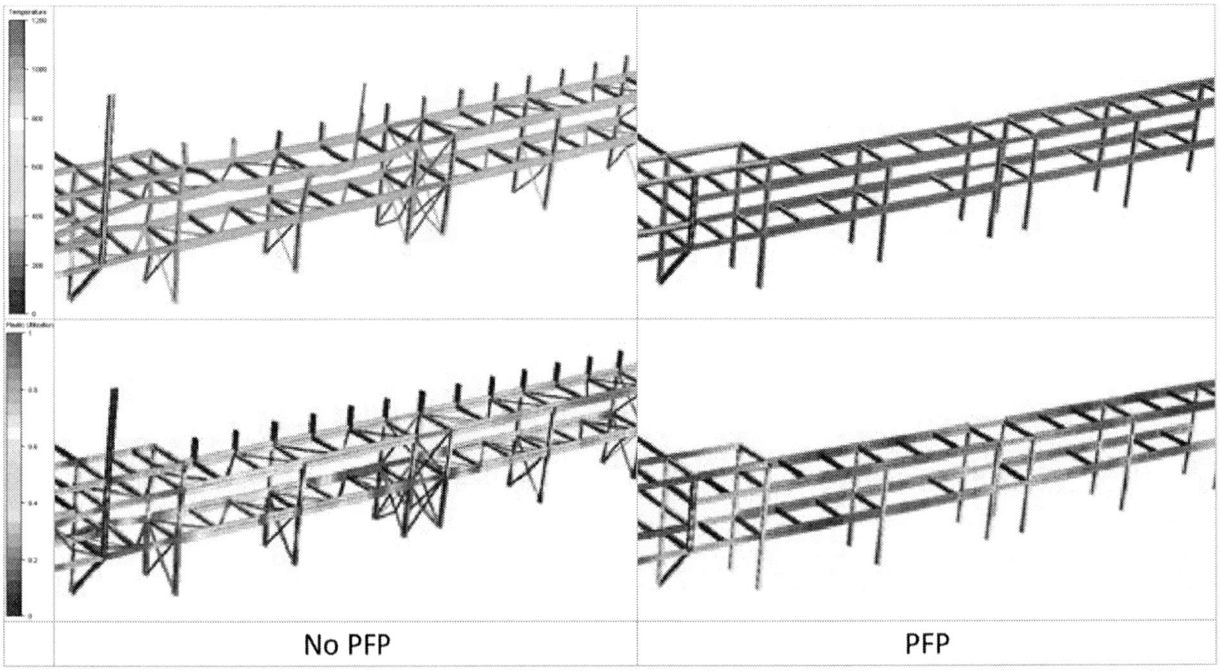

Figure 7: Example Performance Verification of Critical Structural Survival with and without PFP (showing CCT top, plastic utilization bottom)

The process flow diagram, including these elements and where in the process they are considered is shown below.

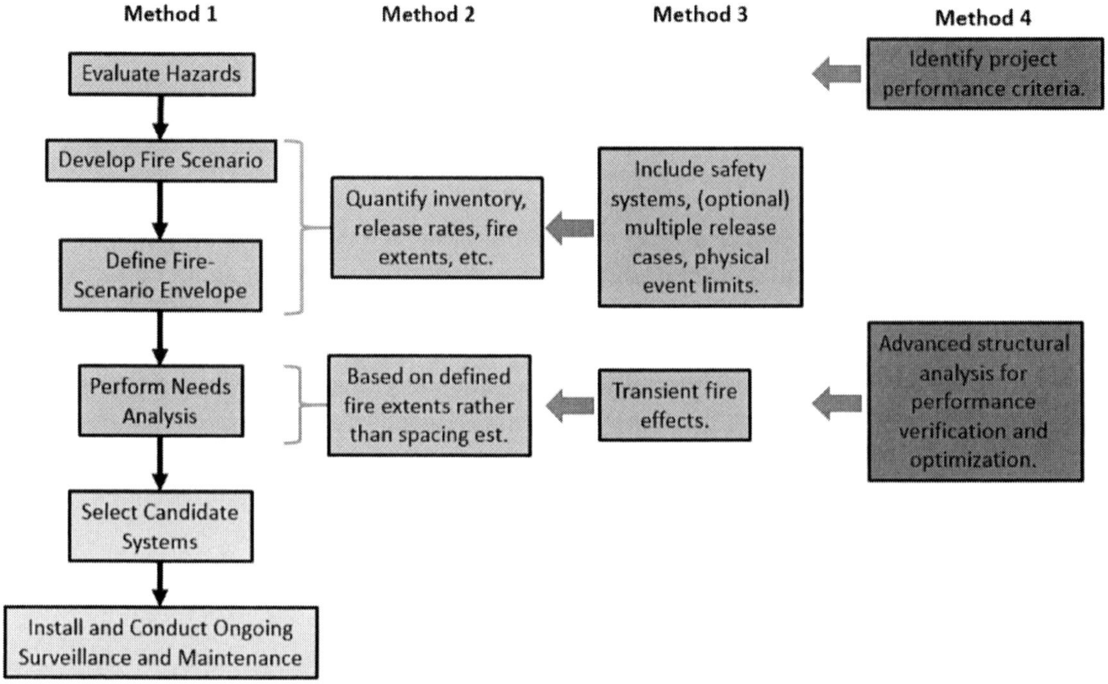

Figure 8: Performance Based PFP Specification Method Flow Chart

The result is this method is a quantitatively optimized scheme, dependent on the structural performance requirements chosen. When conducting this effort, it is critical to ensure that appropriate considerations to piping displacement, movement of cabling, or any other critical equipment is reviewed prior to selecting a performance criteria. Additionally the structural analysis must include all loads expected in the final design (during early design stages allowances must be made for unknown or unspecified equipment to be added later).

2.5 Method-5 (M-5): Risk Based Specification

The final method, risk based, combines the elements of consequence and frequency (likelihood). In this method a full probabilistic approach is conducted to identify a single (or set) of design fire cases against which structural analysis and optimization can be completed. This requires the selection or development of a risk criteria against which the design will be assessed.

The ultimate objective of this method is to understand the risk of the facility (or to specific critical structures) and design a protection scheme to acceptable tolerance criteria. The following elements should be included:

- selection or identification of risk criteria;

- individual equipment counts based on scenario;
- development of base release frequency curves for each scenario based on equipment counts and historical release data;
- development of appropriate event trees and identification of probabilities can include: failure or success of safety systems, directionality, environmental, ignition, etc.;
- risk calculations and identification of design fire case(s); and
- selection of PFP requirements based on design fire case and needs based method (can be conducted based on the process identified in Method-3 or 4).

The culmination of any risk based method for use in design specification is to develop design accident loads (DALs) from which protection schemes can be derived. This is generally and best done through the development of exceedance curves. Exceedance curves are generated from risk based scenario data representing a variation of potential release cases and their historical leak frequencies with probabilistic modeling accounting for variation in outcomes (event trees, ignition probabilities, directional probabilities, potential success and failure of safety systems, etc.). The outcome are cumulative frequency vs consequence (or any end event condition), as shown in the figure below.

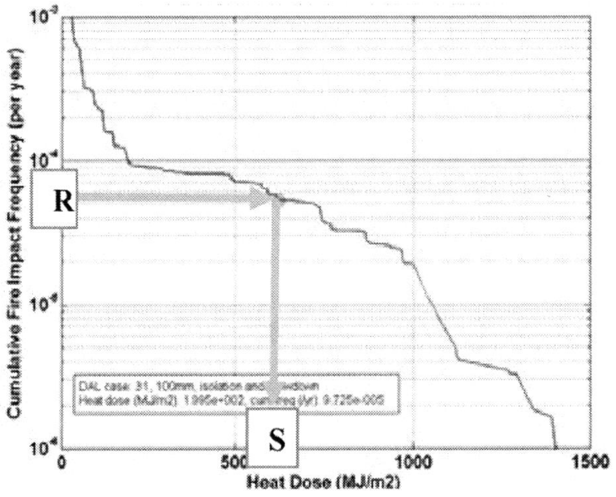

Figure 9: Example Exceedance Curve used to select the Design Fire Case

For fire protection this can be represented as a heat flux (or dose). Based on risk criteria, a design frequency is identified on the exceedance curve (R – in figure above) which leads to identification of the DAL scenario (S – in figure above). In many cases, due to analysis assumptions more than one event may be representative of the DAL selection. In this case it is recommended that careful selection be made of which event is carried forward or for better results, a set of scenarios be carried forward as the DALs. Note, that for LNG facilities, additional exceedance curves must be developed to consider cryogenic events and care should be taken to ensure that their DAL characteristics are contained within the fire specific DALs. Depending on design aspects of the facility, numerous exceedance curves may be developed representing effects to specific critical structures or areas of the facility.

The process flow diagram, including these elements and where in the process they are considered is shown below.

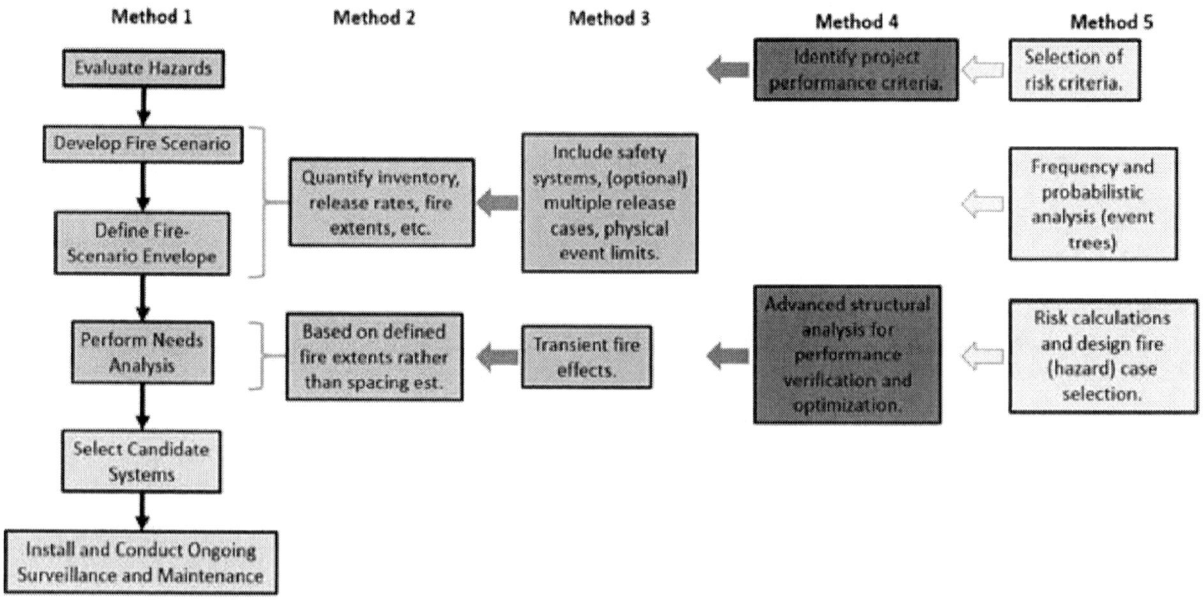

Figure 10: Risk Based PFP Specification Method Flow Chart

Note, that a detailed risk based method has not been outlined within the paper as it is not the focus of the effort and maybe considered for subsequent work. The goal of this paper is to outline the methods that can and are used to specify passive protection schemes. For specific guidance on risk assessments there are numerous references such as CCPS and UK HSE.

3 Current Work

Recently, several PFP specification analyses have been conducted as part of the design of small to medium scale LNG facilities. The most recent assessment included the specific use of Methods 2, 3, and 4 above as part of tiered efforts to reduce project costs related to PFP on critical structural elements (which additionally reduces operational and maintenance costs). The focus of this effort was to provide protection from both thermal and cryogenic events resulting from the potential accidental release of LNG and other flammable materials. The results of the initial work conducted for the LNG facility showed significant reduction in overall PFP weight and area coverage requirements from start to finish.

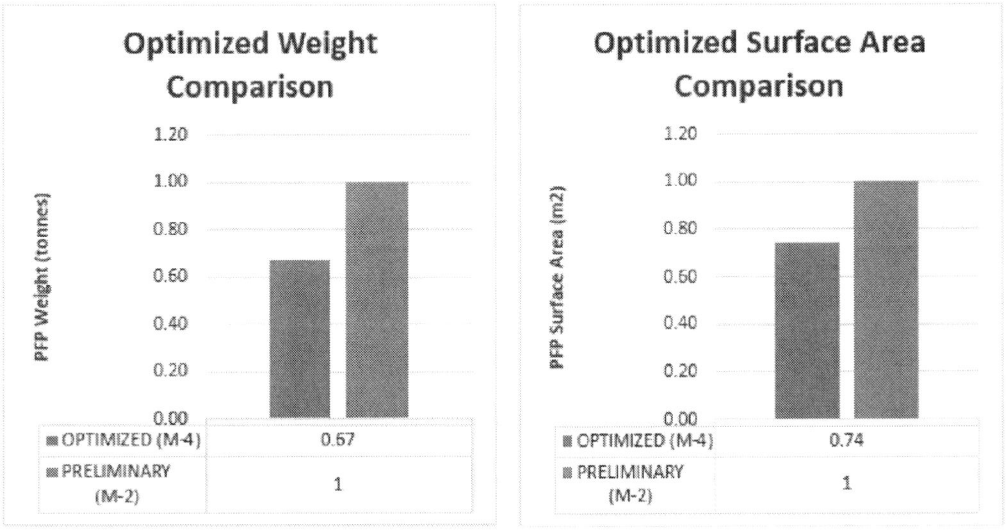

Figure 11: Comparison of weight and area coverage from analysis conducted for small scale LNG facility.

Following the completion of these assessments and development of each PFP scheme, additional analysis and calculations were completed to quantify the costs related to each method, allow a comparison of, and potentially to provide insight on relative effort costs versus PFP costs. The costs to conduct each method were plotted against calculated PFP costs for each method, along with the relative risk ranking of the selected PFP scheme (see Figure 12 below). Costs have been normalized on a basis of one (1) to allow for more simplified comparison.

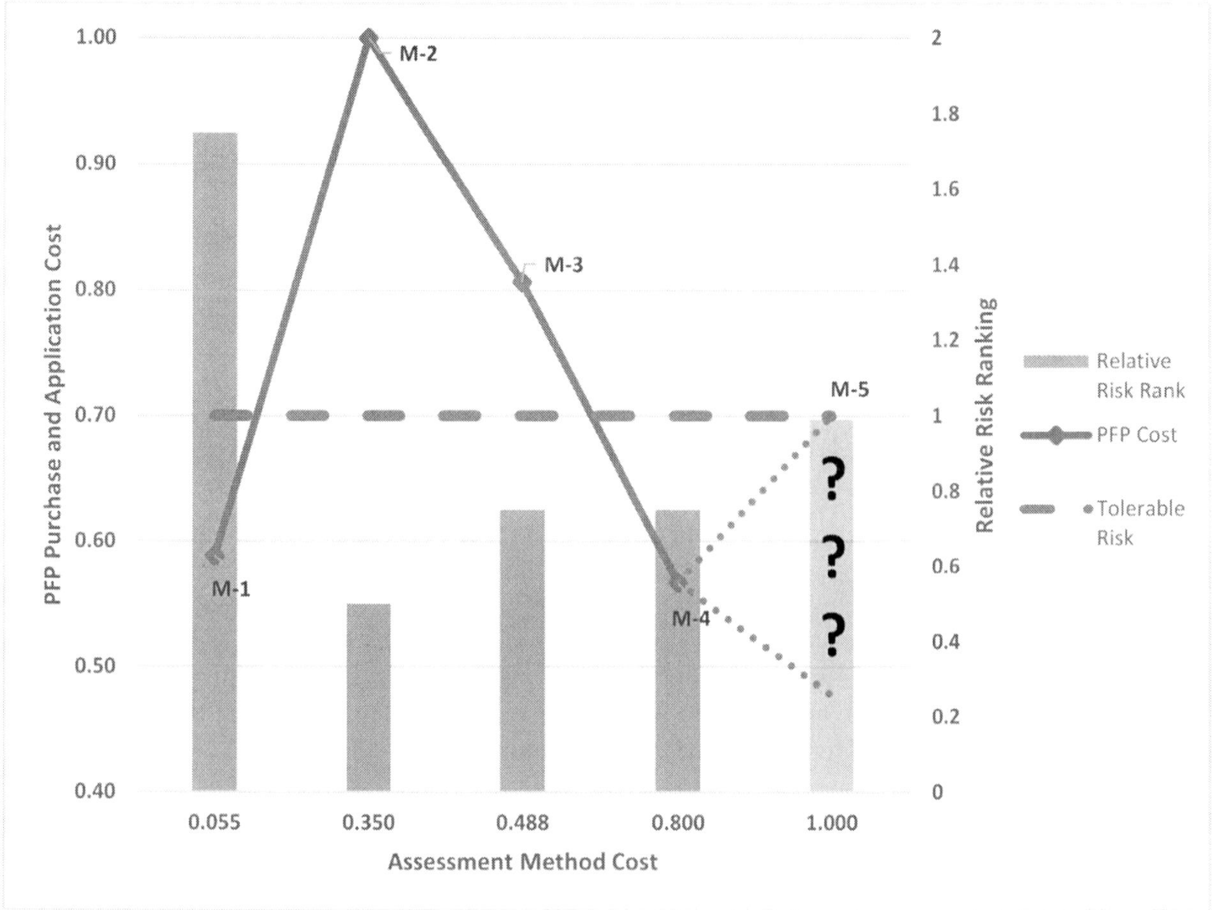

Figure 12: Comparison of PFP Costs and Relative Risk against the various Assessment Method Costs

It is important to note that:

- The costs for PFP represent product purchase and application costs. Cost related to continued maintenance, inspection, and repair throughout the life of the facility are not included.
- The facility assessed included cryogenic protection requirements as part of the design. As such, a PFP product capable of providing both cryogenic and fire protection was used in the optimization of PFP schemes.

3.1 Observations in Method Comparison

In review the comparison of methods and schemes shown in the figure above, we can make several notable observations about each approach and the effect is has on the facility risk as well as costs.

Method-1: While common in practice this method neglects jet fires, cryogenic jets and limits the hazards that are protected by the scheme. This results in a lower cost PFP option, ignores possible hazard scenarios and thus sacrifices protection by significantly increasing the risk of exposure to fire and cryogenic events.

Method-2: This method provides conservative consideration for the greatest set of potential hazards, but doesn't allow for refinement of PFP by accounting for safety systems nor the transient details of releases. With this approach crude fire models are typically employed which cannot appropriately account for geometric effects. The results are the lowest risk protection scheme, but this comes as a significant increase in PFP costs. Increase from Method-1 is simply due to inclusion of all possible hazard scenarios and ultimately may burden operations with maintenance, inspection, and upkeep.

Method-3: Improving on the hazards analysis with the use of transients and consideration for safety systems such as shutdown, isolation, and blowdown represents the fire scenarios close to reality in the amount of release and duration. These improvements result in a reduction of PFP costs over M-2 by 20% with an increase in effort costs of 30%. In terms of savings, the cost of PFP is on the order of 10x more that the effort costs, thus the overall cost reduction factor is 6x. This also comes, with a more refined placement of PFP, which can increase relative risk, but maintaining it below a tolerable level.

Method-4: Adding structural analysis and performance verification to the detailed hazard analysis allows for a formal application of PFP optimization. The selection of load bearing and critical members can be made as with specific survival durations required. Depending on the analysis type, specification of 3-sided vs 4-sided protection can also be made, allowing for reduced PFP and potential construction difficulties. This approach results in no relative change in risk but large reductions in PFP requirements.

For the facility in this study a reduction in PFP costs over M-2 was found to be 44% and resulted in PFP costs slightly less than that prescribed by M-1. This is an overall cost reduction factor of 11x, from M-2.

Method-5: The full risk based method was not conducted for this facility, but the general trend and potential outcome range is known. This method allows a facility to understand (quantitatively) their risk profile and target PFP schemes to meet that desired risk. The result is a targeted scheme which takes the approach of providing a balance between risk and costs. A company can choose to minimize risk and increase protection through PFP or drive risk to tolerable levels and reduce PFP requirements. Depending on the risk criteria selected, the costs for the assessment are typically offset by the PFP savings and general risk to other elements of the facility.

4 Conclusions and Recommendations

The goal of this paper was to take a look at various methods used to assess hazards and specify passive fire protection for small and medium LNG facilities in the US and attempt to quantify the costs of assessment versus PFP material and application. With an aim to use this information to answer: Why do anything? What do you get from each method? What if any, is the best approach? Each method has pros and cons associated with it, but the work shows that definitive savings with respect to costs and risk, can be realized with increased detailed assessment and methods in determining required PFP specification.

For small and medium scale LNG facilities, it is recommended that at a minimum Method-3: Improved Fire Hazards Analysis Specification should be carried out during the design. With minimal assessment costs, significant refinement on PFP specification can be revealed and the risk to the site can be tailored to a tolerable level. This approach provides a relative balance between effort and scheme costs.

For this particular site, the efforts of Method-4 where shown to provide additional significant reduction in PFP costs while maintaining risk levels. It is important though, to not associate these gains with all sites. This facility in particular contained high, long-run pipe racks which provides opportunities to optimize PFP against quick decaying fire events. If a facility contains tightly arranged ground level modules and pipe on sleepers, this method is likely not going to result in the gain realized in this paper. Thus, a review of the facility layout and some preliminary review of fire transient and sizes should be conducted to identify if savings is likely. This, is recommended to be conducted by someone knowledgeable with hazards quantification, structural analysis, and PFP optimization experience.

Ultimately, risk based methods (M-5) are recommended for facilities and projects that have already chosen to utilize a risk informed design for the entire facility. This allows easy integration of PFP optimization methods against risk criteria and allows for the risk assessment costs to be spread across other design influences (cost reductions and mitigations). For example, a Quantitative Risk Analysis (QRA) has been performed in a project, information from that analysis can be utilized in risk based method, M-5. Alternatively, if a particular facility or design has a driving need for further refinement of fire protection (e.g. a facility with limited water supply issues for active systems), then this approach may be the best.

In the future, efforts may include use and inclusion of results specific to risk based methods and the development of a similar method comparison plot for a large scale facility. The authors note, with further investigation and project data, a set of method selection guidelines could be developed to aid in early selection of approach method.

5 References

[1] NFPA 59A, Standard for the Production, Storage, and Handling of Liquefied Natural Gas (LNG), edition 2001, 2006, 2013 and 2016.

[2] API RECOMMENDED PRACTICE 2218, Fireproofing Practices in Petroleum and Petrochemical Processing Plants, Third edition, July 2013

511347 Investigating the Gas Dispersion from a Subsea Gas Release in Stratified Shallow Waters

Abstract:

Investigating the Gas Dispersion from a Subsea Gas Release in Stratified Shallow Waters

Mary Kay O'Connor Process Safety Center – Qatar

Texas A&M University at Qatar, PO Box 23874, Education City, Doha, Qatar

Corresponding author: Moustafa Ali [moustafa.ali@qatar.tamu.edu]

Presenting author: Moustafa Ali

Objective:

Investigate the effects of thermal or saline stratification and high gas rates on the consequence modeling of subsea gas releases with numerical modelling. Phenomena which need better understanding like plume turbulence, bubble size distribution and hydrates formation are addressed. In addition, the potential hazards associated with the toxicity and flammability of surface reaching gas are analyzed.

Abstract

Although the repetitive subsea gas releases incidents occurring in the offshore oil and gas industry, the attention is limited on a case-by-case study; leading to quantitative risk assessment approaches that are limited to similar cases. A subsea gas release can result from a range of different causes including drilling operations; failures in flow lines, gas export lines, and subsea equipment. Such releases can have catastrophic impacts on the environment, offshore platforms and human lives (Olsen & Skjetne, 2016). Natural gas, in general and in middle east, covers a large portion of the world supply and, consequently, this type of incidents could pose significant risk for the related and nearby facilities, like Hydrogen Sulfide (H_2S). For example, 40% of the natural gas reserves in the world are sour gas fields in which middle east holds the highest reserves of sour gas. (Total E&P, 2007). While many approaches have been proposed for the description of underwater/subsea releases, these are not universal and still include deficiencies concerning plume turbulence, hydrates formation, high flowrates, stratified (thermally or owe to salinity) water bodies. According to (Rew, Gallagher, & Deaves, 1995), experimental data for underwater releases exist but are also limited to small and medium scale (compared to actual depth and flowrate).

In this study, we investigated the available computational methods for subsea gas releases applicable for "hot" (e.g. stratified) and saline water bodies like the ones of the offshore facilities at the State of Qatar. An Eulerian based computational fluid dynamics (CFD) model was configured in order to study representative gas release scenarios. Earlier, the CFD model was validated successfully against experimental data from SINTEF and Statoil (Engebresten, Northug, Sjoen, & Fannelop, 1997). It also demonstrated the required sensitivity for critical parameters such as the radial and centerline distribution of velocities and void-fraction. Then, the model was applied for conditions specific to Qatar's offshore industry as a representation of shallow water and a sensitivity analysis was conducted against parameters of local interest such as high-flowrates, salinity, bubble distribution and stratified flows. These parameters are needed to understand the behavior of the gas in shallow depths and above the sea surface. Finally, a discussion on the potential risk level is given concerning fire & explosion and toxicity hazards like H_2S.

Keywords:

Offshore, subsea gas releases, shallow water, bubble plume, CFD, toxic gas, flammable, H_2S, QRA

References

Engebresten, T., Northug, T., Sjoen, K., & Fannelop, T. K. (1997). Surface Flow and Gas Dispersion from a Subsea Release of Natural Gas. In *Proceedings of the Seventh (1997) International Offshore and Polar Engineering Conference* (Vol. I). USA.

Olsen, J. E., & Skjetne, P. (2016). Current understanding of subsea gas release: A review. *Canadian Journal of Chemical Engineering*, *94*(2), 209–219. https://doi.org/10.1002/cjce.22345

Rew, P. J., Gallagher, P., & Deaves, D. M. (1995). *Dispersion of Subsea Releases. Review of prediction methodologies.*

Total E&P. (2007). *Sour Gas - A History of Expertise.*

COMBUSTIBLE-DUST FLASH FIRES: FLAME-RESISTANT CLOTHING

A presentation to the 14th Global Congress on Process Safety
April 22 to 26, 2018
at the Orlando World Center Marriott
by

Richard W. Prugh, PE, CSP, of Dekra Process Safety

Abstract

A life-threatening flash fire can result from "accidental" ignition of a dense cloud of combustible dust, where the concentration of dust in the cloud exceeds the Minimum Explosible Concentration. Dense dust clouds can arise from several sources, including (1) emptying bags, drums, and boxes of dusty materials into mixers, reactors, or product-recovery vessels; (2) cleaning ductwork; (3) emptying dust collectors; (4) emptying dryers; (5) release of dust accumulations from elevated surfaces, following a primary explosion or other "mechanical shock" event; (6) vigorous sweeping of large accumulations of dust on floors; (7) emptying vacuum cleaners, and (8) rupture of dust-conveying ducts, piping, boots, or sleeves.

Where operators, mechanics, and contractors could be exposed to relatively-large dense clouds of combustible dust, they should be protected from possible flash fire. In addition to the eye, head, foot, and hand protection that is typical for chemical plants, they should be wearing clothing that would resist ignition from a flash fire.

Data are available for the intensity of flammable-vapor flash fires, as functions of the weight of vapor the heat of combustion, in terms of kilowatts per square meter. Similar correlations can be developed for combustible-dust flash fires. Also, there is a correlation between housekeeping – to limit the accumulation of combustible dusts on floors and elevated surfaces – and the appropriate flame resistance of clothing.

Good guidance is given for protection from flash fire for flame-resistant ["FR"] clothing, particularly where the hazard is "arc flash" and where compliance with the NFPA 70E standard is involved. With the available data, the degree of protection needed against thermal radiation for a wide variety of combustible-dust flash-fire scenarios, is given in this presentation.

A. Introduction

Combustible dusts are defined as small particles of materials that can be ignited at temperatures of 1,500°F [816°C] or less [Reference 1]. A recent description of the maximum size of dusts is [Reference 2]:

Any solid material composed of distinct particles or pieces, regardless of size, shape, or chemical composition that, when processed, stored, or handled in the facility, has the potential to produce a combustible dust.

Dusts traditionally were defined as material 420 microns [µm] or smaller (capable of passing through a U.S. No. 40 standard sieve). For consistency with other standards, 500 µm (capable of passing through a U.S. No. 35 standard sieve) is now considered to be an appropriate size criterion. If the minimum dimension of the particulate is greater than 500 µm, it is unlikely that the material would be a combustible dust.

Dusts are produced in a wide variety of industrial processes. However, in most processes, the dusts are contained within equipment, and spills and releases of dusts are relatively rare events. Dusts can be generated by size-reduction processes such as grinding and milling and, also, in the conveying of friable powders. In addition, some processes create fibrous materials that do not have the appearance of dust but which can burn very rapidly.

Among the causes for dust releases are:

(1) spillage while manually [or mechanically] emptying bags, drums, and boxes of dusty materials into mixers, reactors, or product-recovery vessels;

(2) manual removal of accumulations of dust in exhaust-ventilation ductwork, usually the result of a conveying-air velocity that is too low;

(3) emptying dust collectors where normal discharge has failed;

(4) manual removal of material from dryers, with product change-over;

(5) release of dust accumulations from elevated surfaces, following a primary explosion or other "mechanical shock" event;

(6) vigorous manual or vehicle sweeping of large accumulations of dust on floors;

(7) emptying vacuum cleaners; and

(8) rupture of dust-conveying ducts, piping, boots, or sleeves.

If the result of spillage or release is a dust cloud, there is a flash-fire or fireball hazard if the concentration of dust in the cloud were to exceed the Minimum Explosible Concentration [MEC]. For a flash fire, the shape of a "spillage" dust cloud is likely to be a hemisphere or a "pancake", as contrasted with a spherical fireball that might issue from ruptured equipment. In either case, it is likely that ignition would be near the edge of a dust cloud, rather than centrally.

There may be many types of ignition sources in the areas where dense dust clouds might occur. They include electrostatic discharges from personnel or equipment that are insulated from ground; electric-power sparks from motors, relay-type controllers, and receptacles; hot surfaces such as ovens; open flames as in boilers and heating-media fire boxes; hot spots in bearings and at other sources of frictional heating; and embers that result from spontaneous combustion.

The increase in pressure in a room or building is not addressed in this document. However, the increase in room temperature could be calculated from the weight of dust that has burned, the heat of combustion of the dust, the size of the room [weight of air], and the heat capacities of the air in the room and of the combustion gases. The increase in pressure then could be determined by application of the Ideal Gas Law.

B. Pertinent Legislation and Regulations

Requirements for employers to determine the appropriate personal protection for employees have been the subject of several regulations and consensus guidelines:

From the Occupational Safety and Health Act [OSHA]:

29 CFR 1910.132(d)(1)

The employer shall assess the workplace to determine if hazards are present, or are likely to be present, which necessitate the use of personal protective equipment (PPE). If such hazards are present, or likely to be present, the employer shall select, and have each affected employee use, the types of PPE that will protect the affected employee from the hazards identified in the hazard assessment.

From the National Fire Protection Association, NFPA 652

8.6 Personal Protective Equipment.
8.6.1 Workplace Hazard Assessment.
8.6.1.1* *An assessment of workplace hazards shall be conducted as described in NFPA 2113.*
8.6.1.2 *When the assessment in 8.6.1.1 has determined that flame-resistant garments are needed, personnel shall be provided with and wear flame-resistant garments.*
8.6.1.3* *When flame-resistant clothing is required for protecting personnel from flash fires, it shall comply with the requirements of NFPA 2112.*

A.8.6.1.1 *A specific evaluation of the work environment to determine the requirement for the wearing of flame-resistant garments should be based on the potential hazards that workers are exposed to as part of their work duties.*
A.8.6.1.3 *It is important to distinguish between the different PPE requirements in NFPA 2112 and NFPA 70E for different exposure hazards. The PPE requirements in NFPA 2112 are not the same requirements in NFPA 70E and might not be sufficient protection for electric arc.*

From the National Fire Protection Association, NFPA 2113

4.1 The organization's selection process for flame-resistant garments shall be based on the conduct of a hazard analysis of the workplace to determine the need for the wearing of flame-resistant garments. The general workplace hazard analysis process shall include a determination of the type of hazards present in the workplace and the potential magnitude, thermal intensity, and duration of the hazard, and the adverse effects of unprotected exposure to the hazards identified.

The purpose of this document is to provide the information and a basis for an analysis of the hazards of combustible-dust fireballs and flash fires.

C. Properties of Combustible-Dust Fireballs and Flash Fires

In order for combustion to propagate through a dust cloud – either as a fireball or as a flash fire – the concentration of dust in the cloud must be above the Minimum Explosible Concentration [MEC]. This corresponds to the Lower Explosive Limit [LEL] for gases and vapors, and the values of MEC and LEL for most materials are similar, at about 30 grams per cubic meter.

A rough estimate of dust concentration can be the lack of light transmission through the cloud. A "rule of thumb" is that a concentration at or above the MEC prevents vision or visibility through about 3 to 6 feet of a cloud of dust [Reference 3].

The result of dust-cloud ignition would be a flash fire or a fireball and, if the combustion is confined, the result would be an explosion, with significant pressure effects.

Important combustible properties of fireballs and flash fires are the Deflagration Index and the Maximum Explosion Pressure. The Deflagration Index is related to the combustion rate and is Kg for flammable gases and vapors, and Kst for combustible dusts. The following Table presents a summary of Kg, Kst, and Pmax data that appear in NFPA 68, 2013 edition [Reference 4].

The data were obtained in a 5-liter test vessel for gases and vapors, and the summary in the following Table exclude acetylene, hydrogen, ammonia, and halogenated compounds, which have unusual properties. The data for dusts were obtained in a 1-cubic-meter test vessel.

Table 1
Summary of Combustibility Properties of "Organic" Dusts

"Organic" Combustible Dusts						
NFPA 68 Table	DEFLAGRATION INDEX Kg [bar-meters per second], FOR GASES AND VAPORS			MAXIMUM EXPLOSION PRESSURE Pmax [barg] FOR GASES AND VAPORS		
	Min.	Avg.	Max.	Min.	Avg.	Max.
D.1	36	88	116	4.4	7.43	8.4
D.2	55	98	148	6.7	7.32	7.9
Averages	-	93	-	-	7.37	-
	DEFLAGRATION INDEX Kst [bar-meters per second], FOR DUSTS			MAXIMUM EXPLOSION PRESSURE Pmax [barg] FOR DUSTS		
E.1(a)	28	116	229	5.8	9.11	10.5
E.1(c)	21	136	364	6.5	8.68	10.6
E.1(e)	12	122	269	5.9	8.67	10.2
Averages	-	125	-	-	8.82	-

As indicated by the above Table for "organic" dusts – such as carbohydrates, hydrocarbons, and other carbon and hydrogen-containing chemicals – there is a wide range of Deflagration Indexes and a relatively-narrow range of Maximum Explosion Pressures. Also, there are many "inorganic" dusts – such as sulfates and carbonates – and halogen-containing dusts that are non-combustible, with zero or near-zero values of these characteristics.

Although there are wide ranges in properties, it is apparent that values of Kg and Pmax for gases and vapors are significantly below those of combustible dusts.

The Maximum Explosion Pressure for gases and vapors are lower because a significant part of the volume in mixtures of fuel and air is occupied by the gas or vapor. In contrast, the volume that is occupied by dust in mixtures with air is almost insignificant.

It is noteworthy that the low values of Kg and Kst correspond to low values of Pmax, and high values of Kg and Kst correspond to high values of Pmax.

Table 2

Summary of Combustibility Properties of Metal Dusts

Combustible Metal Dusts						
NFPA68 Table	DEFLAGRATION INDEX Kst [bar-meters per second], FOR DUSTS			MAXIMUM EXPLOSION PRESSURE Pmax [barg] FOR DUSTS		
	Min.	Avg.	Max.	Min.	Avg.	Max.
E.1(d)	31	228	508	4.1	8.0	17.5

As indicated in the above Table 2 for metal dusts, there are very wide ranges of values for Deflagration Index and Maximum Explosion Pressure. Also, there are several metal dusts that are not explosible, including gold, lead, and zinc, with zero or near-zero values of these characteristics.

From the above data, the following relationship between values of Deflagration Index, K, and Maximum Explosion Pressure, Pmax [in bara], can be obtained:

$$K = 690\, e^{-16.8 / P_{max,a}} \quad \text{bar-meters per second, for Pmax in bara.} \qquad (1)$$

where 16.8 is the "Activation Energy" for explosive combustion of gases, vapors, and dusts, in bar, absolute. Thus, approximate values of burning rate – as expressed by the Deflagration Index - could be determined from the pressure that is developed by combustion in a confined space.

A relationship between "burning velocity" k_r, maximum explosion pressure P [in psig], and rate of pressure increase dP/dt [in psi/second] in test vessels has been developed [Reference 5] for "organic" dusts. The following is a "simplified" form [Reference 6] which assumes only that the initial pressure is atmospheric pressure, and the initial temperature is room temperature:

$$k_r = K1\,(dP/dt) \,/\, [\, P\,(\, P + K2\,)\,] \quad \text{inches per second} \qquad (2)$$

In the Reference:

K1 was given as 3.72;
dP/dt was in terms of psi per second;
P was in psig; and
K2 was +18.3.

However, the values of P in this equation must be in terms of psia, rather than psig, as can be shown by a derivation of this equation.

Further, study of the basic relationship [Reference 5] demonstrated that the value of K1 is a function of the ratio of the internal surface to the volume of the test equipment [S/V]. Also:

K2 should be 3.0;
dP/dt is in terms of psi per second;
P should be in terms of psia; and
K2 should be -14.7.

Thus, the referenced equation would be (for a 1.2-liter cylindrical test vessel):

$$k_r = 3.0\,(dP/dt) / [\,Pa\,(\,Pa - 14.7\,)\,] \qquad \text{inches per second} \qquad (3)$$

This equation was used to obtain values of the burning velocity in the 1.2-liter Hartmann test tube. However, this relationship could be used for confined combustion of dust clouds in equipment having other volumes. Thus, for a 20-liter sphere – where the ratio of internal surface area to volume would be about 5.4 foot $^{-1}$ – the relationship would be:

$$k_r = 10.9(dP/dt)/[Pa(Pa-14.7)] = 10.9(dP/dt)/[(Pg+14.7)(Pg)] \quad \text{in./sec.} \qquad (4)$$

and, for a 1-cubic-meter sphere – where the S/V ratio would be about 1.47 foot $^{-1}$ – the relationship would be:

$$k_r = 40\,(dP/dt)/[Pa(Pa-14.7)] = 40\,(dP/dt)/[(Pg+14.7)(Pg)] \quad \text{in./sec.} \qquad (5)$$

For combustion of "worst-case" near-stoichiometric dust concentrations in a 20-liter sphere, this equation could have the following form:

$$k_r = 0.75\,Kst / \{\,(\,Pmax,g +1\,)(\,Pmax,g\,)\,]\,\} \qquad \text{inches per second} \qquad (6)$$

and, for a 1-cubic-meter sphere:

$$k_r = 2.8\,Kst / \{\,(\,Pmax,g +1\,)(\,Pmax,g\,)\,]\,\} \qquad \text{inches per second} \qquad (7)$$

Based on statements in this reference – that values of k_r are for "non-turbulent" mixtures of dust and air, and "the burning velocities are appreciably lower than those of most flammable gases" – it appears that "burning velocity" corresponds to the "Fundamental Burning Velocity" [FBV].

The Flame Speed for mixtures of combustible materials and air is related to the Fundamental Burning Velocity by [Reference 7]:

$$S_{FLAME} = S_{FBV} [\rho_{BURNED} / \rho_{UNBURNED}] \quad (8)$$

where ρ_{BURNED} is the density of the combustion products, and
$\rho_{UNBURNED}$ is the density of the unburned mixture.

For combustible dusts, the density of the unburned mixture would be that of air, at about 0.075 pound per cubic foot [about 1,200 grams per cubic meter]. The density of stoichiometric burned mixtures – at flame temperatures near 2,250°C [as limited by dissociation of CO_2 and H_2O; Reference 8] – would be about 0.008 pound per cubic foot [about 130 grams per cubic meter]. Thus, the flame-speed density factor would be about 9 for the combustion of dusts. For the combustion of dust/air mixtures near the Minimum Explosible Concentration – where the flame temperature would be near 1,025°C [Reference 9], the density factor would be about 5. In contrast, for most flammable gases and vapors, a significant fraction of the confining volume is occupied by gases and vapors that are denser than air, so the stoichiometric density factor could be as low as 6 [Reference 10], and the LFL density factor would be about 4.

The following Table presents calculated "maximum burning velocities" [k_r] in inches per second [and flame speeds, in feet per second] for a range of Pmax and dP/dt [and Kst] values, as based on values that were obtained for explosions in a 20-liter ["standard"; Reference 11] test vessel:

Table 3

Derivation of Flame Speed from Deflagration Index and Maximum Explosion Pressure

Pmax	Burning Velocities (inches/second) and Flame Speeds [feet/second] for values of Pressure-Increase Rates, in psi/second [bar/second]						
	5,000 [345] Kst=35	10,000 [690] Kst=75	15,000 [1,030] Kst=110	20,000 [1,380] Kst=150	25,000 [1,720] Kst=185	30,000 [2,070] Kst=220	35,000 [2,400] Kst=260
60 psig 4.1 barg	3.96* [3.0]*	7.92 [6.0]	11.88 [8.9]	15.84 [11.9]	19.80 [14.9]	23.75 [17.8]	27.71 [20.8]
80 psig 5.5 barg	2.37 [1.8]	4.73 [3.6]	7.10 [5.3]	9.46 [7.1]	11.83 [8.9]	14.19 [10.6]	16.56 [12.4]
100 psig 6.9 barg	1.57 [1.2]	3.14 [2.4]	4.72 [3.5]	6.29 [4.7]	7.86 [5.9]	9.43 [7.1]	11.01 [8.3]
120 psig 8.3 barg	1.12 [0.8]	2.24 [1.7]	3.36 [2.5]	4.48 [3.4]	5.60 [4.2]	6.72 [5.0]	7.85 [5.9]
140 psig 9.7 barg	0.84 [0.6]	1.68 [1.3]	2.52 [1.9]	3.36 [2.5]	4.20 [3.2]	5.04 [3.8]	5.87 [4.4]
160 psig 11.0 barg	0.65 [0.50]	1.30 [1.0]	1.96 [1.5]	2.61 [2.0]	3.26 [2.4]	3.91 [2.9]	4.56 [3.4]

* Dusts with Kst less than 45 and Pmax less than 4 barg probably are not explosible [Reference 12]

The values that are shown in green, above, are likely combinations of Deflagration Index and Maximum Explosion Pressure for most combustible dusts, as indicated by application of Equation (1), above. The values that are shown in red are unlikely combinations of Kst and Pmax. Thus, a likely flame speed through a cloud of combustible dust would be about 2.5 feet per second.

The data for a Fundamental Burning Velocity [FBV] for dusts are limited. However, the following data are available for a few combustible dusts, as shown in the following Table. The data in the U. S. Bureau of Mines Report [Reference 13] are based on the results that were obtained in 1-liter test equipment, whereas the data in the NFPA publication are based on the results that were obtained in 20-liter test equipment. The values for k_r that are based on the NFPA data were calculated from an equation that is similar to that which was used for the USBM data, but with values of P in psia. The values of flame speed Sf were obtained by multiplying the NFPA value for k_r by the density factor of 9, for "worst-case" near-stoichiometric concentrations.

Table 4

Flame Speeds for Near-Stoichiometric Combustible-Dust Clouds

DUST	USBM RI 7507		k_r (in./ sec.)	NFPA 68,Table E.1		k_r (in./ sec.)	Kst/k_r (bar)	Calculations	
	Pmax (barg)	Kst (barm/s)		Pmax (barg)	Kst (barm/s)			Sf [9X] (ft./sec.)	Sf/Kst
Corn Starch	7.9	66	2.2	10.3	202	4.8	1,650	3.6	0.018
Pittsburgh Coal	5.7	17	1.0	9.2	129	3.8	1,350	2.8	0.022
Poly Ethylene	5.7	18	1.1	8.0	156	6.0	1,025	4.5	0.029
Poly Propylene	4.8	21	1.7	8.4	101	3.5	1,150	2.6	0.026
Stearic Acid	6.0	31	1.7	7.6	126	5.3	950	4.0	0.032
Sulfur	3.9	23	2.8	6.8	151	7.8	750	5.9	0.039
Averages	-	-	1.8	-	-	5.2	1,150	3.9	0.028

As indicated in the preceding Table, the flame speed through a dust cloud could be estimated by multiplying the average value of 0.028 by the Deflagration Index. Thus, for sugar [Kst = 138; Pmax = 8.5 barg], the flame speed through a dust cloud would be about 3.8 feet per second.

Based on the above data, the burning duration for a near-stoichiometric sugar-dust cloud having a diameter of 8 feet – with central ignition – would be about one second, or about two seconds with more-likely "edge" ignition.

It is a "curiosity" that the ratio of Deflagration Index to burning speed corresponds to a value of about 1,150 bar, or about 16,500 psig, although the meaning of this value is not known. However, this value could be used to estimate the burning velocity of other dusts, when the Deflagration Index [Kst] is known [Reference 14].

Note: In the following sections, estimates are provided for the sizes and thermal-radiation intensities for dry 75-micron cornstarch at a "worst credible" weight of 25 kilograms [about 50 pounds] of cornstarch. This weight corresponds to the maximum "Recommended Weight Limit" as established by the National Institute of Occupational Safety and Health [Reference 15]. It would be possible for a dust cloud to result from dropping a 25-kilogram bag of micron-size cornstarch from a working-platform height that would be sufficient to burst the bag, with about 27% of the dusty contents forming a cloud [Reference 16]. At this extent of dispersion, the expanding of the as-yet-unignited dust cloud would have about 6 seconds to encounter an ignition source, to give a fireball or flash fire. Among the possible ignition sources are electrostatic or electric-power sparking, a hot surface, or welding or cutting sparks. Examples are given for a near-spherical fireball and for near-hemispherical flash fires for clouds containing 6.75 kilograms [about 15 pounds] of dusty cornstarch.

D. Thermal Behavior of Combustible-Dust Fireballs

The amount of energy that would be released as a result of fire-ball combustion is the product of the weight of dust and the heat of combustion of the dust [Reference 17]:

$$E = W H_C \quad \text{calories} \tag{9}$$

for W in grams, and
H_C in calories per gram

For example, the combustion energy for a 1-kilogram cornstarch/air fireball would be given by:

$$E = 1{,}000\ (4{,}220) = 4{,}220{,}000 \text{ calories, or about 17.6 megaJoules} \tag{10}$$

This energy is "absorbed" by heating the gases [and steam] that result from the combustion and the nitrogen in the air that provides the oxidant for combustion. For estimation purposes, an average heat capacity of the products of hydrocarbon and carbohydrate combustion in air is about 0.3 calorie per gram of fuel per °C, or about 1.25 Joules/gram-°C, and about 1.15 Joules/g.-°C for air. However, the maximum temperature that could be attained by cornstarch/air combustion is about 2,250°C, as limited by dissociation of carbon dioxide and steam.

a. Fireballs from "Stoichiometric" Dust Clouds

Stoichiometry is defined as the "proportion of reactants such that there is no surplus of any reactants after the reaction is completed." [Reference 18] Thus, for a combustion reaction of a hydrocarbon or carbohydrate with air, all of the carbon in the fuel would be oxidized to carbon dioxide, and all of the hydrogen would be oxidized to water vapor or steam.

The dust concentration that yields the highest values of Pmax and Kst in test equipment can be significantly greater than the calculated stoichiometric concentration. For example, the stoichiometric concentration for cornstarch is about 235 grams per cubic meter, but the "worst-case" concentration is about 500 grams per cubic meter [Reference 19]. For the following discussion, it is assumed that the stoichiometric concentration Cst is the worst-case concentration.

The volume of a dust cloud that has a stoichiometric concentration of W grams of fuel in air would be given by:

$$V_{UNBURNED} = W / Cst \quad \text{cubic meters} \qquad (11)$$

The stoichiometric concentration of a dust dispersed in air is given by:

$$Cst = M / [\, n_{O2} (4.76)(0.0244)] \quad \text{grams per cubic meter} \qquad (12)$$

where M is the molecular weight of the dust;

n_{O2} is the number of gram-moles of oxygen that would be required to oxidize all of the carbon to carbon dioxide and to oxidize all of the hydrogen to steam;

4.76 is the reciprocal of the oxygen concentration in air; and

0.0244 is the number of cubic meters of air in one gram-mole of air.

For example, a value of n_{O2} for cornstarch would be 6, as determined from an oxygen balance:

$$C_6H_{10}O_5 + 6\,O_2 + 18.8\,N_2 \rightarrow 6\,CO_2 + 5\,H_2O + 18.8\,N_2 \qquad (13)$$

Thus, the stoichiometric concentration of cornstarch dust in air is given by:

$$Cst = 162 / [\, 6\,(4.76)(0.0244)] = 235 \quad \text{grams per cubic meter} \qquad (14)$$

The volume of a dust cloud that has a stoichiometric concentration and contains 6.75 kilograms of cornstarch dispersed in air would be given by:

$$V_{UNBURNED} = 6750 / Cst = 6750 / 235 = 29 \quad \text{cubic meters} \qquad (15)$$

The initial diameter of a stoichiometric spherical fireball – prior to ignition – is given by:

$$D_{UNBURNED} = [(6 * W) / (\pi * Cst)]^{1/3} \quad \text{meters} \qquad (16)$$

Thus, the initial diameter of stoichiometric spherical dust cloud containing 6.75 kilograms of cornstarch would be given by:

$$D_{UNBURNED} = [(6 * 6750) / (\pi * 235)]^{1/3} = 3.8 \text{ meters, or about 12.5 feet} \qquad (17)$$

Upon ignition, a cloud of burning dust would expand approximately in proportion to the absolute increase in pressure [or to the increase in temperature] when the combustion is confined, according to the Ideal Gas Law. As shown in Table 4, above, the value of Pmax for cornstarch is 10.3 barg, or 11.3 bar, absolute.

Thus, ignition of a spherical stoichiometric dust cloud containing 6.75 kilograms of cornstarch would result in a fireball having a volume that would be given by:

$$V_{BURNED} = (W/Cst)(Pmax +1)/1 = (6750/235)(10.3 + 1)/1 = 325 \text{ cu.m.} \qquad (18)$$

and the diameter of the fireball would be given by:

$$D_{BURNED} = [(6 * 325) / \pi]^{1/3} = 8.5 \text{ meters, or about 28 feet} \qquad (19)$$

Such a large fireball could occur only in a very large building, with a 28-foot ceiling, so it is likely that the shape of the combustion cloud would be more-nearly a hemisphere [as described later] or a cylindrical "pancake".

The surface area of a fireball would be given by:

$$S = \pi D^2 \quad \text{square meters, for D in meters.} \qquad (20)$$

Thus, the maximum surface area of a stoichiometric 6.75-kilogram fireball would be about:

$$S = \pi\, 8.5^2 = 225 \text{ square meters, or about 2,500 square feet} \qquad (21)$$

The combustion energy for a fireball containing 6.75 kilograms of cornstarch – about 120 megaJoules [at 4,220 calories per gram; Reference 20] minus about 5% for the "PV" work of fireball expansion – over the 225 square meters would give the following thermal-radiation energy:

$$E = 114{,}000{,}000 / 225 = 500 \text{ kiloJoules/sq.m. at the surface, or about 50 J/sq.cm.} \quad (22)$$

As will be shown later, this energy density would be sufficient to cause third-degree burns and to most ordinary clothing if exposure was prolonged for more than a few seconds.

From Table 4, the flame speed would be about 3.6 feet per second. Thus, the approximate duration of burning of a 45-foot-diameter stoichiometric fireball – for central ignition – would then be given by:

$$t = (28 / 2) / 3.6 = 14 / 3.6 = 3.9 \quad \text{seconds} \quad (23)$$

The thermal-radiation intensity at the surface of the fully-developed fireball would be given by:

$$P = E / t = 500{,}000{,}000 / 3.9 = 130 \text{ kilowatts/sq.m., or about 13 watts/sq.cm.} \quad (24)$$

For comparison, the maximum thermal-radiation intensity of sunlight is about 1 kilowatt per square meter [Reference 21]. Thus, this would be hazardous exposure to heat.

With the above similar combustion behaviors for gases, vapors, and dusts, many of the well-established fireball properties of flammable-vapor fireballs [Reference 22] could be applied to fireballs of combustible dusts.

For flammable vapors and gases, the approximate duration of a fireball is related to the weight of fuel in the fireball by [Reference 23]:

$$t = 0.852\, W^{0.26} \quad \text{seconds, for the weight in kilograms.} \quad (25)$$

Thus, for a fireball containing 6.75 kilograms of cornstarch "vapor", the duration would be about 1.4 seconds, as compared to the above-estimated 3.9 seconds for a dust fireball.

It may be of interest that a cornstarch-dust burning velocity and a flame speed were measured at 0.27 and 0.64 meter per second – 10 in./sec. and 2.1 ft./sec., respectively – in vertical 8"x8"x2-meter tubes [Reference 24], supporting the above calculations.

b. Fireballs from "MEC" Dust Clouds

The initial [unburned] volume of a dust cloud in which the concentration would be the Minimum Explosible Concentration would be given by:

$$V_{UNBURNED} = W / MEC \qquad \text{cubic meters} \qquad (26)$$

The MEC for a dust dispersed in air is determined by test, but an approximate MEC is about 20% of the stoichiometric concentration. Thus:

$$MEC = 0.2\, Cst = 0.2\, M / [\, n_{O2}\, (4.76)(0.0244)] \qquad \text{grams per cubic meter} \qquad (27)$$

where M is the molecular weight of the dust;

n_{O2} is the number of gram-moles of oxygen that would be required to oxidize all of the carbon to carbon dioxide and to oxidize all of the hydrogen to steam;

4.76 is the reciprocal of the oxygen concentration in air; and

0.0244 is the number of cubic meters of air in one gram-mole of air.

For example, the MEC for cornstarch dust dispersed in air would be given by:

$$MEC = 0.2\, (235) = 47 \qquad \text{grams per cubic meter} \qquad (28)$$

This compares favorably with MEC values of 40, 30 to 60, and 60 g./cu.m. for cornstarch, in the open literature [References 25, 26, and 27].

The volume of a dust cloud that has a near-MEC concentration and contains 6.75 kilograms of cornstarch dispersed in air would be given by:

$$V_{UNBURNED} = W / MEC = 6750 / 47 = 145 \qquad \text{cubic meters} \qquad (29)$$

The initial diameter of a spherical fireball – prior to ignition – is given by:

$$D_{UNBURNED} = [\, (6 * W) / (\pi * MEC)\,]^{1/3} \qquad \text{meters} \qquad (30)$$

Thus, the initial diameter of spherical MEC dust cloud containing 1 kilogram of cornstarch would be given by:

$$D_{UNBURNED} = [\, (6 * 6750) / (\pi * 47)\,]^{1/3} = 6.5 \text{ meters, or about 21 feet} \qquad (31)$$

Upon ignition, a cloud of burning dust would expand in proportion to the absolute increase in pressure when the combustion is confined, according to the Ideal Gas Law. However, the MEC is defined as "the minimum concentration that is capable of propagating a deflagration." Thus, there would be minimal increase in pressure as a result of ignition of a MEC dust cloud.

Typically, the MEC is determined when the pressure resulting from combustion attains a Pressure Ratio of 0.5 in a confining test vessel. This corresponds to an absolute pressure near 30 psia [References 28 and 29] and, also, to a flame temperature near 350°C and, further, near the Minimum Ignition Temperature of the dust being tested. The expansion of the burning dust cloud would be in proportion to the absolute increase in temperature, according to the Ideal Gas Law.

Thus, ignition of a spherical MEC dust cloud containing 6.75 kilograms of cornstarch would result in a fireball having a volume that would be given by:

$$V_{BURNED} = (W/MEC)(T_{MEC}+273)/(25+273) = (6750/47)(623/298) = 300 \text{ cu.m.} \quad (32)$$

and the diameter of the fireball would be given by:

$$D_{BURNED} = [(6 * 300)/\pi]^{1/3} = 8.3 \text{ meters, or about 27 feet} \quad (33)$$

This compares with the 28-foot diameter for a stoichiometric cornstarch fireball. Thus, it is very likely that the "final" diameters of fireballs for the range of dust concentrations from the MEC to the stoichiometric concentration Cst would be about 28 feet for a fireball involving the combustion of 6.75 kilograms of cornstarch. Cloud volumes and diameters could be estimated through use of the above equations.

The surface area of a fireball would be given by:

$$S = \pi D^2 \quad \text{square meters, for D in meters.} \quad (34)$$

Thus, the maximum surface area of the near-MEC 6.75-kilogram fireball would be about:

$$S = \pi\, 8.3^2 = 215 \text{ square meters, or about 2,300 square feet} \quad (35)$$

The combustion energy for a fireball containing 6.75 kilograms of cornstarch – about 120 megaJoules [Reference 30] minus about 5% for the "PV" work of fireball expansion – over the 215 square meters would give the following thermal-radiation energy:

$$E = 114{,}000{,}000 / 215 = 530 \text{ kiloJoules/sq.m. at the surface, or about 53 J/sq.cm.} \quad (36)$$

As will be shown later, this energy density would be sufficient to cause third-degree burns and ignition of ordinary clothing if exposure was prolonged.

Graphical relationships between pressure and time for closed-vessel tests indicate the pressure-increase rate at the MEC "Go/NoGo" pressure of 1.1 barg would be about 15 bar per second [about 220 psi/second; Reference 31]. Then, from Equation (1) above, the "burning rate" at a pressure of 1.1 barg would be given by:

$$k_r = 40(dP/dt)/[Pg\ (Pg+14.7)] = 40(220)/[16(16+14.7)] = 18 \text{ in./sec. or } 1.5 \text{ ft./sec.} \qquad (37)$$

At the Minimum Explosible Concentration, the flame temperature for most combustible dusts would be near 350°C [Reference 32]. Thus, the density ratio for a MEC fire ball would be near 2.0 because of the excess of air, and the flame speed would be about 3.0 feet per second.

The approximate duration of burning of a 27-foot-diameter near-MEC fireball – for central ignition – would then be given by:

$$t = (27 / 2) / 3 = 13.5 / 3 = 4.5 \quad \text{seconds} \qquad (38)$$

The thermal-radiation intensity at the surface of the fully-developed fireball would be given by:

$$P = E / t = 530{,}000{,}000 / 4.5 = 120 \quad \text{kilowatts/sq.m., or } 12 \text{ watts/sq.cm.} \qquad (39)$$

This is about 90% of the estimated intensity of the stoichiometric cornstarch dust cloud.

c. Fireballs from "Rich" Dust Clouds

For most combustible dusts, there is no clear Upper Explosible Concentration [UEC], but the UEC could be as high as thirty times the stoichiometric concentration Cst [Reference 33]. For example, the UEC for cornstarch would then be about 2,500 g./cu.m. [Reference 34]. For the following discussion of UEC fireballs and flash fires, a value of 2,500 g./cu.m. is assumed, for which the UEC/Cst ratio would be about 11.

The volume of a dust cloud that has a concentration equal to the UEC and a weight of W grams of fuel in air would be given by:

$$V_{UNBURNED} = W / UEC \quad \text{cubic meters} \qquad (40)$$

The volume of a dust cloud that has a UEC of 2,500 g./cu.m. and contains 6.75 kilograms of cornstarch dispersed in air would be given by:

$$V_{UNBURNED} = W / UEC = 6750 / 2{,}500 = 2.7 \text{ cubic meters} \quad (41)$$

The initial diameter of a spherical fireball – prior to ignition – is given by:

$$D_{UNBURNED} = [(6 * W) / (\pi * UEC)]^{1/3} \text{ meters} \quad (42)$$

Thus, the initial diameter of UEC spherical dust cloud containing 6.75 kilograms of cornstarch would be given by:

$$D_{UNBURNED} = [(6 * 6750) / (\pi * 2500)]^{1/3} = 1.7 \text{ meters, or about } 5.7 \text{ feet} \quad (43)$$

In contrast to the preceding discussion, the combustion of dust in a cloud in which the concentration was far above the stoichiometric concentration would behave very differently. Because of the absence of sufficient oxygen for complete combustion, ignition of a UEC dust cloud would result in expulsion of unburned dust into the surrounding oxygen-rich air. The resulting combustion of the expelled dust would result in a fireball that was much larger than the original UEC dust cloud.

An example of such combustion is the venting of unburned dust from a dust collector, in which combustion had been initiated. Based on study of photographs of fireballs from explosion vents, it appears that the volume of such fireballs is about 40 times the volume of the protected dust collector. This is roughly equivalent to the ratio of Pmax to Pstat, where Pstat is the pressure at which the explosion vent opens. Also, it appears that there is no unburned dust outside the fireball; that is, the temperature of the fireball is above the MEC-limit temperature: the temperature – about 1,250°C – at which the MEC becomes zero [Reference 35].

Based on the above considerations, it could be concluded that a UEC fireball would expand until the dust "encountered" sufficient oxygen in the surrounding air to fully burn. That is, the volume of the fireball would be determined from the MEC.

For a cornstarch MEC of about 47, a dust weight of 6.75 kilograms, and a MEC flame temperature near 1,250°C, the volume of the fireball would be given by:

$$V_{BURNED} = (6750 / 47) ([1250 + 273] / 298) = 735 \text{ cubic meters} \quad (44)$$

That is, burning of the dust would result in an expansion of the cloud from 2.7 cubic meters to 735 cubic meters, or about a factor of 270. This is almost an order of magnitude greater than the photographed explosion-venting expansions.

The diameter of the fully-developed fireball would be given by:

$$D_{BURNED} = [(6 * 735) / \pi]^{1/3} = 11 \text{ meters, or about 37 feet} \qquad (45)$$

The surface area of a fireball would be given by:

$$S = \pi D^2 \quad \text{square meters, for D in meters.} \qquad (46)$$

Thus, the maximum surface area of the stoichiometric 6.75-kilogram fireball would be about:

$$S = \pi 11^2 = 380 \text{ square meters, or about 4,000 square feet} \qquad (47)$$

The combustion energy for a fireball containing 6.75 kilograms of cornstarch – about 120 megaJoules [Reference 36] minus about 5% for the "PV" work of fireball expansion – over the 113 square meters would give the following thermal-radiation energy:

$$E = 114{,}000{,}000 / 380 = 300 \text{ kiloJoules/sq.m. at the surface, or about 30 J/sq.cm.} \qquad (48)$$

As will be shown later, this energy density would be sufficient to cause third-degree burns and ignition of ordinary clothing for prolonged exposure.

The flame speed through the UEC cloud would vary considerably, from a low value at the high initial concentration, to a maximum value as the dust/air mixture nears stoichiometry [near 3.8 ft./sec.], and to a low value [near 3.0 ft./sec.] as the concentration nears the MEC. Thus – in the absence of better data – an "average" flame speed for a UEC dust cloud would be about 3.4 feet per second for a UEC cornstarch fireball.

The approximate duration of burning of a 37-foot-diameter of a UEC fireball – for central ignition – would then be given by:

$$t = (37 / 2) / 3.4 = 18.5 / 3.4 = 5.5 \quad \text{seconds} \qquad (49)$$

The thermal-radiation intensity at the surface of the fully-developed fireball would be given by:

$$P = E / t = 300{,}000{,}000 / 5.5 = 55 \quad \text{kilowatts/sq.m., or 5.5 watts/sq.cm.} \qquad (50)$$

This is about 40% of the estimated intensity of the stoichiometric cornstarch dust cloud. However, an important feature of fireballs that have initial concentrations much greater than stoichiometric is the significantly-greater distances – about 35% greater – that are attained by expansion of the fireballs.

E. Thermal Behavior of Combustible-Dust Flash Fires

a. Combustible-Dust Flash Fires

For the purpose of this study, it is assumed that the shape of a combustible-dust flash fire would be a hemisphere, as shown by the electrical-classification diagram for a dense dust cloud [Reference 37]. A "pancake" flash fire could have the same volume for a given dust weight and concentration in air, but the diameter of the flash fire would depend on the height of the "pancake", such as the height of a restraining ceiling. However, several of the properties of a "pancake" flash fire – such as the volume and the thermal-radiation intensity, in terms of Joules per square meter – would be the same as those of a hemispherical flash fire.

The volumes of hemispherical flash fires would be the same as the volumes of the spherical fireballs, as described above, for a given weight of dust and dust concentration in air. However, the diameter of the hemisphere would be greater than that of the sphere by the cube root of two, or about 26%. The "exposed" surface area of the hemispherical flash fire would be about 80% of the total surface of a spherical fireball [$0.5 \times 2^{2/3}$]. About 67% of the total surface area of the hemispherical flash fire would be above the ground or surface level [0.5 / 0.75], with about 33% of the total surface area in contact with the ground or floor.

For the purpose of this study, it is assumed that the ground or floor is "insulating"; that is, thermal energy is not absorbed into the floor or ground. A wooden floor or epoxy-coated concrete could provide a thermal barrier. For example, the "thermal diffusivity" for a wood floor would be about 0.003 square feet per hour; for bare concrete, 0.018; for epoxy-coated concrete, about 0.022 [for epoxy at 0.028]; and for soil at 0.039; as compared to thermally-conductive steel at 0.5 square feet per hour.

b. Flash Fires from "Stoichiometric" Dust Clouds

Ignition of a hemispherical dust cloud containing 6.75 kilograms of cornstarch would result in a flash fire having a volume that would be given by:

$$V_{BURNED} = (W/Cst)(Pmax + 1)/1 = (6750/235)(10.3 + 1)/1 = 325 \text{ cu.m.} \qquad (51)$$

and the diameter of the hemisphere would be given by:

$$D_{BURNED} = [(12 * 325) / \pi]^{1/3} = 10.7 \text{ meters, or about 35 feet} \qquad (52)$$

The "exposed" surface area of a hemispherical flash fire would be given by:

$$S = \pi D^2 / 2 = 1.57 D^2 \quad \text{square meters, for D in meters.} \qquad (53)$$

Thus, the exposed surface area of a stoichiometric 1-kilogram fireball would be about:

$$S = \pi\, 10.7^2 / 2 = 180 \text{ square meters, or about 1,900 square feet} \qquad (54)$$

The area of the hemispherical flash fire that is in contact with the floor would be about:

$$S = (\pi / 4)\, 10.7^2 = 90 \text{ square meters, or about 970 square feet} \qquad (55)$$

However, it is assumed that heat is not conducted into the floor.

The combustion energy for a fireball containing 6.75 kilograms of cornstarch – about 120 megaJoules [Reference 38] minus about 5% for the "PV" work of expansion of the flash fire – over the 90 square meters would give the following thermal-radiation energy:

$$E = 114{,}000{,}000 / 180 = 630 \text{ kiloJoules/sq.m., or 63 Joules/sq.cm.} \qquad (56)$$

As will be shown later, this energy density would be sufficient to cause fatal injury and to ignite most ordinary clothing, for prolonged exposure.

From the preceding discussion, the flame speed would be about 3.6 feet per second. Thus, the approximate duration of burning of a 35-foot-diameter flash fire – for central ignition – would then be given by:

$$t = (35 / 2) / 3.6 = 17.5 / 3.6 = 4.9 \quad \text{seconds} \qquad (57)$$

The thermal-radiation intensity at the surface of the flash fire would be given by:

$$P = E / t = 630{,}000 / 4.9 = 130 \quad \text{kilowatts/sq.m., or 13 watts/sq.cm.} \qquad (58)$$

For comparison, the thermal-radiation intensity of a 6.75-kilogram fireball – entirely above ground level – would be about the same, at 13 watts/sq.cm., but the distance to the outer edge of the flash fire [the radius] would be about 20% greater than that of a fireball [17 feet versus 14 feet, for this example] for a given weight of combustible dust in an ignited cloud.

c. Flash Fires from "MEC" Dust Clouds

As described above for MEC fireballs, the expansion of the burning dust cloud would be in proportion to the absolute increase in temperature, according to the Ideal Gas Law. The temperature of the flame at the MEC for most combustible dusts is about 1,225°C [Reference 39]. Thus, the volume of a 6.75-kilogram MEC cloud of cornstarch dust would increase by about a factor of 2, from 145 cubic meters to about 300 cubic meters, and the diameter of the hemispherical flash fire would be given by:

$$D_{BURNED} = [(12 * 300) / \pi]^{1/3} = 10.5 \text{ meters, or about 34 feet} \qquad (59)$$

This compares with the 35-foot diameter for a stoichiometric cornstarch flash fire. Thus, it is very likely that the "final" volumes and diameters of flash fires for the range of dust concentrations from the MEC to the stoichiometric concentration Cst would be about the same, and about 35 feet for dust clouds containing 6.75 kilograms of cornstarch.

The "exposed" surface area of a flash fire would be given by:

$$S = \pi D^2 / 2 \qquad \text{square meters, for D in meters.} \qquad (60)$$

Thus, the maximum exposed surface area of the near-MEC 6.75-kilogram flash fire would be about:

$$S = \pi\, 10.5^2 / 2 = 175 \text{ square meters, or about 1,850 square feet} \qquad (61)$$

The area of the hemispherical flash fire that is in contact with the floor would be about:

$$S = (\pi / 4)\, 10.5^2 = 86 \text{ square meters, or about 950 square feet} \qquad (62)$$

However, it is assumed that heat is not conducted into the floor.

The combustion energy for a flash fire containing 6.75 kilograms of cornstarch – about 120 megaJoules [Reference 40] minus about 5% for the "PV" work of fireball expansion – over the 175 square meters would give a thermal-radiation energy of about 650 kiloJoules per square meter at the surface, or about 65 Joules per square centimeter.

As will be shown later, this energy density would be sufficient to cause fatal injury and to ignite most ordinary clothing, for prolonged exposure.

It appears that there may be a near-linear relationship between pressure and pressure-increase rate near the MEC [Reference 41], as given by:

$$dP/dt = 118 / 3.9 = 30\ P \quad \text{bar per second, for P in barg, or} \qquad (63)$$

$$dP/dt = 1710 / 57 = 30\ P \quad \text{psi per second, for P in psig.} \qquad (64)$$

Thus, the rate of pressure increase near a pressure of 0.5 barg [7.25 psig] would be about 220 psi/second. Then, from Equation (1) above, the "burning rate" at a pressure of 0.5 barg [about 7.3 psig] – where the pressure-increase rate would be about 15 psi/second – would be given by:

$$k_r = 40(dP/dt)/[Pg\ (Pg+14.7)] = 40(15)/[7.3(7.3+14.7)] = 3.7 \text{ in./sec. or } 0.3 \text{ ft./sec.} \qquad (65)$$

At the Minimum Explosible Concentration, the flame temperature for most combustible dusts is near 1,225°C [Reference 42]. Thus, the density ratio for a MEC flash fire would be near 2.0 because of the excess of air, and the flame speed would be about 3 feet per second.

The approximate duration of burning of a 34-foot-diameter near-MEC fireball – for central ignition – would then be given by:

$$t = (34 / 2) / 3 = 17 / 3 = 5.7 \quad \text{seconds} \qquad (66)$$

The thermal-radiation intensity at the surface of the fully-developed fireball would be given by:

$$P = E / t = 650{,}000 / 5.7 = 115 \quad \text{kilowatts/sq.m., or 15 watts/sq.cm.} \qquad (67)$$

This is about the same as the estimated intensity of the stoichiometric cornstarch flash fire and could cause first-degree burns [similar to sunburns].

For comparison, the thermal-radiation intensity of a near-MEC 6.75-kilogram fireball – entirely above ground level – would be about 120 kilowatts per square meter, as described above. Thus, the thermal-radiation intensity for a fireball and a flash fire would be about the same, but the distance to the outer edge of the flash fire would be about 30% greater for the flash fire [17 feet versus 13 feet].

d. <u>Flash Fires from "Rich" Dust Clouds</u>

As discussed above, it could be concluded that a UEC fireball would expand until the dust "encountered" sufficient oxygen in the surrounding air to fully burn. That is, the volume of the fireball would be determined from the MEC.

For a cornstarch MEC of about 47, a dust weight of 6.75 kilograms, and a MEC flame temperature near 1,250°C, the volume of a flash fire would be given by:

$$V_{BURNED} = (6750 / 47) ([1250 + 273] / 298) = 735 \text{ cubic meters} \qquad (68)$$

The diameter of the fully-developed flash fire would be given by:

$$D_{BURNED} = [(12 * 735) / \pi]^{1/3} = 14 \text{ meters, or about 45 feet} \qquad (69)$$

The "exposed" surface area of a flash fire would be given by:

$$S = \pi D^2 / 2 \quad \text{square meters, for D in meters.} \qquad (70)$$

Thus, the maximum "exposed" surface area of the stoichiometric 6.75-kilogram flash fire would be about:

$$S = \pi\, 14^2 / 2 = 310 \text{ square meters, or about 3,300 square feet} \qquad (71)$$

The area of the hemispherical flash fire that is in contact with the floor would be about:

$$S = (\pi / 4)\, 14^2 = 150 \text{ square meters, or about 1,650 square feet} \qquad (72)$$

However, it is assumed that heat is not conducted into the floor.

The combustion energy for a fireball containing 6.75 kilograms of cornstarch – about 120 megaJoules [Reference 43] minus about 5% for the "PV" work of fireball expansion – over the 310 square meters would give a thermal-radiation energy of about 370 kiloJoules per square meter at the surface.

The flame speed through the UEC cloud would vary considerably, from a low value at the high initial concentration, to a maximum value as the dust/air mixture nears stoichiometry [near 3.8 ft./sec.], and to a low value [near 3.0 ft./sec.] as the concentration nears the MEC. Thus – in the absence of better data – an "average" flame speed for a UEC dust cloud would be about 3.4 feet per second for a UEC cornstarch fireball.

The approximate duration of burning of a 25-foot-diameter UEC flash fire – for central ignition – would then be given by:

$$t = (45 / 2) / 3.4 = 22.5 / 3.4 = 6.6 \quad \text{seconds} \qquad (73)$$

The thermal-radiation intensity at the surface of the flash fire would be given by:

$$P = E / t = 370{,}000 / 6.6 = 55 \quad \text{kW/sq.m., or about 5.5 kW/sq.cm.} \qquad (74)$$

This is about 40% of the estimated intensity of the stoichiometric cornstarch flash fire.

For comparison, the thermal-radiation intensity of a UEC 6.75-kilogram fireball – entirely above ground level – would be about 130 kilowatts per square meter, as described above. Thus, the thermal-radiation intensity for a flash fire would be about the same as that of a fireball, but the distance to the outer edge of the flash fire would be about 30% greater for the flash fire [22 feet versus 17 feet].

F. Literature Data

An alternative method for estimating the emissive power of fireballs and flash fires is via the Stefan-Boltzmann Law:

$$P = \varepsilon \sigma T^4 \quad \text{watts per square meter} \qquad (75)$$

for ε [emissivity], dimensionless;
σ [Stefan-Boltzmann constant] = 5.67×10^{-8} watts per square meter – °K ^{4}; and
T in °K.

Measured values of the emissive power for fireballs [Reference 44] and the back-calculated temperatures – assuming an emissivity of 1.0 – are as follows, with the likely color of the flame [Reference 45]:

Table 5
Measured Emissive Powers for Fireballs

EMISSIVE POWER	TEMPERATURE	COLOR
120 kW/sq.m.	1,210°K [935°C]	Orange
150 kW/sq.m. [avg.]	1,275°K [1,000°C]	Yellow
300 kW/sq.m. [max.]	1,515°K [1,245°C]	White

These data provide support for the above calculations for combustion of stoichiometric and MEC dust clouds.

G. Conclusion for Combustible-Dust Fireballs and Flash Fires

The results of the above calculations can be summarized as presented in the following Table.

Table 6
Properties for a 6.75-Kilogram Burning Dust Cloud of Cornstarch

HAZARD	INITIAL CONC.	PROPERTIES FOR 6.75-KG. CORNSTARCH DUST CLOUD				
		VOLUME (cu.m.)	DIAMETER (m. / ft.)	ENERGY (kJ./sq.m.)	DURATION (seconds)	INTENSITY (kW./sq.cm.)
Fireball	Stoichiometric	325	8.5 / 28	500	3.9	13
	MEC	300	8.3 / 27	530	4.5	12
	Rich	735	11 / 37	300	5.5	5.5
Flash Fire	Stoichiometric	325	10.7 / 35	630	4.9	13
	MEC	300	10.5 / 34	650	5.6	12
	Rich	735	14 / 45	370	6.6	5.5

The fireball and flash fire properties for other combustible dusts can be estimated from the procedures that are provided above.

H. Effects of Human Exposures to Fireballs and Flash Fires

a. Effects of Distance on Thermal Radiation

Thermal-radiation hazards decrease with the square of the distance from the surface of a fireball or flash fire, as given by [References 46 and 47]:

$$P_{RECEPTOR} = P_{FLAME} / [\, R_{RECEPTOR} / R_{FLAME} \,]^2 \text{ kilowatts per square meter} \quad (76)$$

where P_{FLAME} is the thermal-radiation intensity at the surface of the fireball or flash fire;
$R_{RECEPTOR}$ is the distance to the exposed person from the center of the fireball or flash fire, in feet; and
R_{FLAME} is the radius of the fireball or flash fire, in feet

In the above equation, it is assumed that the transmissivity of the atmosphere is 1.0 [that is, low humidity, and/or short distances] and the view factor is 1.0 [that is, the person is exposed to the whole visible part of the fireball or flash fire]. As can be seen from the above relationship, the intensity of thermal-radiation exposure to a receptor at the outer surface of a fireball or flash fire would be equal to the intensity of the fireball or flash fire at the surface of the fireball or flash fire.

The following Table can be constructed from the above equation, for a fireball or flash fire involving the combustion of 6.75 kilograms [about 15 pounds] of cornstarch, for several distance from the surface of the burning dust cloud. In the Table, correlations are presented between energy intensity [in Joules per square centimeter] and the degree of injury to persons outside the fireball or flash fire, where flame-resistant clothing is not worn. Values of energy intensity that could result in first-degree injury [sunburn] are shown in yellow; second-degree injury [blisters] in orange; and third-degree injury or clothing ignition [possible fatal injury] in red, with a high probability in bold red. There would be a very high probability of severe or fatal injury for anyone within the fireball, particularly if clothing were to be ignited. Other entries in this Table indicate painful exposure.

Table 7
Hazard Ranges at a Range of Distances

Hazard	Radius	Surface Energy (J/sq.cm.)	Thermal-Radiation Dose at Given Distances from the Origin of the Fireball or Flash Fire for a 6.75-Kilogram Dust Cloud of Cornstarch					
			15 ft.	20 ft.	25 ft.	30 ft.	40 ft.	50 ft.
Stoichiometric Fireball	14 ft.	50 [3.9 sec.]	45 J./sq.cm.	25 J./sq.cm.	16 J./sq.cm.	[illegible] J./sq.cm.	6 J./sq.cm.	4 J./sq.cm.
MEC Fireball	13 ft.	53 [4.5 sec.]	**40** J./sq.cm.	22 J./sq.cm.	14 J./sq.cm.	[illegible] J./sq.cm.	6 J./sq.cm.	4 J./sq.cm.
Rich Fireball	19 ft.	30 [5.5 sec.]	**30** J./sq.cm.	27 J./sq.cm.	17 J./sq.cm.	[illegible] J./sq.cm.	7 J./sq.cm.	4 J./sq.cm.
Stoichiometric Flash Fire	18 ft.	63 [4.9 sec.]	**63** J./sq.cm.	**51** J./sq.cm.	33 J./sq.cm.	23 J./sq.cm.	[illegible] J./sq.cm.	**8** J./sq.cm.
MEC Flash Fire	17 ft.	65 [5.6 sec.]	**65** J./sq.cm.	47 J./sq.cm.	30 J./sq.cm.	21 J./sq.cm.	[illegible] J./sq.cm.	**8** J./sq.cm.
Rich Flash Fire	22 ft.	37 [6.6 sec.]	**37** J./sq.cm.	**37** J./sq.cm.	29 J./sq.cm.	20 J./sq.cm.	[illegible] J./sq.cm.	**7** J./sq.cm.

As mentioned above, where the diameters of the fireballs exceed the floor-to-ceiling height, the values in the above table for flash fires should be used; for example, to estimate the properties of a "pancake" or cylindrical flash fire. As shown in the above Table, the most-serious exposures would be to flash fires within a distance of 30 feet from the origin, such as the ignition source.

The combustion durations in the above Table indicate the "escape time" that might be available to avoid full exposure to a fireball or flash fire.

b. Consequences of Personnel Exposures to Fireballs and Flash Fires

The following Table 1 presents the consequences of exposure to thermal radiation of a range of energy densities, for exposures of 20 seconds or less [Reference 48] that were used in the preceding Table:

Table 8

Consequences of Exposure to Fireballs and Flash Fires

Consequences Of Exposure	Pain	Sunburn 1st Degree	Blisters 2nd Degree	Cotton Ignition	Fatal 1%	Fatal 50%	Fatal 99%
Energy Density (Joules/sq.cm.)	4	10	17	20	23	50	120

Additional data concerning the ignition of clothing are [Reference 49]:

Table 9

Properties of Clothing Exposed to Thermal Radiation

FABRIC	MELTING TEMP.	AUTOIGNITION TEMPERATURE	AUTOIGNITION ENERGY (J/sq.cm.)	BURN POWER TO SKIN (W/sq.cm.)	FIGURE OF MERIT (seconds)
Acrylic	235 to 320°C	465°C	80 [est.]	3.6 to 5.8	17
Cotton	[No Melt]	280 to 443°C	29 to 55 *	0.44 to 3.9	10
Rayon	255 to 293°C	490°C	40 *	2.7	15
Polyester	252 to 292°C	485°C	75	0.4 to 0.57	150
Nylon	160 to 260°C	425°C	98	0.1 to 0.18	700
Kynol	[No Melt]	614°C	98 [est.]	0.1 to 0.18	700
Wool	[No Melt]	463 to 590°C	83	0.07 to 0.13	830
FR Cotton	[No Melt]	230°C	17 * to 165	0.18 to 1.9	90
FR Wool	[No Melt]	570 to 590°C	17 * to 165	0.12 to 0.16	150 to 1000
Nomex	[No Melt]	675°C	200 [est.]	0.15 [est.]	1250
Teflon	[No Melt]	>600°C	[Not Ignited]	0.15 [est.]	1500

In the preceding Table, the fabrics are arranged in an order that had been suggested by other investigators [Reference 50]. The estimated values were obtained by comparison with pertinent data for other materials. The designers and fabricators of flame-resistant [FR] fabrics can adjust the properties, primarily by controlling the thickness of the material, in terms of dozens of grams per square meter. This can limit (1) the flammability of the fabric and (2) the transfer of heat to the skin – as second-degree burn – from the burning fabric. The ratio of these factors is shown as an approximate "Figure of Merit" (in "allowable" seconds of exposure) in the above Table.

Also as indicated by an asterisk in the preceding Table, several of the above fabrics – including flame-resistant clothing with a low degree of protection – could be ignited by exposure to a fireball or flash fire containing 6.75 kilograms of cornstarch.

It is important to note that materials that melt or drip during the flame-resistance testing cannot be considered to be flame resistant, according to the ASTM standard [Reference 51]. Also, flame-resistant garments must have no "afterflame" [after the thermal-radiation source is removed] for more than 5 seconds. Further, the garments must be designed for easy and rapid removal.

I. Protection Against Combustible-Dust Fireballs and Flash Fires

Whenever there is a hazard involved in handling a combustible material, a hazard assessment should be performed [References 52 and 53]. Depending on the results of the assessment, personal protective equipment [PPE] could be specified for the persons who perform the material handling and, perhaps, the persons nearby. The PPE would be intended to provide protection against fireballs and flash fires, but it would not be practical to provide protection against explosion-caused equipment rupture or structural damage.

a. Minimal Protection Against Fireballs and Flash Fires Involving "Organic" Powders

Where relatively coarse combustible powders are handled, both the Ignition Sensitivity and the Explosion Severity [References 54 and 55] are likely to be low; that is, it is unlikely that a fireball or flash fire would result from mishandling the powder, that a dense cloud would be formed, and/or that the cloud of particles would be ignited by typical ignition sources. The following data for fructose [References: 56 and 57] are likely to apply generally to organic dusts [Reference 58]:

Table 10
Effect of Particle Size on Explosibility Index

Particle Diameter	Min. Ignition Energy	Min. Ignition Temp.	Min. Explos. Conc.	Ignition * Sensitivity	Max. Explos. Press.	Deflag. Index, Kst	Explos.* Severity	Explos. Index
150 μ	1 milliJ.	430 °C	60 g./m³	80	9.0 barg	102 bm/s	1.1	86
200 μ	180 mJ.	440 °C	60 g./m³	0.42	7.0 barg	28 bm/s	0.23	0.10
400 μ	3 J	530 °C	125 g./m³	0.01	6.5 barg	25 bm/s	0.19	0.002
	Standard for Ignition Sensitivity and Explosion Severity: Pittsburgh [Bituminous] Coal Dust							
75 μ	60 mJ.	610 °C	55 g./m³	1.00	8.1 barg	103 bm/s	1.00	1.00

In the above Table, the Ignition Sensitivity is the product of the Minimum Ignition Energy, the Minimum Ignition Temperature, and the Minimum Explosible Concentration, divided into the similar product – about 2,000,000 – for Pittsburgh Coal Dust. The Explosion Severity is the product of the Maximum Explosion Pressure and the Deflagration Index divided by the similar product – about 840 – for Pittsburgh Coal Dust.

The above data support a conclusion that powders that have particle sizes greater than 500 microns do not present a significant deflagration [explosion, fireball, or flash fire] hazard [References 59, 60, and 61].

However small the likelihood of deflagration hazard, persons handling "loose" [unpackaged] combustible powders should be protected by wearing *flame resistant clothing*. Activities where exposure to combustion of powders could occur include the charging of mixers, blenders, and reactors; the unloading of product filter-collectors, dryers, and product conveyors; the recovery of powder product from bags and boxes; the cleanup of powder spills; and changing bags and cartridges in dust collectors.

b. Improved Protection Against Fireballs and Flash Fires Involving "Organic" Dusts

For dusts and powders having particle sizes below 500 microns, better protection should be provided against fireballs and flash fires for persons who could be exposed to dense dust clouds. This would apply particularly for particle sizes below 75 microns, where the Ignition Severity [IS] and Explosion Severity [ES] would be below 0.2 and 0.5, respectively [Reference 62]. Values of IS and ES are available for about 400 materials [Reference 63].

Aluminized flash suits provide protection against thermal radiation for durations of the order of minutes and protection against flame for durations of the order of seconds. A pertinent comments is the following [Reference 64]:

This [reflective] *clothing provides protection against flame lick and intense radiant heat and is worn for short periods only, to enable fire fighters to enter specific high-risk fire-fighting and fire-rescue situations.*

The major difference in thermal resistance is the extent of insulation inside the metalized film and fabric that forms the structure of the suit. In the chemicals and similar industries, the purpose of wearing *flash suits* would be for escape from the hazard, rather than attempting to extinguish or control and hazard. Activities where exposure to combustion of dusts and fine-particle powders could occur include the emptying of dust collectors, cleaning of exhaust-ventilation ductwork, and the cleanup of dust spills.

J. Evaluating the Need for Flame-Resistant Clothing

a. Legislated Requirements

As described above [References 65 and 66]:

The employer shall assess the workplace to determine if hazards are present, or are likely to be present, which necessitate the use of personal protective equipment (PPE).

and

An assessment of workplace hazards shall be conducted [and] *When the assessment has determined that flame-resistant garments are needed, personnel shall be provided with and wear flame-resistant garments.*

The quantitative relationships between the sizes of potential dust clouds and the possible thermal-radiation hazards can provide the input to a PPE-requirements assessment.

For a 6.75-kilogram cornstarch dust cloud, the following might be recommendations from the PPE assessment:

1. Persons handling 25-kilogram bags or boxes of small-particle cornstarch should wear flame-resistant clothing that could withstand exposure to thermal-radiation intensity of 65 Joules per square centimeter.

2. Persons within 30 feet of such operations should wear non-melting flame-resistant clothing that could withstand exposure to thermal-radiation intensity of 25 Joules per square centimeter.

3. Persons within 40 feet of such operations should have clear unobstructed exit paths to a place of refuge or outside the building – without ascending or descending stairs or ladders – in event of a flash fire.

4. Safety showers should be provided within 50 feet of such operations, with frequent tests by the persons who might be exposed to flash fires.

b. <u>Protection Against Fireballs and Flash Fires Involving Metal Powders and Dusts</u>

For metal powders and dusts, protection should be provided against fireballs and flash fires for persons who could be exposed to dense dust clouds. This would apply particularly for particle sizes below 75 microns, where the Ignition Severity [IS] and Explosion Severity [ES] would be below 0.2 and 0.5, respectively [Reference 67]. Values of IS and ES are available for about 400 materials [Reference 68].

K. Reporting of "Near Misses"

Unexpected releases of combustible small-particle powders or dusts which result in a dense dust cloud – with no ignition – should be considered to be a "near miss", with investigation into the cause(s) [Reference 69]. The investigation should include the adequacy of the personal protective equipment and escape paths which would have been needed if ignition had occurred.

L. References

1. National Fire Protection Association, "Standard System for the Identification of the Hazards of Materials for Emergency Response", NFPA 704, Table 2 (2001).

2. National Fire Protection Association, "Standard for the Prevention of Fire and Dust Explosions from the Manufacturing, Processing, and Handling of Combustible Particulate Solids", NFPA 654, Paragraph 3.3.7 (2017).

3. Eckhoff, R. K., "Dust Explosions in the Process Industries", 2nd Ed., Pages 9, 72, and 73 (1999).

4. National Fire Protection Association, "Guide for Venting of Deflagrations", NFPA 68, Tables F.1(a), (b), (c), and (d) (2013).

5. Nagy, J., et al. [U. S. Bureau of Mines], "Explosion Development in Closed Vessels". Report of Investigation 7507, Page 8 (1971).

6. Nagy, J., et al. [U. S. Bureau of Mines], "Explosion Development in Closed Vessels", Report of Investigation 7507, Pages 29 to 32 and Table 4 (1971).

7. Babrauskas, V., "Ignition Handbook", Page 34 (1992).

8. Perry, J. H., "Chemical Engineers' Handbook", 4th Ed., Page 9-40 (1963).

9. Babrauskas, V., "Ignition Handbook", Page 321 (1992).

10. Center for Chemical Process Safety [AIChE], "Guidelines for Vapor Cloud Explosion, Pressure Vessel Burst, BLEVE, and Flash Fire Hazards", 2nd Ed., Pages 3, 88, and 92 (2010).

11. ASTM International, "Standard Test Method for Explosibility of Dust Clouds", E1226 – 12a (December 2012).

12. Frank, W. L., and S. A. Rodgers, "NFPA Guide to Combustible Dusts", Page 92 (2012).

13. Nagy, J., et al. [U. S. Bureau of Mines], "Explosion Development in Closed Vessels", Report of Investigation 7507, Pages 29 to 32 and Table 4 (1971).

14. National Fire Protection Association, "Standard on Explosion Protection by Deflagration Venting", NFPA 68, Equation (D.1) (2002).

15. National Institute for Occupational Safety and Health, "Application Manual for the Revised NIOSH Lifting Equation", Page 13 (1994).

16. Ural, E. A., "Toward Estimating Entrainment Fractions for Dust Layers", in "Fire Research", Table A-1 (June 2011).

17. National Fire Protection Association, "Fire Protection Handbook", 18th Ed., Page A-6 (1997).

18. Babrauskas, V., "Ignition Handbook", Pages 20 and 29 (1992).

19. Eckhoff, R. K., "Dust Explosions in the Process Industries", 2nd Ed., Page 34 (1999).

20. National Fire Protection Association, "Fire Protection Handbook", 18th Ed., Page A-8 (1997).

21. Avallone,E. A., et al., "Marks' Standard Handbook for Mechanical Engineers", 11th Ed., Page 9-13 (2007).

22. Prugh, R. W. [Chilworth Technology, Inc.], "Quantitative Evaluation of Fireball Hazards", Page 83 in "Process Safety Progress", Vol. 13 (2) (April 1994).

23. Center for Chemical Process Safety [AIChE], "Guidelines for Vapor Cloud Explosion, Pressure Vessel Burst, BLEVE, and Flash Fire Hazards", 2nd Ed., Page 338 (2010).

24. Eckhoff, R. K., "Dust Explosions in the Process Industries", 2nd Ed., Page 284 (1999).

25. Grossel, S., "Safety Considerations in Conveying of Bulk Solids and Powders", Page 62 in the "Journal of Loss Prevention in the Process Industries", Vol. 1, No. 2, Table 2 (April 1988).

26. Dechema, et al., "GESTIS-DUST-EX – Tables of Combustion and Explosion Characteristics of Dust", Pages 74 and 75, and 334 and 335 (February 2001).

27. Eckhoff, R. K., "Dust Explosions in the Process Industries", 2nd Ed., Page 609 (1999).

28. ASTM International, "Standard Performance Specification for Flame Resistant and Electric Arc Rated Protective Clothing Worn by Workers Exposed to Flames and Electric Arcs", ASTM F1506 - 17b (2018).

29. Frank, W. L., and S. A. Rodgers, "NFPA Guide to Combustible Dusts", Page 92 (2012).

30. National Fire Protection Association, "Fire Protection Handbook", 18th Ed., Page A-6 (1997).

31. Eckhoff, R. K., "Dust Explosions in the Process Industries", 2nd Ed., Pages 308 and 345 (1999).

32. Eckhoff, R. K., "Dust Explosions in the Process Industries", 2nd Ed., Page 287, Figure 4.17 (1992).

33. Babrauskas, V., "Ignition Handbook", Page 148 (1992).

34. Eckhoff, R. K., "Dust Explosions in the Process Industries", 2nd Ed., Pages 34, 36, and 345 (1999).

35. Babrauskas, V., "Ignition Handbook", Pages 145 and 158 (1992).

36. National Fire Protection Association, "Fire Protection Handbook", 18th Ed., Page A-6 (1997).

37. National Fire Protection Association, "Recommended Practice for the Classification of Flammable Liquids, Gases, or Vapors and of Hazardous (Classified) Locations for Electrical Installations in Chemical Process Areas", NFPA 497, Fig. 6.10(a) (2017).

38. National Fire Protection Association, "Fire Protection Handbook", 18th Ed., Page A-8 (1997).

39. Babrauskas, V., "Ignition Handbook", Page145 (1992).

40. National Fire Protection Association, "Fire Protection Handbook", 18th Ed., Page A-6 (1997).

41. Frank, W. L., and S. A. Rodgers, "NFPA Guide to Combustible Dusts", Page 92 (2012).

42. Babrauskas, V., "Ignition Handbook", Page145 (1992).

43. National Fire Protection Association, "Fire Protection Handbook", 18th Ed., Page A-6 (1997).

44. Frank, W. L., and S. A. Rodgers, "NFPA Guide to Combustible Dusts", Page 92 (2012).

45. Avallone,E. A., et al., "Marks' Standard Handbook for Mechanical Engineers", 11th Ed., Page 4-58 (2007).

46. Society of Fire Protection Engineers, "SFPE Handbook of Fire Protection Engineering", Page 2-54 (1988).

47. Center for Chemical Process Safety [AIChE], "Guidelines for Vapor Cloud Explosion, Pressure Vessel Burst, BLEVE, and Flash Fire Hazards", 2nd Ed., Pages 58 and 397 (2010).

48. Prugh, R. W., "Quantitative Evaluation of Fireball Hazards", in Process Safety Progress, Vol. 13 (2): 83 (April 1994).

49. Babrauskas, V., "Ignition Handbook", Page 815 and Tables103, 104, 105, 113, 114, 116, 117, and 120 (1992).

50. Babrauskas, V., "Ignition Handbook", Tables 111, 112, and 113 (1992).

51. National Institute for Occupational Safety and Health, ""Application Manual for the Revised NIOSH Lifting Equation", Page 13 (1994).

52. U. S. Department of Labor, Occupational Safety and Health Administration, "Safety and Health Standards", 29 CFR 1910.132(d)(1) (November 2016).

53. National Fire Protection Association, "Standard on Selection, Care, Use, and Maintenance of Flame-Resistant Garments for Protection of Industrial Personnel Against Short-Duration Thermal Exposures from Fire", NFPA 2113, Paragraphs 4.2 and A.4.3 (2015).

54. National Fire Protection Association, "Standard on the Fundamentals of Combustible Dust", NFPA 652, Table A.5.2.2(j) (2016).

55. Frank, W. L., and S. A. Rodgers, "NFPA Guide to Combustible Dusts", Page 92 (2012).

56. Eckhoff, R. K., "Dust Explosions in the Process Industries", 2nd Ed., Page 607 (1999).

57. Dechema, et al., "GESTIS-DUST-EX – Tables of Combustion and Explosion Characteristics of Dust", Pages 74 and 75, and 334 and 335 (February 2001).

58. National Fire Protection Association, "Standard on Explosion Protection by Deflagration Venting", Paragraph F.4.3.4.1 (2002).

59. National Fire Protection Association, "Recommended Practice for the Classification of Combustible Dusts and of Hazardous (Classified) Locations for Electrical Installations in Chemical Process Areas", Paragraphs 3.3.3, A.3.3.3, 4.2.1, and A.5.2.3 (2017).

60. National Fire Protection Association, "Standard on the Fundamentals of Combustible Dust", NFPA 652, Paragraph A.5.2 (2016).

61. Grossel, S., "Safety Considerations in Conveying of Bulk Solids and Powders", Page 62 in the "Journal of Loss Prevention in the Process Industries", Vol. 1, No. 2, Table 2 (April 1988).

62. Frank, W. L., and S. A. Rodgers, "NFPA Guide to Combustible Dusts", Page 92 (2012).

63. U. S. National Materials Advisory Board, National Academy of Sciences, "Classification of Dusts relative to Electrical Equipment in Class II Hazardous Locations", NMAB 353-4, Table 4-1 and Appendix A (1982).

64. British Standard, "Protective clothing for fire-fighters - Test methods and requirements for reflective clothing for specialised fire-fighting", EN 1486 (2008).

65. U. S. Department of Labor, Occupational Safety and Health Administration, "Occupational Safety and Health Standards", 29 CFR 1910.132(d)(1) (Nov. 2016).

66. National Fire Protection Association, "Standard on the Fundamentals of Combustible Dust", NFPA 652, Paragraph 8.6 (2016).

67. Frank, W. L., and S. A. Rodgers, "NFPA Guide to Combustible Dusts", Page 92 (2012).

68. U. S. National Materials Advisory Board, National Academy of Sciences, "Classification of Dusts relative to Electrical Equipment in Class II Hazardous Locations", NMAB 353-4, Table 4-1 and Appendix A (1982).

69. Center for Chemical Process Safety [AIChE], "Guidelines for Investigating Chemical Process Incidents", 2nd Ed., Pages 19, 40, 61 to 74, and 437 (2003).

511352 Quantitative Evaluation of "Unconfined" Dust-Cloud Explosion Hazards

Abstract:

<u>QUANTITATIVE EVALUATION</u>

-

<u>OF "UNCONFINED" DUST-CLOUD EXPLOSION HAZARDS</u>

-

A presentation to the 14th Global Congress on Process Safety

April 22 to 26, 2018

at the Orlando World Center Marriott

<u>Abstract</u>

When performing an Occupied Building Risk Assessment, it may be necessary to determine the blast pressure and blast impulse resulting from explosion of a mixture of combustible dust and air in a nearby building or – possibly – outside a building. These scenarios would be similar to a confined explosion of a mixture of air and a flammable vapor [or gas] in a building or an unconfined vapor cloud explosion.

Ignition of dense clouds of combustible dusts with low values of Deflagration Index [Kst] are unlikely to yield significant blast pressures. However, tests involving ignition of dusts with values of Kst above 300 bar-meters per second show that blast effects could be hazardous.

Although more-sophisticated methods for evaluating explosion hazards are available, the "TNT-Equivalence" method – with a "virtual distance" to a TNT source – is simple and provides a good measure of blast hazards. The heat of combustion of the dust would be required, together with a value for the Deflagration Index. Also, if the dust explosion is partially confined within a building or other structure, the pressure at which the structure would "burst" would be needed.

This presentation provides a calculation method for estimating blast pressure and blast impulse, as based on the TNT equivalence of the combustible-dust explosion.

510777 Investigation of Site Elevation on Simple Dispersion and VCE Modeling

Abstract:

Facility siting typically utilizes release modeling to estimate hazards associated with flammable clouds. The results from the release modeling is subsequently used to perform explosion modeling. Release conditions are based on various process conditions including pressure, temperature, and composition. It is also common to address typical meteorological conditions (e.g., wind speed, humidity, temperature, and stability class). A potentially less common adjustment is accounting for a facilityâ€™s elevation (above sea level). The elevation of a facility defines the atmospheric pressure which is used directly in both the dispersion and explosion modeling. This analysis evaluates the effect of elevation on dispersions and explosions independently as well as the cumulative effect. It also examines at which point, if any, special interest should be taken to include facility elevation in facility siting hazard assessments.

511131 *CFD Modeling of Liquefied Gases Discharging through a Pipeline Full Bore Rupture*

Abstract:

Large amounts of substances are transported in pipelines worldwide. This activity represents a hazard that needs to be quantitatively assessed through discharge models that are capable of accurately predict the outflow when a pipeline ruptures. The calculation of the discharge rate of pipelines transporting gases or liquids is well understood and accurate models exist for this purpose; less well understood is the discharge of flashing liquids. When a pipeline transporting a liquefied gas ruptures (e.g., CO_2, LNG or LPG pipelines), the expansion generates a phase transition which results in a two-phase release.

In general, there are two different scenarios when a pipeline ruptures: the full bore rupture, and the puncture case. A full bore rupture is a complete rupture of the pipeline, so the substance flows out through the pipeline cross-sectional area. On the other hand, a puncture is represented by the substance flowing through an orifice or crack in the pipeline walls. A full bore rupture is associated with a quicker depressurization of the system in comparison to the puncture case. Even though a full bore rupture is usually more catastrophic, a puncture is more common in a real scenario.

Numerous researchers have used one-dimensional conservation equations to describe the discharge of liquefied gases when a pipeline full bore rupture occurs. However, a systematic study on how the accuracy of different equations of state affects the depressurization prediction is lacking for this case. The main objective of this research is to propose a two-dimensional discharge model using Computational Fluid Dynamic (CFD) tools in order to predict the pressure and temperature profiles along the pipeline, as well as the discharge rate and phase transition; while investigating the effect of different equations of state on the predictions for the full bore rupture case. In order to validate the 2-D discharge model, full bore rupture experiments with dense-phase CO_2 pipelines have been used. The validation step evaluates the assumptions that have been implemented to model the physics of the transient phenomenon. This study will help to improve the formulation of current discharge models for liquefied gases. Once the full bore rupture study is completed, the subsequent step will be to propose a three-dimensional model to predict the behavior of liquefied gases discharging through a puncture. The previous scenario will complete the picture for predicting the discharge of flashing liquids when a pipelines ruptures, especially for the puncture case, where the physics is less well understood than for the full bore rupture scenario.

Keywords: Consequence Modeling, Source Term, Pipelines, Carbon Dioxide, Computational Fluid Dynamics (CFD)

511210 Understanding the Stability of High Expansion Foam Used for LNG Vapor Risk Mitigation

Abstract:

Natural gas consumption has increased as it is a clean source of energy, and advances in technology which have enabled its extraction from sources previously considered to be economically infeasible. Natural gas occupies a large volume in its vapor form and its liquefaction allows both ease of storage and transportation. Despite these advantages, a leak of Liquefied Natural Gas (LNG) can result in the formation of a cryogenic vapor cloud that has the potential to ignite, which also presents an asphyxiation hazard. NFPA 11 recommends the use of high expansion foam to mitigate the vapor risk of LNG spills. High expansion foam can not only help reduce the heat transfer to LNG through convection and radiation for reduced vaporization, but also heats up LNG vapor that passes through the foam layers, enabling ease of dispersal for risk mitigation. However, foam drains liquid over time which can increase the vaporization of LNG. Therefore, it is important to understand and model these mechanisms, so that the amount of foam required can be accurately quantified. A lab scale foam generator was used to produce the high expansion foam. The effects of several factors like convection, radiation, coalescence, coarsening, evaporation and liquid drainage on foam stability were studied. The factors that are crucial in determining the rate of foam breakage and the amount of liquid drained were identified, which helps determine the amount of foam that needs to be applied for effective vapor risk mitigation.

Keywords: *Liquefied Natural Gas (LNG); vapor cloud; high expansion foam; mitigation; foam stability*

511245 Experience with Conducting a Series of Building Siting Studies per API 752/753

Abstract:

Over the past several years, Plains All American Pipeline, LP (Plains) and Quest Consultants Inc. (Quest) have collaborated on conducting building siting studies per API RP 752 and 753 for a range of facilities in the United States. This paper summarizes the approach developed (from consequence analysis to risk analysis), the types of facilities evaluated, results generated, and the mitigation measures applied to those facilities when required.

The 15 building siting studies that are the subjects of this paper include Plains facilities which range from remote LPG transfer terminals to fully functioning gas plants. As the intent of API RP 752 is on occupants of permanent buildings, the technical approach to identify the hazards and risks associated with the occupancy of such buildings is considerably different than the approach often used to evaluate the hazards and risks to persons outdoors.

This paper also describes how the studies were developed to include potential hazards and risks to occupants of permanent buildings (API RP 752) and portable buildings (API RP 753) that are not addressed in the API documents. While the API recommended practices may not address the full range of hazards a building occupant could experience, other codes such as 29 CFR 1910 do require their evaluation under the â€œgeneral duty clauseâ€ and the Process Safety Management (PSM) program.

510466 Accelerating Safety in the Permian: Implementing Risk Management in Chevron's "Factory Model" Developments

Abstract:

Chevron has turned to a new development approach in the Permian Basin to take advantage of the rapid growth of shale oil and tight gas opportunities. Known as the Factory Model, this consistent and repeatable process has succeeded in tripling the pace of the Permian drilling program using standard designs for tank batteries, compressor stations, and saltwater disposal facilities. Modification of the traditional risk management process was required to align with the repeatable, manufacturing-style pace of the Factory Model; thus, ensuring that Chevron's risk criteria are being met for all new developments.

Each design (i.e., tank battery, compressor station, saltwater disposal facility, etc.) was subjected to consequence analysis to determine potential impacts due to releases from various pieces of equipment. These potential impacts represent the worst-case predicted consequences from each type of facility and are used to:

1. Identify safe distances to build Factory Model developments from public receptors
2. Determine safe placement of field personnel occupied buildings
3. Quantify potential severity rankings in the PHAs performed for Factory Model developments

Achieving a fit for purpose risk management process ensures that Chevron's risk criteria are continually met for Factory Model developments amid evolving designs and process conditions via evergreen review cycles with engineering and execution teams.

509155 Dispersion Study: An Alternative Technique for Determining Classified Area of Electrical Installations

Abstract:

Fire and explosion incident is an inherent risk in process facility handling hydrocarbon fluids. Any release of hydrocarbon to environment may create flammable cloud. However, fire or explosion will only occur if the cloud engulfs an ignition source.

Electrical installation is one potential source of ignition, and API RP 500 and its family are renowned recommended practices applied in the petroleum industry in regard to classifying areas in which electrical installation shall be designed to minimize ignition. They collect decades of practical experience to determine the areas where flammable cloud is normally and might occasionally present. Three-dimensional area is depicted encircling the equipment containing hydrocarbon, and is called classified area. Electrical installation shall suit its classified area or else located away outside of the area.

This paper describes an alternative technique to define the classified area for a case where API RP 500 recommendation is considered as excessive, especially for a facility having space limitation. Performing dispersion analysis to confirm the classified area will be beneficial to ensure equal level of safety assurance while optimizing the requirement for classified electrical installations.

According to API RP 500, classified area for an oily water separator pond, which is part of waste water treatment package, should extend as far as 85 feet distance horizontally. By performing dispersion analysis, the extent of classified area can be optimized down to 6 feet from the pondâ€™s perimeter. The result is based on credible operating conditions and scenarios of the pond. By utilizing this technique, investment cost can subsequently be optimized as the number of classified electrical installation is reduced without compromising safety.

511294 Proven Practices for Communicating Facility Siting Risk to Plant Personnel and Management

Abstract:

Process safety is important. Few people would argue this point, but the implementation of best practices and the breadth, depth, and precision required are constantly being debated. Even when the techniques for performing process safety studies are well established and accepted, disagreements may arise.

Several factors can contribute to this conflict when performing these process safety studies. First, the history and science of process safety is continuously advancing. As a result, there may be more known hazards to review, a larger number of scenarios that can be efficiently analyzed via automation, and a desire to constantly drive down risk. Additionally, at a given site, there may be competing interests between safety and the economics and timing of performing safety analyses. In a setting where time pressures and high dollar mitigation projects are being discussed, there may be an increasing challenge to the basis and accuracy of process safety studies. Finally, there is the issue of regular turnover of plant and project staff. Even though process safety is acknowledged as important, there may be a constant need to educate new and existing staff on the proper process safety techniques to adopt and to convince management of the accuracy and significance of process safety findings.

As a result of these competing factors, the process safety professional must be not only a technical expert, but also an excellent communicator. Especially when prescriptive standards do not currently exist at a facility, the process safety professional may need to champion a study change or mitigation plan to management. Through the use of specific examples in facility siting, the authors will present key lessons learned and offer suggestions to improve process safety communications between plant staff and management. A practical set of communication tools will also be shown.

Improved Along-Wind Dispersion Coefficient Model from Field and Wind Tunnel Data

Jessica M. Morris
Tom Spicer, Ph.D., P.E.
Ralph E. Martin Department of Chemical Engineering
University of Arkansas, Fayetteville, AR
Jmm065@uark.edu

Prepared for Presentation at
American Institute of Chemical Engineers
2018 Spring Meeting and 14th Global Congress on Process Safety
Orlando, Florida
April 22 – 25, 2018

Improved Along-Wind Dispersion Coefficient Model from Field and Wind Tunnel Data

Jessica M. Morris
Tom Spicer, Ph.D., P.E.
Ralph E. Martin Department of Chemical Engineering
University of Arkansas, Fayetteville, AR
Jmm065@uark.edu

Keywords: along-wind, dispersion, finite duration, time-varying, atmosphere, releases, modeling

Abstract

Modeling the effects of along-wind dispersion is particularly important for instantaneous, finite-duration, and time-varying releases of hazardous, toxic, or flammable materials. Current models can under or over predict the along-wind dispersion coefficient σ_x , by a factor of 100, resulting in poor assessment of the consequences of a hazardous chemical release. Past field trial and wind tunnel data have been analyzed to improve the parameterization of σ_x . An ongoing experimental program studying finite duration releases of neutrally buoyant gasses released from a ground-level area source in a wind tunnel is underway to generate data in order to help improve how finite duration releases are modeled.

1 Introduction

Many chemicals used in industry are hazardous due to their toxic and/or flammable properties. When released into the atmosphere through accidental or intentional releases, these hazardous chemicals can have devastating effects on large populations (e.g., Bhopal[1], Flixborough[2], Seveso[3]). The ability to model these air-borne hazardous chemicals is vital to mitigating their adverse effects on the public. The use, storage, and transport of hazardous chemicals play a role in the potential impact of a hazardous chemical release. Accurate modeling of hazardous chemicals released into the atmosphere is imperative in emergency response settings.

Current models exist to predict the concentration and effects of hazardous gasses released into the atmosphere. Instantaneous and finite-duration releases strongly depend of the elongation of the cloud in the along-wind direction (along-wind dispersion). Current models use theoretical or empirical derivations to account for the along-wind dispersion coefficient, σ_x. This paper incorporates theoretical derivations (together with their assumptions), and experimental data to appropriately model the effects of along-wind dispersion.

2 Theoretical Description

An overarching equation to determine the along-wind dispersion coefficient is complex. Varying conditions such as advection wind speed, friction velocity, release height, stability class, time, and other parameters all have effects on the along-wind dispersion coefficient. Established theories account for these different parameters such as the wind speed *u*, friction velocity u_*, time *t*, distance *x*, stability class, and more. The most predominant theories for along-wind dispersion are previously discussed by Morris and Spicer in their paper in 2017[4]. The different assumptions or criteria made by the scientists result in diverse equations for the same dispersion coefficient. Due to these varying assumptions, an overarching equation incorporating all parameters is formidable. Table 1 summarizes the theoretical derivations discussed in a previous paper[4], derived for the overall along-wind dispersion equation, shear component, and turbulent component for each derivation.

Table 1. Comparison of shear and turbulence in relation to along-wind dispersion equation

Author	Overall Equation for σ_x	Shear Component	Turbulent Component
Batchelor[5]	$\sigma_x \, \alpha \, u_* t$	Not explicitly solved	Not explicitly solved
Chatwin[6]	$\sigma_x = 0.803 \frac{u_* t}{\kappa}$	$\sigma_{xs} = 0.596 \frac{u_* t}{\kappa}$	Not explicitly solved
Van Ulden[7]	Neutral: $\sigma_x = 0.84 \frac{u_* t}{\kappa}$ Stable: $\sigma_x = \{(\frac{3}{4} - \frac{16}{9\pi}) \frac{5 u_*^3 t^3}{kL}\}^{1/2} \simeq 0.43 (\frac{5 u_*^3 t^3}{kL})^{1/2}$	$\sigma_s = 0.5 \frac{u_* t}{\kappa}$ $\sigma_c \simeq \sigma_s = 0.5 \frac{u_* t}{\kappa}$	$\sigma_d = 0.67 \frac{u_* t}{\kappa}.$

Theoretically, Batchelor, Chatwin, and van Ulden have different assumptions on parameters that influence along-wind dispersion. All of these derived equations result in an along-wind dispersion coefficient σ_x, as a function of friction velocity u_*, travel time *t*, and a constant of $\sim 0.8/\kappa$ or 2 in neutral stability. Theoretical derivations are predominately determined for neutral stability. The spreading of the contaminant cloud in the along-wind direction is observed under all stability classes. The more stable the atmosphere, the greater effect along-wind dispersion plays. This is shown with van Ulden's derivations and experimental releases in the atmosphere[7].

3 Analysis of Available Data

Available data was accumulated and analyzed for an overarching non-dimensionalized equation for along-wind dispersion. The data was standardized for wind speed reference height and determination of stability class to provide a more uniform data set for analysis. The standardized data was analyzed with a multivariable regression analysis to determine the constants for the non-dimensionalized equation. Geometric mean bias (MG) and geometric mean variance (VG) was reported on the tests for a statistical comparison of the model to the original data.

3.1 Data Available

Data over varying conditions, totaling 446 data sets, was analyzed by Hanna and Franzese from 11 field test programs and 1 wind tunnel test program in 1999[8]. Tests selected from the original 12 programs were analyzed by Hanna and Franzese in an effort to provide a parameterization of along-wind dispersion for ground level releases with an overall equation of $\sigma_x = 2u_*t$. Of the original 452 data sets, six data sets were not used in the analysis due to lack of recorded friction velocity or wind speed. Test programs in the original analysis included DTRA Phase 1, Dipole Pride, KitFox (ERP, URA, and SSR), Hanford, LROD, OLAD, Marchwood Wind Tunnel, Oceanside, Fort Wayne, SRDES, and Victoria. A detailed description of each of these tests are described by Morris and Spicer in their paper in 2017[Error! Bookmark not defined.].

3.2 Standardization of Data

The test programs mentioned above provide a multitude of available data. Majority of the test programs were analyzed and published in journals. Some test programs had the raw data available in addition to journals (if a report was published). If the report and journals did not provide the data needed, the authors or facility were contacted for the original data. Once all of this data was acquired, it was analyzed and standardized for a more uniform analysis. Analysis of the original data was performed to fully understand which methods were chosen to derive specific values, such as time, stability class, along-wind dispersion coefficient, etc. For the along-wind dispersion coefficient, the most common method derives σ_x from σ_t, the standard deviation of the concentration time history, and u_e, the effective cloud wind speed. Some of the older trials report σ_x without any indication of how σ_x was derived. All derivations were clearly understood before analysis was performed. Standardization of available data was performed for model development, uniform wind speed reference height, and uniform determination of stability class described below.

3.2.1 Data Used for Model Development

Not all data are of the same caliber to develop an overarching equation for along-wind dispersion. For adequate analysis, the distance (m), time (s), wind speed (m/s), friction velocity (m/s), and measured σ_x (m) need to be reported or available. Data from DTRA

Phase 1[9], Dipole Pride[10,11], KitFox (ERP and URA) [12,13,14,15], OLAD - van[16], Marchwood[17], and Hanford[18] test cases all have sufficient data recorded for the data analysis conducted. Other tests series used in Hanna and Franzese's analysis included the Short Range Diffusion Experiment Series (SRDES) [19], Fort Wayne[20], Oceanside[21], Victoria[8], and the LROD[22] tests. The SRDES tests could not be used because the wind speed for those tests were not recorded. As well, the use of Oceanside test data was hampered due to lack of records of the friction velocity. Furthermore, major parameters, such as σ_x values, were found to be missing in the Kit Fox SSR and Fort Wayne data collection, thus they could not be used in the present assessment.. Overall, there were 446 recorded data sets with sufficient data reported to be considered for analysis for an overarching equation for along-wind dispersion.

The previous description of the tests show that some of the data sets are directly applicable to addressing the description of along-wind dispersion for ground level releases while some data sets are not. The tests performed in the upper boundary layer with plane releases, were determined to not have as much control as the ground level releases. The LROD and OLAD-plane tests were not used in this analysis due to their elevation. With majority of hazardous releases dispersed before 10,000 m downwind, data with distances less than 10,000 m are emphasized for model data. This excludes the Victoria tests that were used in Hanna and Franzese's analysis. Under these tight constraints, only the most quality data was used for model development resulting in only 89 data points. An overall summary of the tests used for model development are shown below in Table 2.

Table 2. Summary of data sets used for non-dimensionalized equation model development

Name	Location	Release (Chemical)	Duration of Test	Unique Conditions
DTRA – Phase 1	Dugway Proving Ground (z_o = 0.3 mm)	Ground-level Puff (Propylene)	27 – 271 seconds	Ensemble averages
Dipole Pride	Nevada Test Site (z_o = 32 mm)	Ground-level Puff (SF_6)	1,692 – 5,112 seconds	Day and Night
KitFox URA	Nevada Test Site (z_o = 20 mm)	Ground-level Puff (Carbon Dioxide)	65 – 95 seconds	Artificial roughness
KitFox ERP	Nevada Test Site (z_o = 200 mm)	Ground-level Puff (Carbon Dioxide)	72 – 91 seconds	Artificial roughness
OLAD - van	Dugway Proving Ground (z_o = 30 mm)	Ground-level line source (SF_6)	Van: 180 – 11,820 seconds	Truck release
Marchwood	Marchwood Wind Tunnel (z_o = 0.3 mm)	Ground-level Puff (Propane and Helium)	2.1 – 11 seconds	Wind Tunnel
Hanford	Battelle Pacific Northwest Laboratory (z_o = 30 mm)	Ground-level puff (Krypton 85)	39 – 778 seconds	Guillotine like releases of crushed vials

From Table 2, the different test conditions are able to be compared and contrasted. A non-dimensionalized equation developed from the multiple different tests of quality data spanning a variety of conditions will be most beneficial when predicting the along-wind dispersion coefficient for a variety of releases.

3.2.2 Standardization of Wind Speed Reference Height

For a more standardized data analysis, all parameters need to be comparable. Original reference heights varied from 1.5 m, 2 m, to 10 m depending on the test. A standardized reference height of 10 m was chosen to better fit current models. The power law wind speed profile shown below in equation (1) was used to standardize the reference height for all ambient wind speeds.

$$u_x = u_o \left(\frac{z}{z_R}\right)^{\alpha} \tag{1}$$

Where u_o is the reported velocity from the test, z_R is the reference height of wind velocity profile specification in the test, z is the standardized reference height of 10 m, u_x is the standardized ambient velocity at a reference height of 10 m, and α is a constant. The α constant was determined from a weighted least-squares fit with the logarithmic wind speed profile that includes the Monin-Obukhov length λ, shown below in equation (2) calculated in the program DEGADIS.

$$u_x = \frac{u_*}{k}\left[\ln\left(\frac{z+z_o}{z_0}\right)\psi\left(\frac{z}{\lambda}\right)\right] \tag{2}$$

z_0 is the surface roughness, ψ is a function of the surface roughness and Monin-Obukhov length, and k is the Von Karmen constant which is equal to 0.35 within DEGADIS.

Standardized wind speeds calculated at a height of 10 m were used in place of the reported wind speeds in the original data files. This standardized height allows for a more accurate comparison of the effect of wind speed on the along-wind dispersion coefficient.

3.2.3 Standardization of Stability Class

Atmospheric stability is an important parameterization of data due to the parameter being a function of solar insolation, cloud coverage, and wind speed. Pasquill's stability classes are the most widely used parameterization of atmospheric stability[23]. The stability classification of a test is a debatable topic. However, Golder's plot is widely accepted as one of the best methods to determine Pasquill's stability class from surface roughness and Monin-Obukhov length[24]. Golder's plot is shown below in Figure 1 and was used to classify the stability class of each data point.

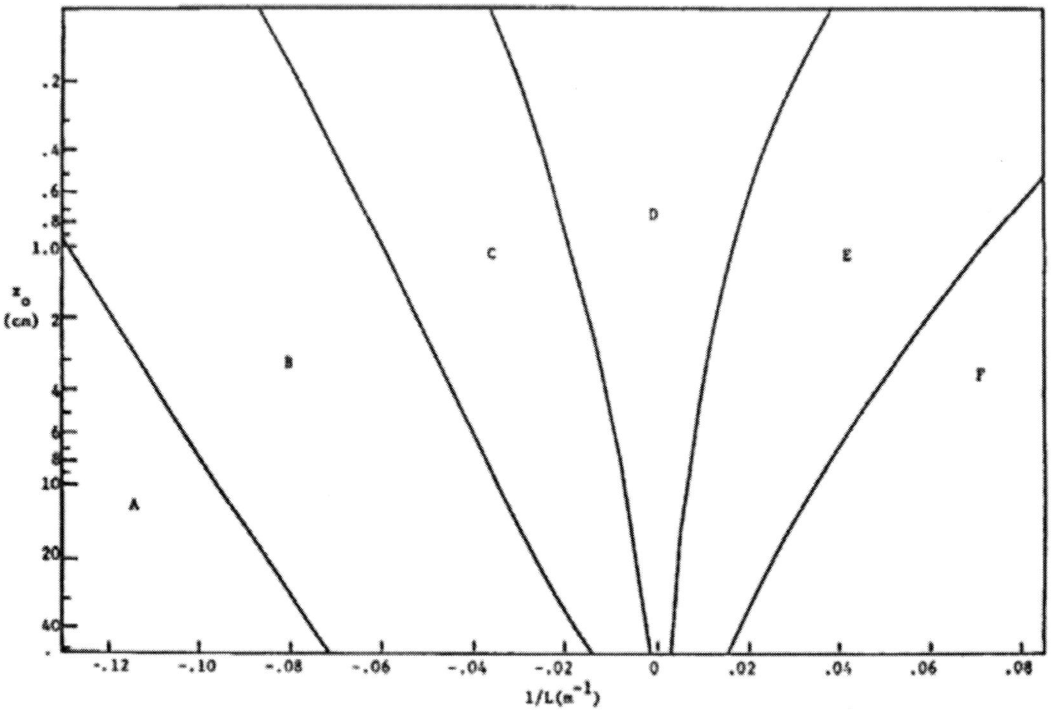

Figure 1. Inverse of the Monin-Obukhov length as a function of surface roughness to determine Pasquill stability class[24]

Figure 1 shows how the inverse Monin-Obukhov length and surface roughness can be plotted resulting in categories to determine Pasquill's stability class. An image analysis software, Enguage, was used to recreate Golder's plot. Data points were then plotted in the recreated plot to determine stability class as a function of surface roughness and Monin-Obukhov length. A total of 307 data points had both the surface roughness and Monin-Obukhov length available for analysis. As mentioned prior, not all tests will be used for model development. Only 89 data points were considered of high quality with all data available to be used for the model development. The classification of these quality data points by stability class are shown below in Table 3.

Table 3. Quality tests arranged by stability class from Golder's plot

Stability Class	A/B	C	D	E	F
Total Tests	5	24	40	15	5

As presented in Table 3, with only the quality data, there are about 29 unstable tests, 40 neutral tests, and 20 stable tests to be used for the model development. This provides a decent spread for stability classes to be used in the model development.

To visually see the effect of stability class on the along-wind dispersion coefficient, data was plotted by stability class. Overall trends in the data by stability class are shown below in Figure 2.

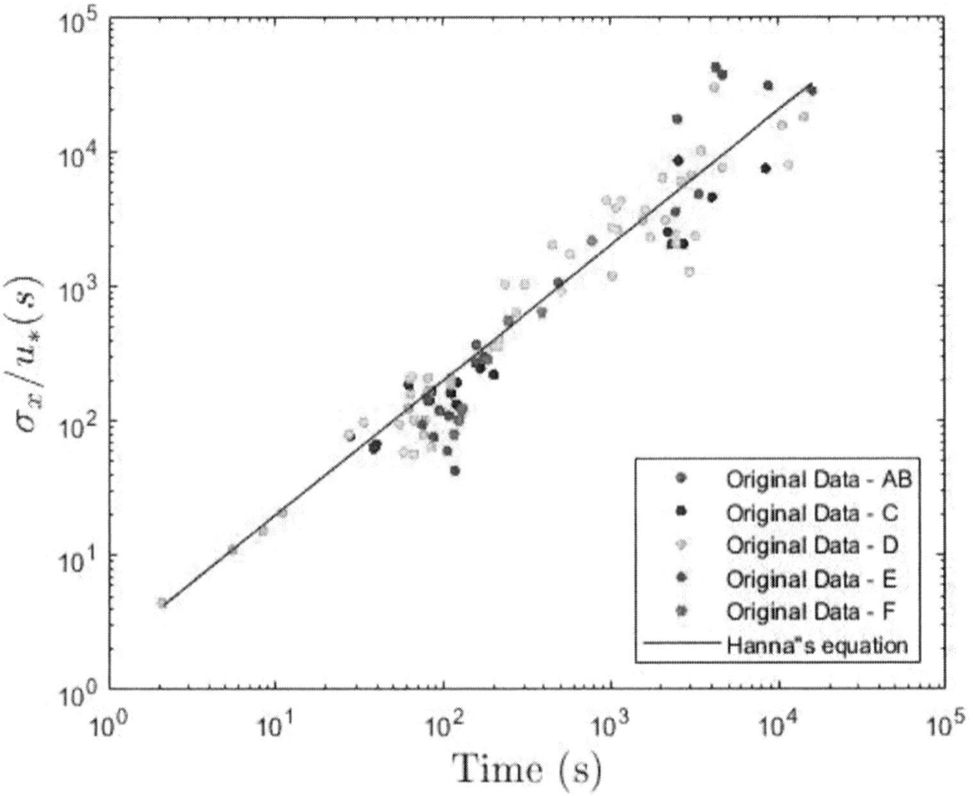

Figure 2. Time (s) vs. σ_x/u_* (s) as a function of the original data by stability class and Hanna's predictions.

From Figure 2, noticeable trends are seen between the different stability classes. Stability Classes AB and C do not have as steep of a slope as the equation predicted by Hanna. Stability Class D spreads out both above and below the equation predicted by Hanna. Stability classes E and F have a slightly steeper slope than the equation predicted by Hanna. These trends show that stability class has an effect on the overall along-wind dispersion coefficient. A non-dimensionalized equation taking into account stability class was developed to capture these trends described in this paper.

3.3 Parameterization and Data Analysis

The along-wind dispersion coefficient depends on multiple different parameters as shown above in Table 1. The best parameterization of data needs to be determined for future models. Experimental data has been visually represented as the along-wind dispersion coefficient as a function of time and friction velocity to compare the model to the original data. This visual comparison will be used in visual comparisons. A geometric mean bias (MG) and geometric mean variance (VG) will be used for a statistical comparison of the model and the original data. A multi-variable regression analysis was used to determine the constants for the non-dimensionalized equation.

3.3.1 Geometric Mean Bias and Geometric Mean Variance

A comparison of MG and VG can be used to mathematically compare predicted data to the original data. This method of comparing MG VG values was developed by Hanna, Chang, and Strimaitis in 1993 when comparing hazardous gas models[25]. The geometric mean bias (MG) equation for an individual point is expressed as:

$$MG_1 = exp[Ln(X_{o1}) - Ln(X_{p1})] \quad (3)$$

where MG_1 is the geometric mean bias for a single point, X_{o1} is the original or observed data, and X_{p1} is the predicted data. For a set of n pairs of data, the following equation can be used to predict the geometric mean bias for a set of points

$$MG_{set} = (MG_1 * MG_2 * \ldots MG_n)^{1/n} \quad (4)$$

where MG_{set} corresponds to the geometric mean bias for the set of data. The MG value shows the numerical value of the predicted value(s) compared to the original. The VG value shows that variance of the predicted data from the original data. The geometric mean variance (VG) can be calculated for a single point using :

$$VG_1 = exp[\{Ln(X_{o1}) - Ln(X_{p1})\}^2] \quad (5)$$

Where VG_1 is the geometric mean variance observed from a single point. The geometric mean variance for a set of data can be determined using :

$$VG_{set} = (VG_1 * VG_2 * \ldots VG_n)^{1/n} \quad (6)$$

Where VG_{set} corresponds to the geometric mean variance observed from the set of data. Values closest to 1 for MG and VG values are most comparable to the original values. Values further from 1 for MG and VG values are further away the original data. If the MG value is less than 1, then the σ_x value is over predicted. If the MG value is greater than 1, then the σ_x value is under predicted. With MG VG for $\sigma_x < 1$, the overall predicted concentration would be less than the observed concentration. Similarly, if MG VG values for $\sigma_x > 1$, the overall predicted concentration would be greater than the observed concentration. Overall, this MG VG method is able to provide a statistical comparison between individual data points and data sets for predicted data compared to the original data.

3.3.2 Multivariable Regression Analysis

A multivariable regression analysis, also known as multiple regression, is an extension of linear regression[26]. A regression analysis determines the relationships among variables using data. With a multivariable regression analysis, the relationships between multiple independent variables can be determined. A regression analysis is a useful tool in understanding the changes of one variable with respect to others. The form of a typical multi-variable regression analysis is given by

$$Y = a + b_1x_1 + b_2x_2 + \cdots + b_nx_n \quad (7)$$

Where the x values are the different independent variables and the Y value is the dependent variable. The values of a and b are constants. A multi-variable regression analysis can be applied to the non-dimensionalized equation to solve for the constants using a least-squares fit. A multivariable regression analysis with known data was performed on the developed non-dimensionalized equations to determine appropriate constants. This analysis was done for the overall data and categorized for analysis using Matlab[27].

4 Results of Non-Dimensionalized Equations (NDEs)

To better predict the along-wind dispersion coefficient, non-dimensionalized equations (NDEs) were developed. The parameters of the non-dimensionalized equations proposed here, were taken from established theory and assumptions regarding trends seen with atmospheric dispersion shown in Table 1. Starting with Batchelor's Similarity Theory[5], parameters were incorporated to allows for the along-wind dispersion coefficient to incorporate wind speed, friction velocity, and time for each test.

4.1 NDE by Stability Class (no Monin-Obukhov Length)

Noticeable trends in the stability class were shown in Figure 2. Stability class is taken into account with the $(\frac{u_*}{u})$ term on the right hand side of the NDE. Friction velocity u_*, and time t, are taken into account on the left hand side of the NDE shown below

$$\frac{\sigma_x}{u_* t} = C_1(\frac{u_*}{u})^{p} \quad (8)$$

where C_1 and p are dimensionless constants. Available data was used to determine the constants for this non-dimensionalized equation with a multi-variable regression analysis. An overall value for p was determined to be -0.5 due to a weighted average of the constants solved by stability class and weighted by quantity of tests per stability class. This new equation is shown below in equation (9)

$$\frac{\sigma_x}{u_* t} = C_1(\frac{u_*}{u})^{-0.5} \quad (9)$$

where values of C_1 are determined for each stability class using a single variable regression analysis when p is set to -0.5. These values are shown below in Table 4.

Table 4. Values for C_1 with p = -0.5

	AB	C	D	E	F	All
C_1	0.311	0.339	0.408	0.429	0.576	0.408
Tests	**5**	**24**	**40**	**15**	**5**	**89**

Table 4 shows clear trends in the C_1 values as a function of stability class when the *p* constant is set to a value of -0.5. The trend observed with the C_1 constant shows that as the stability class transitions from the convective boundary layer to the stable boundary layer, the C_1 value increases. The average value of all of the tests for C_1 is 0.408. This value is consistent with a neutral stability class. A visual representation of the NDE classified by stability class is shown below in Figure 3.

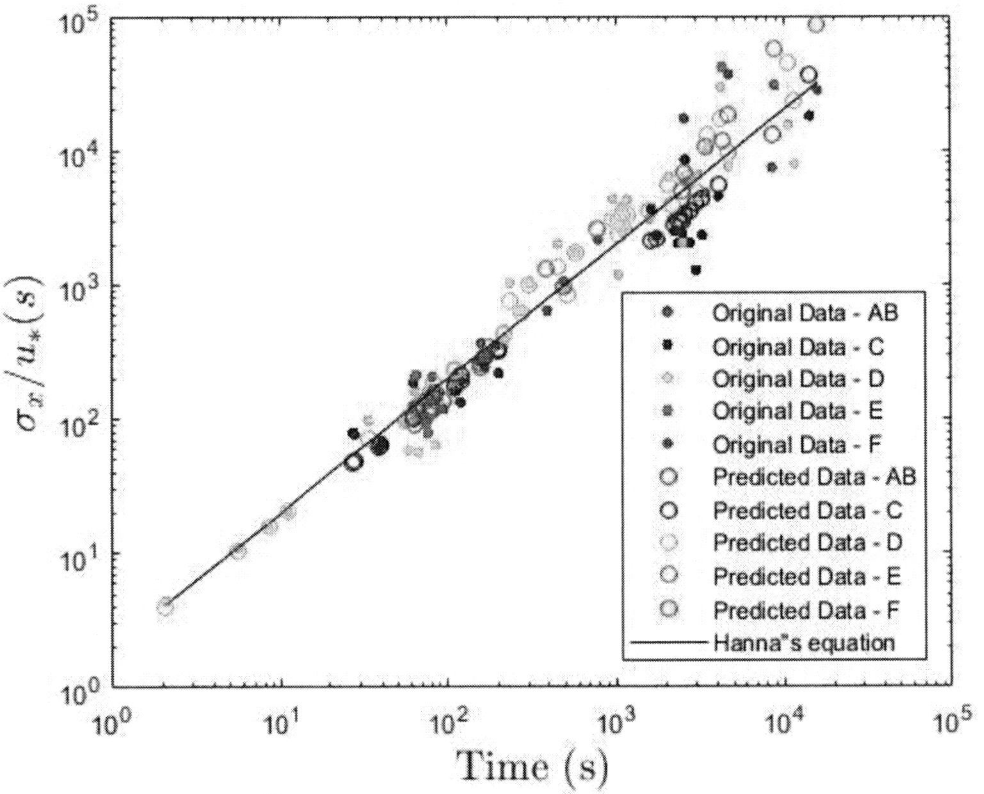

Figure 3. Time (s) vs. σ_x/u_* (s) as a function of the original data by stability class, NDE by stability class, and Hanna's predictions.

Figure 3 shows the original data classified by stability class, predicted data by the non-dimensionalized equation classified by stability class, and Hanna's prediction. Figure 3 visually shows that this non-dimensionalized equation is able to capture the trends by stability class evidenced in Figure 2 appropriately. A statistical comparison of how the predicted data with the NDE and Hanna's equation are shown below with MG VG values in Table 5.

Table 5. MG VG values for NDE by stability class with p = -0.5

	AB	C	D	EF	All
MG – NDE	0.939	0.925	0.889	1.024	**0.931**
VG – NDE	1.096	1.240	1.178	1.458	**1.248**
MG - Hanna	0.728	0.713	1.011	1.217	**0.942**
VG - Hanna	1.234	1.418	1.296	1.484	**1.365**

Overall the MG VG values for all data points are similar for both the NDE MG VG values and Hanna's MG VG in Table 5. Classifying the data points into their respective stability classes shows the MG VG values for the NDE by stability class is able to model the data more appropriately than Hanna's equation. Overall, the NDE by stability class is able to model the data better than Hanna's equation based on the MG VG values closer to 1 by stability class in Table 5 and the visual representation in Figure 3.

4.2 NDE by "Stability Class" (with Monin-Obukhov Length)

While the NDE by stability class is a better fit than the previous along-wind dispersion equation[4], noticeable trends in the C_l constants were observed in Table 4. These trends show that as the stability class transitions from the convective boundary layer to the stable boundary layer, the C_l value increases. An NDE that takes into account the Monin-Obukhov length λ, to model these trends was developed. The previous NDE was used as the basis for the NDE with Monin-Obukhov Length and is shown again in equation (9). An equation for C_l to incorporate the Monin-Obukhov length is shown below in equation (10)

$$\frac{\sigma_x}{u_* t} = C_1 (\frac{u_*}{u})^{-0.5} \quad (9)$$

$$C_1 = C_{s/c} - \left(C_{s/c} - C_d\right) * [\exp\left({}^{\lambda_{s/c}}/_{\lambda}\right)] \quad (10)$$

where $C_{s/c}$ and $\lambda_{s/c}$ are dimensionless constants for the convective and stable boundary layer. Equation (10) is able to make the C_1 values a function of Monin-Obukhov length which directly relates to stability class. Equation (10) is arranged so that at a neutral boundary layer with a Monin-Obukhov length of 10,000, the C_1 value is equal to 0.408 which is consistent with the C_l value for stability class D from Table 5. This arrangement allows for a smooth transition between the convection boundary layer, with negative Monin-Obukhov lengths, and stable boundary layer, with positive Monin-Obukhov lengths. The values for the constants determined from equation (10) with a multi-variable regression analysis are shown below in Table 6.

Table 6. Values for $C_{s/c}$ and $\lambda_{S/C}$ constants with p = -0.5

	Negative L (Convective)	Positive L (Stable)	D (from Table 4)
$C_{s/c}$	0.313	0.408	*0.408*
$\lambda_{S/C}$	186.360	-0.008	
tests	**32**	**51**	***40***

Table 6 shows the constants for $C_{s/c}$ and $\lambda_{S/C}$ for convective and stable boundary layers. There are no obvious trends in the data. The total tests for the NDE with Monin-Obukhov length is slightly less than the previous NDE. Only 83 data points out of the 89 data points reported Monin-Obukhov length which account for this slight decrease. A visual representation of the predicted data is shown below in Figure 4.

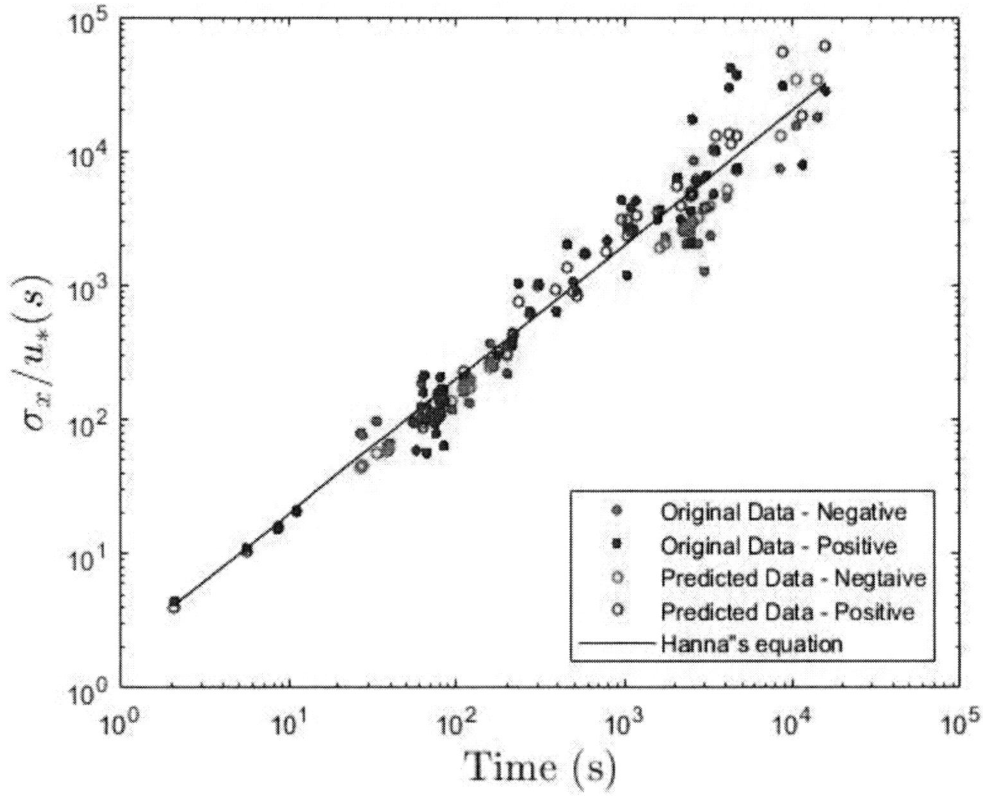

Figure 4. Time (s) vs. σ_x/u_* (s) as a function of the original data, NDE by Monin-Obukhov Length, and Hanna's predictions.

Figure 4 shows the original data, predicted data by the NDE classified by Monin-Obukhov Length, and Hanna's prediction. The slope of the NDE for a negative Monin-Obukhov length is less than Hanna's prediction which aligns with Figure 2 for unstable stability classes. The slope of the NDE for a positive Monin-Obukhov length is greater than Hanna's prediction which also aligns with Figure 2 for more stable stability classes. A statistical comparison of how the predicted data with the NDE and Hanna's equation are shown with MG VG values below in Table 7.

Table 7. MG VG values for NDE by Monin-Obukhov Length

	Negative λ	Positive λ	All
MG – NDE	0.978	1.019	**1.001**
VG – NDE	1.247	1.318	**1.286**
MG - Hanna	0.785	1.117	**0.957**
VG - Hanna	1.446	1.424	**1.433**

The MG VG values for all of the data points in Table 7 are improved for the NDE with Monin-Obukhov Length compared to Hanna's equation. Classifying the Monin-Obukhov length by negative and positive values shows that the NDE has improved means and variances compared to Hanna's equations for both the convective and stable boundary layers. Similar to the NDE by stability class, the NDE with Monin-Obukhov length is an improved model based on the visual representation in Figure 4 and MG VG values closer to 1 in Table 7.

5 Summary and Future Work

Toxic, flammable, or hazardous chemical releases into the atmosphere are critical to model appropriately. An accurate along-wind dispersion equation is needed to ensure safety of the public. Multiple non-dimensionalized equations have been developed in the past through theory and experimental data. Two NDEs were developed in this paper, where both of these equations proved better than the current best equation used for along-wind dispersion. The main difference in the utility of the NDE equations lie in the availability of Monin-Obukhov lengths to incorporate stability class into the model. If the Monin-Obukhov length is available, equations (9) and (10) are suggested to be used with values from Table 6. These equations provide one continuous NDE throughout all conditions. If Monin-Obukhov length is not available, equation (9) is suggested with values from Table 4. Both of the non-dimensionalized equations are able to capture the spread of data better than the current equation for along-wind dispersion due to the additional parameters of friction velocity and wind speed raised to a power which simulates stability class. The NDEs developed here will improve modeling for atmospheric releases due to the effects of along-wind dispersion.

Experimental work is in the process for further validation of these equations through original test releases performed at the Chemical Hazards Research Center in Fayetteville, AR. Finite duration releases from an area source will be performed in an ultra-low speed wind tunnel to provide additional data in a controlled environment to validate the NDEs discussed here.

6 Acknowledgments

We gratefully thank Steven R. Hanna and Pasquale Franzese for sharing data accumulated for this analysis. We also thank Henk Witlox and Maria Fernandez from DNV GL who provided insight and expertise that assisted with the research.

7 References

[1] Boybeyi, Zafer, Sethu Raman, and Paolo Zannetti. "Numerical investigation of possible role of local meteorology in Bhopal gas accident." *Atmospheric Environment* 29.4 (1995): 479-496.
[2] Høiset, S., et al. "Flixborough revisited—an explosion simulation approach." *Journal of hazardous materials* 77.1 (2000): 1-9.
[3] De Marchi, Bruna, Silvio Funtowicz, and Jerome Ravetz. "Seveso: A paradoxical classic disaster." *The long road to recovery: Community responses to industrial disaster* (1996): 86-120.
[4] Morris, J. & Spicer, T.O. (2017). Generalization of Modeling Along-Wind Dispersion. 13th Global Conference
on Process Safety (GCPS), March 2017. San Antonio, TX.
[5] Batchelor, G. K. "Diffusion from sources in a turbulent boundary layer." Arch. Mech. Stosow 3.6 (1964): 661-670.
[6] Chatwin, P. C. "The dispersion of a puff of passive contaminant in the constant stress region." *Quarterly Journal of the Royal Meteorological Society* 94.401 (1968): 350-360.
[7] Van Ulden, A. P. "A surface-layer similarity model for the dispersion of a skewed passive puff near the ground." *Atmospheric Environment. Part A. General Topics* 26.4 (1992): 681-692.
[8] Hanna, Steven R., and Pasquale Franzese. "Alongwind dispersion—a simple similarity formula compared with observations at 11 field sites and in one wind tunnel." *Journal of Applied Meteorology* 39.10 (2000): 1700-1714.
[9] Biltoft, C., Phase I of DSWA transport and dispersion model validation study. DPG-FR-97-058, (1997): 156.
[10] Biltoft, C., Dipole Pride 26: Phase II of DSWA transport and dispersion model validation study. DPG-FR-98-001, (1998): 130.
[11] Watson, T. B, R. E. Keislar, B. Reese, D. H. George, and C. A. Biltoft, The DSWA Dipole Pride 26 field experiment. NOAA Tech. Memo. ERL ARL-225, (1998): 99.
[12] King, S. Bruce. The Kit Fox Field Demonstration Project and Data Set. *International Conference and Workshop on Modeling the Consequences of Accidental Releases of Hazardous Materials.* (1999).
[13] Spicer, T. and Havens, J. Description and Analysis of Atmospheric Dispersion Tests Conducted by EPA. *International Conference and Workshop on Modeling the Consequences of Accidental Releases of Hazardous Materials.* (1999).
[14] Sheesley, D. et al.. Chemical Dispersion Modeling Project, 1995 Kit Fox Field Demonstration DOE Nevada Hazardous Spill Center. Volume I – Experiment Description and Data Processing. (1998).
[15] Hanna, S. R., and K. W. Steinberg, Studies of dense gas dispersion from short-duration transient releases over rough surfaces during stable conditions. *Air Pollution Modeling and Its Applications XI,*Plenum Press, (1996): 481–490.
[16] Biltoft, C., S. D. Turley, T. B. Watson, G. H. Crescenti, and R. G. Carter, Over-Land Atmospheric Dispersion (OLAD) test summary and analysis. WDTC Doc. WDTC/JCP-99/048, (1999): 38.

[17] Robins, A. G., and J. E. Fackrell, An experimental study of the dispersion of short-duration emissions in a turbulent boundary layer. *Air Pollution VI,* C. A. Brebbia, C. F. Ratto, and H. Powers, Eds., WIT Press, (1998): 697–707.
[18] Nickola, P. W., J. D. Ludwick, and J. V. Ramsdell Jr., An inert gas tracer system for monitoring the real-time history of a diffusing plume or puff. *J. Appl. Meteor.,*9, (1970): 621–626.
[19] Sato, J., F. Kimura, and T. Yoshikawa, The longitudinal spread of puff in the short range diffusion experiment. *Pap. Meteor. Geophys.,*32, (1981): 155–162.
[20] Hilst, G. R., and N. E. Bowne, A study of the diffusion of aerosols released from aerial line sources upwind of an urban complex. Vols. I and II. Rep. DA-42-007-AMC-38®, (1966): 231 and 583.
[21] Smith, and B. L. Niemann, Shoreline diffusion program, Oceanside, CA. Vols. I and II. MRI-64-FR-860, (1969): 230 and 456.
[22] Bowers, J. F., G. E. Start, R. G. Carter, T. B. Watson, K. L. Clawson, and T. L. Crawford, Experimental design and results for the Long-Range Overwater Diffusion (LROD) Experiment. DPG/JCP-94/012, (1994): 531.
[23] Pasquill, F. (1961). *The estimation of the dispersion of windborne material*, The Meteorological Magazine, vol 90, No. 1063, pp 33-49.
[24] Golder, Donald. "Relations among stability parameters in the surface layer." *Boundary-Layer Meteorology* 3.1 (1972): 47-58.
[25] Hanna, S. R., J. C. Chang, and D. G. Strimaitis. "Hazardous gas model evaluation with field observations." *Atmospheric Environment. Part A. General Topics* 27.15 (1993): 2265-2285.
[26] Draper, Norman Richard, Harry Smith, and Elizabeth Pownell. *Applied regression analysis*. Vol. 3. New York: Wiley, (1966).
[27] Davis, Richard A. "Practical Numerical Methods for Chemical Engineers: Using Excel with VBA". Lexington, KY, (2015).

The VOCs Emission Source Intensity and Safety Protection Distance Analysis of Urban Refinery Based on QRA

Jiang Xue
SINOPEC Research Institute of Safety Engineering,
QingDao,China
Jiangxue.qday@sinopec.com

Dang Wenyi, Wan Gujun
SINOPEC Research Institute of Safety Engineering,
QingDao,China

Prepared for Presentation at
American Institute of Chemical Engineers
2018 Spring Meeting and 14th Global Congress on Process Safety
Orlando, Florida
April 22 – 25, 2018

The VOCs Emission Source Intensity and Safety Protection Distance Analysis of Urban Refinery Based on QRA

Jiang Xue
SINOPEC Research Institute of Safety Engineering,
QingDao,China
Jiangxue.qday@sinopec.com

Dang Wenyi, Wan Gujun
SINOPEC Research Institute of Safety Engineering,
QingDao,China

Keywords: VOC source intensity analysis, Safety Protection Distance

Abstract

with the "urban refinery" problem becoming increasingly prominent, the identification of VOCs emission source of urban refinery has been the problem to be solved mostly. In addition to the widely used monitoring methods, we combined with the wind tunnel experiment and numerical simulation technology to confirm the model coefficient, and established more universal and quantitative emission source intensity analysis model, with the goal to form a suitable assessment technology for domestic "urban refinery "VOCs emissions problem, as well as to provide scientific data support for the complete sets of environmental and security solution programs.

1 Introduction

With the aggravating process of urbanization in China in recent years, the issue of "urban refinery" has drawn increasing attention. The VOCs we are talking about now mostly refer to organic compounds with atmospheric boiling point below 500℉ or saturated vapor pressure above 71Pa at surrounding temperature[1]. According to the findings of Kyriaki Maria Fameli and others in 2006-2012[2], VOC sources are generally divided into civil fuels, vehicle exhaust and industrial emissions. However, the contribution rates of these three types of emission sources to the overall VOC pollution in the region varies according to regions. For East Asia, where our country is located, the main source is industrial emissions.

At present, the main VOCs emission characteristics of chemical plants and refinery in our country performance at: from the device point of view, in the production and storage process, the connection of the device and pipeline is prone to leakage, and not easy to trace; From the whole scale point of view, if there are electronic components factories or other chemical companies in the area, these enterprises may have complicated gas sources in the plant boundary. The emission of VOCs will result in multiple diffuse plumes. Due to the influence of time, topography, building layout and wind direction, these different diffuser plumes will produce heterogeneous mixing[3], which makes it difficult to measure the surrounding flow field and concentration field, resulting in regional complex air pollution problems. This will have a direct impact on our assessment of the emissions contribution of different sources and the health influence on surrounding community.

Currently the conflicts between "urban refinery" and the development of the surrounding area tend to the relocation of refineries to the industrial zone in countryside, which can only benefit the safe development in urban areas in the short term, but obviously be bad for immigratory area, in which the phenomenon of "urban refinery" is bound to occur again in the long run. On the other hand, due to many years' construction and development of existing refineries, there exist problems such as cross-docking of auxiliary piping and urban infrastructure networks, making the relocation unrealistic. Therefore, to fundamentally solve the problem of urban refinery, it is necessary to use intrinsically safe technology to reinforce the refinery process safety; besides, with quantitative risk assessment technology, conduct the risk assessment for individual risk of surrounding population and environment to establish China's assessment technology for the refinery boundary zone and the surrounding population and business distribution to provide scientific guidance.

2 VOCs Assessment Technique For Urban Refinery in China

Faced with such a situation, primary refinery and petrochemical groups in China have actively developed VOC control technology in process of absorption, desulfurization, catalytic combustion, regenerative combustion and condensation, to purify gas in line with China's current standards [1]. In addition, source-level and plant-scale VOC emissions analysis have also drawn the attention of local governments and relevant enterprises. However, no mature VOC assessment technology is currently developed in China.

Existing methods include regular air sampling, component analysis, and quantitative analysis based on the emission factor index developed in EU and U.S. to evaluate the annual VOC emission levels in the region. The basic formula is[2]

$$E_{i,x} = E_i \times \frac{value_x}{value_{tot}} \qquad (1)$$

where

$E_{i,x}$= annual total emission value of component i in local emissions X.

E_i = annual total emission value of component i in the region.

$value_x$ = the emission factor of X.

$value_{tot}$ = the emission factor of whole region.

However, this method has two defects. First, the component will be affected by the weather, time period and season. Therefore, the sample can not objectively reflect the average condition of the regional air. Second, the factor index is calculated by the European and American countries who are based on their air pollutants composition and historical data, along with unaccustomed to Chinese Refinery. On the whole, this method does not provide a scientific explanation of the level of regional VOC emissions in China, nor can it be used as the main source of data to support our analysis.

3 Establishment of emission source intensity analysis model based on ISCST-3 model

3.1 emission source intensity analysis model

Combined with the complex gas composition of smoke plume diffused in refinery in our country, we need to find a more universal source-strength analysis model based on on-site monitoring and simulation experiments, which can be widely applied to different urban conditions.

According to the characteristics of diversified emission sources and multiple gas components, combined with Gaussian plume model and ISCST-3 model (Complex Industrial Source Intensity Analysis Model) [3], the target refineries and chemical plants in urban areas can be regarded as surface source for the source

$$C = \frac{QK}{2\pi\mu_S}\int_X \frac{VD}{\sigma_y\sigma_z}\left\{\int_Y exp\left[-0.5\left(\frac{y}{\sigma_y}\right)^2 dY\right]\right\}dx \qquad (2)$$

where

C= concentration, g/m^3

Q= surface source emission rate, $g/m^2 \cdot s$

μ_S=wind speed, m/s.

K,V, σ_y, σ_z,= parameter

X=The horizontal distance between monitoring point and center of surface source.

Y=The longitudinal distance between monitoring point and center of surface source.

Z=The vertical distance between monitoring point and center of surface source.

Through this model, all units can be regarded as point sources which make whole plant a surface source. By comparing of the VOCs concentrations of major surface sources in a region, It is possible to assess the contribution of the gas components and the main sources of pollution that contribute to the overall VOCs emissions in the region.

For different emission targets, we first need to determine different coefficients. The determination of these coefficients needs to be combined with the data of the monitoring sampling. At the same time, after the model is determined, the wind tunnel simulation experiment needs to be carried out to confirm and correct, so as to ensure the coincidence of the model with the real diffusing plume and the scientific analysis.

3.2 coefficient determination.

To determine the model parameters, some can be obtained by site monitoring, such as μ_s, x, y; the others can be obtained by wind tunnel test, using monitoring sample as background points, and calculate by wind tunnel diffusion coefficient deviation formula to get these parameters.

3.2.1 site monitoring determination.

By monitoring sampling in same period, we can determine the value of μ_s, x and y. Especially, the monitoring sampling needs to be collected in both horizontal direction (X-Y axis section) and longitudinal direction (XY-Z axis section) in accordance with the wind tunnel experiment, summing up for more than four monitoring points and ensure that at least four points are on the same radius circle centered on the diffusion point source, that is, the same X value is guaranteed. According to the sampling data, we can get the source intensity calculation formula of four sets of background points

$$C_i = \frac{QK}{2\pi\mu_S}\int_{X_i}\frac{VD}{\sigma_y\sigma_z}f(\sigma_y)dx \qquad (3)$$

with

$$f(\sigma_y) = \int_Y exp\left[-0.5\left(\frac{y}{\sigma_y}\right)^2 dY\right] \qquad (4)$$

Since the gas diffusivity Q and the ambient wind speed u_s,,f (σ_y) are the same at the same lateral distance in the same time period, they can be considered as constants. The formula can be regarded as the function of g (V, D, σ_y, σ_z).

3.2.2 tunnel experiment determination.

The existing model calculation methods mostly combine CFD simulation with on-site monitoring. In China, some researchers have used FLUENT model to qualitatively study the influence of buildings' layout on the diffusion of pollutants. Some also used the results of the FLUENT model to verify the results of wind tunnel tests on the influence of pollutant diffusion, but lacking relevant research on the diffusion of pollutants when the sources of pollution are in different locations of buildings.

However, the CFD simulation has obvious shortcomings.1. The diffusional plumes can only simulate the effect of diffusion under single wind direction. The simulation of regional air flow field don't include the influence of air turbulence, as well as idealized diffusion range[4].2. The gas components that directly affect the non-uniform mixing of the diffused plume by different VOC emission sources have a gap with the diffusive plume composition[5].

In this case, wind tunnel experiments can determine the other coefficients of the model and the flow field around the surface area more scientifically and reasonably. The wind tunnel can lay out the concentration point of VOCs emission by using the Reynolds coefficient to simulate the regional wind field in surrounding area and carry out the gas diffusion experiment, taking samples of concentration by different lateral or longitudinal distance from the factory boundaries, to establish the smoke axis trajectory.

The description of plume's axis helps to grasp the movement of plume as a whole. The axis of plume shows the impact of topography on the transport of contaminants. The horizontal trajectory of the plume can give an intuitive indication of the area affected by the plume, such as the flow of plume along the valley. The trajectory of the vertical plume can give a change in the terrain that may occur during transport and spread of the plume. These trajectories can get more accurate parameters to the actual description of smoke plume behavior.

According to the diffusion parameters in ISCST-3 model, the results of wind tunnel can be calculated by MATLAB software. Diffusion parameters reflect the size of the spread of pollutants. In general, the diffusion parameters and the atmospheric stability, surface roughness and so on. After the numerical simulation and calculation of software such as MATLAB, we can get a more objective and reasonable source of VOCs emission analysis results.

As long the experimental data be got by wind tunnel, the data processing can be used to correct the deviation of these data[6-7], the formulas are as follows:

$$y_0 = \frac{\sum_{i=1}^{N} y_i C_i}{\sum_{i=1}^{N} C_i} \qquad (6)$$

$$\sigma_y = \sqrt{\frac{\sum_{i=1}^{N} C_i (y_i - y_0)^2}{\sum_{i=1}^{N} C_i}} \qquad (7)$$

$$z_0 = \frac{\sum_{i=1}^{N} Z_i C_i}{\sum_{i=1}^{N} C_i} \qquad (8)$$

$$\sigma_z = \sqrt{\frac{\sum_{i=1}^{N} C_i (Z_i - Z_0)^2}{\sum_{i=1}^{N} C_i}} \qquad (9)$$

where

C_i=represents the object concentration;

y_i,z_i=measuring point i for y-axis direction and the z-axis direction coordinates;

y_0, z_0=the distance weighted average density;

σ_y, σ_z=target measuring point of the lateral and vertical diffusion parameters .

According to the obtained σy, σz values, and then take the relevant data of the measurement points into the calculation of V, D. Summarizing the values of the various coefficients and bringing them into the source strength analysis model, the total diffusion concentration C of the surface source can be inversely calculated, that is, the diffusion concentration of the VOCs emission source in the unit of the plant boundary. Similarly, source intensity analysis and calculation can be performed on the surrounding VOCs emission enterprises, and then the overall VOCs emission concentration in a specific industrial area of a city can be obtained.

4. Safety Distance Based on Resource Intensity Analysis

To establish a reasonable and scientific safety distance, the scope and normal discharge of the refinery should be considered at the same time. The scope of environmental impact under the conditions is twofold.

4.1 health protection distance

In China, we do can calculate the Health protection distance under the premise of knowing the diffusion concentration of VOCs from various companies (sources), that is,

$$\frac{Q}{Cm} = \frac{1}{A}(BL^C + 0.25r^2)^{0.5} L^D \quad (10)$$

where,

Q= the VOCs emission concentration, $g/m^2 \cdot s$;

C_m=the standard concentration limit, mg/m^3.

L=the health protection distance, m;

r=the diffuse surface source radius of VOCs emission, m;

A, B, C, D are dimensionless coefficients.

with

$$r = \left(\frac{S}{\pi}\right)^{0.5} \quad (11)$$

S=the area of the whole refinery or plant, m^2.

The health protection distance L can be calculated by putting the Q band calculated by the wind tunnel experiment into the above formula.

4.2 the safety distance in an accident situation:

Based on refinery's common types of hazardous chemicals, as well the four dimensions of personnel, property, society and environment impact of accidents, the safety distance should focus on the types of accidents such as explosions, fires, and leakage of toxic substances.

4.2.1 calculation based on the impact of explosions and fire accidents

For explosions and fire accidents, CFD software can be used to simulate the impact of the most serious consequences.

Through the use of CFD software, the plant boundary can be regarded as the major source of leakage of hazardous chemicals within the plant boundary. The surrounding areas of interest are listed as simulation objects, and the results are simulated by the data of the previous on-site monitoring and the local climate and geographical factors. The scope of impact and the overpressure curve of the explosion shock wave are analyzed to obtain the scope of the impact of the explosion; the scope of the impact of the heat radiation from the fire and the flammable gas cloud is analyzed to obtain the scope of the impact of the fire accident.

4.2.2 calculation based on the impact of toxic leakage accidents:

Leakage of Toxic Substances Through wind tunnel tests, the types of toxic hazardous chemicals in the plant boundary can be selected, and the proportion of gas in equal concentrations can be used to simulate leakage of toxic substances at the plant's annual average downward wind speed, and when the lowest limit for human mortality is reached The scope of the diffusion, as a range of toxic leakage accidents.

4.2.3 safety distance

According to the analysis of the health protection distance, the maximum impact range of the accident state, a comparative analysis, you can take the larger of the two as a reference value for the safety protection distance.

5 Conclusion

At present, the contradiction between the development of domestic urban refineries and surrounding communities is outstanding. The main manifestations are: dangerous chemical materials produced and stored in refineries. Once an explosion, fire, or leakage of toxic substances occurs, it will cause serious consequences for the safety of surrounding personnel and property; At the same time, due to the fugitive emissions of refinery VOCs, it will also have adverse effects on the environment and human health of the surrounding areas. In order to fundamentally solve the phenomenon of "urban encircling factories," we can focus on exploration and analysis of the methods for calculating the source emission intensity of VOCs and the quantitative analysis of safety protection distances.

Based on the ISCST-3 model based on the Gaussian plume diffusion principle, an analysis system for the source intensity of VOC emissions was established. An analysis system combining on-site monitoring, wind tunnel testing, and MATLAB numerical simulation was established to calculate multiple sets of data based on the input and integration of on-site monitoring data. The corresponding coefficients, and then the weighted average of the coefficient values of each group, can obtain the preliminarily feasible source intensity analysis necessary coefficients, forming a source intensity analysis system with universal and generalized significance.

6 References (examples given to show format).

[1]Fameli K M, Assimakopoulos V D. The new open Flexible Emission Inventory for Greece and the Greater Athens Area (FEI-GREGAA): Account of pollutant sources and their importance from 2006 to 2012[J]. Atmospheric Environment, 2016, 137:17-37.

[2]Zheng J Y, Shao M, Che W W, et al. Speciated VOC emission inventory and spatial patterns of ozone formation potential in the Pearl River Delta, China.[J]. Environmental Science & Technology, 2017, 43(22):8580.

[3]Fagbeja M A, Hill J L, Chatterton T J, et al. Challenges and opportunities in the design and construction of a GIS-based emission inventory infrastructure for the Niger Delta region of Nigeria.[J]. Environmental Science & Pollution Research, 2017:1-21.

[4]Melas D, Liora N, Poupkou A, et al. An Up-to-Date Assessment of the Air Quality in Greece with the Use of Modeling Tools[M]// Perspectives on Atmospheric Sciences. Springer International Publishing, 2017.

[5]Tsilingiridis G, Zachariadis T, Samaras Z. Spatial and temporal characteristics of air pollutant emissions in Thessaloniki, Greece: investigation of emission abatement measures[J]. Science of The Total Environment, 2002, 300(1–3):99-113.

[6]Zugravu-Soilita N. How does Foreign Direct Investment Affect Pollution? Toward a Better Understanding of the Direct and Conditional Effects[J]. Environmental & Resource Economics, 2017:1-46.

[7] Kwak K H, Baik J J.A CFD modeling study of the impacts of NO x, and VOC emissions on reactive pollutant dispersion in and above a street canyon[J].Atmospheric Environment, 2012, 46(1):71–80.

512911 Leakage Process of Refined Oil in Different Layered Soils

Abstract:

Pipeline transportation as the most important way of refined oil transport. As the refined oil is flammableã€explosive and toxic. In the event of a leakage accident is easy to causing fire, explosion, poisoning and other accidents. If the accident occurred next to the inland river may cause serious impact to the water and its biological. This paper analyzed the leakage process of refined oil in different saturated soils of riverbed. Summarizes the diffusion mechanism and regular pattern of the refined oil in the soil of the river inland and the soil contact with the river. Combined with the leakage characteristics of the pipe hole leakage and its effect on the process of leakage diffusion. The two-dimensional sand box model experiment has established to Analysis and conclusions of the corresponding leakage diffusion.

511277 Management Role in the Building Siting Evaluation

Abstract:

Facility Siting Studies can be done using two approaches, one is the "Consequence-Based" and the other is the "Risk-Based". Both approaches have their own limitations.

The "Consequence-Based" approach generally lead to conservative results that might require action such as building relocation or retrofit. Both of which can be expensive. Where-as "Risk-Based" approach takes into consideration numerical values for both the consequence and the frequency (likelihood) for a range of credible scenarios affecting the buildings. Results could then be compared with the acceptable risk criteria to provide a more realistic view of the possible impact to occupants within the buildings. This approach, as opposed to others, can minimize subjectivity while providing a more rational decision making process for the management. "Risk-Based" approach also provide an additional level of analysis for the management to validate the HAZOPs or What-If studies.

The results from the "Risk-Based" facility siting could be used by management in:

1. Better understand and identify the major contributors to risk and focus the spending on them allowing the company to minimize the risk to an acceptable level;
2. Risk transfer (insurance), allowing the company to determine and work on minimizing insurance premiums;
3. Incident Management, determining the right emergency response plans and exercises for example;
4. Land-use restrictions.
5. Better understanding for the stakeholders.

511308 Bacterial Foraging Optimization Algorithm for Safety Based-Process Plant Layouts

Abstract:

The following article approaches application of inherent safety principles to a process plant layout design based on a bacterial-foraging optimization algorithm. Our approach seeks to maximize the inherent safety level, quantifying it with previously proposed inherent safety frameworks. A computational tool is produced for risk assessment, based on the Inherent Safety Index Module (ISIM) and the Integrated Inherent Safety Index (I2SI). Both approached indices employ process information, such as operating conditions and equipment sizing to develop quantification of damage for fire and explosion cases, although the former includes risk calculation based on consequence and probability estimation modules. Layout-based considerations, such as equipment life cycle, ease of escape and firefighting, wind direction and location of off-site facilities are also considered. The algorithm, inspired on bacterial foraging, places a set of main process units within a specified area and evaluates the aforementioned aspects in order to assess the suitability of a given layout in economic terms. The proposed solution approach provides an alternative to hard-optimization methods, by allowing greater flexibility in accounting for both safety and economic aspects, while providing high quality solutions in a limited computation time. The aim of our proposed solution approach is to provide support to expert decision-making during the early plant layout design stages. A case study based on an acrylic-acid production plant, which has been used by several other articles in the literature, serves the purposes of showing the appropriateness and effectiveness of the method.

511315 Facility Siting, Closing the Gaps

Abstract:

History has seen numerous process safety disasters with catastrophic damage to nearby buildings, resulting in loss of life and property, documented in the literature with some still under investigation. In these disasters, the poor arrangement of equipment within a facility contributed to the magnitude of the disaster.

Despite stern lessons from the past, PSM regulations provide uncertain requirements. PSM require acknowledging or considering facility siting (FS) on their process hazard analysis (i.e. HAZOP), but does not provide a clear guidance on what is a proper approach to consideration and what is adequate. FS can be easily overlooked or even not considered during design; application of corporate siting decisions can be made by compliance exposing operators to unnecessary risk. In many cases FS decisions are considered during late project stages, even after the intended permanently occupied buildings are already in place and mitigation measures are extemporaneous or budget prohibitive.

The American Petroleum Institute (API) and the Center for Chemical Process Safety (CCPS) has invested significant effort in develop API RP-752/753 guidelines and literature. Although these resources provide a good starting point and a sound process via practical flowcharts, the recommended practices leave a lot of room in the selection of adequate technical methods needed to complete the assessment; this is a gap particularly when required answers call for quantitative event simulation and frequency calculation.

For example, once the API screening flowchart is applied, and qualitative assessment is completed; but the building does not meet siting criteria for explosion or toxic, what next? How detailed should the detailed analysis be? Can radiation and blast loads be determined without a full blown quantitative risk analysis (QRA) that is in most cases budget prohibitive? How much of sound analysis can be can be done with a constrained budget? In absence of benchmark public industrial development individual/risk criteria, are there local/corporate siting decision criteria for decision making?

An overall optimized quantitative method is discussed in this paper including four stages. It can be applied for a variety of process and multiple facilities and scope can be managed (i.e. when to stop and what to evaluate). This approach is flexible and cost-effective as it allows screening-out what does not contribute to risk and makes the most of the qualitative assessment, reducing the scope of computational stages.

510455 Leadership Keeping the Balance to Create a High Performance Process Safety Culture

Abstract:

14th Global Congress on Process Safety

22nd-26th April 2018

Orlando World Center Marriott

Call for Papers

Title:

Leadership keeping the balance to create a high-performance process safety culture

Authors:

Coen van Driel **Robert Kauer**

TÜV SÜD Schweiz AG TÜV SÜD Industrie Service GmbH

Mattenstrasse 24 Westendstr. 199

CH-4002 Basel 80686 Munich

Switzerland Germany

coen.vandriel@tuev-sued.ch robert.kauer@tuev-sued.de

Abstract:

In assessing a safety culture there are two aspects for evaluation which is on the one hand the process safety management system and on the other hand the human factor. Both elements are mutually dependent in the realization of a high performance process safety culture.

Why are those elements mutually dependent?

A good example for illustrating the mutual dependency is the challenge of climbing the Mont Blanc or any other challenging mountain. For the safety of the climbers rules and regulations have been developed and converted into management systems. The guides that lead groups of climbers to the summit have to follow the rules and regulation. As in the industry, if something happens with one of the climbers the guide has to go to court. Despite rigorous systems in place incidents and casualties have to be accounted for. The culture of climbing the Mont Blanc is changed driven by money or operational performance of the guides and the perception of the climbers. This trend cannot be changed by focusing just on the system but also requires attention to the mindset of the climbers (culture).

-

How to realize the process safety cultural transformation?

One of the success factors of a process safety cultural transformation is that people are from the start aware of their improvement potential in a way that they admit themselves that they can improve. Something like an "ah" moment. How to get to that moment? Fact based visualization of the improvement potential together with the people is one of the methods that support the process from the beginning, looking at the situation from different perspectives like the technical and management system and the behavior/culture point of view.

Different tools, from brown paper workshops about the processes to case studies support the process in realizing the visualization of mutually understanding the gaps in the system and the individual usage of the system by those people involved indicate the process safety culture within a company.

To understand the culture, it is definitely necessary to get an understanding of the attitudes, values and beliefs of the people in the organization from leadership to the shop floor. This information can be gathered by standard methods such as questionnaires but also with advanced tools like Katalyst anonymous online workshops in which the approach unveils the attitudes, values and beliefs behind the façade people tend to show.

All the perspectives at the end give a realistic picture of the current safety culture by visualizing the blind spots and creating an environment in which it is safe to become conscious incompetent (the "AH" moment) as an organization and person.

Then in cooperation with leadership in the organization developing a transformation roadmap to realize the vision or in other words the picture of the future. It is a roadmap which is owned by the people and leadership in the organization because they have developed it themselves only facilitated by the consultants.

Specific Execution Method for safety culture transformation

To achieve high performance process safety culture, a comprehensive method for diagnosing safety culture is necessary. First, we have to conduct a survey and in-depth interview of managers and employees, including the management team to diagnose the safety awareness of the organization. The survey items consist of management's safety commitment, regulations and procedures, safety atmosphere, etc. The safety awareness level of the members can be measured on a Likert 5 point scale. Second, we also have to conduct site audits, interviews, and identify the company's PDCA operation of Occupational Safety and Health Management System to diagnose the safety management system. After completing such procedures, we can diagnose the problem and provide the solution. Finally, in order to diagnose the safety behaviors, we have to develop critical behavior checklist through identifying accident history, near miss reporting, safety manuals, etc. Behavior observation can be implemented by consultants and supervisors with the checklist. After the observation, the manager should provide the workers with feedback many times during the intervention session. On the basis of the observations, consultants or supervisors should analyze the safety behavior change between baseline and intervention sessions.

What is the role of leadership in realizing a high performance process safety culture?

The key role for leadership is to understand that the road to success is to keep the balance between a system driven and human driven performance improvement to realize the results. Leadership adapts to facilitative leadership competencies, like results focus, managing the boundaries, motivational learning, challenging, etc., that enables them to realize the results.

People have the feeling and motivation that they can improve and they are in the starting position to take the steps for improvement.

Leadership use the assessment to create a clear picture of what Process Safety Excellence mean for the company = creating the beacon for making the transformation plan/ roadmap.

-

The result

What is a high performance process safety culture?

A high performance process safety culture in which people are actively aware of where they are related to best practices and where there is a continuous challenge of the current procedures and behaviors by trying to identify blind spots. This committed culture has risk thinking at its core.

Fig. 1 System and human related elements. Both are mutually dependent. If one fails, the other might not be able to compensate. Leadership has to ensure the right balance.

SARS, Safe, Alert & Resilient Behavior in Social Technical System

Salvador Ávila Filho
Federal University of Bahia, Researcher and Consultant
Rua Graciliano de Freitas 43 Roma CEP 40444340 Salvador BA Brazil
avilasalva@gmail.com

Ivone Cerqueira
Federal University of Bahia, Graduate student
ivonecscerqueira@hotmail.com

Carine Nogueira Santino
Federal University of Bahia, Graduate student
k_santino2006@hotmail.com

Edmara Drigo
Federal Institute of Alagoas, Graduate student
edmarasd@gmail.com

Prepared for Presentation at
American Institute of Chemical Engineers
2018 Spring Meeting and 14th Global Congress on Process Safety
Orlando, Florida
April 22 – 25, 2018

SARS, Safe, Alert & Resilient Behavior in Social Technical System

Salvador Ávila Filho
Federal University of Bahia, Researcher and Consultant
Rua Graciliano de Freitas 43 Roma CEP 40444340 Salvador BA Brazil
avilasalva@gmail.com

Ivone Cerqueira
Federal University of Bahia, Graduate student
ivonecscerqueira@hotmail.com

Carine Nogueira Santino
Federal University of Bahia, Graduate student
k_santino2006@hotmail.com

Edmara Drigo
Federal Institute of Alagoas, Graduate student
edmarasd@gmail.com

Keywords: behavior, competence, communication, cooperation, commitment

Abstract

The organization's Scenario without due resilience to environmental challenges stems from improper behavior. Team development methods, when unsuitable, indicate the preparation of people who are not compatible with the challenges. To motivate safe behavior, it is important to identify the relationship of the organization and the worker to the risks in society and in the family. In an initial project are simulated routine situations to analyze a risk of having inappropriate profiles in the leadership or mental map not suitable for micro decisions in the work or society. the SARS tool, Safe, Alert and Resilient Behavior, depends on movements that generate behaviors following the concepts related to C4t (Competence, Communication, Cooperation, Commitment). The SARS technique that intends to apply an initial strong vaccine from the negotiation of the strategy and the data collection with the training of the front staff, INITIAL SARS, which results in an Initial Behavior. This training that addresses the routine situations of the worker inside and outside the organization is rooted in cultural aspects that can indicate human failures or errors. The implementation of this new behavior brings a new environment where human errors tend to decrease as well as the likelihood of an accident.

Background and Context

The Brazilian industry and in particular the transformation industry live the problem of recurrence of work accidents and processes. In this context, the international industry (SHELL, CHEVRON, AUSTIN POWDER) has been speaking at conferences on the importance of including the analysis of human factors in industrial risks, to the point of making organizational modifications to include non-technical subjects (culture and people) in the level of attention of technical-economic subjects (production).
According to Figure 1, in the Brazilian manufacturing industry, workers are injured at a rate of 2% per year in 2013 with annual growth of 1.2%, which indicates that occupational patterns are not satisfactory [1]. Brazil is the fourth country in the world in Accidents of Work, registering more than 700 thousand cases a year, after China, India and Indonesia [2].
On the other hand, it is noted that the recurrence of process safety events in the oil industry is high. The National Petroleum Agency monitors the events of fire and explosion and despite the task forces, these serious events with a high probability of fatality are repeated, with at least one case per year [3].

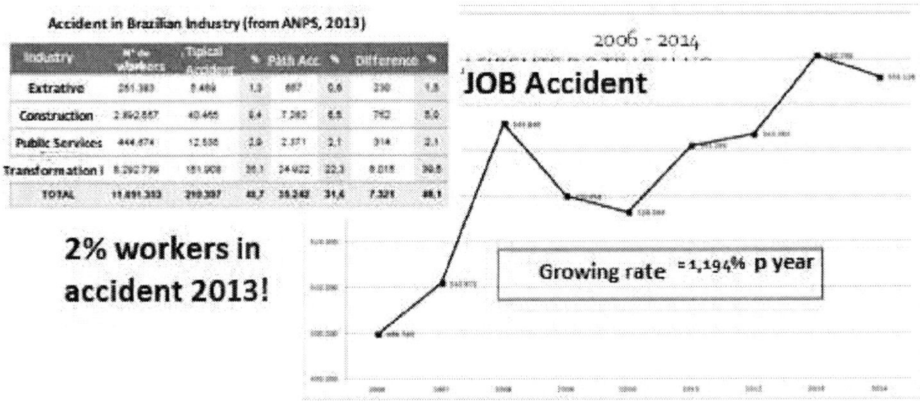

Figure 1 – Accident rate in Brazil

In 2015, the FPSO São Mateus production ship accident occurred in Espírito Santo, where nine fatalities [4] were recorded. The gas alarm at the pump house of the production ship was fired at 11:30 p.m., and after an hour the unit exploded, causing 9 fatalities and 26 workers with moderate to severe injuries. Considering the complexity of the process safety, the possible gaps in the decision making of production leadership in this FPSO are observed. In the period of 1 hour important decisions were taken quickly in the stage of high stress, which influenced the contingency action to isolate the danger in this event. It is in this scenario that it is justified to seek new methodologies based on the routine to correct bad habits, human errors and incorrect decisions of the operator and the management knowing of the cultural influences and of leaderships.
According to [5] the root cause of deviations, failures, incidents and accidents occurred in the chemical industry are related to: economic bias; static risk management; inadequate task planning; design criteria without discussing cognitive constraints; local or global cultural traits (by changing time and form in the decision-making process); and conflicts between policies (top) and practices (bottom) generating an environment conducive to the accident. This paper suggests new methods

of analysis and new intervention strategies in the oil production industry. The system is based on human and organizational behavior and termed as SARS.

1. Introduction

The SARS Management System aims to break the paradigm that human factors have a secondary relationship with production. Understanding the importance of behavior to the organizational culture and safety ensures the sustainability of the Company. Thus, the search for management systems that guarantee a safe, alert and resilient behavior demands bottom-up strategies and interdisciplinary relations in the activities of industrial control.

Learning and organizational analysis depend on the organization's position in relation to nature and business history in relation to society, they also depend on the way in which the affective bonds are established internally so that there is a commitment in the work environment at all levels. The decision process and cognitive action will be favorable to organizational resilience if the balance of these dimensions occurs.

As indicated in Figure 2, the primary matrix in which the organization is built respects local and global history through intuitive processes. Sensations and habits are the result of social beliefs and archetypes that dominate attitudes in the workplace. A second matrix that is inserted in the primary is the intuitive-historical matrix, where the group bond is developed and the leaderships are established. The activities developed have physical and natural limitations that are part of the risk of the task. At this point, the organizational unit emerges with local beliefs and archetypes, with the history of the company and its leaders, and with the format of a decision model.

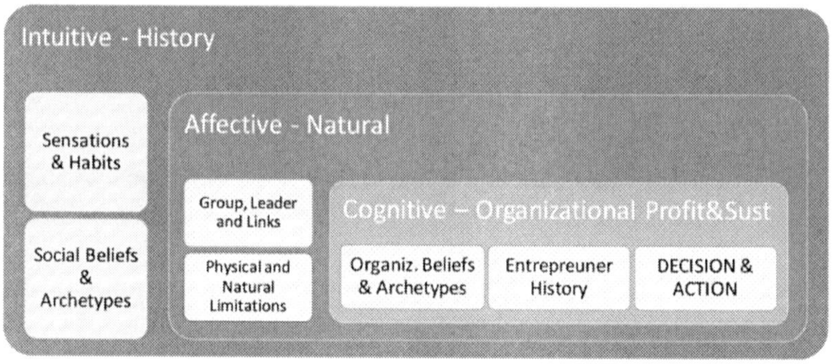

Figure 2 – Organizational Learning Analysis for the task

Decisions about routine problems and repetitions of activities to accomplish the task depend on aspects that make up operational culture from local relationships and global influences. Beliefs, learning about physical and social perception and archetypes will turn into habits that can increase the risk of accidents on the task. The composition of the working memory depends on the previous construction of theoretical and practical knowledge. This knowledge allows one to perceive what is normal or abnormal in the process. Thus, the behavior and behaviors of the production team, together with the support of management and engineering staff, will induce the worker to activate cognitive processing within the appropriate level of attention. The organizational commitment defines the state of alertness and the urgency with which the actions are developed.

2. Discussion

2.1 Multidisciplinary and Behavior

The research on the behavior and habit that can trigger the human error that is the initiating event for the accident in the industry of risk, depends on different dimensions considering that the production is a sociotechnical system [6].
SARS, a proposed method for behavior adjustment, was built in a multidisciplinary environment that involves different subjects in specific professionals in its investigation, who eventually assumed the role of graduate students in an operational risk management group at UFBA, Brazil. While engineers were concerned about the motivation for human error in performing the task [7], some health professionals experienced in high-stress activities [10] worked on the body and task [11,12] or physical discomfort [8,9] in the activities of the industry.
At the same time, engineers, managers and administrators discuss what are the requirements, indicators and standards to consider that the critical activities will be carried out in a normal situation. A great challenge is to enter into the subjectivity of culture and to bring the discussion about the elements that cause guilt in the worker [13] reducing the quality of communication [14] and generating profiles that "command" instead of leading activities.
The communication discussed in linguistics is intertwined with culture metrics [15], and with task control [7] so that, from indicators of commitment and cooperation [6, 16] it can be predicted that safe behavior will occur. Finally, multidisciplinary is a necessity for the prediction of future behavior, the construction of appropriate leadership [17] and the correction of the task and the human performance factors to avoid the accident. SARS is built taking these aspects into account in organizational design.

2.2 Behavior and Culture

According to [18], a Just Culture depends on learning, reviewing concepts and rituals to increase commitment. The technical control of the task results from the relations between communication, competence, cooperation and commitment (C4t).
In a Just Culture the following aspects are installed or installed: (1) communication (standards and feedback) in the horizontal and vertical sense is clear and appropriate to local language schemes; (2) the skills installed and applied to the task are adequate to the demand, where there is resilience to the environmental challenges; (3) a climate of cooperation involves the team where fear is not in place; (4) the cognitive functions of task executors and controllers are aligned with the priorities directing the best decision alternatives.

2.3 Environmental Threats to the workstation

The threats that hamper the stability of cognitive decision-making processes that bring crisis situations to be managed are due to climate change, geopolitical relations and social changes (Figure 3) that impact on the coexistence in work and society.
In the environment of globalization with cost reduction it is necessary to repeat standards to achieve results. The workplace that suffers these threats has the negative result in behavior. Aggression, egocentricity, intolerance, and the conflict between tribal and organizational priority hamper work activity. There are still other social threats in the future behavior such as: diffuse attention to the non-recognition of reality, beyond generational conflicts bringing difficulties

previously not worked by leadership. In addition, water scarcity and energy constraints create an environment of difficult alternatives in the quest for sustainability solutions in cities and industries.

2.4 Bottom Up Research Method

According [19], shop floor workers can add value of new knowledge to their company, they can discover while performing their daily tasks. The transfer of knowledge to the measure manages can better the process efficiencies. Researchers and practitioners perceive value in understanding the range of strategies and initiatives in knowledge transfer, although there is empirical evidence for the discovery of factory floor knowledge in the manufacturing environment.
Identifying where to improve to avoid a failure process when discussing the sociotechnical system becomes more difficult because of the subjectivities involved in behavioral deviations. Thus, picking the signs of the failure in the task and raising the assumptions about the emerging problem requires a shop floor survey to correct some incorrect assumptions in risk management.

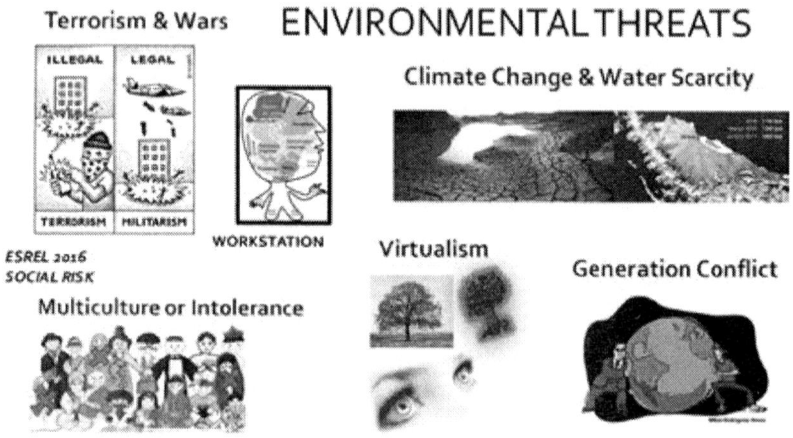

Figure 3 – Threats to workstation

2.5 Level 5 Leader

A Level 5 Leader has extreme humility to the point of assuming for himself certain responsibilities of not communicating a fault, a deviation, a near accident, attributing value to the initiatives that have arisen from the group of operators, for the contingency of the wind. This leadership has the determination and the fierce will to achieve goals, along with its work team, as well as the technical competence that leads to trust in the group, as an agent capable of developing quality safety activities [20].

2.6 Resilience and Organizational Culture

The behavior of workers without proper resilience to environmental threats leads to unsustainable scenarios. Team development methods, when inadequate, indicate the preparation of people who are not compatible with the challenges. Thus, it is necessary to analyze the activities that manage the people in the company: selection, training, integration and allocation of people in the routine.

To motivate safe behavior, it is important to identify the relationship between the organization and the worker with the risks to society and the family [21].
Safe behavior depends on "negotiated culture", technology features, routine tasks requirements, and the level of acceptance of leadership. The subject of operational discipline [22] is not so easy to analyze due, for example, to the behavior of normalization of deviations [23]. In the initial project of the organization, it is sought to simulate routine situations to analyze the risk of having inadequate profiles in the leadership or to have a mental map not suitable for micro decisions in the work and / or society.
The term "negotiated culture" in this paper indicates that there is a movement towards the position of the "average" worker when comparing the bottom values with the organizational and social values (top and out) to control the routine task and the emergency situations .
The variables are many, they are dynamic and it is not appropriate to delimit static rules without analyzing the changes that are occurring in the organization or in society. Thus, the SARS tool depends on the behaviors that follow the concepts and metrics related to C4t [24, 25].
The purpose of this paper is to present the SARS method, which includes tools in the steps of changing habits, adapting competence and adjusting behavior to achieve a just culture. With this method, it will be possible to face the threats and unleash the nodes resulting from organizational diseases.

3. Principles

The principles that guide the SARS methodology for safe, alert and resilient behavior (Figure 4) are based on: the form of learning, understanding about cognitive processing, identifying major hazards and risks, and the relationship between stress and decision-making.

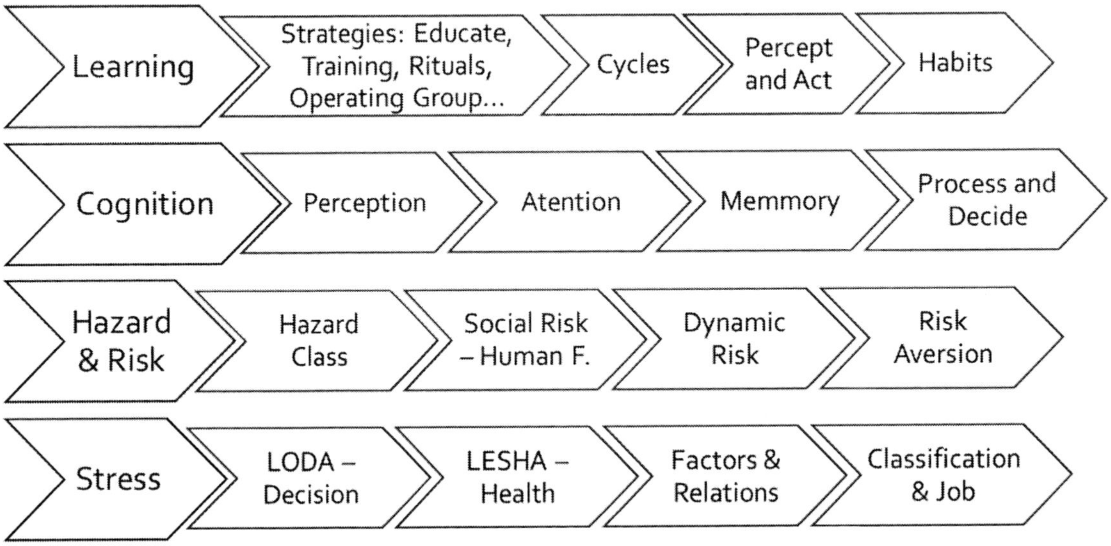

Figure 4 – Principles to Implementation of SARS

3.1 Learning

The SARS method works with learning cycles ranging from habit to behavioral competence leading to an organizational culture with excellence in operations. After the diagnostic steps, interventions are made to correct the various behavioral gaps that cause bias and human error. The employee's previous background needs to be properly analyzed to avoid difficulties or resistance in the learning phase. Thus, experience allows us to increase the power of perception regarding the existence of states of abnormality. In learning it is important to establish the causality of good and bad habits, in addition to the vices so that at the time of intervention, the vaccine is adjusted to organizational disease.

3.2 Cognition

A key step in initiating cognitive processing is the perception of external cues [26]. The status of the equipment or process is recognized by the trained and experienced operator. The use of this ability to perceive depends on the level of commitment of the worker to the business objectives. At other times the lack of experience does not recognize the scenario as an abnormal state delaying the corrective actions and causing the loss of process.
On the other hand, the information flow in the current workstation [27] requires attention filters for interpreting problems. What decreases concentration power making it difficult to focus on activities? How to discern the level of impact to prioritize or define corrective and preventive actions?
Realizing and having attention is the gateway to decision. Memory operations are part of interpretations for pattern recognition authorizing immediate action or decision making [28, 26].

3.3 Identification of Hazards and Risk Estimative

Initially it is important to differentiate risk from risk, accident event, safeguards from difficulties in implementation. Thus, the element that initiates the problem with occupational, environmental or social consequences is called danger. Risk is the way of assessing the level of impact and frequency to define the types and priorities of corrective-preventive action. Social risk that leads to undue behavior is usually not considered. This risk may be embedded in culture or individual personality. The great difficulty is to establish the appropriate safeguards for subjective questions. Today, a well-conducted risk analysis depends on the incorporation of the man or the group as a hazard to the event of the accident. Thus, topics such as: complexity in the task; quality of communication resulting from social and functional relationships and feedback; behavior of workers; characteristics that lead to risk aversion.

3.4 Stress at Social Environment

SARS uses group stress analysis to indicate the best safeguard for each situation. [8] discusses the influence of the stress level in the work place on the decision of the task and the impacts on the worker's body.

4. Methodology

4.1 General

The SARS method aims to apply a strong vaccine against organizational disease that promotes inappropriate behavior in performing critical tasks in cities and industry. Among these behaviors we can mention: (1) passion for the excess that increases the consumption; (2) inertia for change that hinders transformation; (3) avoid conflicts! That promotes non-negotiation; (4) rigidity that hampers transformation, flexibility and diminishing guilt; (5) difficulties in voluntary participation; (6) self-centeredness that does not aid individual resilience; (7) fear and guilt that provoke underreporting; (8) lack of group participation that disrupts the individual in the action of the team; (9) jealousy that hinders trust among colleagues in teamwork; (10) lack of strong internal values in the team which prevents group unity.

The work environment is analyzed to prepare the initial working rules from the history of decisions and errors and then adjusted from the initial perception of the workers in training. This prior survey and interaction in training will shape the leader's and operator's perception of human error and critical decisions, bringing new cultural information to what is happening in the specific workplace.

The habit is influenced by the level of knowledge and depends on the type of labor and social communication adopted by the leadership. The habit depends on the level of cooperation or commitment of the team.

The C4t technique [24, 25] uses the information collected from the initial stage of SARS (INITIAL SARS) from the data collection. SARS addresses subjective issues easily, with simple application and due to the variability of human behavior cannot be based on common sense, without which will not be located emergent social situations that change behavior.

Indicators related to behavior, task, and culture are needed to investigate control or lack of control in the timeline and establish a robust safety culture. Each training is applied to result in safe behavior and at the same time confirm the hypotheses raised from the routine.

The alignment of C4t techniques [24, 25] and the SARS method in addition to the intervention strategies requires previous seminars for the Organization's leaderships, where the phases of change of habit, competence adjustment and change are presented behavior and culture.

The measurement and transformation of culture (C) and behavior (b) promoted by the SARS method follow the implantation processes described in figure 5 and 6. In Figure 5 are the discussion on how to map and investigate the relations between behavior and human factors (PSFs). In Figure 6 the deployment processes are described: mapping the environment (1); analyze Behavior (2 and 3); and measure for adjustment and intervention (4 and 5).

MAPPING AND PRIORITIZING (1). Past behavior, defined by bottom-up techniques: analysis of the operator's and manager's discourse, mapping routine abnormalities, analyzing human error in the executive function, and analyzing rituals and archetypes indicates difficulties in the transformation of culture. By mapping the Sociotechnical Systems and identifying the characteristics of the tasks it is possible to prepare the risk analysis including human factors. Early-stage SARS identifies operational context, operational risks, and sociotechnical environment, perceives knowledge inventory and rituals in shift work. Only then can action be prioritized.

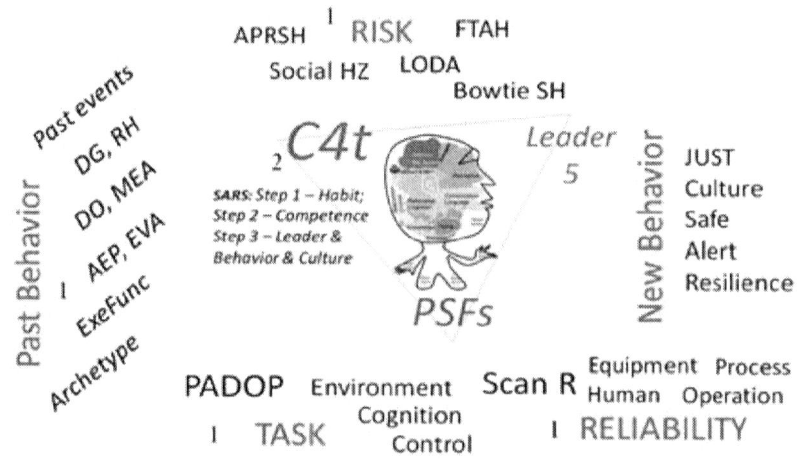

Figure 5 – Map Labor Environment- investigate relation between behavior & PSF

RELATIONSHIPS BETWEEN BEHAVIOR AND HUMAN FACTORS (2, 3). This mapping of past behavior, task, and risk enables us to identify the relationships between C4 (cooperation, commitment, communication, and competence) and human factors (PSFs) to plan interventions for the future. The social and human typology allows establishing the heuristic rules of behavior in the routine (operational culture) and how human errors occur.
MEASUREMENT & ADJUSTMENT, INTERVENTION ON BEHAVIOR AND CULTURE (4, 5). The study of the relations between behavior and human factors that induces the desired culture allows to quantify and intervene to control C4. From then on, investments are planned to carry out the intervention, monitoring and control of behavior and culture. This intervention depends on appropriate training of level 5 leaders to act in the control of process safety.
The chronological steps for implementing SARS are described in Figure 7. In the first step, if you combat bad habit, in the second you adjust your competence and in the third you transform your behavior to achieve a just culture in safety and energy.

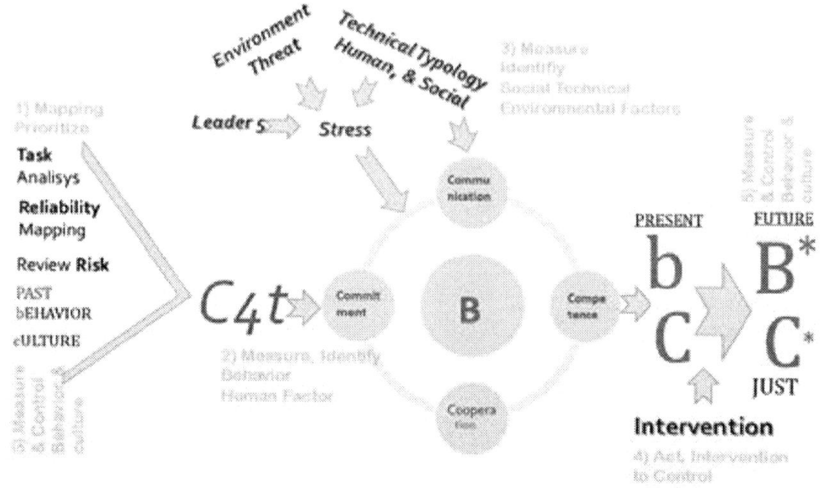

Figure 6 – Processes of SARS Implementation **(b behavior, C culture, * after transformation)**

4.2 1st Step – Data collection, strategy and Habits changing

In the first step, a strategy is negotiated between the company and the researchers as to the techniques, where data are collected to indicate what the bad habits are. This diagnostic serves to prepare the team's training for new habits, INITIAL SARS. The behavior of the data indicates a diagnosis of the average behavior to construct the training (prognosis). This training addresses the routine situations of the worker inside and outside the organization rooted in cultural aspects that may indicate failures or human errors.

The negotiated strategy addresses the management's difficulty in improving the team's profile for process safety. This strategy is intended to include culture analysis and its relationship to the routine for a negotiated culture that is compatible with technology and the tasks indicating criteria for shaping the "new" leadership.

The SARS method aims to review the procedures that involve the management of people in the organization, including the competency building program where, according to [29], the achievement does not reach the projected one. The aim of this method is to install the operational discipline without falling into the abyss of normalization of deviations [23]. It is intended to simulate the routine for analysis of inappropriate profiles in leadership and decisions! Simulate the emergency situation using the technique LODA [12] that develops scenario with progressive stress. It is intended to study the task environment and its design to achieve efficiency with effectiveness [7].

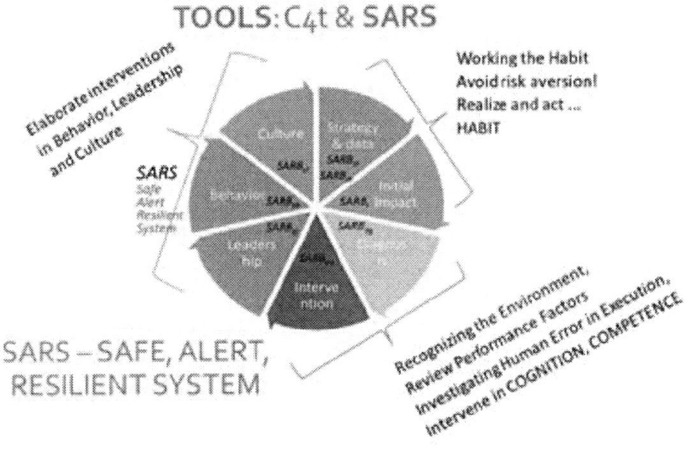

Figure 7 – Steps of SARS Implementation

The SARS method does not only deal with the current aspects identified in the company culture, it also deals with the tendency and the projection of the culture. SARS extends the discussion of organizational culture to understand the influences of regional (local) or global culture on behavior and what factors lead to human error in execution, diagnosis and critical decision. The analysis of socio-economic-affective utility - USEA [30] is performed to identify the reason for occurrence of human error from inappropriate behavior.

Figure 8 shows the identification procedure for changing bad habits, dealing with human errors and incorrect decisions. Routines are repeated for recognition and follow-up as well as establishing procedures for habit control, error and decision.

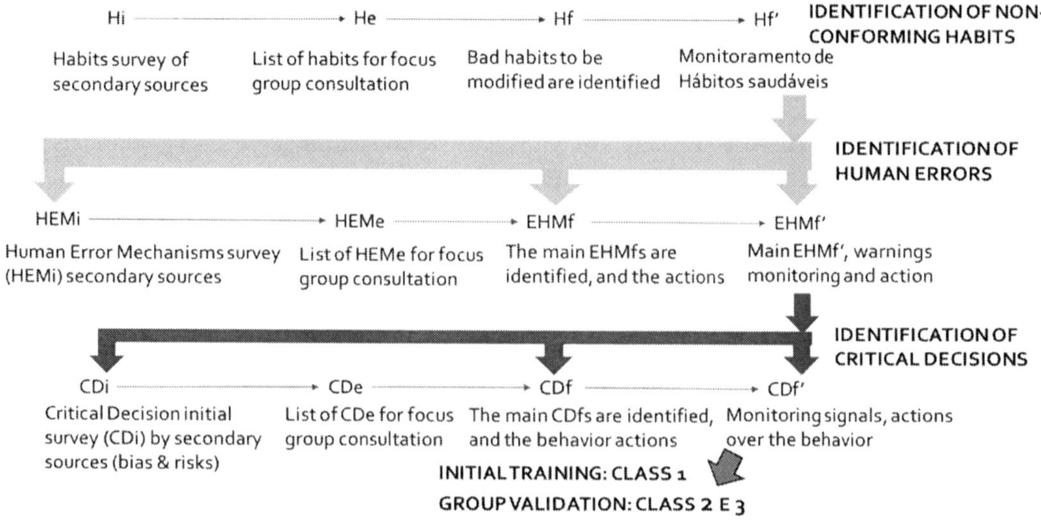

Figure 8– Habit Changing

4.3 2nd Step – Competency Analysis

[31] discuss diagnosis built based on the perception of the operators (Clean Routines), supervisors (LNT - Training Need), company results (Manager's Speech), job observation (Visit), technology (Reading), Book of Occurrence (Operator's Speech - DO). The diagnosis also considers the response of the strategic team to the API770 questionnaire [32]. This diagnosis intends to propose interventions in the following areas: training for concepts, training to avoid mistakes and failures based on human factors and experiences, educational programs and alteration of rituals. These trainings will be scheduled for frontline employees, error-failure investigators, team leaders, and the strategic team.

SARS in the Diagnostic Phase is a method implemented to investigate behavioral oscillations at the workstation. It aims to define the training and interventions needed to transform behavior. Trains the agents that will participate in the transformation. The C4t diagnosis uses operator discourse to map human errors that indicate knowledge gaps and incorrect decisions. C4t directs interventions to improve concepts, understanding mechanisms of human error, cognition deficit, failure mode. C4t proposes new programs for a suitable SKR level [24, 25].

SARS in the intervention phase focuses on conducting training, educational programs, and ritualizing unconscious processes (eg, daily safety dialogue) in an attempt to create new beliefs in the organization that will be corroborated by practice. From this diagnosis and interventions, the leadership-level 5 program for safety and energy can be built. This program aims to develop leaders and review their profile to achieve the goal of reducing accidents and recovering losses, being implemented in the third stage. With the interventions, it becomes possible to validate the preliminary diagnosis in INITIAL SARS and corroborated by C4t. Thus, leadership needs are delimited. The establishment of formal and informal leadership groups allows constructive movement and identification of the potential to adapt to challenges.

4.4 3rd Step – Behavior adjust to Just Culture

At this stage, a diagnosis is made about the quality of communication, and the ability to apply the skills-knowledge-rules, as well as the level of aggregation and commitment of the team. After understanding the cognitive and behavioral gaps, a group of multiplier agents will be prepared to implement the new culture throughout the Company. After the formation of these multipliers for the diagnosis-intervention and after the process of building the level 5 leader, it becomes possible to certify the procedures for establishing a safe, alert and resilient behavior. This behavior will have measures, goals and procedures to be followed by the personnel department in the activities of selection, development and review of interventions in human resources inside and outside the work environment.
SARS in Step 3 works with methods to prepare team leaders in order to reduce human error and gain a resilient vision for organizational and social goals. At this stage, Level 5 leaders are prepared for challenges based on prior-stage diagnostics.
The indicators program initiated in phase 1 (habits) and in phase 2 (competence) is continued for commitment adjustment, inclusion feeling that reflects on cooperation and quality of communication and feedback in formal (standard) or informal processes.
SARS aims to adjust behavior through methods, courses, and interventions to monitor and adjust-negotiate behaviors more appropriate to organizational resilience. At this point, the preliminary diagnosis of C4t will be validated. Behavior metrics will also be validated.
Some subjective-objective risks are analyzed for future projects [33, 34]. The needs for operational control in the routine [7] and in emergency situations [12] are confirmed. A procedure is created for the analysis of equipment failures that includes human factors [35], as well as the analysis of archetypes [36] and Executive Function [37] that support the understanding and the correction of human error. The intention is to create environment for Fair Culture in Security and Energy.

5. Test to Indicators Implementation (C4t tool)

The Scenario for the implementation of SARS and in particular the C4t technique was practiced in the liquefied petroleum gas (LPG) processing industry. The mapping of human and social abnormalities and typology was performed to identify past behavior and schedule changes for future behavior as described in the steps of formulating the indicators in Figure 9. Among the mapping activities are: 4-month analysis of occurrences of the shift book (operator's speech); the analysis of critical process variables (during one year as the gas pressure after compression); maintenance information (such as number of interventions on motorized valves during the 6-month period); occupational and environmental records (such as gas leaks and falls in the working environment for one year); history of fold percentage, deviations by type and quantity processed per shift. In addition, data were collected on human resources (time in the company, in the function and labor information).
The processes involving LPG are described in Figure 9. The LPG generated at the refinery with the possibility of having moisture and presence of sulfur as contaminants is cooled and compressed to be transferred to tank and ship. The critical activities involved are: transfer logistics, thermal cycle and compression, pumping and removal of moisture. Critical equipment is compressors, pumps, ball and tanks. The operational bottlenecks: availability of materials, presence of ethane and capacity, neighboring community and quality of LPG.

Figure 10 shows strategic deliveries that enable the elaboration and analysis of the technical or operational culture installed on the factory floor. By identifying the culture based on the history of human errors it is possible to calculate the indicators as a result of the application of the examination of the technical culture on the group of operators in the areas of competence and communication.

As indicated in Figure 10, the strategic deliverables that are products of the first and second stages of SARS are: anatomy of the fault or map of operational abnormalities; competency map for gap analysis; failure of communication, human errors, decision errors indicated in the occurrences of the shift book and observation; diagnosis of standard quality and feedback.

These indicators are tracked in the shift group and the administrative group, therefore 6 results are recalculated once every 6 months, by sampling the data and the tests performed. Emotional balance analyzes are based on self-response surveys in the cognitive, cultural and health areas (Stage 3 of SARS) and indicate how close the team is to what is considered normal (Figure 10). In addition, another poll attempts to measure the aggregation or disaggregation level of the team with a sampling of 20%, every 6 months. To finalize the impairment indicator is based on empirical equation that has as a result the percentage of the team committed to the critical tasks.

Only after the diagnosis based on technical culture (competence and communication) and group characteristics (cooperation and commitment) can it be possible to define an intervention program to improve behavior. It is also important to analyze the profile of formal and informal leadership to establish intervention strategies. The indicators guide the behavior changing program and culture.

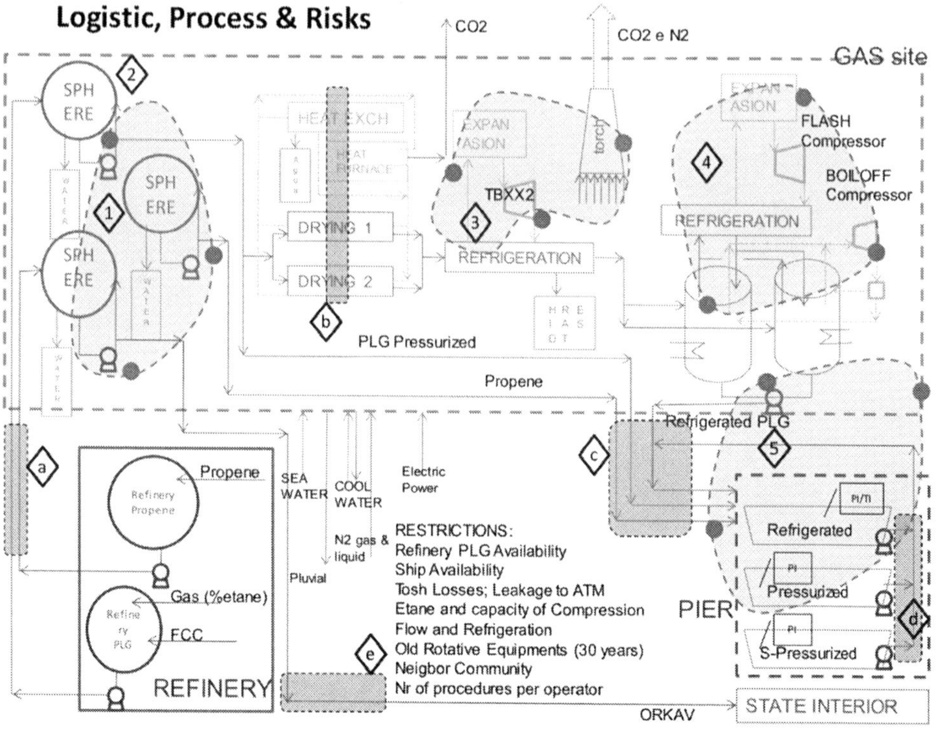

Figure 9 – Process and Logistic in PLG

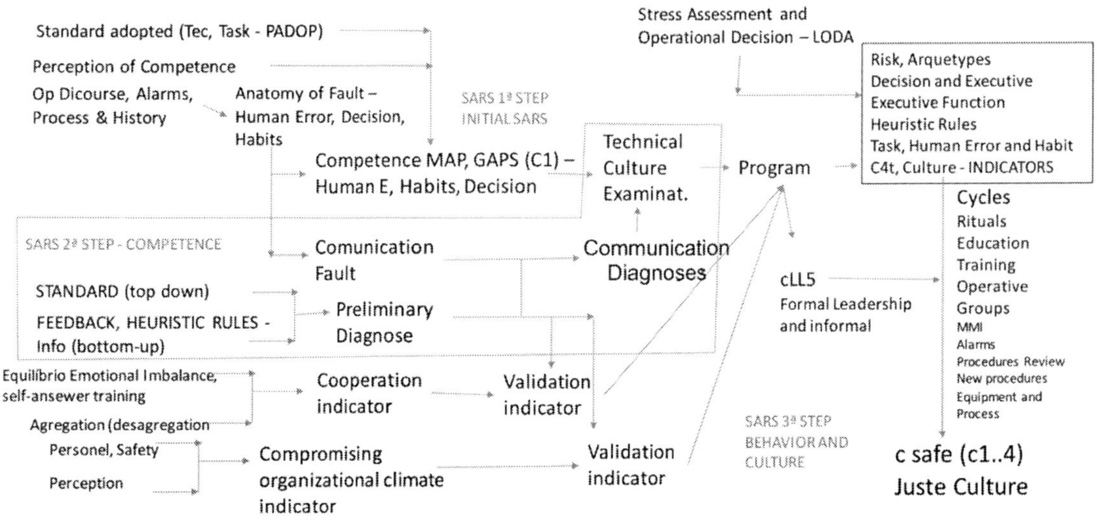

Figure 10 – Roadmap calculating indicators C4t and targeting the SARS Program

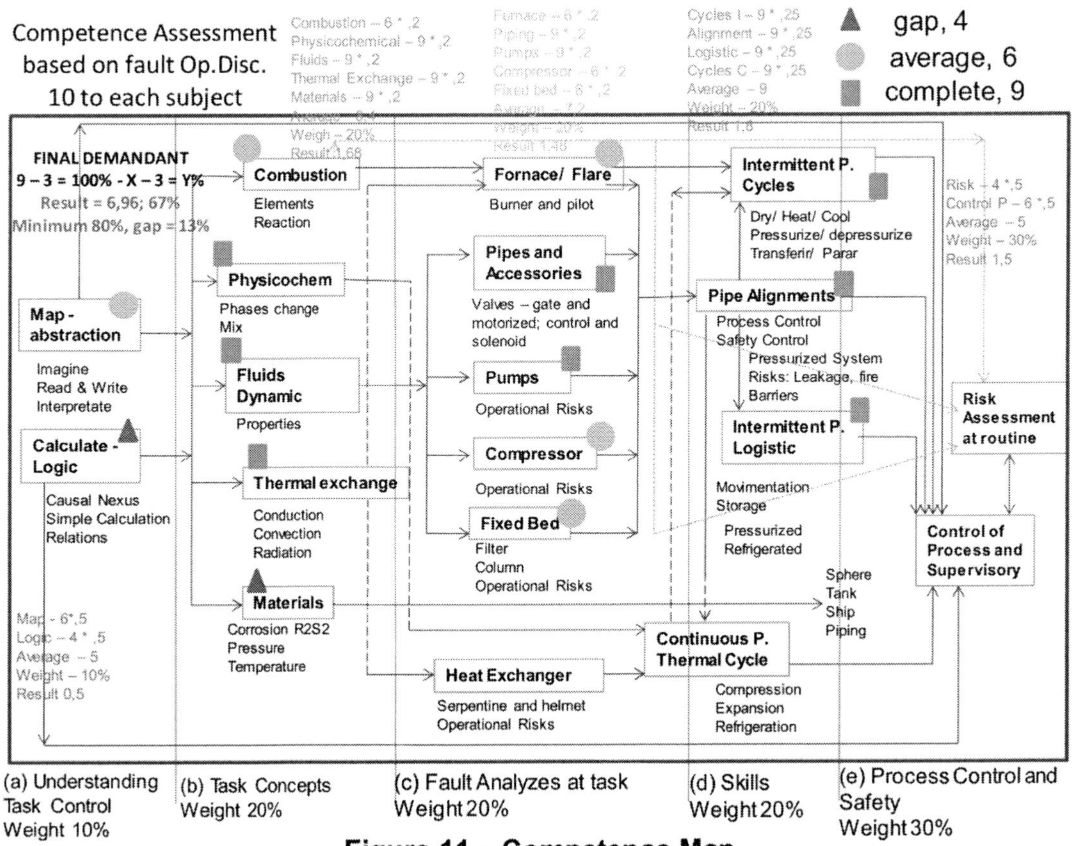

Figure 11 – Competence Map

Competence. The map of competencies indicated in Figure 11 based on the analysis of the human errors of the routine resulting from the mapping of abnormalities shows that regions of knowledge indicated by triangle are missing in relation to other regions that are present. It is therefore formulated a hypothesis that 13% of knowledge is missing, especially in the areas of planning and security that was validated by an average (7) resulting from the examination of the technical culture (10 of the 16 questions applied to 9 operators) indicated in Figure 12. This method was applied to 9 operators which is equivalent to 33% of the shift team.

Cooperation. Emotional balance and the level of aggregation-disaggregation indicate whether there is an appropriate environment for cooperation [25] The resulting indicator is distributed by weight as follows: the emotional balance indicator (EqE +) for group activity has 50% of the weight on the cooperation, 60% equivalent to the level of aggregation (NA +), 20% equivalent to the level of disaggregation (ND-), and 10% in the case of small emergencies (PE-). I Co-operation = 0.5 * EqE + 0.6 * NA - 0.2 * ND - 0.1 * PE. In the case of the application to the LPG plant [6] and according to Figure 11, we have: I cooperation = 0.5 * 63% + 0.6 * 75% - 0.2 * 90% 0.1 * 60% = 52.5%. Compared to maximum cooperation where, I coopemx = 0.5 * 80 + 0.6 * 90 - 0.2 * 40 - 0.1 * 30 = 40 + 54-8-3 = 83%. Compared to minimum cooperation where, I coopemn = 0.5 * 40 + 0.6 * 50 - 0.2 * 60 - 0.1 * 50 = 20 + 30-12-5 = 33%. With a range of 100% equals the difference (Icoopemx - Icoopemn) = 50%. I final cooperation = 19.5 * 100/50 = 39% where the goal is to reach above 60%.

	weight	subject	C4 TOOL	average	minimum	diference	weight
3	2	etane and	comunicat	4,5	8	-3,5	-7
7	3	leakage	comunicat	6	8	-2	-6
8	2	piping	comunicat	5,5	8	-2,5	-5
12	3	refinery	comunicat	6,6	8	-1,4	-4,2
13	2	ship	comunicat	8	8	0	0
14	2	conclusior	comunicat	5	8	-3	-6
							-28,2
				comunication		5,3	-4,7

	weight	subject	C4 TOOL	average	minimum	difference	weight
1	2	corrosion	competence	7	8	-1	-2
2	1	solids	competence	5	8	-3	-3
4	3	spark routine	competence	5	8	-3	-9
5	3	fire concept	competence	7	8	-1	-3
6	3	leakage	competence	6	8	-2	-6
9	3	bleeve	competence	8	8	0	0
10	3	risk	competence	8	8	0	0
11	2	logistic	competence	8	8	0	0
15	2	refrigeration fluid	competence	9	8	1	2
16	2	vapour pressure	competence	2	8	-6	-12
							-33
			competence		7		-3,3

Figure 12 – Technical Culture Examination: competence & communication

Commitment. In the case of impairment, we follow the equation 1 indicated by [6] and the proportion of operators with medium and high commitment, taken out of the norm, defines the group commitment indicator.

$$0{,}8 * \left(\frac{age * levels}{time\ accid * no\ dependent}\right) \quad \text{Equation 1}$$

Thus, high commitment = 5 operators. Low commitment = 3 operators. Out of the rule = 3 operators. Indicator of commitment = 5/8 = 62.5%. Thus, Competence = 70%, Communication = 53%, Cooperation = 39% and Commitment = 62.5%.

Leadership and Stress. Leadership mapping is based on the FIROB indicators [38] regarding social movements of inclusion, affection and control. In the case of LPG, the diagnosis indicates that... The application of LODA [12] for the analysis of stress and impact on the operational decision indicates a behavior of the teams that corresponds to the stress rule [8]. Thus, a graph of the indicators is constructed (Figure 13) identifying regions of risk for loss of process and / or accident.

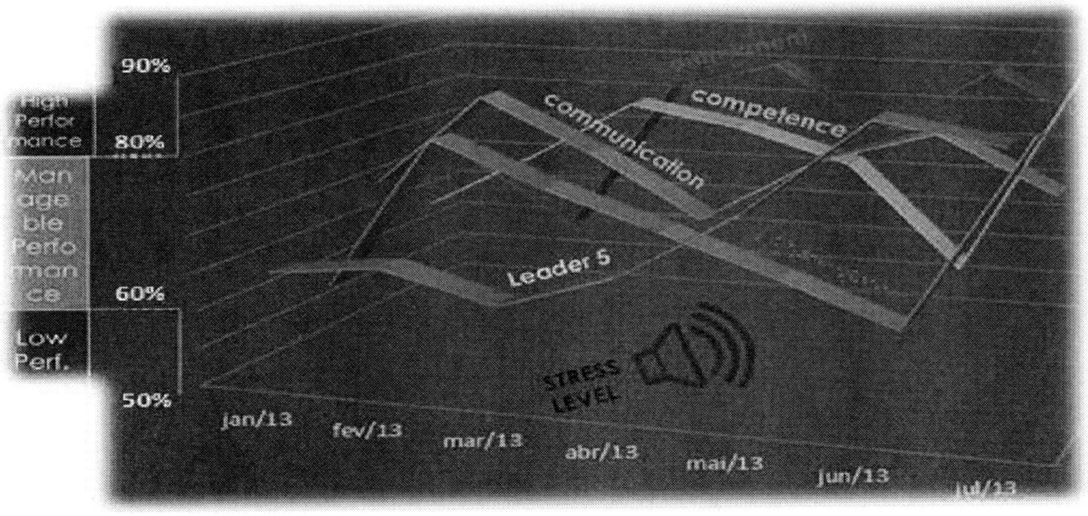

Figure 13 – Indicators C4t in PLG [24]

6. CONCLUSION

The implementation of this behavior brings a new environment where human errors tend to decrease as well as the likelihood of an accident. Power losses are minimized and team productivity is increased. Self-management becomes a hallmark of new behavior.
This methodology is robust and based on operator discourse but it is not very simple to perform because it admits that without subjectivity we do not have answers about latent movements. The metrics or measurements will be stablished in the Organizational Culture project from bottom to up and Company safety is reviewed. Those interested should know this movement!

7. REFERENCE

[1] AEAT 2015. Anuário Estatístico de Acidentes do Trabalho. Secretaria de Políticas de Previdência Social

[2] CORREIO Brasiliense. No Brasil, 700 mil pessoas sofrem acidente de trabalho a cada ano http://www.correiobraziliense.com.br/app/noticia/economia/2017/06/05/internas_economia,600125/acidente-de-trabalho-no-brasil.shtml.

[3] GRANADA G. L. Projeto ANP de Prevenção de Incêndio e Explosões. IV SOMA Workshop de Segurança Operacional e Meio Ambiente 2016. Rio de Janeiro

[4] ALMEIDA A. G., SILVA B.F., MORAIS C.P.M., FERREIRA N.N., PIRES T.S. Relatório de Investigação do incidente de explosão ocorrido em 11/02/2015 no FPSO Cidade de São Mateus. Superintendência de Segurança Operacional e Meio Ambiente. ANP. 2015.
[5] ÁVILA S.F. Assessment of Human Elements to Avoid Accidents and Failures in Task Perform, Cognitive and Intuitive Schemes. Proceedings of 7th Global Congress on Process Safety – GCPS. Chicago, 2011a.
[6] ÁVILA, S. F. (2010). Etiologia das Anormalidades Operacionais na Indústria: Modelagem para Aprendizagem. Tese de doutorado. UFRJ, Rio de Janeiro.
[7] AVILA, S. F, CARVALHO, A. C., SOUZA, C. R. O. (2015). Assessment of complexity in the task to define safeguards against dynamic risks. In: 6th International Conference on Applied Human Factors and Ergonomics. AHFE 2015 Conference Proceedings (International Conference on Applied Human Factors and Ergonomics). Las Vegas.
[8] ÁVILA S. F., GUIMARAES G., NUNES K., SILVA R.C., MAGALHAES R. Protocol to evaluate the influence of stress on the task involving human types, behavior and health: An exercise proposal. Proceedings of European Safety and Reliability Conference – ESREL. Zurich, 2015a.
[9] ÁVILA, S. F., NUNES, K. R. (2015). A influência do estresse na tarefa, comportamento e saúde: uma proposta de protocolo. In: ABRISCO PSAM 2015, 2015, Rio de Janeiro. Anais.
[10] NUNES, K. R., ÁVILA, S. F. (2015) Jornadas excessivas de trabalho e sua influência sobre os acidentes, a produtividade e tomada de decisão dos técnicos de enfermagem em uma UTI. In: ABRISCO PSAM 2015, 2015, Rio de Janeiro. Procedings
[11] CARVALHO, A. C.; AVILA, S. F. (2011). Análise de Fatores Humanos no planejamento e na execução da tarefa. In: Fórum da Operação da DTO-2011 TRANSPETRO, Rio de Janeiro. Resumos do Fórum DTO-2011.
[12] AVILA, S. F. (2011). Dependent Layer of Operation decision analyzes (LODA) to calculate Human Factor, a simulated case with PLG. In: 7th Global Congress on Process Safety CCPS. Chicago.
[13] FRAGOSO, C., ÁVILA, S. F., SOUSA, R., MASSOLINO, C. PIMENTEL, R., CERQUEIRA, I. (2017). Blame Culture in Workplace Accidents Investigation: Current Model Discussion and Shift Requirements for a Collaborative Model. In: P. Arezes (ed.), Advances in Safety Management and Human Factors, Advances in Intelligent Systems and Computing 604, DOI 10.1007/978-3-319-60525-8_33. Springer International Publishing.
[14] DRIGO, E. S. (2016). Análise do discurso do operador e seu instrumento de comunicação entre turnos como ferramenta para o Sistema de gestão. Dissertação de mestrado. Salvador.
[15] SALES, M. D. (2011). Tipologia de cultura organizacional: Proposta de processo de taxionomia da cultura organizacional de uma empresa com base na orientação empreendedora dos líderes/Marcelo Dourado Sales; Campo Limpo Paulista-SP: FACCAMP.
[16] MEYER, J. P.; ALLEN, N. J. (1991). A three-component conceptualization of organizational commitment. Human Resources Management Review,(1), 61-89.
[17] CERQUEIRA I., AVILA FILHO, S., SILVA, J. N. ; MUSSOLINO C. ; PIMENTEL, R. ; FRAGOSO, C. M. (2017). Analysis of Competencies for Decision Crisis Management in Actions Under Progressive Stress: A Simulated Case. In: 8th International Conference on Applied Human Factors and Ergonomics. AHFE 2017, Los Angeles.
[18] DEMOSTHENOUS V. and DEMOSTHENOUS S. Just Culture. SEAHORSE. Glasgow. 2016

[19] RIEGE A., ZULPO M. Knowledge Transfer Process Cycle: Between Factory Floor and Middle Management. Australian Journal of Management, Vol. 32, No. 2 December 2007, © The University of New South Wales.
[20] COLLINS, Jim. Discussion Guide. JIM COLLINS.COM. 2002.
[21] PROGRAMA SAFESTART. http://br.safestart.com/o-que-e-safestart/. 2018.
[22] CENTER OF CHEMICAL PROCESS SAFETY - CCPS. Conduct of Operations and Operational Discipline: For Improving Process Safety in Industry. Editor Willey. 2011.
[23] CENTER OF CHEMICAL PROCESS SAFETY - CCPS. Recognizing and Responding to Normalization of Deviance. Editor Willey. 2017.
[24] ÀVILA, S. F., BITTENCOURT, E. S. (2017). Operator Risk Assessment in the Oil and Gas Industry: Dynamic Tools to Treat Human and Organization Factors. In: OLSON, K. F. Petroleum refining and oil well drilling. New York: New science Publishers.
[25] AVILA FILHO, S., DRIGO, E. S.: Evaluation of Human Factors Through Behavior, C4t. In: III Congress Brazilian Association for Risk Analysis, Process Safety and Reliability – ABRISCO. Rio de Janeiro, 2017
[26] HOLLNAGEL, E. Human reliability analysis context and control. Computers and People Series. San Diego, CA: Academic Press Inc. 1993.
[27] SENNETT, R. A corrosão do caráter. Rio de Janeiro: Record, 2005. 204 p.
[28] STENBERG JUNIOR, R. Psicologia Cognitiva. Porto Alegre: Artmed, 2008. 582 P.
[29] MUCHINSKY, P.M. Psicologia organizacional. 7. ed. São Paulo: Pioneira Thomson Learning, 2004. 508 p.
[30] ÁVILA FILHO S., FONSECA M.N.E., SANTOS A.L.A., SANTINO C.N. Analysis of Cognitive Gaps: Training Program in the Sulfuric Acid Plant. European Safety and Reliability Conference ESREL. Glasgow. 2016.
[31] AVILA FILHO, S., CERQUEIRA, I.C.S.: Training Program based on operator Perception. In: III Congress Brazilian Association for Risk Analysis, Process Safety and Reliability – ABRISCO. Rio de Janeiro, 2017
[32] LORENZO D.K. API770 – A manager's guide to reducing human errors, improving human in the process industries. API Publishing Services, Washington, 2001.
[33] AVILA, S. F., BARROSO, M. P. (2012). Análise preliminar de risco sócio-humano: exercício de aplicação indústria química. In: XXXIII Congresso Interamericano de Engenharia Sanitária e Ambiental - AIDIS, 2012, Salvador. AIDIS.
[34] ÁVILA, S.F., PESSOA, F.L.P. , ANDRADE, J.C.S.(2013). Social HAZOP at an Oil Refinery. Process Safety Progress , v. 32, p. 17-21.
[35] AVILA, S. F., CARVALHO, V. S., AMARAL, J., MENDES, P. C. F. (2012). Analise de fatores humanos em falha de equipamento FMEAH de turbocompressor em planta química. In: 27º Congresso Brasileiro de Manutenção. Rio de Janeiro. ANAIS DO CBM.
[36] ÁVILA S.F., MENEZES M.L.A. Influence of Local Archetypes on the Operability & Usability of Instruments in Control Rooms. Industrial Production. Proceedings of European Safety and Reliability Conference – ESREL. Zurich, 2015.
[37] ÁVILA S. F. and COSTA C. Analysis of Cognitive Deficit in Routine Task: as a Strategy to Reduce Accidents and Increase Industrial Production. Proceedings of European Safety and Reliability Conference – ESREL. Zurich, 2015.
[38] SCHUTZ, W. The interpersonal underworld. Palo Alto: Science & Behavior Books, 1966. 168 p.

510866 Incorporating Safety and Best Practices into Your Process

Abstract:

Examining the benefits and value of state based control. This strategy changes the classic board operator to a process manager that approves steps in the automated procedure. Knowledge capture is embedded in the control system so that the best operating knowledge is always on the board.

511118 Validation of Resilience Model of Multiagency Emergency Management System through Hurricane Harvey and Irma

Abstract:

Modern socio-technical systems necessitate not only safe operations but also resilient performance to adapt to emerging changes and to maintain its control against complex and unexpected disruptions (Hollnagel, Woods, & Leveson, 2007). Such demand for engineering a resilient system becomes increasingly apparent in emergency response operations as seen in multiple catastrophic disasters in past few decades (Comfort, 2002, 2007). To that end, Incident Command Post (ICP), the major operational component of the U.S. National Incident Management System (NIMS), has been modeled as a Joint Cognitive System (JCS) based on literature (Hollnagel & Woods, 2005). This model represents how the Incident Command Post (ICP) as a distributed cognitive system perceives, evaluates input from external events and make decisions to control the abnormalities. To validate this model against the real cases of extreme events, interviews of emergency responders involved in Hurricanes Harvey and Irma are conducted and analyzed to elicit the sources of resilient or brittle performance in the responses to the disasters. The qualitative data analysis (Corbin & Strauss, 2008) of the interview highlights examples and counter-examples of adaptation to varying external inputs, temporally expedited response and recovery, flexible and improvisational resource utilization, organizational sensemaking of the rapidly evolving situation, and changes in organizational structure and personal behavior to satisfy goals of operations. In addition, effects of technological artifacts (e.g., information and communication technology) on such performance are examined.

Comfort, L. K. (2002). Managing Intergovernmental Responses to Terrorism and Other Extreme Events. *Publius, 32*(4), 29-49.
Comfort, L. K. (2007). Crisis Management in Hindsight: Cognition, Communication, Coordination, and Control. *Public Administration Review, 67*, 189-197.
Corbin, J., & Strauss, A. (2008). Basics of qualitative research: Techniques and procedures for developing grounded theory. *Thousand Oaks*.
Hollnagel, E., & Woods, D. D. (2005). *Joint cognitive systems: Foundations of cognitive systems engineering*: CRC Press.
Hollnagel, E., Woods, D. D., & Leveson, N. (2007). *Resilience engineering: Concepts and precepts*: Ashgate Publishing, Ltd.

The Definition of Feasibility as it Applies to Inherent Safety

Michael J. Hazzan, P.E., CCPSC, CPSA
AcuTech Consulting Group Inc.
1919 Gallows Rd, Suite 900, Vienna, VA 22182
mhazzan@acutech-consulting.com

Prepared for Presentation at
American Institute of Chemical Engineers
2018 Spring Meeting and 14th Global Congress on Process Safety
Orlando, FL
April 22 – 25, 2018

The Definition of Feasibility as it Applies to Inherent Safety

Michael J. Hazzan, P.E., CCPSC, CPSA
AcuTech Consulting Group Inc.
1919 Gallows Rd, Suite 900, Vienna, VA 22182
mhazzan@acutech-consulting.com

Keywords: Inherent Safety, IST, process safety management, PSM

Abstract

One of the most difficult technical concepts facing industry and regulators in the implementation of inherent safety is that while inherent safety may be theoretically possible in many cases, it may not be feasible. The term "feasible" can have different meanings and be widely interpreted, and has been a matter of debate since inherent safety became a formal requirement in some process safety (and security) programs approximately ten years ago. In inherent safety regulations in three jurisdictions: Contra Costa County, CA, the State of New Jersey, and most recently in the State of California for oil refineries, this term has been used as a qualitative measure to help determine whether a particular inherent safety measure should be implemented. Underlying its use is the principle that a claim of infeasibility of an inherently safer measure shall not be based solely on evidence of reduced profits or increased costs.

Definition of "Feasible"

The American Heritage dictionary defines "feasible" as (Ref 2. American Heritage):

> 1. Capable of being accomplished or brought about; possible. 2. Capable of being utilized or dealt with successfully; suitable. 3. Logical; likely.

While this general definition encompasses the use of the word feasible in the context of inherent safety, the application in inherent safety is somewhat narrower in that it applies to the feasibility of prospective inherent safety improvements in facilities where the process safety risks are significant.

Use of the Term "Feasible" by OSHA

OSHA has long had criteria by which their safety and health standards are required to be evaluated before adoption. These originate in the Occupational Safety and Health (OSH) Act of 1970, which is the legislative basis for the agency's existence and all of OSHA's standards and how they are created and enforced. Section 6(b) of the OSH Act requires technological and economic feasibility analyses to adopt workplace safety and health standards. In general, as this provision has been construed by the courts, any workplace health standard adopted by OSHA must meet the following requirements:

1) The standard must substantially reduce a significant risk of material harm.
2) Compliance with the standard must be technically feasible. This means that the protective measures required by the standard currently exist, can be brought into existence with available technology, or can be created with technology that can reasonably be developed.
3) Compliance with the standard must be economically feasible. This means that the standard will not threaten the industry's long-term profitability or substantially alter its competitive structure. The preamble to the PSM Standard states that Congress does not appear to have intended to protect employees by putting their employers out of business (Ref 11. PSM Preamble). A standard is economically feasible even if it portends "disaster for some marginal firms," but it is economically infeasible if it "threaten[s] massive dislocation to, or imperil[s] the existence of," the industry. This conclusion about economic feasibility is supported by the case law. (Ref 1. ATMI)
4) It must reduce risk of adverse health to workers to the extent feasible.
5) The standard must be supported by substantial evidence in the record, consistent with prior agency practice or is supported by some justification for departing from that practice.

Regarding the PSM Standard, the OSH Act requirements for technological and economic feasibility analyses were met during the rulemaking process for the Standard, which is derived from the Clean Air Amendments of 1990.

> In promulgating the Clear Air Act Amendments of 1990, Congress conclusively determined that "a process safety standard designed to protect employees from hazards associated with accidental releases of highly hazardous chemicals in the workplace" is necessary and that the standard must, at a minimum, require employers to adopt fourteen specified planning, procedure and training safety measures. Pub.L. 101-549 (Nov. 15, 1990), reprinted at 29 U.S.C.A. 655 note (Supp. 1991). For the reasons explained in detail throughout this statement of findings and conclusions, the standard's fourteen planning, procedure and training requirements, when fully implemented, reduce the risk of catastrophic fire and explosion (330 fatalities and 1,917 injuries/illnesses annually) by 80 percent. This constitutes a substantial reduction of significant risk of material harm. Compliance is technologically feasible because the standard's requirements are already being implemented to some extent. Compliance is economically feasible because all regulated sectors can readily absorb or pass on compliance costs during the standard's first five years, and economic benefits will exceed compliance costs thereafter. The standard's costs, benefits, and compliance requirements are consistent with the Clean Air Act Amendments, as well as with other OSHA safety standards. OSHA considered and responded to all substantive comments on their merits; OSHA evaluated all suggestions for their impact on worker safety, their

feasibility, their cost effectiveness, and their consonance with the OSH Act and the Clean Air Act Amendments.

> OSHA reviewed the process safety management practices currently in place across industry as well as the recommended practices of industry trade associations and standards-setting organizations. On the basis of substantial current compliance found by OSHA and its consultants, widespread familiarity with the concepts and procedures of PSM, and the availability of technical consultation within and outside the affected sectors, OSHA has determined that the final rule for managing process hazards is technologically feasible. (Ref 11. PSM Preamble)

Following the adoption of the PSM Standard, OSHA established criteria for rejecting PHA and incident investigation recommendations in the Compliance Directive for the PSM Standard that was published in September 1994 and includes the consideration of feasibility of the recommendations (Ref 12. CPL):

> An employer can justifiably decline to adopt a recommendation where the employer can document, in writing and based upon adequate evidence, that one or more of the following conditions is true:
>
> 1. The analysis upon which the recommendation is based contains material factual errors;
> 2. The recommendation is not necessary to protect the health and safety of the employer's own employees, or the employees of contractors;
> 3. An alternative measure would provide a sufficient level of protection; or
> 4. The recommendation is <u>infeasible</u>. (emphasis added)

Although OSHA did not define "feasible" in the Compliance Directive, the common practice of applying "feasible" in the context of evaluating recommendations for possible rejection is that: 1) the recommendations is technically possible and doesn't require the facility/company to invent something new; 2) that the laws of physics and chemistry will support the recommendation; 3) that the recommendation is physically possible, e.g., moving a maintenance shop to avoid facility siting risk does not exceed the property boundary of the facility; 4) that the risk does not increase as a result of implementing the recommendation; and 5) the economic impact of implementing the recommendation is not extreme, e.g., the facility will not go out of business.

<u>Use of the Term "Feasible" by EPA</u>

Unlike the OSH Act, which created OSHA and includes OSHA's requirement to include an explicit technological and economic feasibility analysis, the National Environmental Policy Act of 1969, which created EPA and forms the basis for the nation's environmental protection program, does not contain such a provision. Neither does the Chemical Release Prevention provisions of the Clean Air Act Amendments of 1990, which authorized both the PSM Standard and the RMP Rule. However, EPA has also used the term "feasible" in the development of its regulations, including the Risk Management Program (RMP) rule. The preamble to the original RMP Rule contains several examples of the use of the term "feasible" in the development of the regulation: (Ref 5. RMP 1996)

- EPA believes that assessing the likelihood of a release at most sites for site-specific individualized Program-level determinations is neither technically feasible nor cost-effective. In most cases, the data do not exist to conduct a meaningful analysis; where they do exist, the cost of developing a defensible analysis and overseeing it could well exceed the cost of compliance with the rule. Such an approach would resemble a permit program, which would be resource-intensive for sources and implementing agencies. EPA determined that the simpler approach for assigning sources to Program 1 would provide regulatory relief for those sources that could not affect the public while allowing other sources to devote their resources to prevention activities rather than to analyses that would be subject to legal challenges.
- EPA proposed in the draft RMP rule to define the worst-case release as the "loss of all of the regulated substance from the process ... that leads to the worst offsite consequences" and that the scenario should assume "instantaneous release." During the public comment on the draft rule, many commenters stated that instantaneous loss of the total process contents is not technically feasible for complex systems and, therefore, represents a non-credible worst case that would provide no useful information to the public or the source for risk communication, accident prevention, and emergency preparedness. In response to these comments, EPA redefined a worst-case scenario as the release, over a 10-minute period, of the largest quantity of a regulated substance resulting from a vessel or process piping failure.

EPA Interpretation of "Feasible" in Revised RMP Rule of January 13, 2017

In the revision to the RMP Rule published in the Federal Register on January 13, 2017, EPA included a new requirement in the PHA element to perform a Safer Technology Alternative Analysis (STAA) (Ref 7. EPA RMP 2017). Although the implementation of this revised RMP Rule has been delayed, the rulemaking process for it provides valuable insight by one of the two main federal regulators of process safety in the U.S. into the meaning and interpretation of the term "feasible" as it applies to inherent safety. In the draft Rule, the new requirement for a STAA, which is a requirement for an inherent safety review under a different name, also included the concept of "feasible." In the draft rule, EPA proposed "feasible" to mean capable of being successfully accomplished within a reasonable time, accounting for economic, environmental, legal, social, and technological factors. EPA further clarified in the definition that environmental factors would include consideration of potential transferred risks for new risk reduction measures. Following the rulemaking process, EPA changed the term "feasible" to "practicable" in the final rule. This was done to avoid confusion over different definitions of the term "feasible" between OSHA and EPA as OSHA was considering a similar requirement in their revisions to the PSM Standard at the same time. However, EPA chose to define the term "practicability" as follows:

> *Practicability* means the capability of being successfully accomplished within a reasonable time, accounting for economic, environmental, legal, social, and

technological factors. Environmental factors would include consideration of potential transferred risks for new risk reduction measures. (Ref 6. EPA Docket)

This definition is consistent with the definition of "feasible" during the rulemaking process for the 2017 RMP Rule revision. It is also consistent with the definition of feasible used by Contra Costa Health Services as described below. EPA did not define the various factors, such as ''economic'' or ''social'' used in the proposed definition of ''feasible'' nor in the revised term ''practicability.'' The examples in the proposed rulemaking preamble by EPA are taken from the guidelines provided by CCPS in the 2nd Ed. of their Inherent Safety Guidelines (Ref 3. CCPS), and are not exclusive of other situations. EPA believes that the definition of ''practicability'' in the final rule provides sufficient flexibility for the owner or operator to determine whether an IST or ISD considered could be successfully accomplished. EPA does not believe that ''economic or social factors'' should be further defined in the rule because further specificity of these terms would likely be too prescriptive and would not encompass all the possible conditions and outcomes that might be encountered when determining the practicability of an IST or ISD considered in the STAA.

EPA expects that facility owners and operators will use their expertise and make reasonable judgements when considering the appropriate meaning of economic or social factors so that any decisions regarding possible implementation of IST is not driven towards changes that would cause unintended adverse consequences. Finally, EPA disagreed with assertions by rulemaking commenters that accounting for the factors in the definition of ''practicability'' could be used as an excuse to avoid necessary implementation measures. EPA did not require IST or ISD implementation in the final rule and, therefore, further clarifying the practicability definition did not impact IST or ISD implementation. EPA believed that the same tools and methods that facilities currently use for their PHAs can be used to identify and measure hazards and risks of any safer alternative options. While EPA decided not to define economic, environmental, legal, social and technological factors included in the ''practicability'' definition of the final RMP, further explanation of these terms can be found in NJDEP's Guidance for Toxic Catastrophe Prevention Act (TCPA) - Inherently Safer Technology (IST) Review Feasibility Guidance described below.

In responding to comments who argued for mandating the adoption of IST measures for all facilities wherever feasible, EPA acknowledged that non-IST methods for preventing accidents may be reasonable in some circumstances. EPA also addressed the notion that the regulations have an overriding requirement to be reasonable:

> "While it is true that similar quantities of chemicals under the same conditions present similar hazards regardless of sector, various sectors present different likelihood of release. Some sectors handle chemicals differently under conditions that are more likely to lead to severe releases. The record reflects that the likelihood of severe accidents is greater in the sectors that must conduct STAA analysis under this final rule. Thus, it is reasonable to have different requirements for these sectors than for others. Independent of whether any new IST/ISD is adopted, there is a cost to conducting an STAA analysis. EPA has reasonably limited STAA analysis requirements to sectors

that we view as most likely to likely to have more frequent, severe releases that are most likely to be benefit from STAA review. Inherent in our approach is distinguishing among classes and types of facilities. We expect that the adoption of STAA analysis requirements in this final rule will advance IST not only in the sectors targeted by the rule, but also more generally as experience is gained and opportunities for technology transfer are developed." (Ref 6. EPA Docket)

Regarding the consideration of cost as a criterion for evaluating inherent safety measures for implementation, EPA concluded that cost is a valid consideration when determining whether a risk management measure can be successfully accomplished. Because EPA did not require implementation of any IST, they saw no reason to exclude this factor from a practicability determination. EPA also disagreed with the suggestion to limit consideration of reduced profits when assessing a risk management measure because they believed that cost is a valid consideration for practicability. Identifying an amount of an allowable cost for an IST was not something that can be prescribed in the regulation because cost decisions are highly dependent on the economics involving a particular process, facility and industry. (Ref 6. EPA Docket)

In the RMP rulemaking process EPA also stated that the STAA requirement, as written will not lead to an overall reduction in safety measures. The current rule already requires the PHA to consider active, passive and procedural risk management measures, however, the requirements do not prescribe exactly which type or exactly what engineering and administrative controls must be implemented. The regulations allow facilities to use their specific knowledge and expertise of the process to meet the PHA requirement to ''identify, evaluate and *control* the hazard'' [emphasis added]. In the revised RMP Rule EPA added a requirement for certain industry sectors to conduct a STAA that also considers IST in the hierarchy of controls. However, EPA recognized that requiring facilities to implement IST instead of using passive, active or procedural safeguards can involve extensive and very expensive changes to a facility's process, depending on the IST, especially if it involves substitution of alternative chemicals and/or major process redesign. EPA believed that a practicability consideration should address whether an IST or ISD can be accomplished technologically, is economically possible, does not result in an increase in hazards or other risks that cannot be controlled, or cannot be successfully accomplished because of other considerations. (Ref 6. EPA Docket)

EPA's insights described above on how and why they finalized the STAA provisions in the 2017 revised RMP Rule, and they reflect their intent on how the term "feasible" should be interpreted in the context of inherent safety (and how they might enforce the STAA provisions). However, the January 13, 2017 revised RMP Rule is not yet effective. On June 9, 2017 the EPA Administrator delayed the implementation of the final 2017 RMP Rule until February 19, 2019.

Consensus evaluation methods that examine all of the relevant factors in feasibility are currently not available. New Jersey's inherent safety guidance provides the most detailed criteria published thus far, but these criteria, or variants of them have yet to be adopted in other jurisdictions or voluntarily by industry. However, it is clear that feasibility is a multi-faceted examination that is specific to each proposed inherently safety measure, specific to each facility and its risk profile, and is also specific to other industrial sectors that might be affected by potential risk transfer of the proposed measure. Feasibility also includes economic feasibility. Methods to estimate the value of inherent safety measures and to quantitatively assess whether a given process is "as inherently safe as is practicable" are currently unavailable or unproven. It will be difficult to reach consensus on truly objective financial criteria regarding proposed inherent safety measures as assessing the feasibility of the environmental and public health and safety impacts will require assigning monetary values to the lives of the those affected by the process as well as the monetary value of damage to the environment. This can and has been done but it is controversial in the public domain, at least in the U.S. It is less controversial in some other countries, e.g., The Netherlands.

Both the public and regulators may have higher risk reduction expectations of inherent approaches than what is ultimately found to be feasible in a given situation. The public and regulators expect to see a realistic effort to apply first- and second-order inherent safety to achieve major risk reduction, as defined in the CCPS inherent safety guidelines (Ref 3. CCPS). However, only considering first- or second-order IS measures may be at odds with other practical and cost-effective risk reduction options, especially for existing processes. This difference between public and industry perceptions of risk - and a misunderstanding or lack of appreciation of the impacts of some IS modifications - may create a barrier to both common expectations, and to an accepted definition of "feasibility" between the public and industry. Any move in an inherently safer direction is likely to be a positive risk reduction move, and should be encouraged.

Contra Costa Health Services Definition of "Feasible"

Contra Costa Health Services (CCHS) defines "feasible" as "capable of being accomplished in a successful manner within a reasonable period of time, taking into account economic, environmental, legal, social, and technological factors." However, a claim of financial infeasibility "shall not be based solely on evidence of reduced profits or increased costs, but rather shall include evidence that the financial impacts would be sufficiently severe to render the inherently safer system as impractical." This contention shall be supported by sufficient evidence to demonstrate that implementing this inherently safer system is impractical, and the reasons for this conclusion. This definition incorporates the notion of possibility and successful accomplishment which are part of the general definition of the word, but adds the concepts of reasonable time and the financial impact of accomplishment. Specifically, the use of reduced profits or increased costs shall not be the sole evidence used to claim that an inherent safety is infeasible. (Ref 4. CCHS)

For new processes, the CCHS guidance identifies the following different project phases when reviews should be performed to confirm if IS is applicable:

- during the chemistry forming (synthesis) phase for product/process research and development to focus on the chemistry and process;
- during the facilities design scoping and development prior to completion of the design basis to focus on equipment and configuration;
- and during the basic design phase of the project.

The guidance contains more detailed questions and review items to apply during each of these project phases. The guidance does recognize that IS may not be applicable for all facilities and project phases in the same way or in the same sequence. IS should be examined as early as possible to minimize the impact on the project.

For existing processes, the CCHS guidance notes that the consideration of IS is limited to scenarios where a Major Chemical Accident or Release (MCAR) could reasonably occur. This can be accomplished and documented by denoting which scenarios identified in PHAs have consequences (without crediting any active safeguards) that rise to that level. In most chemical/processing industry PHAs that are part of formal PSM programs, this is documented in PHA worksheets by describing the consequences so that offsite and onsite impacts that result from major releases are clearly identified. This is also usually supplemented by a qualitative or quantitative consequence or severity ranking that clearly indicates the severe consequences that are the hallmark of a MCAR. Facilities can perform one of the following methods to ensure that ISS are considered and documented for the covered processes:

- By conducting an independent ISS analysis that is done in addition to a PHA. Checklists or guideword analysis contained in the guidance document that incorporates ISS can be used to accomplish this analysis.
- By conducting an ISS analysis that is incorporated into the existing PHA review process. This would require that the PHA for each process be re-done in its entirety to include an initial ISS analysis. Revalidated PHAs that examine only portions of the process may not be sufficient to satisfy the initial ISS review if the whole process is not evaluated. Checklists or guideword analysis contained in the guidance document that incorporates ISS can be used to accomplish this analysis.

Whichever type of ISS analysis is implemented by the facility the CCHS guidance specifies that the analysis be conducted and documented to 1) document the qualifications of the team facilitator/leader and team makeup, 2) document the ISS's considered as well as those implemented, document the method used for the analysis if independent ISS analysis was performed, 3) if a checklist for the ISS was used, for items that were not considered, document why those items were not considered (i.e., not applicable or were already considered in previous consideration), 4) document the grounds that were used to make the feasibility determination, 5) document the ISS considered, the ISS implemented, and the ISS not implemented, 6) revalidate the ISS at least once every five years. (Ref 4. CCHS)

New Jersey TCPA Definition of "Feasible"

In the original NJ TCPA regulation, NJDEP included a requirement to perform a "state-of-the-art" (SOA) evaluation when the likelihood of release of hazard scenarios identified in the PHA and risk assessment that exceeded the acute toxicity concentration (ATC, the NJ limit of concern for consequences of releases and similar to AEGLs and ERPGs in the RMP Rule) at the site boundary. The definition of "state-of-the-art" in usage was:

> Current technology that, when applied to an owner or operator's EHS equipment and procedures will result in a significant reduction of risk. The technology represents an advancement in reduction of risk and shall have been demonstrated at a similar referenced facility to be reliable in commercial operation or in a pilot operation on a scale large enough to be translated into commercial operation. The technology shall be in the public domain or otherwise available at reasonable cost commensurate with the reduction of risk achieved. (Ref 9. NJ TCPA 2008)

In 2003, NJDEP added IS review provisions to the TCPA but only for new processes, and amended them in 2008 to extend the review requirements to new and existing processes (Ref 9. TCPA 2008). In 2009, the TCPA was further amended, in included in this set of changes was the deletion of the state-of-the-art definition and evaluation requirement (Ref 10. TCPA 2009). With the extension of the IS provisions, the SOA provisions were seen as redundant. Although the IS and SOA provisions of TCPA required owners/users to periodically evaluate the status of their covered processes to current technology, IS and SOA are not identical. Inherent safety, as currently defined and included in several regulations, is more comprehensive than the previous SOA requirements, and is more detailed in its requirements, e.g., defining and applying the four IS strategies of Substitution, Moderation, Minimization, and Simplification.

In 2006, NJDEP published guidance on how to perform IS reviews. This guidance has been updated several times since then, most recently in 2017. The guidance contained provisions for the qualifications of the persons performing the IST reviews, how they were to be documented, and a definition of feasibility. The guidance specified that the technology should be available at reasonable cost commensurate with the anticipated reduction or elimination of the hazard, that it was generally not expected that research into new technologies would be conducted, or that a major process modification or replacement be instituted to implement an inherently safer alternative. (Ref 8. NJ IST)

The New Jersey Department of Environmental Protection (NJDEP) guidance for the Toxic Catastrophe Prevention Act (TCPA) regulations contains more detailed criteria than CCHS for determining the feasibility or infeasibility of a prospective inherent safety measure. The 2017 detailed guidance includes criteria for: environmental and public health and safety feasibility, legal feasibility, technological feasibility, and economic feasibility. The criteria published by NJDEP are not intended to be all-inclusive. Also, NJDEP has provided a clarification for whether a qualitative and/or quantitative justification is expected. To date this guidance on the feasibility of IS options is the most detailed that has been published by a regulator or organization.

Environmental and Public Health and Safety Feasibility

- There would be a significant negative environmental impact (Consideration of water resources, water pollution, air pollution, solid and hazardous wastes, noise, etc.)
- The IST could decrease the hazard but would increase the overall risk.
- The risk would be shifted to another location where the risk would be the same or higher (If infeasibility is claimed because of risk, document the difference in frequency and/or consequences with and without the IST).
- All of the above require both qualitative and quantitative justification.

Legal Feasibility

- The IST could result in a conflict with existing federal, state, or local laws.
- The IST could violate a license agreement and the license agreement cannot be modified and must remain in effect
- Qualitative justification is required for both.

Technological Feasibility

- Is in conflict with Recognized and Generally Accepted Good Engineering Practices (qualitative, cite RAGAGEPs)
- Product quality specifications cannot be met (qualitative and quantitative).
- Availability of materials (qualitative and quantitative).
- Space restrictions (qualitative and quantitative).
- Impact on production rate (qualitative and quantitative).
- Commercially available or not (qualitative).

Economic Feasibility

If a given IST is claimed to be infeasible due to economic reasons the infeasibility justification should address the following elements:

- Life Cycle Analysis
- Capital investment, including design and implementation
- Net operating costs
- Change in the cost of materials including transportation and handling related costs
- Change in energy consumption
- Change in human costs such as number of operators, training
- Any other direct manufacturing costs
- Net regulatory compliance cost, change in fees
- Demolition and future cleanup and disposal cost
- All of the above criteria require a quantitative justification (cost/benefit analysis)

Generally, an IS measure is feasible if it has been successfully applied to similar processes or similar situations unless there are unique circumstances at the facility. Quantitative and/or qualitative justification is required to use these criteria. The justifications should highlight those unique circumstances and how they relate to the feasibility factors. (Ref 8. NJ IST)

Conclusions

The regulatory history of the use of the term "feasible" by OSHA, EPA, and state regulators is varied. The term has been used both formally in regulations themselves, and informally in published guidance and in actual practice. The following conclusions with respect to "feasible" and inherent safety can be drawn from this history and industry practice:

Regulatory Definition. Except for the New Jersey DEP, regulators have refrained from providing detailed definitions of "feasible" or "practicable" and its supporting terms, i.e., economic, environmental, legal, social, and technological. They seem to rely, at least currently, on the judgement of facility personnel who evaluate inherent safety measures for their feasibility. They do expect that when a measure is found to be infeasible that this decision is supported by a reasonable written rationale. New Jersey DEP has provided additional more detailed guidance defining these terms and where they expect quantitative and qualitative evidence to support a finding of infeasible.

Methods & Guidance. Consensus industry or regulatory evaluation methods that examine all of the relevant factors in feasibility are currently not available. New Jersey's inherent safety guidance provides the most detailed criteria published thus far, but these criteria, or variants of them have yet to be adopted in other jurisdictions or voluntarily by industry in a widespread manner. However, it is clear that feasibility is a multi-faceted examination that is specific to each proposed inherently safety measure, specific to each facility and its risk profile, and is also specific to other industrial sectors that might be affected by potential risk transfer of the proposed measure. Therefore, consensus evaluation methods for feasibility of inherent safety measures will be very difficult to achieve.

Economic Considerations. Feasibility does include economic considerations. Regulators have consistently stated that economic considerations play a role in defining feasibility, although they have also stated consistently that it is not the only factor, nor is it the most important factor. Methods to estimate the value of inherent safety measures and to quantitatively assess whether a given process is "as inherently safe as is practicable" are currently unavailable or unproven. It will be difficult to reach consensus on truly objective financial criteria regarding proposed inherent safety measures. This is because assessing the feasibility of the environmental and public health and safety impacts will require assigning monetary values to the lives of the those affected by the process as well as the monetary value of damage to the environment. This can and has been done but it is controversial in the public domain, at least in the U.S. It is less controversial in some other countries, e.g., The Netherlands, where quantitative risk analyses are required by law and are publicly disseminated.

Both the public and regulators may have higher risk reduction expectations of inherent approaches than what is ultimately found to be feasible in a given situation. The public and regulators expect to see a realistic effort to apply first- and second-order inherent safety to achieve major risk reduction, i.e., eliminate the hazard in its entirety, or use the four strategies of Substitution, Moderation, Minimization, and Simplification where possible to reduce the risk substantially. However, only considering first- or second-order IS measures may be at odds with other practical and cost-effective risk reduction options, especially for existing processes. This difference between public and industry perceptions of risk - and a misunderstanding or lack of appreciation of the impacts of some IS modifications - may create a barrier to an accepted definition of "feasibility" between the public and industry. However, in a overall sense, any improvement made that is an inherently safer measure is likely to be a positive risk reduction move, and should be encouraged.

References

1. American Textile Manufacturers. Institute (ATMI), 452 U.S. at 513 n. 31; American Iron and Steel Institute. v. OSHA, 939 F.2d 975, 980 (D.C. Cir. 1991).

2. American Heritage Dictionary of the English Language, New College Edition, 1976.
3. Center for Chemical Process Safety, *Inherently Safer Chemical Processes: A Life Cycle Approach, American Institute of Chemical Engineers, 2nd Ed.*, 2009.
4. Contra Costa Health Services (CCHS), California, *Industrial Safety Ordinance Guidance Document, Section D*, June 2011.
5. Environmental Protection Agency (EPA) 40 CFR §68, *Risk Management Programs for Chemical Accidental Release Prevention; Final Rule - Preamble* (RMP), 1996.
6. Environmental Protection Agency (EPA), *Response to Comments on the 2017 Proposed Rule Further Delaying the Effective Date of EPA's Risk Management Program Amendments (April 3, 2017; 82 FR 16146)*, Docket EPA-HQ-OEM-2015-0725, June 8, 2017.
7. Environmental Protection Agency (EPA) 40 CFR §68, *Risk Management Programs for Chemical Accidental Release Prevention; Final Rule - Preamble* (RMP), 2017.
8. State of New Jersey Department of Environmental Protection, Bureau of Release Prevention, *Guidance on Inherently Safer Technology (IST)*, September 25, 2017.
9. State of New Jersey, *Toxic Catastrophe Prevention Act (TCPA) Program*, N.J.A.C. 7:31 et. seq., 2008.
10. State of New Jersey, *Toxic Catastrophe Prevention Act (TCPA) Program*, N.J.A.C. 7:31 et. seq., 2009.
11. Occupational Safety and Health Administration (OSHA), 29 CFR §1910.119, *Process Safety Management of Highly Hazardous Chemicals, Explosives and Blasting Agents; Final Rule - Preamble*, 1992.
12. Occupational Safety and Health Administration (OSHA), CPL 02-02-045, *Process Safety Management of Highly Hazardous Chemicals -- Compliance Guidelines and Enforcement Procedures*, 1992.

512896 Inherently Safer Design Applications and Case Study of Thermally Coupled Distillation for C3-Alkyne Hydrogenation

Abstract:

Inherently Safer Design Applications and Case Study of Thermally Coupled Distillation for C3-alkyne Hydrogenation

Meng Qi[1,2*], Dongfeng Zhao[1,2], Yi Liu[1,2]

E-mail: qimeng166@163.com

[1] Center for Safety, Environmental, and Energy Conservation Technology

[2] College of Chemical Engineering

China University of Petroleum (East China), Qingdao 266580, P R China

Inherently safer design eliminates hazards from the source of process, reduces the risk of chemical plants, and makes the process more economical and safer. But at present, ISD for the chemical process is still in the theoretical stage, and there is no better application practice. The emergence of reactive distillation technology has greatly simplified the manufacturing process of methyl acetate, reaction and separation process has integrated in one column, reducing the residence time of the reactant in the plant and improving the safety level. Therefore, reactive distillation process of methyl acetate is the most outstanding embodiment of the application of inherent safety principle in chemical process design.

In recent years, thanks to the development of chemical process integration technology, reactive distillation, thermal coupling distillation, divided wall column distillation and other new technologies not only reduce energy consumption, simplify the process, but also make the process more inherently safer. In this paper, taking catalytic hydrogenation of C3 alkynes as an example, using ISD principle at the early stage of process design to make the safety and techno-economic analysis of conventional, thermally coupled and divided column distillation. Based on the kinetic and process flow data from literature to establish the steady simulation of three different processes, extracting results such as operating conditions, heating & cooling requirements and chemical characteristics to carry on the calculation of the operating and fixed costs.

The results show that the integrated distillation technology not only improves the conversion rate of the reaction, greatly reduces the energy consumption and enhances the economic, but also improves the inherent safety level of the process. The inherent safety index was used to compare the inherent safety level of three processes and analyze how the process inherently safer design affects the actual operating cost of the plant. It is verified that the application of inherent safety principle in the initial stage of the process design greatly improve the overall safety level of the plant, as well as reduce the energy consumption of the whole plant and the actual operating cost.

References

[1] Gupta, J. P., D. C. Hendershot, and M. S. Mannan. "The Real Cost of Process Safety¡ªA Clear Case for Inherent Safety." *Process Safety & Environmental Protection* 81.6(2003):406-413.

[2] Gupta, J. P., and D. W. Edwards. "A simple graphical method for measuring inherent safety." *Journal of Hazardous Materials* 104.1¨C3(2003):15-30.

[3] Y Wang, Y Luo, X Qian, X Yuan. "Simulation of thermally coupled catalytic distillation flowsheets for C3 alkyne selective hydrogenation." *CIESC Journal* 67.2(2016):580-587.

[4] YU Z Q, GAO B L, XU J M, et al. Kinetic study of C3 cut liquid-phase selective hydrogenation over supported metal catalyst [J]. *Industrial Catalysis*, 2002, 10(5): 11-16.

18AIChE Spring Meeting
14TH GLOBAL CONGRESS ON PROCESS SAFETY
ORLANDO FLORIDA 2018

Use of Various High-Nickel Alloys for Supercritical Water Reactions at Severe Temperature and Pressure Conditions

Matthew Conley
Sustainable Energy and Advanced Materials Laboratory
Russ College of Engineering and Technology
Ohio University
Athens, OH 45701
mc339810@ohio.edu

Matthew Factor
8031 Watson Road
St. Louis, MO 63119
mjfact@gmail.com

Sunggyu Lee
Sustainable Energy and Advanced Materials Laboratory
Russ College of Engineering and Technology
Ohio University
Athens, OH 45701
lees1@ohio.edu

Prepared for Presentation at
American Institute of Chemical Engineers
2018 Spring Meeting and 14th Global Congress on Process Safety
Orlando, Florida
April 22 – 25, 2018

Use of Various High-Nickel Alloys for Supercritical Water Reactions at Severe Temperature and Pressure Conditions

Matthew Conley
Sustainable Energy and Advanced Materials Laboratory
Russ College of Engineering and Technology
Ohio University
Athens, OH 45701
mc339810@ohio.edu

Matthew Factor
8031 Watson Road
St. Louis, MO 63119
mjfact@gmail.com

Sunggyu Lee
Sustainable Energy and Advanced Materials Laboratory
Russ College of Engineering and Technology
Ohio University
Athens, OH 45701
lees1@ohio.edu

Keywords: Supercritical Water, High Pressure, High Temperature, High-Nickel Alloys

Abstract

Supercritical water has increased in popularity as a reaction medium for hydrocarbons due to its greatly enhanced solvent capabilities towards these substances in the supercritical fluid region. While a supercritical condition could be achieved for water as long as it is subjected to both high temperature (T>374°C) and high pressure (P>22.1MPa), the realistic process conditions that are required for some of the supercritical water reactions far exceed the water's critical point barrier. Such reactions potentially include selective oxidation, gasification, reformation, and more. These elevated temperatures and pressures, along with corrosion considerations, require the use of special materials in the reactor hardware, pipes, and fittings utilized in these reaction systems. Some commonly used materials are various high-nickel alloys due to their great mechanical properties and corrosion resistance at elevated temperature and pressure conditions. This paper will focus on a comparison of five commonly used high-nickel alloys, viz., Hastelloy C276, Inconel 625 (Grades 1 and 2), Haynes 230, and Haynes 282, based on prior experimental reformation studies conducted in our laboratory on the reactors made of these specialty alloys and material data sheets. This study is intended to provide a comprehensive comparison of the yield and tensile strengths, creep and rupture strengths, corrosion resistance, and varying degrees of catalytic properties of the individual alloy materials. The comparisons of the different strengths will be made over temperature ranges that are commonly encountered in supercritical water reformation reactions, thereby establishing the safe operational limits and ranges for the system.

1 Introduction

Supercritical water has increased in popularity as a reaction medium for hydrocarbons and hydrocarbon containing materials as it is able to convert these hydrocarbons into their highest oxidized form.[1] Because of this dual capability as reaction medium and principal reactant, supercritical water reactions, including selective oxidation, partial oxidation, gasification, reformation, and more, can be used to destructively decompose harmful wastes and also convert hydrocarbons into synthesis gas, which is a mixture of mostly hydrogen, carbon dioxide, and carbon monoxide with some lower chain hydrocarbons such as methane and ethane also present.

For water to reach a supercritical state, its temperature and pressure must exceed 374 °C (705 °F) and 22.064 MPa (3200 psia) respectively. However, realistic operating conditions for many supercritical water reactions far exceed water's critical point. Being in the supercritical region greatly alters many properties of water including the density, dielectric constant, solubility, viscosity, and the ion product. Water has a relatively high dielectric constant of 78 at ambient temperatures, but it decreases to less than 5 when water reaches supercritical conditions.[2] Because of the high dielectric constant, ambient water is a very good polar solvent. As the temperature approaches supercritical conditions for water, the dielectric constant decreases continuously and drastically decreases once water becomes a supercritical fluid, which leads to supercritical water being a poor polar solvent but a good solvent for nonpolar substances such as hydrocarbons due to high miscibility with organics.[3] This change in dielectric constant, especially when very close to the critical point, can create a corrosive environment by changing the pH of the system.[2] Because of the corrosive and oxidative tendencies of supercritical water as well as the high temperatures and pressures required for supercritical water reactions, special materials are required for the reactor hardware, pipes, and fittings utilized in these reaction systems.

Some commonly used materials are various high-nickel alloys due to their great mechanical properties and corrosion resistance at elevated temperature and pressure conditions as well as their excellent fabricability. These alloys include Hastelloy C276, Inconel 625 (both Grade 1 and Grade 2), Haynes 230, and Haynes 282. Hastelloy C276 is a good material for reactions up to 538 °C and has very good resistance to stress corrosion cracking in chloride-bearing solutions.[4] Hastelloy C276 has been used as a supercritical water reactor in the supercritical water partial oxidation of polystyrene to depolymerize the polystyrene into its monomer, oligomer, and other hydrocarbons at 382 °C.[5]

Inconel 625 comes in two grades – Grade 1 and Grade 2 with Grade 2 being created by solution treating Grade 1. Inconel 625 Grade 1 has been utilized as a first generation reactor for use by Lee's research group for the reformation of JP-8 jet fuel at temperatures of 650 °C and pressures up to 5000 psi.[6]

In order to increase the operating conditions for temperatures closer to 800°C, a Haynes 230 alloy second generation reactor was designed to reform both military grade JP-8 jet fuel and civilian Jet-A jet fuel at a temperature of 767 °C and 3500 psi.[7] Haynes 230 is a nickel rich alloy that is designed to perform in high temperature environments up to 1149 °C.[8]

Also utilized as a high-alloy material is Haynes 282, which was developed for use in high temperature structural applications including in aero and industrial gas turbine engines.[9]

This paper will focus on a comparison of these five commonly used high-nickel alloys based on prior experimental reformation studies conducted in our laboratory on the reactors made of these specialty alloys and material data sheets of these materials.

2 Tensile and Yield Strength

Two important properties for materials utilized in high temperature and pressure situations are tensile strength and yield strength. Tensile strength is the stress that a material can withstand while being pulled or stretched before breaking whereas the yield strength is the stress applied to plastically, or non-reversibly, deform a material. For all five high-nickel alloys, the tensile strength is relatively constant for a wide range of temperatures up to approximately 600 °C (Hastelloy C276 is only suitable up to 538 °C[4]) where a sharp decrease results in the strengths. Haynes 230 and Inconel 625 Grade 2 have relatively similar strengths until approximately 800°C when Haynes 230 shows a slightly higher tensile strength while Inconel 625 Grade 1 is actually better in terms of tensile strength up until about 700 °C when it becomes comparable to Grade 2. Haynes 282 consistently has the highest tensile strength of all the high-nickel alloys. The tensile strengths of all of the materials for temperatures up to 1000 °C can be seen in Figure 1.

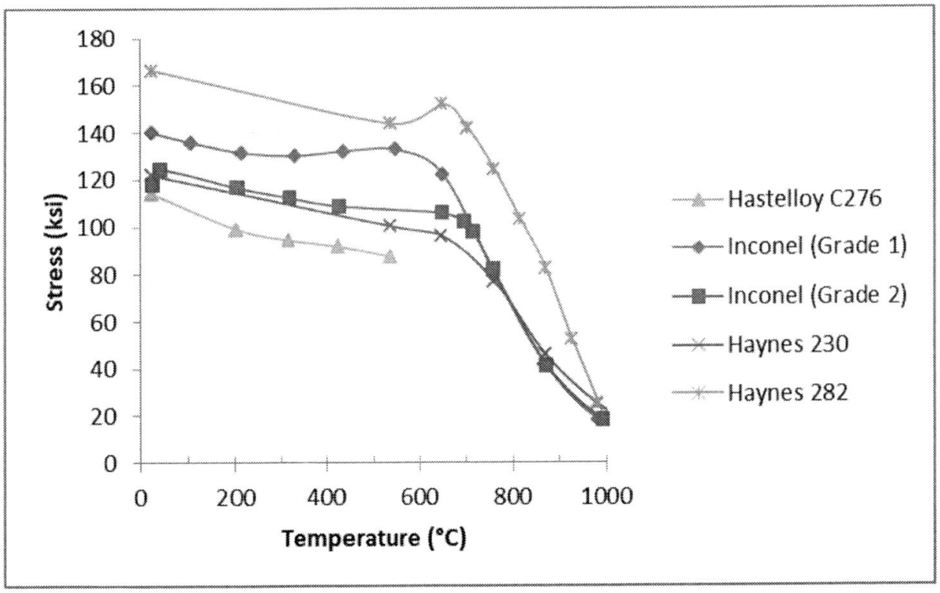

Figure 1. Tensile strengths of Hastelloy C276[4], Inconel 625 (Grades 1 and 2)[10], Haynes 230[8], and Haynes 282[9] for temperatures up to 1000 °C

Figure 2 shows the yield strength for all of the high-nickel alloys. The yield strengths for Hastelloy C276, Haynes 230, and Inconel 625 Grade 2 are nearly the same for temperatures up to 538 °C (data for Hastelloy C276 is lacking above this point) with Haynes 230 and Inconel 625 Grade 2 continuing this trend up to 1000 °C. Inconel 625 Grade 1 actually has higher yield strength than Inconel 625 Grade 2, while Haynes 282 is superior to all of the other alloys in terms of yield strength. However, in terms of tubular reactor design, creep and rupture are more important material properties as these occur at lower stresses and can render the reactor perilous or prone to failure if a deformation or rupture occurs.

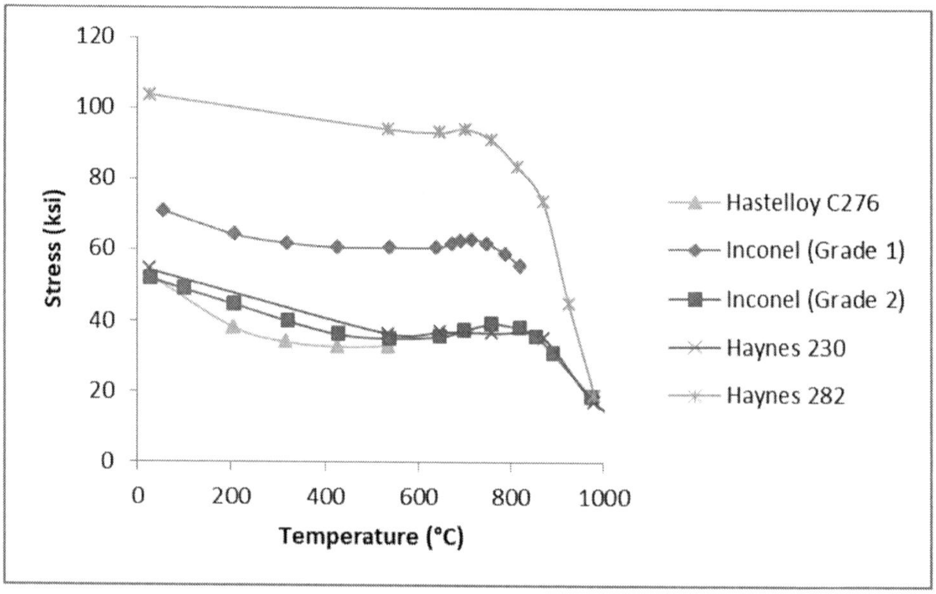

Figure 2. Yield strengths of Hastelloy C276[4], Inconel 625 (Grades 1 and 2)[10], Haynes 230[8], and Haynes 282[9] for temperatures up to 1000 °C

3 Creep and Rupture Strength

In terms material properties, creep, or permanent deformation due to stress, and rupture, or complete material failure, are very important in that they are indicators of when a piece of equipment, namely a reactor, will fail as a deformation or rupture will cause the reactor to become unreliable for practical applications. To determine which material is better suitable as a supercritical water reactor for temperatures up to 800 °C, creep and rupture need to be compared and minimized.

Even though Inconel 625 Grade 1 is superior in terms of yield strength and tensile strength than Inconel 625 Grade 2, Grade 2 has substantially better resistances to both creep and rupture at temperatures around 800 °C, which can be seen in Figures 3 and 4.

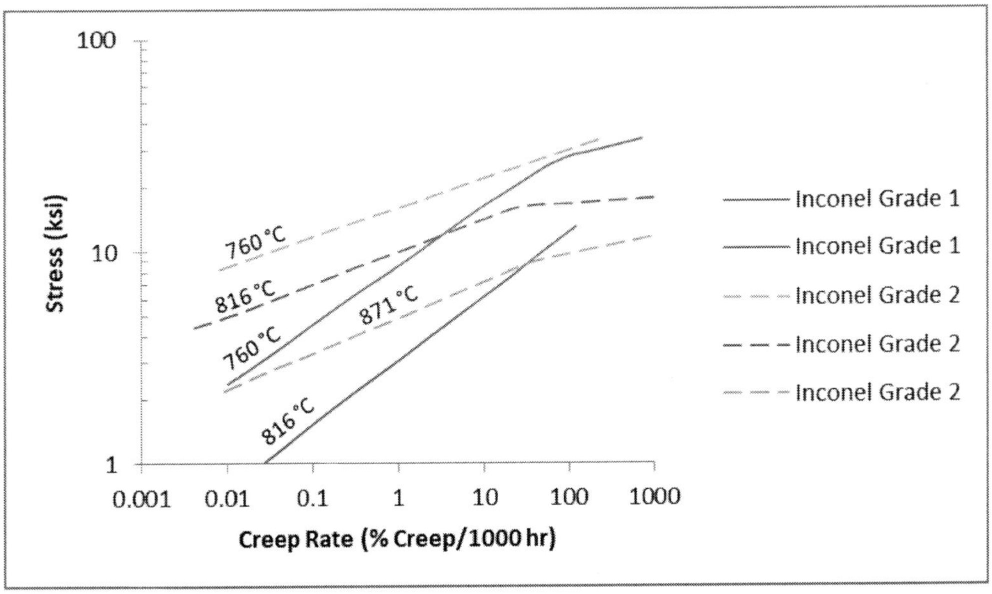

Figure 3. Creep rate for Inconel 625 Grades 1 and 2 for temperatures around 800 °C[10]

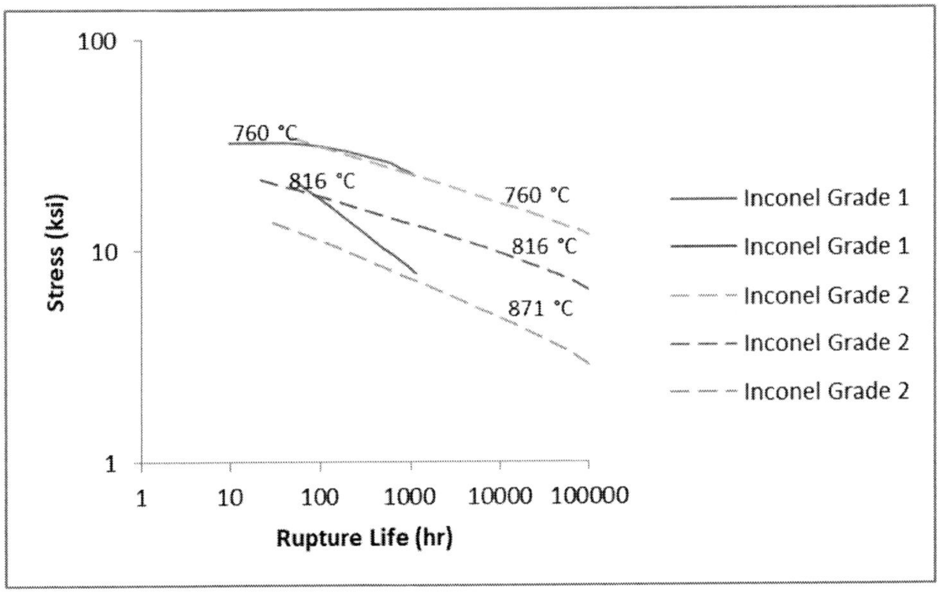

Figure 4. Rupture life for Inconel 625 Grades 1 and 2 for temperatures around 800 °C[10]

When deciding which grade of Inconel to use, as Grade 1 is superior in tensile and yield strength while Grade 2 is superior in creep and rupture, the American Society of Mechanical Engineers (ASME) pressure vessel guidelines are important to follow. Inconel 625 Grade 1 is approved by the ASME Boiler and Pressure Vessel Code up to 650 °C while Grade 2 is approved up to 871 °C making Inconel 625 Grade 2 the only option for Inconel at temperatures of 800 °C.[1]

When looking at rupture strength and 1% creep strength at 1,000 and 10,000 hours for Inconel 625 Grade 2, Haynes 230, and Haynes 282, it is found that Inconel 625 Grade 2 and Haynes 230 have similar values for both (no information was available for 1% creep for Inconel 625 Grade 2 at 10,000 hours) with Inconel 625 Grade 2 having slightly higher values. Haynes 282, however, has significantly higher values for both rupture and 1% creep, which can be seen in Table 1.

Table 1. Initial stress at 1% creep and rupture stress at 1,000 hours of operation and 10,000 hours of operation for Inconel 625 Grade 2[10], Haynes 230[8], and Haynes 282[9] with the values at 800 °C being interpolated values

		1,000 hrs		10,000 hrs	
Alloy	**Temp (°C)**	**1% Creep (ksi)**	**Rupture (ksi)**	**1% Creep (ksi)**	**Rupture (ksi)**
Inconel 625 Grade 2	760	16.1	22.8	-	16.7
	800*	11.9	16.7	-	12.1
	816	10.0	13.3	-	9.8
	871	4.9	7.3	-	4.8
Haynes 230	760	11.5	20.0	8.0	14.2
	800*	9.7	16.0	6.8	11.2
	816	9.2	14.0	6.5	9.8
	871	6.0	9.5	4.4	6.2
Haynes 282	760	36.0	41.0	-	27.0
	800*	26.7	30.5	-	19.0
	816	22.0	25.0	-	14.0
	871	12.0	14.0	-	8.0

While Haynes 282 has superior rupture strength than that of Haynes 230 for the temperatures and lifetimes investigated, Haynes 230 should have higher rupture strength after continued use due to a shift in the slope of strength versus rupture life as seen in Figure 5.

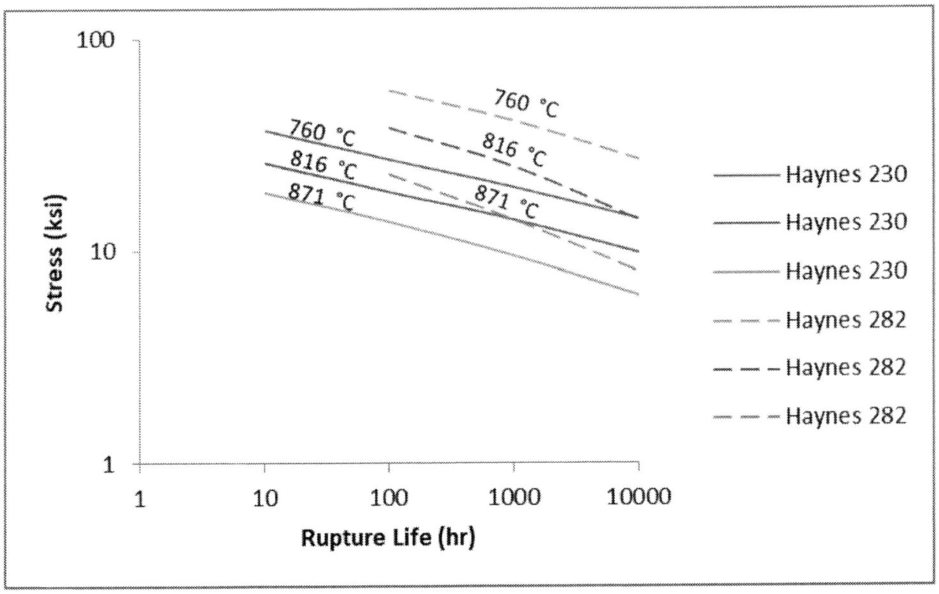

Figure 5. Rupture life for Haynes 230[8] and Haynes 282[9] for temperatures around 800 °C

When examining the rupture life for the three alloys at temperatures of 816 °C and 871 °C and a pressure of 4.5 ksi, which is a typical pressure for supercritical water reactions, Haynes 282 is the superior alloy at 816 °C while Haynes 230 is slightly better at the higher temperature with Inconel 625 being the worst of the three compared at both conditions. This information can be seen in Table 2.

Table 2. Operational lifetimes for Inconel 625 Grade 2[10], Haynes 230[8], and Haynes 282[9] for 816 °C and 871°C at a stress of 4.5 ksi

Alloy	Temp (°C)	Liftetime (hrs)
Inconel 625 Grade 2	816	1.8 million
	871	14,000
Haynes 230	816	2.3 million
	871	88,000
Haynes 282	816	2.1 million
	871	128,000

This information shows that Haynes 282 has a similar lifetime to that of Haynes 230 with a superior lifetime to that of Inconel 625 while having superior creep and rupture strengths to both. Another important consideration for material selection is corrosion resistance as supercritical water is a corrosive reaction medium.

4 Corrosion Resistance

Corrosion of metals is a major issue in material selection for any piece of equipment, but especially so in the corrosive medium of supercritical water. If a reactor is used for fuel reformation, knowing the oxidation patterns for the materials in question in the environments created by fuel reformation is beneficial. These oxidative environments associated with fuel reformation at high temperatures include air, wet hydrogen, hydrocarbon fuels, and oxygenated intermediates.

The oxidation resistances of Inconel 625 and Haynes 230 are very well-known and easily comparable, but, according to Meyer and Deodeshmukh, Haynes 282 "was developed to have excellent creep resistance while maintaining good fabricability and thermal stability," but "its long-term oxidation resistance has not been fully established."[11] This lack of data inhibits some comparison between Haynes 282 and the other alloys, although there is some overlap in certain conditions.

Parabolic oxidation rate constants (k_g) explain how quickly a material will oxidize in the environment that the constant was calculated for, with a smaller constant meaning that the oxidation will be a slower process than that with a larger constant. In a study performed by England et al., Haynes 230 has a smaller oxidation constant than that of Inconel 625 in both flowing air and humidified hydrogen at 800 °C.[12,13] The parabolic rate constant for Haynes 230 oxidized in air at 800 °C is about 22 times smaller than that of Inconel 625, while the rate constant for Haynes 230 oxidized in humidified hydrogen is about 9 times smaller. The activation energy (Q) for Inconel 625 for each condition is also higher than that of Haynes 230, meaning that the oxidation of Haynes 230 is less likely to happen. These values can be seen in Table 3.

Table 3. Parabolic oxidation rate constants for Haynes 230 and Inconel 625 at 800 °C oxidized in air[12] and in humidified hydrogen[13]

	Air		H_2	
Alloy	**kg (g^2/cm^4h)**	**Q (kJ/mol)**	**kg (g^2/cm^4h)**	**Q (kJ/mol)**
Haynes 230	4.64x10^-12	340	1.20x10^-10	180
Inconel 625	1.01x10-10	314	1.06x10^-9	165

For oxidation in a hydrocarbon fuel, data from a burner rig oxidation test was examined.[8] In this setup, samples were exposed to the products of the combustion of 2 parts Number 1 fuel oil and 1 part Number 2 fuel oil burned at an air to fuel ratio of about 50:1 with the samples being removed from the gas stream every 30 minutes and cooled to ambient temperatures before being reinserted into the combustion stream. As seen in Table 4, Haynes 230 has better oxidation resistance in terms of metal loss than Inconel 625 at both 870 °C and 980 °C, while Inconel 625 is completely consumed at higher temperatures.[8] At lower temperatures, Inconel 625 has less average affected metal than Haynes 230, which shows that the oxidation is more localized under these conditions but becomes more widespread than that of Haynes 230 at higher conditions. Although there is no data for 800 °C, similar trends of oxidation resistances should be present at lower temperatures, leading Haynes 230 to be the better metal over Inconel 625 in terms of oxidation resistance.

Table 4. Burner rig metal loss for Inconel 625 and Haynes 230 at 870 °C and 980 °C for 1000 hours of exposure[8]

	870 °C, 1000 hr		980°C, 1000 hr	
Alloy	Metal Loss (μm)	Average Metal Affected (μm)	Metal Loss (μm)	Average Metal Affected (μm)
Haynes 230	23	99	71	142
Inconel 625	30	56	94	152

As stated earlier, there is little oxidation resistance data currently available for Haynes 282 as it is a relatively new material. There has been some overlapping data obtained in comparison to Haynes 230 in regards to oxidation at 871 °C for 360 days in flowing air while cycled to room temperature every 30 days. This comparison is shown in Table 5.

Table 5. Oxidation data for Haynes 230[8] and Haynes 282[11] at 871 °C for 360 days in flowing air

Alloy	Metal Loss (μm)	Average Metal Affected (μm)
Haynes 230	5	43
Haynes 282	28	71

Based on the above data, Haynes 230 is actually a better material in terms of oxidation resistance than Haynes 282 over long periods as its metal loss and average affected metal were both less than that of Haynes 282. Other short term (1000 hour)

oxidation tests have been run in a burner rig similar to what was mentioned earlier except the samples were only cooled to roughly 260 °C instead of ambient temperatures as was the case with the Haynes 230 and Inconel 625, making the tests not absolutely comparable. However, at 870 °C and 980 °C in the burner rig setup, Haynes 282 had values of metal loss of 46 μm and 203 μm. At lower temperatures, the value was similar to that of Inconel 625 and Haynes 230 even though the higher temperature exposure (only cooled to 260 °C instead of ambient temperature) would lead to more oxidation, making Haynes 282 an acceptable corrosion resistant replacement material for either Haynes 230 or Inconel 625. Also, Meyer et al. showed that Haynes 282 had similar oxidation behavior to Haynes 263 and Haynes R-41 and better oxidation resistance than Waspaloy; Haynes 263, Haynes R-41, and Waspaloy are other nickel-based alloys commonly used for gas turbine engines[11]. More data on long term oxidation and other oxidative environments is needed for a full understanding of the oxidation resistance of Haynes 282.

5 Catalytic Properties

The high-nickel alloys, due to the presence of nickel and other catalytic metals such as chromium, molybdenum, and cobalt among others, can show some catalytic effects towards different reactions, including the important water-gas shift reaction that occurs in the supercritical water gasification process.

According to Boukis et al., Inconel 625 played an important role in catalyzing the gasification of methanol at pressures ranging from 25 to 45 MPa, temperatures ranging from 400 °C – 600 °C, feed concentrations ranging from 5 wt% to 64 wt%, and residence times ranging from 3 seconds to 100 seconds. The inner surface of the Inconel 625 reactor wall was able to catalyze the reaction to a point of 99.9% methanol conversion without the need of other added catalysts.[14]

The water-gas shift reaction, favored in the forward direction at lower temperatures, accounts for the relative amounts of hydrogen, carbon dioxide, and carbon monoxide gases in the syngas product stream of supercritical water gasification.

$$CO + H_2O \leftrightarrow CO_2 + H_2 \tag{1}$$

Various studies have shown that this reaction is indeed catalyzed by nickel alloys such as Hastelloy C276 and Inconel 625, with Inconel 625 being especially favorable. Yu et al. showed that during glucose gasification with a 0.2 M glucose solution at 600 °C, 34.5 MPa, and 34 second residence time, Inconel 625 catalyzes the water-gas shift reaction and produces a gas that is very highly rich in hydrogen and carbon dioxide, where Hastelloy produces a gas stream rich in carbon monoxide.[15] This conclusion, that Inconel 625 catalyzes the water-gas shift reaction, is also supported by Taylor et al. in the gasification of methanol at 26.7 MPa and temperatures ranging from 550 °C to 700 °C with the reaction being kinetically limited at temperatures below 700 °C.[16]

There is a general dearth of literature information with regard to the monolithic catalytic effects of Haynes 230 and Haynes 282 either on the water gas shift reaction or the hydrocarbon gasification reaction, even though some preliminary experimental evidence of their distinct catalytic behaviors from that of Inconel 625 have been obtained by the current authors' team. It is particularly noteworthy that Haynes 282 has the highest cobalt composition (10.0%) among all the high-nickel alloys considered in the current study, namely Haynes 232 (≤5.0%), Hastelloy C276 (2.5%), and Inconel 625 (≤1.0%). While nickel has been established as a primary fuel reformation catalyst, cobalt, along with molybdenum, has been found as a major ingredient in effective formulations of hydrocracking catalysts[17]. The observed differences may be at least partially explained by the fact that hydrogen is a primary product of the fuel reformation reaction, whereas hydrogen is a primary reactant of the fuel hydrocracking reaction.

6 Conclusion

High nickel alloys are strong and tough materials suitable for supercritical water gasification reactor components as they can withstand high temperatures; have strong creep, rupture, and tensile strengths; and are corrosion resistant to the environments seen in supercritical water. The material best suitable for these applications for temperatures around 800 °C is Haynes 282 as it is superior with respect to creep and rupture strengths, yield and tensile strengths, and lifetime compared to the other investigated alloys while having similar corrosion properties. Due to this, Haynes 282 was selected as the material for the newest generation of reactor as it is able to withstand high temperatures and pressures. Likewise, for reactions under 650 °C, an Inconel 625 thermowell was added to not only ensure temperature consistency throughout operation, but also to act as a catalyst to ensure a hydrogen rich product stream. Moving forward, more corrosion data, especially long term corrosion data, is needed to verify that the corrosion resistance of Haynes 282 is adequate for long-term use of the reactor.

7 References

1) M.J. Factor, H.B. Lanterman, J.E. Wenzel, and S. Lee. Use of Haynes Alloy 230 for Supercritical Water Reactors, AIChE Annual Meeting (2007).

2) R. Gupta. "Supercritical Water Oxidation" in *Encyclopedia of Chemical Processing*. Taylor and Francis Group, New York, NY, 2006.

3) O. Yakaboylu, J. Harinck, K.G. Smit, and W. de Jong. Supercritical Water Gasification of Biomass: A Literature and Technology Overview, Energies 8, (2015), 859-894.

4) Haynes International. *HASTELLOY® C-276 alloy*; Brochure H-2002E; Haynes International, Inc.: Kokomo, IN, 2017.

5) D.W. Lilac and S. Lee. Kinetics and mechanisms of styrene monomer recovery from waste polystyrene by supercritical water partial oxidation, Adv. Environ. Res. 6, (2001), 9-16.

6) S. Lee, H.B. Lanterman, J.E. Wenzel, N. Edwards, A. Adams, J. Wootton, and A. Garcia. Noncatalytic Direct Reformation of JP-8 Fuel in Supercritical Water, Natl. Meet. Book Abstr. 2006.

7) J.W. Picou, J.E. Wenzel, H.B. Lanterman, and S. Lee. Hydrogen Production by Noncatalytic Autothermal Reformation of Aviation Fuel Using Supercritical Water, Energy Fuels 23, (2009), 6089-6094.

8) Haynes International. *HAYNES® 230® alloy*; Brochure H-3000K; Haynes International, Inc.: Kokomo, IN, 2017.

9) Haynes International. *HAYNES® 282® alloy*; Brochure H-3173B; Haynes International, Inc.: Kokomo, IN, 2017.

10) Special Metals Corporation. *INCONEL® alloy 625*; Brochure; New Hartford, NY, 2013.

11) J. Meyer and V. Deodeshmukh. Long-Term Oxidation Resistance Of Several Precipitation Strengthened Ni-Based Superalloys, ASME Turbo Expo 2017, ASME. Charlotte, NC. 2017.

12) D.M. England and A.V. Virkar. Oxidation Kinetics of Some Nickel-Based Superalloy Foils and Electronic Resistance of the Oxide Scale Formed in Air Part I, J. Electrochem. Soc. 146, (1999), 3196-3202.

13) D.M. England and A.V. Virkar. Oxidation Kinetics of Some Nickel-Based Superalloy Foils in Humidified Hydrogen and Electronic Resistance of the Oxide Scale Formed Part II, J. Electrochem. Soc. 148, (2001), A330-A338.

14) N. Boukis, V. Diem, W. Habicht, and E. Dinjus. Methanol Reforming in Supercritical Water, Ind. Eng. Chem. Res. 42, (2003), 728-735.

15) D. Yu, M. Aihara, and M.J. Antal. Hydrogen production by steam reforming glucose in supercritical water, J. Energy Fuels 7, (1993), 574-577.

16) J. Taylor. Hydrogen production in a compact supercritical water reformer, Int. J. Hydrog. Energy 28, (2003), 1171-1178.

17) S. Lee. "Clean Liquid Fuels from Coal" in *Handbook of Alternative Fuel Technologies*. CRC Press, Boca Raton, FL, 2014.

510248 Optimization of Blowout Preventer Design for Optimal Cost and Reliability

Abstract:

Abstract for 2018 AIChE Spring Meeting and 14th Global Congress on Process Safety April 22-25, 2019 Orlando, FL

Session Selection: 52nd Annual Loss Prevention Symposium (LPS) (T1C)

T1C00 Asset Integrity and Reliability - Oral Session

Optimization of Blowout Preventer Design for Optimal Cost and Reliability

Denis Su-Feher, Nilesh Ade, Bin Zhang, Yogesh Koirala, and M. Sam Mannan

Mary Kay O'Connor Process Safety Center, Artie McFerrin Department of Chemical Engineering, Texas A&M University, College Station, TX, 77843-3122, USA

Abstract

Incidents such as the Deepwater Horizon blowout highlight the necessity of reliable, well designed blowout preventers (BOP) on offshore drilling platforms. Even if a blowout does not occur, extensive economic loss can occur when a BOP fails due to the expenses of performing corrective maintenance. Therefore, a well-designed, well maintained BOP is crucial to safer operation of a drilling platform. On the other hand, it is not practicable to overdesign the BOP as this will result in increased cost with little change in reliability. An optimized BOP design is necessary to balance the capital and operating cost of a blowout preventer with its risk of failure.

The capital cost and operating cost must be considered when designing a BOP, along with its reliability over time. A previous study has shown that a model based on fault tree analysis can relate the failure probability of each component in a BOP design to its overall failure probability and provide a globally optimized preventive maintenance schedule with minimum maintenance cost subject to a minimum reliability threshold. A number of maintenance schedules can be generated from this model, as well as a Pareto-optimality frontier that demonstrates the optimal maintenance cost vs reliability of a particular design.

This study presents a number of BOP designs and generates a corresponding Pareto-optimality frontier for each design. The reliability of each design is compared, which can be used to optimize the design of a BOP by considering the capital and maintenance cost for a number of different designs. A Pareto-optimality frontier is derived for all defined BOP designs, which simultaneously determines the optimal design and preventative maintenance schedule for the BOP.

Keywords: *Safe Design, Asset Integrity and Reliability, Upstream, Blowout Preventer, Preventative Maintenance, Multi-Objective Optimization, Simultaneous Design and Scheduling*

Corporate Memory Loss Ongoing

– Except CCPS 4th Pillar (Learn from Experience) as Espionage

Long(Steven) Zhang
No affiliation since Jul 2014 off-hiring from ConocoPhillips
11152 Westheimer Road APT #524, Houston TX 77042
longz.houston@gmail.com

Dr Wahed Wasel
AEYAAA International, LLC
2929 Hayes Road APT#1103, Houston TX 77082
wahed.wasel@hotmail.com

Prepared for Presentation at
American Institute of Chemical Engineers
2018 Spring Meeting and 14th Global Congress on Process Safety
Orlando, Florida
April 22 – 25, 2018

Corporate Memory Loss Ongoing

- Except CCPS 4th Pillar (Learn from Experience) as Espionage

Long(Steven) Zhang
No affiliation since Jul 2014 off-hiring from COP
11152 Westheimer Road APT #524, Houston TX 77042
longz.houston@gmail.com

Dr Wahed Wasel
AEYAAA International, LLC
2929 Hayes Road APT#1103, Houston TX 77082
wahed.wasel@hotmail.com

"By broadening our conception of the forces that make up and control our lives, average people can never again be completely caught in the grip of patterned behavior of which they have no awareness."

- Anthropologist Edward T. Hall

Without activated Process Safety Socio Cultural Communication (PSSCC), Process Safety Techno Cultural Communication (PSTCC) is some kind of meaningful work and simultaneously some kind of meaningless work i.e. mass brainwashing about the true big picture.

- the author, on RBPS Socio-Technical Systems & Leadership Media Universe

Keywords:

Memory Loss, Learn from Experience, Espionage, Diagnosis, Foreknowledge, Epistemology

Abstract

In this paper, the authors first would introduce the relationship between Process Safety Culture and RBPS Core Value. Then the authors would rationalize Process Safety Culture's Chronic Malnutrition and its 28 Congenital Infra (Strategic) Cultural Activities. At end, the authors would characterize CCPS 4th Pillar (Learn from Experience) as Chemical Engineering (RBPS) Espionage/Diagnosis/Knowledge in light of reflections both of RBPS Architecture & Ethical Dimension and of Epistemology of Engineering Knowledge so that the potentials of CCPS 4th Pillar (Learn from Experience) and its 4 Elements could be fully unlocked.

1 Introduction

The issue of "Corporate Memory Loss" might have been one of the critical topics which have not yet attracted due attention in process industry till very recently. In the authors' opinions, "corporate memory loss" in process industry is of some patterns of the universal challenges of "Bridging Engineering to the Humanities". And in 2007, Center for Chemical Process Safety (CCPS) published its Guidelines for Risk Based Process Safety (RBPS), within which the issue of "corporate memory loss" was also speculated as some ultimate topic concerning future process safety management [1].

In reality, since 1st Global Congress on Process Safety (GCPS), there have been a series of attempts /efforts to bridge between Chemical Engineering/RBPS and the Humanities:

At 1st GCPS (2005), Process Safety Culture was mentioned first time [2], and then as one of those 10 recommendations of BP Texas Refinery Explosion Report [3], and then as one Element within CCPS Risk Based Process Safety Temple/4 Pillars/20 Elements (2007).

At 9th GCPS (2013), 21st Element (Process Safety Cultural Communication (PSCC)) and 22nd Element (RBPS Core Value) were identified as necessary further efforts complementing Process Safety Culture and resulting in Egg model (refer to Scheme 1st). And one of the author's opinions was: the temple model is a good building with a correct design intent, but there should be a landlord i.e. a core value to live within and PSCC be his livelihood; the temple may still look like a good building outwardly but without an alive core value, it could become a tomb in reality. [4]

At 1st CCPS Asia-Pacific Conference on Process Safety (2013), some efforts of benchmarking Egg model against Sun Tzu's ***The Art of War***, and against **Traditional Chinese Medicine** as well [5], prove, along with those subsequent further bridging efforts [6][7] (e.g. An Inevitable Ethical Dimension of Vision 20/20 & its correlation with Operational Discipline (OD), A New Vocabulary of Process Safety Culture, RBPS Socio-Technical Systems & Leadership Media Universe etc.), to be so rewarding that both Sun Tzu's principle of foreknowledge /espionage of war (results) and the principle of diagnosis of Chinese Traditional Medicine have been jointly shedding more and more light on the unique role CCPS 4th Pillar (Learn from Experience) would

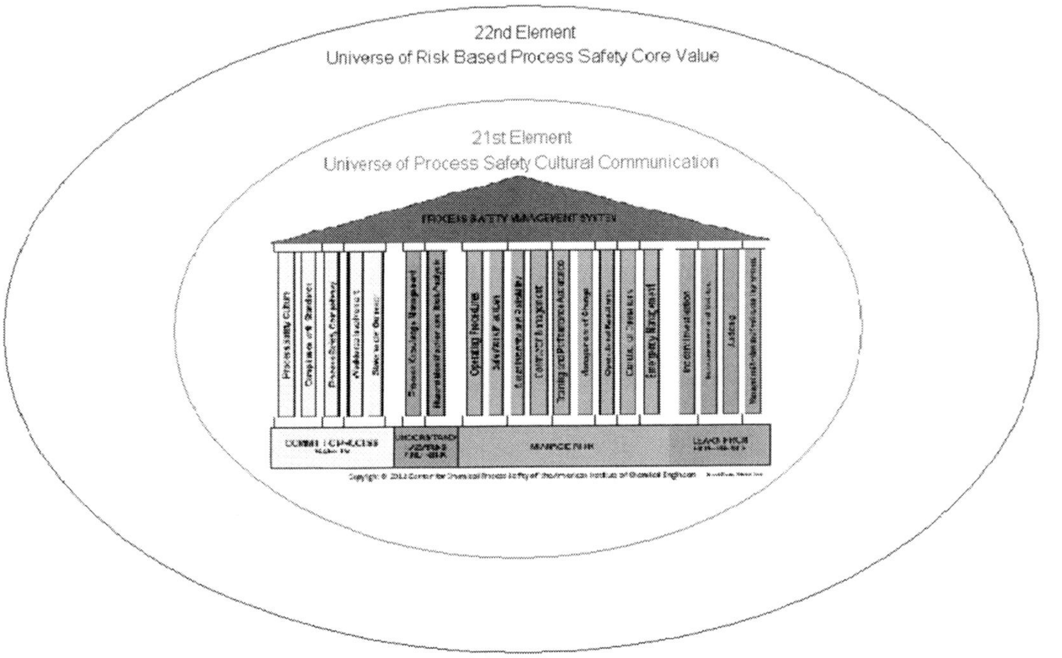

The Temple (4 pillars) Figure Source: CCPS, Process Safety Boot Camp, contributed by David Guss, Nexen Inc. Copyright 2010 AIChE and used with permission

Scheme 1st

<table>
<tr><td colspan="2">RBPS Egg</td><td>The Art of War</td><td>Philosophy of Process Safety</td><td></td></tr>
<tr><td colspan="4">RBPS Core Value - Tao - RBPS Worldview</td><td>Worldview</td></tr>
<tr><td rowspan="4">Commit to Process Safety</td><td>Process Safety Competency</td><td rowspan="2">Leadership</td><td rowspan="2">People</td><td rowspan="7">Process Safety Cultural Communication (Process Safety Culture) RBPS Methodology</td></tr>
<tr><td>Workforce Involvement</td></tr>
<tr><td>Compliance with Standards</td><td rowspan="2">Weather/Climate</td><td rowspan="3">Plant</td></tr>
<tr><td>Stakeholder Outreach</td></tr>
<tr><td>Understand Hazards & Risk</td><td></td><td>Terrain</td></tr>
<tr><td>Manage Risk</td><td></td><td>Discipline</td><td>Procedure</td></tr>
<tr><td>Learn from Experience</td><td></td><td>Espionage</td><td>Engineering & Humanities</td></tr>
</table>

Table 1st

likely play in future Chemical Engineering and Risk Based Process Safety framework / corridor (Refer to Table 1st).

2. Process Safety Culture's Congenital 28 Infra (Strategic) Cultural Activities

2.1 Process Safety Culture is a Prison unless we have RBPS Core Value to unlock it

Since the first time being mentioned at 1st GCPS (Kadri & Jones, 2015) and its subsequent inclusion both as one of those 10 recommendations of BP Texas Refinery Explosion Report (2007) and as one of those 20 Elements of CCPS Risk Based Process Safety Temple/4 Pillars (2007), Process Safety Culture might have been universally perceived as the biggest breakthrough in the history of process safety management. In authors' opinions, however, Process Safety Culture is just an Element but with a great potential, which could be fully unlocked across process industry through a series of further efforts of bridging (Chemical) Engineering /RBPS to the Humanities [8].

At 9th GCPS, 21st Element (PSCC) and 22nd Element (RBPS Core Value) were identified and CCPS 4 Pillars/Temple upgraded to Egg Model. Within Egg Model, Process Safety Culture was rationalized as a tree of organizational culture and RBPS Core Value the root of the tree; the ultimate purpose of Process Safety Cultural Communication (PSCC) is to convey the message of RBPS Core Value, i.e. PSCC functions not only the nourishment from the root but also the photosynthesis for the root; and PSCC is the future form of Process Safety Culture.

In recent process industry, we may occasionally hear the saying "we need to build a Process Safety Culture". However, in spite of the fact that Process Safety Culture just has been an Element of CCPS 4 Pillars since 2007 it might had already been over there since the very moment process industry was first initiated. And at 1st Asia-Pacific Conference on Process Safety (2013), some new perspective even was presented like: "**Process Safety Culture is a prison unless we have RBPS Core Value as a key to unlock it.**" And the original saying by American anthropologist Edward T. Hall was "It (culture) is in fact a prison unless one knows that there is a key to unlock it" [9], which seems still brand new to Chemical Engineering /RBPS, even to Engineering. Hence it seems not so reasonable for us only saying "We need to build a Process Safety Culture" without mentioning at same time that "the ultimate purpose of Process Safety Culture /PSCC is to convey the message of RBPS Core Value".

2.2 Process Safety Culture's Congenital 28 Infra (Strategic) Cultural Activities

Without RBPS Core Value and PSCC, what we likely could only achieve might be just a chronic mal-nutritional Process Safety Culture, toward which at 10th GCPS a 7 Mega Elements' map of Process Safety Culture /PSCC was drawn and presented to the process industry [10]. And the brief background is as follows:

The author treats these 7 Mega Elements as 7 Primary Message Systems as anthropologist Edward T. Hall did for the culture study so that a similar two-dimensional grid could be constructed (refer to Table 2nd). And in this manner, we have totally possible 49 Infra (Strategic) Cultural Activities resulting from the various combination of the 7 Mega Elements.

And each Infra (Strategic) Cultural Activity was defined as the critical verbal statement & behaviors related to the interactions between Mega Element A and Mega Element B. And in special cases, i.e. these 7 ones in the diagonal from upper left to lower right are in fact of the critical verbal statement & behaviors related to the definition of the Mega Element concerned. Hence all those 21 Infra (Strategic) Cultural Activities above the diagonal, which are exactly the same with the ones below the diagonal, along with those 7 ones in the diagonal, could make up a full map of Process Safety Culture /PSCC and be collectively deemed as those Congenital Infra (strategic) Cultural Activities of a Process Safety Culture.

	RBPS Core Value 0	**P.S.C.C. (Process Safety Culture) 1**	**Leadership 2**	**Weather/ Climate 3**	**Espionage 4**	**Terrain 5**	**Discipline 6**
RBPS Core Value 0	**00**	**01**	**02**	**03**	**04**	**05**	**06**
P.S.C.C. (Process Safety Culture) 1	**10**	**11**	**12**	**13**	**14**	**15**	**16**
Leadership 2	**20**	**21**	**Leadership with 4 new dimensions 22**	**23**	**24**	**25**	**26**
Weather/ Climate 3	**30**	**31**	**32**	**33**	**34**	**35**	**36**
Espionage 4	**40**	**41**	**42**	**43**	44	**45**	**46**
Terrain 5	**50**	**51**	**52**	**53**	**54**	**55**	**56**
Discipline 6	**60**	**61**	**62**	**63**	**64**	**65**	**66**

A Map of Process Safety Culture /PSCC

Table 2nd

At this stage it seems obvious enough that **Process Safety Practices, which might be far broader and more diverse than universally perceived, are more likely to come from those 28 Congenital Infra (Strategic) Cultural Activities, rather than exactly from CCPS 4 Pillars /20 Elements.**

2.3 *Why Chronic Malnutrition of Process Safety Culture?*

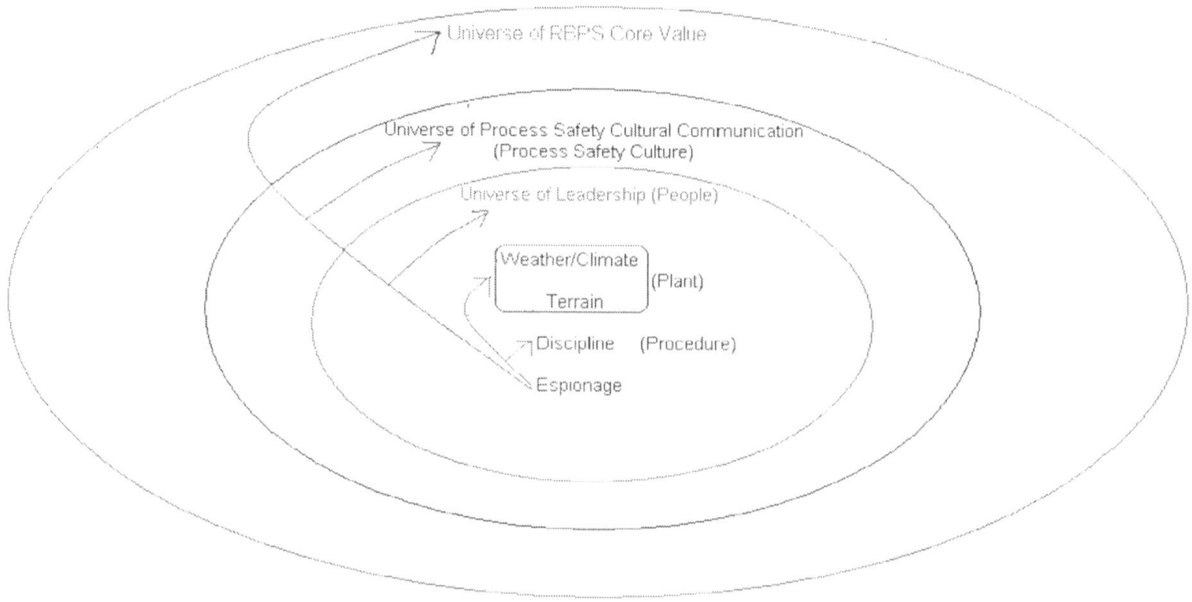

Scheme 2nd

Sun Tzu's *The Art of War* has told us 2500 years ago that war results/outcome could even be foreknown through 5 criteria/facets: Tao (RBPS Core Value), Weather/Climate, Terrain, Leadership and Discipline, let alone post-war root cause analysis. In the author's opinion, only through Espionage could the intelligence of those 5 criteria/facets be reliably acquired, and the author added another one i.e. PSCC, which is in fact an advanced form of Process Safety Culture and whose ultimate purpose is to convey the message of RBPS Core Value. So totally, we have 6 strategic assessment criteria/facets, along with Espionage, within RBPS Egg.

As 21st Element (PSCC) & 22nd Element (RBPS Core Value) were identified (9th GCPS), Leadership upgraded with those 4 new Process Safety (Cultural Communication) Competencies as dictated by PSCC & RBPS Core Value (1st Asia-Pacific Conference on Process Safety) and Weather/Climate newly defined & migrated from post-accident remedy measures to the Root Category of RBPS Core Value, PSCC & Leadership (10th GCPS) Egg model could be further simplified to the scheme 2nd and the rationale of the current actual color code of those 28 Congenital Infra (Strategic) Cultural Activities be as simplified as below:

Whenever creative minority (RBPS Core Value inspired process safety engineers /practitioners) is absent in PSCC Leadership, the Leadership would probably become a prison isolating whatever within from the 2 universes of RBPS Core Value and PSCC as well as from Leadership itself so that all the 18 Infra (Strategic) Cultural Activities in yellow background would be vacant, i.e. those corresponding 18 categories of **Process Safety Cultural Vitamins (PSCVs)** would be missing. And similarly as an incorrect Leadership dictates strategic failure-bonded

Weather/Climate those 4 Infra (Strategic) Cultural Activities /PSCVs in blue background (33, 34, 35 & 36) missing as well. And in process industry we might find that Espionage currently only could be carried out toward Symptoms (Terrain & Discipline); and it seems making sense that as there should be no espionage to Roots (RBPS Core Value, PSCC, Leadership & Weather / Climate) due to the necessity of "state security /counter-intelligence" of an organization the actual Infra (Strategic) Cultural Activity /PSCV (44) would be half missing of a Mega Element Square Unit in isolation. Totally Process Safety Culture, however, loses its 22.5 categories of Infra (Strategic) Cultural Activities among its 28 congenital ones, or, we could say: **the current process industry average Process Safety Culture has a serious chronic malnutrition (with 80.4% due nourishment missing). And the current process industry average Process Safety Culture Health Index (PSCHI) could be deemed at 19.6% or about 1/5.**

Nourishment Missing Ratio: (18+4+0.5)/28 = 80.4%

Process Safety Culture Health Index (PSCHI): 19.6%

3. Chemical Engineering/RBPS Diagnosis /Espionage /Foreknowledge

- CCPS 4th Pillar in light of Bridging Engineering to Humanities

From the analysis above, we know there might be unprecedentedly great potential with Process Safety Culture to be unlocked across the process industry if we really have some key to unlock the prison of Process Safety Culture. And today we really have had 22nd Element (RPS Core Value) as the key, whose brightness is also assured by 21st Element (PSCC). And as dictated by RBPS Core Value & PSCC, we upgraded Leadership with those 4 new Process Safety (Cultural Communication) competencies as below:

Competency in embracing RBPS Core Value;

Competency in upward Process Safety Cultural Communication;

Competency in downward Process Safety Cultural Communication; and

Competency in Process Safety (Cultural Communication) leadership planning, assessment & intervention

And based on the author's rationale that Weather/Climate (external public communication with stakeholders, or Corporate Social Responsibility (CSR)) should not only be deemed as post-accident remedy measures for the damaged public image and reputation, but also, and more importantly, as a contributing (even foreknowledge) factor to the catastrophic accidents, Weather/Climate has immigrated from Symptom category to Root (cause) category, so that finally we arrived at RBPS Socio-Technical Systems and Leadership Media Universe, both of which could be deemed as nicknames of Process Safety Culture.

RBPS Socio-Technical Systems - Leadership Media Universe														
	PS Socio Cultural Communication (PSSCC) Sub-Socio System (7 Elements + Espionage 4 Elements)													
								PS Techno Cultural Communication (PSTCC) (Espionage 4 Elements + 11 Elements) Sub-Technical System						
22 Sub-elements	RBPS Core Value	Process Safety Cultural Communication	(Process Safety Culture)	Process Safety Competency with 4 new dimensions	Workforce Involvement	Compliance with Standards	Stakeholder Outreach	Incident Investigation	Measurement & Metrics	Auditing	Management Review & Continuous Improvement	Process Knowledge Management	Hazard identification & Risk Analysis	9 Sub-elements (in light of organizational effectiveness, *inherently safer chain of command* & logistical support)
7 Mega Elements	Tao/PS Worldview	P.S. Methodology		Leadership		Weather/ Climate		Espionage (RBPS 4th Pillar Learn from Experience)				Terrain (RBPS 2nd Pillar Understand Hazards & Risk)		Discipline (RBPS 3rd Pillar Manage Risk)
CTM*	One Root (4 dimensions)							Traditional Chinese Medical Diagnosis				One Symptom (2 dimensions)		
AIC**	Formal Culture							Informal Culture				Technical Culture		

CTM*: Chinese Traditional Medicine
AIC**: Anthropology & Intercultural Communication

Table 3rd

Another major discovery throughout these efforts of Socio-Technical Systems & Leadership Media Universe is: Process Safety Techno Cultural Communication (PSTCC) is just one ingredient of PSCC; the other **Process Safety Socio Cultural Communication (PSSCC), whose ultimate purpose is to "Bridge between Chemical Engineering and the Humanities".**

3.1 Espionage /Diagnosis /Foreknowledge

– CCPS 4th Pillar in Context of Architecture, Ethical Dimension & Epistemology

Espionage & Foreknowledge From Sun Tzu's perspective, war results and outcome could be strategically learned from and even foretold in terms of 5 critical criteria, i.e. the Tao, the Weather, the Terrain, the Leadership and the Discipline. And Sun Tzu even had a saying like "Thus the reason the farsighted ruler and his superior commander conquer the enemy at every move, and achieve successes far beyond the reach of the common crowd, is foreknowledge. (Roger T. Ames's translation) [11]", from which we could have a strong rationale that foreknowledge is so critical for winning wars or avoiding a war. However, where does high quality foreknowledge come from? It comes from the actual and accurate data of these 5 criteria, which only could be acquired through Espionage. **Sun Tzu really made a great contribution to Military Socio Cultural Communication (MSCC) and achieved 1st successful effort of bridging (military) Engineering to the Humanities.** Similarly, we could say: only through PSCC (PSSCC) could the results and outcomes of the "war" against catastrophic process safety accidents in process industry be truly learned from and reliably foretold.

Traditional Chinese Medical (TCM) Diagnosis Within TCM, there are 4 diagnostic methods, i.e. inspection, listening & olfaction, inquiry, and pulse-taking & palpation, all of which collectively function as medical "espionage" /foreknowledge to the results/outcomes of "war" against diseases so that symptoms (syndromes) and roots (diseases) could be identified and strategically addressed in a wise manner. We even could simplify these methods as a Symptom-Diagnosis-Root approach (refer to Table 3rd), which exactly matches not only with the Egg Architecture of 7 Mega Elements (22 Elements) but also with the mechanism among PSSCC, CCPS 4th Pillar & PSTCC. Put another way, TCM Diagnosis handles not only Symptom but also Root, the latter of which functions as Medical Socio Cultural Communication (Med.SCC); and similarly, **CCPS 4th Pillar's congenital purpose not only for Terrain & Discipline but also for RBPS Core Value, PSCC, Leadership and Weather/Climate.**

Chemical Engineering (RBPS) Diagnosis – Architecture, Ethical Dimension & Epistemology

Toward Bridging Chemical Engineering to the Humanities, Architecture, Ethical Dimension and Epistemology might collectively accommodate as rationale for some unique role CCPS 4th Pillar (Learn from Experience) would probably play in future Chemical Engineering and Risk Based Process Safety framework /Corridor. Since 2007 when CCPS 4 Pillars (20 Elements) were presented to process industry, 21st Element (PSCC), 22nd Element (RBPS Core Value) and their Egg model at 9th GCPS might have been 1st important reflection on the architecture of CCPS 4 Pillars (20 Elements). And an Inevitable Ethical Dimension of CCPS Vision 20/20 and its inherent correlation with Operational Discipline (OD) at 12th GCPS could be deemed as further efforts of exploration with architecture redesigning or upgradation.

And all of these efforts and attempts mentioned above actually concern one philosophical branch "Epistemology of Engineering knowledge", which "studies the nature of knowledge, justification, and the rationality of belief." [12]. In another words, reflections on RBPS Architecture & Ethical Dimension and Epistemology of Chemical Engineering knowledge would lay down the principles for the congenital purpose of CCPS 4th Pillar and for how truly effective each of those 4 Elements of CCPS 4 Pillar should be, rather than CCPS 4th Pillar and its 4 Elements themselves. At a couple of occasions, the author made some statement as below: "Without activated Process Safety Socio Cultural Communication (PSSCC), Process Safety Techno Cultural Communication (PSTCC) is some kind of meaningful work and simultaneously some kind of meaningless work i.e. mass brainwashing about the true big picture.". To put another way: **without reflections on RBPS Architecture & Ethical Dimension and Epistemology of Knowledge of Chemical Engineering, CCPS 4th Pillar (Learn from Experience) would be the diagnosis only to Symptoms (Terrain & Discipline), ultimately covering up those true root causes (RBPS Core Value, PSCC, Leadership & Weather/Climate).**

3.2 Investigation, Metrics, Auditing & Management Review – Unlocking the Potentials

With due reflections /commitment on Architecture, Ethical Dimension & Epistemology of Knowledge of Chemical Engineering, it is optimistic that those Congenital 28 Infra (Strategic) Cultural Activities of Process Safety Culture would all be activated and Process Safety Culture completely recover from its chronic malnutrition.

Those 16 questions within Table 4th below could be traced back to those 4 Infra (Strategic) Cultural Activities: 04, 14, 24 & 34. At 12th GCPS, the author made statement like: those 16 questions with "Yes", especially those 4 ones within the column of "Incident Investigation", would be of the true characteristics of companies with great process safety performance or Inherently Safer Operational Discipline (ISOD) in the future, and should be bravely faced to by both senior executives and board of directors.

	Chemical Engineering Diagnosis (CCPS 4th Pillar - Learn from Experience)			
	Incident Investigation	**Measurements & Metrics**	**Auditing**	**Management Review**
RBPS C.V.	**If RBPS Core Value is examined as a contributing factor to incident?** YES	**If RBPS Core Value is an item within process safety KPI?** YES	**If RBPS Core Value is an item within auditing?** YES	**If RBPS Core Value is an item within management review?** YES
P.S.C.C. *	**If PSCC is examined as a contributing factor to incident?** YES	**If PSCC is an item within process safety KPI?** YES	**If PSCC is an item within auditing?** YES	**If PSCC is an item within management review?** YES
Leadership with 4 new PSCC competencies	**If Leadership is examined as a contributing factor to incident?** YES	**If Leadership is an item within process safety KPI?** YES	**If Leadership is an item within auditing?** YES	**If Leadership is an item within management review?** YES
Weather/Climate	**If Weather/Climate is examined as a contributing factor to incident?** YES	**If Weather/Climate is an item within process safety KPI?** YES	**If Weather/Climate is an item within auditing?** YES	**If Weather/Climate is an item within management review?** YES

*** P.S.C.C. is the author's addition in light of anthropology and intercultural communication**

Table 4th

Incident Investigation

Currently in process industry, many corporations have their own process safety management system, of which nearly each Element would have a champion & coordinator assigned for the corresponding program. Put Element Incident Investigation as example: having a champion and coordinator for the Incident Investigation Program would enhance consistency. However, who would be the champion and coordinator for the consistency of due reflections on Architecture & Ethical Dimension of RBPS and on Epistemology of Chemical Engineering Knowledge? In the author's opinion, for corporations, the answer probably should be CEO; for CSB, the answer probably Chairperson. And not only should Table 4th be included as an integral part of future Incident Investigation report template, but also Table 2nd, Table 3rd & Scheme 2nd referred as a couple of versions of true big pictures, in context of which the potential of Element Incident Investigation (actually whole CCPS 4th Pillar) could be fully unlocked.

The process industry and most process safety practitioners could speculate how meaningful the 7 Mega Elements Framework would be if all of those catastrophic (process safety) accidents that occurred such as Bhopal, Piper Alpha, BP Macondo Disaster, Tianjin Port Big Explosion etc. are "re-investigated" within the framework of 7 Mega Elements and its 28 Infra Strategic Cultural Activities. And then we would be optimistic that the lessons with most catastrophic accidents that occurred could be truly learned from and corporate memory loss prevented.

Measurement & Metrics

In the latest years, CCPS and API published their relevant Process Safety Metrics /Leading & Lagging Indicators [13], which might be assumed the best way to provide a complete picture of effectiveness of process safety management system, and which might have contributed to some degree to the process safety performance improvement in chemical and petroleum industries.

In process industry for many years there has been a popular saying "You don't improve what you don't measure". However, the problem now is: do we really want to measure and hence to improve RBPS Core value, PSCC, Leadership, Weather/Climate and their relevant 22.5 Infra (Strategic) Cultural Activities? Or if we really want to diagnose out the root(s) and ultimately cure the chronic malnutrition of Process Safety Culture? Hence in author's opinion, CCPS (and API) Process Safety Metrics / Leading & Lagging Indicators should be complemented by those new Process Safety Metrics (all Leading Indicators) of 22.5 Infra (Strategic) Cultural Activities.

In future process industry, these 7 Mega Elements might function as an unprecedentedly powerful vocabulary of Metrics for strategic communication, improvement, diagnosis & foreknowledge among BoD, Executive Management, Legal, HR, CSR, Operation Managers, process safety engineers, frontline supervisor and operators etc. And this strategic communication and diagnosis probably need to start with new definitions of those 5 Mega Elements (RBPS Core Value, PSCC, Leadership, Weather/Climate & Espionage), i.e. those Infra Strategic Cultural Activities: 00, 11, 22, 33 & 44.

Auditing

In process industry, we understand the process safety audit's purpose is for periodic evaluations of all RBPS Elements and managing the audit findings and resolutions. However, it seems we have been neglecting the evaluation of the current RBPS's Architecture since 2007 when CCPS 4 Pillars were first time presented. In the author's opinion, during either 1st party, 2nd party or 3rd party audit, we need to monitor not the degree of maturity not of each element of RBPS 4 Pillars but also, more importantly, the degree of maturity of current RBPS Architecture. In another word, should RBPS Core Value, PSCC, Leadership & Weather/Climate be included as auditing objects or items within future Process Safety Audit template.

In future process industry, these 7 Mega Elements might function as an unprecedentedly powerful vocabulary of Auditing for strategic communication, improvement, diagnosis & foreknowledge among BoD, Executive Management, Legal, HR, CSR, Operation Managers, process safety engineers, frontline supervisor and operators etc. And this strategic communication and diagnosis probably need to start with new definitions of those 5 Mega Elements (RBPS Core Value, PSCC, Leadership, Weather/Climate & Espionage), i.e. those Infra Strategic Cultural Activities: 00, 11, 22, 33 & 44.

Management Review & Continuous Improvement

Similarly we could say: Architecture & Ethical Dimension of CCPS 4 Pillars and Epistemology of Chemical Engineering Knowledge shall be reviewed first, without which whatever Continuous Improvement (CI) would remain meaningless or even worse.

In future process industry, these 7 Mega Elements might function as an unprecedentedly powerful vocabulary of CI for strategic communication, diagnosis & foreknowledge among BoD, Executive Management, Legal, HR, CSR, Operation Managers, process safety engineers, frontline supervisor and operators etc. And this strategic communication and diagnosis probably need to start with new definitions of those 5 Mega Elements (RBPS Core Value, PSCC, Leadership, Weather/Climate & Espionage), i.e. those Infra (Strategic) Cultural Activities: 00, 11, 22, 33 & 44.

4 Dimensions of Chemical Engineering Diagnosis

From the above analysis, we may be aware that **those 4 Elements of CCPS 4th Pillar collectively play the role of espionage /diagnosis /foreknowledge of Chemical Engineering /RBPS and might be deemed as 4 dimensions of Chemical Engineering Diagnosis, whose ultimate purpose is to diagnose the Roots (those 4 Mega Elements or Process Safety Socio Cultural Communication). And the RBPS Architecture & Ethical Dimension and the Epistemology of Chemical Engineering Knowledge make up the context, in which the purpose and potential of CCPS 4th Pillar be revived and unlocked.**

4. Conclusion

In this paper, the author first gave a brief background introduction of Egg model and RBPS Core Value. Then the author rationalized why Process Safety Culture's Chronic Malnutrition and a map of those 28 Congenital Infra (Strategic) Cultural Activities. At end, the author characterized CCPS 4th Pillar (Learn from Experience) as Chemical Engineering Diagnosis & Knowledge through reflections both on RBPS Architecture & Ethical Dimension and on Epistemology of Engineering Knowledge so that those 4 Elements of CCPS 4th Pillar (Learn from Experience) become 4 dimensions of Chemical Engineering Diagnosis and their potentials be unlocked.

5. References

[1] Center for Chemical Process Safety, Guidelines for Risk Based Process Safety, John Wiley & Sons, Inc., 2007

[2] Kadri, Shakeel H. and Jones, David W., Nurturing a Strong Process Safety Culture, 20th Annual CCPS International Conference, AIChE Spring Meeting, Atla, April 2005

[3] B.P. U.S. Refineries Independent Safety Review Panel, The Report of the B.P. U.S. Refineries Independent Safety Review Panel, 2007

[4] Zhang, L., An Upgraded Definition of Risk Based Process Safety Core Value and Its Impact in Light of Anthropology & Intercultural Communication on the Future Evolution of Process Safety Culture & Management, 2013 AIChE Spring Meeting and 9th Global Congress on Process Safety, San Antonio, April 2013

[5] Zhang, L., Benchmarking Risk Based Process Safety Egg (RBPS Temple+2 Universes) against The Art of War, 1st CCPS Asia-Pacific Conference on Process Safety, Qingdao, Sept 2013

[6] Zhang, L., How Committed to Process Safety is Committed Enough? – Doing Process Safety Cultural Communication (PSCC) as the Inevitable Ethical Dimension for Vision 20/20, 2016 AIChE Spring Meeting and 12th Global Congress on Process Safety, Houston, April 2016

[7] Zhang, L., Redesigning Architecture of Risk Based Process Safety in Context of Emerging Ethical Dimension of Vision 20/20 – from 4 Pillars to 7 Mega Elements, 2017 AIChE Spring Meeting and 13th Global Congress on Process Safety, San Antonio , Mar 2017

[8] Bucciarelli, Louis L., Engineering Philosophy, Delft: DUP Satellite, 2003

[9] Hall, Edward T., The Silent Language, Anchor Books, 1990

[10] Zhang, L., People Means Leadership, NOT Simply mean People – 4 New Dimensions of Process Safety Competency, 2014 AIChE Spring Meeting and 10th Global Congress on Process Safety, New Orleans, April 2014

[11] Sun-Tzu and Ames, Roger T., Sun-tzu: The Art of Warfare, Ballatime Books, 1993

[12] https://en.m.wikipedia.org/wiki/Epistemology

[13] CCPS, Process Safety Leading and Lagging Metrics, 2011

506204 Human Factors Engineering Checklist for Shale Hydrocarbon Owner-Operators

Abstract:

Shale rock formations lie more than a mile beneath the earth's surface and contain natural gas deposits trapped inside tiny spaces within the rock. The hydrocarbons are extracted from the shale rock using a process known as hydraulic fracturing or "fracking." This involves first drilling both vertically and then horizontally to the depth of the formation to expose more of the shale. Once the hole is drilled and the formation is reached, gallons of water, sand, and manmade chemicals are injected into the well under high pressure. The combination of the high pressure fluids and materials create small factures in the formation releasing the hydrocarbons.

Fracking is decades old technology that has gained attention from the public eye in recent years due to environmental and safety concerns. The environmental challenges range from groundwater contamination and air pollution to increasing traffic. As a result, Oil and Gas companies follow Best Management Practices (BMPs) for construction and operation of the drilling site including roads for the transport of heavy equipment, leveling of the site, structures for erosion control, the excavation of pits to hold drilling fluids, and the placement of racks to hold the drill pipe and casings. The dangers present for hydraulic fracturing workers include vehicle accidents; respirable silica sand; explosions and fires; and exposure to chemicals. While United States government organizations are working with shale producers to develop and implement environmental and health regulations for activities at shale hydrocarbon exploration and production sites, the safety culture of the worker in some companies tends to be overshadowed by the pressures to capitalize on the opportunities. The pressure to capitalize on the shale production opportunities is in part driven by the break-even production costs, which has decreased from $80/bbl to $35/bbl since 2013. Therefore the need for shale producers to stay vigilant and address corporate safety culture is ever more important to avoid unnecessary injuries and accidents.

The National Institute of Occupational Safety and Health (NIOSH) reported between 2003 and 2009 that in the oil and gas extraction industry there is an annual occupational fatality rate of 27.5 per 100,000 workers, which is seven times higher for all other workers in the United States. Additionally, in 2013, the US Bureau of Labor and Statistics reported a 23 percent annual rise in Oil and Gas fatalities. The Occupational Safety and Health Administration (OSHA), a division of the United States Department of Labor, assures safe and healthful working conditions for working men and women by setting and enforcing standards. These standards require employers to provide a safe working environment, including but not limited to shale production sites. One component of these standards is 29 CFR 1910 which set the minimum human factors design requirements related to the typical operations and maintenance activities conducted by any Owner of shale production companies.

Owners of shale exploration and production companies are mostly unfamiliar with Human Factors Engineering requirements because Shale Exploration and Operation opportunities are often fast tracked to increase profitability. Consequently, issues surrounding worker safety may not be realized until it's too late or may be overlooked all together. Therefore, the need to develop new approaches or improve on existing ones and consider mechanisms to incorporate human factors requirements to avoid negative effects associated with fracking is even more important.

A checklist is presented whereby the Owners and Operators of Shale exploration and production facilities can ensure that the minimum human factors requirements are integrated into their machinery units before they are put into operation. The checklist is intended to provide a uniform and economical approach to identify safety concerns in remote worksites based on, and supported by, various industry codes and standards. When well-implemented and managed, the checklist allows for a systematic identification of issues, after which the Owner-Operators can use the results to make trade-off decisions or find alternative solutions. An added benefit from using this checklist is that it can aids in early identification of risks that can potentially compromise the safety and integrity of personnel and the overall achievement of enterprise objectives and goals.

18 AIChE Spring Meeting
14TH GLOBAL CONGRESS ON PROCESS SAFETY
ORLANDO FLORIDA 2018

Research on Emergency Response Evaluation of Major Process Safety Accidents

Presenter/Principal Author
PhD. Candidate:Wei Zheng
College of Mechanical and Transportation Engineering,
China University of Petroleum (Beijing), Beijing 102249, PR China
E-mail Address: safetyzhengwei@163.com

Other author :Professor Jian Shuai
College of Mechanical and Transportation Engineering,
China University of Petroleum (Beijing), Beijing 102249, PR China

Prepared for Presentation at
American Institute of Chemical Engineers
2018 Spring Meeting and 14th Global Congress on Process Safety
Orlando, Florida
April 22 – 25, 2018

Research on Emergency Response Evaluation of Major Process Safety Accidents

Wei Zheng*, Jian Shuai
College of Mechanical and Transportation Engineering, China University of Petroleum (Beijing), Beijing 102249, PR China
(Corresponding author PhD. Candidate *Tel: +86 13466733814. Fax: +861089733772. E-mail Address: safetyzhengwei@163.com)

Keywords: process safety; emergency response; AHP; fuzzy comprehensive evaluation.

Abstract: The major accident emergency is the last line of defense for process safety control. In this paper, an evaluation index system for emergency response capacity of chemical accidents based analytic hierarchy process(AHP) method was established, according to the four different stages of major accident emergency in chemical process including precaution phase, preparation phase, response phase and recovery phase. By analyzing the weight of different index in evaluation index system, fuzzy comprehensive evaluation method was used to solve the quantification of the emergency response ability of major accidents as well as the protection analysis more completely of the chemical process. Through the study of these comprehensive safety technologies and the implementation of relevant technical measures, the emergency response capability of major chemical accidents can be improved and the safety risks during the production of hazardous chemicals can be reduced.

The major accident emergency system is the last defence of process safety control.To improve the ability to respond to major accidents and establish a quick and efficient emergency response system is helpful to prevent and control accidents and to reduce the risk of the process safety accidents[1].

When enterprises face a variety of emergency situations, different emergency services lead to very different consequences for the enterprises[2]. At present, one of the important reasons for the frequent accidents and the serious consequences of casualties in China is that the emergency response system is not perfect and the emergency response capability of enterprises is insufficient.

The purpose of the emergency capability assessment is to find out the problems and weaknesses in the normal work of emergency management and to improve and perfect the enterprise emergency management system[3]. In order to strengthen and perfect the emergency management work, enterprises should focus on improving the ability to prevent and handle unexpected incidents and effectively supervise, inspect and promote the daily emergency management work according to the characteristics of internal and external assessment so as to enhance emergency response management capabilities, and comprehensively strengthen the emergency management system[4].

1. Evaluation index system of major accident emergency response

Emergency response refers to the ability of enterprises to deal with emergency prevention and control of unexpected incidents. Major accident emergency response refers to the ability of enterprises to deal with major accident prevention and control[5]. A major accident emergency work is a long-term and continuous process, including four stages of prevention, preparation, response and recovery.

According to the characteristics and content of the emergency system[6], the fuzzy comprehensive evaluation based analytic hierarchy process(AHP) is used to evaluate the emergency response capability of major accidents in the chemical industry and the effect of the emergency system on the chemical process hazard control.

According to the main contents of major accident emergency, the analysis of the major factors that affect the emergency ability of major accidents in each of the four stages was performed[7]. Then, the evaluation index system of major accident emergency response based on these analysis results was established, shown in table 1.

Table 1 evaluation index system of major process safety emergency response capability

Target layer	**evaluation stage (Guidelines layer)**	**Evaluation stage (Indicator layer)**
Major Process Safety Incident Emergency Ability Evaluation Index System A	Prevention phase B_1	Collect regulations and standards B_{11}
		Effective safety management B_{12}
		Technical measures for safety protection B_{13}
		Establish monitoring and early warning B_{14}
		Implement of disaster insurance B_{15}
		Conduct safety research B_{16}
	Preparation Phase B_2	Complete emergency response agencies B_{21}
		Scientific and reasonable emergency plan B_{22}
		Emergency notification and alarm system B_{23}
		Adequate and comprehensive emergency resources B_{24}
		Sign mutual aid agreement B_{25}
		Regular emergency training and exercises B_{26}
	Response phase B_3	Enterprises' quick and efficient implementation B_{31}
		Personnel correct, quick emergency response B_{32}
		Promptly report to government agencies B_{33}
		Effective emergency assistance B_{34}
		Promptly issue an emergency announcement B_{35}
		Effective protection measures B_{36}
	Recovery phase B_4	Restore normal order as soon as possible B_{41}
		Conduct damage assessment B_{42}
		Quick disinfection and decontamination B_{43}
		Start post-disaster reconstruction B_{44}
		Insurance compensation B_{45}
		Plan review and experiences summary B_{46}

2. Theories AHP and fuzzy comprehensive evaluation

AHP is an easy way to make decisions on the more ambiguous or more complex decision-making problems by means of combination of qualitative and quantitative analysis[8]. In particular, it can qualify the experience and judgment of decision-makers and show people's thinking process hierarchically through comparing the relevant factors and testing the rationality of the comparison results one by one so as to provide a more convincing evidence. Many decision-making problems usually appear as scheduling problems of a series of programs, which can be solved by AHP[9]. The basic steps of AHP are as follows: (1) establishing the hierarchical structure model; (2)constructing the judgment matrix; (3) checking the consistency of the matrix; (4) calculating the weight vector; (5) calculating the combination weight vector; (6) Overall consistency test. For major decision-making issues, the general consistency test can be considered acceptable only if the overall agreement ratio CR was small enough [10].

Fuzzy comprehensive evaluation makes use of the single-factor evaluation results related to evaluation objects to form the corresponding evaluation matrix, and uses the weight factors which determine the degree of importance of each factor to do fuzzy transformation to finally obtain the evaluation results of the evaluation objects[11].

Fuzzy comprehensive evaluation involves the following three factors[12]:

(1) factor set $U=\{u_1,u_2,\cdots u_n\}$

(2) decision-making set $V=\{v_1,v_2\cdots v_m\}$, such as { higher, high··· lower}, also called comment set

(3) Single-factor evaluation, which is a fuzzy mapping from U to V, $f:U\rightarrow V$, that is

$$u_i \rightarrow f(u_i)=(r_{i1},r_{i2}\cdots,r_{in})\in F(V)$$

fuzzy mapping f can determine a fuzzy relationship:

$$R_f \in F(U\times V)$$

$$R_f(u_i,v_i)=f(u_i)(v_i)=r_{ij}$$

Thus the fuzzy matrix R_f can be represented by a fuzzy matrix $R\in M_{n\times m}$

$$R=\begin{vmatrix} r_{11} & r_{12} & \cdots & r_{1m} \\ r_{21} & r_{22} & \cdots & r_{2m} \\ \cdots & \cdots & \cdots & \cdots \\ r_{n1} & r_{n2} & \cdots & r_{nm} \end{vmatrix}$$

Thus (U,V,R) forms a comprehensive evaluation model, in which a given subset of $A\in u_{i\times n}$ is a fuzzy subset of v (the weight of the factor), and the comprehensive evaluation $B=A\cdot R\in u_{i\times n}$, which is obtained from the weigh of various factors according to A, is a subset of V.

3. Fuzzy comprehensive evaluation application in major process safety accident emergency response capability

3.1 Set up factor set

Determine the enterprise emergency management capability evaluation index system A, B_1, B_2, B_3, B_4 in the table 1.

3.2 Set up comment set

In accordance with the size of emergency management capabilities, from high to low rating for the five levels.

The comment set is: $V = \{V_1, V_2, V_3, V_4, V_5\}$

Where V_1 = higher, V_2 = high, V_3 = medium, V_4 = low, V_5 = lower, for all kinds of levels of enterprises emergency response capability from high to low.

3.3 Set up judgment criteria of evaluation index

The judgment criteria of the various evaluation indices are shown in Table 2, Table 3, Table 4 and Table 5.

Table 2 Prevention stage indices B_1

Indices	Description
Collect regulations and standards B_{11}	The more comprehensive collection and utilization of relevant laws, regulations and standards, the higher the indicator value
Effective security management B_{12}	The more comprehensive and effective security management, the greater the impact and the higher the indicator value
Technical measures for Safety Protection B_{13}	The more comprehensive and detailed safety precautions, the higher the indicator value
Establish monitoring and early warning B_{14}	Establishing effective and good monitoring and early warning system leads to a higher indicator value
Implementation of disaster inspection B_{15}	Implementation of disaster insurance to transfer risk, the higher the index value
Conduct safety research B_{16}	The deeper the safety research work has been done, the higher the indicator value

Table 3 Preparation stage indices B_2

Indices	Description
Complete emergency response agencies B_{21}	The more reasonably the emergency agencies is settled, the higher the index value
Scientific and reasonable emergency plan B_{22}	The more scientifically, reasonably and practically the emergency plan is prepared, the higher the index value
Emergency notification and alarm system B_{23}	The more effective emergency and alarm system, the higher the indicator value
Adequate and comprehensive emergency resources B_{24}	The more adequate and effective resources, personnel, technology and other emergency resources, the higher the index value
Sign mutual aid agreement B_{25}	If mutual aid units signed an agreement to improve emergency response capacity, the indicator value becomes higher
Regular emergency training and exercises B_{26}	If regularly organize emergency training and exercises has been performed, the indicator value becomes higher

Table 4 Preparation stage indies B_3

Indices	Description
Quickly and efficiently execute the emergency B_{31}	The indicator value becomes higher, when the enterprises take a quick and efficient emergency action at the first time of the accident
Correct and quick emergency response B_{32}	The indicator value becomes higher, when the personnel at the first scene of accident take a correct and quick emergency response
Promptly report to government agencies B_{33}	The indicator value becomes higher, when the accident is reported to government agencies at the very first time
Effective emergency assistance B_{34}	At the time of the accident, obtaining quick and effective emergency assistance leads to a higher index value
Promptly issue an emergency announcement B_{35}	Timely release of emergency bulletins is conducive to emergency response and to a higher index value
Effective protection measures B_{36}	The more effective individual protection, evacuation and asylum, search and rescue measures, the higher the indicator value

Table 5 Recovery stage indices B_4

Indices	Description
Restore normal order as soon as possible B_{41}	After the accident is eliminated, the sooner the return to normal order, the higher the indicator value
Conduct a damage assessment B_{42}	After the accident is eliminated, the more promptly the damage evaluation is conducted, the higher the index value
Quick disinfection and decontamination of water B_{43}	The indicator value becomes higher, when decontamination is completed as soon as possible
Start post-disaster reconstruction B_{44}	The sooner the post-disaster reconstruction is started, the higher the indicator value
Insurance compensation B_{45}	Quick and full insurance compensation leads to a higher index value
Plan review and experiences summary B_{46}	Improving the plan and summing up experiences in time lead to a higher index value

3.4 Weight calculation and consistency test

The invited expects construct a judgment matrix according to the importance of each index. the results of calculating the weight vector and conducting the consistency index (CI) test through the constructed judgment matrix are shown in Table 6.

In order to verify the rationality of the judgment matrix, it is necessary to test whether there is a serious inconsistency in the judgment matrix from the maximum eigenvector $\lambda_{\max}$. The consistency test can be completed by solving the consistency ratio (CR).

$$\mathrm{CI} = (\lambda_{\max} - n)/(n-1)$$

$$\mathrm{CR} = \mathrm{CI}/RI$$

Table 6 Consistency index test of judgment matrix

Judgment matrix	n	$\lambda_{\max}$	CI	RI	CR	consistency test result
A	4	4.2299	0.0766	0.9000	0.0851	Pass
B1	6	6.4355	0.0871	1.2400	0.0702	Pass
B2	6	6.3779	0.0756	1.2400	0.0609	Pass
B3	6	6.5694	0.1139	1.2400	0.0918	Pass
B4	6	6.4606	0.0921	1.2400	0.0743	Pass

As can be seen from the above table 6, the CR of all the judgment matrixes are less than 0.1, all of which pass the consistency test.

The weight vector of each indicator is shown in Table 7, which can be used to calculate the B1, B2, B3, B4 index layer combination weight vector. The calculation results represent that CI (3) is 0.1023, RI (3) is 1.2398, CR (3) is 0.0825, and CR (2) is 0.0851, all of which passed the consistency test. The CR * is 0.1676, which is acceptable.

Table 7 Index weight Vector of each layer of evaluation index system

Judgment matrix	Weight vector w					
A	0.0558	0.1533	0.5976	0.1931		
B1	0.0703	0.3990	0.2782	0.1577	0.0536	0.0409
B2	0.1287	0.2471	0.0574	0.4404	0.0324	0.0938
B3	0.2095	0.4262	0.1130	0.1341	0.0354	0.0815
B4	0.4182	0.0871	0.2356	0.1124	0.0918	0.0546

3.5 Evaluation of the benchmark value and the determination of the results

In the evaluation index system of emergency response capability of major accidents, the defined range of evaluation index is from 0 to100. The determination and description of the standard values are shown in Table 8.

Table 8 The standard values of evaluation index

Evaluation index value	**emergency management capacity**	**emergency level**
90-100	higher	Very high emergency capacity, maintain this level
80-90	high	High emergency capacity, several projects need to be optimized
70-80	medium	Medium emergency capacity, needs to be enhanced
60-70	low	low emergency capacity, need to be improved urgently
low Less than 60	lower	very low emergency capacity, need to be restructured immediately

3.5 Case study

In a case of chemical plant, several experts were invited to assign each of the indicators in Table1 according to the actual situation of emergency in the chemical plant, evaluate the emergency response capability of major accidents through the established evaluation model and give a evaluation score of this enterprise by fuzzy comprehensive evaluation based AHP. The emergency response evaluation score was 85.10, which indicated that this enterprise's emergency response capability was high. However, it was also suggested that the enterprise should strengthen the training of emergency personnel and organize the emergency drills regularly.

4. Conclusion

The evaluation index system for emergency response capacity of chemical accidents was established based on Analytic Hierarchy Process (AHP), which included various indictors in four different stages named precaution phase, preparation phase, response phase and recovery phase. According to the characteristics and content of the emergency system, the fuzzy comprehensive evaluation was used to evaluate the emergency response capability of major accidents in the in process safety. A case of chemical plate was studied using this AHP based fuzzy comprehensive evaluation method and its emergency response evaluation score was 85.10, which indicated that this enterprise's emergency response capability was high. The AHP based fuzzy comprehensive evaluation method can solve the problem of quantifying the protective functions of major accidents and strengthen the control of process safety more completely.

Reference

[1] Ken Bingham, layer of Protection Analysis for Rigorous Risk Evaluation[J], Process Safety,2004, 9:26-38.

[2] Georage D Haddow, Jane A Bullock.Introduce to Emergency Management[M],TX:butter worth-Heniemann,2003.

[3] Fang JS, Mannan MS. Value at risk perspective on layers of protection analysis[J], Process Safety and Environmental Protection. Vol.85(2007):85-87.

[4] Mostafizur Rahman, Anna-Mari heikkila, Markku hurme. Comparison of inherent safety indices in process concept evaluation [J], Journal of Loss Prevention in the process industries, 2005, 18:327-334.

[5] Roy E Sanders.Practicing chemical process safety; a look at the layer of protection[J], Journal of Hazardous Materials, 2004,145:141-147.

[6] Thomas L.Saaty. A note on the AHP and expected value theory[J], Socio-Economic Planning Sciences,1986,6(20):397-398.

[7] Saaty Thomas L,Tran Liem T.On the invalidity of fuzzifying numerical judgments in the Analytic Hierarchy Process[J],Mathematical and computer modelling, 2007m, 46(10):962-975.

[8] C.Cacciabue, Human factors impact on risks analysis of complex system, J.Hazard Mater.71(2000)101-116.

[9] CCPS, layers of protection analysis: Simplified Process Risk Assessment [M],AIChE, New York, 2001.

[10] Chunyang wei, William J, Rogers M,Sam Mannan. Layer of protection analysis for reactive chemical risk assessment[J], Journal of Hazardous Materials, 2008,159:19-24.

[11] Tseng J M, Liu M Y, Chang R H. Emergency response plan of chlorine gas for process plants in Taiwan[J], Journal of loss prevention in the Process Industries, 2008, 21:393-399.

[12] Saaty Thomas L. Rank from comparisons and from ratings in the analytic hierarchy/network processes[J],European Journal of Operational Research,2006,1(2):557-570.

515622 Exploration and Research on Safety Management of Scientific Research Laboratories in Colleges and Universities

Abstract:

Scientific research laboratories usually have the characteristics of higher risk of experimental environment and conditions,more complicated experimental items and contents,more unknown factors in the experimental process,and most students independently carry out experiments. Therefore, potential laboratories safety issues are even more pronounced. Based on the operational characteristics of scientific research laboratories and the essential causes of accidents in recent years, the paper analyzes the causes, starting conditions and control measures that may cause laboratory accidents and puts forward the establishment of a systematic safety management system and an operational safety standards system of scientific research laboratories,which strives to be intrinsically safe and prevents laboratories accidents in the first place.

The influence of initial pressure on the explosion characteristics of 2, 3, 3, 3 – tetrafluoropropene

Yuxiao Li [a], Yi Liu [b, c]

[a] Center for Safety, Environmental Protection and Energy-Conservation Technology, China University of Petroleum, Qingdao 266580, China;

[b] College of Chemical Engineering, China University of Petroleum (East China), Qingdao, 266580, China;

[c] State Key Laboratory of Heavy Oil Processing, China University of Petroleum (East China), Qingdao 266580, China;

Keywords: refrigerant;r1234yf;maximum explosion pressure;maximum rate of explosion pressure rise

Abstract

Refrigerants are used more and more widespread, but there are potential safety hazards in the use of refrigerants. In particular, the possible explosion of refrigerants caused the world's refrigeration industry attach great importance. In this paper, 20L ball was used to study the explosive properties of refrigerant 2, 3, 3, 3 – tetrafluoropropene at different initial pressures. The experimental results show that the maximum explosion pressure increases with the increase of the pressure. When the pressure is constant, the maximum explosion pressure reaches the maximum value of 7.6%. In addition, this article also calculated the explosive index at various concentrations to indicate the size of the explosive power.

1. Introduction

With the continuous improvement of society and the improvement of living standards, air conditioners and refrigerators have become very common in our life. According to the statistics, the total output of air conditioners and refrigerators has been increasing year by year. In 2015, the total output of air conditioners and refrigerators reached 221.93 million. The continuous increase in output of refrigeration systems will inevitably lead to a ever-increasing demand for refrigerants. In addition, due to people's growing awareness of environmental protection, protect the ozone layer and curb global warming has attracted attention worldwide, which also contributed to the constant upgrading of refrigerants, the international community has banned the use of CFCs, while gradually reducing the use of HCFCs.

Ideal refrigerant should have zero ozone depletion potential, a small potential greenhouse effect, good thermal performance, operating pressure suitable, non-toxic, non-flammable, good material compatibility, good solvent resistance, low cost. However, no such refrigerant has been found up to now, and most of the refrigerants are flammable. When mixed with air, it is possible to form explosive mixture gas. When an ignition source, a high-temperature substance or an oxidant is encountered, it has the risk of explosion, so many studies has made to research the characteristics of the explosive and combustion of the refrigerant [1].

Zhang et al. test different concentrations of R32 and R290 explosion pressure and the maximum explosion pressure rise rate through 20L ball [2]. Razus et al. studied the variation of the explosion pressure of propane air mixture with initial temperature and initial pressure in a closed container [3]. Kondo et al. studied the effect of humidity on the combustion characteristics of the refrigerant in a 12L spherical glass flask with ASHRAE method [4]. In addition, he studied the explosion limits of ammonia with R1234yf, R1234ze, R134a and R125 [5], while the influence of pressures on the explosion limits of R1234yf, R32 and methane was also studied [6]. Chen et al. studied the combustion characteristics of the binary mixture DME/R134a, while calculated the explosion limits and the critical explosion suppression curve of the ternary mixture DME/R1234yf/R134a [7]. Papas et al. determined the laminar flame speeds of R32/air at stoichiometric concentrations and a binary mixture of R32/R1234yf and

R290/R1234yf with air [8]. Moghaddas et al. studied laminar flame speeds and flame structures of R32/air and R152a/air mixtures at different pressure and temperature conditions in spherical and cylindrical vessels, demonstrating that laminar flame speeds of these two refrigerants have a significant dependence on temperature and pressure [9]. Pagliaro et al. determined the laminar flame speeds of premixed CH4-air and C3H8-air flame by adding C6F12O and compared with the flame speed of premixed flame with CF3Br-suppressed[10]. Igarashi et al. tested the flame speeds of R32/air and R32/R134a / air using vertical-tube method [11]and closed-vessel method[12].

R1234yf has zero ODP, low GWP as new refrigeration. This paper study the explosion characteristics of R1234yf through 20L ball under different concentration and pressure conditions.

2. Experimental

The experimental apparatus is suitable for the explosion parameters determination of gas or dust under special conditions. It includes microcomputer control system, explosion vessel (20L ball), vacuum system, automatic ignition device, automatic gas distribution system, wireless transmission system and computer monitoring system.

The pressure measurement range of experimental equipment is as follows. Explosion pressure: 0 ~ 4 MPa (absolute pressure); Sample chamber pressure: 0 ~ 4 MPa (absolute pressure); Gas distribution pressure: 0 ~ 1 MPa (absolute pressure), accuracy: 0.1% FS; The range of temperature measurement and control: Ambient temperature ~ 200 °C; Resolution: 0.1 °C; The interval of explosive pressure sampling: 0.2 ms; Mixing method: 100 ms rapid stirring; Cooling method: Spontaneous combustion cooling; The volume of explosive chamber: 20L (double stainless steel ball); Ignition method: Chemical ignition and High-pressure pulse ignition; Gas distribution number: two vacuum proportional gas distribution; Gas distribution accuracy: ± 0.1%.

In this paper, we chose nine concentration points of 6.8%, 7.0%, 7.2%, 7.4%, 7.6%, 7.8%, 8.0%, 9.0% and 10.0%. The initial pressures were set to 0.1MPa, 0.13MPa, 0.15MPa and 0.2MPa. The explosion pressure and explosion pressure rise rate of R1234yf at different initial pressures and concentrations were measured at 25 °C.

3. Experiment analysis

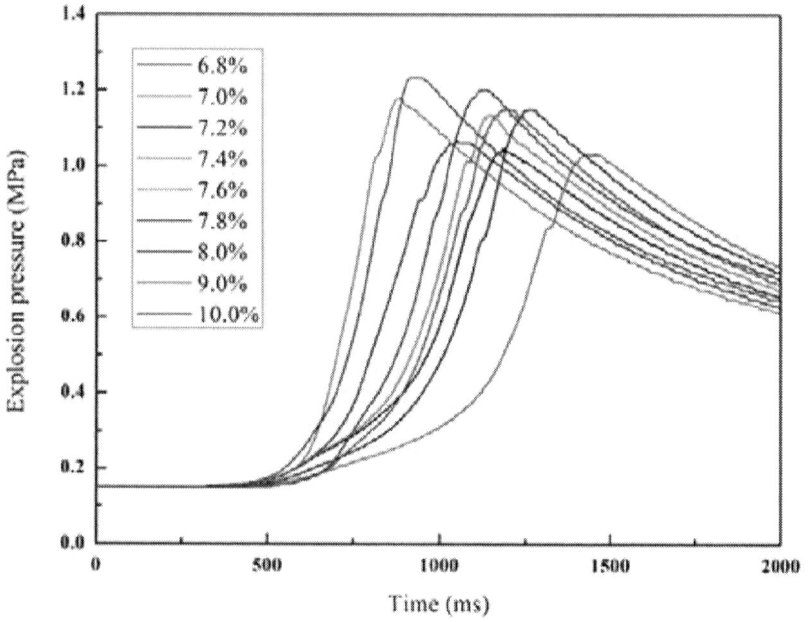

Fig. 1 Explosion pressure with time at atmosphere pressure and 25 °C

Fig. 1 shows the explosion pressure change with time at atmosphere pressure and 25 °C. The explosion pressure increased sharply and then decreased. The changes of maximum rate of explosion pressure rise with time at atmosphere pressure and 25 °C is shown in Fig. 2. The rate of explosion pressure rise increases firstly and then decreases, and the maximum rate of explosion pressure rise can obtain from Fig. 2. Fig.3 shows the explosion pressure of R1234yf at different initial pressures and concentrations. It can be seen that the explosion pressure increases and then decreases, and the explosion pressure reaches the maximum of 0.742MPa at 7.6%. The stoichiometric concentration of R1234yf is 7.4%. According to the previous study, the maximum explosion pressure appears at a position where the concentration of chemical equivalent is little large, which is consistent with previous literature studies. At the same concentration, as the initial pressure increases, the explosion pressure also increases. Because the initial pressure increases resulting in the gas being compressed, the amount of flammable refrigerant contained in the explosion vessel increases to the enlargement of the explosion pressure. However, under the condition of pressurization, the explosion pressure of R1234yf at the same initial pressure does not reach the maximum at 7.6%. At the volume concentration of 6.8% to 7.8%, the explosion pressure increases firstly and then decreases, and the explosion pressure is a peak at 7.6%. The trend from 7.8% to 10.0% increased and then decreased, and also appeared a peak. Under the conditions of pressurization, the explosion characteristics of R1234yf change. The trend of maximum explosion pressure at same initial presents M-type. It may be because at high pressure and high concentration, the intermolecular spacing decreases, while R1234yf has low flammability. The narrowing of the intermolecular spacing increases the degree of combustion reaction at high concentrations.

Fig. 4 shows the maximum rate of explosion pressure rise of R1234yf at different initial pressures and different concentrations. It can be seen that at atmosphere pressure, the maximum rate of explosion pressure rise increases firstly and then decreases. At high pressure, the maximum rate of explosion pressure rise under the same pressure also shows a M-type trend, which consistent with the trend of the maximum explosion pressure. As the initial pressure increases, the maximum rate of explosion pressure rise at the same concentration increases firstly, then decreases and then increases and decrease finally. The trend of the maximum explosion pressure is inconsistent with the maximum explosion pressure. This may be because the initial pressure has an influence on the flame propagation velocity. The flame propagation velocity of R1234yf is 1.5 cm/s and the flame propagation velocity decreases with the increase of the initial pressure. However, the narrowing of the intermolecular spacing increases the rate of combustion reaction at high concentrations.

In addition, when the pressure reach to 0.25MPa above, the explosive gas mixture in the container does not explode at any concentration. For R1234yf, when the pressure is greater than 0.25MPa, the substance is more secure. In this paper, the initial pressure of 0.25MPa is defined as the critical explosion pressure of R1234yf, which can also be explained by the effect of pressure on flame propagation. Therefore, when the pressure rises to some extent, the flame will not propagate and the explosion can not proceed.

Some errors also exist in experiment, which are mainly due to the following reasons.

(1) There is no stirred system in the experimental device. In the experiment, the instantaneous impact force generated by pressing the sample chamber to 2.1MPa can achieve the stirring effect. Therefore, it is difficult to uniformly mix the gas in the explosion container.

(2) There is a certain error in the accuracy of the gas injection volume and initial pressure.

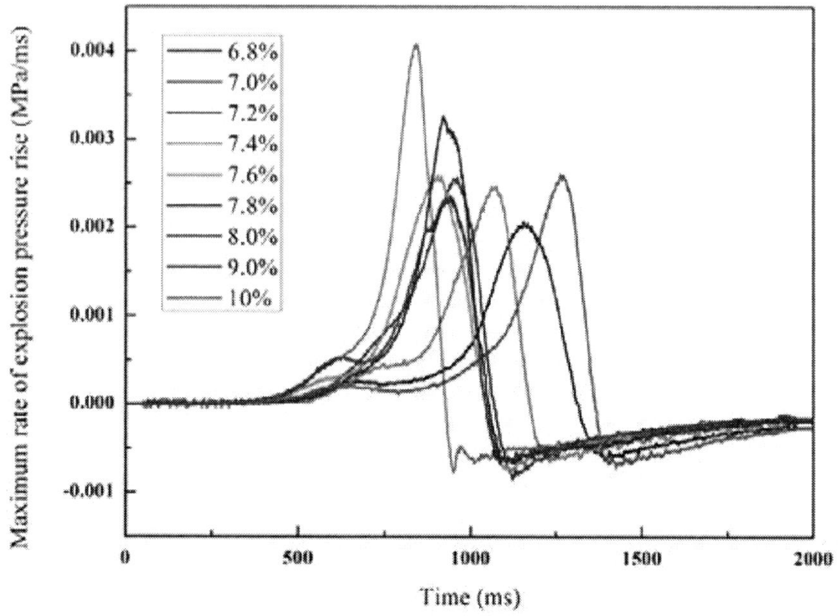

Fig. 2 Maximum rate of explosion pressure rise with time at atmosphere pressure and 25 °C

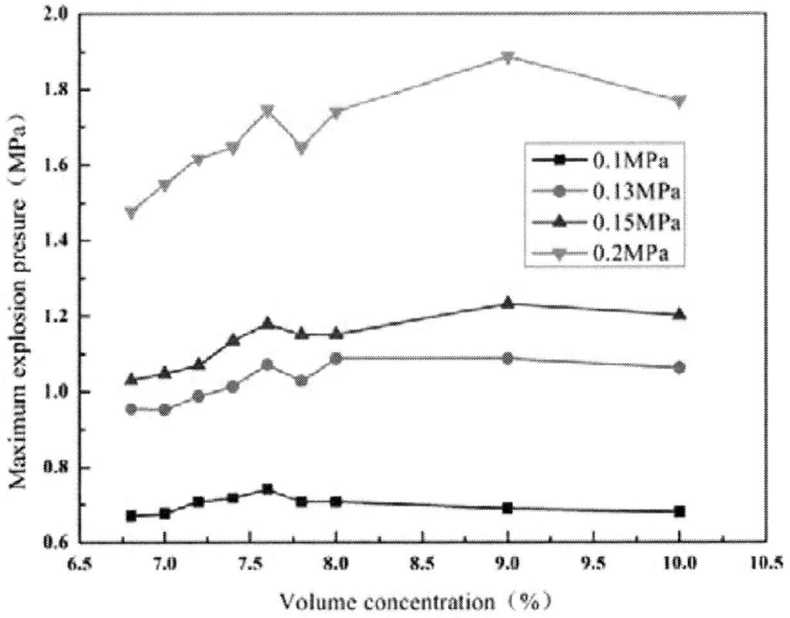

Fig. 3 Maximum explosion pressure at different initial pressures and concentrations

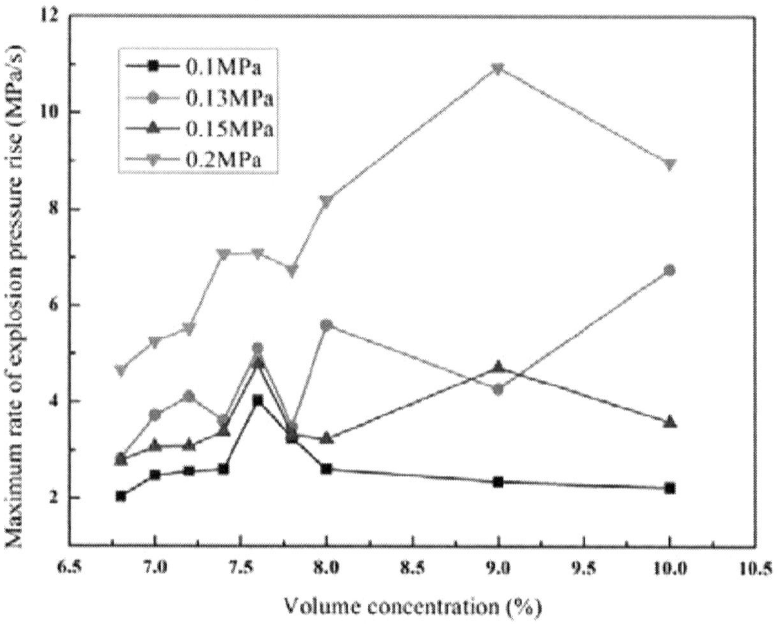

Fig. 4 Maximum rate of explosion pressure rise at different initial pressures and concentrations

Table 1. Comparison of explosion parameters for R290, R32 and R1234yf

	Maximum explosion pressure (MPa)	Maximum rate of explosion pressure rise (MPa/s)
R290	0.875 4%	60.38 4%
R32	1.02 20%	74.76 20%
R1234yf	0.742 7.6%	4.04 7.6%

Table 2. Flammability classification of typical refrigerants

Class Ⅰ: flame propagation	Class 2L: lower flammability and low	Class 2: lower falmmability	Class 3: higher

	burning velocity(≤10 cm/s)		flammability
R744	R1234yf/R1234ze	R152a	R290
R410A	R32	R143a	R600a
R134a	R717		

Similar studys of R290 and R32 had been done at atmosphere temperature and pressure, the experimental results shown in Table 1. Table 2 shows the flammability rating of some typical refrigerants. It can be seen from Table 1 - 2 that under the atmosphere temperature and pressure, the maximum explosion pressure of R1234yf and R290 is basically the same, which is less than that of R32. This is because the volume concentration is 20% when R32 reaches the maximum explosion pressure, then the amount of combustible material more than the other two substances more. The flammability of R1234yf is similar to R32 but lower than that of R290, so the maximum explosion pressure of R1234yf at 7.6% less than R290 at 4%. The maximum rate of explosion pressure rise of R1234yf is significantly less than that of R290 and R32. Comparing with R290, it is easy to understand that R290 is a flammable substance, so the maximum rate of explosive pressure rise is bigger. The combustion grade of R1234yf is same to R32, but the maximum rate of explosion pressure rise of R1234yf is less than that of R32. It may be because the flame propagation velocity of R1234yf is less than that of R32. The flame propagation velocity of R1234yf is 1.5 cm/s and that of R32 is 6.5 cm/s[13], so R1234yf is safer than R32 in spite of the same flammability classification.

4. Conclusion

In this paper, a large number of experiments have been done to determine the maximum explosion pressure and the maximum rate of explosion pressure rise of R1234yf under different concentrations and pressures. The following conclusions were obtained:

- The maximum explosion pressure and the maximum rate of explosion pressure rise of R1234yf at the optimum concentration is less than that of R290 and R32.

- The maximum explosion pressure of R1234yf increases firstly and then decreases at atmospheric pressure, and is M-type under the condition of pressurization.

- The maximum rate of explosion pressure rise of R1234yf present M-type at atmospheric pressure. As the initial pressure increases, the maximum rate of explosion pressure rise at the same concentration increases firstly, then decreases and then increases and decrease finally.

- R1234yf is safer than R32 in spite of the same flammability classification.

References:

[1]. Xin, W., et al.,Development trend of refrigerant substitute and the present refrigerants production status in China. Refrigeration and Air-conditioning, 2011(01): p. 110-115.

[2]. Zhang, W., et al., Study of the explosion characteristics and combustion products of air conditioner using flammable refrigerants. Journal of Fire Sciences, 2015. 33(5): p. 405-424.

[3]. Razus, D., et al., Temperature and pressure influence on explosion pressures of closed vessel propane–air deflagrations. Journal of Hazardous Materials, 2010. 174(1-3): p. 548-555.

[4]. Kondo, S., K. Takizawa and K. Tokuhashi, Effect of high humidity on flammability property of a few non-flammable refrigerants. Journal of Fluorine

Chemistry, 2014. 161: p. 29-33.

[5]. Kondo, S., K. Takizawa and K. Tokuhashi, Flammability limits of binary mixtures of ammonia with HFO-1234yf, HFO-1234ze, HFC-134a, and HFC-125. Journal of Fluorine Chemistry, 2013. 149: p. 18-23.

[6]. Kondo, S., et al., On the pressure dependence of flammability limits of CH2=CFCF3, CH2F2 and methane. Fire Safety Journal, 2011. 46(5): p. 289-293.

[7]. Qi, C., et al, Experimental study on the flammability of a new type of refrigerant R1234yf/R134a. Acta Energiae Solaris Sinica, 2016. 37(1): p. 152-156.

[8]. Papas, P., et al., Laminar flame speeds of 2,3,3,3-tetrafluoropropene mixtures. Proceedings of the Combustion Institute, 2017. 36(1): p. 1145-1154.

[9]. Moghaddas, A., et al., Laminar burning speeds and flame structures of mixtures of difluoromethane (HFC-32) and 1, 1-difluoroethane (HFC-152a) with air at elevated temperatures and pressures. HVAC&R Research, 2014. 20(1): p. 42-50.

[10]. Pagliaro, J.L. and G.T. Linteris, Hydrocarbon flame inhibition by C6F12O (Novec 1230): Unstretched burning velocity measurements and predictions. Fire Safety Journal, 2017. 87: p. 10-17.

[11]. Igarashi, N. and K. Tokuhashi, Assessment of burning velocity test methods for mildly flammable refrigerants, Part 2: Vertical-tube method. ASHRAE Transactions, 2013. 119: p. 255.

[12]. Takagi, S. and K. Tokuhashi, Assessment of burning velocity test methods for mildly flammable refrigerants, Part 1: Closed-vessel method. ASHRAE Transactions, 2013. 119: p. 243.

[13]. Minor, B.H., D. Herrmann and R. Gravell, Flammability characteristics of HFO - 1234yf. Process Safety Progress, 2010. 29(2): p. 150-154.

515620 Promoting the Production Safety Responsibility System in an Orderly Management System

Abstract:

This paper starts with the key factors that affect the responsibility system for production safety and systematically analyzes the inherent relationships among the various factors. On the basis of this, aiming at the typical problems existing in the concrete work at present, this paper put forward the viewpoint of "orderly management" with team as the main body of security and put forward suggestions and countermeasures on how to implement orderly management and implement the responsibility system of work safety .

515617 Spontaneous Polymer Risk and Control of C5 Hydrogenatedpetroleum Resin Plant

Abstract:

C5 hydrogenated petroleum resin device due to the process of their own reasons, since the polymer in the pretreatment processwill produce a certain amount of, regular cleaning, self polymerin air and oxygen oxidation has thespontaneous combustioncharacteristics, device has repeatedly because ofself polymer processing and storage of fire accidents caused by improper. On the risk from the polymer is analyzed, on the basisof summing up experiences, in strengthening management and technical transformation of two has takeneffective measures to control risk, since the polymer spontaneous, put an endto the spontaneous combustion accident occurring again.

515616 Analysis on Safety Management of Three Additives for Ethylene Plant

Abstract:

Because of the need of technology and equipment, different additives is injected in different procedures of Ethylene production process. These additives have different physical and chemical properties, and have negative effects on corrosion, toxicity or adverse effects on the environment. These additivesâ€™ management is an important part of the safety management of chemical production process, this paper carries on the discussion of safety management from the additivesâ€™ transportation, on-site storage and filling process of the additives, to standardize management better, and better protect the environment and personnel safety.

515613 Special Equipment Inspection Strategy Research and Application Analysis

Abstract:

In this paper, combining with the relevant legal requirements, special equipment management and inspection by the special equipment in service inspection method and the analysis of failure mode, absorb the risk-based inspection (RBI) core idea, in the regular inspection rules on the basis of introducing the concept of personalized inspection, will use the unit risk warning and professional testing unit of targeted testing, the combination of in order to improve the accuracy and effectiveness of special equipment inspection, formed through sum up and "four inspection ichigo management innovation work method". In full with the daily management of special equipment, regular inspection, risk prevention and control, special operation strategy analysis and practical research as the main line, management of special equipment, especially the special equipment of petrochemical industry has a practical significance.

515612 Corrosion Protection System Controls in the Application of the Refining Equipment Management

Abstract:

On face of inferior crude oil of the refining units, equipment corrosion intensifies the plight of device due to frequent equipment corrosion problems seriously affect the safe, stable and long period running, and the management of the corrosion protection management does not reach the designated position, anticorrosion control ability is insufficient, in combination with devices using the environment of the refining units, processing factors such as raw materials, process conditions, technical requirements, make a comprehensive analysis, set up the mechanism of corrosion protection system control, through the establishment and application of control mechanism of corrosion protection system, improve the device corrosion resistant ability, the effective protection of the refining unit safe and stable production.

Effect of the Content of Carbon Oxide in Ethylene on the Production Process of Gas Phase Polyethylene

Fangang JI
Jilin Petrochemical Company, Jinlin, China
Keywords: polyethylene; gas phase; carbon oxide content; control parameters

Abstract

The gas phase fluidized bed technology of PE is widely used in the chemical industry, its character is the requirement of very pure raw material .And the individual impurity content is less than 0.1mg/kg, total content is not more than 1mg/kg. According to the device running state when the carbon oxide (CO and CO2) content in ethylene is out of the standard,we do the analysis and summarization about the process of device parameters and actual operating condition.It is concluded that the problem of raw ethylene which the carbon oxide is out of standard cause the following questions.For example, the catalyst activity decreases significantly, reaction temperature fluctuates frequently, reactor electrostatic increases, the hydrogen modulation and copolymerization performance decrease and the fraction of small particles increases. Therefore, the corresponding measures have been made to solve the similar problem of the fluctuation of the impurity content of raw materials.

The LDPE plant in ethylene factory of Jilin Petrochemical Company adopts gas phase fluidized bed technology which is developed and Industrialized by the UCC company, it has the advantages of simple process flow, low investment , low energy consumption, large flexible operation and stable product quality, a large span of the product, etc. Meanwhile , it can produce a full density linear polyethylene product with a melting index of 0.01g/min to 150g/min and a density of 915kg/m3 to 970kg/m3. In September 1996, it begins to put into operation formally. The original annual design of production capacity is 100kt/a. After three expansion, the production capacity of the device upgrades to 274kt/a.

1. The process of ethylene supply and the change of carbon oxide content

1.1 The brief introduction of ethylene supply process

The ethylene supplyed by the ethylene unit is first heated in the ethylene heater (E-2001) and then goes into the acetylene (C-2104) to remove acetylene. And the ethylene which remove acetylene comes into the heat exchanger (E-2105A/B) and makes heat exchange with the ethylene from deoxidizing bed (C-2109). Then it is preheated by the ethylene preheater (E-2106), and then enters the bed for removing carbon monoxide (C-2107), and then enters the deoxidizing bed (C-2109) to remove the trace oxygen. After cooling from the ethylene heat exchanger (E-2105A/B), the ethylene is dehydrated by ethylene dryer (C-2112/2113) at last. A small part of the ethylene is used for the butene system of the bed preload, and the vast majority is used to polymerize in the reactor system.

1.2 The actual change of the content of carbon oxide in ethylene

1.2.1 Changes of carbon oxide content before the refining of ethylene

The mass fraction of carbon monoxide and carbon dioxide in the ethylene before purification is less than 1*10-6and 1 5 *10-6. But as you can see from Figure 1, carbon monoxide and carbon dioxide content is always near the upper limit of index or exceed the standard index during the period from January 26 to February 10 of 2017 ,and which reach the maximum in February 4th, 1.8 * 10-6 and 4.4 * 10-6, significantly higher than normal control index.

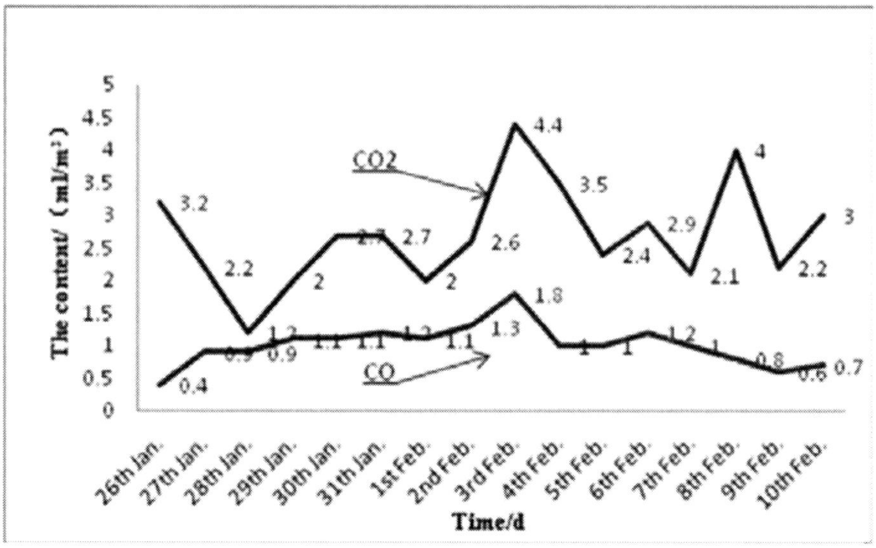

Fig.1 The change diagram of carbon oxide content in refined ethylene

1.2.2 The Change of carbon oxide content in refined ethylene

The carbon oxide content of refined ethylene keeps below the instrument detection limit of 0.1 mL/m³and is in a relatively stable state. During the period from January 26 to February 10th, the content of carbon oxide in ethylene is fluctuated and higher than in the stable period. Owing to the device has a carbon monoxide bed, the data of the carbon oxide content after the refining of ethylene are collected and analyzed. The specific detection data are shown in Table 1. It can be seen from table 1, The content of the carbon oxide is at the maximum in the period from February 3rd to February 4th, and in the other time is not clear because of the limited precision of the instrument. In addition, the sampling analysis can only represent the data of one time point, and can not be monitored in time. Therefore, the data in Table 1 provide only a reference, which does not fully reflect the true change in the content of carbon oxide.

Table 1. The change of carbon oxide content in refined ethylene

Date	Content of carbon oxide/ ($mL \cdot m^{-3}$)					
	A	B	C	D	E	F
26th Jan.	<0.1	2.3	<0.1	2.2	<0.1	<0.1
27th Jan.	<0.1	<0.1	<0.1	1.1	<0.1	<0.1
28th Jan.	<0.1	5.0	<0.1	3.5	<0.1	<0.1
29th Jan.	<0.1	3.0	<0.1	1.3	<0.1	<0.1
30th Jan.	<0.1	4.6	<0.1	1.1	<0.1	<0.1
31th Jan.	<0.1	2.0	<0.1	2.0	<0.1	<0.1
1st Feb.	<0.1	2.6	<0.1	1.0	0.4	<0.1
2nd Feb.	<0.1	0.7	<0.1	2.8	<0.1	<0.1
3rd Feb.	0.5	4.4	0.5	4.6	<0.1	<0.1
4th Feb.	0.4	3.6	0.5	3.1	<0.1	<0.1
5th Feb.	<0.1	4.4	<0.1	4.6	<0.1	<0.1
6th Feb.	<0.1	2.4	<0.1	2.4	0.5	<0.1
7th Feb.	<0.1	2.0	0.4	<0.1	<0.1	<0.1
8th Feb.	<0.1	3.6	<0.1	2.2	<0.1	<0.1
9th Feb.	<0.1	3.4	<0.1	2.8	<0.1	<0.1
10th Feb.	<0.1	1.2	<0.1	4.2	<0.1	<0.1

Notes: A is the content of CO in refined ethylene of the old system

B is the content of CO_2 in refined ethylene of the old system

C is the content of CO in refined ethylene of the new system

D is the content of CO_2in refined ethylene of the new system

E is the content of CO in circulating gas system

F is the content of CO_2in circulating gas system

2. The change of the main control parameters of the device

2.1 The change of catalyst activity

The polyethylene titanium catalyst is very sensitive to the impurities in the raw materials, even if the content of the catalyst is very low, it also has a great effect on its activity. The carbon oxide (carbon monoxide and carbon dioxide) which is described in this paper, is a toxic substance of the titanium catalyst, which can lead to a significant decrease in its activity. As can be seen from Figure 2, the average activity of the catalyst is obviously lower than the usual control index (20 000kg/kg) during the period from January 26th to February 10th. In particular, on February 6th, the average activity of the catalyst decreased greatly, only about 15000 kg/kg. It can be concluded that when other conditions are not adjusted, the increase of carbon oxide content in ethylene feedstock will directly lead to the decrease of catalyst activity, which will further increase the consumption of catalyst [1].

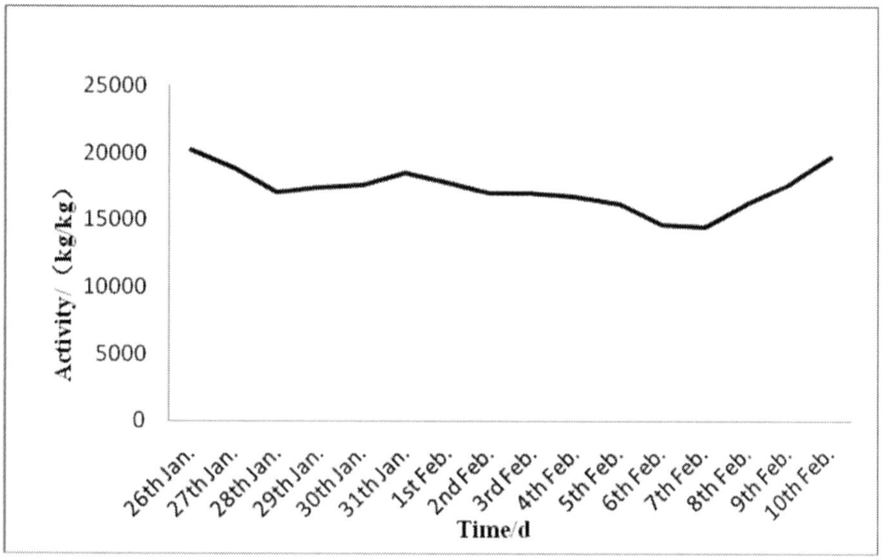

Fig.2 The change diagram of catalyst activity

2.2 The reaction temperature control

The reaction temperature is an important parameter to measure the reaction state. Too high is easy to cause the resin adhesion and caking, and block up the discharging system. Too low not only reduces the reaction time and space efficiency, but also causes the product index to exceed the standard. As you can see from Figure 3, in normal production, the fluctuation range of reaction temperature is very small, and oscillates up and down in the range of 86~87 degrees, so the operator can hardly operate smoothly without interference. From the figure 4, when the content of carbon oxide is over standard, the reaction temperature is not easy to control, and the fluctuation range is extended to 85~88 degrees. The operator needs to intervene repeatedly to avoid the phenomenon of "runaway temperature". The reason is that the concentration of carbon oxide in the reactor is fluctuating, when the content of carbon oxide is low, the catalyst will release a lot of effective activity instantly, and so the reaction temperature raises;when the content of carbon oxide is high, the catalyst in the reactor is consumed by poison, the activity is inhibited, and so the reaction temperature is greatly reduced. To sum up, the excessive carbon oxide content brings great inconvenience to the reaction temperature control, which directly affects the stability of the normal production operation.

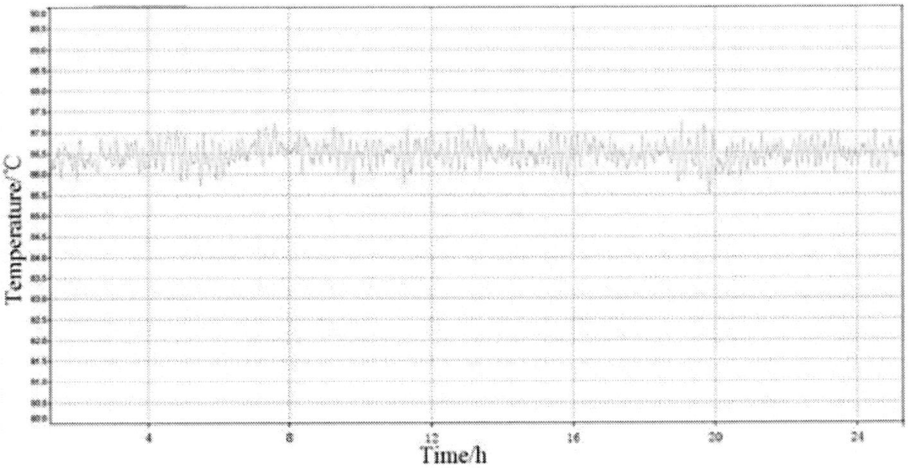

Fig .3 The normal temperature control chart

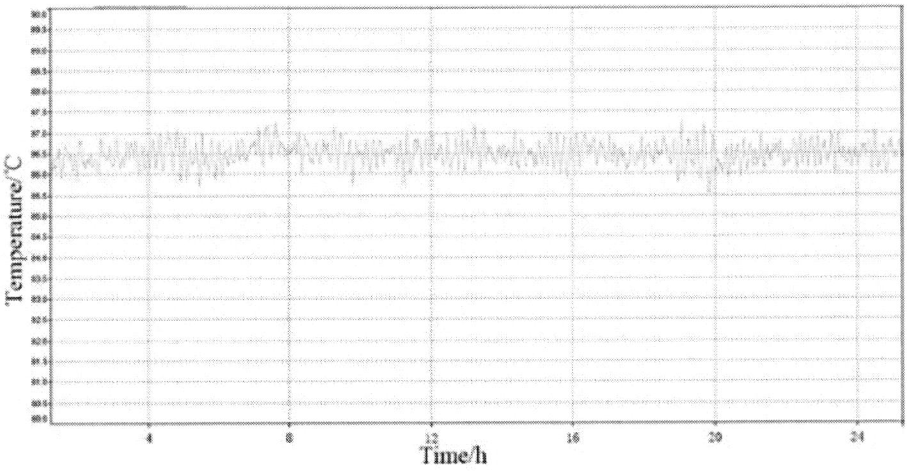

Fig.4 The temperature control chart when the carbon oxide content is fluctuated

2.3 The static change of the reactor

Sudden growth impurity content in raw gas will cause a sharp rise in the electrostatic. During the normal production, the electrostatic fluctuation in the reactor keeps-0.100kV around. From Figure 5, we can see that during January 26th to February 10th, the static electricity of reactor increases significantly.especially in February 3rd, reaching the maximum value of -0.250kV, and this is coincided with the sudden increase of carbon oxide content in February 3rd (table 1). It is known,as the reactor static accumulates,as the electric field intensity increases, so the electric field force of the catalyst particles and resin particles increases, while the resin maintaining flow state tension decreases gradually, especially the "dead zone" minimum resin is gradually absorbed to the wall and causes serious,agglomeration phenomenon [2]. Therefore, it is necessary to pay close attention to the content of ethylene impurities and using antistatic system to prevent the occurrence of blockage of distribution plates, the occurrence of forced parking.

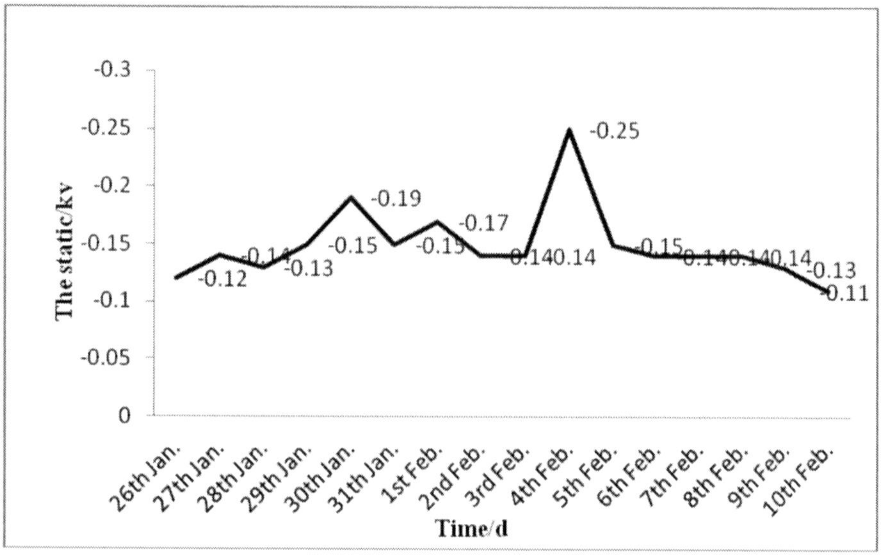

Fig.5 The static change diagram of the reactor

2.4 The change of hydrogen adjustmen and copolymerization

The ratio of hydrogen to ethylene and the ratio of butylene to ethylene is an important indicator which directly reflects the performance of hydrogen adjustment and polymerization, and it can indirectly reflect the unit consumption of the device. Therefore, it is necessary to analyze and summarize the performance of the reactive hydrogen adjustment and copolymerization when the content of carbon oxide is fluctuating. From Figure 6, we can see that the ratio of hydrogen to ethylene and the ratio of butylene to ethylene are first increased and then decreased, and the value increased by about 3% compared with the normal period. Thus, when the content of carbon oxide in the raw gas is increased, the hydrogen adjustment and copolymerization property of the reaction unit decreases and the single consumption increases. The main reason for this phenomenon is that the surface active point of the catalyst is occupied by the poison, and the degree and depth of the polyethylene chain is decreased[3].

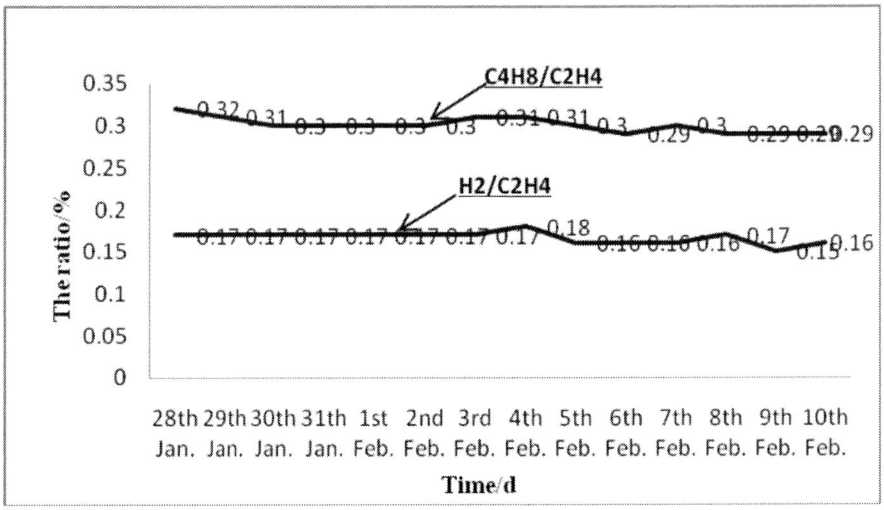

Fig.6 The change diagram of hydrogen adjustment and copolymerization

2.5 The size change of powder particles

The size distribution of the powder resin particles has an important influence on the feeding efficiency and the fluidization state of the reaction unit. From the data analysis in the table 2, during January 26th to February 10th, the average size of resin particles decreases significantly and the particle size shifted to a small offset as a whole. The proportion of powder with a size of 75 μ m and 124 μ m is significantly improved. This indicates that the increase of carbon oxide content not only affects the activity of the catalyst, but also affects the surface structure of the catalyst itself, which makes the resin particles more easily broken and smaller during the friction process. In addition, if the content of fine powder in the reactor increases, the entrainment in the circulating gas will increase.And the inlet filter of compressor and the tube bundle of heat exchanger (circulating water) will be blocked for a long time, which is not conducive to the long term operation of the device.

Table 2. The change of mass percentage of different size particles in powder resin

Date	The change of mass percentage of different size /%						
	1700 μm	880 μm	425 μm	250 μm	124 μm	75 μm	Average size /μm
26th Jan.	1.81	20.76	47.36	26.16	3.84	0.04	0.788
27th Jan.	1.79	22.37	47.80	24.69	3.17	0.05	0.806
28th Jan.	1.13	23.24	49.16	23.09	3.02	0.05	0.802
29th Jan.	1.39	22.85	45.38	25.42	4.38	0.06	0.789
30th Jan.	1.25	19.75	49.04	25.66	4.06	0.05	0.768
31th Jan.	1.25	19.73	49.04	25.65	4.07	0.06	0.767
1st Feb.	1.35	16.89	50.39	26.46	4.12	0.04	0.743
2nd Feb.	1.46	18.05	49.71	26.39	4.12	0.04	0.757
3rd Feb.	1.33	17.92	46.89	28.52	4.95	0.07	0.741

4th Feb.	1.52	20.87	46.11	26.6	4.54	0.1	0.775
5th Feb.	1.56	21.2	45.52	27.12	4.38	0.04	0.778
6th Feb.	2.20	19.46	48.49	25.55	3.48	0.03	0.787
7th Feb.	1.89	14.32	49.55	29.19	4.43	0.02	0.728
8th Feb.	1.56	21.2	45.52	27.12	4.37	0.04	0.778
9th Feb.	1.15	17.12	52.52	25.58	3.37	0.01	0.750
10th Feb.	1.90	24.64	50.82	19.83	2.49	0.01	0.844

2.6 Change of the foam product quantity

In the normal production period, the number of foam product is less than 30 bags per day. From the table 3, it can be seen that during the period from January 26th to February 10th, the number of foam product rises sharply and reaches 185 bags on January 29th. This shows that the reaction temperature fluctuates very often and the reaction temperature is not brought in time, and so more bulk product is formed[4].

Table 3. The quantity change of foam product

Date	Foam product/（bag • d−1）	Date	Foam product /（bag • d−1）
26th Jan.	56	3rd Feb.	94
27th Jan.	109	4th Feb.	67
28th Jan.	138	5th Feb.	52
29th Jan.	185	6th Feb.	51
30th Jan.	101	7th Feb.	61
31th Jan.	87	8th Feb.	49
1st Feb.	53	9th Feb.	46
2nd Feb.	38	10th Feb.	10

3. The consumption of catalyst and auxiliary

As the content of carbon oxide in ethylene exceeds the standard, the consumption of catalysts and auxiliaries increases. Now we make the following calculation for the main material consumption of the unit from January 26th to February 10th.

3.1 The consumption of main catalyst

From January 26th to February 10, the production of PE is 12163.52t. According to the catalyst consumption index 0.05kg/kg, the weight of 0.608t catalyst is needed, while the actual catalyst consumption is 0.708t. Therefore, the event causes the catalyst consume 0.1t more. The main reason is that the carbon oxide causes the poisoning of the catalyst, which leads to the decrease of the activity and the increase of the consumption.

3.2 The cocatalyst consumption

3.2.1 The consumption of aluminum tri-n-hexylammonium(T3)

According to the consumption plan of tri-n-hexylammonium aluminum 0.025kg/kg, the cocatalyst weight of 0.304t is needed, while the actual consumption is 0.313t. Therefore, this incident causes more consumption of 0.009t. The main reason is that T3 reacts with the catalyst in a certain proportion. When the consumption of the catalyst increases,the consumption of T3 also increases.

3.2.2 The consumption of the diethylaluminum chloride(DC)

According to the consumption plan of diethylaluminum chloride 0.075kg/kg, the catalyst weight of 0.912t is needed, while the actual consumption is 1.12t. Therefore, this incident causes more consumption of 0.208t. The main reason is that DC reacts with the catalyst in a certain proportion. When the consumption of the catalyst increases,the consumption of DC also increases.

3.3 The consumption of pentane oil for condensing agent

According to the pentane oil consumption plan 3.2kg/kg,the pentane oil weight of 38.92t is needed.But a total weight of 44.13t is actually consumed. Therefore, this incident causes pentane oil more consumption of 5.29t. The specific reason is that the reaction rate is affected by the activity of catalyst,and resulting in the high and low level of reactor and unstable discharge frequency.Thus, a large loss of circulating gas is happened and the consumption of pentane oil increases.

4. The main measures taken

4.1 Increasing frequency of analysis and strengthening communication with related workshops

Daily sampling and analysis are carried out and the increase frequency of analysis is needed. The online analysis instrument is used to monitor the change trend of the composition of the circulating gas in real time. Strengthen the communication with the ethylene workshop and improve the mastery degree of ethylene impurity replacement.

4.2 To improve the level of the index control with coupe operation

The double duty supervision system should be adopted in the reaction position, and each of them should perform their duties to improve the browsing frequency of the DCS screen. We should always pay attention to the change of the parameters of the reaction unit, do well in tracking analysis and adjustment work, so as to ensure the smooth operation of the device.

4.3 Pay attention to the index of bed layer to improve the effect of impurity removal

According to the ethylene oxide content of the boundary area and reaction load, the optimal regeneration frequency of each ethylene carbon monoxide bed is calculated. And the strict regeneration frequency ensure that the refining bed is in the optimal working condition at any time. Thus, it is necessary to ensure smooth removal of carbon monoxide, carbon dioxide and other light components,and to reduce the impurities as far as possible in the ethylene gas.

5. Concluding remarks

The superstandard carbon oxide in ethylene will cause the catalyst activity decrease significantly; the reaction temperature fluctuates frequently and hard to control; the increase of static reactor and serious caking phenomenon; the hydrogen modulation and copolymerization performance decrease obviously; the increase content of fine powder, the unstable discharge frequency and the increase of raw material consumption. Therefore, when the device happened the incident similar to the situation, we should be based on the composition and content of the impurities , and take corresponding measures to move impurity and retreat that ensure long period smooth running.

6. Reference

[1] Guo Z.F., " Progress of Polyethylene Ziegler-Natta Catalyst for Slurry Processes,"*Petrochemical Technology*, Volume 4, No.453-457,2009

[2] Cai X.J., "Prevention of Static Electricity Risk in Gas Phase Polyethylene Reactor," *Zhejiang Chemical Industry*, Volume 4, No.27-30,2010

[3] Xiao S.J.,Yu F.S., "Olefin Coordination Polymerization Catalyst and Polyolefin,"Beijing: Beijing University of Technology, 2002

[4] Wang H.T., "Reasons and Preventive Measures of Coking in Vapor Phase Polyethylene Reactor ," *China Synthetic Resin and Plastics*, Volume 2, No.52－55,2005

The research of Infrared thermal imaging technology in preventing fire explosions in petrochemical enterprise

Peng Guilin
Petrochina Liaoyang Petrochemical company Safety and Environmental Department, Postcode 111003

Key words: infrared thermal imaging detection; fire; explosion

Abstract

Most of the raw material, intermediate and product of petrochemical enterprise are flammable, explosive, toxic and corrosive substances; many different kinds of equipment, units spread relatively concentrated. During petrochemical producing process there are high probability of fire explosions and other accidents, mass death and casualty. It might cause many chain reaction if fire or explosion happens, the risk is high, with high loss. As many equipment on site are covered by insulation layer, it is difficult by traditional equipment checking or manual inspecting to find out hidden danger inside. This might cause hidden danger could not be troubleshoot for very long time and even worse, some might lead to big loss at last. Infrared thermal imaging detection technology is a new effectively detecting technology, with the advantages of remote detection, image is intuitive, detection area is large, high sensitivity and high resolution, etc. It is very important in preventing fire explosion. By studying present situation of production and the application of Infrared thermal imaging detection technology, to find a way that can prevent fire explosion reasonably in petrochemical enterprise.

1. Induction

Along with the continuous develop and progress of science and technology, in order to find potential accidents effectively, to prevent fire explosions and other accidents in advance, many kinds of NDT technology have been well developed. Among them, the representative conventional NDT technologies are Ultrasonic testing technology, Gamma ray detection technology, Magnetic particle detection technology, Infrared detection and infrared diagnosis technology etc. Infrared thermal imaging detection technology is a new method to detect defective material and stress check. By the way of analyzing the thermal imaging of devices to diagnose the thermal safe state and hidden defects, to provide evidence for making pointed improvement measure, it has got wide attention for its special advantage. But since its popularizing scope is small, low frequency of use, the advantages and characteristics of thermal imaging detection technology has not been put in good use in petrochemical enterprise daily safety management. Therefore there is still big room for the technology to be used in preventing fire explosions and other accidents in petrochemical enterprise, to supply better service for safety management.

2. Working principle and characteristics of Infrared thermal imaging detection

2.1 Principle of Infrared thermal imaging detection

Infrared thermal imaging detection using infrared radiation theory, to receive the infrared radiation energy from the surface of the target object by different kinds of detectors, the temperature distribution on the surface of the object will be transformed into a thermal image as showed in fig.1.

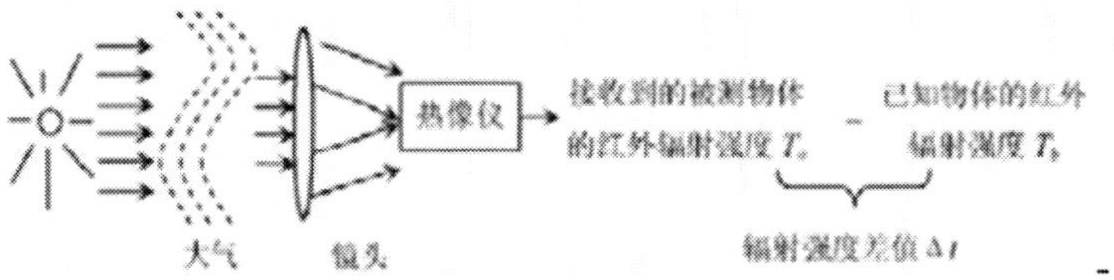

Fig.1 Infrared thermal image temperature measurement schematic diagram

Infrared thermal image use the influence of discontinuity defects on the thermal conductivity of the target object, and then reflect on the temperature of the surface of the object, there will be a temperature gradient on the port of the surface of the object, which will result in a difference of the infrared radiation energy, by this difference it can draw a conclusion whether there is any defect and the degree of the defect.

2.2 The characteristics of Infrared thermal imaging detection technology

Compare with traditional detecting technology, Infrared thermal imaging detection technology has the characteristics as below.

2.2.1 Imaging detection, image visualization

At present, the main thermal imager instrument is Non - cooling focal plane infrared thermograph, the image is clear with high quality. During detecting process, whenever there is any object enter the scope of the instrument's visual field, the image will be showed on the screen.

2.2.2 High sensitivity, high resolution

All of the thermal imagers have high sensitivity, even when the temperature difference is less than 1 degree they can also tell. Therefore they can find the very small change of the thermal state on equipment to some extent. This is the reason that they can detect equipment in detail parts and also do the remote detecting at the same time.

2.2.3 High detecting efficiency

When it works normally, the instrument will aim at the target equipment, the thermal distribution will show on the screen immediately. If there is any defect point, the temperature of the point will be detected afterwords.

2.2.4 Non-touching detection, detecting area is big

When detect the thermal defect point of electrical equipment using infrared thermal imager, the distance can be several meters, several decameters or even more than hundred meters. So even the electrical equipment with big area it can be detected by the infrared thermal imager. The instrument will observe the whole situation, if there is any defect point they would be detected carefully.

2.2.5 Immunity from electromagnetic interference

The infrared thermal imager works in the range of 8~14 μm band, which is far from the electromagnetic spectrum. Therefore even in the field with strong electromagnetic such as transmission power station, high pressure line and so on, the infrared thermal imager will not be affected, thermal images will always be reliable.

2.2.6 The scope of the application is wide, the benefit is high

There are many test methods used in the current preventive test of electrical equipment, such as chromatographic analysis, partial discharge test, leakage current and resistance current test, direct resistance measurement, pressure test. Each of them is applicable to all kinds of fault detection of electrical equipment. But, as to test method, infrared imaging detection technology can be applied to the detection of most faults in high voltage electrical equipment.

2.2.7 Computer analysis can be used

At present most of the infrared thermal imager can be connected with computer. Using image processing software to analyze thermal processing and stored in the PC card, and also can analyze failure point and make a report.

3. The application of infrared thermal imager in petrochemical enterprise and room of improvement

3.1 The application of infrared thermal imager in petrochemical enterprise

3.1.1 Detect the inner lining of high temperature equipment by Infrared thermal imaging technology

Normally, to maintain equipment with lining of high temperature resistance or/and corrosion resistance, it needs to shutdown, open the vessel, check and then to repair if necessary. This method could not make sure the real situation during the equipment is running, and the shutdown will affect the operation if not necessary. Using Infrared thermal imaging technology, temperature of different part of the equipment at same time and same situation can be detected, as well as temperature difference of same part at different time, these can be used to judge whether the lining is damaged.

The degree of damage can be diagnosed according to the temperature limit of the lining damage, and then, based on the temperature upper and lower limit criterion, based on the actual situation of the unit, the reference standard of the damage criterion is summarized to diagnose the extent of the damage in the lining. For example, using the infrared technology application of the sinopec group recommended by catalytic equipment lining damage degree temperature limit criterion, combined with the test of heavy oil unit for the actual situation, put forward a temperature limit equipment lining damage criterion. See table1.

Table1. The External Surface Temperature Contrasts to The Lining Damage Degree

Damage degree	good	slightly	Below average	Middle	Above average	Serious	Very bad
External surface temperature	70~90	90~120	120~160	160~200	200~250	250~300	330~450

In figure 2.a, The temperature of the overtemperature zone is moderate, therefore it is judged as middle-range hollowed out. In figure 2.b, The temperature of the overtemperature zone is above-average, therefore it is judged as penetrating crack. Through a large number of infrared thermal images and the actual inspection of lining damage, each kind of injury has the corresponding regular infrared thermal image. It is not necessary to shutdown, to open the vessel, and not need to contact directly. It is easy to operate and can find out the damage position of the lining of the vessel and the extent of the damage. For its features of fast, non-contact and so on, it is especially suitable for real-time monitoring and on-line detection of serving equipment, which provides a scientific basis for precognitive maintenance.

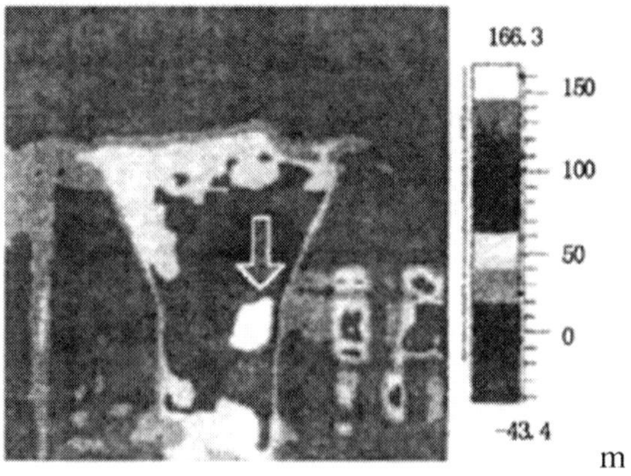

(a)Thermal image of parts of wear resistant material of external heat exchanger fall off

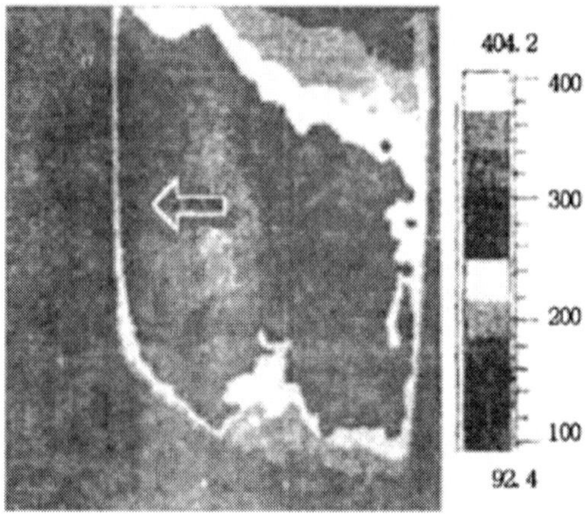

(b) Thermal image of cracks of the internal lining of the compressed plate

Fig.2 Thermal image of extent of the lining damage

3.1.2 Detecing insulation layer of pipeline by Infrared thermal imaging technology

In most enterprises, the pipeline is widely using insulation covered structure. The pipeline is too long to be welded in sections, and the polyurethane is used in the welding area, and galvanized sheath is used to insulate the insulation. The sealing performance of the casing is decreased with time, which can be easily seeped into water and corrosion, which will lead to pipe defects and even leakage accidents.

The picture of gas pipeline in visible light and its infrared thermal image see figure 3. It can be seen clearly that there is a significant temperature rise in the bottom part of the pipeline. The maintenance result showed that the insulation layer of the area has been infiltrated.

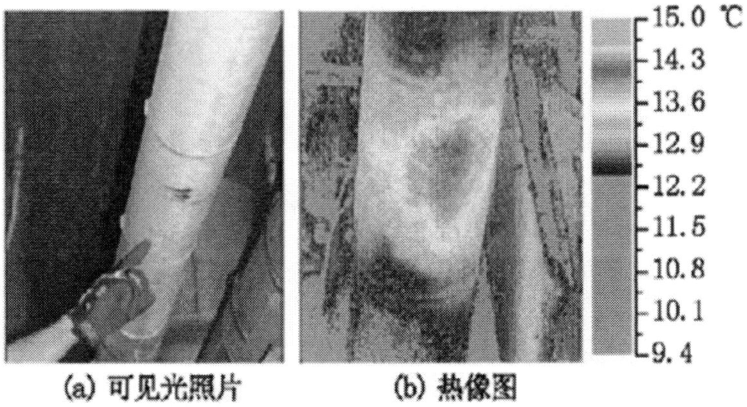

Visible photograph thermal imaging picture

Fig.3 Photograph of gas line and its thermal imaging picture

Using Infrared thermal imaging technology to detect defects of the insulation layer has the advantages of low cost, safe, reliable operation, convenient and quick detection, which is not available for routine testing methods. The technology can also be used to evaluate thermal insulation effect. Accurately and systematically evaluate the thermal insulation effect, which is conducive to the economical and reasonable implementation of the insulation process transformation

3.1.3 Infrared thermal imaging technology for gas leakage detection

Liquefied natural gas, liquefied petroleum gas and liquefied ethylene are colorless and tasteless and often require special instruments to detect, the infrared thermograph can be used to quickly determine the leakage of the gas through the temperature change, so as to locate the leak point and prevent the fire, explosion and other accidents. For example, the gas leakage infrared thermograph can quickly detect the leakage of methane and other volatile organic compounds (VOC) gases, and the gas leakage thermal image can be seen in the picture 4. Compared with the methods of concentration sensor, Infrared thermal imaging technology for monitoring of chemical gas leak in chemical industry, has the advantages of long distance, large area and fast, and is not affected by environmental conditions such as wind power or others.

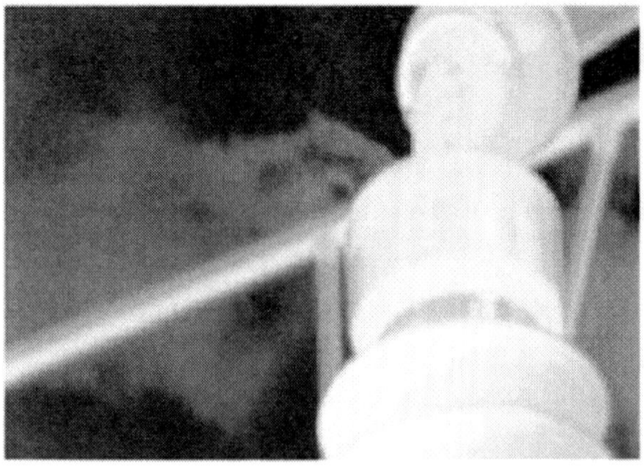

Fig.4 Gas leakage thermal image

3.2 The improving room for the application of Infrared thermal imaging technology in petrochemical industry

The temperature of most part of the equipment with failure will change obviously when operating in petrochemical enterprises. We can monitor the equipment with Infrared thermal imaging technology to check out the failures and fix them in time. Infrared thermal imaging detective devices have many advantages such as long-distance, non-contacting operation, fast, high sensitivity and so on. Especially compared with other devices they have many unmatched advantages in high temperature, high pressure, poisonous and electrical environment. So it is safe and have high economical value. Yet, using frequency of the Infrared thermal imaging Technology in daily operating is not high. We often use them only when we do non-destructive testing once a year or half a year. They are seldom used in daily safety management.

Taking Liaoyang Petrochemical Company's application of Infrared thermal imaging technology as an example, their application of the technology is lack. Liaoyang Petrochemical Company was started up formally at 1974. After 40 years' developing, it has been developed into one industrial structure with refining, aromatic and olefins as its main industrial part. There are 67 big refining units now. Recent years, the company has actively promoted the HSE system construction continuously, take the risk management as the core and localization management as the starting point to continuously promote various process risk evaluation methods; call on all staff to identify job risks and find hidden danger around. And plenty of funds were put to encourage and reward for employee. Even though there are some improvement on safety management and the amount of accidents were largely declined as well, yet there were still large number of processing safety accidents happened every year . Among of them fire and explosion accidents were the most. By studying the accidents and summarizing in the meetings, we found out the main reasons that resulted in these accidents were the equipment facilities have the safety hazard of wear, corrosion, crack and leakage. Normally these defects could not be identified directly by operators.

At present the company to find safety hidden dangers mainly depends on traditional manual inspection. While the main hardware equipment being used in manual inspection were Vibration Meter, Combustible Gas Alarm, Oxygen Content Alarm and wrench and so on. During manual inspecting operation it often depends on the operators to see, to smell, to listen and touch these chemical equipment closely to tell whether there is any safety hidden danger. But this kind of method often has big deviation as it is vary from person to person, and also be affected by the weather, experience and even the mental state. And when the operators inspect the equipment regularly they are directly exposed into the unseen and unpredictable dangerous chemical material environment that injured possibility is high.

To detect the equipment with Infrared thermal imaging technology only when the main equipment and valves need be tested professionally during plants' maintenance or when the special inspection for special equipment need to be taken by the third party security testing group. The advanced technology is rarely applied in daily safety management. The main reason why this technology is not widely applied in the company probably are:

3.2.1 Infrared thermal imaging technology is dependent on hardware too much

The infrared thermal imager used by enterprises is not easy to carry, using conditions are higher and the price is more expensive, which is not conducive to the promotion.

3.2.2 Infrared thermal imaging technology has not been recognized as an advanced technology and has great value for security management

At present, Infrared thermal imaging technology has not been incorporated into corporate technology, also a lack of knowledge in the daily training process, which results in that infrared thermal imaging technology is unknown in enterprises. No matter the company's decision-makers, management, and grassroots technicians and operators all do not know much about the technology.

3.2.3 The infrared thermal image technology is not integrated into the safety management consciousness of enterprises

The main means of preventing fire explosion are the on-site combustible gas alarm and fire alarm device at present. But often when these devices send out the alarm, to the personnel to respond and deal with, it is too late, the fire explosion accident has already happened.

Therefore there is still big room for Infrared thermal imaging technology in preventing fire explosion in petrochemical enterprise. The safety and superiority of the technology can be widely used in the daily inspection, weekly inspection, season inspection, year inspection and daily inspecting and repair quality inspection. By increasing the using frequency of Infrared thermal imaging technology, we can make full use of this technology to help find all kinds of security risks, by analyzing the causes of hidden dangers and provide a powerful image basis for how to deal with the hidden dangers. It is helpful to develop effective measures to prevent fire explosion, thus effectively reducing the occurrence of accidents.

4. conclusion

The Infrared thermal imaging technology is a new detecting technology. It has the advantages such as fast detecting temperature, high sensitivity, wide temperature-testing range, non-touching, visible imaging and non-interference testing. It has been used in various aspects of petrochemical industry. With the continuous development of scientific technology, the Infrared thermal imager will also be developed into miniaturize and portable and low price devices. This will make the infrared thermal imaging technology more easily accepted and promoted by enterprises. Enterprises can also use this technology as a project to solve the problem, combine the characteristics of the company, and cooperate with infrared thermal imaging equipment to research and develop the equipment applicable to the production site,; integrate the advanced science and technology with the actual production process to better serve the fire explosion prevention work of petrochemical enterprises; continuously improve the safety management level of enterprises and ensure the safe and stable development of enterprises.

5. References

[1] Li Y.H., Sun X.G., Lian J.H.,"Research on Measurement Temperature Technology and Application of Infrared thermal imaging,"Modern Electronics Technique, Volume 1, No.112-113.,2009

[2] Jin G.X., "Application of Infrared thermal imaging Technology to Internal Corrosion Detection of Petrochemical Equipment ," Petro-Chemical Equipment, Volume 1, No. 83,2013

[3] Xue X.Y., "Application of Infrared Thermography Detection Technology in Petrochemical Industry," Management & Technology of SME， Volume 1, No. 267-268, 2014

[4] Liu H. "Application of Infrared Thermography Detection technology in Petrochemical Industry ," Petro-Chemical Equipment, Volume 1, No. 48-52, 2010

[5] Huo R.,Yang Z.,H.,Liu J.X., "Fire and Explosion Prevention and Control Engineering,"Beijing: China Machine Press，2007

Evaluate the HAZOP analysis quality from Technical View

Zhang Xue[a], Ding Fangfang[a], Qian Cheng[b]

[a] **CNPC Research Institute of Safety & Environment Technology, Beijing China**

[b] **China University of Petroleum-Beijing, Beijing China**

Keywords: HAZOP; evaluation; analysis quality

Abstract

With the increasing security concept in the petrochemical industry, Hazard and Operability Study (HAZOP) is more and more used in the safety management of petrochemical industry. As a result of HAZOP analysis has a strong dependence on people's experience, and it's a relatively high requirement for analyzing the overall quality of the team, HAZOP analysis team cannot meet the huge demand of HAZOP analysis market. HAZOP analysis project inevitably exist low bidding, low level of analysis process, and poor quality of analysis results and other issues. If these problems are not solved in time, the market will misunderstand the HAZOP method, and even abandon this method. The author has been engaged in HAZOP analysis for many years, Through a large number of investigation and research, This paper summarizes several aspects of evaluating the quality of HAZOP analysis from the technical view, hoping to provide effective means for enterprises to quality control of HAZOP analysis.

1. Whether the HAZOP analysis report is complete or not

The completeness of the HAZOP analysis report is the most essential element for evaluating the quality of the HAZOP report. At a minimum, the report should include an overview of the project, analysis data information, analyst information, analysis node information, analysis meeting record information and analysis node graph.

2. Whether the overview is comprehensive or not

Usually, there is a chapter in the HAZOP analysis report that explains the source of the project, the execution of the project, the purpose, object, scope, etc. of the HAZOP analysis. As a result, the report readers can understand at a glance, the number of drawings, the number of nodes and the time spent in the report. The timing of the HAZOP analysis is preliminary design or detailed design? The device of the HAZOP analysis is new or in service? The purpose of the HAZOP analysis is to identify safety and operational issues in the process or to determine the impact of the installation site on public safety.

3. Whether the team members are configured properly or not

The HAZOP analysis report should include the names, the professional or technical positions of all members of the HAZOP analysis team, and even the signature page for all members of the group or the HAZOP meeting sign-in form.

It is well-known that the essence of the HAZOP analysis method lies in "brainstorming": The analysis group composed of people with different knowledge backgrounds work more creatively and systematically than they work alone, and can identify more problems [1]. A team of people configured properly is the basis for a high-quality HAZOP analysis. Analysis team should have 4-6 members. At a minimum, hosts, recorders, designers who are familiar with processes and operations and the owner should be included. If the device is new, designer is best to attend. Some HAZOP analysis team shave only three members, namely the host, the recorder and a process engineer from the owner. Obviously there is no way to achieve a "brainstorm". HAZOP analysis team with too few people will not get high-quality analysis results due to the limited knowledge and experience of participants. Although one person can complete the entire HAZOP analysis, this analysis cannot be called a HAZOP analysis at all.

4. Whether the analysis data is complete and accurate or not

The HAZOP analysis report should include the data used in the analysis. PID drawings, PFD drawings, as well as operating procedures, instrument control charts, logic diagrams and so on are commonly used. PID chart is the most important. The data used in the analysis should be uniquely identified. Especially HAZOP analysis of in-service installations, process safety information kept by the enterprises is often not complete, and the drawings are of poor quality. The information provided by field technicians and operators involved in the analysis and the operational experience on the site are all important basis for the analysis. All these should be explained. Only in this way can we effectively verify whether the follow-up rectification work has implemented the recommended measures or improvement proposals proposed by the HAZOP analysis and that the HAZOP report is readable and traceable.

Divided nodes are best identified in different colors on the PID drawings, which are an integral part of the HAZOP analysis. If the node is too small or too divided, there will be a lot of duplication of work. If the nodes are too large, the results of the HAZOP analysis will be greatly deviated, and some results will even be missed. It can be seen from this that the smaller nodes are much better than the larger nodes if time allows.

5. Whether the analysis deviation is comprehensive or not

There are some different ways to identify deviations in the HAZOP analysis team. The most commonly used are "lead-based methods" and "process-based methods." The guideline-based approach is to use all of the process parameters sequentially for each guideword, and the method based on the process parameters is to use all the guidewords sequentially for each process parameter. Usually nodes include a series of interlinked process parameters, it is best to use the guidewords for each process parameters, such the guidewords for flow are no, too low and too high, and then the guidewords for liquid level is no, too low and too high.

Commonly there are seven guidewords used in the HAZOP process, respectively, no, too much, too little, with, part, reverse and anomalies, refer to IEC 61882-2001. If each node involves five process parameters on average, it will produce 7 * 5 = 35 deviations, if there are 20 nodes in this device, it will produce 7 * 5 * 20 = 700 deviations, the number of deviations is obviously Huge, even though many of these deviations may be meaningless. Occasionally, in order to improve efficiency, hosts or recorders tend to screen deviations before the HAZOP analysis meeting to determine the deviations suitable for this node. This screening process may lead to some deviations due to the lack of experience or knowledge. Therefore, if screening is performed, it is necessary to ensure the reliability and completeness of the selected deviation. As a rule of thumb, the equipment or procedures in the nodes are a key part of the HAZOP analysis and should include, as a minimum, a corresponding deviation analysis based on the type of equipment (eg, reactor, tower, pump, tank, heat exchanger, etc.) It has been analyzed either upstream or downstream of this node that it can no longer be considered.

6. Whether the record of the meeting are clear or not

The record of the HAZOP analysis meeting is the most important part of the HAZOP report and is the outcome document of the entire HAZOP project that reflects the level and workload of the HAZOP analysis team best. Many companies and agencies have established different HAZOP analysis results file methods. Below are some common ways to prepare HAZOP meeting records.

(1) All causes, consequences, protective measures and recommended measures are associated with a deviation. This method of presentation is relatively simple when analyzing records, but every reason listed in this deviation does not necessarily have all the consequences. In other words, there is no correspondence between causes, consequences, protective measures and recommended measures relationship. With such records, those who did not participate in the analysis need to re-infer the causes, consequences, protective measures and the relationship between the proposed measures, or restore the scene of the analysis at that time, its readability is not strong.

(2) Only unusual meeting records. In such records, only those deviations that the analysis team considered to be reliable and of serious consequence were recorded and deviations that did not have serious consequences were not recorded. In this record, business managers simply cannot judge whether the HAZOP analysis traverses all possible deviations or whether the missing records are considered safe to be analyzed or simply missed and not analyzed.

(3) Records of meetings with suggested measures only. Such records are considered to minimize the length of the document and only record the recommendations of the analysis team to enhance safety. While such a record could provide a risk management strategy, the quality of the analysis cannot be demonstrated at all.

(4) In summary, the reason-based recording method now is the best, as shown below. In the records, there is an accurate correspondence between the causes, consequences, protective measures and recommended measures. Whenever readers read this report, they can accurately find a variety of reasons for a deviation, the corresponding consequences of each reason and protective measures. Therefore, this method of recording HAZOP meetings is the most accurate and least ambiguous.

Figure: HAZOP analysis records indicate intent

			Date	2017.11.1	
			Members	Host : ZHANG Xue Secretary : DING Fang Fang Participants : ###	
Parameter	Guideword	Reason	Consequence	Existing protective measure	Suggested action
		Oil from the system into the device flow is low	Oil pump P-1/2 pumping failure, raw material leakage, leakage heat exchanger unilateral heat, heating furnace raw material interruption, the tube burned through, the fire explosion	FIC1101 ; PG1002 ; Combustible gas alarm	
Flow	too low/none	Pre-pump filter plugged	Oil pump P-1/2 pumping failure, raw material leakage, leakage heat exchanger unilateral heat, heating furnace raw material interruption, the tube burned through, the fire explosion	FIC1101 ; FIC1102 ; PG1002 ; PG1003 ; Combustible gas alarm	1, it is recommended that theoil pump P-1/2 before and after the filter additional pressure gauge 2, the proposed Marui oil pump P-1/2 before and after the filter additional heat
		Oil pump P-1/2 failure to stop	Oil pump P-1/2 pumping failure, raw material leakage, leakage heat exchanger unilateral heat, heating furnace raw material interruption, the tube burned through, the fire explosion	Spare pump ; FIC1101 ; FIC1102 ; PG1002 ; PG1003 ; Combustible gas alarm	3, the overall proposal: the pump operating status signal and other parameters into the DOS remote display
		FIC1101 circuit failure , FV1101too big	Oil pump P-1/2 pumping failure, raw material leakage, leakage heat exchanger unilateral heat, heating furnace raw material interruption, the tube burned through, the fire	Off before the crude oil heat exchanger outlet temperature	

explosion

Figure: HAZOP analysis records indicate intent

7. Whether the proposed measures are feasible or not

The HAZOP analysis focuses on the problems discovery, not on the solution. So the recommendations made by the HAZOP are often easy-to-implement solutions, not "design solutions." In order to avoid inadvertent changes and introduce new dangers, the modification of the process should be done by professionals. Measures proposed by HAZOP analysis mainly come from three aspects, one is how to avoid or detection of the causes of deviation early, such as monitoring methods like controlling valves and increasing the alarm, isolation methods like using blinds and checking valves, verifying whether the design parameters are suitable for on-site conditions and so on. Another aspect is how to reduce the severity of the consequences of deviations, such as safety valves and cofferdam. The last is to increase the workability of workers, such as automatic control, boosting agencies, security, operational space and other issues.

Enterprises should evaluate the proposed measures in the report further and strike a balance between cost control and risk reduction according to the principle of reasonable and feasible risk management so as to accept the proposal, accept it in part, propose better measures and further Research, or even refuse to do so.

8. Conclusion

As can be seen from the above description, the key to complete a high-quality HAZOP analysis is people. But the accuracy of the HAZOP analysis data; the experience and expertise of the HAZOP team; the time taken for the HAZOP analysis as this may affect the ability of the team to traverse deviations, envision the cause of the consequences; and the integrity of HAZOP logger records, etc. All of the factors have a significant impact on the HAZOP analysis quality.

References

[1] Liao Xuepin Hazard analysis of chemical process [M] Beijing：Chemical Industry Press，2005,32-44

Study on Safety Performance of Nitrogen Sealing in Internal Floating Roof Tank

Wu Zuoxiang , Li Jiayi
(CNPC Research Institute of Safety and Environment Technology, Beijing)

Keywords: internal floating roof tank; safety performance of nitrogen sealing; ventilation volume

Abstract

Nitrogen sealing technology used in internal floating roof tank is a common means of VOC treatment. The internal floating roof tank adopts nitrogen sealing technology, which reduces the breathing loss, guarantees the quality of oil storage, and reduces the possibility of fire and explosion caused by oil and gas accumulation. However, the safety problems caused by nitrogen sealing technology are rarely considered, which may occur in practical applications, especially the design of nitrogen sealing ventilation volume. In this paper, nitrogen sealing system should consider the ventilation volume under various working conditions and other safety problems.

Under the present condition of vigorously promoting the energy saving and emission reduction, the control and reduction of VOC emission become the key mission in the petrochemical industry and the nitrogen sealing technology used in internal floating roof tank is one of the important means for VOC governance. The internal floating roof storage tank is made by adding the floating roof inside the vault storage tank, the floating roof added inside the tank can reduce the vaporization loss of medium and the external vault can prevent the rainwater, snow and dust entering into the tank to keep the medium inside the tank clean. However, the breather valve of internal floating roof tank is directly connected to the atmosphere, the pressure of storage tank is lower than the atmospheric pressure, the air enters into the storage tank and the oil product contacts with air to react and deteriorate, thus the quality of oil product will be affected. If the pressure of storage tank is higher than the atmospheric pressure, lots of oil product will volatilize into atmosphere, the fire and explosion accidents may occurs. So, as for the internal floating roof storage tank storing the combustible liquid of low flash point, the nitrogen sealing system is usually equipped in the top of storage tank and the nitrogen is filled into the gas-phase space in the top of storage tank to isolate the air; meanwhile, the nitrogen is filled inside the storage tank to reduce the vaporization of oil and gas and improve the storage safety of oil product.

1. Principle of nitrogen sealing technology

The nitrogen sealing technology is to use the nitrogen to supplement the gas space inside the tank. Because the nitrogen is lighter than oil and gas, the nitrogen floats on the surfaces of oil and gas. When the breathe-out course occurs, the nitrogen rather than the vapor of oil product is breathed out the tank. Breathe-in course occurs when the pressure inside the tank lowers, the nitrogen automatically enters into the tank to supplement the gas space, reduces the vaporization loss and prevents the oil product to contacting and oxidizing. Take the internal floating roof storage tank storing light oil of one company as an example and the process flow diagram of its nitrogen sealing system is shown as Figure 1.

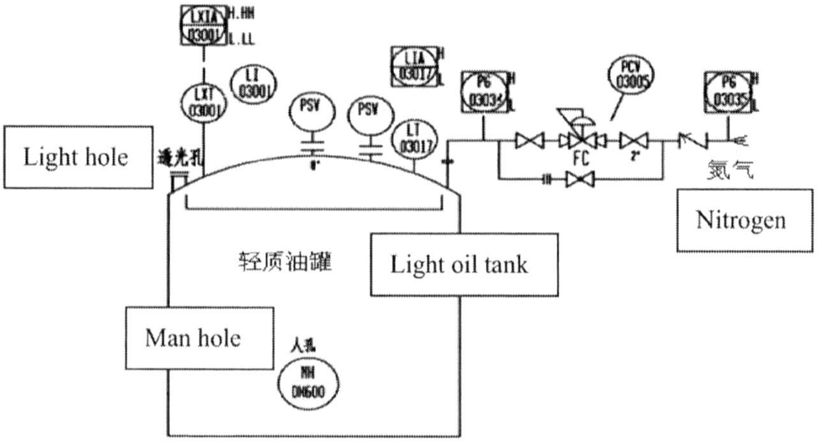

Fig. 1 Process flow diagram of nitrogen sealing system

Set the nitrogen sealing system, set the pilot-operated type nitrogen sealing valve group and restriction orifice bypass in the storage tank, the gas-phase space pressure inside the tank is maintained at about 1.2KPa during the normal use condition of nitrogen sealing valve group. When the gas-phase space pressure is higher than 1.4KPa, the nitrogen sealing valve is closed and the nitrogen supply stops; when the gas-phase space pressure is lower than 0.8KPa, the nitrogen sealing valve is opened and the nitrogen supply starts to ensure the air will not be breathed in during the normal operation process of storage tank and prevent the formation of explosive gases. The purity of nitrogen used by storage tank nitrogen sealing system should not be lower than 99.96% and the pressure of nitrogen is desirable to be 0.5-0.6MPa.

If the nitrogen sealing valve is malfunctioned and cannot be timely closed which causes the pressure inside the tank to exceed 1.5KPa, vent the nitrogen outwards through the breather valve with flame arrestor; if the nitrogen sealing valve is malfunctioned and cannot be timely opened which cause the pressure inside the tank to lower to -0.3KPa, supplement the air into the tank through the breather valve with flame arrestor to ensure the pressure inside the tank will not be lower than the low limit of design pressure (-0.5Kpa).

2. Ventilator capacity calculation

The designed nitrogen sealing system reduces the vaporization loss and prevents the oil product deterioration, but the nitrogen sealing system comes with its own risks. When the nitrogen sealing valve is malfunctioned and fully opened, the pressure inside the tank is extremely high, the physical explosion of storage tank may occur to cause accidents once the venting capacity of breather valve cannot meet the requirement. So the ventilator capacity calculation of breather valve is the most important thing of nitrogen sealing system, the specification of breather valve in the tank top needs to be checked to confirm whether it can satisfy the requirements after set the nitrogen sealing. In addition to satisfy the big and small breathes of storage tank, the ventilator capacity of breather valve also should take other factors into consideration such as the inflow capacity in the case of the nitrogen sealing valve cannot be closed.

The big breath means that the liquid flows into and out the storage tank, the liquid level rises and falls to cause the increase and decrease of gas volume which will cause the change of static pressure difference, this breath function formed due to the change of liquid level inside the tank is referred to as big breath process, its breathing air capacity is independent of storage tank type; the small breath means that the change of environment temperature causes the oil product vapor full of storage tank gas-phase space to rise temperature and expand (breath out) or to cold and coagulate (breath in). In summer, strong oil and gas smells may be around the tank during feeding. This phenomenon is jointly caused by big and small breathes, the oil and gas vented from the breather valve in the tank top volatilize into atmosphere.

2.1 Ventilator capacity calculation of big breath

The ventilator capacity vented by oil tank in the big breath can be calculated according to empirical formula (1).

$$V_1 = \frac{G(P_1 - P_2)}{P_2} \qquad \text{[Eq. 1]}$$

which:
V1—Ventilator capacity vented by oil tank in the big breath (m3)
G—Oil product amount pumped out from oil tank (Kg)
P1—Annual average atmospheric pressure (KPa)
P2—Oil product vapor pressure (KPa)

2.2 Ventilator capacity calculation of small breath

The ventilator capacity vented by oil tank in the small breath can be calculated according to empirical formula (2).

$$V_2 = \frac{1.9\times10^5 \rho(507+9/5t_2)P_1 t_1^{0.5} D^{1.73} H^{0.51}}{MP_2^{0.32}(P_1 - P_2)^{0.68}} \quad \text{[Eq. 2]}$$

In which:
V2—Ventilator capacity vented by oil tank in the small breath (m3)
ρ— Density of oil product inside the tank(Kg/ m3)
t1—Maximum diurnal average temperature difference (K)
t2—Temperature of oil product of inside the tank (K)
P1—Annual average atmospheric pressure (KPa)
P2—Vapor pressure of oil product inside the tank at the moment of t2 (KPa)
D—Oil tank diameter (m)
H—Average height of oil and gas space (m)
M—Molecular weight of oil product vapor

2.3 Ventilator capacity calculation in the case of nitrogen regulating valve malfunction

In the normal operating condition, the ventilator capacity required by nitrogen sealing is the sum of big and small breathes. In the case of nitrogen sealing valve is malfunctioned and cannot be timely closed or even fully opened, the ventilator capacity in this operating condition is larger than the sum of big and small breathes and the ventilator capacity is equal to the maximum flow of nitrogen regulating valve. If the ventilator capacity of breather valve is smaller than the maximum flow of nitrogen regulating valve, then the breather valve cannot timely vent the nitrogen flow in the case of regulating valve malfunction, the storage tank is to explode due to extremely high pressure. So the ventilator capacity of breather valve in the case of nitrogen regulating valve malfunction should be taken into consideration in the design process of storage tank nitrogen sealing, the ventilator capacity in this operating condition should not be smaller than the maximum flow of nitrogen regulating valve. In this way, even the nitrogen regulating valve malfunctions, the air can be timely vented through the breather valve and the storage tank will not be overpressure, the nitrogen sealing of internal floating roof tank can achieve the essential safety.

3. Safety issues during the nitrogen sealing using

According to the application situations of nitrogen sealing system, the following issues should be taken into consideration when design the nitrogen sealing system. (1) The filter should be equipped in the front of regulating valve to prevent the rusts and impurities of nitrogen pipeline entering into the nitrogen regulating valve to cause the nitrogen regulating valve blocked and plugged. (2) The position of connection pipe of nitrogen supply pipe should be away from the connector of breather valve and it should insert in the storage tank to a certain depth from the tank top, so that the nitrogen will not be directly vented after the nitrogen enter into tank, then the purpose of nitrogen sealing can be achieved. (3) The pressure remote transmission indicator and alarm should be equipped in the front and back of nitrogen pressure regulating valve, so that the causes of malfunction can be judged based on various pressure indication situations in the case of nitrogen sealing system malfunction. (4) In order to prevent the sudden interruption of nitrogen supplied from public works and the oil and gas inside the storage tank flow back into the nitrogen system, the one-way valve should be equipped in the nitrogen supply pipe.

4. Conclusions

The internal floating roof storage tank adopts the nitrogen sealing technology to supplement the clean nitrogen into the tank which avoids inhaling air and prevents the oxidization of oil product. At same time, it greatly reduces the breathing loss which can effectively restrain the vaporization and breathing of light component in the oil product and ensure the storage quality of oil product. By adopting the nitrogen sealing technology, the gas space in the storage tank is gas mixture of flammable liquid gas phase vapor and nitrogen, it can lower the flammable liquid concentration to keep it out the gas explosion limit range, so the tank does not have the conditions for the combustion explosion. Due to the breathing loss of oil product has been greatly reduced and the nitrogen is breathed out, the poisonous and harmful oil product vapor in the operating space has been effectively reduced, then the environmental pollution caused by oil and gas can be avoided.

The completeness of internal floating roof storage tank structure and its safety facilities relates to the essential safety of storage tank, so the ventilatory capacity of breather valve should be correctly calculated in the nitrogen sealing design process and the safety issues in the using process should be noted.

5. References

[1] Zheng X.Z., "Application of Oil Tank Nitrogen Sealing System," *Petroleum Products Application Research*,Volume 3 ,No.41,1999

[2] Xiao C., "Important Meaning of Nitrogen Sealing Technology Applied in Light Oil Tank," *Petrochemical Safety Technology*, Volume 6, No.19-21,2004

[3] Wen G.M., Wu D.Z., "Technology Improvement of Vertical Vault Oil Tank Nitrogen Sealing System,"*Oil and gas storage and transportation,* Volume 9, No.9-11,1997

[4] Qiao Z.L., "Liquid Sealing Height Calculation of Storage tank hydraulic relief valve,"*Oil and gas storage and transportation*, Volume 5, No.24-26,2003

[5] SH/ T3007 - 2007 Specification for design of tank farms for petrochemical storage and transportation system

[6] HG/T 20570.16-95 Gas sealing setting

Safety Control in Test Process of Coal Chemical Plant

Yang zhigang

Shenhua Ningxia Coal Industry Group, Yinchuan , Ningxia

Wang zhiquan,Sun yong,Ding yulin,Wangpenghong

Shenhua Ningxia Coal Industry Group, Yinchuan , Ningxia

Key words: coal chemical industry; safety in production; safety process control

Abstract

In recent years, some areas in our country relying on abundant coal resources, have launched coal chemical projects. While solving the energy shortage in the traditional petrochemical industry, coal chemical industry also conceals many potential safety hazards due to its long production process and complicated equipment, which can easily lead to accidents and seriously affect the safety of its production and the personal safety of related personnel, especially in the process of device test, which is the stage frequent accidents occur. Accordingly, we should attach importance to and implement the safety management of coal chemical production to improve the safety control in the process of trial production, reduce the safety risk and improve the production efficiency, laying a solid material foundation for the social and economic development of our country. This paper mainly expounds the software and hardware investment, the construction of safety management system, the analysis of unsafe factors, the establishment of the system, the identification of risks, the management of hidden dangers and the establishment of the emergency system in the process of trial running, and explores the new ideas and new direction of safety control and management in the process of commissioning of the coal chemical industry.

Preface

The production of coal chemical enterprises has its remarkable particularity. During the process of production test, there are many hidden dangers, which seriously affect the safety of production safety and related personnel. It is of great practical significance to strengthen the production safety management of coal chemical enterprises in order to control the safety hidden dangers in test process of coal chemical production, as well as reduce its safety risks and raise production efficiency.

1.Analysis of unsafe factors in test process of coal chemical production

1.1 Flammable, explosive, high temperature and high pressure

Most of the raw materials, intermediate products, products and by-products in various coal chemical production processes are flammable and explosive media, and many materials are in high temperature and high pressure in the production process. From the input of raw materials up to the output of the products, as well as in the process of storage of raw materials and products, poor sealing of equipment, facilities and pipes, misoperation of operators, abnormal fluctuation of process parameters of high temperature and high pressure equipment, may lead to leakage of process media and trigger fire and explosion accident.

1.2 Poisoning, asphyxiation and corrosiveness

During the trial run of coal chemical plant, toxic, corrosive materials such as carbon monoxide, ammonia, benzene, toluene, methanol, hydrogen sulfide and sodium hydroxide will be generated. At the same time in the process of equipment on and off, the need to replace the process gas in the equipment and pipes with nitrogen, high concentrations of asphyxiating materials in the equipment, or entering the equipment before detection of the oxygen content or failing to take protective measures will cause poisoning, suffocation and other dangerous if entering the device.

1.3 Other factors

In addition, during the trial operation of coal chemical plant, dust, noise, vibration, crashes, mechanical injury and radioactive radiation may also cause danger and harm to workers.

According to the above, there are many unsafe factors in the trial production of coal chemical industry, which determines that the requirements for coal chemical enterprises are getting higher and higher in terms of safety technology, safety production and management.

2. Effective strategy of safety management in the trial production of coal chemical industry

2.1 Improving staff safety awareness

Before the commissioning of coal chemical plant, all managers and operators must firmly establish safety awareness. Coal chemical enterprises should systematically educate employees so that employees always pay attention to safety during their work and develop good safety habits.

In the trial production process, operators should strictly follow the safety norms to eliminate "three violations" to ensure that the implementation of all safety measures in place. At the same time, safety education and training should be strengthened. Through the "three-level" safety education and various types of safety publicity and education activities, the safety awareness of workers should be continuously improved so that employees' awareness should be gradually changed from 'asking me for safety' to 'I want safety'.

2.2 Establishing and improving the responsibility system for safety production

The coal chemical industry should set up the responsibility system of production safety for the first person in charge before the test run. The safety production responsibility system must cover all the positions and personnel in the enterprise, clarify the division of safety responsibilities, and implement the responsibility management at different levels, so as to put safety responsibility in place.

2.3 Establishing a long-term safety management system

Managers at all levels of coal chemical industry should lead by example and change the wrong concept that whether the accident happens or not as an evaluation model to measure the quality of the enterprise's safety work, advance the safety management and establish and improve the safety management system with risk pre-control as the core, and put the production safety management rules and regulations into the specific work. At the same time, the enterprises should establish and improve the appropriate safety awards and assessment methods, fixed inspection assessment, implement strict assessment system, regularly organize the assessment of enterprise safety management and control, formulate improvement measures and gradually establish a long-term safety mechanism.

2.4 Strengthening the risk identification in the process of coal chemical production trial production

HAZOP analysis should carried out from the design phase of the process package design, basic engineering design, detailed engineering design and commissioning. It is not only conducive to the construction of intrinsically safe coal chemical enterprises, but also saves investment costs. With the large-scale production of modern coal chemical industry and complexity of technology and product, along with the probability of occurrence of accidents and greatly increased degree of harm, equipment failure, design flaws, operating conditions failure, hazard control failure, human error and so on all easily lead to major production safety accidents and disastrous consequences. Using HAZOP analysis technology, risk identification is conducted before the design and production operations to identify potential hazards and then take effective preventive measures to ultimately reduce risks and eliminate accidents.

2.5 Increasing the hardware and software input in the safety production of coal chemical industry

During the operation of coal and chemical enterprises, unscrupulous factors such as equipment aging and wear of parts may appear, which will inevitably bring about new hidden dangers and dangers to employees in their work safety. Therefore, we must attach great importance to the safety of production equipment and facilities. Coal chemical industry enterprises shall, according to the requirements of the relevant laws and regulations, systematically withdraw the safety production costs in a planned way and make full withdrawal and earmarking only. In addition, it is necessary to systematically reorganize and reform existing production equipment every year, and ensure the safe and reliable operation of production equipment and facilities from technical measures. The enterprise should strengthen the introduction of research and development and utilization of safety technology, adopt advanced and reliable monitoring and early warning systems and equipment, increase investment in hidden danger management, minimize the loss and impact of accidents, prevent the equipment from running with sickness, optimize the operation of the process, constantly upgrade the equipment and facilities of enterprises safety level, and eventually provide material assurance for long-term safe and stable operation.

2.6 Strengthening supervision and inspection of safety production

Coal chemical industry should further strengthen supervision and inspection of work safety and carry out supervision and inspection of work safety in all aspects of production. By using routine inspection, special inspection, quarterly inspection and comprehensive inspection, the enterprises will promptly detect hidden troubles and eliminate hidden dangers and eliminate accidents in the bud, and ensure the safety of production operations.

2.7 Actively and effectively carrying out safety standardization

Active and effective safety standardization work can help enterprises effectively implement the main responsibility for safety in production. At the same time, to carry out safety standardization work is also an effective means for enterprises to improve their safety management level and improve the conditions for safe production. Enterprises should continue to do a good job of safety management, in accordance with the requirements of standardization, to maximize the protection of staff safety and health. In this way, the coal chemical production enterprises can achieve a sustained and healthy development and make greater contributions to the progress of the economy and society.

2.8 Establishing and improving the emergency rescue system of coal chemical industry

Coal chemical enterprises should strengthen the construction of contingency plan system and established a three-level contingency plan system consisting of comprehensive emergency plans, special plans and site disposal plans. Emergency plans should be feasible and can be revised in time in accordance with the actual situations that have taken place. At the same time, emergency supplies should be included in the emergency management, and sufficient number of emergency supplies may ensure they can meet the requirements in the accident. Plans for emergency drills and organizing emergency drills on a regular basis are also necessary. Through drills, emergency response and handling of accidents at all levels of personnel would be enhanced.

3. Conclusion

In summary, the coal chemical test process has a significant risk. The presence of a variety of risk factors seriously affect the safety of its production. Therefore, we should start from the management level, pay attention to and implement the safety management in coal chemical test process, strengthen personnel safety education and establish a long-term mechanism for safe production management, so as to effectively prevent unnecessary losses of property and casualties caused by safety hazards in the production process and promote the stable and sustainable development of the coal chemical industry.

4.References

[1] Zhang Honghui, "Research on strengthening safety production of coal chemical enterprises," Chemical management, Volume 27, 2016

[1] Li Huiying, "Analysis of safety production of coal chemical enterprises," China's science and Technology Expo, Volume 34, 2015

Based on the maintenance operation package mode coal chemical inspection and maintenance safety management practices

Tang Xuexiang
Shenhua Ningxia Coal Industry Group, Yinchuan , Ningxia

Key words: Overhaul homework package ; PDCA Closed-loop mode ; Process safety management ; maintenance site ; Troubleshooting

Abstract

The safety management about Inspection and maintenance for coal to oil chemical is a process safety management system. Maintenance work package is based on PDCA closed-loop mode, include four-step (plan–do–check–act). the inspection and maintenance operations process management provide a set of maintenance work package as the core of a series of process control methods. Maintenance work package provides a set of perfect and standardized inspection and maintenance safety operations procedures, standardizing equipment maintenance processes, operating methods prescribed for workers. Following the principle of who major who confirm , confirm the key points about quality and safety step by step . Further ensure the safe operation of the process controlled and improve inspection and maintenance quality and efficiency, so as to achieve the standard staff on duty, the standard of living standards to ensure that the overall inspection and maintenance process management is safe, high quality and progress- controlled.

1. Maintenance package implementation background and significance Content

Maintenance safety management is a huge systematic project, involving many factors: people, machines (machinery and equipment), materials (materials), method (maintenance work methods), the ring (working environment), due to the above element control Impact, maintenance safety management is often difficult to achieve the intended goal, process safety management can not be completely controlled.

Maintenance work package to the maintenance staff to provide complete, standardized and standardized maintenance operations procedures, to ensure the safety of the maintenance process under the premise of improving maintenance quality and efficiency, centralized management, focus on the entire process of maintenance projects, maintenance operations package is maintenance Operating procedures for the overall procedural, standardized and standardized management, the formation of a unified and standardized safety management system.

1.1 In order to standardize the operating procedures, to achieve the standard staff posts, dry standard live

The accident occurred mostly because of human's unsafe operation, so it is very important to regulate people's operation method in the equipment inspection and maintenance process. Overhaul of work packages from the preparation and implementation of the process have reached the program, standardization and standardization, in accordance with national and industry standards for the development of safe operating procedures and maintenance standards, so that each process has provisions, standards, there are checks, for each Equipment compilation to form a systematic inspection and maintenance of safety management standards, to achieve the standard staff on the job, the standard requirements of dry work.

1.2 In order to improve the quality and efficiency of maintenance, inspection and maintenance process to achieve safe and controlled

In the equipment inspection and maintenance process to deal with the relationship between safety, quality and progress in order to maximize the benefit of the enterprise, maintenance work package to equipment as the object, based on human skills, process management as the basis, through the research equipment Run the process of the various parameters, accurate positioning of equipment operating defects, maintenance inspection standards based on repair carried out by key maintenance and repair of the key points of quality control step by step, to achieve safety inspection and maintenance, quality inspection standards to improve Maintenance efficiency, and thus efficient and reasonable completion of inspection and maintenance activities of the organization and planning activities and resource allocation.

1.3 In order to regulate the maintenance operation procedure, set up inspection and maintenance safety management mode

Maintenance work before the overhaul of the inspection package will often have some problems, no clues to the job, messy workplace, the environment makes the work of workers is not motivated, delayed overhauling delay ... These issues to the equipment maintenance and safety management Bring a great impact, so that work efficiency greatly reduced. Maintenance work package to a large extent maintenance operations provide a comfortable working environment and safe place of work to eliminate potential safety problems and further enhance staff enthusiasm for work and improve work efficiency to ensure the quality of equipment overhaul and improve equipment operation cycle , From the equipment maintenance preparation, implementation, acceptance and evaluation to improve the establishment of a set of system maintenance operation safety management mode.

First, the preparatory work before the full sufficiency, from the staff preparation, safety preparation, equipment and material preparation, technical preparation, on-site preparation, training positions prepared for the deployment of systematic arrangements to ensure the smooth implementation of the maintenance process.

Second, before the maintenance of maintenance personnel on the safety skills, inspection and maintenance level assessment, organization of safety knowledge and maintenance training methods, equipment maintenance process to carry out desktop exercises to eliminate accidents caused by overhauling blind operation.

Third, standardize the security conditions before equipment maintenance check. In accordance with the safety management requirements of "one single five cards" (ie, maintenance orders, risk identification cards, risk control cards, energy isolation cards, quality acceptance cards and emergency disposal cards), the content of the overhaul should be clearly identified and the dangerous hazard factors identified according to the code Standards and management system to develop a risk control measures to implement the table, by the "operating person in charge" one by one implementation, "guardian" in each process before the re-confirmation, to ensure that all risks are controllable; secondly, according to the equipment may accidentally release energy to formulate energy Isolation list and the equipment operator to implement energy isolation, the operator isolation effect test to ensure that all energy isolation; Finally, according to the job site environment by drawing the emergency escape route map, the development of emergency measures and to inform all staff involved in the operation , Improve safety awareness.

Fourth, standardize the implementation of equipment maintenance phase of the dynamic identification of risk management. Before the operation of equipment during daily operations, the "responsible person" should lead the whole staff to carry out a comprehensive risk identification and assessment and implement the safety measures one by one. The safety personnel should carry out safety experience sharing, and the person in charge of work arrange the contents of the day and inform each link of existence Risk and preventive measures, the guardian of the whole process of supervision of safety risk prevention and personnel insecurity, and enhance safety risk control.

Fifth, the entire equipment maintenance process are strictly in accordance with standard procedures. Maintenance work site a variety of industrial equipment, safety equipment, maintenance supplies, spare parts arranged in an orderly manner to ensure that the working environment in good order; disassembled parts in accordance with the "flip down" the order of the orderly display to prevent due to work The program is complicated and leaks, misplaced phenomenon, avoid rework and increase the workload.

Sixth, standardize equipment maintenance methods to enhance the quality of work. Carry out detailed measurement, inspection and repair on each part, and record the inspection conditions and measurement data in detail. According to the principle of establishing a record card for each part, check the contents, measurement data, measurement methods used, judgment results Recorded in the record for the future safety and stability of equipment to provide data analysis basis. In addition, the measured data of some equipment parts and components do not match the international, domestic and industrial standards, but it is applicable and feasible for the safe and stable operation of this equipment. After repeated measurements and verification, we will be included in the maintenance operation package to form the Taiwan equipment maintenance standards, the use of online or offline testing equipment in the operation of continuous verification of the accuracy of maintenance standards, the overhaul of the technical standards package is more instructive, thus reducing the workload of each maintenance, and ensure the safety of personnel and equipment to enhance Maintenance quality and efficiency.

Seventh, strengthen the supervision of safety and quality. In the overhaul, the key point of the safety and the quality of the key points of the implementation of program-based management, according to the critical stage of the risk level to determine the acceptance of responsibility, from the operator, responsible person, plant-level confirmation and company-level verification of the implementation of four levels of safety, Quality "four" to confirm the management, clear responsibility of each level of acceptance, to ensure the safety of personnel and equipment control.

Eight is to focus on the training of young staff skills. According to the contents of the overhaul prior to the maintenance of training program for the preparation of posts, training plan to develop by the "master" every day overhaul repair process in the field of practical operation training, each student based on what they learned and grasp the situation to write learning Summarize and deepen understanding to further enhance the skill level.

Nine is the implementation of equipment maintenance operations information and network management, so that each time the equipment maintenance data can be consulted at any time to establish a history of each equipment maintenance database for scientific equipment maintenance basis.

2. Maintenance work package implementation of the main approach

Stable operation of large units is an important part to ensure chemical plant "safety, stability, length, full, excellent", in order to do a good job in unit maintenance management to ensure the smooth progress of maintenance work, installation and maintenance branch in the unit overhaul full implementation of the operation package management. According to the PDCA mode, comprehensive control over the safety, quality and progress management of inspection and maintenance can be achieved from the aspects of overall preparation, overhaul implementation, inspection and acceptance, and evaluation improvement. Equipment maintenance operations to achieve the information management, the establishment of the history of each equipment maintenance database for the scientific maintenance equipment provided a basis for creating a sustainable improvement of maintenance inspection and maintenance management.

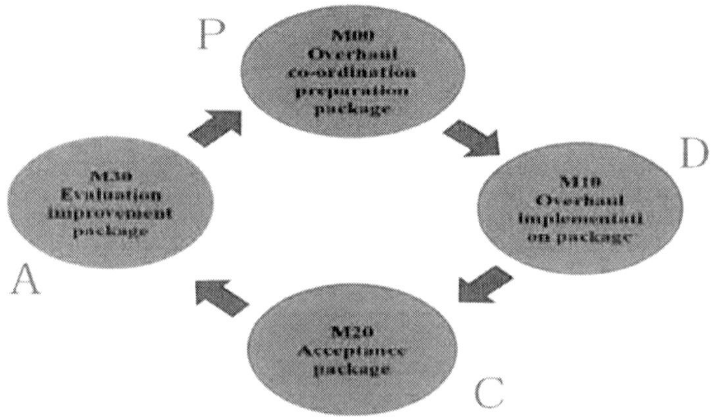
P
M00
Overhaul
co-ordination
preparation
package
D
M10
Overhaul
implementati
on package
C
M20
Acceptance
package
A
M30
Evaluation
improvement
package

Maintenance work package frame

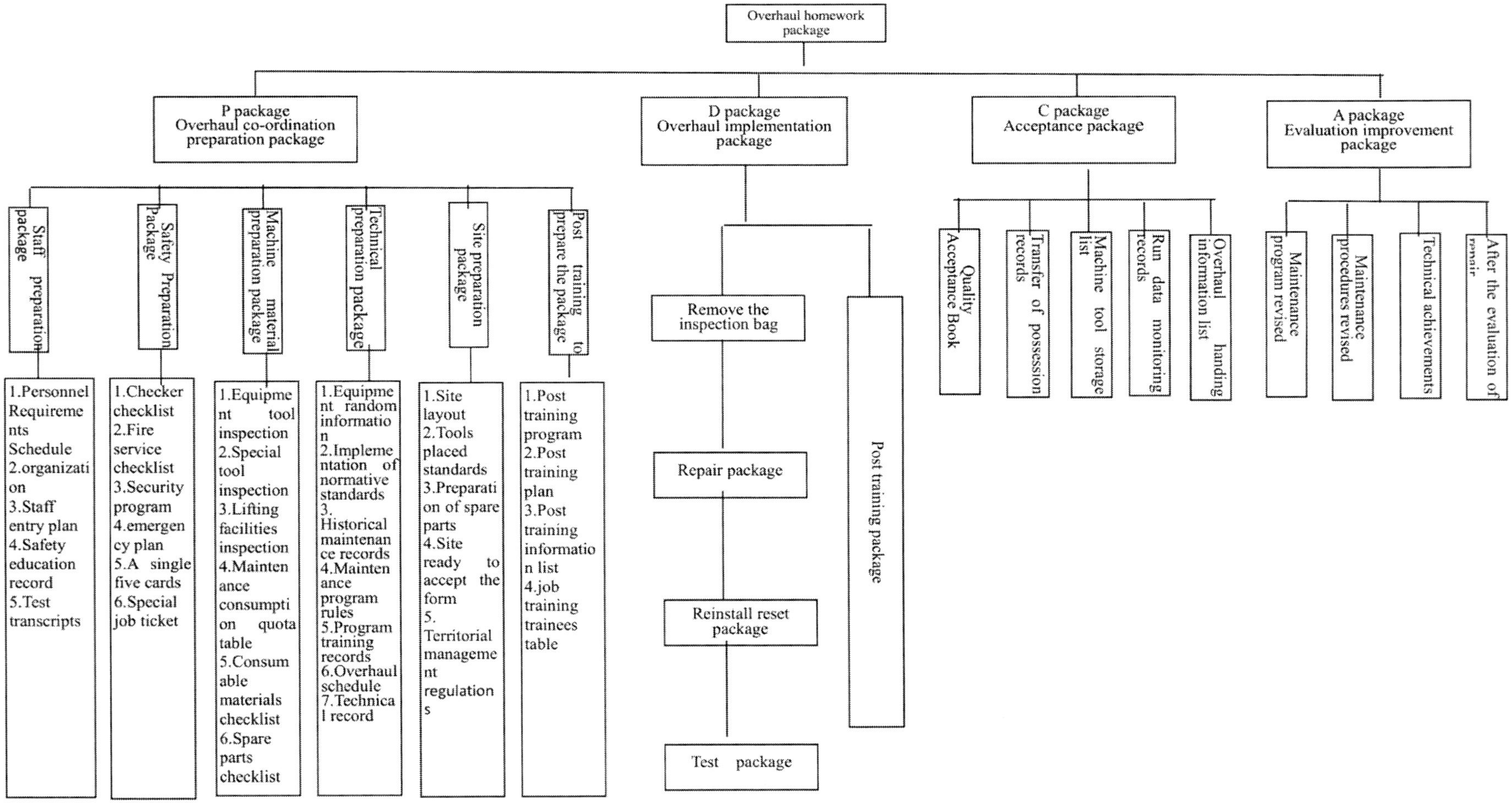

2.1 Overhaul co-ordination preparation package

Overhaul preparation package is a guide document that guides overhaul personnel to carry out various preparatory work for overhauling operation. According to the preparation contents, it is divided into personnel, safety, machine materials, technology, on-site and post training package as shown in Figure 2.

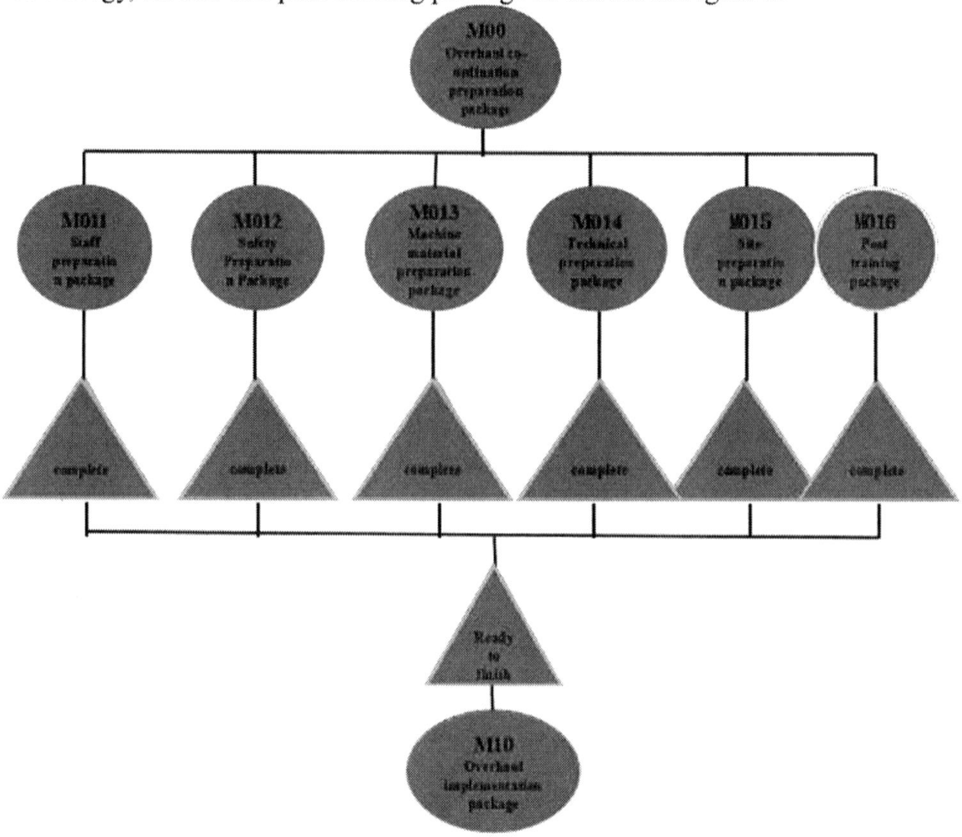

Fig. 2 Maintenance Preparation Package

2.1.1 Personnel preparation package

In accordance with the preparation of maintenance package package content (Figure 3), the development of personnel requirements plan, the establishment of organizational units, according to the maintenance time to develop personnel admission plans, organize safety education and examinations, payment of labor protection supplies and documents.

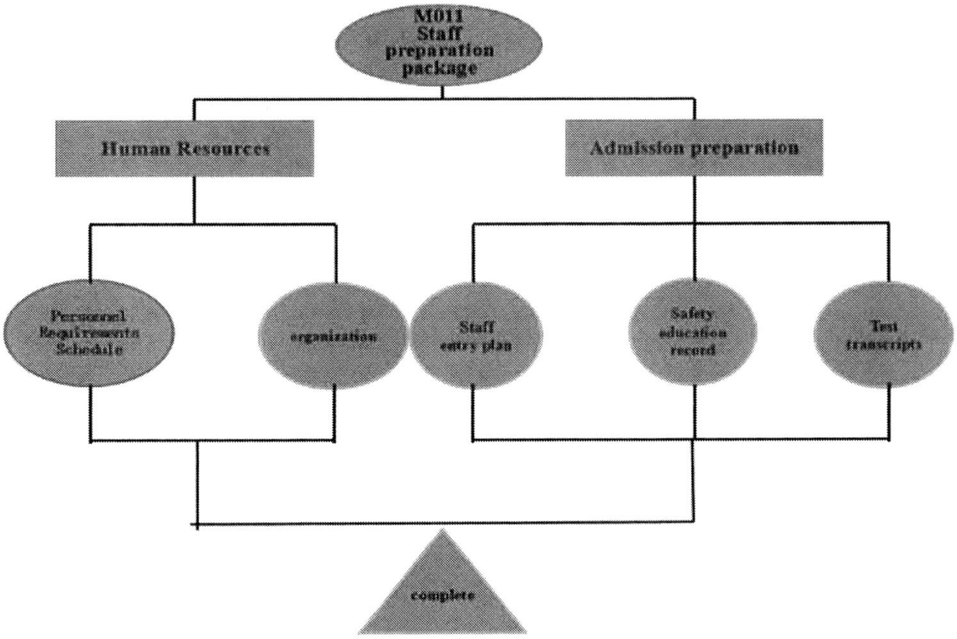

Fig. 3 Personnel preparation package

(1) Human Resources: According to the inspection items and contents, according to the maintenance man-hours quotas and the definite human resource requirements, we formulate the demand plan for the configuration of operators (see Figure 4), and set up the maintenance organizations according to the division of work, so as to clarify the safety of personnel at all levels Duties (see Figure 5).

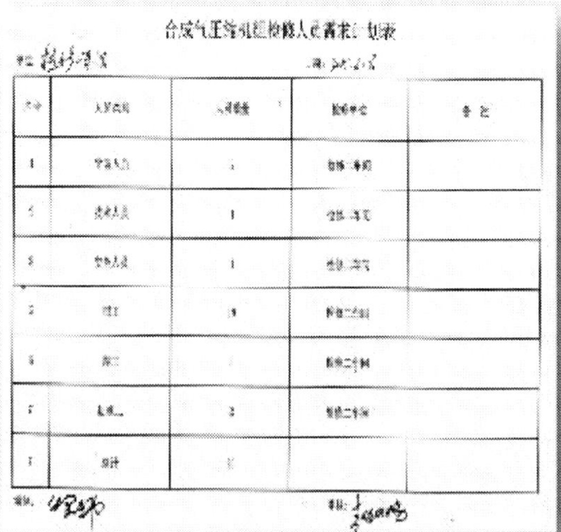

Fig. 4 Personnel Requirements Planning Chart

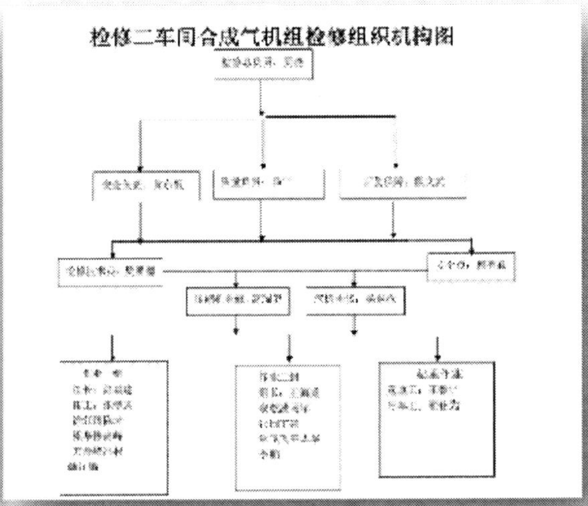

Fig. 5 Maintenance Organization

(2) Admission Preparation: After the human resources are in place, according to the inspection time, formulate the entry schedule of the personnel (see Figure 6), clear the approach time, and organize the safety education for the operators before the inspection, and issue the assignments after the education is qualified Document.

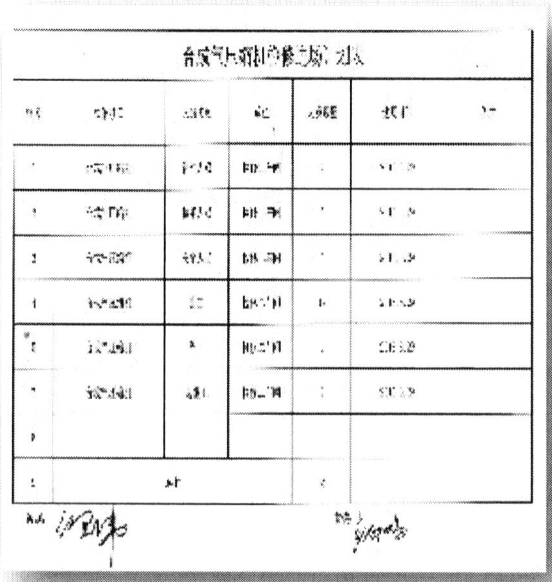

Fig.6 Staff entry plan

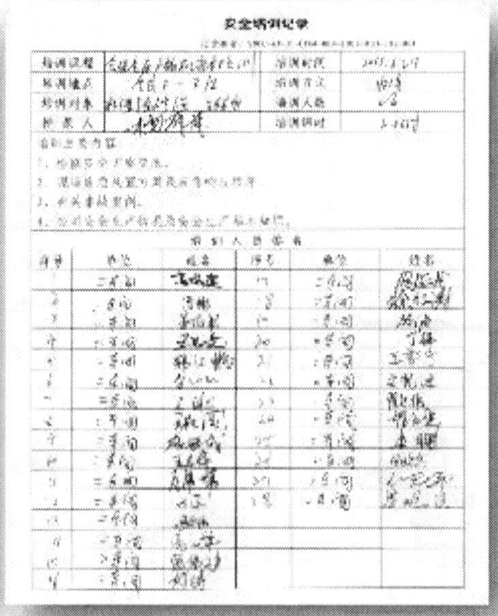

Fig.7 Safety training record table

2.1.2 Safety Preparation Package

The safety work kit (shown in Figure 8) during the preparation and implementation process is regulated by the preparation and implementation of safety equipment protective equipment, safety plan development, ticket handling and risk identification before maintenance. The safety preparation package During the operation, the existing safety risks are dynamically identified, and the control measures are formulated and implemented by the responsible person. The guardian confirms the safety and control of the overhaul operation.

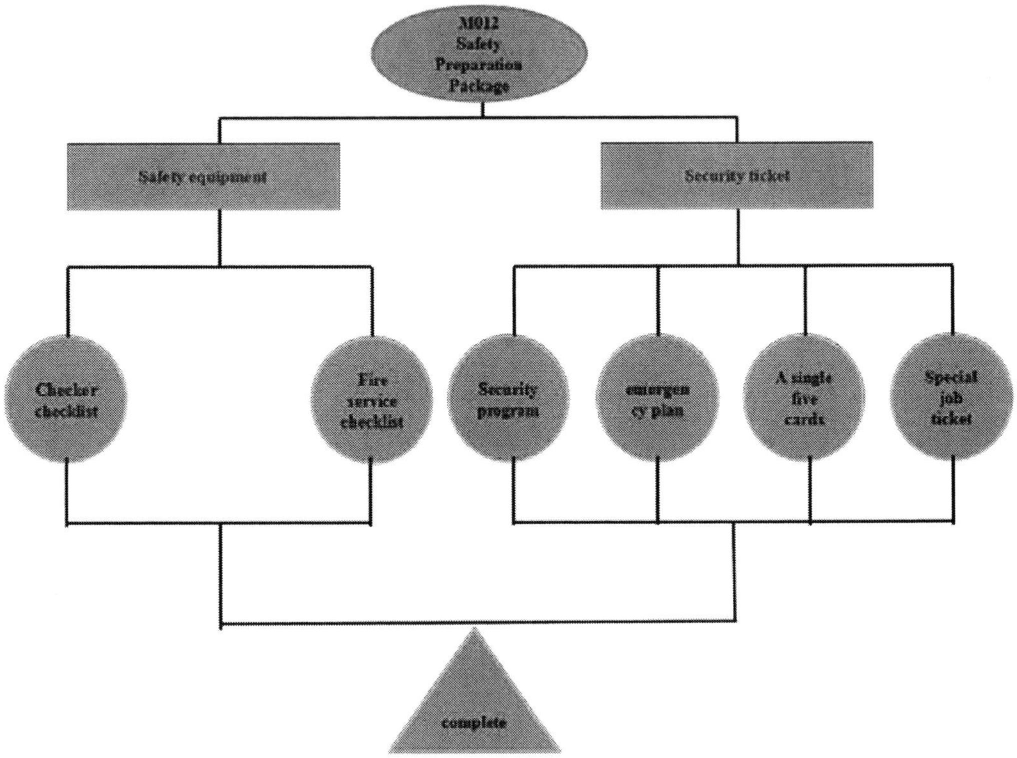

Fig. 8 Security Preparation File Package

(1) Safety equipment: Equipped with safety testing and protective equipment according to the working environment. Before use, it shall be inspected and recorded by a special person (see Figure 9) to ensure the complete use.

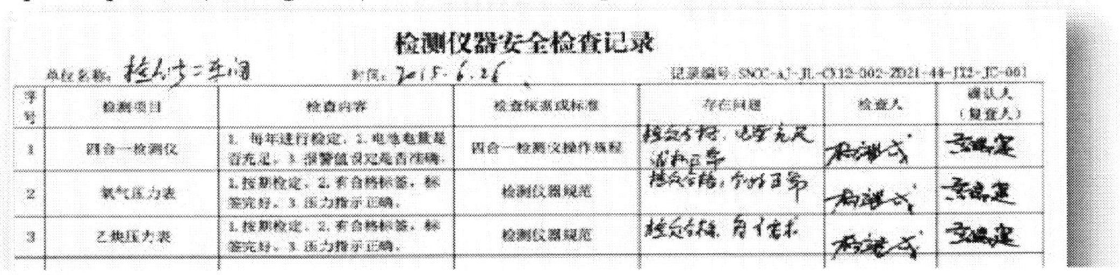

检测仪器安全检查记录

单位名称：[illegible]　时间：[illegible]　记录编号：SNCC-AJ-JL-C012-002-2021-44-JT2-JC-001

序号	检测项目	检查内容	检查依据或标准	存在问题	检查人	确认人（复查人）
1	四合一检测仪	1. 每年进行检定，2. 电池电量是否充足，3. 报警值设定是否准确。	四合一检测仪操作规程	[illegible]	[illegible]	[illegible]
2	氧气压力表	1.按期检定，2.有合格标签，标签完好，3.压力指示正确。	检测仪器规范	[illegible]	[illegible]	[illegible]
3	乙炔压力表	1.按期检定，2.有合格标签，标签完好，3.压力指示正确。	检测仪器规范	[illegible]	[illegible]	[illegible]

Fig. 9 Safety equipment testing record sheet

(2) Safety ticket: formulate safety management and control plan from aspects of equipment safety surrender, risk identification and Risk control management, process supervision and inspection so that the objectives and methods of process safety management are clear and responsibilities are up to the person; Emergency response plan and make timely response and disposal to emergencies. According to the requirements of "one sheet of five cards" and special operation management, various safety measures before operation are confirmed one by one to ensure the safety of the process.

2.1.3 Machine material preparation package

The adequacy of equipment materials preparation, a direct impact on maintenance progress. Before the repair according to the provisions of the contents of the kit material package (see Figure 10), the general maintenance tools required for maintenance, special equipment and lifting appliances to check and confirm, to ensure safe and reliable. Consumption according to maintenance costs fixed supplies needed to ensure that various types of maintenance consumables to meet the demand. Operators on the maintenance required spare parts for inspection to confirm, to ensure its quality standards.

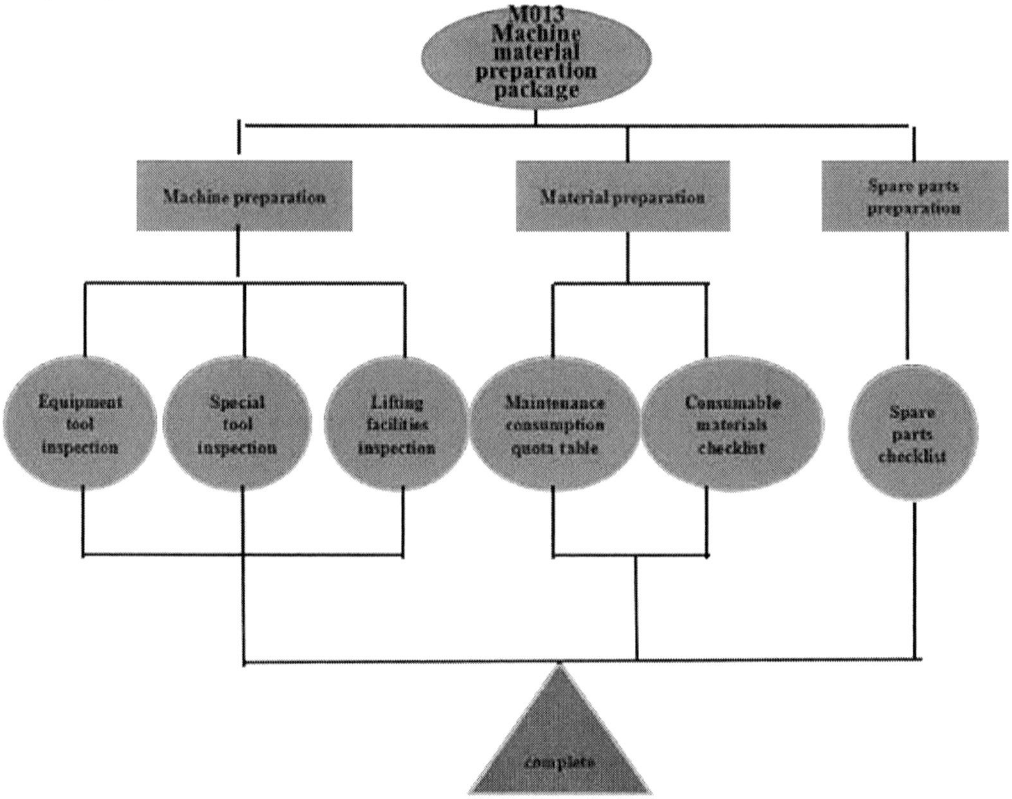

Fig. 10 Machine Materials Preparation Package

(1) Machine Preparation: In order to ensure the safety and reliability of the operation and meet the requirements of the operation, the person in charge and the worker should check the tool before operation and the special tools required for overhauling (as shown in Figure 11) People and technicians to verify the confirmation.

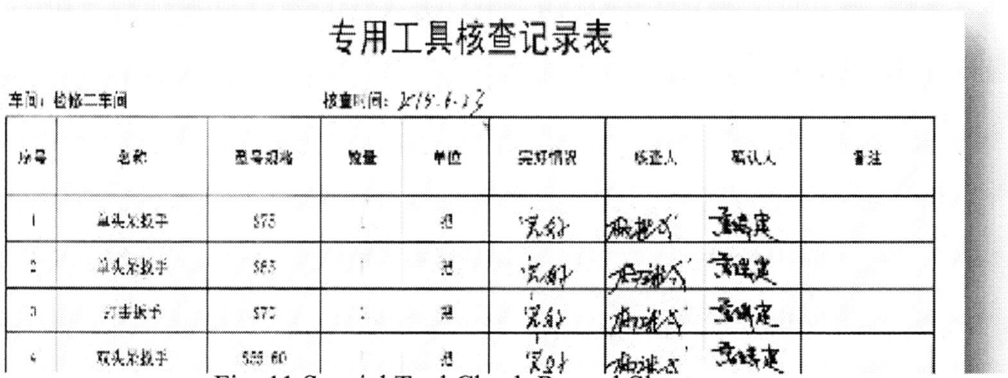

专用工具核查记录表

车间：检修二车间　　核查时间：[illegible]

序号	名称	型号规格	数量	单位	完好情况	核查人	确认人	备注
1	单头呆扳手	S75	1	把	完好	[illegible]	[illegible]	
2	单头呆扳手	S65	1	把	完好	[illegible]	[illegible]	
3	打击扳手	S72	1	把	完好	[illegible]	[illegible]	
4	双头呆扳手	S55-60		把	完好	[illegible]	[illegible]	

Fig. 11 Special Tool Check Record Sheet

As the hoisting work safety control is particularly important, the integrity of the hoisting equipment and facilities affect the safety of the whole maintenance work directly. Therefore, the inspection of the hoisting equipment is jointly confirmed by the professional lifting personnel and the person in charge (as shown in Figure 12), To ensure the safety and reliability of lifting equipment and facilities.

起重设施检查表

序号	名称	规格型号	单位	数量	完好情况	检查人	确认人	备注
1	[illegible]	[illegible]	[illegible]	4	[illegible]	[illegible]	[illegible]	
2	[illegible]	[illegible]	[illegible]	4	[illegible]	[illegible]	[illegible]	
3	[illegible]	[illegible]	[illegible]	4	[illegible]	[illegible]	[illegible]	
4	[illegible]	[illegible]	[illegible]	[illegible]	[illegible]	[illegible]	[illegible]	
5	[illegible]	[illegible]	[illegible]	3	[illegible]	[illegible]	[illegible]	
6	[illegible]	[illegible]	[illegible]	3	[illegible]	[illegible]	[illegible]	
7	[illegible]	[illegible]	[illegible]	3	[illegible]	[illegible]	[illegible]	

Fig. 12 Lifting facilities inspection record sheet

(2) Material Preparation: Before maintenance, the operation team shall provide inspection supplies according to consumable consumption quotas (as shown in Figure 13) to ensure the supply of inspection supplies and reduce the waste of consumables.

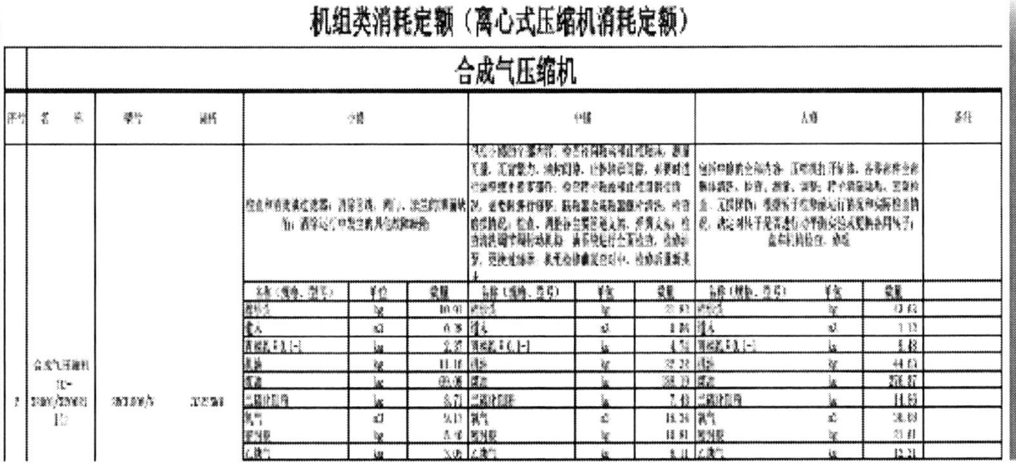

机组类消耗定额（离心式压缩机消耗定额）

合成气压缩机

Fig. 13 Consumption Quota Chart

(2) Spare parts preparation: Before the repair, the responsible person in charge and the technician check and confirm the spare parts, to ensure the spare parts quality and technical requirements.

2.1.4 Technical preparation package

Technical preparation is an important guarantee for the quality control of maintenance process. Through the technical preparation of the package (see Figure 14), the maintenance technical information, historical maintenance records and videos, the implementation of norms and standards involved in the maintenance process are collected and collated. During the maintenance process, standards and records can be inspected at any time to ensure Evidence based on the maintenance process.

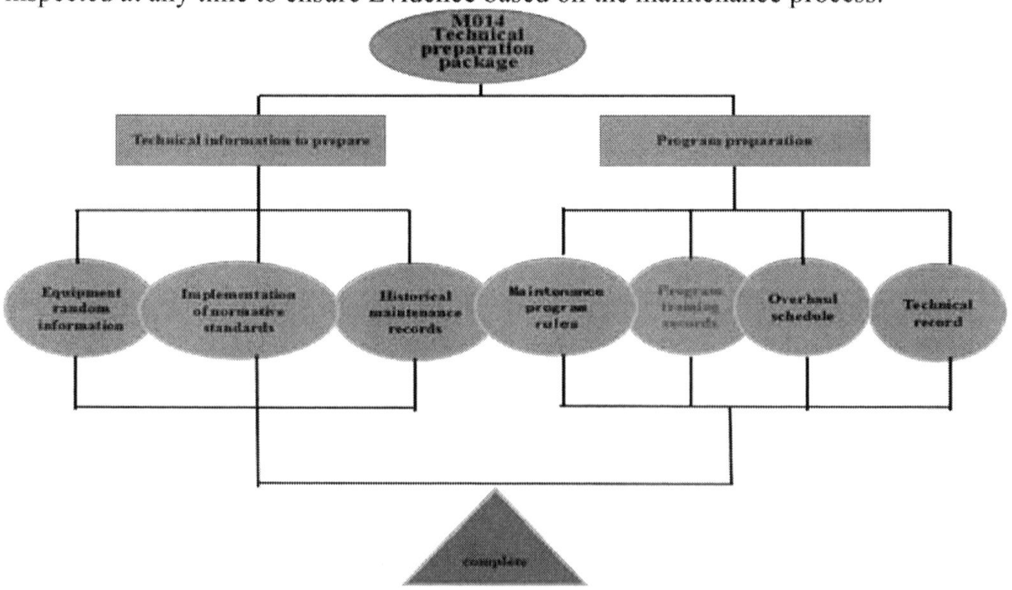

Fig. 14 Technical Preparation Package

Maintenance programs and operating procedures is the basis for operations guidance, based on standard specifications of the equipment overhaul process scientifically and reasonably to develop the best overhaul programs and operating procedures, and prior to the operation of the operating staff to overhaul the maintenance program and procedures (see Figure 15), Organize the crew members to carry out maintenance process desktop exercises(see Figure 16), clarify the inspection procedures, the safe operation of the special tools, the data measurement methods and the technical requirements for overhauls to ensure the quality and progress of the overhaul process.

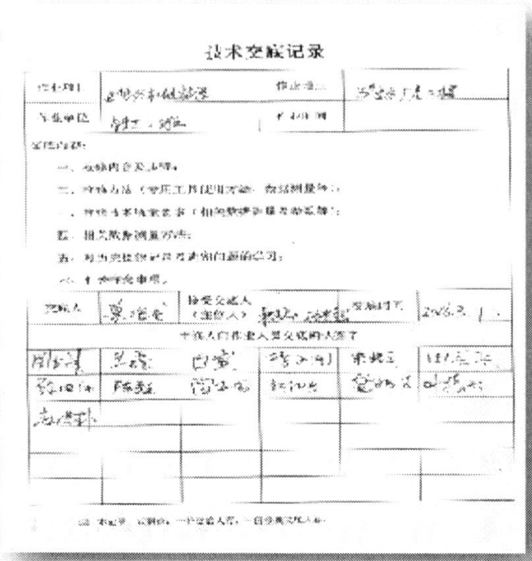
技术交底记录

Fig. 15 Technical submission record

Fig.16 Desktop Walkthrough

2.1.5 Site preparation package

Reasonable and standardized layout of the job site, in accordance with the requirements of the field package (see Figure 117), according to "unified management of standardized maintenance" to set up maintenance operations belonging to the ground (as shown in Figure 18), the development of tools, spare parts exhibition panels Put the standard. To implement various security measures in the overhaul area, clearly overhaul the operation boundary and civilized production boundary and implement the localized safety management according to the principle of "who is in charge and who is responsible for overhauling", set up the overhaul service and regulate the customization within the territory Placed on the maintenance site to implement standardized management, maintenance order in an orderly manner to improve the enthusiasm of the working staff.

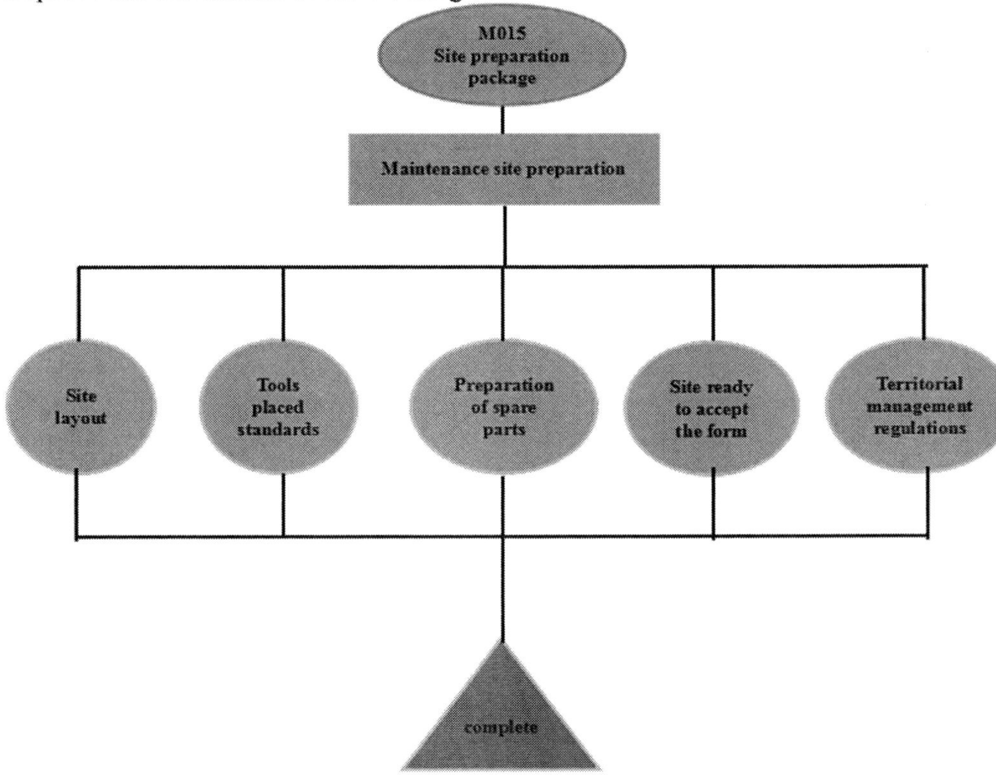

Fig. 17 Site preparation package

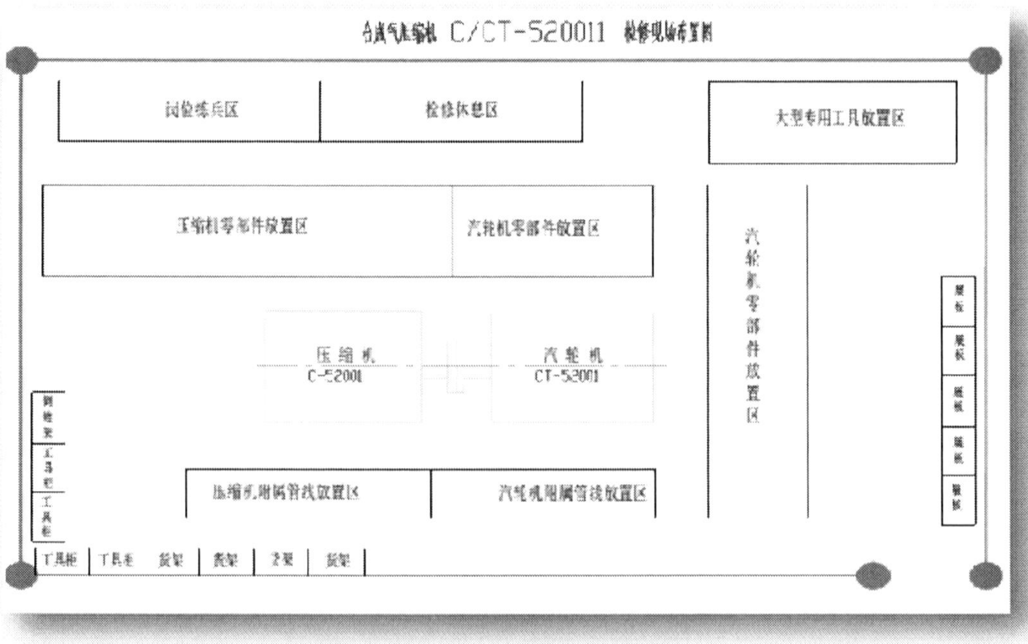

Fig. 18 Site ownership layout

2.1.6 Post training to prepare the package

In actual combat training, improve the level of operational skills, and promote the maintenance process safety management. Carry out various types of theoretical hands-on training for young staff members such as "teachers and students only" and on-site training in positions and posts to enhance the skills of young staff. In accordance with the requirements of the training packs (as shown in Figure 19), the training program for posts is formulated, the contents of the training programs for the posts (as shown in Figure 20) and personnel are defined, and the training is conducted for the operators.

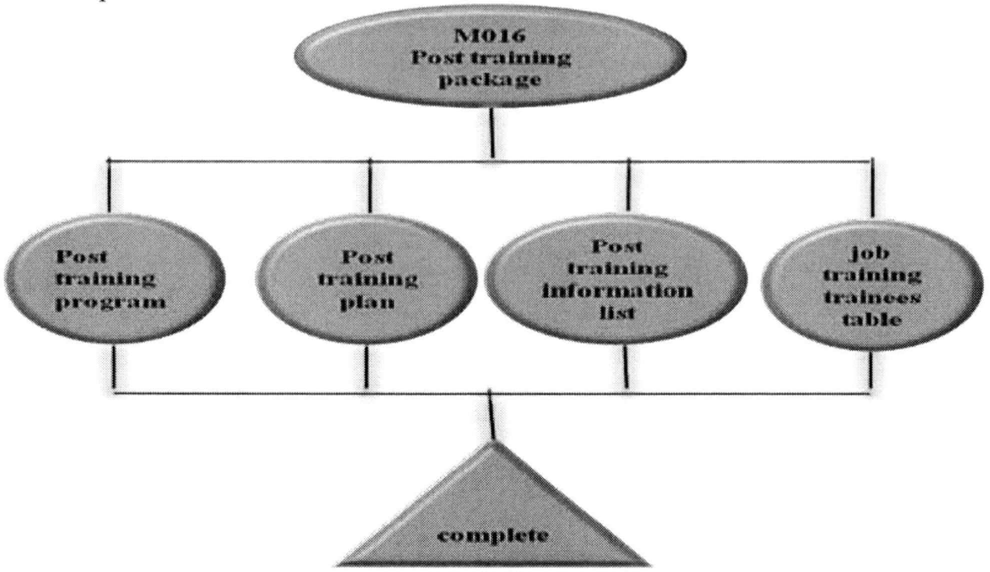

Fig. 19 Post training package

附件1

安装检修公司烯烃一套合成气机组岗位练兵计划（透平检修）

序号	岗位练兵内容	练兵时间	练兵对象	实操指导	技术指导	负责人	备注
1	了解汽轮机的基本结构和组成部分	检修第一天	参与合成气压缩机透平检修练兵学员	杨继成	赵旭鹏 吴鹏德	陈建平	
2	掌握轴承描述结构原理及轴承间隙及紧力检查方法	检修第一天	参与合成气压缩机透平检修练兵学员	杨继成	赵旭鹏 吴鹏德	陈建平	
3	螺栓加热器使用方法	检修第二天	参与合成气压缩机透平检修练兵学员	杨继成	赵旭鹏 吴鹏德	陈建平	
4	转子结构原理学习	检修第四天	参与合成气压缩机透平检修练兵学员	杨继成	赵旭鹏 吴鹏德	陈建军	
5	石墨填料更换方法	检修第五天	参与合成气压缩机透平检修练兵学员	杨继成	赵旭鹏 吴鹏德	陈建军	
6	掌握滑动轴承间隙的测量方法	检修第六天	参与合成气压缩机透平检修练兵学员	杨继成	赵旭鹏 吴鹏德	陈建军	
7	掌握转子推力间隙测量和调整方法	检修第七天	参与合成气压缩机透平检修练兵学员	杨继成	赵旭鹏 吴鹏德	陈建军	
8	掌握迷宫间隙的测量方法	检修第八天	参与合成气压缩机透平检修练兵学员	杨继成	赵旭鹏 吴鹏德	陈建军	
9	轴承间隙及紧力检查方法	检修第九天	参与合成气压缩机透平检修练兵学员	杨继成	赵旭鹏 吴鹏德	陈建军	
10	复习螺栓加热器使用方法	检修第十天	参与合成气压缩机透平检修练兵学员	杨继成	赵旭鹏 吴鹏德	陈建军	
11	掌握轴承箱结构原理及轴承间隙及紧力检查方法	检修第十一天	参与合成气压缩机透平检修练兵学员	杨继成	赵旭鹏 吴鹏德	陈建军	
12	机组三表找正方法	检修第十二天	参与合成气压缩机透平检修练兵学员	杨继成	赵旭鹏 吴鹏德	陈建军	
13	岗位练兵总结	检修第十三天	参与合成气压缩机透平检修练兵学员	杨继成	赵旭鹏 吴鹏德	陈建军	

Fig. 20 Post training plan

2.2 Overhaul implementation package

During the overhaul process, the operator performs the dismantling inspection, repair process, back-pack reset, commissioning and acceptance procedures in accordance with the maintenance implementation package (shown in Figure 21), and carries out maintenance work step by step. Through the steps of safety critical point and quality key point Step confirmation, standardize the operation of the operating personnel to ensure that the entire maintenance process controlled safely.

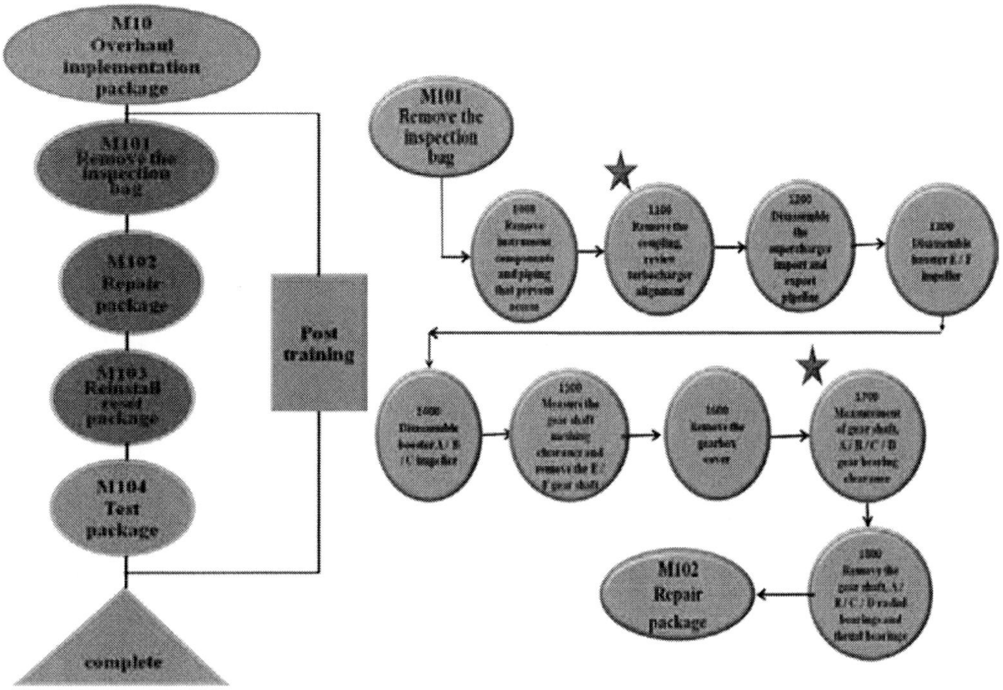

Fig. 21 Maintenance Implementation Package

Job implementation package will be the entire maintenance process in sequence into the various maintenance status, each state contains specific maintenance activities, such as demolition of the turbine coupling shield and the middle spacer, re-check the turbine center, by the master of pairs When each step of disassembly is confirmed and the safety inspection (as shown in Figure 22), when the safety measures of the safety critical steps of the maintenance work are confirmed, it is necessary to confirm the safety inspection by the person in charge, the maintenance technician and the workshop level Process safety controlled.

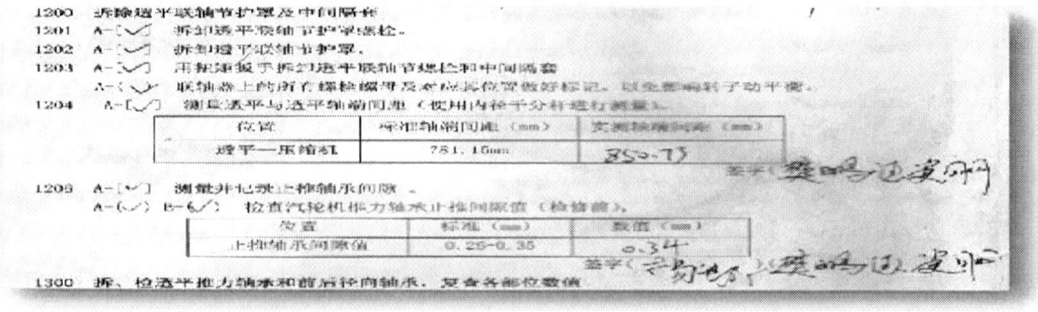

Fig. 22 Maintenance Implementation Package

In addition, the maintenance safety, quality key points and maintenance methods, inspection (test) methods, test results were clearly described (marked with a red asterisk) (Figure, to achieve the maintenance of safety and maintenance of quality standardization (see Figure 23).

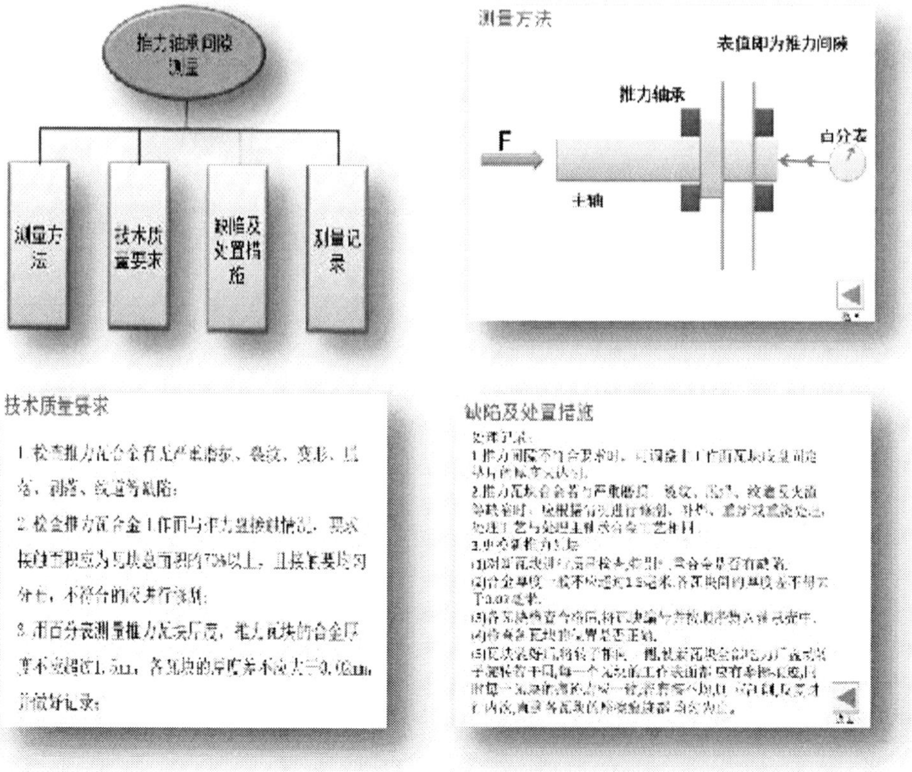

Fig.23 Key point measurements

Throughout the course of overhauling, the training unit of the unit overhaul the work package as a platform (see Figure 24), through the desktop deduction, on-site practice, writing study notes, so that young staff quickly grasp the maintenance difficulties, key points, and effectively enhance Overhaul skills of young employees, from the technical norms of safe operation.

Fig. 24 Field practice and training notes

2.3 Check the acceptance package

From the quality of acceptance, transfer of possession, storage of machinery and equipment, operational data monitoring, delivery of information on five aspects of inspection and acceptance (Figure 25). Standardize the quality of acceptance documents, and to track the implementation of the test run to carry out.

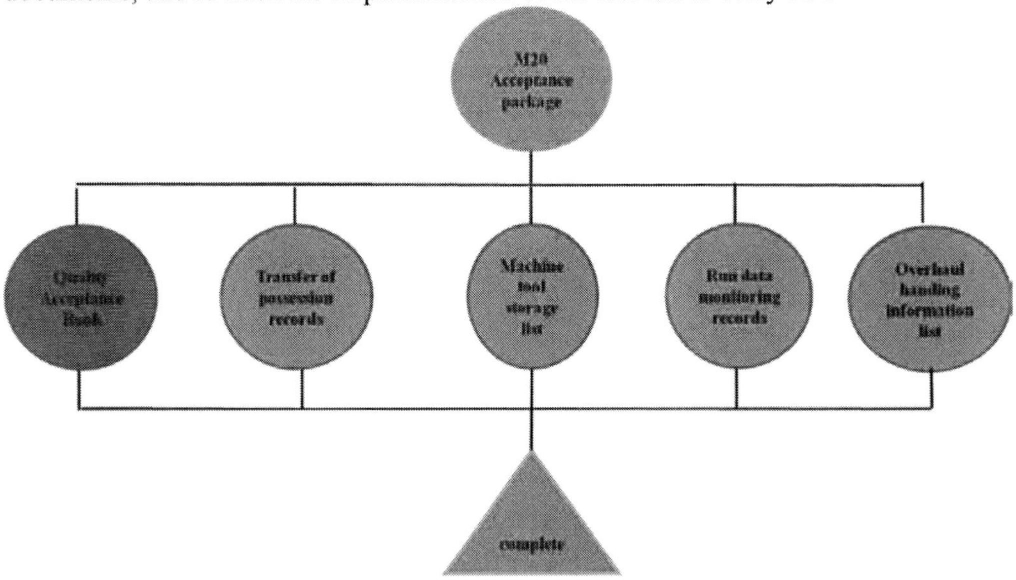

Fig. 25 Check the acceptance package

Quality Acceptance Book. The overhaul of the safety, quality key points data collected, and signed step by step to confirm, to ensure accurate and reliable data.

Territorial transfer records. The overhaul shall be handed over to the equipment user to ensure that all the tools, consumables and rubbish used have been cleaned and cleaned.

Machine tool storage list. The equipment list will be repaired and returned to the warehouse.

Run data monitoring records. The maintenance of equipment operation data continued monitoring, analysis of equipment operation.

Overhaul handing information list. This equipment maintenance process safety, quality and progress of the data to conduct a comprehensive card sorting, organized into volumes for the next overhaul to provide data reference.

2.4 Evaluation improvement package

The evaluation improvement package (Figure 26) summarizes the inspection from four aspects: inspection plan, operation rules, technical achievements and post-inspection evaluation, and revises and improves the safety, quality and schedule management methods and safety operation standards for workers. According to the maintenance of video, photos and other video data to produce training courseware, organize training to further enhance the level of safety and skill of workers.

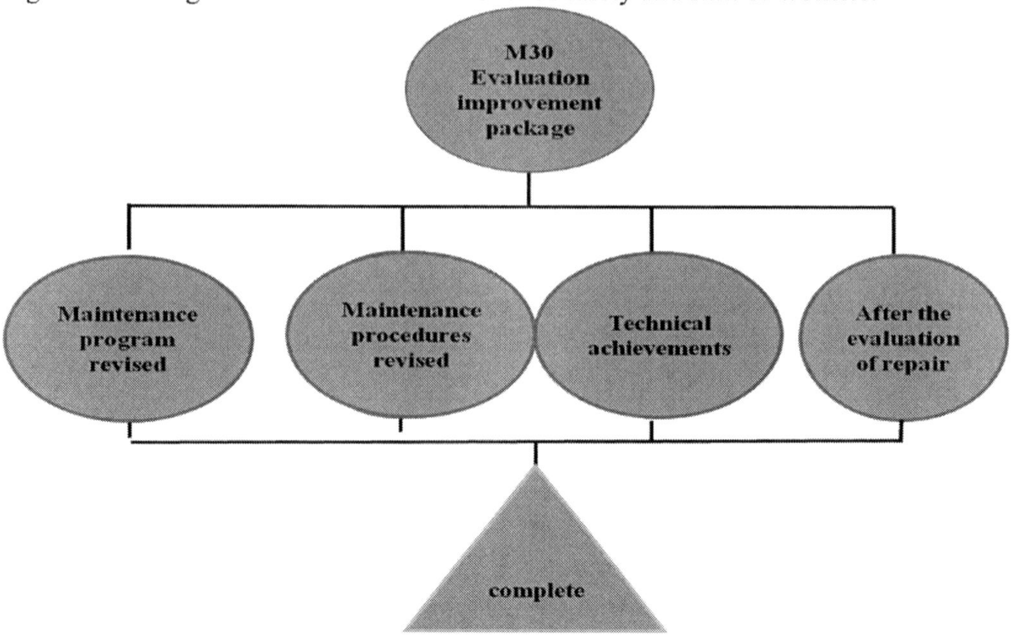

Fig. 26 Evaluation Improvement Package

Equipment maintenance within one month after the end of maintenance organizations to carry out maintenance evaluation, is to implement continuous improvement of specific measures to the maintenance of safety management to give a true and objective evaluation. Through the post-maintenance evaluation, it can evaluate the technical and economic indexes after the overhaul of the unit. More importantly, it can re-examine all aspects or parts of the safety management in the whole process of maintenance and repair, find out the existing problems and gaps and make improvements to achieve continuous improvement Overhaul the whole process of safety management purposes.

3 Overhaul operation package implementation effect

Through the implementation of safety standardization in the overhaul process, the operation process was safely controlled, the quality was qualitative, the responsibilities were defined and the records were clearly recorded. The operators truly met the requirements of standard posts and standards and achieved three improvements overall.

3.1 Improve the safety control level of maintenance work

The management of overhaul work packages is carried out in the overhaul process to stipulate the preparations before maintenance, ticket handling, risk identification, safety protection articles, etc. The dynamic risks of every action in the work process are dynamically identified and the control measures And supervise the implementation. Through the implementation of the standard process safety management, making the maintenance tasks in place, risk identification in place, the implementation of on-site safety measures, safety checks in place. The final realization of safe and controlled maintenance operations, to prevent accidents.

3.2 Enhance the level of operational staff

Through the maintenance of job kits to carry out job training, improve the level of maintenance personnel skills, adhere to the "training in actual combat, in the hard work to enhance" concept, the practical training into the inspection and maintenance to the large unit maintenance project training content for the post, Using on-site training module and student short board, the need to combine work, learning content and the scene in kind, a combination of learning and answering questions on the spot, a total of training 10 crew members, 10 crew maintenance technicians, Making maintenance personnel technical skills level has been effectively improved.

3.3 Improve the quality of equipment maintenance

Overhaul of the operation package In the process of overhauling and implementing, the overhauling operation will be disassembled, repaired, reset and rebuilt. The technical requirements for each overhaul will be standardized according to international, national and industry standards . In the implementation of the process of strict quality control of key points and overhaul methods, inspection (test) methods, test results for the maintenance person in charge, technicians, plant-level and company-level "four" to achieve the quality of maintenance over standardization.

Application of safety instrument system in the process of safety management of coal oil chemical process

Liu baoyi
GShenhua Ningxia Coal Industry Group， Yinchuan , Ningxia

Key words: SIS, coal chemical industry, control system

Abstract

This paper briefly introduces the composition, function and characteristics of the safety instrument system, and briefly introduces the application of Tricon, the SIS system of Triconex company, in the project of methanol branch of shenhuining coal group. Both gasification, synthesis and sulfur recovery of methanol branch use the SIS system Tricon. Tricon safety instrument system all the hardware and software are from German tuv certification, with a triple redundant I/O bus and triple redundant communication bus, separated from each other, no bus management module, to ensure the safe and stable operation of the equipment. Tricon has a scanning cycle of up to 25ms, with a diagnostic coverage rate of 99.9% and 99.999% of the system. The Tricon safety instrument system greatly improves the safety and reliability of the device and reduces the safety hazard.

In the field of coal chemical industry, the safety instrument system is mainly used for complicated interlock logic control unit, used to implement the accident situation interlocking emergency stop and various open parking safety system logic control, through the safety instrument system can effectively ensure the safe and stable operation of the plant. The project of methanol branch of shenhuining coal group includes 250,000 tons per year of methanol, 600,000 tons per year of methanol, 60,000 tons per year of formaldehyde. In the process of methanol gasification of 250,000-ton/year coal methanol gasification process, the methanol gasification process of methanol in 600,000 tons/year is selected from the four nozzled coal slurry developed by east China university of science and technology and yanmine.

1. Summary of safety instrument system

1.1 definition of safety instrumentation system

Safety instrumentation System, is used to perform one or more safety instrumented functions,mainly for the factory control system alarm and interlocking parts, practise alarm action or stop control according to the test results in the control system, is an important part of automatic control of factory。

1.2 level of safety instrumentation system

SIL is short for safety integrity level, the Chinese expression is the comprehensive security level,is the classification of the safety instrumentation system in the process industry in the S84.01 standard by the American institute of instrumentation (ISA), is divided into 1, 2, 3 three levels。

The average probability of failure risk of SIL1 level is 0.10 ~ 0.01 per year.

The average probability of sil2-level failure risk is 0.01 ~ 0.001 per year.

The average probability of failure of SIL3 level is 0.001 ~ 0.0001 per year.

SIL has a corresponding relationship to the probability of failure.so, SIL reflects the extent of the effect of security functions on the overall level of risk reduction.Assess and determine the level of safety integrity can be by qualified specialized agencies, but also by the enterprise based on their own practical experience to determine, According to the characteristics of the production plant, the danger of the process, the safety level required by the relevant laws and regulations of the country and the enterprise. The level of safety typically required for the large chemical, petrochemical and power industries is SIL3.

2. Selection of safety instrument system for coal chemical industry

2.1 Safety grade of safety instrument system in coal chemical industry

The process of gasification adopts the technology of coal slurry gasification. The gasification process is carried out under high temperature and pressure, the highest temperature is 1500 degrees C, and the highest pressure level is ANSI 1500LB. The medium of coal water slurry is a high viscosity non Newtonian fluid with high abrasiveness. Oxygen is a medium with strong oxidizing properties. Hydrogen gas in gasification gas is highly combustible and explosive, and carbon monoxide flammable and explosive is moderately toxic. Therefore, the coal gasification process is highly flammable and explosive continuous production process, and PDP developers locate the security level of tuv5-6, The production line is arranged in an open frame, and the explosion-proof level is Zone II Group C T4 group. The SIS system USES three redundant systems certified by international structure to ensure reliable operation.

The sulfur recovery process equipment is advanced, the automation level is high, the process production process continuity is strong, flammable, explosive, have corrosion,
In addition, the production equipment will cause the chaos of continuous production during the sudden power failure, and may cause major accidents such as fire, explosion, personal injury and equipment damage. The main production device adopts centralized control mode, in the central control room in DCS system for testing and automatic control, and USES the independent TUV5 ~ 6 levels of safety levels of SIS system interlock protection of the device, According to the scale of production, process, characteristics of processing equipment, process operation requirements, and refer to the domestic automation level of the same kind of device.

The safety interlocks of coal gasification and whole installations are determined by an SIS system that conforms to the safety grade of tuv5-6. The SIS system is in accordance with the international security level, the system can distinguish the first accident, and the sound and light alarm. The system has the function of accident recall, and after the interlocking, the data scene of the automatic high-speed memory accident before and after the accident is printed out in order to analyze the cause of the accident. The SIS system trigger signal generally adopts hard wiring, and one component, logic design and actuator are all fail-safe.

2.2 Field instrument for safety instrument system of coal chemical industry

According to the relevant data, 40 percent of the fault of the SIS system belongs to the sensor, and 50% belongs to the actuator. Improving the reliability of the sensor and actuator can greatly reduce the probability of failure of the safety instrument system and improve the availability of the system. The detection component of the safety instrument system selects the imported transmitter product, and the important parameters are used in the triple detection to ensure that the input link generally reaches the level of SIL3 safety integrity, In order to reduce the failure rate of detection change.The control system adopts UPS and municipal power supply, and the valve in the field USES the gas shut-off valve, which can prevent the electromagnetic valve from being refused. the valve switch position feedback is collected to the SIS system, and the interlocking SIS system stops when the control instructions are inconsistent with the site valve position feedback, In order to ensure the system timely action after the valve action.

3. The characteristics of Tricon system

TRICONEX products are continuously certified by iso---- TUV, CSA, FM, IEC, UL, CE. Tricon provides fault tolerance through the triple modular redundancy structure (TMR). Triple modular redundancy (TMR) structure (see figure 1-1) ensures the tolerance of the equipment, and can appear in the components, hardware failures or transient fault from internal or external sources provide uninterrupted control of intact.Each I/O module contains three separate sub-circuits. Each circuit on the input module reads the process data and sends the information to its corresponding main processor. The three main processors communicate through a dedicated high-speed bus system called TriBus.

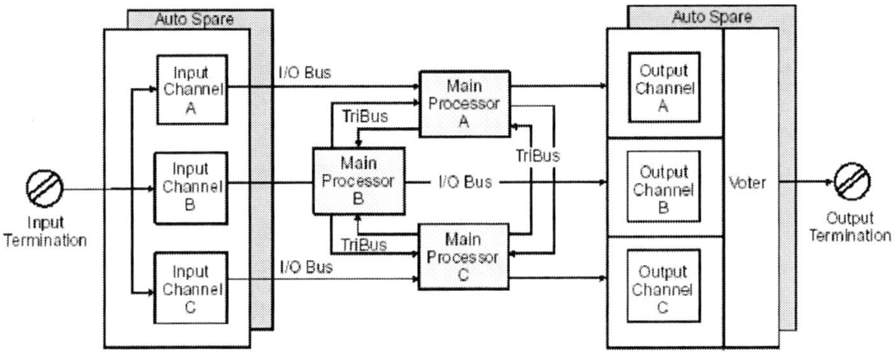

Fig. 1 Tricon controller's triple structure

The system is composed of three identical system channels (except the power module, which is doubly redundant). Each system channel executes the control independently and works in parallel with the other two channels. The hardware voting mechanism votes and diagnoses all digital inputs and outputs from the site. The simulated input is processed with the median value. Because each of the dividers is separated from the other two, any fault in any of the sub-circuits will not be passed to the other two sub-circuits. Maintenance work, including removal and replacement of fault modules with sub-circuit faults can be performed online without interruption of process control. (in case of a hot standby card, and confirm that the hot standby card is in working condition.) A wide range of diagnostic work for each circuit, module and functional circuit can detect the fault in operation in time and instruct or alarm.

A Tricon system can contain at most 15 racks, to accommodate the installation of input, output, and hot standby module, and the communication module and other appropriate combination, card a total of up to 118 pieces. There are three types of racks: main frame; Extension frame (the distance from the main frame is as far as 30M); And the remote frame (as far as 12KM from the main frame).

4. Application of Tricon system in the project of Ningxia coal and methanol

The safety instrument system of ning coal methanol is designed and constructed by Beijing kangji automation equipment technology co., LTD., which adopts the method of separating and setting between the central control room and the field cabinet. Field instrument signals through the cable into the scene between the cabinet, the cabinet room to the central control room with dual redundant communication links, SIS system controller Settings in the corresponding field between cabinets, auxiliary work station and operator station, SOE station set up in the central control room, GPS clock is set in the central control room.

4.1 Power supply configuration

Each rack is equipped with two power modules with a power supply voltage of 220VAC, which is powered by UPS and municipal or dual UPS. Dc power supply adopts the redundant power supply of dc power supply, and the output voltage of power supply is 24VDC, and different power supply is used in the on-site power supply and internal power supply of the system. The field instrument, safety grid and terminal board are all supplied with dc 24V, and the solenoid valve power supply circuit sets 1A ceramic fuse. All field input meters are connected to the system by the safety grid, and all output signals are isolated by relay. All dc power supply and ac power supply state access system. When any power supply fails, the system will alarm to ensure reliable operation of power supply.

4.2 The system configuration

Each device sets up the SIS controller independently, and sets the controller independently for the multi-series device (gasifier and air compressor unit). System network configuration redundancy, can be with honeywell DCS system, this TSI state monitoring system redundancy RS485 communication. The interlocking output signal of the TSI state monitoring system and the signal of the auxiliary operating station are connected to the SIS system by hard wiring. For data to be displayed in DCS system and SIS system, the signal is sent to DCS system and SIS system respectively by using one or two security gate ICC312. The SIS system of each unit is equipped with operator station, SOE station (and engineer station). Gasification device, sulfur recovery unit and generator auxiliary operating station, auxiliary operating station power supply adopts dc 24V, with emergency stop button and stop alarm instruction.

4.3 Clock synchronization

The system configuration communication network card is the TCM card, the supported clock synchronization mode has GPS, SNTP and Tricon master node. The clock synchronization of the coal methanol project is SNTP, and the clock source is from a GPS in the central control room, and the external benchmark adopts GPS/ beidou double clock source. The DCS system is the PKS of honeywell, the clock synchronization mode is NTP, and the clock source is consistent with the SIS system. Each operator station, SOE station, engineer station, server and so on also pass NTP update time, the maximum degree ensures the clock unity. When the SIS system's clock source is offline, the system will generate alarm information.

4.4 System diagnosis and alarm

System by triple modular redundancy Yu Jiegou (TMR) provides fault tolerance, is composed of three identical system channel, the diagnosis and maintenance, online identification and error compensation control system, and to allow in does not affect the normal production of on-line maintenance. The diagnosis and alarm system of the system is quite rich, including CPU fault, I/O module fault, communication failure, power module fault, battery power voltage, and power module temperature too high alarm. All alarms are displayed simultaneously on the frame and on the rack. The common LOAD alarm of AO module is the output line opening, and the unused AO output channel is short to eliminate the alarm. After the I/O module has a FAULT, it needs to be processed and replaced online.

5. Summary

Through the debugging test before driving, the safety instrument system Tricon is successfully applied in the ning coal methanol device, which further verifies the reliability and safety of the system. The system has perfect diagnosis and alarm, which can provide predictive maintenance for users, greatly reducing the downtime caused by equipment failure.

Study and Practice of "11311" Safety Management and Control model in Modern Mega CTL and Chemicals Industry Base

Yao Min

Shenhua Ningxia Coal Industry Group, Yinchuan , Ningxia

Key words: Safety Culture, Safety System, 3 Basic construction, talents team safety vision

Abstract

With a mission to achieve " zero accident, zero fatality", and to build the long term & effective safety development mechanism , the thesis is focused on "11311" safety management and control model , covering productive study on building "11311" safety management and control philosophy , the cultivation of the special safety cultures, the risk precaution and management, "3 basic "works construction as well as the talent team construction etc.

Through the grand execution of upgrading and transition strategy undertaken by Shenhua Ningxia Coal Industry Group (SNCG)., the CTL & Chemical Industrial Base has become the new leading cluster industry among all its clusters.Since 2004, the accumulated investment in the base has been standing at nearly 100 billion RMB, with 8 modern coal to chemical pilot plants being completed, namely, CTL 4 million tons/y , coal to olefins 2 million tons/y ,etc. The capacity of coal based chemical production is 10million tons per year. Converting 34 million tons coal every year. A native largest and word-leading mega CTL and chemical base has been initially formulated. As a mega state-owned enterprises, SNCG attaches great importance on safety production. During its construction in the base, based on the fact with giant scope,new tech,complex process and great safety management difficulty, SNCG put the target on building long term effective safety management system and set the mission to achieve “ zero accident, zero fatality”, through active study and research, it formulated “11311” safety management and control model, which covers the aspects of culture construction, risk prevention, “3 basic” construction as well as talent team construction,etc.

1. The necessity of establishing SNCG “11311” safety management and control system

At the beginning of chemical base construction, SNCG followed the safety principle of coal production industry. It was difficult to meet the chemical character with “ high pressure, high temperature, easy combustion and poison and hazardous” . Specially the staff of coal based chemical had different background with great different culture. In order to enhance the safety consciousness and streamline the realization on safety, through years and rounds of practice, knowledge and practice, draw lesson from local petro chemical companies' method and learning, finally comes up a workable safety management and control model.

1.1 the structure of the employee is one of the key element

Modern coal to liquids and chemical industry is of high technology , talents and resource intensive, a high quality talents team shall be prerequisite to deliver good support. As a traditional coal exploration enterprise, SNCG lacks the talents in coal chemical area. To break this bottleneck, by adopting concept of “ seeking talents, request talents, absorb talents and cultivating talents” , through outbound attraction inbound cultivation, currently SNCG has formulated a high talent team specialized in coal chemical industries with more than 8000 people. While , still the following phenomenon exists:

Since the source of the the talents coming from various areas, their safety knowledge ,behavior, safety consciousness and safety management model is quite different, which leads to non-unified standard and action. A streamlined safety culture must be built to integrate and merger the individual differences;

Since the scale of CTL and Chemical projects are mega projects, most of its core technology is commercially utilized for the first time, there are tremendous renovate process and more creations, it is quite necessary to enhance team building with the focus on base construction, basic work and basic skills ,so-called " 3 basic" constructions.

1.2 the high danger of the industry, another one key element

Coal to liquids and Chemicals industry is of high danger industry. Due to its inherent dangerous specialty, accidents are more likely to occur , such as fire, blast, leakage,explosion,poisoning,and such accidents could cause very bad damage. Once if it is not controlled properly, the Domino effect could probably happen, which could cause big disasters,such as heavy fatalities, ecology damages and even harm to national security. Therefore, building "11311" Safety management and control system with the integration of inclusiveness, guidance , developing and with its adoption in all rounds is the effective way to constructing long term effective safety system.

1.3 The company's developing mission.

On 19th,Jul 2016, when General Secretary Mr.Xi visited 400kta CTL project issued a call that socialism depending on hard working. In the same year, on 28th Dec, when the project worked out qualified end production, Mr Xi asked the companies must keep on going, constantly, secure the operation with safety, stable and clean. Enhance the advantage of coal process and conversion of China. Speed up the revolution of energy producing and consumption. In order to put the mission into real, it is necessary to build up a safe, stable and clean safety management module.

1.4 National "safety development" strategy , one key element

President Xi Jinping commented: safety production is closely related to the well being, and related to social development. Life is the most priority, development can not be achieved via the sacrifice of the people' s life, this this the red line that is impassable. To build one moderately society and to realize the rejuvenation of the Chinese Nation, it needs a long-term steady and secure production environment. Human-being orientated and safety development persistence has been taken as the national strategy and been written in the updated < safety production law> , it is the basic symbol for constructing a harmonies society. As a basic solution, building safety management and control model is of significant importance to achieve peace and propriety for the enterprises.

2. The management concept that 11311 safety management & control mode based on

2.1 People-oriented concept. On the one hand, life is priceless

The development of enterprises must take health and safety as first priority, and to achieve sound and fast growth, employees' health and safety must be assured. Thus, the key of safety concept and safety value is to highlight people-oriented concept, and to show the right life values. On the other hand, Individual is the most active, unstable and decisive factor in many factors of productivity. Among people, equipment, environment, and management these four factors, the behavior of people rank first and have deeper and wider impact on safety production. According to Herbert William Heinrich, the famous safety engineer of America, 85-90% of industry accidents result from unsafe act. Therefore, people factor is prior and important. To achieve intrinsic safety shall first think highly of training employees to work and think safely. The training on consciousness,activity and quality is prior to other factors.

2.2 Pre causion & prevention concept

According to the accident causal chain theory, the process of the occurrence and development of industrial injury accidents is linked to a certain causal chain. The fatality is the result of accident. The cause of accident is unsafe activity of human-being or unsafe state of equipment. The shortage of human-being cause the unsafe activity of human-being and unsafe state of equipment. The shortage is caused by Adverse environment. So when building up the safety management model, we must analyze the risk source and method of causing accident and possibility from human-being, machine,environment and management.

2.3 Systematic management concept

During production, the accidents could usually result from various factors rather than single factor, Thus, safety management must abide by the integrated, comprehensive, and systematic concept. Realizing the target of 'Zero injury, Zero fatality', to build up long-lead safety management mechanic. By cultivating safety culture, building risk precaution system, strengthen safety base and training staff, building up a suitable safety management model for human-being, machine,environment and management.

2.4 PDCA closed-loop management & non-stop improvement concept

Enterprises development is a dynamic process, the old problems solved, the new issues prop up. Therefore, process control is important. The circulation of planning, practice, inspection, action continuously makes risk identification and management optimized in PDCA operation, which eliminate the potential risks at the very beginning, and realize closed-loop management and optimization. Meanwhile, the system itself could also get improved and updated during circulation.

3 The Exploration and Practice on"11311"Safety Control Mode

Based on mentioned above, "11311" Safety Control Mode is to foster a culture, build up a system, lay three solid bases, create a workforce and deliver a vision.

3.1 Foster Distinctive Safety Culture and Create sound atmosphere for safe development

Safety culture is the soul of the safe operation of an enterprise. The safe behaviors of people are the primary elements among all the elements of safe operation. The improvement of safety culture is regarded as the priority by the enterprise, wide domestic and foreign advanced safety culture essences are positively learned from the strong points of others. Throughout starting with the concept culture, system culture, behavior culture, material culture, after exploration, summary, extraction from others, the safety culture with the distinctive "5+2" main concept of the enterprise has been built up, which has become the strategic measures for forging the soul, fostering talents and modeling image of the enterprise.

Concept Culture highlights the concept of "5+2. ①The five-core concepts that "employee' s life and health is above all" , any operational accident can be prevented", "all the hidden hazards can be controlled", "any leakage is an accident", "excessive emission/discharge is and accident"have been firmly established. ②"Two Rules" are held. With reference to Iceberg Theory and Heinrich' s Rules, the quantitative change will cause qualitative change, the prevention of serious accident must start from elimination of minor accidents and hidden hazards and must eliminate the root causes of the accidents and hidden dangers. Safety concept culture is a special and unique safety concept system that was derived and summarized from the past safe operation practices, which has been embedded into the enterprise safety value standards through the publicity, guidance and introduction to different channels and multilevel.

System Culture is the link to convert safety concept culture into safety behavior culture and material culture, is the actual reflection on safety concept culture incorporated into system, and is also the culture to extend an influence on the standardability and disciplinary of the behaviors of the enterprise and the employees. SNCG is formulating system based on national safety laws and regulations, industry code and advanced management method, combined with "5+2"concept. SNCG have made more than 5200 operation standard,management standard and technology standard. They are the basis for employee to operate accordingly.

Behavior Culture: Human-being's unsafe activity is the main reason to accident.The widely recognized and accepted enterprise safety behavior culture system has been established in accordance with recognition of safety behaviors, (including safety commitment, safety oath, demonstration of leaders), safety behavior rules and regulations(including safety prohibitions, paramilitary management, employee behavior rules, eight-prohibition and eight-must, nine-non-repairing), safety behaviors incentives (including goals, incentives, risk mortgage, promotion, recognition of excellence, performance assessment), etc. to achieve the change from things and material-oreinted to people-oriented, and create the "self-control" atmosphere for the employee behaviors, which changed the requirement of the systems and procedures into the voluntary behaviors of the employees and eliminates the unsafe factors.

Material Culture:SNCG established her own unique" 6+2" material culture with the guidance of concept culture and the normalization of system culture, in the support of new technology and informatization, and based on equipment modernization to set up six large systems, including safety interlock, fire & gas detection and alarm, emergency and rescue, inspection patrol tracking, board operation and field operation training, portable testing and monitoring, and two big platforms of an integrated GCP(goal-Conduction) performance management and a safety management informatization.

3.2 To provide effective measures to safe process control through construction of risk prevention system

With the thoughts of prevention in advance, SNCG has set up a risk prevention management system based on risk identification, taking risk pre-control as principles and taking the elimination of unsafe behavior of the employees to cut off the cause and effects chains of the accidents as measures.

*3.2.1The five relations for system establishment ar*e that duties create operation which make risks that create systems which make records that support duties, which are both cause and effects relations and system linkage. “Five-relation” is the general framework for the establishment of recognition and construction of the systems.

3.2.2 Exploration is the effective carriers for the implementation of series of risk prevention system

“One schedule and five cards” they are work task schedule, risk identification card, risk prevention card, energy insulation card, quality acceptance card and emergency response card. According to “One schedule and five cards”, the risk factors for each task are identified to make sure the hazard degree of the unsafe factors and to prepare control standards and measures. During the working and operating, putting the safety responsibility into reality, the risk source is under control.

“ 16 Standard Systems are implemented, they are “ Dynamization of risk identification, classification of risk management & control, work permit to operation, normalization of emergency response, sharing of safety experiences, standard indexes for the implementation of measures, standardization of safe technology examination, demonstration of the leaders, lock up of energy insulation, standardization of daily work completion, well establishment of jobs, unification of contracting works, localization of responsibilities, standardization of maintenance site, standardization of work site, visualization of operation site.

Environmental protection insists on the principles for “control of emission deduction, ensuring to meet the standards and recycling” . Through the application to, research and development of the new technology, zero discharge of waste water and maximum recycle of the waste gas and solids have been strived for.

Safety training ensures “training, warning and reflections”. Safety training for all the employees are accomplished by means of training packages, passing test, post duty training. In the safety warning and training base, variety of safety accident cases collected and sorted out every year are used for accident cases warning orientation. Those who violated the three violation behaviors shall be retrained for warning cautions. All the employees shall discuss the reflection against the warning training to improve the safety qualification. All the staff processed “Nine Question” reflection and discussion to strength the safety consciousness.

Hidden hazards are managed through the process of “ check, correction and supervision” . All the employees are encouraged to check hidden hazards to strictly correct the hazards according to PDCA closed-loop principle. Any hidden hazard shall be supervised in the workshop level.

“Five-Star & Five-type” team is promoted, and “monthly inspection, quarterly assessment and annual top-ten excellence and “2020 eight-step method” are created. The construction of “Five-Star & Five-type” team is promoted according to five-type of intrinsic-safety, quality-performance, innovation-driven, energy-conservation and harmonious -development types and five levels, monthly inspection and quarterly performance assessment are delivered, and the top ten team leaders will be recognized to upgrade the level of the team construction.

The shift exchange is carried out strictly according to “2020 eight steps” , including 20 minutes for inspection, 20 minutes for shift exchange, inspection, roll-call, checking, shift handover and takeover, learning, task assignment, work summary and oath of work.

3.3 The consolidation of safety management provided a solid basis for safe

development

It is difficult to succeed in doing anything without a solid basis. Primary construction, basic management and basic skill are the footstone of safety management.

Primary construction: it is primarily focused on the construction of the Party branches based on the construction of the Party and leadership to establish basic organization and work teams to strive for "five-excellence Party branches", civilized team, safety team, "five-star & five-type" team, the construction of trade union community and "Youth" branding by the Youth League.

Basic Management: Having taken the post responsibilities as the principles, the quality, measuring, standardization, systems and working procedures as the main work description, to specifically define the operation processes and standards, breakdown the post duties to make applicable, simple, effective rules and regulations, systematic manuals, procedures and records.

Basic skill training: The aim is to behavior the employee by increasing safety and professional skills. The safety consciousness is improved by tests including safety, fire and gas prevention,process and parameter, interlock and operation regulation,equipment knowledge,start and stop procedure, accident learning, regulation, profession training, skill competition, Package training, Teacher and apprentice, simulation training, 3 class 4 step, point system. The employee's safety consciousness has changed from asked safety to need safety.

3.4 Building up a high quality team, creating the personnel guarantee for the safe development

Human-being is the determine factor for companies.Human is the crucial factor to safety. human is the main part of management executing, meanwhile it is the main part of implementing of various of safety and operation regulations, it is the first and crucial factor for establishing safety management. The importance of team building is obvious for safety developing.

Step 1:Draw up a employee developing plan:From the perspective of employee career development, construction of professional development channel for operators, technicians and managers. For operators ,skilled worker, skilled master and other 6 level are adopted. For technology staffs, the 9 levels of first,second,third,expert are adopted. For managers, preliminary, middle and advanced are adopted. By managing promotion, helping employees to design their career. Employees can check their future clearly. Meanwhile combining the individual development with the developing of company, win-win will realize.

Step 2: Operators will accept 3 round pre-training before working about company's culture, technology. Before new process, new tech ,new material and new machine are put into use, employees will adopt relative training. Then they are capable with skills,safety knowledge and positive attitude. Technicians will be improved by having tech lecture, education training, professional knowledge study, advanced practice learning, tech workshop, learning of new process, tech and material. The aim is to make them have strong capacity on tech practice and R&D. Managers will be improved by workshop,lecture,practice exchanging with companies and registry MBA, EMBA education. They focus on HR, production management, cost managing, performance appraisal, motivation and communication. To improve the managers' political literacy, management ability, innovation ability, interpersonal communication ability and professional ability and other comprehensive qualities.

Step 3: Establishing employee's performance appraisal file. Building up Target conducting performance management system. In this system, the main substances are: Company's strategy is leading. Aiming management is guide. Entities and employees' performance appraisal management is content. Salary incentive is means. Based on the analyze of daily working target and performance of each employee, aware employee at the end of year. All of these will be part of employee's file. They are the base of employee's performance test.

Step 4: Building up employee's performance test system. Building up employee's comprehensive performance test system based on the file. For operators, the test focuses on detail evaluation, skill evaluation. For technicians, the test focuses on detail evaluation, position,democratic evaluation, scientific and technological achievements (thesis, patent,awards). For managers, the test focuses on performance file, democratic evaluation, competition for a post. The result of test will be adopt on promotion, the development of employee is fair.

3.5 Establishing a long-term mechanism for safety development, realize 'Double Zero' target

3.5.1 Company's vision

Following the 3 steps of natural instincts, strict supervision, self-management and team management to make the top-level design of safety development. Making safety regulations, clarifying the details and contents of each step. Establishing a long-term safety mechanism on the basis of gradual progress. Implementing 'Zero fatality Zero injury'.

Strict supervision:①By many safety activities like the Year of Safety Base, the Year of Safety Base optimizing, the Year of Safety base Consolidate, the basic works are tamped.②implementing risk pre-caution system, launching the Year of System Construction, the Year of System Optimizing,the Year of System Qualified. Making the system into real.③implementing standard compliance activity. Launching the Year of Standard Creation,the Year of Standard Qualified, the Year of Standard Proven, the Year of Standard progressing. The employee knows how to work properly. By the strategy of three steps, pushing the change of safety method from 'ake me be safe' to 'I ask for safe'. The safety management is changed from 'ask me be safe' to 'I ask for safe'.

Self-management. The potential capacity of each employee is regarded. Encouraging all staff show their value. Employee regard safety as their own achievement. Everyone eager to take their own responsibility. They have high safety willing with whole knowledge and skill and good behavior. They implement self-lifting,self-innovation,self-exceed. They form a method of 'we need safety'.

Team management.It is mainly embodied in the team building high quality, unity and stability of the team. All staff shares unique target. The management system are accumulated. Safety culture is winning support among the people. Employees regard safety as team honor. During daily work, they not only watch our their own job with following regulations and standard, but also get aware of team member's behavior. Reminding member to operate safely. Sharing their own safety knowledge and experiences to others.

3.5.2 building up a safety evaluation system with check+review+evaluation

Following the method of combination of daily check, special review, stage evaluation, increasing the safety management level step by step.

Daily check--solving key issue. Outstanding the check of key operation and process. Solving individual risks and key issues by daily check, professional check, seasoning check, holiday check.

Special review--solving series issue. Focusing on single theme or factor, the review is much deeper on a special issue. The issue will be analyzing from very beginning to the end based on linear method.

Stage review--solving systematic issue. Building up Safety Assessment Standard, evaluating safety qualitative and quantity state by chart and data to define the shortage and defect. Defining the stage of safety developing on and main target and content. Implementing progressing activities, promote the healthy and orderly development of enterprise safety orderly.

4.Conclusion

Since the '11311' safety management model forming and promoting, the company's safety management has great innovation and earn achievement periodically. ①the staffs from all over China share unique safety consciousness. The safety method is unified. During operating, staffs has changed from 'asked to be safe' to 'asking for safe', from obeying negatively to preventing positively. Employees deem safety as a responsibility, an achievement, a team honor.②the preferential risk of employee's unsafe behavior decrease obviously. The quality standards have covered all aspects. Everyone, everything, every time, everywhere have standard. The safety management level increased obviously.③the safety regulation and practice are combined tightly. The action rules are easy to get, remember and implement. The accident in base has been decreasing 20-30% every year. All companies accord with Class 2 and above of National Danger Chemical Safety Code. The safety management makes the human-being,equipment and environment harmonized. Practice has approved that '11311' Safety Management Model is a suitable and effective safety management model for coal based chemical companies.

5.References

[1] Chen Weimin, Xu Lian, "Safety Risk Pre-caution Management,"Press of China University of Mining and Technology, 2006

[2] Qi Youhong, Qi Youjin, "Fine Safety Management, ", XInhua Press 2009

[3] Luo Yun, "The Management of Employee's Safe Behavior, " Chemical Industry Press, 2012

[4] Luo Yun, "Systematic Management of Safety Production, " Chemical Industry Press. 2014

[5] Zhang Guoshun, Cai Junhui , "Treasury Book for Safety Production, " Coal Industry Press, 2015

[6] Sun Huashan , "Safety Production Rish Management, " Chemical Industry Press, 2012

[7] Xu Deshu, "Study and Survey on China's Safety Culture Construction, " Sichuan Science and Technology Press, 1994

[8] Shenhua Group , "Research and Practice on Safety Developing of Coal Mine of Shenhua Group , " China Coal Industry, Volume 08, 2018

[9] Wang Jian , "Research and Practice on Building Essential Security Coal Mine, " Journal of Xian Science and Technology University , Volume28, NO.1, 2008

[10] Heinrich H W. , " Industrial accident prevention , " .New York:Mcgraw-Hill,1931

[11] Bird JR., Frank E., "Management guide to loss control, " .Atlanta:Institute Press,1974

Study on the corrosion mechanism and behavior of large crude storage tank bottom plate

Luo Fangwei, Li Mosong
CNPC Research Institute of Safety & Environment Technology, No. 1 building, Huanghe north road, Changping District, Beijing, China

Abstract: As the major storage facility, large crude storage tank plays an important role in the whole product line of oil and gas industry. Corrosion has, until recently, become the major reason of storage tank failure with the increase of crude importation in China. Sulphuric, acidic as well as salt substances in varied types of crude bring more unfavorable factors to the safety of storage tank. In this work, the results obtained from the electrochemical experiments of carbon steel in Russian crude and Daqing crude were reported respectively. The mechanism and rate of both general corrosion and localize corrosion were revealed to support the further study on risk assessment and corrosion control technology of storage tank.
Keywords: Large crude oil storage tank, electrochemical corrosion, carbon steel

The large vertical crude oil storage tank of normal pressure which is the main storage facility of crude oil is widely used in oil field, storage, transportation, refining, chemical and other industries. The use of the tank has increased significantly with the continuous expansion of production capacity, especially in the petrochemical enterprises. The storage of crude oil is in large quantities, moreover, the medium is inflammable, explosive, toxic and environmental polluting. The leakage of tank is much more likely to lead to a major fire and explosion accident. The accidents of East chemical plant "6.27", Huangdao oil depot "8.12", Dalian petrochemical "8.29" and Shandong Shida science and technology "7.16" show that the oil storage tank accidents are still in multiple status. The failure mechanism of crude oil storage tank is different due to the differences of design and manufacture level, operation conditions, medium conditions, external environment, overhaul and maintenance conditions. Without considering the operation causes and accessory failures, the main failure mechanisms of tank body include corrosion thinning, stress corrosion cracking, brittle fracture and mechanical fatigue. The corrosion thinning is the most important reason [1] among these above. The United States statistical analysis on 242 cases of storage tank accidents in the past 40 years shows that 72 cases (about 30%) of accidents were caused by corrosion [2]. China's crude oil imports has been increasing and the sources of imports has been diversifying recently. The corrosion problem brought by high sulfur, high acid and high salt crude oil to storage tanks is more and more prominent [3].

1. Experiment research

Daqing and Russian crude oil storage tanks were selected to collect tank bottom water as corrosion medium to study the corrosion law of common storage tank steel Q235 in these two mediums. Firstly, the bottom water of 40000m^3 crude oil storage tank in the storage and transportation workshop of a refining and chemical enterprise was collected to analysis. The differences between these two kinds of water samples are list in Table 1:

Table 1 Composition of bottom water in storage tank

compositions / samples	PH	conductivity	SO_4^{2-}	Cl^-	CO_3^{2-}	HCO_3^-	Mg^{2-}	Ca^{2-}	S^{2-}	Fe^{2-}	Total salt
		μS/cm	mg/L	mg/L	mg/L	mg/L	mg/L	mg/L	mg/L	mg/L	mg/L
Daqing	8.2	1.22×10^4	33.8	2.39×10^3	132	1.32×10^3	3.12	612	1.83	0.37	6.94×10^3
Russion	8.8	3.68×10^4	36.1	489	117	372	1.92	87.2	3.38	0.12	1.19×10^3

It can be seen that the salinity of Daqing's bottom water is obviously greater than that of Russian, while the S^{2-} content of Russian oil is higher. The composition of the selected Q235 steel in the experiment is shown in Table 2.

Table 2 Q235 steel composition

C (%)	Si (%)	Mn (%)	P (%)	S (%)	Cr (%)	Ni (%)	Cu (%)
0.22	0.35	1.40	0.05	0.05	0.30	0.30	0.30

Since both of the bottom waters contained anions such as sulfate ion, chloride ion and sulfur ion, the following electrochemical corrosions inevitably occurred in the base plate:

$$Fe^{2+} + S^{2-} \rightarrow FeS$$

$$Fe^{2+} + 2Cl^- \rightarrow FeCl_2$$

$$Fe^{2+} + SO_4^{2-} \rightarrow FeSO_4$$

The collected bottom water was placed in the experimental device, as shown in Figure 1, to carry out corrosion hanging and electrochemical experiments. The surface of metal hanging pieces were sanded by 320, 600, 800, 1000 and 1200 mesh sandpaper ,cleaned up with distilled water, absolute ethyl alcohol and propanol ethyl carbinol (analytical pure), then placed into drying utensil for 24 hours, and weighed by electronic balance (accuracy 0.1mg). Metal hanging pieces were arranged on the hanging rack and immersed into the bottom of the storage tank to test. The circular surface of 15.80mm diameter (1.96cm^2 area) emerged out of water. The experiment period was one month. Then the surface morphology of test pieces were observed by OLYMPUS BX61 metallographic microscope with the magnification of 200x. The corrosion outcomes were removed and the rest test pieces were weighed to calculate corrosion weightlessness.

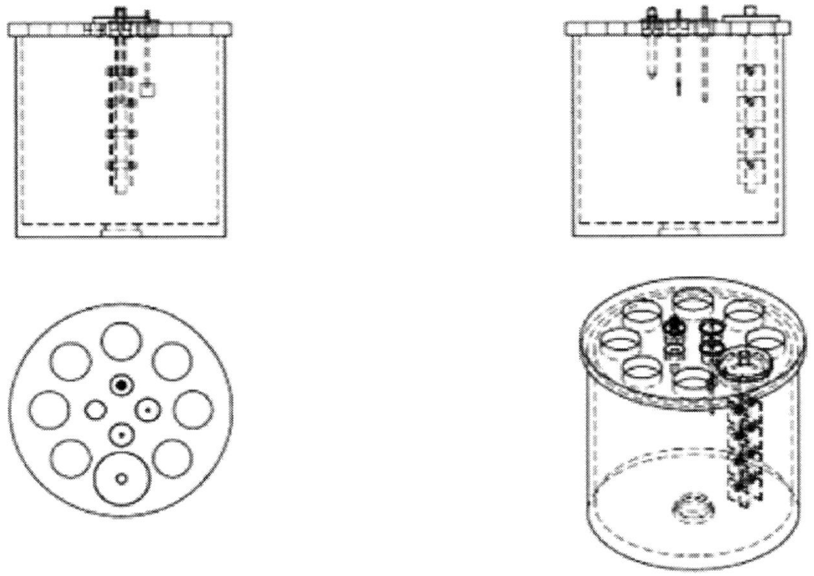

Figure 1 The device diagram of corrosion simulation experiment

The electrochemical working electrode adopted the processed 1cm*1cm*3mm Q235 steel pieces, welded copper wire, sealed with epoxy resin, shown in Figure 2. The surfaces were also sanded by 320, 600, 800, 1000 and 1200 mesh sandpaper, cleaned up with distilled water, absolute ethyl alcohol and propanol ethyl carbinol (analytical pure), then placed into drying utensil for 24 hours. The electrochemical testing instrument was IM6 electrochemical workstation, the auxiliary electrode was platinum black electrode, and the reference electrode was Ag-AgCl electrode. The working electrode was immersed into the solution then measured the electrode potential for 48 hours.

At the same time, the linear polarization, polarization curve and AC impedance test were carried out one hour after the electrode immersing in the solution. The AC impedance spectrum was measured again after 48 hours.

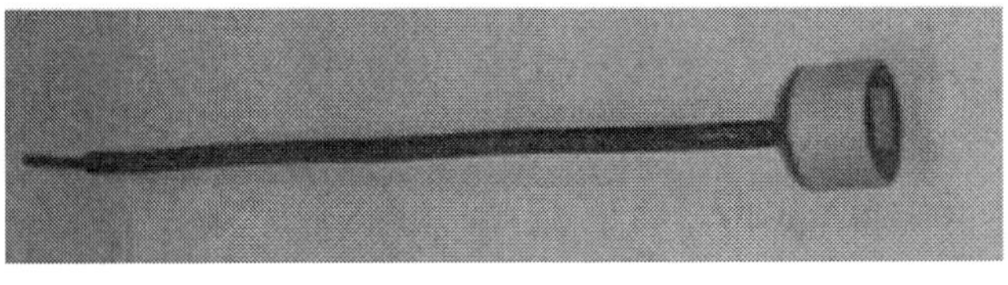

Figure 2 Working electrode of experiment

2. Results and discussion

After one month's immersion, the weight loss of metal hanging piece in Daqing crude oil bottom water was 2.5mg and the corrosion rate was 0.02mm/a. The weight loss of metal hanging piece in Russian's was 1.6mg and the corrosion rate was 0.012mm/a. The corrosion morphologies are shown in Figure 3 and 4 separately. It can be seen that after a month's immersion, the corrosions are uniform corrosion and both of them have partial corrosion. However, the partial corrosion development of Daqing crude oil is more serious than Russian oil. Moreover, obvious pitting corrosion pits have occurred.

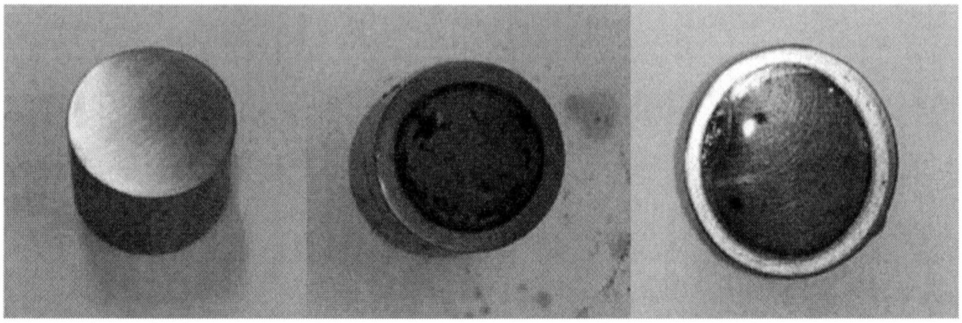

Figure 3 Original hanging piece, piece has immersed in bottom water of Daqing and Russian crude oil for a month

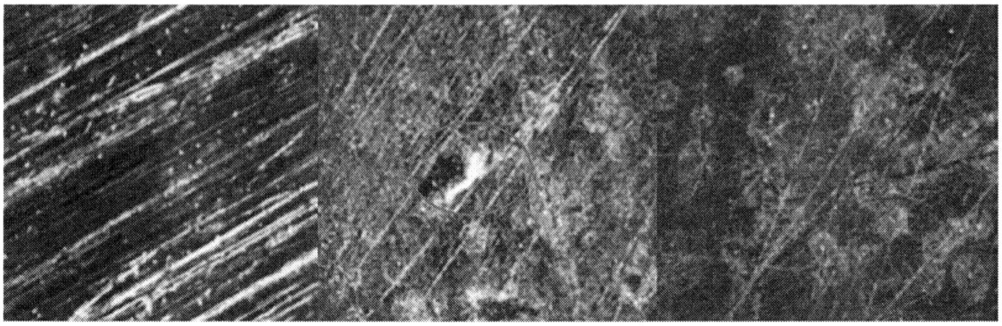

Figure 4 Morphology of original hanging piece, piece has immersed in bottom water of Daqing and Russian crude oil for a month

The electrode potential changes after 48 hours' immersion are shown in Figure 5. The both potentials of these two test pieces showed negative shift rapidly then positive shift. The electrode potentials became basically stable after about 3 hours. The positive shift of test pieces in Daqing bottom water was about 30mV, that was -0.530V (Ag-AgCl). A large scale positive shift in Russian's was 230mV, that was -0.280V (Ag-AgCl). The electrode potential difference was 250mV. It is probably due to the high content of S^{2-} in bottom water of Russian oil, it's reactant of iron sulfide film can sufficiently prevent the corrosion process, while that of Daqing is too less to form a complete protective film. The same appearance is also reflected in the surface morphology of the hanging sheet after a month's immersion.

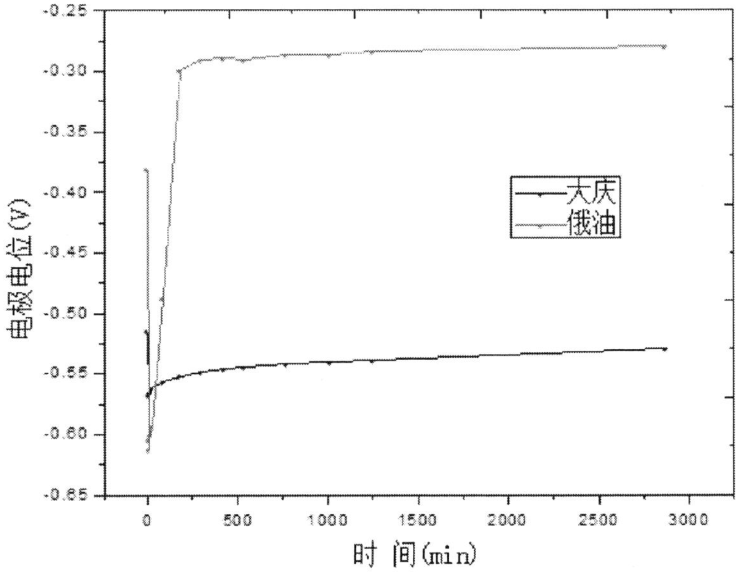

Figure 5 The electrode and potential
Red for Russian, black for Daqing

The test conditions of polarization curve are that the scanning range is relative electrode potential from 250mV to -250mV and the scanning speed is 5mV/S. The results are shown in figure 6.

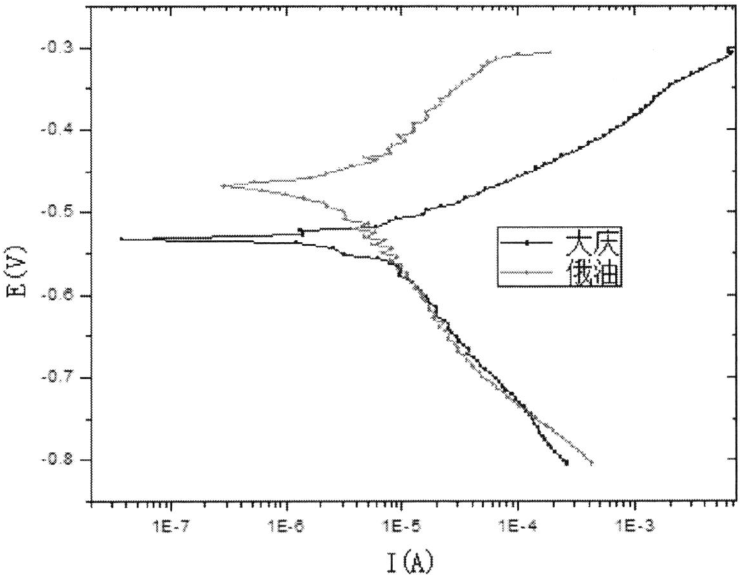

Figure 6 Polarization curve
Red for Russian, black for Daqing

The calculation results based on polarization curve are list in the following table 3:

Table 3 The calculation results based on polarization curve

Water samples	Ecorr (V vs Ag/AgCl)	Icorr (μA • cm^{-2})	Corrosion rate v(mm • a)
Daqing Water	-0.533	6.1	0.07
Russian water	-0.469	2.9	0.03

The test conditions of AC impedance are that AC amplitude is + 10mV and frequency range is 10KHZ-100KHZ-10mHZ. The test results are shown in Figure 7 and 8.

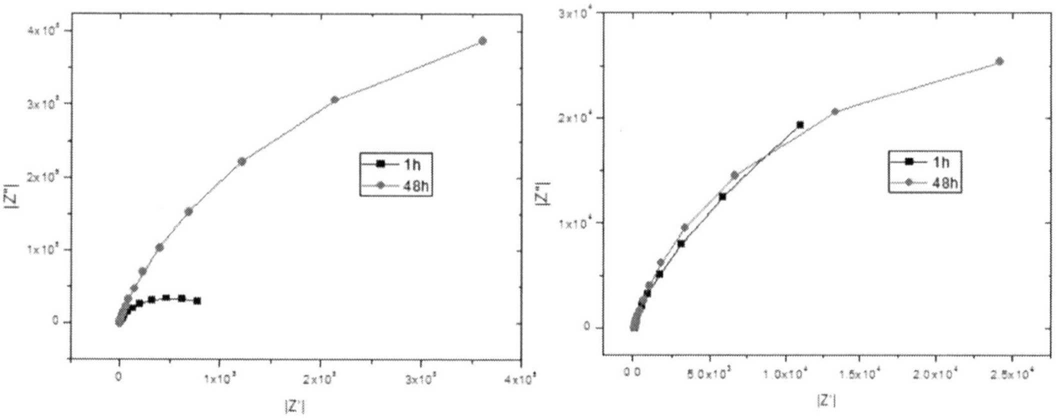

Figure 7 AC impedance Nyqusit figure of the test pieces in bottom water of Daqing and Russian crude oil

The impedance equivalent circuit is shown in Figure 8. The parameters of the equivalent circuit are shown in Table 4.

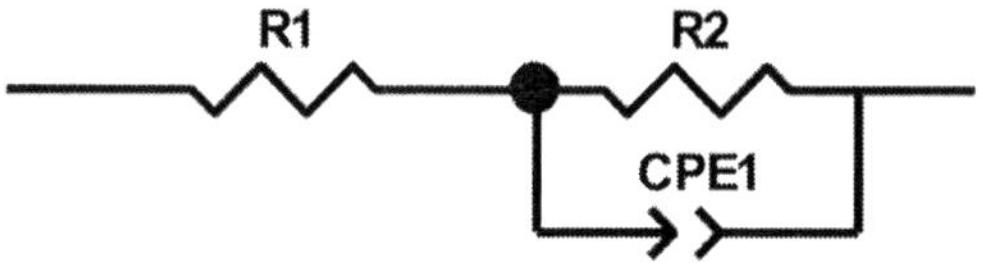

Figure 8 AC impedance equivalent circuit

Table 4 Fitting parameters of impedance spectrum

Test pieces	R1(Ω)	R2(Ω)	CPE-T(F)	CPE-P
Daqing1h	38.15	84603	3.7763E-5	0.86362
Daqing48h	39.71	974480	1.2678E-7	0.0016
Russian1h	107.4	59703	0.00048	0.8838
Russian48h	114	69706	0.00025	0.86892

3. Conclusion:

Through the experiments above, the internal corrosion conditions of baseplate of Russian and Daqing crude oil tank can be concluded as follows:

1) There are a variety of corrosive anions in both bottom waters of Daqing and Russian crude oil storage tank, and there are many kinds of corrosion mechanisms on the internal tank baseplate.

2) The studies on hanging pieces and electrochemical corrosion show that the corrosivity of Daqing's bottom water is higher than that of Russian. Both of them have uniform corrosion and local corrosion at the same time, but the localized corrosion tendency of Daqing is more obvious.

3) The electrode potential of steel in the bottom water of Russian crude oil is obviously more positive than that of Daqing oil by the changes of electrode potential. It may be due to that the steel surface in Russian crude oil bottom water forms a more complete corrosion protective film, but it need further surface composition analysis to confirm.

Reference:

[1] Guo Bing. Research on risk assessment of large atmospheric storage tank group [D]. Hebei University doctoral thesis. 2010:13

[2] Li Chunshu. Research on corrosion detection and residual life prediction technology of normal pressure tank bottom plate [D]. East China University of Science and Technology doctoral thesis. 2009:1-2

[3] Liu Weiming, Zhang Hanqian, Qu Zhaoxia, etc. Research progress on corrosion properties of steel for large crude oil storage tank [J]. Corrosion science and protection technology. 2008, 20 (6): 448-453.

Research projects: This paper is funded by the national key research and development plan, "Safety monitoring, early warning, protection integration and Engineering demonstration of storage facilities for typical hazardous chemicals" (2016YFC0801205).

Author: Luo Fangwei, male, born in Oct. 1976, graduated from China University of Petroleum (Beijing) in 2008, industrial Ph.D., works in CNPC Research Institute of Safety & Environment Technology as senior engineer, engaged in corrosion and protection technology research.

Contact: No. 1 building, Huanghe north road, Changping District, Beijing, China

Zip code: 102206

Tel: 010-80169771

E-mail: luofangwei@cnpc.com.cn

515208 Study on the Integrated Protection Technology of Lightning and Static Electricity in Large Oil Storage Tank

Abstract:

In recent years, there have been many petroleum and petrochemical enterprises caused by thunder and lightning static electricity caused by the explosion of oil storage tank fire accidents, accident prevention and anti-static research, oil storage tank in the anti-static anti-static issues still exist; storage Canister anti-static anti-static mainly take two measures, one anti-static lightning grounding, the second is anti-tank anti-lightning installation of lightning receptors; but these two measures did not solve the fundamental problems, resulting in electrostatic lightning caused The flash explosion fire hidden danger has not been solved. In response to the accidental risk of electrostatic discharge of oil storage tanks, we conducted systematic research on the causes of the formation of electrostatic precipitations of lightning cans, the integrated lightning protection technology and the corresponding technical standards in the 11th, 12th and 13th five-year plan period. According to the research results, the cause of lightning electrostatic discharge of oil storage tank is introduced, and the lightning static electrostatic integrated protection technology for oil storage tank is explored.

Using Aloha to Establish Potential Risk Areas and Verified with Actual Incidents for Toxic Chemical Operation in Taiwan—a Case Study in Middle Taiwan

Liao Kuang-Yu / Horng, Jao-Jia
National Yunlin University of Science and Technology
123 University Road, Section 3,Douliou, Yunlin 64002, Taiwan, R.O.C.
liaoky@gmail.com/Horngjj@gmail.com

Li, Min-Chang 、
Chang, Chih-Wei
National Yunlin University of Science and Technology

minchangs@gmail.com
ccw0812@gmail.com

Prepared for Presentation at
American Institute of Chemical Engineers
2018 Spring Meeting and 14th Global Congress on Process Safety
Orlando, Florida
April 22 – 25, 2018

Using Aloha to Establish Potential Risk Areas and Verified with Actual Incidents for Toxic Chemical Operation in Taiwan—a Case Study in Middle Taiwan

Liao Kuang-Yu / Horng, Jao-Jia

National Yunlin University of Science and Technology
123 University Road, Section 3,Douliou, Yunlin 64002, Taiwan, R.O.C.

liaoky@gmail.com/Horngjj@gmail.com

Li, Min-Chang /Chang, Chih-Wei
National Yunlin University of Science and Technology

Keywords: ALOHA, Emergency response, Super GIS

Abstract

ALOHA had been used to simulate the influencing areas after one-hour dispersion and consequence analysis for the prevention measures and response plan in operating toxic chemical substances in Taiwan. This study aimed at those factories operating toxic chemical substances in the central of Taiwan, Taichung, Changhua, and Nantou counties. The data combination of ALOHA simulation and Super GIS established a hazard analysis graph that could evaluate the potential hazards and impacts in leaking incidents of toxic chemical substance. Furthermore, this evaluation was verified by 2012 to 2016 hazardous material incidents and statistically analyzed. Those results were drawn on the GIS analysis graph to examine the potential areas with high and middle risks and to evaluate the emergency response and preparedness measures in order to comprehensively understanding the effectiveness of consequence analysis and the practical uses.

1 Introduction

Recently, in the field of industry and high technology growing rapidly year by year in Taiwan. Especially the industry related to chemicals have account for 31% and up to 2.7 trillion dollars of manufacturing's total value in Taiwan, the most important is that chemicals related products closely related to daily life. However, with the continuous development of technology, it need more various materials and energy sources, various petrochemical processes and technology industries have become increasingly complex and their scale has gradually expanded. The types of chemicals use have also gradually increased. Since chemicals have hazard characteristic of toxic, flammability, corrosive, reactivity etc. No matter operate in the process of manufacture, use, storage, transportation, disposal etc. If you didn't noticed, may cause serious accidental accidents.

Environmental Protection Administration, ROC (Taiwan) statistics of 394 chemical disaster cases about types of materials from 1997 to 2006 were as follow:

(1) Liquid gas tank leaking by turns over and fire accidents for 87 cases, account for 18.4%, coming in first. (2) Toxic high pressure gas and substances (included Ammonia chlorine gas) for 82 cases, account for 17.3%, coming in second. (3) Flammable liquid for 72 cases, account for 15.2%, coming in third. (4) Class 5 explosives for 59 cases, account for 12.5%, coming in fourth. (5) Flammable high pressure gas for 51 cases, account for 10.8%, coming in fifth. (6) Class 6 strong acids and strong alkalis for 43 cases, account for 10.1%, coming in sixth.

Types of disasters as follows: Fire:138 cases (35%), explosion:48 cases (12%), leak:168 cases (43%). Therefore, we should bring suitable protect equipment depend on the types of chemical disaster incidents in non-fire rescue. But, if it had already exploded, should change into fireproof-clothes.

Environmental Protection system receive 1442 cases of chemicals disaster incidents announced from all over the country in 2007-2011, fire, leak, explosion, etc. were mostly happened in the types of chemical disasters in Taiwan. From 2007 to 2010 (November), caused incidents were factory accidents, traffic accidents, others accidents, storage accidents, school accidents, etc. Factory accidents for 905 sessions, account for 63%, coming in first, the survey also indicate the announcement of factory accidents cases were risen in the past two year, should directly relate to the increasing chemicals operation. The types of incidents occurred, fire (44%), leak (22%), explosion (10%), and transportation (11%) were the most.

Following to population and the development of industries, the use of toxic chemical materials will be increasingly impact on megalopolitan area, so chemical disaster prevention and public disaster reduction system of response need to be planned completely. Environmental Protection Administration, Executive Yuan apply the spatial analysis ability of GIS to integrate the database of chemical disaster prevention information, complement simulate pattern to simulate the spread of toxic chemical materials, as the investigate analysis of chemical disaster prevention, countermeasures, and support the necessary decision, also can bring into the normal release control operation of toxic chemical materials, set up a chemical disaster risk assessment system to analysis short and long term of risk location in megalopolitan area. Due to major chemicals disasters occurred, such as fire, explosion or toxic gas spread, etc. May caused unpredictable results. If can simulate the impact on range before disasters occurred and assess the possible attack from environment, also address countermeasures of possible deaths, damage, etc. They can control timely when the accidents occurred, and reduce the damage of personnel, property, and environment. Environmental Protection Agency's Risk Management Program apply air pollution pattern to consider different situation of factory and the spread of storage tank leaking heavy gas, factories need to address consider response, reduce risks accordingly and complement emergency response. Following to deal with major chemical disasters in Taiwan, limit to response equipment and resources insufficient, should complete disaster prevention in advance, avoid wasting resources and unnecessary deaths.

This survey is aimed at the toxic materials operation places in Central Taiwan (Taichung, Changhua, and Nantou), perform ALOHA spread simulate analysis, analysis potential damage range of operation places, and combine with Super GIS to draw layers in damage range, also discuss simulate analysis results, as references for operation factories or related rescue units to prevent disaster and emergency response.

2 Research methods

2.1 Introduction of software application

2.1.1 ALOHA derives from Steady–State Gussian Plume Spread continuously, updated version 5.4.7 now. Since Environmental Protection Agency (EPA), Chemical Emergency Preparedness and Prevention Office (CEPPO), National Oceanic and Atmospheric Administration Office of Response and Restoration (NOAA) are the assessment tools to apply the range of chemicals toxic gas spread suggest by EPA; ALOHA need to apply information as table 1, leakage and simulate spread process as figure 1.

ALOHA build up 900 various chemical materials database, including terror attack relate to toxic industry chemicals, including TABUN, sarin, V agent, Mustard gas, lewisite, Hydrogen Cyanide, Cyanogen chloride, Chlorine, phosgene, Tear gas, chloroacetophenone, etc. This database main consider about physical characteristics, simulate scenario included toxic gas spread, fire, and explosion, including four types of leak sources (Direct, Puddle, Tank, and Gas Pipeline), can simulate situation as table 2, also can combine with map (or Google Earth) by export graphics, draw damage range and plan the evacuate distance of sensitive environment area.

2.1.2 Geographic Information System (Super GIS): Super GIS provide the function of map drawing, edit, geographical analysis, raster, etc. To predict damage results into potential disaster analysis figures relate to spatial distribution, provide the logistics operation of emergency measures decision support or plan.

Table1. ALOHA software analysis database

geographical location	Chemical materials leaking situation	Spread
➢ Types of near buildings ➢ Simulate the event time ➢ Weather: Atmospheric Environment background information	➢ Choose which chemical materials need to be analysis ➢ Types of leakage: spurt, become puddle, tank leaking or pipeline leaking	➢ Heavy Gas dispersion ➢ Gaussian dispersion

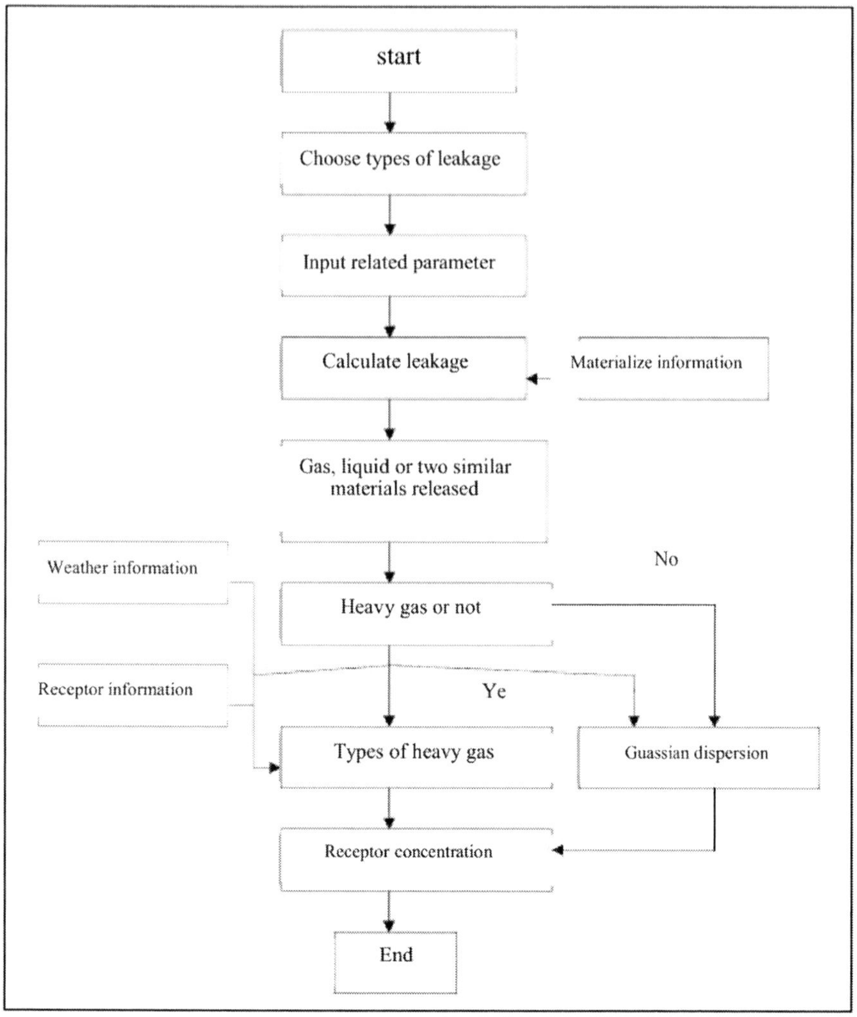

Figure 1. Calculate process of chemical disaster toxic gas spread

Table 2. ALOHA-5.4.7 Types of leaking and Scenario hypothesis

Types of leaking	Types of leaking	Toxic gas situation	Fire situation	Explosion situation
Direct	Direct leaking	Toxic vapor	Flammable area(flash)	Vapor explosion
Puddle	Vapor	Toxic vapor	Flammable area(flash)	Vapor explosion
	Flame (Puddle fire)	Puddle fire	— —	— —
Tank	Non-flame	Toxic vapor	Flammable area(flash)	Vapor explosion
	Flame	— —	Spray blaze or Puddle fire	— —
	BLEVE	— —	BLEVE (fireball or puddle fire)	— —
Gas Pipeline	Non-flame	Toxic vapor	Flammable area(flash)	Vapor explosion
	Flame (Spray blaze)	Spray blaze	— —	— —

2.2 Consequence simulate analysis

2.2.1 Collect background information: ALOHA simulate is major to consider the factor of chemical materials spread effection, such as types of chemical materials, wind velocity, wind direction, temperature, humidity, etc. The factor of damage: types of tank, storage capacity (volume), hole diameter, location, etc. Target: TWA, IDLH, etc. Use Pasquill-GIFFORD and DEGADIS to analysis gas and liquid materials, can get the range of damage effection. First, collect related information, then draw up possible situation and types of accidents, choose which is symbolized, significant potential damage and high probability of occurrence to perform consequence analysis, and understand the affected range of accidents, necessary information as follows:

2.2.1.1 Storage equipment information: including factory location and types of storage equipment standards, SDS in the storage, deal with emergency situation (HAZMAT, prevent and rescue disaster manual) and others.

2.2.1.2 Weather information: including wind speed, probability of wind direction, relative humidity, temperature, atmosphere stability, etc. Summarize and count central area typical weather conditions in past five years as table 3.

Table 3. Central Taiwan weather information statistics in past five years

Scenario simulate hypothesis	A city	B city	C city
Wind speed (m/sec)	1.5	1.84	0.8
Temperature (°C)	29	29.7	30.2
Humidity (%)	71.33	73.06	70.7
Cloud coverage	0	0	0
Atmosphere stability	F	F	F
Wind direction	NNE	NE	WNW

Notes: Weather information: Central Weather Bureau and Environmental Protection Administration air quality detection station

2.2.2 Damage scenario simulate hypothesis: toxic chemicals leaking occurred, liquid evaporate to toxic vapor before spread in the air, simulate the worst scenario hypothesis is the largest capacity of unit chemicals storage tank leaking completely in ten minutes, materials vapor speed reference to Chemical Exposure Index (CEI), ALOHA simulate and assessment the hazard range of toxic gas spread in one hour.

2.2.3 Simulate the assessment of objects and range: This simulate assess the operation places where performed the toxic gas leak simulation by chemicals operation in three cities of Central Taiwan (Taichung city, Changhua county, and Nantou county).

2.2.4 Draw a hazard analysis graphic: This search defined to the toxic affected area of toxic chemicals leaking up to ERPG-2 (personnel expose in toxic gas environment about one hour, the maximum allowable concentration that does not cause the body to cause irreparable hazard), and hazard radius (km) is simulate by ALOHA. In probability of occurrence, the storage depends on different qualities and types, may cause different probability of accidents (hole), reference the probability of hole announced by America chemical industry association, pressure compartment is 9.55×10^{-4} s/y, normal metal storage compartment is 8.63×10^{-3} s/y, barrel trough is 1.03×10^{-2} s/y. The hazard range of toxic vapor accidents are related to wind direction on site, so take wind direction and population into account. Thus, in drawing whole hazard analysis graphic, demonstrate each location of simulate factories in GIS first, the leakage of toxic chemicals spread centered around hazard distance (ERPG-2 affected range), use Super GIS's Nearby to do hazard range (Buffer); statistics the probability of 16 position of wind direction average in the past 5 years, plus hazard range which overlap area to each other, also consider area population density to build up attribution information, drawing population density potential risk situation, output to Suefer program and draw hazard risk contour line overlap with Super GIS hazard analysis graphic. Use color level to represent risk hazard, from red to purple represent high risk area value (0.283-634.305), pink to orange represent medium risk area value (0.021- 0.282), green to deep blue represent low risk area value (0-20).

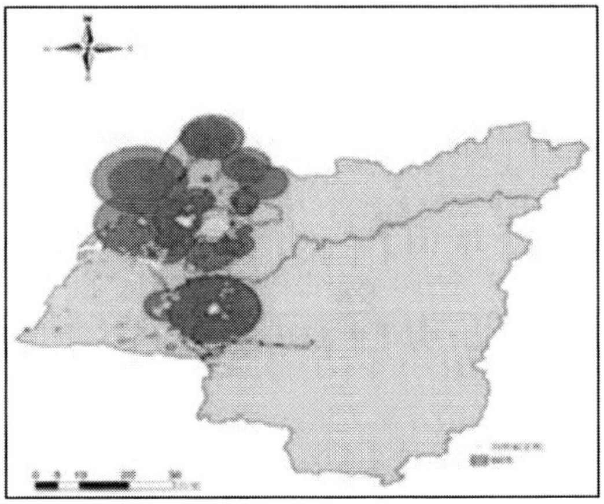

Figure 2. Central Taiwan ALOHA simulate risk analysis

3 Results and discussion

This research major analysis the 535 controlled factories in Taichung city, Changhua county, and Nantou county, in high and medium risk area have 432 factories, account for 80.6%, as table 4, main distribute in Nantou city, Shengang Township, Lukang town, Wuqi District, Xitun District, Houli District, Daya District, Dali District, etc. As figure 2. Table 5 and figure 3 show the location of fire units, the location set up in high and medium risk area only account for 34%, obviously, the energy of fire units in risk area are insufficient, at the present stage, fire units deal with chemical disaster accidents equipment included environment detection and personal protect equipment (e.g. protect clothes, breath protect, etc.), if fireman can't control or protect on site, will often need to wait for Environmental Incidents Specialist Team to support. Which can strengthen the response energy of fire units and industry area factories, simulate abrupt case and exercise emergency response plan through mutual defense in industry area, also strengthen disaster control and response at first moment.

Besides, check the locate distribution of school, health care units, and shelters (as figure 4 to figure 6 and table 4), in high and medium risk area of schools account for 40.6%, health care units account for 59.5%, and shelters account for 50.1%, Although schools and health care units both have emergency evacuate plan now, but most evacuate plans are about natural disaster (figure 7), shelters are suitable for typhoons, floods, earthquakes, mudslide, and tsunamis at present, no special plan for chemical disaster shelters, the concept of chemical disaster shelter on site still insufficient, we should take this analysis graphic to restart the plan of evacuate shelters, consider to increase special evacuate shelters of chemical disaster and guide the evacuate on site education.

This research also search the database of toxic disaster rescue management system in Environmental Protection Administration, Executive Yuan, statistic chemicals disaster accidents in 3 cities of Central Taiwan (Taichung city, Changhua county, and Nantou county) from 2012 to 2016, and analysis types of accidents according to fire, leak, explosion, etc. Following to statistic analysis results indicate that fire account for 73%, leak account for 12%, explosion account for 3%, and others (transportation and unknown chemicals) account for 12%, as figure 8; analysis types of accident places results indicate factory accidents account for 70%, traffic accidents account for 11%, school accidents account for 3%, storage accidents account for 3%, and others accidents account for 13%, as figure 9.

Compare to accident location and chemical disaster potential analysis (as figure 10 to figure 11) from 2012 to 2016, table 5 indicate types of accidents in high and medium risk area, factory accidents account for 73.5%, traffic accidents account for 66.6%, storage accidents account for 83.3%, school accidents account for 63.7%, and others account for 77.3%; table 6 indicate types of disasters in high and medium risk area, fire account for 73.2%, leak account for 80.5%, explosion account for 70%, and others account for 67.8%, the probability of accidents occurred in high and medium risk area are truly high.

Table 4. Risk area sensitive object distribution statistic

Area	High risk area	Medium risk area	Low risk area
Controlled factories	66.8%	13.8%	19.4
Rescue units	24.8%	9.5%	65.7%
Schools	23.8%	16.8%	59.4%
Health care units	38.1%	21.4%	40.5%
Shelters	33.1%	17%	49.9%

Notes: % represent the percentage of factories in the area

Table 5. Types of accidents in risk area distribution statistic from 2012 to 2016

Area	High risk area	Medium risk area	Low risk area
Factory accidents (70%)	49.4%	24.1%	26.5%
Traffic accidents (11%)	48.7%	17.9%	33.4%
Storage accidents (3%)	58.3%	25%	16.7%
School accidents (3%)	45.5%	18.2%	36.3%
Others accidents (13%)	61.4%%	15.9%	22.7%

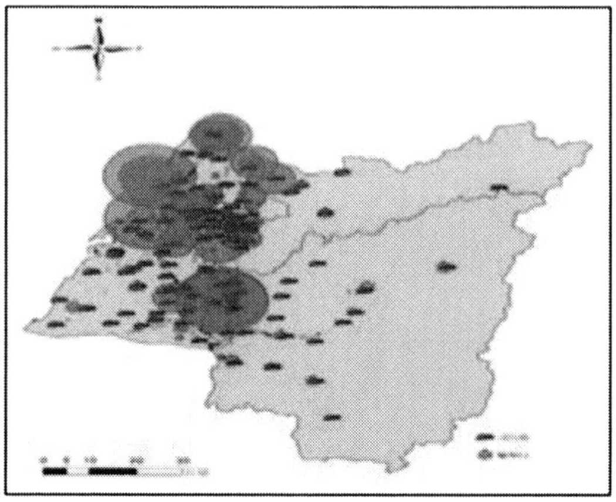

Figure 3. ALOHA consequence simulate results and rescue units distribution in Central area

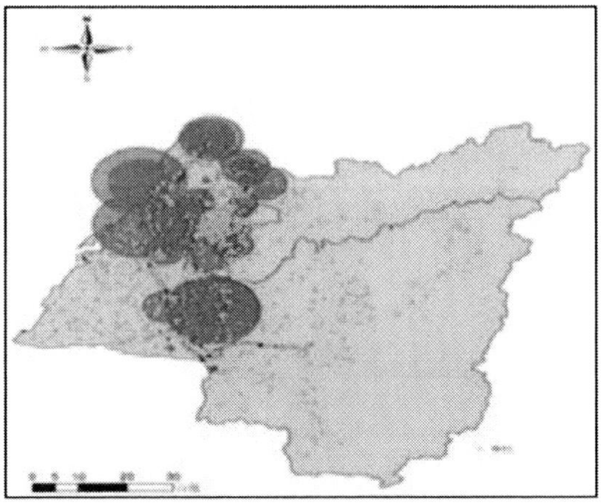

Figure 4. ALOHA consequence simulate results and schools distribution in Central area

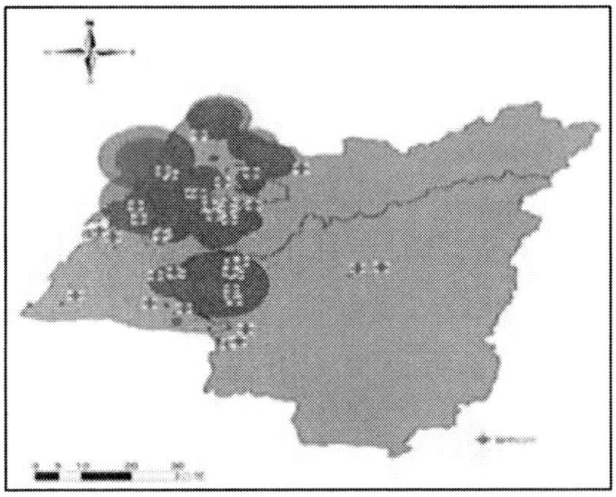

Figure 5. ALOHA consequence simulate results and health care units distribution in Central area

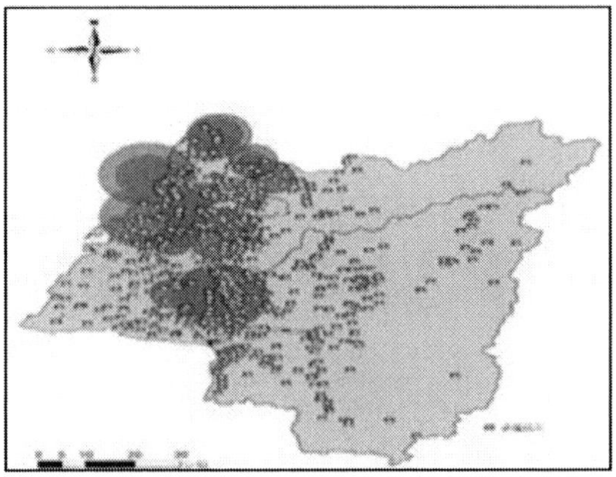

Figure 6. ALOHA consequence simulate results and shelters distribution in Central area

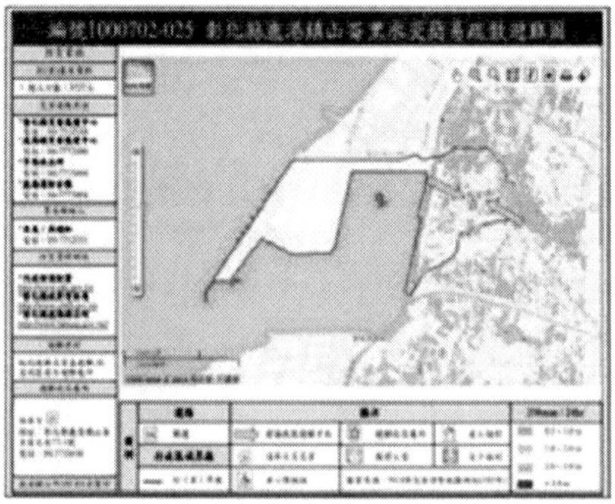

Figure 7. Simply evacuate graphic of floods at SHAN-LUEN LI, Lukang Township, Changhua County

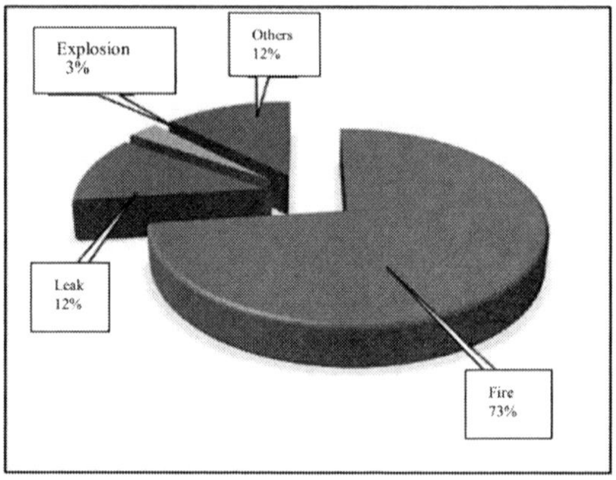

Figure 8. Types of chemicals disasters in Central Taiwan from 2012 to 2016

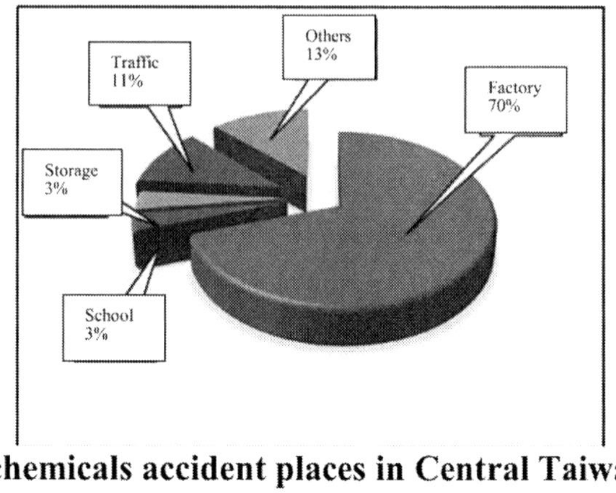

Figure 9. Types of chemicals accident places in Central Taiwan from 2012 to 2016

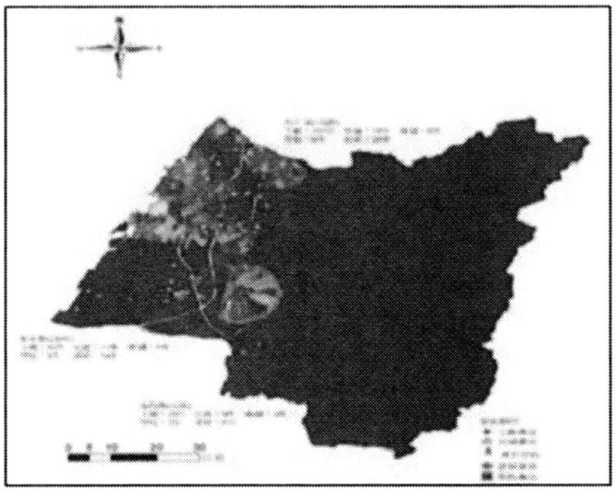

Figure 10. Types of places and hazard analysis in Central area

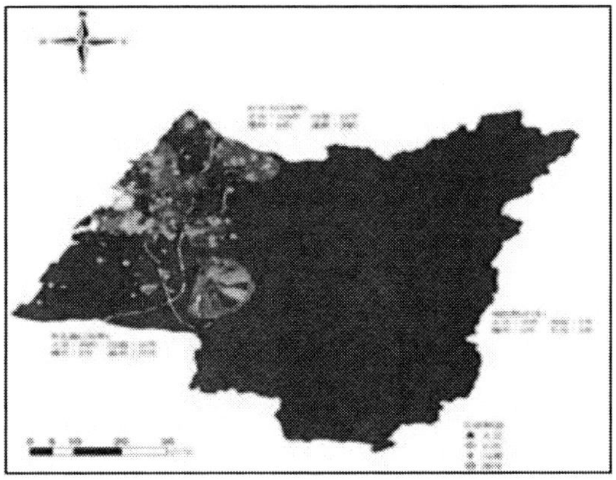

Figure 11. Types of disaster and hazard analysis in Central area

Table 6. Types of disasters in risk area distribution statistic from 2012 to 2016

Area	High risk area	Medium risk area	Low risk area
Fire (73%)	53.1%	20.1%	26.8%
Leak (12%)	68.3%	12.2%	19.5%
Explosion (3%)	50%	20%	30%
Others (12%)	47.7%	20.1%	32.2%

4 Conclusion and suggestion

This research use ALOHA to simulate toxic chemical disasters potential analysis of industry area in Taichung city, Changhua county, and Nantou county, controlled 535 factories are in high and medium risk area account for 80.6%. Fire units distribute in high and medium risk area only account for 34%, obviously, the response energy of fire units in risk area are insufficient. Check with school, health care units, and shelters distribution, schools account for 40.6%, health care units account for 59.5%, shelters account for 50.1% in high and medium risk area. Although schools and health care units both have emergency evacuate guide plan, but most evacuate plans are natural disasters, no special plan for chemical disaster shelters, the concept of chemical disaster shelter on site still insufficient.

Compare to accidents location from 2012 to 2016 and ALOHA chemical disaster shelters indicate types of accidents in high and medium risk area, factory accidents account for 73.5%, traffic accidents account for 66.6%, storage accidents account for 83.3%, school accidents account for 63.7%, and others account for 77.3%; types of disasters in high and medium risk area, fire account for 73.2%, leak account for 80.5%, explosion account for 70%, and others account for 67.8%, the probability of accidents occurred in high and medium risk area are truly high.

When accidents occurred, especially rescue units, health care units, and shelters, how to use response ability? We should plan and prevent in advance, research suggestion as follows:

4.1 Consider more about population density when set up rescue units at present, if not, may affected chemical disasters range, industry area and factories need to develop the energy of save themselves, Environmental Protection Administration and Ministry of Labor both request factories to set up emergency response plan and strengthen mutual defense, hand in the response plan to rescue units can know factories operation of chemicals and set up measures of emergency response to match abrupt case.

4.2 School and health care unit both have plan about natural disaster emergency evacuate guide, and shelters for chemicals disaster are less now, we can restart the plan and plan specific shelters by chemical disaster potential analysis.

4.3 Factories should prepare response resources and the energy of support ordinarily, when disasters occurred, can start mutual defense assistance, energy adjustment depends on different accidents, in order to control the disasters at first moment, cooperation with government and civil groups can fulfill emergency response plan.

5 References

[1] Industrial Development Bureau, Ministry of Economic Affairs：Petrochemical Industry Development and Industrial Security；2015。

[2] National Fire Agency, MOI：Research of chemical disasters scale classification, response pattern and the assessment and integration mechanism of rescue ability. National Fire Agency project；2006。

[3] Hsieh, Yi-Da, Shen, Jia-Jay, Horng, Jao-Jia：Analysis operation disasters of chemicals in Central area recently. Area and environment resources sustainable development seminar；2011。

[4] Chang, Nai-Bin, Tseng, Ti-xing, Kao, Zhi-Yu：Combine with toxic chemical materials spread and Geographic Information System of chemical disasters prevention. Environmental Protection Administration, Executive Yuan project；1996。

[5] Yi-Shu Lin：Research of use different patterns and America risk management plan to assess consequence analysis of chlorine operation plants in Central Taiwan. [master's dissertation]，National Yunlin University of Science and Technology Department and Graduate School of safety Health and Environmental Engineering；2004。

[6] Tsai, Jia-Yi：America risk management cases analysis. Industry pollute prevention 2004；89：107-123。

[7] Fan, Guo-Shu, Chen, Zheng-Ren , Liang, Xiao-Jun , Yang, Hui-Ning：Toxic materials disaster potential analysis and plan assessment. Environmental Protection Administration, Executive Yuan project；2008。

[8] NOAA and U.S. EPA., ALOHA 5.4.7, Office of Response and Restoration of the National Oceanic and Atmospheric Administration and Chemical Emergency Preparedness and Prevention Office of the U.S. Environmental, Seattle, WA; 2016.

[9] Horng, Jao-Jia、Yi-Shu Lin：Use consequence analysis to set up evacuate and shelter plan in factories and industry area. the second edition of International environmental disasters and emergency response technology seminar symposium 2004。

[10] Horng, Jao-Jia , Kuo, Jin-Ying , Yi-Shu Lin：Discuss and plan research to storage problem of reactive chemicals. the third edition of International environmental disasters and emergency response technology seminar 2005；229-239。

[11] Jao-Jia Horng, Yi-Shu Lin, Chi-Min Shu, Eric Tsai, Using Consequence Analysis on Some Chlorine Operation Hazards and their Possible Effects on Neighborhoods in Central Taiwan, Journal of Loss Prevention in the Process Industries, 2005, Vol. 18, No. 4-6: 474-480.

[12] Central Weather Bureau，http://www.cwb.gov.tw/

[13] National Fire Agency evacuate and shelter graphic,http://www.nfa.gov.tw/main/ftplist.aspx

[14] Environmental Protection Administration, Executive Yuan toxic disaster rescue management information system，https://toxicdms.epa.gov.tw/index.aspx?type=p

512508 Using Leading Indicators to Measure Safety Management Performance of Urban Refineries

Abstract:

There is growing interest in advancing knowledge and practice on the use of leading indicators to measure safety management performance in organizations. Meanwhile, with the speed up of the urbanization process, some of the inland refining factories have been surrounded by the cities. In response to the need to improve the effectiveness of safety management in these factories, the aim of the paper is to propose a method to develop leading indicators to assess safety management performance of urban refineries. According to the indicator scores in ISRS, NOSA, API and Basic Norms for Work Safety Standardization(BNWSS), surveys were conducted, safety experts were consulted and invited to express their views, either through interviews or workshops. Based on these surveys above, a leading indicators system was built for assessment of safety management performance. Finally, the practical research showed that the conclusion drawn from the measurement model was credible in that it agreed with those from ISRS assessment, BNWSS audit and safety culture measurement.

511185 Managing Operational Risks in an Integrated Manner—Extracting Value While Enhancing Safety

Abstract:

The Chinese chemical industry is the largest in the world and is a global leader in the production, storage, and transport of chemicals. However, as the accident in the Chinese port city of Tianjin showed last year, it can have a devastating impact on the local community, company, and government reputation. The explosion last year in the vibrant Tianjin economic zone killed 165 people, injured nearly 800, destroyed more than 300 buildings, damaged 12,400 cars and more than 7,500 shipping containers and caused an estimated $1.1B damage to the city and one of China's busiest seaports. Though the government held company executives accountable for poor safety management and local officials for insufficient safety oversight, it must be remembered that the chemical industry is a major source of economic growth and pride in many Chinese provinces. The challenge that the Chinese chemical industry, as well as other companies face, is how to appropriately ensure safety during operations, while still creating expected company growth.

In general, the processing industry is rapidly changing, and we cannot look at process safety management in isolation from all enterprise risks. Industries now operate in a complex environment with constantly changing challenges and limited view of potential risks ahead. There are various macro driving forces that impact a company's safety and business success. Some of them are: economic downturn and cycles creating operational uncertainty; product margin and price volatility resulting in less money for maintenance; increased local and international competition adding pressure to produce faster and cheaper; Evolving regulatory environment generating uncertainty of new requirements; technology breakthroughs such as remote maintenance monitoring through increased SCADA collection introducing new data management challenges; and increased business operating complexity such as management of long international supply chains. Each can have a tremendous impact on process safety and make it difficult for a company to understand how to be successful while still being a leader in process safety management.

Because many companies look at risk (including process safety risk) through silos, their approach is fragmented. For example, workplace health, safety, and environment, are managed to ISO 14000 and the new 45000 standards, chemical process safety management through Decree 591, "Regulations on Safe Management of Hazardous Chemicals" (and its sub-regulations), quality management through ISO 9000, etc. and many other standards such as project management risk, and integrity management risk. Our view is that companies should look at all the operational risks that can impact a company's business objectives in an integrated fashion and understand how to best protect people and assets while still extracting business value from operations, no matter the internal or external cause.

The complex nature of operational risks poses many challenges to proper recognition and effective risk management. Diverse stakeholders such as employees, suppliers, regulators, plant operators, and even shareholders all look at risk differently and separately. This leads to unclear risk ownership and silos across company departments. Because operational risks are many times disconnected from plant operations and front-line workers, it means that the risk impact or mitigation owner is not always obvious. Understanding the relationships and interdependencies is essential to break the risk silos across the organization.

To best protect enterprise value and extract new enterprise benefits, an integrated operational risk management (ORM) approach should be adopted. The ORM approach should be tightly linked to the enterprise strategic planning, take an integrated view of operational risks (including PSM), consider all stakeholders and their needs, manage operational risks while balancing returns and most importantly, be dynamic and adaptive to the changing environment of the world.

Once operational risks are understood and managed, the resilient companies can rapidly adapt and respond to internal or external events—changes, disruptions, opportunities or, threats—and continue operations, sustaining and enhancing business performance. This is done by establishing effective ORM capabilities at every level and across the entire enterprise, tightly linked to decision-making and resource allocation.

There are numerous benefits to implementing an integrated approach to operational risk management. Some of the operational benefits are improved integrated safety performance, enhanced business resilience and agility to manage major disruptions in the supply chain, increased reliability and reduced production losses, and better customer service delivery and quality. Financial benefits include higher profitability and return from optimization of capital allocations and investment, cost reduction due to decreased need for rework and reprocessing, decreased risk perception resulting in easier access to capital and lower cost of debt and optimization of insurance coverage. There are also numerous reputational and compliance benefits including: protection of market share due to high reliability and service quality, strengthened trust among shareholders, improved reputation amongst regulators and increased level of compliance with laws and regulations. And the most important of all—increased and sustainable process safety management!

515879 Research Progress in Olefin Polymerization Technology

Abstract:

Petrochemical Research Institute PRI is a subsidiary of PetroChina Company Limited research institutes, principally engaged in the development of new products such as refining, petrochemical process and catalyst research and development, synthetic resin and synthetic rubber, self-innovation refining technology covering 80% of refining process, independent innovation of chemical technology covering 50% of chemical processes and providing PetroChina refining and petrochemical business Technical Support.

PRI is first research Institute in China to develop alpha - olefin related technology, involved two research directions of series alpha - olefin synthesis and high added-value of alpha-olefin, accumulated rich experience in catalyst, process development and polymerization reaction engineering during 30 years history.PRI have multiple full-process pilot plant such as hexene-1/PAO/ PE ,which is support for technology research and development strongly.

Complete set of hexene-1 technology, which integrates trimerization catalyst, full-process process and high efficiency reactor since 1998, has gone through laboratory research, small test, pilot and industrial test, which have been successfully industrialized for twice. This technology could be characterized as low Matter/energy consumption, high Catalytic efficiency, short process and high quality hexene-1 products, which could meet the most requirements of various polyethylene devices.

Synthesis technology of Poly alpha-olefin (PAO) ,which is known as based oil ,is one of research directions in alpha-olefins application for PRI. PRI's PAO technology adopts the self-developed full-process technology, which has the characteristics of narrow molecular weight distribution, continuous operation, and cleaning process, could produce PAO 4, PAO 6, PAO 8 and other low viscosity serial product.

In the future, PRI will expand the existing technology in alpha-olefin synthesis and application field , strengthen foreign cooperation.

Research on Analysis of Abnormal Emission Sources in Petrochemical Enterprises

Xiaoying Tian1,2*, Dongfeng Zhao1,2, Yi Liu1,2
1 Center for Safety, Environmental, and Energy Conservation Technology
2 College of Chemical Engineering, China University of Petroleum (East China), Qingdao 266580, P R China
tianxiaoying7633@163.com
zhaodf@vip.sina.com
liyx@upc.edu.cn

Keywords: Petrochemical enterprises; Abnormal emission source; Analysis

Abstract

As we all know, petrochemical enterprises are the important pillar of our national economy, but the petrochemical enterprises also belong to the high pollution industry. The abnormal emission is one of the important sources of pollutant emissions in petrochemical enterprises, especially during startup, shutdown and maintenance, petrochemical plants will emit large quantities of waste gas due to process vessel depressurization, purging and venting, which includes both the uncontrolled emission and the organizing was gas that cause serious damage the atmospheric environment in our country. However, there are few researches on petrochemical enterprises under abnormal working situation and the relevant laws and necessary conditions are imperfect. Therefore, it is an urgent problem to analyze abnormal emission sources of petrochemical enterprises.

In view of the above problems, this article reviewed the relevant literature at home and abroad and determined the concept of abnormal emission through a comprehensive comparison of the existing standards between China and the United States. It also summarized and classified the types of pollutants of abnormal emission in petrochemical enterprises, and then introduced the environmental hazards caused by the pollutants. It summarized abnormal emission sources in petrochemical enterprises through an analysis of operating procedures for abnormal situation, and then determined the pollution source strength. Finally, it gave thought to the management and control of the emission source of the petrochemical enterprises in the abnormal situation and put forward some reasonable suggestions.

1. Introduction

The Abnormal Situation of petrochemical enterprises refers to the running state of petrochemical plant in pre-production trial, startup and shutdown, maintenance, production load change, local process conditions change, local equipment failure and periodic special operation [1]. The Abnormal Situation of the petrochemical enterprises can be divided into planned and unplanned abnormal situation. The Planned Abnormal Situation refers to the startup, shutdown and maintenance of the equipment approved by the operator or the mandatory plan. The Unplanned Abnormal Situation is the startup, shutdown and maintenance that are result from equipment or system utilities failure, operation errors or accidents or the irresistible factors in nature.

The abnormal situation involves different petrochemical equipment in petrochemical enterprises with complex operation. During the abnormal situation, the pollution sources will change largely resulting in the unit emissions increased sharply in the short term, which contains many kinds of pollutants, even some toxic and harmful gases, polluting the atmosphere of our country seriously. The research on abnormal emission of oil refining units in western countries is earlier, and the relevant standard system is more perfect. But the abnormal emission sources are not attention of government departments in China, and the current abnormal emission standards is imperfect, even the analysis of abnormal emission sources are still blank. Moreover, it is very difficult for the state to control the emission of exhaust gas from the abnormal situation, so the analysis of the abnormal emission sources in the petrochemical enterprises have great significance.

2. Research status of abnormal situation

2.1 Research status of abnormal situation in China and America

2.1.1 Relevant studies in China

Under abnormal situations, petrochemical enterprises that will not discharge pollutants under normal working conditions will generate emissions due to purging, boiling and other processes, or even exceed the normal emissions. However, the government does not pay attention to the abnormal emission sources of petrochemical enterprises, and the related policies and standards are formulated relatively late and the management system is not perfect.

Since twenty-first Century, more and more attention has been paid to the atmospheric emissions of petrochemical enterprises in China. In 2003, China implemented the "Technical Guidelines for Environmental Impact Assessment-Petrochemical Construction Project", which pointed out that normal and abnormal emissions of various pollutants should be made. The abnormal emission is the discharge of the pollutants under the conditions of startup, shutdown, maintenance and accident leakage during the operation period. In 2008, "Technical Guidelines for Environmental Impact Assessment -Atmospheric Environment" was issued, which asked operator to estimate the abnormal emission of toxic pollutants and introduced that abnormal emission refers to emission of pollutants under abnormal situation, for example, the emission that is result from the ignition furnace, equipment maintenance, pollutant emission control measures fail to achieve the desired efficiency, non-normal operation process equipment. In 2014, "Shanghai City Chemical Plant Startup, Shutdown and Maintenance of VOCs Emission Control Regulations" and "Shanghai City Chemical Plant Startup, Shutdown and Maintenance of VOCs Emission Control Technology Guide" that pointed out that the plant shutdown refers to the shutdown process of the chemical equipment for the planned maintenance, which includes the stopping, emptying and cleaning. It also placed the principles and measures for VOCs emission control in Shanghai petrochemical and chemical enterprises during startup, shutdown and maintenance, as well as the technical requirements for monitoring and reporting. At the same time, enterprises are required to record the pressure and temperature of the container when the device is empty and record the gas components before the inspectors enter into the unit. In 2015, "Emission Standard of Pollutants for Petroleum Refining Industry"(GB 31570) and " Emission Standard of Pollutants for Petroleum Chemistry Industry "(GB 31571) was issued, which pointed out that abnormal situation means the process parameters of a production facility are not planned to exceed the elastic changes in the design of the device. The "Technical Guideline for Environmental Impact Assessment of Construction Project-General Program" (HJ2.1-2016) was implemented in 2016, which is proposed to be classified to analyze and calculate the production and emission of conventional pollutants and characteristic pollutants according to the production, loading, storage, transportation and get the Pollution factors and their production and emission methods, concentrations, quantities and so on.

2.1.2 Relevant studies in America

Historically, the rules of the commission and its predecessor agencies have not specifically required authorization of MSS activities. However, in December 2005, the commission established deadlines for different facility types to submit an application to authorize planned MSS emissions or lose the ability to claim an affirmative defense for unauthorized emissions during those activities. Senate Bill (SB) 1134, 82nd Legislature, 2011, created Texas Health and Safety Code (THSC), §382.051962, which extended the deadline for certain oil and gas facilities from January 5, 2012 to January 5, 2014. This proposed rulemaking will provide applicants a streamlined authorization mechanism for planned MSS to meet the extended statutory deadline.

Texas Health Safety Code (THSC) states planned MSS activity means an activity with emissions or opacity that: (1) is not expressly authorized by commission permit, rule, or order and involves the maintenance, start-up, or shutdown of a facility; (2) is part of normal or routine facility operations; (3) is predictable as to timing; and (4) involves the type of emissions normally authorized by permit [2]. Planned MSS emissions are part of normal or routine facility operations, which include emissions that are predictable as to timing. Examples of planned MSS emissions may include activities such as routine maintenance and other activities such as equipment blowdowns, pipeline pigging, or tank de-gassing. But planned MSS emissions do not include unplanned MSS, upsets, or emission events.

Texas Administrative Code states unplanned MSS activity is an activity with unauthorized emissions that are expected to exceed a reportable quantity or with excess opacity, which means: (1) a startup or shutdown that was not part of normal or routine facility operations, is unpredictable as to timing, and is not the type of event normally authorized by permit; or (2) a maintenance activity that arises from sudden and unforeseeable events beyond the control of the operator that requires the immediate corrective action to minimize or avoid an upset or malfunction [3]. The Division stated that the discharge of pollutants from unplanned MSS activities is not permitted. Maintenance emissions from normal operations can be planned and predicted, but the emissions associated with unplanned maintenance activities are uncertain and unpredictable. No matter how to control and arrange the startup after the shutdown due to unplanned emissions, the frequency of occurrence is unpredictable, and the activity will be accompanied by unpredictable pollutant emissions. In addition, any MSS emissions from enterprises that fail to meet the standards and those that are not controlled and related to the emission related equipment are illegal. In addition, any MSS discharge from equipment that fail to meet the standards and lack control measures is illegal.

2.1.3 Definition

Through the above research and found that the domestic relevant current standards for petrochemical enterprises of abnormal emission definitions are diverse and not uniform, moreover, the abnormal situation in petrochemical enterprises have many types, but study in the abnormal situation abroad mainly focused on startup, shutdown and maintenance. At the same time, startup, shutdown and maintenance is one of the type of abnormal situation that has large emission and occurs frequently. Therefore, the abnormal emission that this paper study refers to the MSS emission in petrochemical enterprise, namely emission from startup, shutdown and maintenance.

2.2 Types and hazards of abnormal emission pollutants

Abnormal production conditions are generally not long that could be a few days or hours [1]. During abnormal situation, the process equipment will be returned material, cleaned and released, and the vaporized material is first discharged to the flare system or other processing facilities. After reaching certain requirements, the vapor will be emptied to the atmosphere [4]. The specific emission of exhaust gases under abnormal situation is mainly related to the factors such as production materials, process flow, and product type. The main pollutants are $S0_2$, NO_2, CO, VOCs, $PM_{2.5}$ and PM_{10}. The abnormal emission of petrochemical enterprises has the following characteristics:

(1) Abnormal exhaust gas is from emission and volatilization of oil materials that mainly caused by abnormal operation of petrochemical plants during startup, shutdown and maintenance, which is the loss of raw materials and products;

(2) Abnormal emission belongs to fugitive emissions, pollutants enter into the atmosphere directly without dilution that will form a concentration exceeding standard area near the emission sources;

(3) Waste gas from abnormal emission of petrochemical enterprises includes various kinds of pollutants, and most of which are highly toxic substances pose a threat to human health;

(4) an abnormal discharge of pollutant has a longer diffusing time, so it has long-term effects on atmospheric environment;

(5) The distribution of abnormal emission sources is wide, which involves many equipment in production and storage, and it becomes irregular surface source pollution, which brings difficulties to monitoring, accounting and management.

Because of the above characteristics, the abnormal emission of petrochemical enterprises enter into the atmosphere will pollute the environment and cause human respiratory disease or acute poisoning, even threaten the health of life. It endangers animals and plants for a long time, even affect the health of the population and then destroy the ecological balance. The acidic or organic pollutants in the exhaust gas will cause the corrosion of the equipment and factory buildings, even cause acid rain or "Greenhouse effect", or change regional climate.

3. Analysis of MSS activity

3.1 MSS operating procedures

3.1.1 startup procedures

Because of the different production technologies, process flows, raw materials and product types of petrochemical plants, the corresponding startup procedures are different.

(1) atmospheric and vacuum distillation unit

The operation rules of atmospheric and vacuum distillation unit include device inspection and acceptance, steam purging, vacuum air tightness test, Installation of crude oil, system circulation and heating up and the start of other systems.

(2) hydrocracking unit

The startup procedures of hydrocracking unit consist of two systems, including reaction system and fractionation system. The reaction system includes nitrogen replacement, air tightness test, catalyst loading and drying, hydrogen replacement and air tightness test, catalyst vulcanization and passivation, and Installation of raw materials. The fractionation system includes steam blowing, cold oil cycle, hot oil cycle, and installation of raw materials.

(3) hydrogen production unit

The operating procedures of the hydrogen production unit include the inspection and acceptance of the device; steam blowing; nitrogen replacement, air tightness test; circulation; the installation of raw materials and the start of PSA system.

Therefore, the discharge of pollutants in the process of petrochemical equipment is mainly due to the operation of the equipment after the introduction of raw materials. The reactor temperature cannot meet the reaction conditions immediately, the process material flow rate is lower than the normal operation conditions, and the operation state is not stable enough, so it will inevitably discharge a large amount of pollutants.

3.1.2 Shutdown and maintenance procedures

The main materials in the petrochemical plant are mainly liquid, gas, solid or gas-liquid two phase. During shutdown and maintenance, the equipment will be returned material, cleaned and released. The device systems have different shutdown and maintenance procedures due to different materials, production conditions and equipment configuration, but the following steps are generally included [4]:

(1) stop work and return the material, and keep normal pressure in equipment and pipeline;

(2) Treatment after shutdown, including water washing and system pipeline treatment;

(3) Blowing, including sealing steam to collection tube, equipment air entraining blowing,

keeping blowing for 48 hours, and stopping blowing;

(4) Water washing and cooling, water diversion into a collection tube, Water diversion into different container through a sweeping line, water injection into each tower with a reflux pump, bottom emptying, and stop washing;

(5) Retreat material completely, purge and release the equipment pipeline thoroughly, separate all pipelines by blind plates, clean equipment to meet the maintenance requirements, and repair equipment by construction companies.

The operation flow introduces that the pollutants mainly come from the original material in the equipment, including the device material in the discharge to the recycling container, headspace gas emissions from recycling containers, the pollutants in headspace gas after the device retreating, liquid materials attached to all kinds of wall inside the device and material remaining in low-lying and corner of equipment because of its inability to clean up.

3.2 Analysis and calculation of MSS emission source

3.2.1Analysis and calculation of MSS emission source in China

China has a relatively late understanding of abnormal emissions from petrochemical enterprises. Overall, studies on abnormal emission sources are still insufficient. Research on the emission sources of MSS situation is only for the discharge of flare and the vessel depressurization. At present, the Ministry of Environmental Protection has only issued accounting methods for VOCs emissions, while other petrochemical equipment emission sources have not been further studied.

The "Guidelines for VOCs Pollution Source Inspection in the Petrochemical Industry" released by the Ministry of Environmental Protection in 2015, which stipulated that total VOCs emissions shall be the sum of the emissions from vessel depressurization and purging during shutdown and maintenance. Normally, the emissions from vessel depressurization and purging are first discharged to the flare system or other processing facilities. After reaching certain requirements, the vapor will be emptied to the atmosphere. Therefore, two parts of VOCs emissions should be considered respectively: the emissions of flare or other processing facilities and the emission of atmosphere. If the emission accounting for flares is based on the material balance method and the calorific value coefficient method, the initial VOCs emissions from pressure relief blowing can be included in flare emissions. If the VOCs emissions accounting for flares is calculated by other methods or the flare is not used due to failure, all emissions be included in the discharge of pressure relief blowing [5].

(1) flare emission

The material balance method is based on the VOCs emission that enters flare and the flare combustion efficiency, which needs to continuously monitor the flow and composition of the gas entering flare (or the composition analysis of artificial sampling are at least 1 times per 3h in the intermittent emission event). The flare combustion efficiency can be test data, or the hypothetical default combustion efficiency.

$$\sum flare,\ i = \sum_{n=1}^{N} [Q_n \times t_n \times C_n \times \frac{M_n}{22.4} \times (1 - F_{eff}) \times 10^{-3}] \qquad (1)$$

Where:

$\sum flare,\ i$= VOCs emissions of flare "i" (t/a or t/event);

n=The number of measurements;

N=The number of times of annual measurement or the number of times the flare is measured at work time;

Q_n=Flare gas flow at the nth time(m3/h);

t_n= Flare working time at the nth time(h);

C_n= VOCs volume fraction at the nth time;

M_n=VOCs molecular weight at the nth time(kg/kmol);

22.4=Molar volume conversion coefficient(m3/kmol);

F_{eff}=Flare combustion efficiency (%).

The flare default combustion efficiency under abnormal situation is obtained as follows:

Table 3-1 The flare default combustion efficiency under abnormal situation

Type of combustion gas	Flare operating condition	Flare combustion efficiency
none	cannot meet the conditions about net heat value of flare gas and the flow outlet velocity	93%
steam	cannot meet the conditions about net heat value of flare gas and the flow outlet velocity	93%
air	cannot meet the conditions about ratio of steam to gas	80%

(2) venting amount

The guide lists "abnormal situation (including MSS activity) emissions" as one of the 12 sources of VOCs pollution in petrochemical industry. It puts forward VOCs accounting method, namely the formula method, which calculates the discharge from the process of pressure relief and blowing during MSS activity.

① Gas container

The method assumes that there is no liquid material in the container and the gas obeying the ideal gas law. The VOCs venting amount formula is as follows:

$$E_{Gas,\ i} = \frac{P_V}{P_0} \times \frac{T_0}{T_V} \times V_V \times (1 - V') \times \frac{M_W}{22.4} \times V_F \quad (2)$$

Where:

$E_{Gas,\ i}$= The VOCs venting amount of gas container "i"(t/event);

P_V =Absolute pressure of container when the pressure gas is discharged into the atmosphere(kPa);

P_0= Average pressure under standard state (101.325 kPa);

T_V =The temperature of the vessel when the pressure gas is discharged into the atmosphere(K);

T_0= The temperature under standard state (273. 15K);

V_V=Volume of container, (m^3);

V'=The volume fraction of material, catalyst, or tray in a container, When there is no internal component in the container, the value is 0;

M_W=Mole mass of VOCs (kg /kmol);

22. 4= Molar volume conversion coefficient(m3/kmol);

V_F= Volume fraction of VOCs in a container。

In the above parameters, V_V and V' are easier to obtain, P_V、T_V、M_W and V_F need extra monitoring to obtain.

② Liquid container

When calculating the VOC emissions of a liquid-containing container during purging, it may be assumed that the mass fraction of residual liquid in the liquid thin layer or pre-purge container, according to the viscosity of the liquid, the internal surface area, the position of the discharge opening. At the same time, it is assumed that the mass of the remaining liquid in the container is far greater than the amount of air pollutants.

When the container has two-phase volatile materials, emissions are the sum of the emissions of gas container and the liquid container. The VOCs venting amount formula is as follows:

$$E_{Liquid,\ i} = V_V \times (1 - V') \times f_1 \times d \times WF \times 10^{-3}[f_2 \times (1 - F_{eff}) + (1 - f_2)] \quad (3)$$

Where:

$E_{Liquid,\ i}$ The VOCs venting amount of gas container "i" (t/event);

V_V=Volume of container, (m^3);

V'=The volume fraction of material, catalyst, or tray in a container, When there is no internal component in the container, the value is 0;

f_1=The volume fraction of a liquid thin layer or a residual liquid before blowing (0.1%~1%);

d=Liquid density, kg/m3;

WF=The mass fraction of VOCs in the container;

f_2=The mass fraction of a liquid thin or residual liquid being blown to a flare or other treatment facility;

F_{eff}=Flare combustion efficiency (%)

The flare efficiency is shown in table 3-1, and the efficiency of the treatment facilities need to be measured.

3.2.2 Analysis and calculation of MSS emission source in America

(1) Texas Administrative Code

Texas Administrative Code pointed out the discharge source of the specific equipment during MSS activity: engine, compressor, turbine, and other combustion facilities maintenance; repair, adjustment, calibration, lubrication, and cleaning of site process equipment; replacement of piping components, pneumatic controllers, boiler refractories, wet and dry seals, meters, instruments, analyzers, screens, and filters; turbine or engine component swaps; piping used to bypass a facility during maintenance; pigging and purging of piping; blowdowns; emptying, purging, degassing, or refilling of process equipment, storage tanks and vessels (except landing floating roof tanks for convenience purposes); abrasive blasting, surface preparation, and surface coating of facilities and structures used at the site in oil and gas handling and production [6]. The regulations require that the total emissions should not exceed 0.25 tpy. For the operators who do not accept the default value, the regulations provide a separate calculation:

①Blowdowns and MSS Pigging and Piping Components:

VOC = ((Pressure of Gas Inside the Unit Before Venting) * (Actual Volume of the Vented Unit)) / (Frequency of events) * (Molecular Weight) * VOC wt%

(4)

②Engine Oil change/Filter change:

Losses from emptying of oil into an open pan can be estimated using AP-42 Loading Equation: $L_L = 12.46\,\frac{\mathrm{SPM}}{\mathrm{T}}$

(5)

Evaporation Losses from open pan:

$$L_E = 4.14 \times 10^{-5} U_S^{0.78} P_V M_W{}^{0.67} A_p{}^{0.94} t \quad (6)$$

Where:

L_L=loading loss, pounds per 1000 gallons (lb/103 gal) of liquid loaded

S=a saturation factor

P=true vapor pressure of liquid loaded, pounds per square inch absolute(psia)

M=molecular weight of vapors, pounds per pound-mole(lb/lb-mole)

T=temperature of bulk liquid loaded, °R (°F+460)

③Changing Solution in Glycol Dehydration Units and Amine Units:

Loading Equation:

$$L_L = 12.46\frac{\mathrm{SPM}}{\mathrm{T}} \quad (5)$$

Clingage Loss Equation:

$$L_{SL\,max} = 0.60\ (PV_V\ /\ \mathrm{RT})\ \mathrm{Mv} \quad (7)$$

④Changing Engine Rod Packings and Wet/Dry Seals and Heater Treaters:

Emissions result from the evaporation of the lubricant adhered to the rod packing and seal casings.

Clingage Loss Equation:

$$L_{SL\,max} = 0.60\ (PV_V\ /\ \mathrm{RT})\ \mathrm{Mv} \quad (7)$$

⑤Aerosol Lubricants: Emissions from the use of aerosol lubricants such as WD-40 can be estimated by assuming that 50% of the contents of a 16 oz can are volatile hydrocarbons that will volatilize.

⑥Calibration: Emissions resulting from the use of calibration gases can be estimated by weighing the calibration gas cylinder every time it is used.

（2）Oil refinery emission estimation protocol

Oil Refinery Emission Estimation Protocol Issued by the US Environmental Protection Agency gives the calculation formula for the gas and liquid vessel depressurization and purging during startup and shutdown. During startup and shutdown, the process operation is abnormal and the primary emission event is expected to be vessel depressurization and purging (or degassing).

During startup, the reactor temperatures may not be sufficient for the necessary reactions to occur or the process flow rate may be well below typical operating conditions. Under these conditions, processes that do not typically have atmospheric emissions may have significant emissions [7]. While most control devices will operate well at lower vent rates, some control devices, such as Venturi scrubbers, require a minimum flow for effective operation. During startup and shutdown, the emission characteristics of the processes may be significantly different than those during normal operations. When the emission estimation methodology relies on normal operating conditions (e.g., most site-specific and default emission factors), it is important to specifically estimate the emissions during startup and shutdown and include these emissions in the facility's annual emission totals. During shutdown, the process vessels need to be degassed and purged. Process unit startup and shutdown of one process unit may also affect the emission characteristics of other units, particularly with respect to the fuel gas system. When a hydrotreater or hydrocracking unit is shutdown, this often leads to excess hydrogen being added to the fuel gas. This leads to higher NOx emissions from process heaters.

① Gaseous Process Vessel Depressurization and Purging

Process vessel depressurization for gaseous processes can be estimated using Equation (8). This equation assumes ideal gas law; for gases that do not follow ideal gas law, compressibility factors can be included. The equation includes a void fraction term to account for inert material, such as packing material, distillation trays, and certain catalyst particles, that reduce the gaseous space. This equation may underestimate the emissions if the solid material in the vessel may adsorb certain gas constituents (e.g., petroleum coke or activated carbon), that can desorb at lower pressures and temperatures.

$$E_i = \{\frac{(P_V+14.7)}{14.7} \times \frac{528°R}{T_V} \times [V_V \times f_{viod}] \times \frac{MW_i}{MVC} \times MF_i\} \quad (8)$$

where:

E_i = Emissions of pollutant "i" during depressurization event (lbs/event);

PV = Gauge pressure of the vessel when depressurization gases are first routed to the atmosphere (pounds per square inch gauge, psig);

14.7 = Assumed atmospheric pressure (pounds per square inch, psi);

TV = Absolute temperature of the vessel when depressurization gases are first routed to the atmosphere (degrees Rankine, °R);

528 = "Standard" temperature used in the calculation (°R);

VV = Volume of the vessel (cubic feet, ft3);

f_{viod} = Volumetric void fraction of vessel. Use 1 unless packing material or trays are present in the vessel;

MW_i = Molecular weight of pollutant "i" (lb/lb-mole);

MVC = Molar volume conversion factor (385.3 scf/ lb-mole);

MF_i = Volume fraction of pollutant "i" in vessel gas (scf pollutant i/scf of gas).

Typically, depressurization and purging occur initially to the refinery fuel gas system or a flare or similar control device. Combustion of gases purged to the refinery fuel gas system (either directly or via a flare gas recovery system) should be accounted for under stationary combustion source emissions regardless of the methodology used for stationary combustion sources. Depending on the methodology used to estimate the emissions from the flare, these gas flows may or may not be included in the flare emission estimates. If the flow rate and composition or heating value of the flare gas is measured, special calculations are not needed because episodic releases to the flare, such as those from depressurization and purging events, are adequately accounted for. If the flare emission methodology does not account for these flows, Equation (8) can be used to estimate the mass of gas sent to the flare or other control device (using initial depressurization temperatures and pressures) and these mass rates can be adjusted by the flare or control device efficiency to estimate the emissions during the controlled venting period.

(2) Liquid Process Vessel Depressurization and Purging

For vessels that contain liquids, depressurization and purging emissions can be estimated based on the "heel" or fraction of the liquid that remains in the vessel prior to purging. This method assumes that the mass of liquid remaining in the tank will overwhelm the quantity of contaminants in the headspace. For systems operating with very volatile material, it may be necessary to consider the initial gaseous emissions for the most volatile components as well as the liquid heel. A heel fraction may be between 0.1 and 1 percent, depending on the viscosity of the liquid, the surface area of internals within the vessel, and the location of the drainage port or pump line. Using the heel method, it is assumed the entire liquid heel is eventually volatilized. Knowledge of the process stream can be used to speciate the emissions. Again, some of the heel may be purged to a flare or other control device and a control efficiency for this portion is warranted, but some fraction will also be purged directly to the atmosphere.

3.3 MSS emission control

With regard to the management and control of pollutant emissions during MSS, there are fewer relevant regulations at home and abroad. The government requires petrochemical enterprises to strictly formulate operating procedures and pollution control measures for abnormal conditions such as opening and stopping, maintenance, and production anomalies in China. At the same time, the operator shall file with the competent department of environmental protection before the implementation of the planned operation, strengthen environmental supervision during the implementation process, and conduct an assessment afterwards. Unplanned operations must strictly control pollution, eliminate accidental emissions, make timely assessment afterwards, and report to the competent department of environmental protection. In order to avoid the formation of secondary pollution, the exhaust gas generated by catalytic combustion and thermal incineration, as well as the organic wastewater generated by adsorption, absorption, and condensation, should be discharged after being treated. Replacement of adsorbent and other processes should be recorded in the operation information, and waste absorbent should be properly disposed of according to relevant requirements.

40 Federal Code of laws and regulations (CFR) 63 stipulates that the liquid materials in the equipment should be retreated as far as possible before the exhaust gas is vented to the atmosphere, and the equipment can be depressurized through a control device, a fuel gas system or a process gas recovery system until one of the following conditions is satisfied:

①The lower explosive limit (LEL) of combustible gas in the maintenance equipment is less than 10%.

②If the lower explosive limit of combustible gas in the equipment cannot be measured due to design problems, the pressure in the repair equipment should be reduced to 34.5 kPa or less. When the maintenance vent is opened, the blowing operation cannot be performed unless the lower explosive limit of combustible gas within the device is less than 10%.

③The quality of VOCs in the maintenance equipment shall not exceed 32.66 kg.

④ If the equipment containing flammable catalyst (such as hydrocracker) is repaired without nitrogen supply, when the equipment is purging and emptying, the explosive limit of flammable gas in the equipment is not more than 35% only allowed one time in the whole year and the remaining must be less than 20%.

The standard also requires that the lower explosion limit or equipment pressure of the vessel should be measured and calibrated using a process instrument or portable instrument. The VOCs content in the equipment should be determined based on equipment size and material balance.

For planned MSS activity, there are clear requirements for the reporting and record keeping of MSS operating conditions in petrochemical enterprises in Texas Administrative Code. The owner or operator of a regulated entity conducting a planned MSS activity shall notify the commission office for the region in which the regulated entity is located and all appropriate local air pollution control agencies with jurisdiction at least ten days prior to any MSS activity that is expected to cause an unauthorized emission that equals or exceeds the reportable quantity (RQ). If notice cannot be given ten days prior to a scheduled maintenance, startup, or shutdown activity, notification must be given as soon as practicable prior to the activity. The notification for a planned MSS activity must identify:

①the name of the owner or operator;

②the commission Regulated Entity Number of the regulated entity, if a Regulated Entity Number and air account number exist(s), or if there is not a Regulated Entity Number, the air number of the regulated entity. If a Regulated Entity Number and air account number do not exist, then identify the location of the release and a contact telephone number;

③the physical location of the points at which emissions from the MSS activity will occur;

④the type of MSS activity and the reason for the activity;

⑤the expected date and time of the MSS activity, and expected duration of any maintenance activity;

⑥ the common name of the process units or areas, the common name and the agency-established facility identification number of the facilities that will be involved in the emissions activity, and the common name and the agency-established emission point numbers where the unauthorized emissions may be released to the atmosphere. Owners or operators of those facilities and emission points that the agency has not established facility identification numbers or emission point numbers for are not required to provide the facility identification numbers and emission point numbers in the report, but are required to provide the common names in the report;

⑦the expected duration of the emissions from the MSS activity;

⑧the basis used for determining the quantity of air contaminants to be emitted;

⑨the actions taken to minimize the emissions from the scheduled maintenance, startup, or shutdown activity.

For unplanned MSS activity, the frequency and duration of operation resulting in unauthorized emissions must be minimized and all possible steps should be taken to minimize the impact of the unauthorized emissions on ambient air quality, because the periods of unauthorized emissions from any unplanned MSS activity cannot be prevented through planning and design, at the same time, all emissions monitoring systems are kept in operation if possible and the owner or operator actions during the period of unauthorized emissions from any unplanned MSS activity should be documented by contemporaneous operating logs or other relevant evidence.

4. Conclusions

(1) Although China has formulated a number of standards concerning the abnormal emission sources of petrochemical enterprises, and defined and proposed the definition and management regulations, it is still not comprehensive enough to compare with the relevant standards issued by the United States.

(2) China has proposed a calculation method for VOCs in abnormal emission sources of

petrochemical enterprises. When calculating VOCs emissions according to the formula, it is necessary to monitor the temperature, pressure and gas components of the vessel before the pressure-relief gases are discharged into the atmosphere. However, there are no specific requirements in the current operating specifications in petrochemical enterprises, which greatly reduces the accuracy of accounting for abnormal emission sources.

(3) China currently only provides calculation formulas for the VOCs emission from the gas containers and liquid containers venting. But the United States also provides calculation formulas for the emission of specific petrochemical equipment, as well as the emission limits.

(4) For the management and control of abnormal emission sources of petrochemical enterprises, compared with the detailed requirements and complete management procedures of the United States, there are fewer relevant regulations in China, and the management and evaluation system are imperfect.

In order to improve the accuracy of MSS emission source calculation in China's petrochemical enterprises and the governance level of atmospheric pollutants in China's petrochemical industry, build a system of pollution control, This article refers to the US refinery emission standards and the related regulations for emission management of MSS conditions in Texas, and puts forward the following recommendations: (1) MSS emissions from petrochemical enterprises are mainly concentrated in the process of startup and shutdown. To achieve environmental standards, it is necessary to summarize experiences and lessons to optimize the operating procedures. (2) In China, there is still a lack of complete laws and regulations for abnormal emissions regulation. We need to systematically sort out the environmental and national level to formulate environmental governance standards according to current status of abnormal emissions and develop advanced pollution control technologies at the same time.(3) To establish a corporate environmental information disclosure system, enterprises should actively disclose relevant environmental information such as discharge of pollutants during MSS activity and the operation of treatment facilities, and accept public supervision.

References

[1] Yu An. Research on the Law Enforcement Management of Flare Emission under the Abnormal Situation in the Petrochemical Industry[A]. Chinese Environmental Sciences Association. Excellent Papers of the 2006 Academic Annual Meeting of the Chinese Environmental Sciences Association (Volume Two) [C]. Chinese Environmental Sciences Association: Chinese Environmental Sciences Association, 2006: 4.

[2] Texas. Texas Health & Safety Code (THSC). TITLE 5. RULE §382.051962. 2017

[3] Texas. Texas Administrative Code (TAC). TITLE 30. RULE §101.1. 2017

[4] JIA Yu-ling, CUI Ji-shan, NIU Hao, WANG He-jing. Discussion on VOCs Emission Estimation during Petrochemical Device Shutdown and Purging Process[J]. Environmental Monitoring Management and Technology,2017,29(03):61-63+68.

[5] Office of Environmental Protection Department. Office of Environmental Protection Department [2015] NO.104. Guidelines for VOCs Pollution Source Inspection in the Petrochemical Industry [S]. Beijing: Environmental Protection Department, 2015.

[6] Texas. Texas Administrative Code (TAC). TITLE 30. RULE §106.359. 2017

[7] RTI. Emission estimation protocol for petroleum refineries (Ver-lion 3) [R]. North Carolina: RTI, 2015.

516183 Why I am Passionate About Process Safety

Abstract:

Louisa Nara joined CCPS in 2010 after 30 years of working in the chemical, petrochemical and pharmaceutical industries. She is passionate about process safety in large part due to the 5 decades of incidents that impacted her on a personal basis. These incidents have served as key motivators in shaping her career in process safety and her passion around sharing incidents, lessons learned and best practices on a global basis.

In 1968, at the age of 10, Louisa was impacted by her first process safety incident as a â€œkey receptorâ€, a member of the community evacuated following a fire and toxic gas release at a chemical plant less than 1 aerial mile away. The incidents discussed have affected Louisa, her family, friends, co-workers and the global community committed to process safety.

516182 That Can't Happen

Abstract:

A lot of attention is given to "near misses" or "near hits". It is my view that these terms downplay the potential for severity of outcome. What if near misses were viewed as just being "lucky"? I think most managers would agree that luck runs out and that it sounds a bit more uncertain as to the outcome. Ken Tague has spent 37 years in operations and corporate safety and is currently the Process Safety Manager for Archer Daniels Midland Co. He shares two incidents with completely different outcomes that shaped his appreciation of and passion for process safety.

516181 The HDAY-4 Unit Catastrophic Fire. Three Decades of Asset Integrity Opportunities

Abstract:

The 1984 Amuay refinery hydrodesulfurization unit 4 (HDAY-4) fire caused the crash shutdown of the 630,000 barrel-per-day refinery. Three hydrogen plants and four HDS units were heavily damaged. Despite the magnitude of emergency and demanding activities to control it, none of the refinery personnel or 133 emergency responders were injured. I was a young fire chief and directed the emergency response operations. The fire caused a loss of $151 million in todayâ€™s money. A hot oil line, known to have excessive vibration, failed in fatigue due to hydrogen embrittlement.

512535 When the House Was a Rockin', Process Safety Came a Knockin'

Abstract:

On June 13, 2013 a catastrophic fire and explosion occurred when a reboiler ruptured at the Williams Olefins Plant located in Geismar, Louisiana. At the time of the incident, the authors were at a neighboring facility leading a risk assessment focused on identifying and managing various business loss scenarios due to internal events and external supply disruptions. As the incident unfolded, the site and community emergency alarm system activated and a mandatory shelter-in-place was initiated by the site emergency response team. This unfortunate incident had minimal impact on our site but resulted in two fatalities, 167 reported personnel injuries, an 18 month plant shutdown, fines, and settlements with injured parties. For the authors, process safety â€œcame knockinâ€ in years prior through responsibility and oversight of high hazard processes, innovation in the development of inherently safer processes, and informal mentorship of young engineers. We intend to share some of our personal experiences, as well as our professional perspectives on how the future of process safety requires more interdisciplinary collaboration to solve complex problems and enhanced business acumen to create and sustain value.

505748 The Day a Three Mile Island Event Did Not Happen

Abstract:

In 1986, while I was a new engineer at a nuclear power plant, the plant experienced a total loss of feedwater event when multiple systems and equipment failed, including safety critical equipment. This event could have led to a meltdown of the reactor core, similar to the Three Mile Island Nuclear Plant event several years earlier, but several operators literally saved the day through their actions. Subsequent reviews of the management system failures that led to the equipment failures provided for a wake-up call in the industry to refocus on process safety. The corollaries to process safety events and process safety management systems in the refining and chemical industry were uncanny. How could simple management system / procedural failures have led to a near catastrophe? How did these failures affect an industy, my colleagues, and I? What more can we learn from this event and others?

Why Are the New People Making the Same Mistakes That the Old-Timers Made 25 Years Ago? Managing Organizational Change

A.M. (Art) Dowell, III
Principal Engineer
Process Improvement Institute
16430 Locke Haven Dr
Houston, TX 77059 USA
adowell@piii.com

Prepared for Presentation at
American Institute of Chemical Engineers
2018 Spring Meeting and 14th Global Congress on Process Safety
Orlando, Florida
April 22 – 25, 2018

Why Are the New People Making the Same Mistakes That the Old-Timers Made 25 Years Ago? Managing Organizational Change

A.M. (Art) Dowell, III
Principal Engineer
Process Improvement Institute
16430 Locke Haven Dr
Houston, TX 77059 USA
adowell@piii.com

Keywords: memory, managing change, personnel, organizational change

Abstract

History repeats itself. The same process safety incidents occur over and over. The same occupational safety incidents occur repeatedly.

Why? "Organizations have no memory" – Trevor Kletz. As the operating, maintenance, and engineering personnel are promoted, retire, take vacation, are absent, OR job duties are reallocated, critical safety information and expertise is lost. The paper gives examples of memory loss and its consequences.

The right critical knowledge and skills must be available on site, all the time, when they are needed. The paper describes a straight-forward Management System to maintain the needed competency as the organization changes. The system is self-documenting and can easily be kept up-to-date. Examples are given from years of experience.

1 Introduction

Many of us have experienced one or more days when process safety hit home. For this author and his colleagues, one such day occurred when we realized that the same process safety incidents were occurring over and over. We also realized that the same occupational safety incidents were occurring over and over. New people are making the same mistakes that the old-timers made years ago. The problem was that history was repeating itself.

Why did history repeat? Trevor Kletz observed "Organizations have no memory" [1]. We asked ourselves **why** do organizations have no memory? It seems that each person was learning from his/her own mistakes, but had not learned from mistakes made by their predecessors.

We recognized that we needed a system to help people have the necessary information, training, and skills to avoid repeating the incidents of the past.

2 The memory issues

The memory issue occurs because operating and maintenance personnel (and support personnel, as well) are promoted, retire, take vacation, are absent for one reason or another, OR their job duties are reallocated. As these changes occur, without a formal system, it is unlikely that the new person will know and understand the same hazards and safeguards that the outgoing person knows. What the new person knows depends on his/her training, background, and experience. Should sufficiency of the incoming knowledge be left to chance?

In 50 years of process safety experience, the author has observed or experienced a six-month operator training a new operator. The new operator simply followed the six-month person around for two weeks and was considered to be trained.

The author has observed production manager changes where there was no overlap whatsoever. The old manager left and the best overlap was that the new manager sat down in the office swivel chair just as it stopped spinning from the old manager's exit. In other cases, there were gaps of days or weeks between the old and the new manager. Well, managers know how to manage, right? That may be true, but does the new manager understand what the hazards are in his/her unit? Does he or she know what is needed to sign off on bypassing a safeguard safely?

The other problem with memory is that people in the same job can forget what they once knew, especially for tasks that are done infrequently. Consider that modern units may run for 1 to 5 years between startups and shutdowns. Therefore, a system is needed to provide timely refresher training or drills so that people know what they need to know when they do a task.

3. Example solutions

3.1 New employee

The author observed a unit in which, within the first six months, almost every operator had an occupational injury (US OSHA recordable) or a first-aid injury from exposure to an organic acid. After six months, no experienced operators had the exposures even though they were continuing to do the same tasks. Why did this occur?

Before an exposure, no one understood that exposure to the organic acid was as bad as they had been told – they just didn't believe it. They treated the job task in a cavalier manner. All the training was done on the job. After an exposure – an injury – they understood the severity of the consequences and the reasons underlying the procedures to avoid exposure. Procedures were appropriate, but new folks believed shortcuts were okay.

The facility began a new operator training program using air and water in the pumps and tanks of an out-of-service unit. The trainees made mistakes and took shortcuts, but they were exposed only to water. The trainers explained the severity of a chemical exposure in similar situations.

The result was the number of injuries in the first six months declined.

After a few years, the training unit was repurposed for another product and was no longer available for training. Because of an economic downturn at the time, the hands-on water and air training was discontinued. Anecdotally, the author understands that the injury rate went up.

W. Edwards Deming said the first of the 14 points for good management principles is "Create constancy of purpose for improving products and services" [2]. We have all seen incident-reducing initiatives that seem to be effective, but for, one reason or another, were discontinued, with a subsequent increase in incidents

3.2 Promotion

When one employee is promoted to another position, the replacement employee must eventually be able to do what the incumbent did. Alternately, the duties must be parceled out to other employees. In either case, **ALL** the job duties must be identified. The training required must be made available and taken **before** the change is made or **before** the new person undertakes the new duties.

It doesn't make sense that management would put a new person into a job and not have a list of the job duties. It doesn't make sense to expect a person to do a new task safely before training is given.

While the principles in this section would seem to be common sense, we have heard some surprising stories.

3.2.1 Promotion example

A unit Day Foreman retired. A Shift Foreman was promoted to Day Foreman. An Operator was promoted to Shift Foreman. The previous Shift Foreman had managed the water quality in the cooling tower. The new Shift Foreman had no knowledge of water quality and why it might be important, and did not manage the cooling tower water. Legionella bacteria developed in the cooling tower basin. The result was a major public relations disaster even though no cases of Legionella were identified.

3.3 Reorganization

Reorganization is similar to the promotion example above, except that it is much more extensive. **ALL** the duties of ALL organizations being reorganized must be identified:

- Are there job descriptions for the jobs being eliminated?
- Do they cover all aspects of the jobs?

The new system must cover **ALL** the duties that were being done by the old organization. It is a massive task to identify exhaustively all the tasks that are done by a worker; it is even more daunting to identify those associated with the whole department. Extensive and intensive debriefings must be done. It is not a trivial task.

The more departments that are being reorganized, the more work is being performed. All the tasks for all the positions must be identified. Then the tasks must be sorted into two categories:

- Not done?
    - Can they be safely eliminated? It takes a meeting and analysis with knowledgeable people to know.
- Can they be done by others who already know how to do them?
    - Who and where are the others who already know how to do them? Are they contractors? Or in other departments? How do we know that they know that they now have the tasks?

3.3.1 Reorganization example

An extensive site reorganization was done. After several years, a process safety audit found that several relief valves had not been inspected and tested as required by the company's mechanical integrity program. The auditors found that all the personnel who might have had the responsibility in the past to ensure the inspections and tests were done were sure that they did not have that responsibility in their current jobs. The auditors could not find anyone who believed the inspections and tests were part of their current jobs. The task had been lost in the reorganization.

- Everyone thought that someone else had the task.
- Everyone was sure that he/she did not have the task.
- Reality: **NO ONE** had the task.

3.4 Temporary absence long-term

A long-term absence is when the incumbent is missing for a day to several days to several weeks (e.g., a vacation or an illness).

How will the replacement be done? Will it be from a pool of equally qualified individuals? Will it be temporary help? Will the tasks of the absent person be added to the workload of others?

Are the duties to be performed known by the new people that are intended to be doing them? Is there a job description? Is it correct and up-to-date? Is the updating of job descriptions part of the management of change process?

Management must clearly communicate to all involved what each individual is being asked to do and how frequently. And of course, what is the training for the replacement people?

3.4.1 Long term example absence

A pumper had the task to gauge a 15 million pounds storage tank to do a weekly check on the inventory. (To do this task, the pumper went to the top of the tank and inserted a wooden stick with measurement lines to measure the depth of the material in the tank.) The tank had a single level indicator, but it was known to be wrong frequently. The pumper went on vacation, and nobody was designated to gauge the tank during this time. After a week of production and barge receipts with no inventory confirmation, the tank overflowed, creating a potentially serious exposure hazard, a difficult cleanup, and potentially serious environmental consequences.

3.5 Temporary absence short-term

A short-term absence is for a few hours, such as a meeting away from the unit, a trip into town, a medical appointment, etc. Are there critical tasks being performed during this absence (e.g., startup, shutdown)? Have minimum coverage standards for these tasks been established (e.g., number of people, experience level)? Will these tasks be adversely affected by this absence? Has there been any opportunity to plan for this absence (e.g., calling in a replacement, deferring critical tasks)?

3.5.1 Short-term example – missing knowledge

As a consultant, the author has been to a number of different facilities where we discussed emergency response procedures for certain tanks of hazardous materials. The operations, technical, and safety personnel in the meeting did not know critical details of the emergency response procedure. "Well, John Doe is the expert on that and he is away today, so we can't answer that question." We asked, "Can we see the written procedure?" "Well, ah, we can't find it right now." Someone in the meeting commented, "I guess it is a good thing that we don't have an emergency involving that tank today!"

3.5.2 Short-term example – a good program

One unit, comprised of multiple trains producing a toxic intermediate, established a clear operating policy for start-up of the trains. They always started up a train on the evening shift, because there were fewer people in the plant at that time. They also ensured that the skill level of operations personnel on shift was adequate during the startup.

Another unit normally would hold over a shift, thereby providing double staffing during a startup, for training and quality assurance. All managers were cognizant of the personnel on shift during startup, and provided coverage by foreman as needed.

5 A management system

What is needed to avoid incidents introduced by personnel change?

The right critical **knowledge** and **skills** must be available:

- Where the task is being done,

- When the task is being done,
- To the person doing the task.

A Personnel Management of Change System is needed to accomplish this objective. The Personnel Management of Change System must document the tasks, and document the needed skills and knowledge to accomplish those tasks safely. The Personnel Management of Change System must institute and maintain a formal management of personnel change process. The Personnel Management of Change System must provide assignments to specific individuals to manage the Personnel Management of Change System.

4.1 Document

We can answer the documentation question by asking (and answering) more questions. What knowledge does the technician (operator or mechanic) need to know? What skills does the technician need to do the task safely, reliably, and with good quality? What are **ALL** the tasks that need to be done? What is the best way to do the task? Institute training, observation, and positive reinforcement to ensure that everyone does the tasks in the best way!

4.2 Formal management of personnel change system

The formal management of personnel change system should identify the knowledge, skills, and tasks for each position. The management system develops an action plan to ensure the new person learns the knowledge, demonstrates his/her skills, and understands his/her task and responsibilities. Further, the management system develops a plan to provide the knowledge, skills, and tasks safely while the new person is in training – that is, it develops a transition plan.

The management of change system should provide a system to manage temporary absences. First, the system must recognize when critical knowledge and skills are missing. Second, it should ensure critical operations are not performed when key people are missing. And, third, it should provide alternate resources to ensure safe operations.

4.3 Managing the management of change system

Managing the management system means that there is a plan in place with identified people to take care of the Personnel Management of Change system to ensure that the system is working properly, to ensure that problems are corrected, and to ensure that improvements are implemented. The Personnel Management of Change system should confirm:

- Who is documenting the changes to task, skills, and knowledge?
- Who is responsible for scheduling and conducting the formal management of change processes?
- Who is responsible to ensure that the necessary training is completed on schedule?

- Who audits the Personnel Management of Change system?

4.4 Auditing management of change system

Remember that we must inspect and test the equipment and procedures for independent protection layers that prevent serious consequences in the operating facility. We test relief valves [3]. We test safety instrumented functions [4]. We should inspect dikes. We should drill our operators on response to alarms. If we find deficiencies, we should correct them (e.g., repair/replace, train, etc.).

Likewise, it is critical to "inspect and test" the management system for organizational change to ensure that it is providing personnel with the right skills and knowledge to do the right tasks.

5 Conclusion

This paper merely skims the surface in examples of process safety incidents that have occurred because of personnel change that was not well managed. A common element of these types of incidents is that people did not have the knowledge and skills to do their job. People did not know certain duties were part of their new job.

This paper gives several examples for improved management of personnel change systems and provides a number of questions to ask. If the questions can be answered, then a strong management of change system can be developed. Building a management of personnel change system will make major strides in providing company memory.

CCPS has a guideline book that can be helpful in developing a management system for personnel change [5].

6 Acknowledgment

The author gratefully acknowledges the contribution of his colleague and dear friend, Stan Anderson, to this paper. The two of us outlined this paper together about a decade ago, but other work requirements prevented us from writing and publishing it. Stan was called away from this life a few years ago and we hope he is singing tenor in heaven.

7 References

1. IChemE. The wisdom of Trevor Kletz – the 'founding father' of inherent safety (Day 353). Available at https://ichemeblog.org/2015/05/15/the-wisdom-of-trevor-kletz-the-founding-father-of-inherent-safety-day-353/ Accessed on February 14, 2018.
2. ASQ. W. Edwards Deming's 14 Points for Total Quality Management. Available at http://asq.org/learn-about-quality/total-quality-management/overview/deming-points.html Accessed on February 14, 2018.
3. ASME Boiler and Pressure Vessel Code, Section VIII, "Pressure Vessels". 2017 Edition.

4. ANSI/ISA 84.00.01-2004 (IEC61511-1 Mod), Functional Safety: Safety Instrumented Systems for the Process Industry Sector Part 1: Framework, Definitions, System, Hardware and Software Requirements, 2004.
5. CCPS. *Guidelines for Managing Process Safety Risk during Organizational Change.* Center for Chemical Process Safety, American Institute of Chemical Engineers, New York, NY, 2013.

518673 Tutorial in Fire Safety & Emergency Management

Abstract:

When you think of Process Safety what comes to your mind? For a lot of people itâ€™s Process Hazard Analysis, Management of Change or even Process Technology. Often times when facilities grow, chemistry changes and people transition to different roles, the area of Fire Safety & Emergency Management can receive less attention than needed to ensure systems and processes are adequate when needed. This oral presentation will provide an overview of why fire safety and emergency response systems are essential in a PSM facility and something that should not be taken for granted. This presentation includes best practices of fire safety and emergency response systems along with real world case studies from two experienced veterans in the high hazard chemical, nuclear and petrochemical industry.

518674 Tutorial in Hazardous Materials Response & Emergency Management

Abstract:

Have you thought about how you are prepared if you have a hazardous materials release? You might think you may have water fog nozzles, secondary containment, etc. These items are good and in many cases necessary to prevent a hazardous materials release from getting bigger than you intended. However, what if you had a large release that you needed to activate a response team or if you are being impacted by a release from a neighboring facility. The organization presenting in this session has trained over 85,000 first responders in 21 countries and has responded to more than 1,000 incidents without a single injury. This oral presentation by two experienced veterans in the high hazard chemical industry will provide an overview on best practices for planning hazardous materials responses and case studies to make your plans more effective.

518676 Tutorial in Mechanical Integrity / Quality Assurance

Abstract:

Mechanical Integrity and Quality Assurance is no doubt an important element of your Process Safety Management Program. When you first think about Mechanical Integrity you may think of piping wall thickness, vessel integrity, relieve device condition, valve contracture, flanges, etcâ€¦ which are all really good things to have as part of your Mechanical Integrity program. When budgets get tight and when focus is on keeping the production running in an ever so challenging business environment do you see any fluctuation of mechanical integrity focus on fire and emergency responses systems? This oral presentation will cover what mechanical integrity is important to your overall PSM program and provide some focused discussion on the importance of maintaining mechanical integrity programs on fire and emergency response systems even during difficult business times. Our presenter is a veteran of the U.S. Navy and also a veteran mechanical integrity professional in the high hazard chemical industry that will provide a very informative discussion on this topic area.

Process Safety Aviation Series:

And… Who Is the Pilot?

Ignacio Alonso/Mike Broadribb
Consejo de Seguridad de Procesos, Latinoamérica
Bogota, Colombia
ialonso@csp-la.org

Mike Broadribb(s)
Consejo de Seguridad de Procesos, Latinoamérica
Bogota, Colombia
mbroadribb@bakerrisk.com

Prepared for Presentation at
American Institute of Chemical Engineers
2018 Spring Meeting and 14th Global Congress on Process Safety
Orlando, Florida
April 22 – 25, 2018

Process Safety Aviation Series: And... Who is the Pilot?

Ignacio Alonso/Mike Broadribb
Consejo de Seguridad de Procesos, Latinoamérica
ialonso@csp-la.org

Mike Broadribb(s)
Consejo de Seguridad de Procesos, Latinoamérica
mbroadribb@bakerrisk.com

Keywords: Culture, Leadership, Organization principles, Human Error, Human Factors

Abstract

The aeronautical industry have developed extensive research in causal factors and methodologies to prevent risks associated with human factor. Pilots work in complex environments that require significant skills and deep knowledge of the way in which the planes and the systems that control them work. However, this may not be sufficient if there are no clear protocols to interact with air traffic controllers and their own crew in crucial moments that could end in a catastrophe.

The environment in which operators and managers of a processing plant are exposed to can be similar to those of an aircraft crew. The paper will refer to these similarities and will find aspects that relate the duties and responsibilities of airline pilots with those of the plant managers, and also relate situations that have involved leadership flaws in incidents from both industries.

The presentation will discuss organizational and cultural aspects that have contributed to catastrophic accidents in both industries and will focus on the role that pilots and plant managers must assume as individuals to deal with hazards. The document also emphasizes the skills and the training required by pilots and plant managers and the relationship that they need to maintain with their crews in order to act effectively as a single team during difficult situations.

The paper discusses human factors associated to cultural issues and relate them to key elements of CCPS Risk Based Process Safety highlighting some requirements used in the aviation industry that could be followed by the chemical process industry.

This paper is a continuation of the work presented by the authors in the 13th Global Congress on Process Safety titled "Human Error, a Myth Eclipsing the Real Causes" that associated issues from the aeronautic industry with Process Safety.

1 Introduction

The commercial aircraft of today are designed to transport people safely in the air. Modern airplanes are provided with advanced instrumentation and alarm systems that warns pilots when an abnormal situation arises during flight or when deviations to the operational parameters established by the manufacturer occur, which can jeopardize the lives of passengers and crew. In addition, sophisticated and comprehensive air traffic control systems exist now to help pilots in flight, providing precise instructions to avoid other traffic and to safely depart and approach airports.

Detailed air routes, procedures and navigation charts have been developed for global use by airline pilots. Recognized official and private bodies have worked together to set the rules for air transportation and the requirements that airlines must comply with in order to be safe and maintain their license to operate. These requirements also include specific inspection and maintenance programs to ensure that the airplane`s integrity is not compromised.

As in the aeronautical business, the chemical and petrochemical industry has developed standards and codes to manage safety during day to day operations. This industry has implemented design measures, fabrication and construction methods, and developed sophisticated equipment and safety instrumented systems to manage the risks inherent in processing hazardous substances. Companies dedicate lots of resources and effort to develop practices and procedures to assure that people perform consistently under normal or abnormal situations. Many countries have also established regulations to enforce the implementation of management systems and protective equipment that act as barriers for preventing or mitigating hazards.

2 The Unavoidable Human Factor

It is recognized that major risks are inherent in air transportation as well as in the petrochemical industry. The protective systems, procedures and regulations mentioned above for both industries have been developed and implemented as an attempt to avoid disasters. However, regardless of all these efforts, accidents still happen.

Levels of automation have been increased in order to minimize human intervention, but when it comes to managing risks and dealing with hazardous situations, the automated systems often fall short, and only the reasoning ability of a human being can successfully resolve situations where safety may be at risk. However, what happens when the ability of the person in charge of dealing with these situations is compromised?

Factors that can affect a person´s ability to take charge can include:

- Cultural issues
- Leadership
- Sense of vulnerability
- Lack of competence

- Improper training
- Poor communication
- Operational discipline
- Corporate policies

The following examples will describe situations involving these factors that ended in catastrophic accidents in the airline industry, with a correlative example from the petrochemical industry involving the same factors.

3 Accidents and common factors

This section describes some accidents within the aerial industry and relates them to common accident factors in the hydrocarbon and chemical industries. The aircraft incidents are summarized from an RBPS (Risk Based Process Safety) perspective to show the association with cultural and organizational aspects that can have an effect on the behaviors of both pilots and plant operators.

3.1 Flight 101. Polish Air Force – President Lech Kaczynski

April 10, 2010. Warsaw to Smolensk

Occupants 96 / Fatalities 96

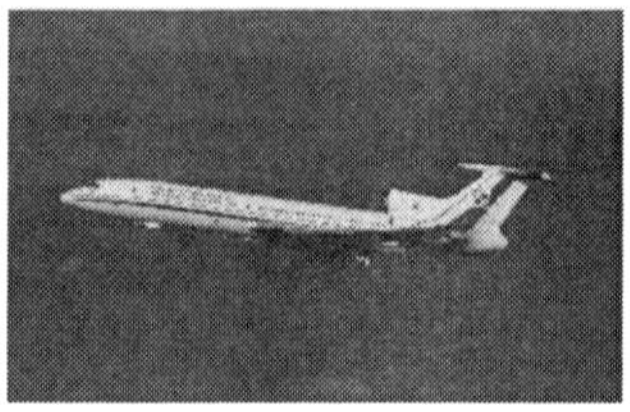

A Tupolev Tu-154 crashed into a hill while approaching the Smolensk airport killing the Polish president, his wife and the principal members of the government. In total 8 members of the crew and 88 passengers lost their lives.

The accident occurred under dense fog while approaching the Smolensk Airport which was not equipped with ILS (Instrument Landing System). It was a single flight scheduled as part of a diplomatic mission (Commemoration of the 70th anniversary of the Katyn massacre) to relieve tensions between the Polish and Russian governments.

The following circumstances contributed to the disaster **(Reference 1):**

1) Failure to monitor altitude by means of a pressure altimeter during a non-precision approach due to intentional manipulation of the altimeter to avoid nuisance alarms;

2) Failure by crew to respond to the PULL UP warning generated by the TAWS (Terrain Awareness and Warning System);

3) Attempt to execute the go-around maneuver under the control of ABSU (automatic go around);

4) Approach Control confirming to the crew the correct position of the airplane in relation

to the runway threshold, glide slope, and course which might have affirmed the crew's belief that the approach was proceeding correctly although the airplane was actually outside the permissible deviation margin;

5) Failure by LZC (Landing Zone Controller) to inform the crew about descending below the glide slope and delayed issuance of a level-out command;

6) Incorrect training of the Tu-154M flight crews by the Polish Air Force.

The RBPS perspective

Contributing circumstances included:

1) Incorrect coordination of the crew's work, which placed an excessive burden on the aircraft commander in the final phase of the flight – Leadership, HIA, MoC, workforce involvement

2) Insufficient flight preparation of the crew – Hazard Identification and Analysis, Management of Change

3) Crew's insufficient knowledge of the airplane's systems and their limitations – Competence, Protective systems,

4) Inadequate cross-monitoring among the crew members and failure to respond to the mistakes committed - Culture, workforce involvement, sense of vulnerability

5) Crew composition inadequate for the task – Competence, culture

6) Ineffective immediate supervision of the Polish Air Force flight training process by the Air Force Command – Leadership, training

7) Failure by the Polish Air Force to develop procedures governing the crew's actions in the event of:
   a) Failure to meet the established approach criteria – procedures, safe work practices, operational discipline;
   b) Using radio altimeter for establishing alarm altitude values for various types of approach – unsafe work practices, conduct of operations;
   c) Distribution of duties in a multi-crew flight – Leadership, workforce involvement

8) Sporadic performance of flight support duties by the LZC over the last 12 months, in particular under difficult weather conditions, and lack of practical experience as an LZC at the Smolensk North airfield – Training, Protective Systems, Culture.

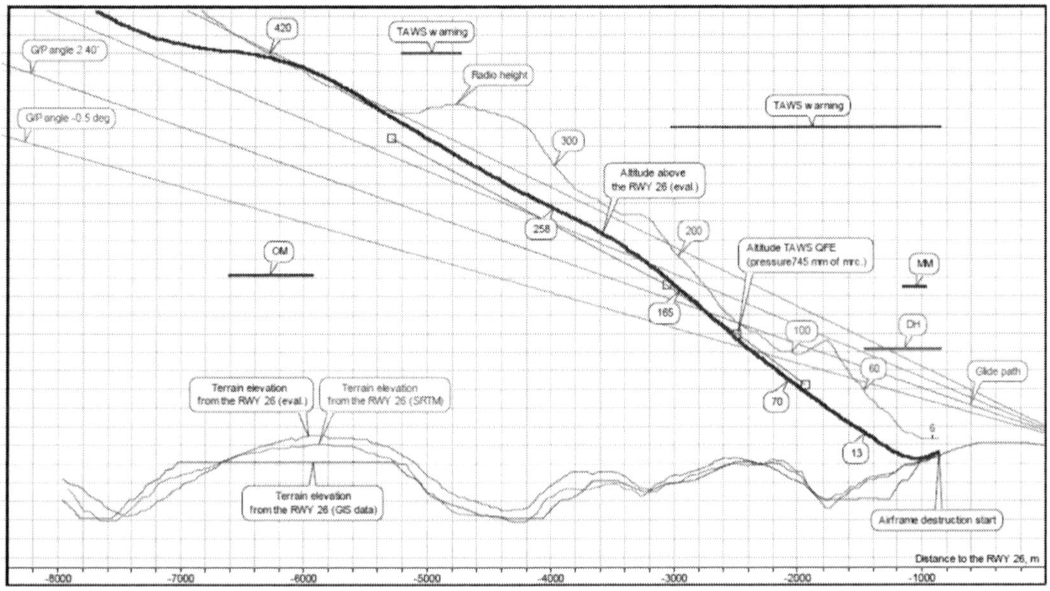

Figure 1. Flight Parameters on the Approach to Smolensk

The flight crew had been alerted that conditions were not adequate for landing. However, no one in the crew tried to convince the pilot to abort and try to land at another airport which could suggest a cultural issue from the military that is traditionally focused on objectives. Preserving the life of the president should have been the first priority.

The altimeter was intentionally misconfigured for the approach to avoid a nuisance alarm that warned about proximity to terrain (TAWS). This was a common practice among the air force pilots. Intentional disregard of safety controls has also been practiced by some plant operators from organizations with a poor process safety culture.

This was a one-off flight to a location of sporadic air traffic. A MoC process should have been conducted by the appropriate authorities and involving the crew in order to identify all risks associated to this operation.

The controller (LZC) did not speak Polish, so all communications had to be held in Russian. The Pilot in Command (PIC) was the only crew member who spoke Russian, so he was handling all communications with LZC and other air traffic in the area while operating the controls of the aircraft. This overloaded his capacity and reduced his level of attention decreasing his ability to assess the risks.

Recordings from the FDR (Flight Data Recorder) indicated that the Protocol director of the government was in the pilot's cabin, which indicates that the cabin was not isolated from passengers. Even when he didn't give any instruction, investigators believe that his presence in the cabin affected decision making and a proper communication between crew members. This implies issues associated with a "power culture" in which adherence to Procedures, Competence and Conduct of Operations are overlooked.

Two years before, in August 2008, the pilot was flying as first officer of the presidential airplane. The pilot of that flight refused to land in Tbilisi due to unsafe conditions at the airport disobeying orders from their superiors. As a result the pilot was removed and never selected again to fly the presidential plane. This episode brings to mind the power culture issues mentioned above and put additional pressure on the pilot that surely wanted to keep his job. It also suggests a lack of sense of vulnerability.

The Smolenk airport was poorly equipped. The ATC (Air Traffic Control) radar did not have the capacity to provide a detailed position and height of the presidential aircraft to guide the pilot under bad weather conditions.

The ATC tried to convince his superiors to redirect the flight to another airport but they did not have authority to modify plans already set by the members of the government who surely didn't have the competence to identify and analyze the risks. Again, a MoC process would have been useful in an environment that suggests a strong "power culture".

This incident brings to mind the 1987 Piper Alpha platform disaster in the North Sea **(Reference 2)**. After an explosion and while major fires were burning, the Claymore and Tartan platforms continued to pump oil to Piper Alpha despite being aware of the emergency. The back pressure from this pumping continued to feed the fires through ruptured pipework, although the fires would have burnt out had Claymore and Tartan shut down when they first became aware of Piper's emergency.

This ultimately condemned Piper when the fires caused the catastrophic failure of two gas risers. Despite losing voice communication with Piper and seeing the glow on the horizon from the major fires, neither the Claymore nor Tartan platform manager had authority to shut down and required permission from the respective leadership onshore. This is another example of a power culture in which the onsite leaders did not have full authority (nor confidence) to make and implement risk management decisions in an emergency.

3.2 Flight 518 – Santa Barbara Airlines

February 21-2011. Mérida to Caracas

Occupants 46 / Fatalities 46

An ATR-42-300 operating Santa Bárbara Airlines Flight 518 crashed in a mountain shortly after takeoff from Mérida-A Carnevalli Airport in Venezuela. All 46 on board were killed.

Mérida Airport is located in a mountain valley at an elevation of 5010 feet. Terrain in the valley rises from 1000 to 19.500 feet. Arrivals and departures have to use the Rio Chama corridor to the west of the airport.

Santa Bárbara Airlines however used an unpublished approach and departure procedure, using a valley to the northeast of the airport. This procedure saved about 15 minutes of flying time on the route to Caracas.

All passengers had already boarded for flight 518 when the pilots arrived at the aircraft. Being late, they immediately started engines while simultaneously asking for permission to do so.

Because another flight was approaching the airport, the crew were told to either expedite their departure or wait for the arrival of the incoming aircraft. The flight crew decided to go and began taxiing only 2 minutes and 40 seconds after startup. Consequently the AHRS (attitude and heading reference) system of the airplane was unusable. The aircraft needed to remain still for 3 minutes after the battery is turned on so that all gyroscopes, accelerometers and magnetometers can be stabilized and synchronized. (Operational discipline, operating procedures

The AHRS provides attitude information for aircraft, including roll, pitch and yaw. The captain possibly expected to re-initialize the system while they were at cruising altitude. After departure from Runway 25 the crew planned to use the unpublished procedure. Climbing through clouds a 180-degree turn was initiated. Using the unreliable magnetic compass, the aircraft made a 270 degree turn, heading towards rising terrain. The captain took over control from the copilot. When visual contact with terrain was regained, the crew noted they were heading for mountains. The captain tried to avoid the rising terrain but the aircraft impacted the side of a mountain at 4100 m (13.400 feet).

The crew made several violations to flight protocols and procedures. They have used the unofficial route before even knowing that it was dangerous due to extremely high mountains. (Normalization of deviation)

In addition, the two pilots were very good friends. They had no one to strongly challenge any decision taken by the other. They were overconfident and clearly lacked a sense of vulnerability. (Leadership, organizational culture)

The safety culture in the airline was poor and the actions of the pilots taking shortcuts revealed behaviors focused primarily on saving costs.

Contributing factors included:

- Unauthorized route
- Missing checklist
- System configuration
- Visual Flight
- Incoming flight
- Overconfidence

Figure 2. Yellow: Approved departure procedure; Blue: Unpublished route; Red: Route to disaster

The RBPS perspective

From the above deviations the investigation determined Human Error as the root cause of the accident. However, the tolerance shown by the airline and the traffic controllers also denoted poor safety culture and lack of operational discipline following procedures, air transportation rules and flight safety standards.

The organizational culture was so relaxed that Air Traffic Controllers (ATC) allowed the passengers to board the plane without a crew member on board and permitted the pilots to take decisions that could also impact safety of other air traffic. Their competence and training can also be questioned as they should know the time required to configure the basic controls of the airplane.

The use of a non-published route (read: non-approved) constitutes a "normalization of deviation" that seemed to be tolerated by both airline and ATC, even knowing that it involved hazards.

The investigation concluded "human error" as the "root cause". However, it is perhaps an immediate cause **(Reference 3).** All the deviations described above conditioned the pilot's behavior and significantly contributed to the disaster. As in many of the Chemical and petrochemical catastrophes the "root cause" should be found into the policies and cultural aspects of the airline and the aeronautical authorities.

In 2005 an explosion a hydrocarbon vapor cloud was ignited and violently exploded at the isomerization process unit at BP's Texas City refinery in Texas City, Texas, killing 15 workers, injuring more than 180 others and severely damaging the refinery. An independent report **(Reference 4)** found that BP has not instilled a common, unifying process safety culture among its five U.S. refineries. Each refinery has its own separate and distinct process safety culture. While some refineries were far more effective than others in promoting process safety, significant process safety culture issues exist at all five U.S. refineries, not just Texas City.

Although the five refineries do not share a unified process safety culture, each exhibits some similar weaknesses. The Panel found instances of a lack of operating discipline, toleration of serious deviations from safe operating practices, and apparent complacency toward serious process safety risks at each refinery.

3.3 Flight 447 – Air France

June 2010. Rio de Janeiro to Paris

Occupants 228 / Fatalities 228

An Air France Airbus A330-200 was destroyed when it crashed into the sea during a transatlantic flight from Rio de Janeiro to Paris-Charles de Gaulle Airport. The airplane carried 12 crew members and 216 passengers.

After contacting the last Brazilian ATC the captain found an area of cumulonimbus clusters typical of the tropical Atlantic in the month of June. Just before entering the turbulence zone the captain left the cockpit to rest and put the two co-pilots in command.

Flying at 35,000 ft into the heavy clouds the Pitot probes that provide air speed input to the flight computers were obstructed by ice crystals giving false information to the pilots and automatically disconnecting the auto trust mode. The PF (Pilot Flying) took the controls of the airplane and some moments later the PNF (Pilot Not Flying) declared "alternate law" which is a kind of manual control in which the pilot provide input to the computer in order that it keeps operating the control surfaces.

The pilots were surprised by the disconnection of the auto mode while flying at cruise altitude and did not understand what was actually happening with the airplane as the PF repeatedly provided "nose up" inputs even after some stall warnings. When the Captain came back to the cabin the plane have entered into a sustained stall and it was too late for a recovery maneuver. The aircraft crashed into the sea.

Several attempts were made to locate the wreckage of the airplane. Finally on April 3, 2011, a search vessel using unmanned submarines located pieces of wreckage including an engine, landing gear, fuselage and wing parts on the Ocean floor.

The accident resulted from the following succession of events **(Reference 5)**:

- Temporary inconsistency between the airspeed measurements, likely following the obstruction of the Pitot probes by ice crystals that, in particular, caused the autopilot disconnection and the reconfiguration to “alternate law”;
- Inappropriate control inputs that destabilized the flight path;
- The lack of the crew to identify any link between the loss of indicated speeds called out and the appropriate procedure;
- The late identification by the PNF of the deviation from the flight path and the improper correction applied by the PF;
- The crew not identifying the approach to stall, their lack of immediate response and the exit from the flight envelope;
- The crew’s failure to diagnose the stall situation and consequently a lack of inputs that would have made it possible to recover from it.

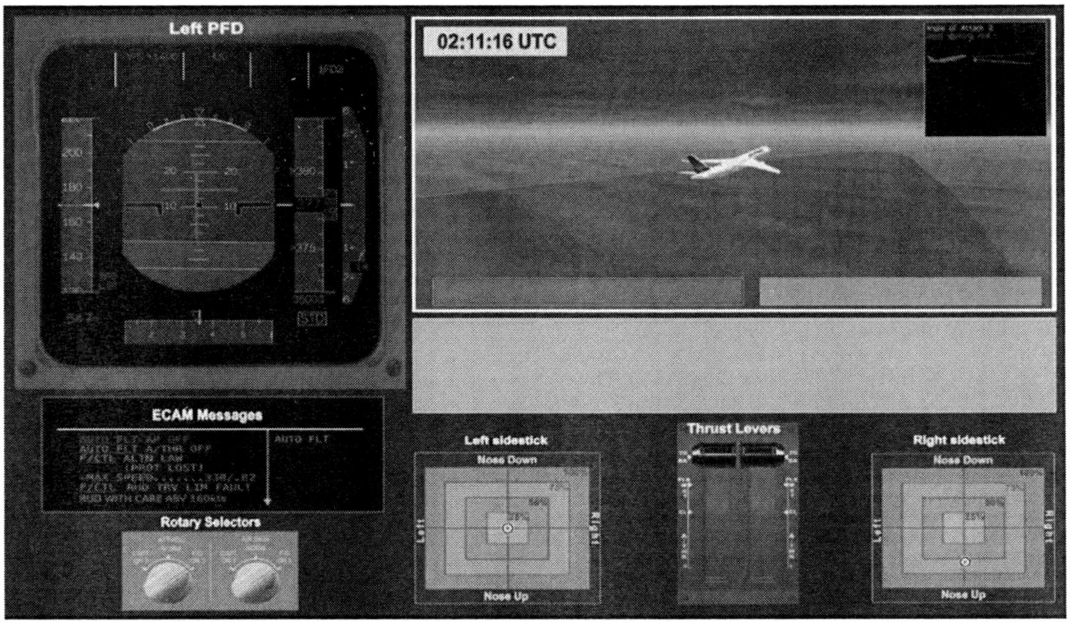

Figure 3. Electronic Centralized Aircraft Monitoring and inputs from sidesticks and throttle

These events can be explained by a combination of the following factors:

- The feedback mechanisms on the part of all those involved that made it impossible to:
  * Identify the repeated non-application of the loss of airspeed information procedure and to remedy this
  * Ensure that the risk model for crews understanding the consequences of icing of the Pitot probes at cruise altitudes.
- The absence of any training, at high altitude, in manual airplane handling and in the procedure for 'Vol avec IAS douteuse' (Flying with Indicated Air Speed in doubt)
- Task-sharing that was weakened by:
  * Incomprehension of the situation when the autopilot disconnection occurred,

* Poor management of the startle effect that generated a highly charged emotional reaction by the two copilots;
- The lack of a clear display in the cockpit of the airspeed inconsistencies identified by the computers;
- The crew not taking into account the stall warning, which could have been due to:
  * A failure to identify the audible warning, due to insufficient training on stall phenomena, stall warnings and buffet,
  * The appearance at the beginning of the event of transient warnings that could be considered as spurious,
  * The absence of any visual information to confirm the approach-to-stall after the loss of the limit speeds,
  * The possible confusion with an overspeed situation in which buffet is also considered as a symptom,
  * Flight Director Indications that may have led the crew to believe that their actions were appropriate, even though they were not,
  * The difficulty in recognizing and understanding the implications of a reconfiguration in alternate law with no angle of attack protection.

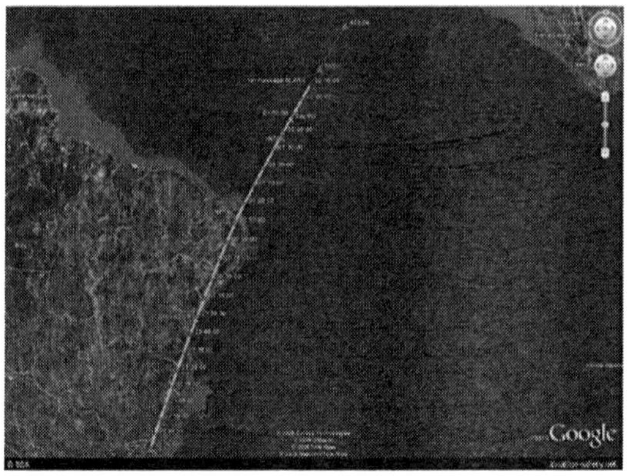

Figure 4. Flight 447 trajectory.

Figure 5. Recovering tail of the Airbus A330

The RBPS perspective:

The obstruction of the Pitot probes by ice crystals during cruise was a phenomenon that was known but not well understood by the Airbus pilots at the time of the accident. After initial reactions that depend upon basic airmanship, it was expected that this phenomenon would be rapidly diagnosed by pilots and managed where necessary by taking precautionary measures on the pitch attitude and the thrust. This identifies a failure in the lesson learning process that should warn all pilots of the potential for this problem and lead to development of a detailed procedure on how to identify and react to the situation.

The occurrence of the failure in the context of flight in cruise completely surprised the pilots. The apparent difficulties with airplane handling at high altitude in turbulence led to excessive handling inputs in roll and a sharp nose-up input by the PF. The destabilization that resulted from the climbing flight path and the evolution in the pitch attitude and vertical speed was added to the erroneous airspeed indications and ECAM (Electronic Centralized Aircraft Monitoring) messages, which did not help with the diagnosis. There is a need to improve in the design of the HMI.

In the minute that followed the autopilot disconnection, the failure of the attempts to understand the situation and the de-structuring of crew cooperation fed on each other until the total loss of cognitive control of the situation. The underlying behavioral hypotheses in classifying the loss of airspeed information as 'major' can`t be used to justify this accident. Confirmation of this classification supposes additional work on operational feedback that would enable improvements in aspects like crew training, the ergonomics of information supplied by the Flight Management System and the development of detailed procedures.

The airplane went into a sustained stall, signaled by the stall warning and strong buffet. Despite these persistent symptoms, the crew never understood that they were stalling and consequently never applied a recovery maneuver. The combination of the ergonomics of the warning design, the conditions in which airline pilots are trained and exposed to stalls during their professional training and the process of recurrent training did not generate the expected behavior in a reliable way. An examination of the training for airline pilots did not provided convincing indications of the building and maintenance of the associated skills.

There are similarities in the reaction of the airline industry to this incident and the reaction of owners of processing plants after major incidents. For example, a change in the type of the pitot probes had been recommended but not enforced until after this accident happened. The absence of training for manually handling the aircraft in certain conditions and the fact that the captain left the cabin just before the airplane was about to enter uncertain meteorological conditions may suggest a high dependability on the systems that control modern aircraft.

As with plant operators, providing procedures and training on how to respond to the signals provided by HMIs in the control rooms is not sufficient. Operators should also be trained on the physical principles associated with the process and the basics of the protective devices designed to maintain control of the operating parameters in order that they can properly identify and deal with any abnormal situation.

An incident also involving poor hazard identification, risk management, poor training and involvement in procedures, and poor decision-making resulted in six fatalities and 50 injuries in February 2010 in a natural gas explosion at the Kleen Energy power plant under construction in Middletown, Connecticut. The explosion resulted from planned work activities involving a large release of flammable natural gas in the presence of workers and ignition sources. The incident occurred during planned cleaning of fuel gas piping, part of the commissioning phase of the project. Workers were conducting a "gas blow" to force natural gas through piping at 650 psig pressure and high volume to remove debris (rust, welding slag, foreign material), which was vented to atmosphere.

On the day of the incident, the cleaning crew did not have a safety meeting that specifically discussed the hazards and procedures associated with gas blows. At the time of the explosion, natural gas was being blown from an open-ended horizontal pipe in a congested area between two large steam generators in an area immediately south of the power generation building. Dispersion of the gas was likely adversely affected by the congestion and orientation of the vent pipe. Although efforts were made to eliminate or control potential ignition sources outside of the power generation building, ignition sources remained outside and inside, such as electrical power, welding, and diesel-fueled heaters. The gas blows may have also created sparks or static accumulation.

The CSB in their investigation determined that feasible alternatives (air, nitrogen, steam, chemical cleaning) exist to the unsafe practices that led to the explosion, and that many companies (though not all) use safer methods for handling or venting natural gas. CSB also noted serious deficiencies that could result in unsafe work practices, including:

- Lack of involvement of workforce in gas safety procedures, training and communication,
- Lack of technical evaluation of vent piping to ensure adequate air mixing and release to a safe location,
- Ill-defined instructions to control or eliminate potential ignition sources,

Failure to recognize that a gas blow itself may provide a source of ignition (static charge accumulation, sparks).

3.4 Flight 1549 - US Airways

January 15th 2009. New York to Charlotte, NC

Occupants 155 / Fatalities 0

In the climb out after takeoff from New York City's LaGuardia Airport the airplane struck a flock of Canada geese just northeast of the George Washington Bridge and consequently lost all engine power. Unable to quickly reach any airport, pilots Chesley Sullenberger and Jeffrey Skiles glided the plane to a ditching in the Hudson River off Midtown Manhattan. All 155 people aboard were rescued by nearby boats and some with few serious injuries.

The investigators of the NTSB used flight simulators to test the possibility that the flight could have returned safely to LaGuardia or diverted to Teterboro; only seven of the thirteen simulated returns to La Guardia succeeded, and only one of the two to Teterboro.

Furthermore, the NTSB report called these simulations unrealistic **(Reference 6)**: "The immediate turn made by the pilots during the simulations did not reflect or account for real-world considerations." A further simulation, conducted with the pilot delayed by 35 seconds resulted in a crash. In testimony before the NTSB, Sullenberger maintained that there had been no time to bring the plane to any airport, and that attempting to do so would likely have killed those onboard and more on the ground.

The Board ultimately ruled that Sullenberger had made the correct decision, reasoning that the checklist for dual-engine failure is designed for higher altitudes, when pilots have more time to deal with the situation. While simulations showed that the plane might have just barely made it back to LaGuardia, those scenarios assumed an instant decision to do so, with no time allowed for assessing the situation.

What went wrong?
Canada geese flocks are common in the New York area this time of the year. There was no way for the airplane to avoid birds coming into the engines.

What went well here?

The RBPS perspective:

The relevant aspects of this incident are focused in positive aspects related to elements of Process Safety.

- Crew was qualified and experienced (Competency)
- Crew rigorously followed communication and command protocols (codes and standards, practices, procedures)
- They properly and timely communicated an emergency situation to ATC (procedures, workforce involvement)
- There were no external pressures or interventions over the pilots in command (Culture, leadership)
- Captain and first officer discussed options, evaluated risks and agreed on the best option in the short time that they had available (MoC, Risk Management, open communication culture)
- Crew did not put money before safety trying to save the plane by attempting to land in an airport (Culture)
- Captain demonstrated great ability maintaining overall integrity of the aircraft and the life of the passengers after performing a ditching maneuver (training, competency)
- Management of emergency, evacuation equipment and procedures by all members of the crew demonstrated adequate training and good Safety Culture of the organization (Emergency management, sound culture, safe operating practices and training)

Figure 6. Rescue operation

Figure 7. The crew of flight 1549

4 Summary

The accidents described show that in working environments where the safety culture is poor, priorities are focused on saving costs or maximizing production. In these situations, no open communication exist, procedures are not consistently followed and a "power culture" reigns, which are all scenarios prompt for a catastrophe.

They also demonstrate that for pilots as well as plant operators, there is little point in taking shortcuts to save costs or to keep their jobs when lives are at stake.

An important learning from disasters is that putting economic interests before safety is far more costly than assigning resources to achieve an adequate level of safety and reliability. Some airlines have disappeared after having a catastrophe and even the giant BP stumbled after the 2010 Gulf of Mexico disaster.

This paper wants to make a call to corporations to pay more attention to the factors that condition the behaviors of the employees and to understand that these cultural issues open the door to the "human error". Actions of employees are highly conditioned by the corporate culture that exists and by the importance placed on corporate directives for a strong process safety culture.

Captain Sullenberger (Sully) is a good example of a well-trained, experienced and competent pilot who used his authority, without additional pressure and followed established standards and procedures to properly interact with the crew and the air controllers. He was able to make a decision to manage an emergency and mitigate the consequences. And even so, the result could have been catastrophic.

But… how many Sullys are out there? Perhaps not many. We can think for a minute what would have happened if he had been the pilot in command of the Polish president? Would he have been able to sort the situation and walk away alive?

As with Sully, organizations need to ensure that personnel in charge of critical activities are skilled, competent, well trained and experienced. However, these attributes are never sufficient if organizations don't create a working environment in which communications are open and principles like respect and value for life are prioritized over business objectives. That is what an effective Process Safety culture is all about.

5 Conclusions

Safety in organizations is almost totally dependent of the principles and values imparted to employees. History shows that best operating management systems, practices, procedures and competence are not sufficient to operate safely when the operational discipline is not properly oriented to obtain the appropriate human behaviors.

Organizations need to have processes in place to deploy experienced personnel, empower operators and establish rules and communication protocols that allow operators to perform correctly not only during emergency situations, but always.

From the learnings of the accidents described in this paper organizations need to support their employees by:

- Establishing corporative policies with principles that define the values of the organizational culture
- Genuinely prioritizing safety before economic interests
- Creating an honest environment of open communication at all levels
- Setting clear rules and monitoring operational discipline within the organization
- Defining clear roles and responsibilities for PSM throughout the organization
- Establishing formal PS training and competence programs to ensure that operators understand the risks from the chemical and physical processes they manage
- Maintaining formal processes to identify and manage risks
- Developing practices and procedures that are consistent with the values of the organization
- Creating a system to appropriately manage and communicate the lessons learned from incidents
- Most importantly: "Lead by example"

6 References

[1] Interstate Aviation Committee, Air Accident Investigation Commission. Final Report - Tupolev Tu-154M tail number 101 of the Republic of Poland.

[2] Broadribb, Michael P., *What Have We REALLY Learned? (25 years after Piper Alpha)*, Process Safety Progress, Vol. 34 Issue 1, March 2015.

[3] Ignacio Alonso / Mike Broadribb. CCPS; *Human Error – A myth eclipsing the real causes. Presented at the* 13th CCPS Global Congress on Process Safety, and published in Aiche, Process Safety Progress, Wiley 2017

[4] Baker Panel. The Report of the BP U.S. Refineries Independent Safety Review Panel. January 2007.

[5] Bureau d'Enquêtes et d'Analyses pour la sécurité de l'aviation civile, Ministère de l'Écologie, du Développement durable, des Transports et du Logement, Final Report - On the accident on 1st June 2009 to the Airbus A330-203 registered F-GZCP operated by Air France flight AF 447 Rio de Janeiro – Paris, Paris 2012

[6] NTSB, Loss of Thrust in Both Engines After Encountering a Flock of Birds and Subsequent Ditching on the Hudson River, US Airways Flight 1549 Airbus A320-214, N106US, May 4th, 2010, Weehawken, New Jersey

Reference in order of first citation

Web resources:

[1] Aviation Safety Network, Flight Safety Foundation, Title, Available at http://aviation-safety.net/database/record. Accessed on March 10, 2018

Preparing for a Successful HAZOP/LOPA (Making or Breaking Quality & Efficiency)

Steven T. Maher, PE CSP
Risk Management Professionals
Steve.Maher@RMPCorp.com

Carine M. Black
Risk Management Professionals
Carine.Black@RMPCorp.com

John M. Johnson
Risk Management Professionals
John.Johnson@RMPCorp.com

Prepared for Presentation at
American Institute of Chemical Engineers
2018 Spring Meeting and 14th Global Congress on Process Safety
Orlando, Florida
April 22-26, 2018

Preparing for a Successful HAZOP/LOPA (Making or Breaking Quality & Efficiency)

Steven T. Maher, PE CSP
Carine M. Black
John M. Johnson

Risk Management Professionals

Keywords: PSM, RMP, CalARP, SMS, Process Safety, PHA, HAZOP, LOPA

Abstract

Although many factors can affect the quality and efficiency of a Hazard and Operability (HAZOP) Study or Layer of Protection Analysis (LOPA), one of the most important influences on the outcome is preparation. Contemporary HAZOP/LOPA guidebooks, such as the ones published by the Center for Chemical Process Safety (CCPS), have long dispelled the notion that a facilitator merely needs to show up and ask "blind" questions. Beyond facilitator preparation, given that each team and each evaluation is unique, thought must be given to ensuring that preparation is focused and tailored to the needs of the particular effort. Failure to properly prepare for the HAZOP/LOPA is one of the easiest ways to compromise results. The objective of this paper is to encapsulate experiences of the authors in facilitating HAZOP/LOPA over a wide-spectrum of industries, owner/operators, and facility-types, and provide the reader with some practical tips for transforming focused preparation activities into a successful HAZOP/LOPA outcome. Key paper topics include:

- Why quality and efficiency are both pivotal to a successful outcome
- Synchronization with organizational and regulatory objectives
- Guidance for ensuring process, operations, and the remainder of the core support team are properly prepared
- Advice for ensuring vendors are properly prepared
- Effective use of part-time "Subject Matter Experts"
- Making good use of a design review
- Master timeline in preparing for a successful HAZOP Study
- Pros and cons for pre-definition of causal events
- Nuisances of preparation for capital projects vs. operating facilities
- Differences in preparing for a new HAZOP/LOPA vs. a revalidation
- Ensuring the right tools are available

1. Focus on Objectives – Why is a Quality Process Hazard Analysis (PHA) Important

Most people attempt to avoid undesirable outcomes and organize their actions to minimize risk; however, very few apply a structured evaluation of potential undesired outcomes to avoid. Whereas this may be a personal decision for the individual, this is unacceptable for a complex process system, where the undesired events can have impacts well-beyond those directly involved. These events may be rare and involve a complex set of initiating events and failure of safety/mitigation systems; thus, it is necessary to perform thorough evaluations.

FIGURE 1.1 – Tragedies to Avoid

The December 2, 1984 Methyl Isocyanate (MIC) release from the Union Carbide Bhopal Facility is considered a pivotal event in catalyzing the application of Safety Management Systems (SMS) approaches to enhance process safety. The MIC release and the magnitude of the tragedy (3,928 fatalities and over 100,000 estimated permanent injuries) [1], drew the attention of industry, the public, and the regulatory community to the potential consequences associated with process safety events (Figure 1.1). Industry's response was swift and definitive. The American Institute of Chemical Engineers (AIChE) founded the Center for Chemical Process Safety (CCPS) in 1985, recognizing that the most effective mechanism for addressing process safety was not the application of additional prescriptive mechanisms, or by addressing any specific action, but by effecting changes in the way business is done (i.e., safety culture and management systems). CCPS Guidebooks are currently considered key references in conveying the technologies needed for process safety, and the very first guidebook ("Guidelines for Technical Management of Chemical Process Safety" [2]) published in 1987 was designed to address this pressing need.

A key part of managing process safety has always been the identification and understanding of potential hazards and their consequences. This need pushed practical techniques for hazard identification to the front lines in the effort to manage process safety (i.e., the Hazard and Operability (HAZOP) Study, developed in the decades prior to the Bhopal tragedy).

2. Brief History of Key PHA Techniques and Regulatory Requirements

Although there are quite a few tools in the PHA toolkit, the team-oriented, patterned-brainstorming sessions associated with HAZOP Studies [3] are generally considered the workhorse of the industry (Figure 2.1). Layer of Protection Analysis (LOPA) [4], frequently integral with the HAZOP Study, is a complimentary tool that is best patterned to integrate with the HAZOP Study. LOPA provides additional insights, some of which can be directly used for identifying appropriate reliability targets for key safety features. Although HAZOP Studies have been a core part of an acceptable hazard evaluation process referenced in various industry guidelines [2,3], as well as regulatory requirements such as Process Safety Management (PSM)[5] and Risk Management Programs (RMP)[6] for onshore facilities in the United States, LOPA is a relatively new tool that simplifies Quantitative Risk Assessment (QRA) techniques to a manageable level to facilitate usage. LOPA applications are gradually becoming best practice for addressing higher consequence/risk events and are especially useful for an initial assessment of the reliability needed for key safety systems. As well as an industry accepted practice for important applications, for California Refineries, PSM [7] and the California Accidental Release Prevention (CalARP) Program [8] require the development of Safeguard Protection Analysis (SPA), with LOPA as an acceptable tool.

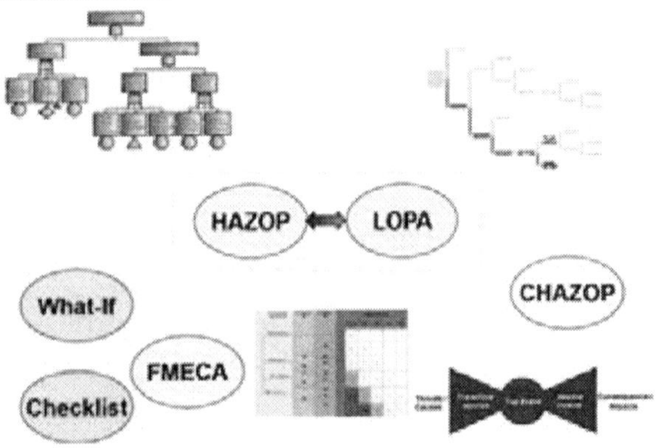

FIGURE 2.1 – HAZOP & LOPA are Core Elements of Hazard Evaluation

Although the core of the HAZOP approach hasn't changed, HAZOP and LOPA applications have been evolving to accommodate higher expectations of both regulators and industry practitioners with respect to quality and level-of-detail. HAZOP/LOPA must also address and integrate some of the evolving regulatory requirements, e.g., California's mandate for the application of SPA, Damage Mechanism Review (DMR), and Hierarchy of Hazard Control Analysis (HCA) to California Refineries.

To achieve this requires careful planning and preparation, but even before that, definition of clear objectives and defining the approach.

3. Defining Objectives and Approach

The core objectives of all hazard identification exercises (e.g., HAZOP/LOPA) are to uncover potential weaknesses/vulnerabilities in system design/operation that could result in an undesired outcome, e.g., injury, environmental impact, equipment damage, operational impact, or compromised company reputation. Additionally, the focus of the HAZOP/LOPA and the approach chosen can vary significantly for:

- Capital Projects (Early Stages)
- Capital Projects (Latter Stages)
- Operating Facility Initial HAZOP/LOPA
- Operating Facility HAZOP/LOPA Revalidation
- Addressing Specific Regulatory Requirements and Issues Associated with the Geographical Location

- Owner-Specific Priorities

When defining the objectives for all HAZOP/LOPA efforts, it is important to ensure that system boundaries are clearly defined. Table 3.1 outlines some general HAZOP/LOPA objectives.

TABLE 3.1 GENERAL HAZOP/LOPA OBJECTIVES
• **Identify causes that could lead to potential safety, environmental, or operability issues.**
• **Recommend changes, or further study, to overcome safety, environmental, or operability problems.**
• **Provide training to inexperienced personnel in the design intentions, expected operation, and potential hazards of the facility.**
• **Provide background and bases for the preparation of plant manuals and operating procedures, mechanical integrity programs, and understanding of facility hazards.**
• **Encapsulate background for recommended actions.**

3.1 Capital Project HAZOP/LOPA

Team dynamics and composition can be substantially different for capital projects compared to an operating process unit at a large industrial facility. [9,10] Capital projects typically:

- may have fundamental vulnerabilities or design options that HAZOP/LOPA can help resolve;
- involve large multi-disciplinary teams with numerous stakeholders who wish to participate in the HAZOP/LOPA – Preparation must accommodate a team that may be widely-separated geographically;
- have knowledge significantly compartmentalized, increasing the importance of the HAZOP/LOPA function to harmonize the design, while increasing the challenge;
- encounter financial pressures to accelerate project schedule;
- use package vendors to accelerate project schedule and compartmentalize design work;
- spread financial project risk over many companies; or
- have significant financial impacts associated with design mistakes that can result in re-work and start-up challenges.

TABLE 3.2 HAZOP/LOPA SUCCESS = PLANNING & PREPARATION
• **Quality**
• **Team Utilization & Effectiveness**
• **Session Time**

Many large projects make increasing use of vendor packages to distribute costs and accelerate progress. This modularized approach can allow multiple elements of the project to proceed in parallel, theoretically streamlining completion and allowing the specialist companies to focus on their areas of expertise. It's important to note that this significantly increases the complexity of the project and increases the importance of planning and preparation for the HAZOP/LOPA. HAZOP/LOPA can be a crucial element for binding these activities and flushing out potential weaknesses in the design or its integration, and previous papers have discussed the need for systematic application of HAZOP/LOPA during the design process.

In the early-stages of a capital project, the HAZOP/LOPA may focus on higher consequence scenarios to support decisions regarding fundamental design issues. Design details (e.g., instrumentation) are likely not readily available, but are generally not necessary to support these fundamental decisions or demonstrate a proof-of-concept for design acceptability. Significant value can be obtained from working out these issues early and ensuring that the various design participants are moving in the right direction. Starting the HAZOP/LOPA at these early stages, and later building on it, can support accelerated project progress. [9] In the latter-stages of a capital project, design details become readily available, and the focus of the HAZOP/LOPA shifts towards finalizing and vetting design details, design optimization, and neutralizing start-up challenges.

During capital project development, significant pressure is placed on the design team for maintaining project progress and doing things right the first time. Additionally, as the design progresses, changes to the design become more difficult and more expensive. Both of these challenges are typically best addressed by a phased approach to HAZOP/LOPA through the design process in an incremental fashion and with periodic sessions to keep pace with the design process. HAZOP/LOPA should synchronize with the availability of design information, and thus, the planning and preparation activities match the design details available, and objectives are adjusted to match the decision-making needs for that portion of the project. It should be noted that, properly structured, the HAZOP/LOPA can be readily updated as the design progresses.

FIGURE 3.1 – Use HAZOP/LOPA to Harmonize Subsystems

3.2 Operating Facility HAZOP/LOPA

Objectives for an Operating Facility HAZOP/LOPA are quite different. The key areas of focus are typically meeting regulatory requirements for the performance of HAZOP/LOPA and ensuring that company risk acceptance thresholds are maintained. There is no great impetus to optimize or simplify the design, as equipment has already been purchased and installed. It is important to treat continual improvement opportunities in a different manner than actions that may be necessary to meet regulatory requirements or the company's risk acceptance thresholds. Continual improvement opportunities include improvements suggested by the team for system configuration and/or operation, improvements to safety margins, and improved risk acceptance thresholds, even if current thresholds are already met.

Although this focus is somewhat simpler, there are other facets that may be required by regulation or company best practices that require additional planning/preparation for an Operating Facility HAZOP/LOPA, including the following.

- Simplification of high-maintenance-cost portions of the design
- Removal of dead-legs
- Incident Investigations
- Management of Change
- Facility Siting
- Damage Mechanism Review
- Safeguard Protection Analysis
- Hierarchy of Hazard Control Analysis
- External Events (e.g., seismic, storm loadings)

Operating Facility HAZOP/LOPA studies typically involve much smaller teams, with individuals typically having a wider-range of knowledge and experience of design and operation, greatly simplifying planning and preparation logistics. Various Subject Matter Experts, however, may be needed to support the above Operating Facility specific objective, along with the relevant documentation.

Defining the objectives for Operating Facility HAZOP/LOPA Revalidations may also involve addressing the perennial dilemma of deciding whether it is more fruitful to take a fresh look with the team or build on the results of the previous HAZOP/LOPA. A proper decision on objectives and approach is best made by accessing the previous HAZOP/LOPA and evaluating the following.

- Previous Technical Participants – If the previous team was highly capable and experienced, updating the previous effort may result in the best product, especially if the current team has limited experience.
- Previous Facilitator – Even with a good team, if the previous facilitator did not drive the team to excel, any use of the previous HAZOP/LOPA should be exercised with caution.
- Previous HAZOP/LOPA Documentation – Even if the team and facilitator were capable, if the document is sparse, its use has the potential for misleading the team into a false sense of comfort.

Researching these items and making good decisions regarding objective and approach in preparation for the HAZOP/LOPA Revalidation can have a huge impact on the quality of the results. To optimize the quality and validity of the revalidated HAZOP/LOPA, it is typical to build on the previous study and carefully drill down into the details. This also minimizes any potential liability associated with the Revalidation Team missing an important risk that may have been documented in the previous effort. If there were significant deficiencies in the facilitation or documentation quality of the previous study, recreating the study may still be beneficial. Proper and thorough documentation in a fresh study will lead to higher quality Revalidations in the future, as long as considerations made in the previous study are still addressed.

4. Planning Tips for the HAZOP/LOPA

A successful HAZOP/LOPA can be measured by its quality, completeness, and effectiveness, among other parameters. Planning and preparation are pivotal to a successful HAZOP/LOPA outcome. Planning activities such as ensuring the attendance of core personnel are essential and must be done well in-advance. The following paragraphs explain essential planning activities for a successful HAZOP/LOPA Session. Figure 4.1 presents a general timeline that is divided into two key types of activities, planning and other preparation activities. Planning and other preparation activities are also detailed in Sections 5 through 7 of this paper. It is important to note that planning may require significant lead time to establish the correct team and align schedules for a successful study, while other preparation tasks may be completed within a shorter timeframe before the session is scheduled to begin.

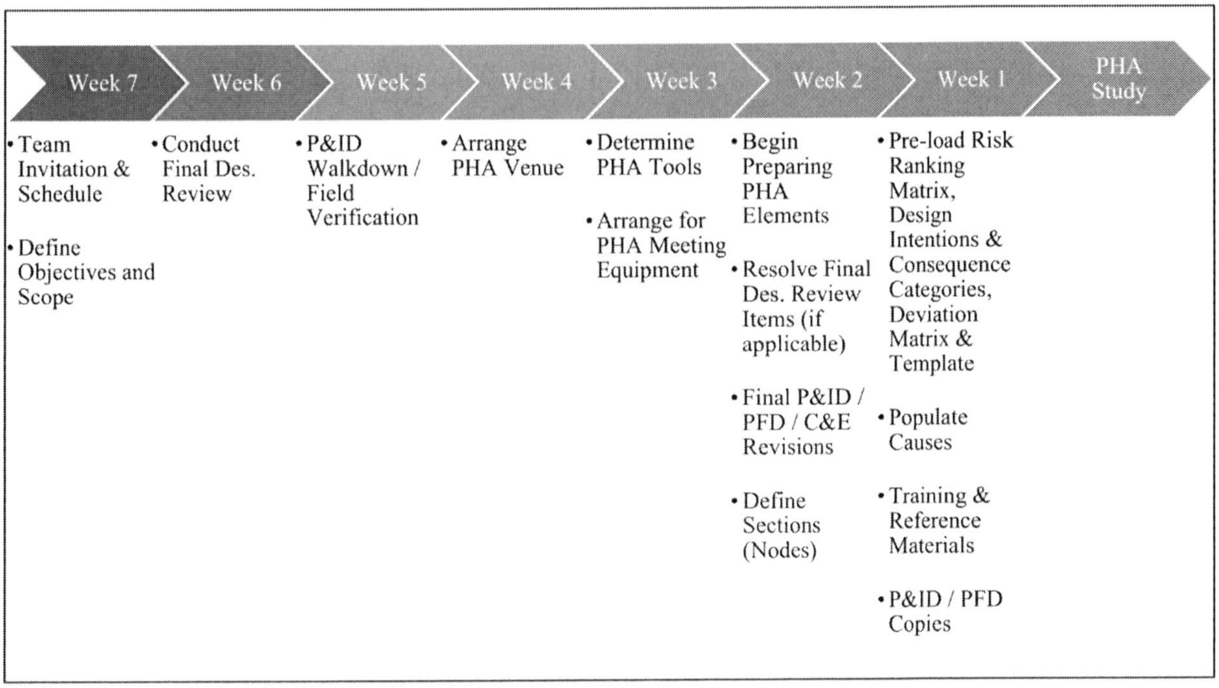

FIGURE 4.1 – Timeline for PHA Planning/Preparation

Whether for an Operating Facility or Capital Project, it is important to recognize that in addition to it being an independent safety evaluation, a HAZOP/LOPA is a technical problem-solving and decision-making exercise. As such, a critical element is the quality of the resources (i.e., design/safety information and personnel) available to the HAZOP/LOPA. Since unlimited attendance is not practical, it is important to understand the technical resources needed for technical problem-solving and decision-making. Table 4.1 below identifies the key disciplines useful for any HAZOP/LOPA Team.

The investment in resources aligned to a HAZOP/LOPA Team can be significant. Channeling these resources and energies can yield significant benefits, and rapid feedback can result in being able to utilize these insights much more quickly, either to progress a Capital Project or address potential risks associated with an Operating Facility. It is essential that the right people participate in the HAZOP/LOPA Study, and that they are motivated to objectively evaluate the design and to openly participate in technical problem-solving. Personnel involved must:

- be knowledgeable in the design of their portion of the system,
- be knowledgeable in the dynamics of system response, and
- have critically considered the response of their portion of the design to upset conditions.

Table 4.1
Key Disciplines Useful for a HAZOP/LOPA
(R = Required by PSM/RMP)

- Facilitation/Leadership (R)
- Team Scribe/Recorder
- Process/Project Engineering (R)
- Operations (R)
- Maintenance (R)
- Control/Protection Instrumentation Engineer
- HSE Engineering
- Rotating Equipment Specialists
- Specialists to Address Unique Requirements, e.g., DMR, SPA, HCA

Note that an optimal team size for an Operating Facility HAZOP/LOPA is 6-8 individuals.

4.1 HAZOP/LOPA Facilitator

Having a competent facilitator can make or break a HAZOP/LOPA study. Their main goal is to motivate the Team and ensure that they are being pushed enough to address all potential hazards. A facilitator should approach the study with a clear understanding of the process, and be prepared to ask the key questions that will inspire active discussion to ensure each scenario is being thoroughly considered. Additionally, it is the responsibility of the facilitator to keep the Team on track, and prevent them from wasting precious session time designing solutions to every scenario. This is a delicate balance between promoting creativity, and knowing when to re-focus the Team. An exceptional facilitator will likely have extensive industry experience and be able to work with a wide range of Team dynamics, and still be able to maintain control of the room.

4.2 Scribe Support

Scribing skills include typing, software usage, and computer interfaces with various types of equipment. The facilitator can be tasked with both facilitation and scribing duties depending on various parameters of the HAZOP/LOPA, such as system complexity and team composition. For a relatively small, manageable group, a facilitator might be able to also accommodate scribe duties. The pivotal decision-making parameter is whether the additional cost of a scribe outweighs the money saved by shortening the HAZOP/LOPA Sessions and the subsequent documentation time. Many professionals, especially in a large company, tend to underestimate the value of their time and that of their peers. The employee-time cost can be substantial and shortening the HAZOP/LOPA sessions could have a significant financial benefit for an individual facility. In most cases, this financial threshold is reached once the team size (excluding facilitator and scribe) exceeds 3 to 4 individuals for local projects and possibly 4 to 5 individuals if travel costs for the facilitator and the lower-cost scribe are considered. As another benchmark, Reference 16 recommends to "use a dedicated Scribe, for meetings longer than 4-total hours" and "Use a well-trained scribe to take the documentation load off of the team. This rule can save 30-50% of meeting time and increases brainstorming (because the team is not daydreaming as they wait for Leader to complete the notes)." Along with tangible savings, the scribe can provide an improvement in the quality of the documentation, and the session itself, by allowing the facilitator to better focus on engineering issues and maintain an optimal pace of the HAZOP/LOPA.

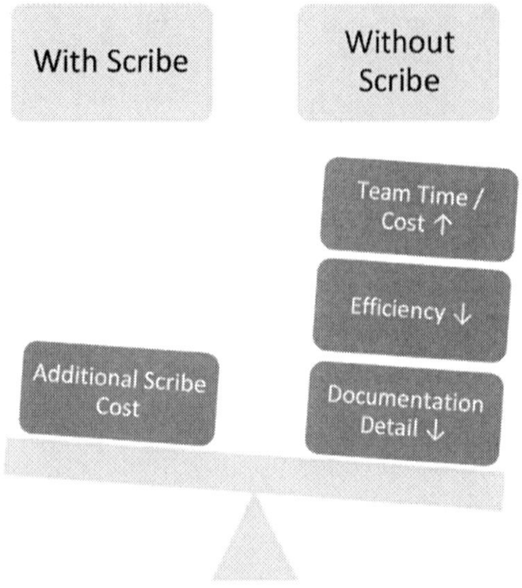

FIGURE 4.2 – Weighing Your Scribe Options

Of course, the decision to involve a scribe pivots on the scribe's capabilities. Some key characteristics to look for in a scribe include familiarity with HAZOP/LOPA techniques, reading engineering drawings, and the HAZOP/LOPA documentation software. For a HAZOP/LOPA, its advised that the facilitator avoid offers from the facility to provide a scribe. Unless properly trained and motivated, a poor scribe can compromise quality and team progress. Typically, a younger engineer that may be undergoing training to become a facilitator in the future is the ideal candidate.

4.3 Regulator

For some studies, it is appropriate to have the regulator's involvement in the HAZOP/LOPA study. This is particularly helpful in areas with more involved regulatory requirements, such as California. Including the regulator in the process will appease them knowing that you are taking regulatory requirements seriously and help develop a solid relationship between the facility and the regulators. Additionally, if the regulator chooses to participate, they can potentially instill some insight from their experiences attending a wide variety of HAZOP/LOPA studies, while also having the opportunity to express their concerns on design and operations at the facility.

5. Preparation Tips for the HAZOP/LOPA

5.1 Key Preparation Needs – Process Safety Information

Section 4 discusses many of the long-term planning requirements which pivot around personnel resources, which may need to be arranged months before the HAZOP/LOPA Sessions. The other key part of a successful HAZOP/LOPA is the quality, completeness, accuracy, and availability of necessary Process Safety Information (PSI). The adage "garbage in, garbage out" applies to the PSI used for the HAZOP/LOPA. Not only can incomplete and inaccurate PSI result in erroneous HAZOP/LOPA conclusions, but the recovered time that would have been lost by the HAZOP/LOPA Team to resolve information discrepancies/inaccuracies or to re-work previously completed scenarios, as information materializes, can pay for accurate PSI. This is particularly important when considering accurate Piping and Instrumentation Diagrams (P&IDs). For operating facilities, it is helpful to have an operator do a walkdown of the system to document any mistakes in the drawings prior to the HAZOP Study.

Table 5.1 illustrates the general information requirements for the performance of a HAZOP/LOPA. HAZOP Studies have been in use since the 1970s, and there are many styles of application. Many Teams have not yet been exposed to a HAZOP/LOPA that thoroughly challenges key elements of the design, and that is imperative for achieving the desired level of safety and operability. Thus, providing clear expectations and helping the individuals responsible for preparing the PSI can be very helpful.

TABLE 5.1 – GENERAL INFORMATION REQUIREMENTS FOR A HAZOP/LOPA

- **Process Flow Diagrams**
- **Piping & Instrumentation Diagrams (with changes identifiable from any previous HAZOP Studies)**
- **Cause & Effect Diagrams**
- **Alarm & PSV Setpoints**
- **Site Layout / Platform Location Drawings**
- **Accident/Incident History & Reports**
- **Management of Change (MOC) & Pre-Startup Safety Review (PSSR) Documentation**
- **Previous HA/PHA Recommendation Status**
- Equipment Data Books
- Operating & Emergency Procedures
- Maintenance Records
- System Descriptions
- Previous HA/PHA Reports
- Toxic, Chemical, and Physical Properties
- Prevention Program Compliance Audits

Bold items in the list above are generally more important.

It is important to note that not all HAZOP/LOPA studies need to be treated with the same level of detail. Some systems that are lower risk, or are unregulated, can utilize less detailed PSI information and still address a majority of the significant hazards. For large Capital Projects, it is often helpful to take a phased approach to HAZOP/LOPA studies. As previously discussed in Section 3.0, incorporating HAZOP/LOPA studies earlier in the design process allows for the Team to correct potential design issues early on. The Team must be cognizant of the level of completeness and accuracy of the PSI information available and must ensure that the areas in the process with limited information is re-HAZOP'd at later phases of the design process, as additional design information becomes available. [9]

5.2 Preparation Timing to Support HAZOP/LOPA Objectives

Section 5.1 stressed the importance of PSI preparation before the HAZOP/LOPA. Especially in support of a capital project, if PSI is poor and/or the team is poorly prepared to support a challenging HAZOP/LOPA, it can lead to an unfortunate waste of time and critical resources. This can result in a huge negative impact on a project. In addition to frequent and quality communications, a key part of a facilitator preparing their team includes verifying that their concept of accuracy, completeness, and detail is consistent with HAZOP/LOPA needs, confirming that the depth-of-challenge that the facilitator is planning for the HAZOP/LOPA is consistent with what they may have been exposed to previously, and emphasizing the need for pacing their preparation activities accordingly.

P&IDs should contain complete information and the key technical support members on the team should be prepped regarding the types of questions that materialize during the HAZOP/LOPA, otherwise the Team will struggle to identify and evaluate potential design/operations weaknesses, which may lead to more time and increased session costs associated with poorly focused resources.

TABLE 5.2
SCHEDULE FOR EFFECTIVE HAZOP/LOPA DESIGN INFORMATION PREPARATION

- **3 Weeks Before HAZOP/LOPA Session (Facilitator Actions)**
    - Review complete Drawing Package (PFD, Material Balance, Process Description, P&ID, and Cause & Effect (C&E) Diagram) to ensure a clear understanding of the process.
    - Transmit to the Process Engineer any questions based on any perceived design information gaps (note that other background information may have been provided as part of the design package beyond the P&IDs).
    - Verify with the Control/Protection Systems Engineer that their design integrates with the rest of the plant and that common functions such as the Emergency Shutdown (ESD), Fire and Gas Detection System, and Power & Other Utilities have been addressed.

- **2 Weeks Before HAZOP Session (Facilitator Actions)**
    - Communicate with Process Engineer and Control/Protection Systems Engineer and ask a few questions to spot-check that they understood and are addressing any gaps.

- **1 Week Before HAZOP Session (Process Engineer Actions)**
    - Provide updated drawings and other design information, in a form suitable for reproduction.
    - Having the ultimate responsibility for the design, verify that gaps have been addressed and the drawings are complete and accurate. In general, Process Engineers are the only ones capable of taking on the role of ensuring that the design and control philosophy addresses the system functional requirements.

- **2 Days Before HAZOP Session (Process Engineer Actions)**
    - Ensure that copies of pertinent information are reproduced for the HAZOP Study Team Members.

Please note:

- If multiple design documents are provided (e.g., P&IDs, PFDs, C&Es), they should all be up-to-date, legible, and consistent.

5.3 Types of Information

Important general design information requirements were specified in Section 5.1. Any information provided should be in a form that is consistent with project specifications, up-to-date, legible, and consistent. Section 5.2 provides a schedule for working with the HAZOP/LOPA Team to ensure that key preparation activities are performed. Table 5.3 and Table 5.4 provide examples of common information gaps in design information, and key HAZOP/LOPA-related issues that the Team should be prepared to address. This is a critical part of preparation for a HAZOP Study.

TABLE 5.3 – COMMON DESIGN INFORMATION GAPS

General P&ID Information Content:

- Key Equipment Design Information Blocks (either top or bottom of P&ID) – Include at least design pressures (e.g., vessels, pumps, filters, strainers, and heat exchangers), design temperatures, material specifications.
- Equipment Tag Numbers
- Some Indication of Piping Specifications (e.g., size, material, design pressure)
- Piping specifications conform to facility standards

Control Valves:

- Valve Failure Positions (e.g., loss of air, loss of power, loss of hydraulic pressure, thermostatic control)
- Actuator Types
- Size
- Setpoints

Relief Valves:

- Setpoints
- Size/Sizing Basis

Pumps:

- Blocked-in Discharge Pressure
- If Stopped, is Flow Reversal possible?
- AutoStart Function - Flow/pressure/electrical fault? Setpoint?

Block Valves:

- "Normal" Position (unless clear from context)
- Hazardous Area Drawings, as required

TABLE 5.4 – ADDITIONAL QUESTIONS/INFORMATION DETAILS THAT THE TEAM SHOULD BE PREPARED TO ADDRESS

Heat Exchangers/Coolers:

- Impact of Loss of Heating/Cooling Medium Flow?
- Impact of Changes in Heating/Cooling Medium Temperature?
- Impact of Loss of Power (total or partial) to Cooler Fans?
- Impact of Fin-Fan Cooler Louver Malfunction Open/Closed (also, what is the "failure position")?

General Process Condition Changes:

- Consequences of loss of flow, excess flow?
- Consequences of temperature excursions high/low?
- Do these consequences violate a) safety limits resulting in near-term damage, b) safety limits resulting in long-term damage or equipment degradation, or c) compromising warrantee or stable operation?

Instrumentation:

- Protection System Setpoints?
- Protection System Actions? If complex, a C&E may be necessary.
- Alarm Setpoints?
- Alarm Annunciation Locations (i.e., local, DCS, remote center)?
- Alarm Effect (i.e., panel indicator, audible alarm, flashing lights)?
- Communication with the DCS and partitioning of instrumentation and control between the local panel and from the DCS?

Electric Immersion Heaters:

- Damaged if Uncovered (without activation of protection features)?
- Will Control/Protection Temperature Transmitters function as-intended, if uncovered?

Other (primarily applicable for lube oil systems):

- For three-way valves, is total isolation possible? Can flow go down both paths if the valve is mispositioned?
- Sensitivity to lube oil composition?

Tanks:

- Vent locations and potential hazards?
- Provisions for tank maintenance, e.g., draining?
- Accessibility / procedures for filling?
- Materials of construction?
- Secondary containment and draining requirements?

6. Venue Tips for the HAZOP/LOPA

There are many factors to consider when deciding on the most efficient venue to perform a HAZOP/LOPA study. It is vital to select an appropriate location, especially because more complex HAZOP/LOPA studies can last several weeks at a time. Additionally, while having shorter session times is desirable because it can result in more active Team participation and focus, this is often not very feasible because of project time constraints and team availability. Thus, session times can often drag out longer than preferable, resulting in long hours. Selecting a venue that is comfortable, convenient, and flexible can optimize the Team's morale and even streamline the study with the appropriate preparation.

6.1 Location

Choosing the right room for the HAZOP/LOPA sessions can help contribute to the success of study. The room should be large enough for all Team members to have plenty of space. Ideally, everyone will have their own set of P&IDs that they will want to spread out and review, so larger tables are preferable. The room should also have plenty of wall space to hang up large drawings as needed, and to project the session notes and electronic versions of the P&IDs to allow the Team to easily follow along. Ensure that the room selected also has extra chairs and table space to accommodate the part-time participants who may need to join the session to offer their expertise periodically.

The actual venue location should also be thoughtfully considered. While it is convenient to select a room that is right near the regular workplace for the majority of the team, this has several potential downsides, as the close proximity to normal work-place areas can lead to major distractions. This includes the potential for Team members' getting preoccupied with day-to-day matters, wandering off during breaks, or session time getting interrupted by coworkers asking for support from Team members for unrelated tasks. Selecting a room location that is slightly removed from the centrally located and highly trafficked areas can mitigate these distractions and will allow the Team to remain focused on the task at hand.

The room selected for the study should also consider proximity to the actual site. Often times, performing a walkdown of the system being assessed helps the Team get a better idea of potential hazards. For currently developed facilities, the Team should assess the need for a walkdown and select a venue accordingly. During design HAZOPs, the focus should be on locations that will maximize Team comfort and engagement. Make arrangements for site access and reserve meeting locations (e.g., conference rooms) well in-advance.

6.2 Team Member Distribution

Some strategy can be put into arranging the Team members to optimize participation. Try to evenly disperse individuals around the table so it is easy for every team member to speak and be heard. the best spot for the facilitator and scribe is at the front of the room next to the displays/screens. This is so that the facilitator can easily guide the scribe, get up to captivate people's attention, or point out important items in the session worksheets or drawings being projected onto the screens.

For more outgoing or key members of the HAZOP/LOPA Team that are bound to talk frequently, it is sometimes helpful to place them on the other side of the room to help draw the rest of the Team members into the discussion. For quieter individuals, do not allow them to sit in the corner answering emails on their phone. Ideally, move them to a more central location, possibly near the facilitator, to force them to engage with the rest of the Team. No one on the HAZOP Team should be there to just to observe. Everyone should have some input, share their insights and expertise, and be encouraged to contribute their viewpoints for each scenario. [11]

6.3 Venue Amenities

The Venue selected should be equipped with several helpful items that the Team may require. This includes large displays with so that the scribe can project the worksheets and relevant P&IDs. White boards should also be available so that individuals can quickly throw together rough drawings, calculations, diagrams, or quick summarizations of main points. Numerous highlighters of varying colors should also be available on the conference table, should the Team need to change up the nodes used for the study.

While it is easy to get caught up in the technicalities of HAZOP/LOPA studies, some consideration should also be put into the more basic preparations for a HAZOP/LOPA. Ensuring that snacks, coffee, and refreshments are available throughout the study will help HAZOP/LOPA Team members maintain focus and keep up Team morale. Additionally, simple arrangements such as ordering lunch to be delivered ahead of time can save the HAZOP/LOPA Team lots of precious study time spent digressing from relevant conversations to discussions about where to eat and when. The duration of lunch breaks can also be minimized by planning ahead. Table 6.1 the Venue Selection Tips to keep in mind.

TABLE 6.1
VENUE SELECTION TIPS

PHYSICAL ACCESS
- **Physical Site Access**
- **Conference Room Availability**
- **Location Removed from Daily Workplace Areas**

ROOM CONFIGURATION / RESOURCES
- **Adequate Table Space**
- **Strategically Distributed Team Members**
- **Sufficient Resources (e.g. computer displays, wall space, white boards, etc.)**

6.4 Including Remote HAZOP/LOPA Team Members

Although it would be ideal to schedule a perfect HAZOP where all key members can be physically present, this is highly unlikely to occur. It is smart to prepare for the inevitable scenario where at least a few individuals will have to attend the sessions remotely. Performing a HAZOP/LOPA remotely is not the right answer for many applications, but for some, it can be a very effective solution that can engage resources (e.g., Vendors or Subject Matter Experts) that wouldn't be available otherwise. All of this can accelerate the progress of efforts such as capital projects, aid in project control, enhance teamwork, and result in tangible savings to the Owner/Operator. [12]

To prepare for including remote Team members, the Venue selected should have internet access and the scribe should anticipate that they will need to share their screen so that the remote Team members can also follow along. This can easily be accomplished if the HAZOP Team shares relevant contact information in advance, and the Team utilizes remote meeting programs to connect with new members quickly. The conference room selected should also have a local phone accessible to contact remote Team members and not worry about spotty service concerns affiliated with cellphones.

Certain preparation efforts are more important for a successfully including remote HAZOP/LOPA Team members due to the potential technical problems associated with additional equipment. The following paragraphs explain essential preparation activities for a successful remote HAZOP/LOPA Session, and for convenience, these are summarized in the form of a checklist in Table 6.2.

Adequate Computers/Displays

To help ensure that remote team members are properly engaged, the scribe should ensure that they have both the HAZOP/LOPA notes and the P&IDs readily accessible on the screen they are sharing. The scribe should switch over to the P&IDs during more complex team discussions to ensure that the remote Team members follow along and can contribute effectively. Alternatively, it is often helpful for a process engineering representative or the "owner" of the P&IDs to also share their screen with remote member and "drive" the display to focus the Team's attention on the portions being evaluated. This additional communication link can also be used to share other information pertinent to the HAZOP/LOPA, e.g., photographs, other engineering information, etc.

TABLE 6.2
REMOTE HAZOP/LOPA PREPARATION CHECKLIST

EQUIPMENT
- **Adequate Computers/Displays**
- **Cameras**
- **Arrange Communications Access and IT Support**
- **Testing**

RESOURCES AND ACCESS
- **Check Power and Communications Connections for All Computer Equipment**
- **Ensure all rooms have adequate internet access, and phone service or landline phone access.**

TECHNICAL PREPARATION
- **Send all necessary documentation to remote Team members ahead of time (e.g., highlighted P&IDs, risk ranking matrix, relevant PSI, etc.)**

Cameras

Some companies have interactive video conferencing between their sites. This can help the Facilitator verify that the remote Team members are present and engaged. Even if a dedicated video system is not available, screens with a video feed of team members can be setup if the Team feels it is necessary.

Pre-Arrange Communications Access and IT Support

If bringing computers with specialized software, make arrangements for any necessary login or network permissions, plan on setting up early, and make arrangements for contingency IT support ahead of time. Also, verify that power connections are sufficient and accessible for all equipment, and if wireless Internet access is insufficient, ensure data communications cables can reach.

7. Final Readiness Tips

When it comes to a successful HAZOP/LOPA, there are few activities more important than planning and preparation. Just as the HAZOP/LOPA is used to identify what can go wrong with the chemical process being evaluated, teams should postulate potential problems that can arise with the HAZOP/LOPA Sessions and how the problems might be addressed. An essential way to mitigate the problems addressed is by sufficiently reviewing the process being studied and formulating a framework for the anticipated study. The following paragraphs explain essential preparation activities for a successful HAZOP/LOPA Session.

7.1 Pre-Define Causes

Pre-defining causes by the facilitator is a contentious issue for some professionals. Although some individuals express concerns of putting boundaries on the team's imagination, for an in-person HAZOP/LOPA there can be a number of benefits that include the following.

- **Completeness** – For a cause-by-cause HAZOP Study, pre-defining causes can help avoid missing an important initiating event. By the facilitator defining the prominant causal events up-front, it frees the team to brainstorm the less-obvious causes and subtle process issues.
- **Future Use** – Rather than random brainstorming, careful patterning of causal events before the interactive chaos of a HAZOP Study can facilitate future use.
- **Quickly Locating Equipment During the Session** – Careful patterning up-front, and including tag numbers and P&ID references, streamlines locating the equipment during the HAZOP/LOPA Session.
- **Grouping Causes** – If carefully patterned, a natural grouping of the initiating events can minimize multiple detailed discussions of the consequences, adding clarity, consistency, and accuracy to the HAZOP/LOPA.

The team can be tasked with reviewing the causes and being prepared to explain the associated process dynamics and credible ultimate consequences. From experience, defining the process dynamics and associated ultimate consequences can represent the single largest segment of session time, so if the team can be aided in converging on these issues team focus is improved and the overall HAZOP/LOPA effort streamlined.

7.2 Pre-Define Questions

In the same way that pre-defining causes can focus the team's efforts, for many industries and for many countries, the patterned-brainstorming approach to HAZOP/LOPA is unfamiliar and not consistent with typical business approaches. Providing examples of HAZOP/LOPA scenario development and creating a list of specific questions (typically regarding the basis-of-design for the equipment that they are responsible for) can be very helpful, especially for members of the team that are inexperienced in HAZOP/LOPA best practices. Pre-defining causes and questions can be beneficial for all HAZOP/LOPA efforts and even more beneficial when involving remote HAZOP/LOPA Team members.

7.3 P&ID Grid Format

During any HAZOP/LOPA, the Facilitator will likely need help in focusing the team's attention to a specific location on the P&ID. If possible, margin grids should be included on all engineering drawings to be used for the Study.

8. Summary

Taking the appropriate steps to be sufficiently prepared for a HAZOP/LOPA can help improve the quality and efficiency of the HAZOP/LOPA study, and put a successful framework in place. This can be accomplished by first acknowledging the most important objectives for the particular project. For Capital Projects, this includes focusing on optimizing the process design and minimizing costs. Operating facilities should focus more on mitigating current risks and vulnerabilities in the process, and adhering to regulatory requirements.

Regardless of the focus, all HAZOP/LOPA studies require basic planning to be successful. This includes scheduling all key members as soon as possible to ensure their attendance. Additionally, the Team needs to obtain and update all necessary PSI, prepare the appropriate technical questions that need to be asked during the study, and secure a venue that will optimize productivity. Overall, properly preparing for a HAZOP/LOPA study can significantly decrease the amount of time and money invested into a project, and improve the quality of the design, while also accomplishing the main objective of mitigating risks in the hazardous process.

9. References

[1] http://en.wikipedia.org/wiki/Bhopal_disaster – Accessed on March 20, 2015.
[2] CCPS "Guidelines for Technical Management of Chemical Process Safety," 1987.
[3] CCPS "Guidelines for Hazard Evaluation Procedures," 1985.
[4] "Layer of Protection Analysis: Simplified Process Risk Assessment, " October 2001.
[5] PSM – 29 CFR 1910.119, "Process Safety Management (PSM) of Highly Hazardous Chemicals, Explosives and Blasting Agents," 1992.
[6] RMP – 40 CFR Part 68, "Risk Management Programs (RMP) for Chemical Accidental Release Prevention," 1996.
[7] California Code of Regulations, Section 5189.1, "Process Safety Management for Petroleum Refineries," October 1, 2017.
[8] "California Accidental Release Prevention (CalARP) Program," October 1, 2017.
[9] Paper 246177, 2012 Global Congress on Process Safety, "Assimilating Design Formulation and Design Review into a HAZOP."
[10] Paper 246192, 2012 Global Congress on Process Safety, "Practical Approach to Vendor Package HAZOP Studies and Preparing the Package Vendor."
[11] http://www.RMPCorp.com/HAZOP-Study-series-module, HAZOP/LOPA Facilitation Best Practices Webinar Series.
[12] Paper 479475, 2018 Global Congress on Process Safety, "Optimizing Your Remote HAZOP/LOPA Experience."

Understanding the Interrelationships Between the PSM Elements For Effective Implementation

Revonda P. Tew
Process Improvement Institute, Inc.
1321 Waterside Lane
Knoxville, TN 37922, USA
rtew@piii.com

Prepared for Presentation at
American Institute of Chemical Engineers
2018 Spring Meeting and 14th Global Congress on Process Safety
Orlando, Florida
April 22 – 25, 2018

Understanding the Interrelationships Between the PSM Elements For Effective Implementation

Revonda P. Tew
Process Improvement Institute, Inc.
1321 Waterside Lane
Knoxville, TN 37922, USA
rtew@piii.com

Keywords: PSM elements, interrelationships, MOC, PSI, effective implementation

Abstract

Every element of PSM has a role in controlling risk. These elements have an interrelationship – each depends on the implementation of one or more elements for effective implementation. Some elements have a greater influence than others. Understanding these relationships is critical for building and sustaining effective PSM programs. This paper will show through figures and discussions, these interrelationships, key attributes of each element that influence other elements and how poor implementation of each element adversely affects the implementation of the other elements.

1 Introduction

The elements of process safety work together to control risk, each providing a level of safety management. Some elements help to ensure processes are designed using Recognized and Generally Good Engineering Practices (RAGAGEP) for process safety, including the proper hazard reviews to ensure processes are safeguarded against human error and equipment failures that can lead to loss of containment and employee injuries and fatalities. Some elements set standards for day-to-day operations – operate within the design specifications, review changes prior to implementation, maintain the safeguards and equipment, investigate near misses and incidents to improve any design or operational shortcomings. Audits provide a periodic assessment to identify overall management system weaknesses and make recommendations to improve program shortcomings and effectiveness.

The governing PSM elements are Process Safety Information (PSI) and Management of Change (MOC). These two elements have the greatest influence on the development and execution of the other PSM elements. Discussions around these elements will show their core relationships with all other elements followed by identification of key relationships between some other elements.

2 Relationship Between Process Safety Information and Management of Change

2.1 MOC Protects PSI

The Process Safety Information element documents the process safety design. It is the foundation for process safety. Ninety-nine percent of changes are changes to PSI (others being organizational changes, some facility changes and stand-alone procedural changes). MOC is the umbrella, protecting the process safety design **(Figure 1)**. Not only is current PSI needed to evaluate the technical basis and evaluate the risk of the change, the PSI that is changing due to the process change must be identified and must be updated to reflect the change.

Figure 1. Relationship Between MOC, PSI and Other PSM Elements

2.2 MOC Protects Other PSM Elements

When a process change is proposed and evaluated, the quality and thoroughness of the MOC review extends beyond the change itself. The implementation of a change will require identifying other elements of PSM that must be changed/updated. For example, if a new instrument is added to a process:

- Operating procedures must be updated with instructions for operating, monitoring, and troubleshooting the instrument, and should have a description of its process safety function
- Mechanical Integrity program must be updated. The instrument data sheets must be filed properly, inspection, testing and preventive maintenance program established and entered into the Computerized Maintenance Management System (CMMS) and spare parts ordered

- Training guides must be updated to include the instrumentation, safe operating limits, consequences of deviations, and troubleshooting guidelines
- All related design and specifications of the instrument must be filed properly in the specific PSI documents such as loop sheets, P&IDs, etc.

2.3 Day-to-Day Use of PSI

On a day-to -day basis plant staff relies on the PSI documents to make decisions concerning process changes. Therefore, if documents are not updated to reflect changes as they occur, eventually the process safety design documents will be out of date and future process change decisions will be made on outdated information. PSI is the basis for implementing all elements of PSM. Some examples where PSI is used:

- Standard Operating Procedures (SOPs) content
- Training materials for operators
- Emergency Response Plans
- Training for Contractors and Mechanics
- Establishing ITPM for Equipment and Instrumentation
- Conducting Process Hazards Analyses (PHAs)
- Conducting Incident Investigations
- Evaluating MOCs

2.4 MOC Procedural Step to Protect PSI

To keep PSI updated and accurate, a critical step in the MOC process is to identify all the documents associated with elements of PSM that require updating, to reflect the process change with respect to the PSI. Many companies' MOC systems lack a mechanism to help the MOC owner identify all documents that contain PSI to be updated. **Table 1, "Documentation Checklist,"** is an example list that identifies all the documents related to PSM implementation that may need updating to reflect a process change. However, identifying the documents is only the first step. This trigger list consists of documents that define PSI, and documents that support the implementation of other PSM elements or proof documents of implementation. [1]

Table 1. Documentation Checklist Used to Identify All Documents Impacted by A Process Change

Processing
- ☐ Operating Manuals
- ☐ Training Material
- ☐ Standard Operating Procedure(SOP)
- ☐ Standard Job Procedure (SJP)
- ☐ Asset Operating Procedure (AOP)
- ☐ Operating Best Practice (OBP)
- ☐ Operating Instruction (OPI)
- ☐ Department Policies
- ☐ Quality Manual
- ☐ Training records
- ☐ Process Description
- ☐ Product Specifications
- ☐ Refinery Energy Model
- ☐ Refinery Utility Capacity Report
- ☐ Loss of Utility Emergency Procedures
- ☐ Shift Log / Turnover Log
- ☐ Shift Schedule for Plant
- ☐ Shift Personnel List

Piping
- ☐ Steam & Trap Docs
- ☐ Pipe Specs
- ☐ Pipe rack drawings
- ☐ LX drawings
- ☐ Piping isometrics

Organizational
- ☐ Corporate and Refining Phone List
- ☐ Organizational Charts
- ☐ Outlook Distribution lists
- ☐ On-Call Schedule
- ☐ Vacation Schedule
- ☐ Cintellate Hierarchy
- ☐ Oracle Hierarchy

Technical-PTS
- ☐ Process Flow Diagrams (PFDs)
- ☐ Piping and Instrument Diagrams (P&ID)
- ☐ Safe Upper and Lower Limits
- ☐ Maximum Intended Inventories
- ☐ Evaluation of the Consequence of Deviations
- ☐ Material and Energy Balances
- ☐ Description of the process materials and process chemistry
- ☐ Hazard Effects of mixing two chemicals
- ☐ Design basis
- ☐ Control Narratives

Technical-PCG
- ☐ Safety systems
- ☐ Cause and Effect Matrices
- ☐ DCS Configurations
- ☐ Process Historical Data (PHD)
- ☐ Event Analysis
- ☐ Process Control Instructions (PCI)
- ☐ Process Control Program
- ☐ Computer Graphics
- ☐ Safety Requirement Specifications (SRS)

Risk Identification
- ☐ Blast Study
- ☐ Toxic Release Study
- ☐ HAZOP
- ☐ LOPA
- ☐ FMEA (RELI-139)
- ☐ PPM Record
- ☐ Safety Integrity Level (SIL)

Health, Safety & Environment
- ☐ Evacuation Routes and Strategies
- ☐ Emergency Response Manual (ER)
- ☐ Health & Safety Instructions (HIS)
- ☐ Process Safety Instructions (PSMI)
- ☐ Environmental Instruction (EMSI)
- ☐ Training database
- ☐ OH Manual
- ☐ C. of A. records
- ☐ Waste Characterization
- ☐ Disposal Permit
- ☐ Process Hazard Analysis Report
- ☐ MSDS

Civil
- ☐ Equipment Plot Plans
- ☐ Underground drawings

Instrumentation
- ☐ SIS Instrument Check Procedures
- ☐ JX,PX drawings
- ☐ Loop diagrams
- ☐ Instrument Datasheets
- ☐ AMS Files

Reliability
- ☐ Materials of Construction Drawings and Datasheets
- ☐ Design Codes and Standards
- ☐ Equipment Datasheets
- ☐ Vessel Drawings
- ☐ Fire water system drawings
- ☐ Underground Drawings
- ☐ Lubrication records
- ☐ Vibration Database Records
- ☐ Spare Parts
- ☐ PD-18
- ☐ Equipment specification sheets
- ☐ CMMS records
- ☐ Vendor information
- ☐ Asset Criticality Assessment (RELI 138)
- ☐ Design Calculations
- ☐ EWO/TWO
- ☐ Reliability Instruction (RELI)

Electrical
- ☐ E-MENU data
- ☐ Single-line drawings
- ☐ Electrical Area Classification Docs
- ☐ Heat Tracing Dwg.
- ☐ Heat Tracing Controller Settings
- ☐ Panel Schedules
- ☐ MCC Schematics
- ☐ Wiring Drawings
- ☐ Protection Device Parameter Update (Multilin)

Mechanical Integrity
- ☐ Radiography Documents
- ☐ Pump Bench Records
- ☐ Notes to Equipment Files
- ☐ PSV Documentation
- ☐ Inspection Piping Isometrics
- ☐ Material Test Report (MTR)
- ☐ Non-Conformance Report (NCR)
- ☐ Mechanical Integrity Instructions (MII)
- ☐ Quality Assurance Manual
- ☐ Vessel (drawings, inspection reports)

Maintenance
- ☐ Repair Procedures
- ☐ Open Issue List
- ☐ Basic Equipment Maintenance Plan
- ☐ Materials Mgt. Instruction (MMI)
- ☐ Maintenance Instructions (MNTCI)
- ☐ Repair Procedures

Document Control/Management
- ☐ *All IOR Instructions, Policies & Standards*

Documentum:
- ☐ Bill of Material
- ☐ Breaker Testing Records
- ☐ Calculations
- ☐ Canadian Registration Number (CRN)
- ☐ Certificate of Compliance
- ☐ Data Sheets
- ☐ Drawing (Engineering Drawing, Isometric Loop, P&ID, PFD, Plot Plan, Sketch)
- ☐ Drawing (Above Grade Piping, Absorber, Boiler, Building, Column, Compressor, Concrete, Contractor, Cooling Tower, Design Data, Drum, Electrical, Exchanger, Fin Fan Coil, Foundation, Furnace, Grading, Insulation, Misc. Mechanical Equipment, Paving, Piping Isometric, Pump & Driver, Reactor, Stripper, Structural Steel, Tank, Vessel, Underground Piping
- ☐ eAM Equipment Data
- ☐ Emergency Procedures
- ☐ Engineering Equipment Summary
- ☐ Engineering Work Order (EWO)
- ☐ Feed Package
- ☐ Forms
- ☐ Inspection Record
- ☐ Inspection Test Plan
- ☐ Install Op Maintenance Manuals
- ☐ Instructions
- ☐ Instrument Index
- ☐ Isometric
- ☐ Line List
- ☐ Manufacturers Data Reports
- ☐ Material Testing Report (MTR)
- ☐ Parts List
- ☐ Performance Curves
- ☐ Principles & Philosophies
- ☐ Process Design Data
- ☐ Project PEP, PPM
- ☐ Quality Plan
- ☐ Refinery Application Manual
- ☐ Sketch
- ☐ Specifications
- ☐ Studies and Technical Reports
- ☐ System Turnover Books
- ☐ Technical Memorandum
- ☐ Unit Operating Procedures
- ☐ Vendor – Weld Procedures, Test Reports, Procedures and Drawings
- ☐ Catalogues, Vendor Data
- ☐ PSMI 201
- ☐ Datasheets – Equipment, Instrument, Material of Construction and Datasheets
- ☐ Drawings – Firewater, Loop, Piping & Instrument, Process Flow, Single Line Electrical, Underground, Vessel
- ☐ Electrical Area Classification
- ☐ Inspection Piping Isometrics
- ☐ MSDS
- ☐ Safety Systems/Cause and Effect Matrices
- ☐ Toxic Release and Blast Studies
- ☐ Vendor Data – Process Material, Process Chemistry, Hazardous Effects of Mixing Two Chemicals

3 Relationship Between Process Safety Information and Other Elements

As mentioned previously, MOC protect the process safety design based on PSI, PSI is the foundation of process safety. Defined as chemical, technology and equipment information, this information defines all aspects of the process design – including process safety. Each category of PSI is utilized in a variety of ways to meet PSM implementation. **Figure 2** shows PSI's strongest relationships.

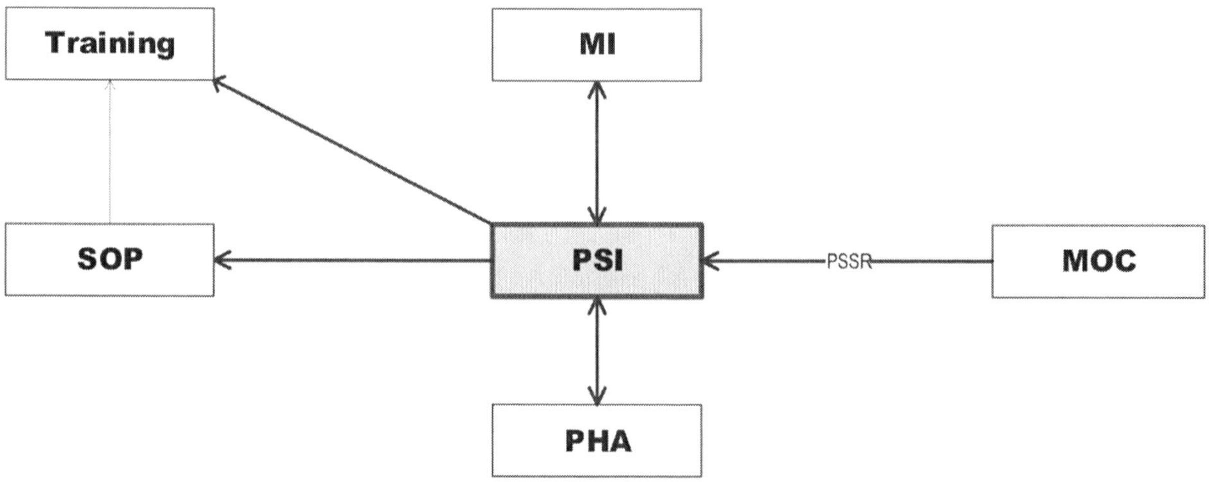

Figure 2. PSI's Strongest Relationships

3.1 Chemical Information

Chemical Information consists of physical properties and safety hazards of the chemicals, chemical reactivity and incompatibility hazards. Understanding the nature of the chemicals in storage or under process conditions (reactivity, corrosivity, thermal and chemical stability) is a critical component for equipment specifications, worker exposure hazards (toxicity, permissible exposure limits) and process safety hazards. The hazards of inadvertent mixing are necessary for managing process safety hazards of certain operational tasks where the incompatibles could possibly make contact. Some of these activities include:

- Raw material storage
- Raw material unloading
- Equipment cleaning
- Operation of multi-use equipment
- Maintenance
- Relief systems
- Berms/Sumps
- Chemical Sewers
- Waste Storage

- Product storage

3.1.1 Chemical Information Used in Implementation of Other Elements

PHAs – Inadvertent mixing is a potential cause of parametric deviations resulting in a process safety consequence of interest, requiring safeguard evaluation such as process design to minimize interactions along with evaluation of human error prevention, and detection and mitigation safeguards of inadvertent mixing. The continuous HAZOP deviation "high concentration of contaminants" typically identifies any incompatible chemicals and typical causes are equipment failures centered around the activities listed above including human errors with equipment operation (skipping steps and performing steps incorrectly). If inadvertent mixing data is missing or outdated, then these scenarios will be missed during the PHA. Incompatibilities may also be identified during parametric deviations such as "high temperature" and "high pressure" as incompatible materials may generate exothermic reactions and increased pressure.

Chemical information is also necessary during PHAs to estimate the severity and type of the consequence based on physical properties, toxicity and flammability.

SOPs and Training - Steps to avoid inadvertent mixing, if a concern, are incorporated into SOPs as warning statements and these steps are evaluated during the PHA of non-routine modes of operation for continuous processes and for batch procedures, to evaluate if steps skipped or performed incorrectly could possible lead to inadvertent mixing with process safety consequences. If so, then steps requiring human intervention are needed and information on how to avoid inadvertent mixing. During the PHA, evaluating how the operator would know they have made a mistake (such as how quickly the process responds to such errors and how the process is instrumented/alarmed to alert the operator of the deviation such as high temperature, high pressure, high level, etc.) is critical and this information must be incorporated into useable information in the SOP and in training guides.

MOCs - Chemical properties and inadvertent mixing information are used in the evaluation of MOCs from a technical basis and a process safety basis to ensure changes do not introduce incompatibility issues (for new or existing chemicals) with the existing or new process equipment, and that equipment configuration changes and SOP changes do not introduce inadvertent mixing potential. Materials of construction and other equipment design information must be reviewed along with the process operating conditions and chemical properties to ensure the process is designed for the interactions between chemicals, operating conditions and equipment specifications.

Mechanical Integrity – As mentioned under MOCs in the previous paragraph, a thorough MOC review will include understanding the impact of any chemical related changes, operating conditions and compatibilities with existing or proposed equipment. Of course, as part of the MOC process the impact of the change on other PSM management systems are identified. For

Mechanical Integrity (MI), evaluating whether increased inspections and testing or different testing protocols should be established based on anticipated corrosion rates, for example, and other mechanical issues that a new chemical or a new chemical in conjunction with process operating conditions can have on equipment must occur, and how the inspection testing and preventive maintenance will be impacted.

Safe Work Practices/Contractors – Contractors perform a variety of maintenance tasks on equipment containing highly hazardous chemicals. Chemical information is used when assessing the hazards associated with maintenance work such as performing Job Safety Analyses, completing work permits such as hot work, confined space entry and lock out/tag out. Maintenance procedures also contain information on chemicals.

Incident Investigations – With incident investigations all aspects of the work being performed when the accident occurred are analyzed. With many process safety incidents highly hazardous chemicals are released resulting in employee exposure, injuries and death, and equipment damage and environmental damage. Chemical information is one part of PSI that is needed to help determine the sequence of events and the potential impacts (in cases of near misses).

Emergency Planning and Response – Emergency responders use chemical information to determine PPE requirements, treatments for employee exposures who may be injured by the chemical release and for how to mitigate and clean up the release.

Audits – Auditing includes analyzing every element of PSM to ensure information and requirements for regulatory, industry and corporate standards are included in the management system governing the element. Auditing Chemical Information includes reviewing documents defining the chemical information such as summary of properties and hazards, SDS availability and hazards of inadvertent mixing documents.

Trade Secrets – Trade secrets can contain chemical information in the descriptions of the process technology, process chemistry descriptions and volumes.

3.2 Technology Information

Technology Information comprises the topics listed below:

Block Flow Diagram – Shows major process equipment, the inlet and outlets of process materials and how they interconnect throughout the process equipment. More complex diagrams, often called Process Flow Diagrams also show process operating conditions.

Process Chemistry – Defines raw materials, intermediates, products, byproducts, any purge gases, chemical equations, exothermic and endothermic conditions.

Maximum intended inventory – Is the maximum amount of PSM chemicals inventoried in storage tanks, feed tanks, in-process equipment and product storage vessels. This quantity is typically expressed in a usable, relative unit (such as pounds, kilograms, or volumes as opposed to a percentage of tank/vessel volume).

Safe Upper and Lower Process Parameter Limits – In PHAs the consequences of deviating from normal operating conditions is documented. The safe upper and lower limits are derived from the equipment design and from process chemistry. Safety systems set another tier of safety limits, establishing alarm set points and safety shutdown set points to keep the process from exceeding the equipment design limits. This information is related to safety systems, including relied system design and design basis, process chemistry and equipment design information.

Evaluation of Consequences of Deviations (CODs) – COD is a summary of quantities and materials released under specific conditions and the resulting impact of the release (health impact on employees and the size of the release (limited to release point area, impacts plant site, or off-site consequences). This information is documented in the PHA corresponding to specific process deviations, causes and safeguards. As part of the PHA, the team identifies how the consequence is detected (detection safeguards) and identifies the administrative and engineering controls to mitigate the potential severity of the consequence.

3.2.1 Technology Information Used in Implementation of Other Elements

PHAs – The PHA team collectively uses chemical, technology and equipment information to identify and evaluate the process hazards. Technology information and the other parts of PSI are needed to identify process deviations, causes, consequences and safeguards. This piece of PSI is dependent on the PHA process and other PSI necessary to conduct the PHA and specific to CODs, equipment design information, process chemistry, maximum intended inventory, operating conditions, safe operating limits, chemical hazards, and safeguards identifiable within safety systems documentation (cause and effect diagrams, firefighting equipment, emergency deluge systems, scrubbers), and P&IDs that show all equipment, instrumentation and the interconnectivity.

SOPs – SOPs contain all Technology Information to some degree. Chemistry Information, chemical amounts, safe operating limits and the consequences of exceeding those limits are collectively incorporated into the action steps to complete the process operations. Warning statements about CODs (exceeding safe operating limits), and additional information concerning safety systems (alarms, operator responses and safety shutdown systems) are important process safety information housed in SOPs, often in the troubleshooting guide section.

Training – Operators' understanding of process parameters, their limits, causes, resulting consequences, equipment design, operator response and automatic responses (safety shutdown systems) is imperative to an effective risk management system.

MOC – Many process changes involve process chemistry, chemical inventories, the resulting safe operating limit changes and the consequences of deviations. The MOC process evaluates these aspects of the process to ensure engineering and administrative controls are adequate or added and all aspects of other PSM elements are updated as a result of the change.

Mechanical Integrity - To establish inspection testing and preventive maintenance program to maintain the ongoing integrity of the equipment, the equipment design information is needed –

design limits, materials of construction; process chemistry and process operating conditions and the design and function of instrumentation and controls and shutdown systems.

Safe Work Practices / Contractors – Applicable Technology information is part of the contractor training program and in applicable procedures they may use on site to perform their work.

Incident Investigation – Incident investigators piece together the sequence of events that actually occurred. The team uses defined PSI to understand how the process deviated from normal to identify potential equipment failures and human errors, that if eliminated, would have stopped the chain of events. Understanding information from PFDs, process chemistry and safe operating limits is necessary for building the sequence.

Emergency Planning and Response - Emergency planners use all of the Technology Information in addition to PHA scenarios (which are developed in part from Technology Information) to develop emergency response scenarios – how to respond based on the chemical released under specific operating conditions and the amounts released. **Evaluation of Consequences of Deviations** information is needed to determine PPE requirements for workers in the area, necessity for escape PPE such as 5-minute escape respirators, shelter in place procedures, evacuation procedures for the plant and/or community and for emergency response procedures – how to respond and mitigate the release.

Auditing – Auditing includes analyzing every element of PSM to ensure information and requirements for regulatory, industry and corporate standards are included in the management system governing the element. Auditing Technology Information includes reviewing process chemistry documents, safe operating limit tables, CODs (often presented together due to their relationship), and chemical information including auditing numerous other PSM elements to ensure consistency, such as reviewing SOPs, where the PSI listed above is required content. Some companies, in their PSI manuals, point to the SOPs for this PSI to eliminate outdated discrepancies between copies of the same information located in two places. PSI auditors look for consistencies and inconsistencies across PSM implementation documents that share the same information. Skilled auditor knows the interrelationships between the PSM elements and uses that information for a thorough audit.

Trade Secrets – Trade secrets can contain Technology Information - the chemicals and volumes, process flows (PFDs), chemistry, and operating conditions.

3.3 Equipment Information

Equipment Information – Equipment information exists in specification sheets, U1As, data sheets for pumps, tanks, pressure vessels, instrumentation and includes materials of construction, thickness of materials, dimensions, volumes, nozzles and other connections, pressure and temperature ratings; for piping information such as schedule, diameter, connection information (flange information), materials of construction, piping support information and for pumps information such as pump type (positive displacement, centrifugal), inlet and outlet size, head pressure, etc. Instrumentation specifications are also part of equipment information.

P&IDs - A piping and instrumentation diagram (P&ID) shows the interconnection of process equipment and the instrumentation used to control the process along with the following information: [2]

- Process piping, sizes and identification, including: pipe classes or piping line numbers, flow directions, interconnections references, permanent start-up, flush and bypass lines
- Mechanical equipment and process control instrumentation and designation (names, numbers, unique tag identifiers), including: valves and their identifications (e.g. isolation, shutoff, relief and safety valves), control inputs and outputs (sensors and final elements, interlocks), miscellaneous - vents, drains, flanges, special fittings, sampling lines, reducers and increasers
- Interfaces for class changes
- Computer control system
- Identification of components and subsystems delivered by others [3]

Electrical Classification - The National Electrical Code (NEC) defines hazardous area classification as those areas "where fire or explosion hazards may exist due to flammable gases or vapors, flammable liquids, combustible dust, or ignitable fibers or flyings." Electrical classification defines the design of electrical and other equipment necessary to prevent ignition sources where flammables or combustibles are present. [4]

Relief System Design and Basis - Describes the pressure relief system whose purpose is to control or relieve excess pressure. Relieving devices include rupture discs, relief valves, conservation vents, and others. The basis for the excess pressure is also documented such as the following scenarios - runaway reaction, blocked flow, external fire, and loss of power. Documents include the calculations defining set relieving pressure and capacity of the device, and orifice information. Design information also includes inlet size and outlet size of the relieving device, the inlet and outlet piping, whether the relief is to an open or closed system (involving knock out vessels and flares).

Ventilation System Design – Ventilation system information refers to ventilation related to process safety such as for buildings containing process equipment, fan sizes, exchange turnover rates, information on systems activated to maintain electrical classification in the event of a release, and also information for positive pressure control panels, etc. Other ventilation equipment includes shelter-in-place buildings and ventilation systems for toxic chemical storage and processing.

Codes and Standards – Refers to the design and construction industry codes and standards used for the process equipment and systems. These are found on equipment specification and data sheets, in design calculations and equipment fabrication drawings.

Safety Systems – Safety systems include for example, location of eye washes, safety showers, emergency stops, emergency scrubbers, deluge systems, emergency flares, relief devices and piping, process parameter alarms, interlocks and emergency shutdown systems, detection systems, and emergency isolation valves.

Trade Secrets - Trade secrets can contain Equipment Information - the equipment design, P&IDs, safety systems and material and energy balances.

4 Other Key Relationships Between the Elements and How to Use Your Understanding of the PSM Element Interrelationships to Strengthen PSM

As described in Section 2, MOC and PSI have the strongest interrelationships with all the elements. Section 3 provided examples of how specific PSI is used in implementing all PSM elements and within that discussion the role of the MOC system was highlighted:

- Up-to-date PSI is needed to evaluate technical basis and evaluate process safety risk for proposed process changes
- MOC process must ensure PSI gets updated as part of change management

The summary below highlights other interrelationships between elements beyond MOC and PSI.

Employee Participation – It is vital that operators and mechanics participate in the development and implementation of the PSM elements. This results in the staff taking ownership of the elements and provides a better understanding of their role in implementing each element.

PSI – Understanding that PSI is used to some degree in the implementation of all elements, provides the framework for ensuring PSI is managed properly. Incomplete and outdated PSI clearly results in weak overall PSM. MOC is the mechanism to ensure PSI is kept updated along with identifying the other elements of PSM that must be updated as a result of PSI changes. Failure to identify process changes will subsequently result in outdated PSI and impact the evaluation of process safety risks during the MOC review and approval process. Some PSI is built on other PSI. Chemistry and equipment information are needed for material and energy balances and this information is used to create P&IDs, establish SOLs and define safety systems, some of which are summarized on Cause and Effect Diagrams. **Figures 3 and 4** show how PSI impacts other parts of PSI, PHAs, MI, SOPs and Training.

PHAs – PHAs should not be performed unless complete, updated/current **PSI** is available. Expending the resources to conduct a PHA on a process that does not exist occurs when P&IDs are not up to date, safety system information, and material and energy balance information are unavailable or inaccurate. This results in scenarios that are poorly defined and risks that are under estimated, for example, taking credit for safeguards that do not exist or that are not designed to protect against the applicable hazard scenario. Using the results from PHAs is necessary to meet required **SOP** content requirements (consequences of deviations, supplements troubleshooting guide information on how to detect, avoid and correct process deviations) and to develop **Training** guides. Furthermore, CODs contain consequences identified and evaluated during the PHA and any updates to alarms and interlocks (human and system responses). Entries from the Cause and Safeguard columns are extracted for troubleshooting guide information. **Mechanical Integrity** program uses PHA results to prioritize maintenance activities, based on the importance of certain safeguards to detect, prevent, intervene and mitigate process safety scenarios. During revalidations **MOCs** and **incident reports** are reviewed. If these are poorly evaluated and/or poorly documented then the PHA revalidation team must spend more time to discuss each and the quality of these reviews will be negatively impacted given lapses in time since their occurrence and changes in staffing, meaning vital information to process safety may be forever lost.

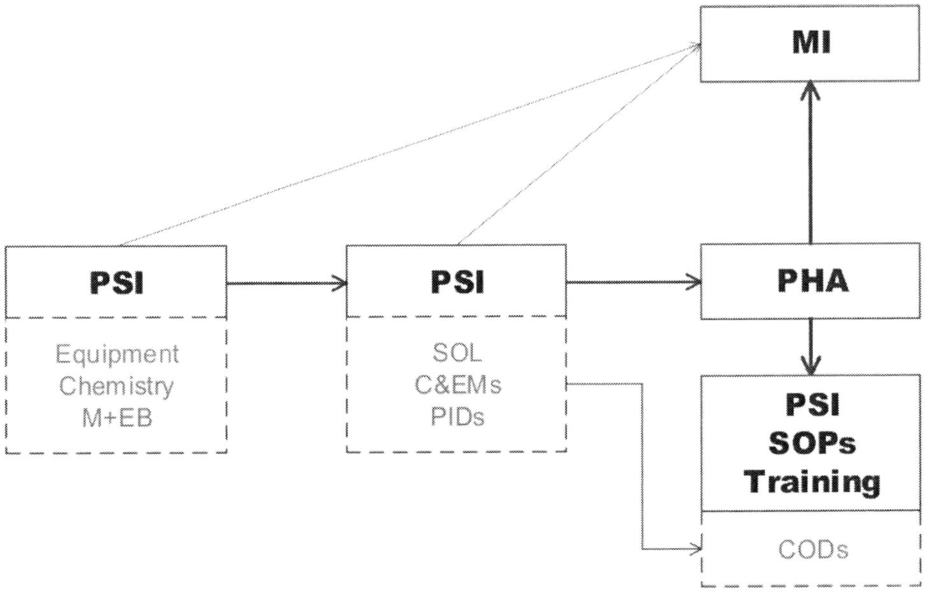

Figure 3. Relationship of Specific PSI to PHAs, MI, SOP and Training

SOPs and Training – As shown in **Figure 4,** SOPs rely on PSI and PHA results to be complete and accurate. SOPs and PSI are the building blocks for **Training.** Weak implementation of PSI, PHA and SOPs ultimately impact Training.

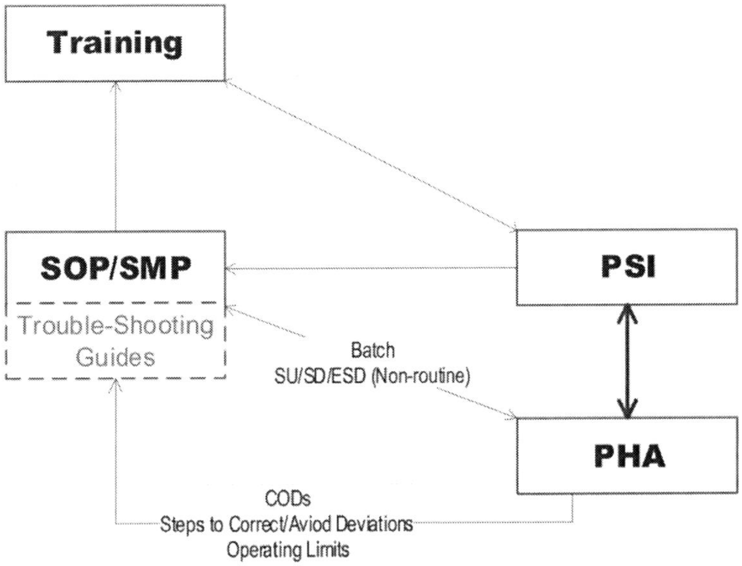

Figure 4. Relationships between PSI, PHA, SOPs and Training

Mechanical Integrity – Mechanical Integrity program exists to protect the design specifications and integrity of the process equipment. Up-to-date, well documented design information on equipment (part of **PSI**) is necessary for evaluating process changes, necessary for the hazard identification and analysis – how equipment can fail, to establish safe operating limits. If equipment is not maintained, then its operability and safety design can be compromised. Based on the equipment type, materials of construction, chemical service and operating conditions, inspections and testing and preventive maintenance plans (ITPMs) are developed. The other part to MI is quality assurance with respect to installing equipment that meets design specifications, and includes equipment designed for the service intended, procuring the correct equipment, fabrication of equipment, receiving the correct equipment, storage and retrieval systems, installation (instructions for maintenance techs, adequate procedures and training, tools and equipment), and operating within safe operating limits. ITPMs' test results and reporting near misses and incidents can identify mechanical integrity deficiencies. Deficiencies must be corrected or process changes (as a result of the deficiencies or continuing to operate with the deficiency) must be evaluated and approved. **Figure 5** summarizes the MI role.

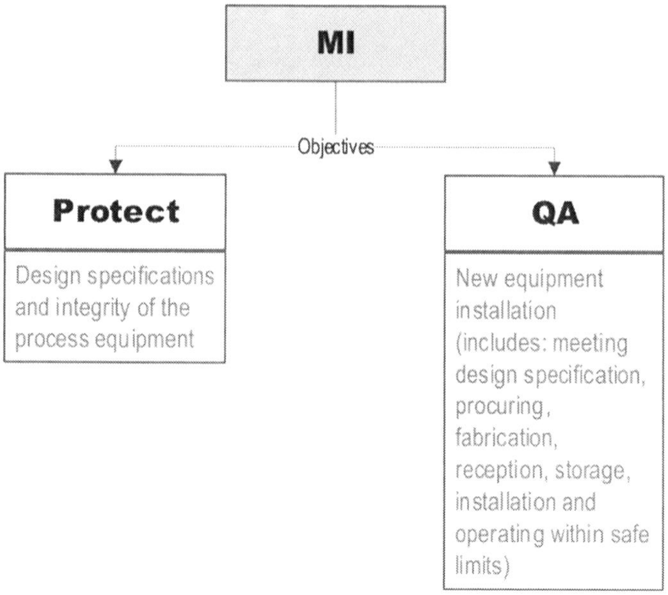

Figure 5. Role of Mechanical Integrity

5 Moving Forward

Figure 6 shows the overall relationships between the elements, and also shows some elements missing from OSHA's PSM program that are not discussed in this paper (leadership and accountability, key performance indicators and human factor optimization). Ensuring PSI is complete and accurate before start up, during the PHAs, and for use in MOC evaluations, and that subsequent changes to PSI and other PSM elements are updated to reflect process changes, are key to an effective PSM program. Understanding the interrelationships between the elements helps identify all information and documents impacted by a change (defining PSI and other implementation documents such as MI plans, SOPs, Training, Emergency response…) in the MOC process. Otherwise, the day-to-day implementation of all the elements will be weakened as all depend on PSI as their foundation. Some elements also strongly depend on other elements to be effective. Poor implementation of PSI, MOC and other inputs results in some degree of weakness in every element, leading to ineffective process safety risk management.

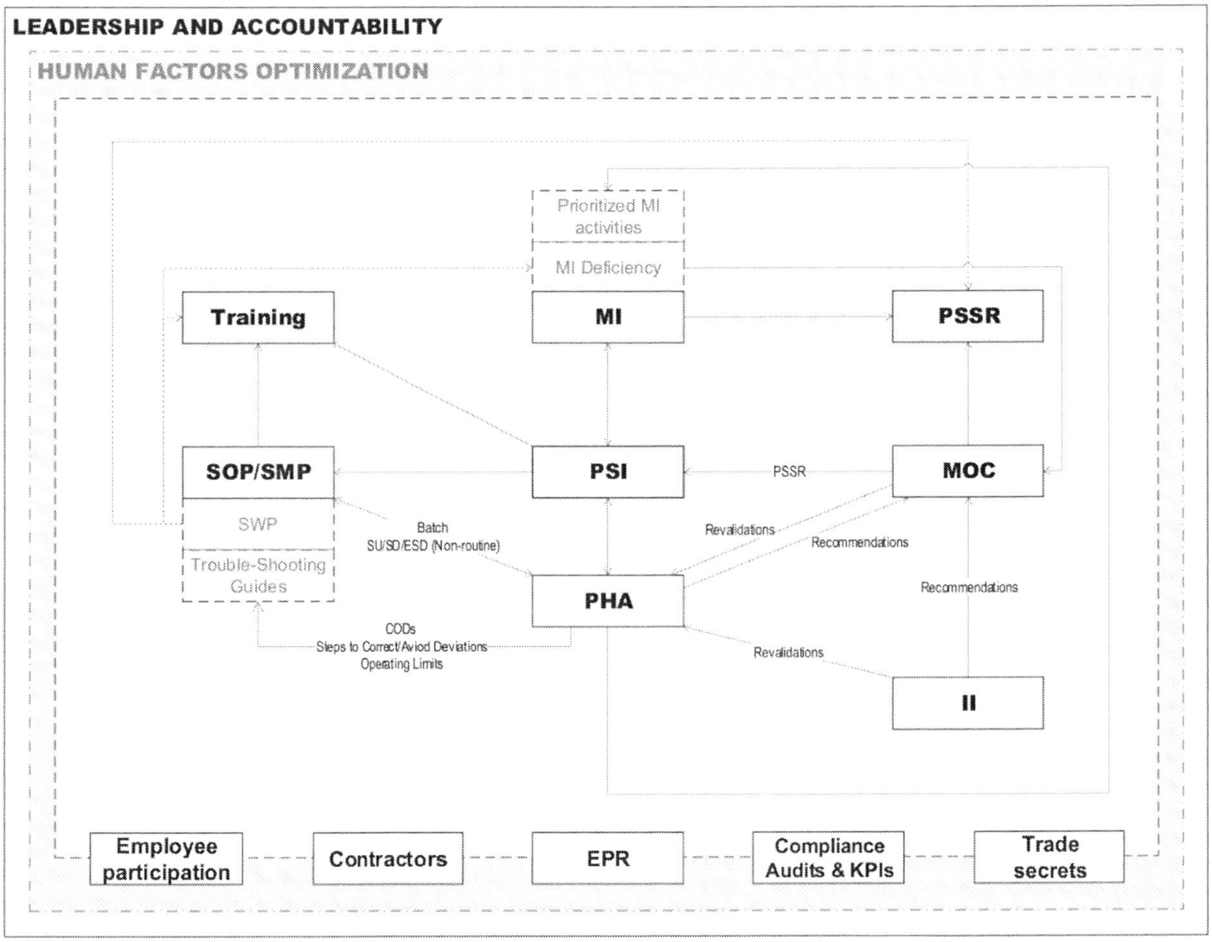

Figure 6. Interrelationships of PSM Elements

6 References

[1] R. Tew, T. Waugh, M. LaPlante and D. Pastirik, *Keys to Avoiding Shortcomings in Your MOC System*, New Orleans, Louisiana: 10th Global Congress on Process Safety, 2014.

[2] Institute of Instrumentation and Control

[3] Piping and Instrument Diagram, Available at

https://en.wikipedia.org/wiki/Piping_and_instrumentation_diagram

[4] CCPS. *Guidelines for Auditing Process Safety Management Systems.* Center for Chemical Process Safety, American Institute of Chemical Engineers, New York, NY, 2011.

512589 Dust Explosion Propagation and Isolation

Abstract:

Industrial processes handling combustible dusts usually comprise numerous pieces of equipment, such as grinders, mixers, dryers, classifiers, dust collectors, and storage hoppers, interconnected by a more or less complex network of pipes and conveying systems.

If a combustible dust cloud is accidentally ignited in a process enclosure, large amounts of heat and high pressures are generated in milliseconds. The resulting dust deflagration may, or may not compromise its structural integrity depending if the vessel is adequately protected or not. Nevertheless, the deflagration pressure will push the fireball through any existing openings, including process interconnections. The flame front will stretch and speed up due to increased turbulence, accentuating the severity of the initial event. Ultimately, a more violent secondary deflagration can take place when the flame reaches the next process enclosure, such that existing protection measures may well fail under these aggravated conditions.

This presentation will review the basic mechanisms of dust explosion propagation, unravel common myths about dust explosion propagation in process pipes and conveying systems, and provide detailed information on how to protect a process against dust explosion propagation.

Analysis of Dust Concentration Measurements Taken during the Filling Process of a Silo

L.L. Lulbadda Waduge[a] *S. Zigan*[a], *L.E. Stone*[c], *A. Belaidi*[b], *P. García-Triñanes*[a]

[a] *The Wolfson Centre for Bulk Solids Handling Technology, University of Greenwich, Central Avenue, Chatham Maritime, Kent ME4 4TB, United Kingdom*

[b] *Department of Mechanical Manufacturing and Design Engineering, University of Greenwich, Central Avenue, Chatham Maritime, Kent ME4 4TB, United Kingdom*

[c] *Faculty of Engineering and Information Sciences, University of Wollongong, Northfields Avenue, Wollongong, NSW 2522, Australia*

Abstract

The consequence of the release of fine particles smaller than 500 microns than can form dust clouds in the handling and storing equipment can be severe. These dust clouds present potential dust explosion hazards because the airborne dust can occur in concentrations within the dust explosion limits of the material, during the filling process of storage silos. In terms of safety, a considerable number of factory workers have lost their lives throughout the years due to severe accidents such as; 2003 explosion at West Pharmaceutical Services in Kinston, North Carolina (6 casualties), the 2008 explosion at Imperial Sugar Plant in Savannah, Georgia (killing 14), and the 2009 explosion in a coal silo at WE Industries power plant in Oak Creek, Wisconsin (injuring 7) (Davis et al., 2011). Hence, dust explosions become a significant concern of industries (e.g. power, chemical, pharmaceutical or food industries) handling organic products, grain and flour, metal products or pulverised coal to mention just a few. Preventing dust explosions and the damage of plant infrastructures require a profound understanding of the particle/ air dynamics in the dust cloud circulating in the storage silo.

This research aims to construct a model which predicts the spatial distribution of dust concentrations during the filling process of a silo. Then passive explosions protection systems such as panels on the silo roof ventilation can be effectively designed to prevent explosions. The dust distribution was measured with an optical method developed by (Lulbadda Waduge et al., 2016). The optical method employed in a small scale experimental rig which is essentially a fluidised bed with fine materials and a single camera together with a laser source was utilised to capture the dust concentration in different areas when the dust is airborne. Then images of repeated runs were taken and post-processed to compare different dust concentration levels in each set.

The hydrodynamics of the process supposed to behave the same at each repeated run of the experiment as experiments were identical to each other regarding materials and the scale. Having the same ranges of concentrations distribution at each repeated test were obtained.

Keywords: Laser; Optical method; Dust explosions; CFD; Experimental; Storage silos; Particle dynamics

Introduction

Predicting the airborne dust concentration in confinement is essential to prevent dust explosions. Most of the models that predict airborne dust concentrations are based on numerical simulations. However, numerical simulations are based on a set of assumptions, and therefore numerical simulations must be validated using empirical data. Simulations platforms such as Fluent CFD provides results that cannot be compared directly with the empirical result as the number of particles simulates with numerical simulations are considerably lower than the empirical data. This paper focused to provide a solution in comparing the simulation results with empirical data for the purpose of validating the simulations.

According to Oxford dictionary of English, the term concentration is defined as; "The relative amount of a particular substance contained within a solution or mixture or in a particular volume of space"[1]. As an example, for application of the term concentration is found in chemistry. In chemistry, the amount of Avogadro number of Hydrochloric acid molecules in a litre of water is known as the concentration of Hydrochloric acid in water and with the standard unit of moles per litre (mol/m^3)[2]. Furthermore, the density of Hydrochloric acid can be calculated by multiplying its molar mass (molar mass of Hydrochloric acid is 36.4600 g/mol) and hence, the term density can be derived as "The quantity of mass per unit volume of a substance"[1].

However, the standard unit for measuring dust concentration is used as grams per cubic meter [3] which is the unit of density. Using the unit of density to measure dust concentration seems to be beneficial for the calculation done for the application of dust concentration such as calculating the limitations according to environmental regulations [3]. Although it is beneficial for calculations, it is essential to keep the constancy with the units that used in other fields to measure concentration for the purpose of avoiding confusion. Hence, in this paper the unit for the dust concentration is used as a number of particles per unit volume and the unit of dust density is used as grams per unit volume.

Methodology

The concentration of airborne fines/dust can be calculated as the number of particles per unit volume. However, in reality, airborne dust is not distributed uniformly in a considerably larger volume, and therefore dust concentration can only be calculated for a smaller volume. Calculating the concentrations of dust particles in a small volume is somewhat questionable since the applications of dust are always related to larger confinement. In order to overcome such a problem, the larger volume that contained dust needs to be broken into smaller hypothetical volumes which are also known as cells in a mesh. Figure 1 illustrates an example for a mesh extracted from a cylindrical silo.

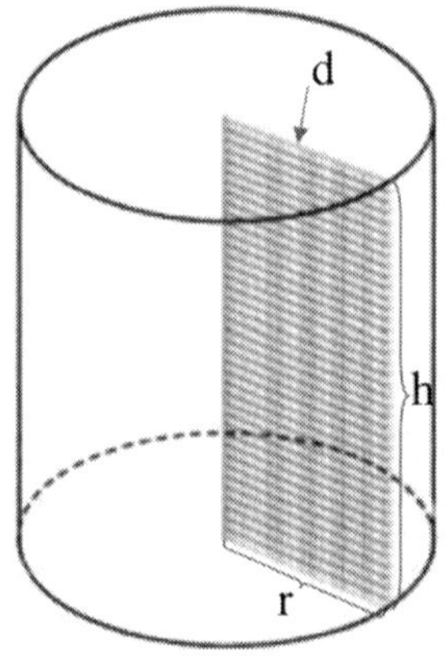

Figure 1: A 3D mesh with cuboid cells. where; 'h' is the height of cylinder (confinement), 'r' is the radius of the cylinder (width of the confinement), 'd' is the depth of the mesh (theoretically this should be equal to size of the largest airborne particle)

Use of a mesh with cells is commonly used in numerical simulations such as Fluent. Extracted simulation results related to a mesh from Fluent includes the locations of each cell and the dust concentration values corresponding to each cell. Typically, this dust concentration values are calculated by counting the number of particles which are also known as 'packets' that are found in each cell.

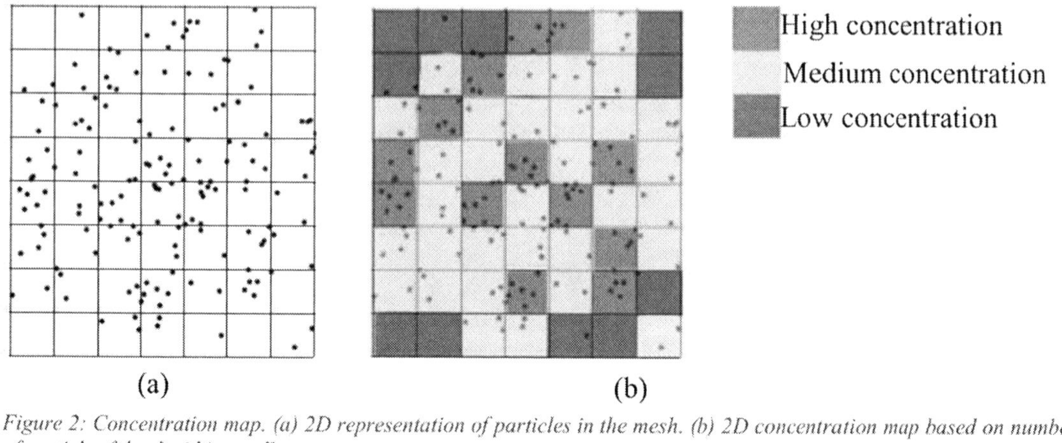

Figure 2: Concentration map. (a) 2D representation of particles in the mesh. (b) 2D concentration map based on number of particles folund within a cell.

The practical use of mapping dust concentration distribution is to understand zones. However, the dust concentration map shown in Figure 2 does not illustrate distinct zones of dust concentrations. Instead, the map shows a scattered distribution of dust concentration.

Processing the dust concentration map shown in Figure 2 through an averaging technique resulted in a map with distinct concentration zones. As the averaging technique, a widow ran through each cell in the concentration map by averaging the number of particles found nearby cells.

n_1	n_2	n_3
n_4	n_5	n_6
n_7	n_8	n_9

Figure 3: Example for an averaging filter. Where: 'n_1 to n_9 are the number of particles found in each cell and 'n_5' is the number of particles on the middle cell which is the filter is based on

Equation 1 shows the calculation to obtain the average concentration of the cell where the filter window is based on. The middle cell of the filter window is aligned with a cell in the map, and the average concertation is calculated with the nearby cells as shown in Figure 2 and Equation 1.

	Average number of particle concertation in the middle cell	=	$\frac{n_1 + n_2 + n_3 + n_4 + n_5 + n_6 + n_7 + n_8 + n_9}{9}$	Equation 1

As such, the middle cell was placed on cell at the top left corner of the map and shifted left up to the cell at top left corner and then continued to the next row. The filter window will exit the colour map from the cell at the bottom right corner after going through each cell in the map.

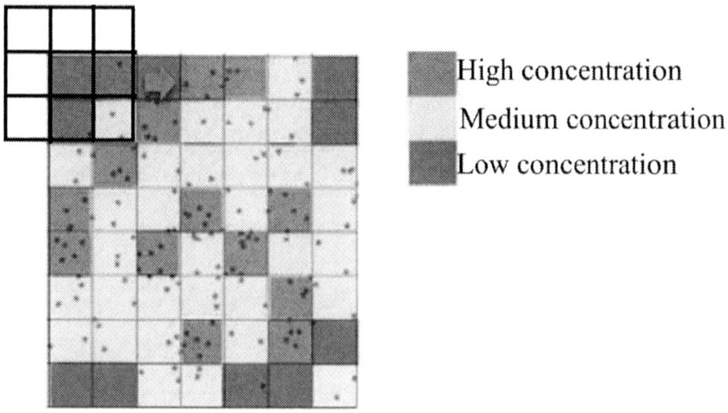

Figure 4: The averaging window is shifted in to each cell starting from top left corner of the concentration map.

The calculated value for the position where the window is positioned in Figure 4 is calculated as 0.55 particles per unit cell ($\because \left(\frac{0+0+0+0+0+1+0+2+2}{9}\right) = 0.5556$). Then the window was shifted to right side cell from the current cell for next calculation.

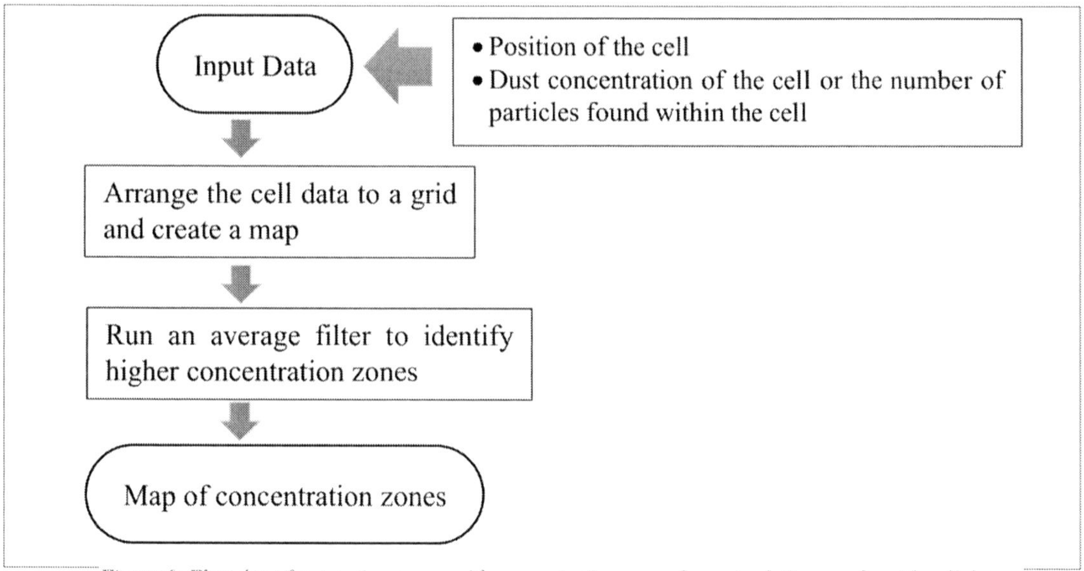

Figure 6: Flowchart for creating a map with concentration zones from simulation results with cell data

Figure 6 shows the overall process of creating a map with simulation results and Figure 5 shows an example for a final map with concentration zones.

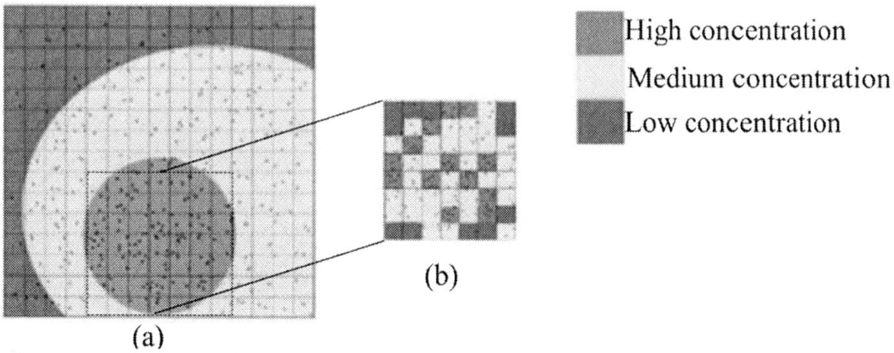

Figure 5: Concentration maps. (a) Final map which shows the concentration zones. (b) An area of the final map before running the average filter.

Furthermore, the practical results obtained from an optical method [4] contains a digital image with particles or an image with dust concentrations. The digital image is a set of pixels arranged in the same way as a cell structure that discussed previously and therefore the same method can be applied in processing the practical data to obtain a map with concentration zones.

Results

Simulation results contained Cartesian locations of airborne particles was obtained as the first step and filtered the data to extract the locations belongs to particles in one plane. Particles were mapped into a mesh as the second step.

Figure 7: Extracted data from a simulation which then mapped in to a particles in a 2d plane.

The mapped mesh in Figure 7 was refined by reducing the mesh size and labelled each cell with the number of particles found in each cell as Figure 8.

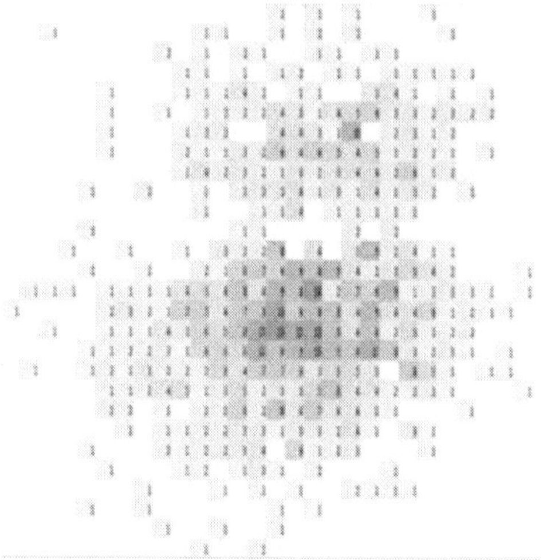

Figure 8: Particles in a refined mesh and each is numbered with the number of particles found.

An average filter was applied to the particle concentration map in Figure 8, and the resultant image is shown in Figure 9(b). The concentration map in Figure 9 is showing a distinguishable map of different concentration zones.

(a) (b)

Figure 9: Colour map processing. (a) The colour map with concentration on each cell. (b) The final colour map after application of an average filter.

Conclusion

The result from a numerical simulation can be converted into a contour map which represents concentration zones. The contour map created from numerical results is comparable with the contour map created from an experiment which used an optical method [4]; hence the result from a numerical simulation can be validated with empirical data.

Reference

[1] A. Stevenson, *Oxford dictionary of English*, 3rd ed. Oxford University Press, 2010.

[2] E. R. Cohen, "IUPAC: Quanities, Units and Symbols in Physical Chemistry," in *Green Book*, M. Quack, J. Stohner, H. L. Strauss, M. Takami, A. J. Thor, E. R. Cohen, T. Cvitas, J. G. Frey, B. Holström, K. Kuchitsu, R. Marquardt, I. Mills, and F. Pavese, Eds. Cambridge: Royal Society of Chemistry, 2008, p. 81.

[3] D. Carruthers, D. Carslaw, R. Colvile, P. Richard Derwent OBE, S. Dorling, P. Bernard Fisher, P. Roy Harrison OBE, M. Heal, P. Duncan Laxen, S. Lindley, I. McCrae, J. Stedman, P. Mike Ashmore, M. Jenkin, P. Woods, S. Moorcroft, T. Murrells, P. Quincey, and K. Stevenson, "Particulate Matter in the United Kingdom," *AQEG Part. Matter UK Summ. Defra, London.*, p. Chapter 5, 2005.

[4] L. L. Lulbadda Waduge, S. Zigan, L. E. Stone, A. Belaidi, and P. García-Triñanes, "Predicting concentrations of fine particles in enclosed vessels using a camera based system and CFD simulations," *Process Saf. Environ. Prot.*, 2016.

513734 The Design and Operation of Multi-Point Dust Collection Systems

Abstract:

It is very common in particulate-producing plants to have a vacuum dust collection system to collect fugitive dust from all manner of unit operations such as container filling, container dumping, size reduction, transfer points, etc. The design methods have been well established in the *Industrial Ventilation* manual published by the American Conference of Governmental Industrial Hygienists [1]. Nevertheless, in practice these systems are not always properly designed or maintained, thereby not providing the dust collection and subsequent worker protection for which they are intended. Following the guidance of the ACGIH manual, we will first outline the philosophy and methodologies to design a multi-point vacuum pick up system. At first glance, this would appear to be an easy task, however when one considers the multiple requirements (saltation velocity, minimum velocity to meet NFPA guidelines, required volumetric flow rate of air for dust collection at each pickup point, etc.), the problem is considerably more difficult. In practice, many systems will either not operate as expected or performance will degrade over time. We will highlight some of the pitfalls in the design and operation of these systems, such as intermittent operation of branches of a system, unintended extension of the system by adding longer lengths of flexible tubing, etc. This highlights the need for the design engineer and the system operator to work together to conceptualize and design a system which provides the level of dust collection needed and can be easily maintained for the life of the process plant.

[1] Industrial Ventilation: A Manual of Recommended Practice, 29th Ed, American Conference of Governmental Industrial Hygienists (ACGIH), 2016.

515912 20 years of Chemical Safety and the Changing Environment

Abstract:

The latter phrase can be changed to "Changing Safety Environment". It is meant to be a dual meaning about the developments in chemical safety generally and in extreme weather issues to highlight Arkema.

18 AIChE Spring Meeting
14TH GLOBAL CONGRESS ON PROCESS SAFETY
ORLANDO FLORIDA 2018

Bonded, Grounded, and Burned to a Crisp: Electrostatic Ignition of Flammable Gases

Brenton L. Cox
Exponent, Inc.,
4580 Weaver Parkway, Warrenville, IL 60555
bcox@exponent.com

Justin A. Bishop, Russell A. Ogle, Nicholas A. Traina, Jay R. Prigmore
Exponent, Inc.
4580 Weaver Parkway, Warrenville, IL 60555
jbishop@exponent.com, rogle@exponent.com, ntraina@exponent.com, jprigmore@exponent.com

Prepared for Presentation at
American Institute of Chemical Engineers
2018 Spring Meeting and 14th Global Congress on Process Safety
Orlando, Florida
April 22 – 25, 2018

Z140010 - 5751

Bonded, Grounded, and Burned to a Crisp: Electrostatic Ignition of Flammable Gases

Brenton L. Cox
Exponent, Inc.,
4580 Weaver Parkway, Warrenville, IL 60555
bcox@exponent.com

Justin A. Bishop, Russell A. Ogle, Nicholas A. Traina, Jay R. Prigmore
Exponent, Inc.

Keywords: Static Electricity, Electrostatic Discharge, Fire, Natural Gas

Abstract

Ignition by electrostatic discharge of clean flammable gas, such as natural gas, being released to the atmosphere during blowdown operations is unlikely. However, the presence of foreign material in the form of solid particles or liquid droplets changes the risk level significantly. The discharge of a flammable gas stream with a cloud of foreign material poses a potential electrostatic hazard, even if the piping system and associated equipment is otherwise bonded and grounded. Foreign material may be solid debris originally present in the system and subsequently suspended and transported by a gas stream. Alternatively, the foreign material may be aerosolized liquid, entrained in the gas stream due to impingement on pooled liquid or condensation from the vapor phase. In either case, the foreign material may be capable of accumulating sufficient electrostatic potential to ignite the flammable gas if an oxidant is present.

There have been several documented incidents where ignition has occurred during a flammable gas blowdown. In some instances, gas lines were being cleaned, while in others, the blowdown was completed as a safety measure for another process. This paper will focus on case studies where a stream of flammable gas was ignited during a blowdown and an obvious ignition source (other than electrostatic discharge) was not present, but foreign material was observed before, during, or after the incident. The potential influence of particle size and material properties is presented. This paper discusses guidance offered by the available literature on the control of electrostatic ignition hazards in flammable gas streams.

1 Introduction

Most process safety professionals will be familiar with the notion of bonding and grounding. Both bonding and grounding are essential for the control of electrical ignition sources if electrostatic discharges would otherwise be likely between conductive objects in flammable atmospheres. Properly bonded objects will maintain approximately the same electrical potential, and properly grounded objects provide a path to dissipate excess charge buildup to earth. Thus, if all conductive objects are either grounded or bonded to ground, then the potential for electrostatic discharge between those objects is greatly reduced. So, does that mean that a well-specified and properly managed bonding and grounding policy will totally eliminate electrostatic ignition? Not necessarily.

After a fire or explosion has occurred, static electricity often appears in the line-up of usual suspects. Fortunately, in many cases, bonding and grounding can be confirmed after the fact. If the conductive components such as pipes and vessels are indeed bonded to ground, then many investigators will rule out electrostatic ignition as a potential explanation. On the other hand, if damage does not allow for confirmation of bonding and grounding, and no other satisfactory ignition source can be identified, a presumptuous investigator may assume that equipment bonding and/or grounding was inadequate and played a role in ignition. However, there are circumstances in which proper bonding and grounding still do not eliminate the potential for electrostatic discharges capable of igniting flammable atmospheres. The following case study is one such circumstance.

2 Incident Summary

A fire occurred at a newly constructed portion of a facility undergoing commissioning. The event occurred during a test of a natural gas compressor, when a fault occurred in the programmable logic controller (PLC) for the compressor system. This fault resulted in a loss of signal to a fail-open valve that served as the vent valve for the compressor system. The high pressure natural gas vented through a blowdown silencer, as intended by design. However, the natural gas ignited upon exiting the silencer. The remainder of the natural gas contained within the facility was released as part of the emergency response. As a result, the adjacent blowdown stacks also ignited. The fire did not cause any injuries, but it did cause equipment damage.

The ignition event was recorded by two security cameras. Still captures from each video can be seen in Figure 1, showing the frame prior to ignition, and the first two frames recorded after ignition. An opaque plume was visible prior to ignition, and the fire was first observed approximately one second after the plume appeared. The location of the fire—when first observed in the video—was in close proximity to the top of the silencer.

Figure 1. Frames captured from the security camera footage from immediately before ignition (labeled a), and the first two (unique) frames post ignition (labeled b and c, respectively).

3 Investigation Methodology

Several ignition hypotheses were developed and tested by the facility's Root Cause Analysis (RCA) team. The RCA team ruled out all but two potential ignition sources: mechanical impact or electrostatic discharge. Exponent was asked to assist in the evaluation of ignition hypotheses involving these two potential sources.

As part of its investigation, Exponent conducted an inspection of the facility. Exponent also reviewed working papers, laboratory test results, and calculations prepared by the facility RCA team. A literature survey quickly revealed that similar incidents had occurred at other facilities while lines were being cleaned or blown out with high pressure natural gas. In several of these incidents, ignition was attributed to one of the two mechanisms mentioned above: mechanical impact or electrostatic discharge.

The fire hazard associated with high-pressure natural gas blowdown activities has been recognized by the U.S. Chemical Safety Board (CSB) [1] and the National Fire Protection Association (NFPA). [2, Chapter 5] [3] An important lesson learned from the fire and explosion incidents documented by the CSB is that, in most cases, an obvious ignition source was not present. In many of these cases, neither foreign material (particle impact) nor static electricity could be eliminated as potential ignition sources. For example, in the case of the Kleen Energy explosion, the cloud of foreign material evident during a gas blow is a compelling observation.

Photograph from CSB Kleen Energy report

Figure 2. Photograph of foreign material exiting the natural gas pipe during a gas blow that led to an explosion.

4 Engineering Analysis of the Ignition Hypotheses

The analysis of relevant ignition hypotheses is structured around the fire triangle, i.e., a fire can only occur if fuel, oxidizer, and a competent ignition source come together simultaneously. Thus, a viable hypothesis must include each component of the fire triangle.

All conductive components of the system were found to be adequately bonded to ground. Thus, electrostatic discharge between components was eliminated as a potential fire cause. However, after the accident, foreign material was observed inside the silencer and in the pipe elbow that connects directly to the silencer. This foreign material included a name plate and other solid debris. Liquid water was also found in the system. While the water observed post incident was likely the result of suppression activities, liquid water was still considered as a possible contaminant to the gas stream, as will be discussed. These two categories of contaminants: foreign materials and potential presence of water droplets, led to the evaluation of two potential ignition sources: impact by solid debris or electrostatic discharge from entrained material (either solid debris or water droplets).

4.1 Fuel and Blowdown Discharge

The natural gas involved in the incident was released at approximately 1100 psig from the vent valve. The facility's data historian indicated the inlet gas temperature was approximately 78°F. Natural gas traveled through an orifice plate, and then through an underground run of piping until it ultimately exited through the blowdown silencer, third in the array of silencers shown in Figure 3. An opaque plume was visible in the surveillance video of the discharge—similar to the plume seen in the Kleen Energy photograph—suggesting the presence of foreign material (i.e., solid debris and/or entrained liquids). This plume ignited within a second of becoming visible in the video footage.

Figure 3. Photograph from the date of the incident showing the array of silencers post-incident. The fire reportedly initiated at the indicated silencer.

A pipeline flow simulation performed by the RCA team indicated the maximum gas flow during the blowdown was approximately 35,700 SCFM (standard cubic feet per minute). This corresponded to a maximum gas velocity in the pipe of 495 feet per second and an exit velocity out of the silencer of 114 feet per second. The gas discharge out of the silencer was a subsonic flow of a buoyant, turbulent jet.

A routine gas composition analysis performed by the facility confirmed the natural gas quality. The gas analysis indicated that the gas was dry, pipeline quality gas (water vapor content below the detection limit) and the energy content was typical of natural gas (nominally 1000 BTU per standard cubic foot).

The silencer was constructed primarily of steel, with some carbon steel components and some stainless steel components. As stated above, the silencer and all of its conductive components were bonded to ground. All metal components were welded together. However, the silencer did contain glass packing. Additionally, an identification tag was found loose within the silencer.

The silencer is shown in Figure 4. Areas of rust show where paint was burned off of the exterior surface of the silencer stack. Lifting lugs can be seen on the front and rear of the stack, near the top.

Figure 4. Exponent images of the involved silencer.

A sealed bottom "bowl" exists below the diffuser, connected to a plugged drain line. It was observed after the incident that the drain line for the silencer was plugged. A representative sketch of this bowl can be seen in Figure 5. Approximately one-and-a-half inches of rain fell in the ten days prior to the incident. Thus, the bowl likely contained some amount of water at the time of the discharge.

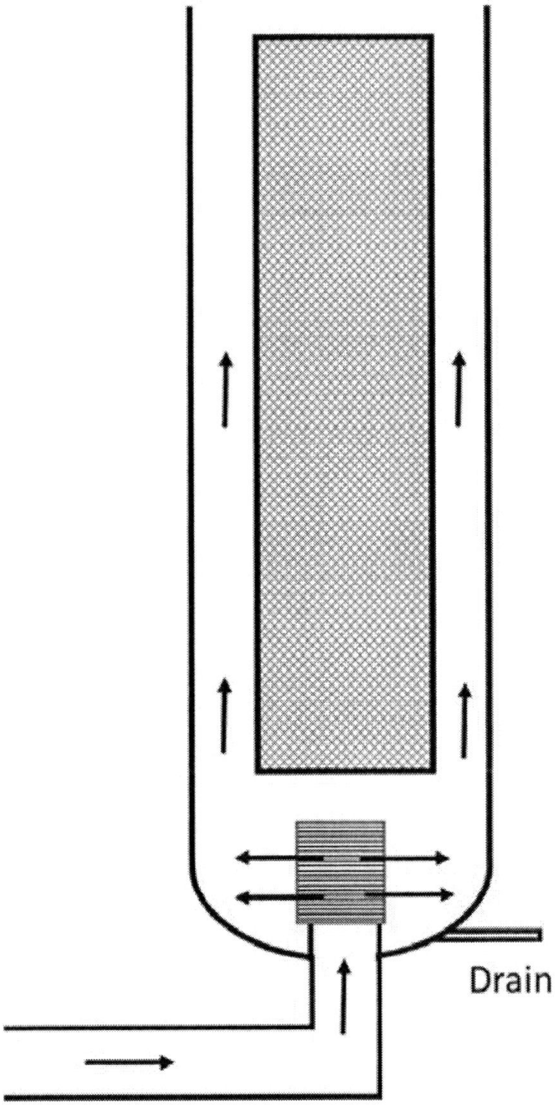

Figure 5. Diagram of the diffuser and "bowl".

4.2 Oxidizer: mixing of the natural gas discharge with the atmosphere

The surveillance video footage shows that the prevailing wind caused the jet to bend away from the vertical axis of the silencer. Because the gas discharge was 100% natural gas, ignition could only occur at the edges of the jet where air was entrained (mixed) with the natural gas discharge. Once ignited, the bending of the jet flame imposed a radiant heat flux onto the ground surface of sufficient magnitude that it ignited the grass.

The ambient temperature at the time of the incident was 68°F. This was slightly warmer than the calculated gas discharge temperature of 61°F (as determined by facility modeling efforts). Due to the low molecular weight of natural gas compared to air (roughly 17 versus 29 gram/mole, respectively), the natural gas was buoyant (less dense than air) and therefore exhibited a tendency to rise. This buoyant behavior inference was borne out in the dispersion behavior documented by the surveillance videos.

4.3 Ignition source: particle impact

Ignition, the initiation of combustion in a flammable gas mixture, requires thermal energy. There are a variety of ways to produce the thermal energy of ignition. The quantity of energy and the temperature of its source vary with the type of energy. Electric spark discharge, static electricity, hot surfaces, small flames, and mechanical sparks are all potential sources of ignition energy. In this blowdown incident, there were no obvious ignition sources active in the vicinity of the natural gas release. Thus, it is necessary to search for ignition sources that would naturally arise during the natural gas blowdown. One such ignition source would be hot particles generated by particle impact. The basic concept for this scenario is that the blowdown flow of natural gas had to create hot particles by causing solid debris to impact the metal surfaces of the piping or silencer. Then, through a combination of fracture and abrasive wear, incandescent (glowing) particles would be created in the impact process and be transported by the natural gas flow through the silencer and out into the natural gas-air cloud forming outside the silencer.

Ignition of a flammable gas mixture requires a deposit of thermal energy large enough to initiate a process called thermal runaway. In a thermal runaway, heat must be generated by combustion faster than it is lost to the environment. If that happens, a flame will propagate from the point of ignition towards the unburnt gas mixture. [4, pp. 132-135] Methane, the principal component of natural gas, is relatively easy to ignite with an electrostatic discharge. [5, pp. 333-361] In contrast, compared to most flammable gases, it is one of the more difficult flammable gases to ignite by a hot particle. [6] Furthermore, the ignition effectiveness of a hot particle depends strongly on the materials of construction of the spark-generating components. [7] The more difficult a gas is to ignite, the larger and hotter the particle must be to achieve ignition. An exception to this statement is if the metal particles are incendiary.

Two types of foreign material were identified during the inspection: a pool of water at the bottom of the silencer and solid debris in both the piping and in the silencer. The pool of water is significant as a potential carrier for electric charge. This is discussed in the electrostatic ignition section. Water is not a potential ignition source due to friction or impact.

Two types of solid debris were identified in the blowdown piping-silencer system. The first type was small particulate matter found in the pipe elbow that attached to the silencer. This material was chemically analyzed and determined to be organic material. The ignition scenario related to particle impact is based on the impact of particulate material against the confining walls of the blowdown piping-silencer system. The confining walls are constructed of either carbon steel or stainless steel alloys. Impact by organic particulate material against carbon steel or stainless steel alloys will not generate sparks or frictional heating. Thus, the organic particulate material does not constitute an ignition hazard due to impact. The organic particulate material does have the capability to accumulate electric charge and is further discussed in the electrostatic ignition section.

The second type of solid debris found in the "bowl" of the silencer was a metallic equipment identification tag. The manufacturer of the silencer indicated that the material was made of stainless steel. However, tests performed by the facility indicated that the tag was at least 97% aluminum. This is a significant finding because it is well known that aluminum particles striking oxidized (rusty) carbon steel surfaces can result in a thermite reaction. The thermite reaction is an exothermic oxidation reaction that can generate very high temperatures causing the aluminum particles to burn. [8] These burning particles are very effective ignition sources for difficult to ignite flammable gas mixtures like natural gas. [9]

4.4 Feasibility of ignition by foreign material

Given the types of foreign material found in the blowdown piping-silencer system, the only credible ignition scenario involving particle impact would be aluminum particles generated by abrasive wear of the aluminum equipment tag. To be a feasible ignition source, the discharge of blowdown gas must lift and project the tag into oxidized carbon steel surfaces. This impact must be sufficiently violent so as to liberate burning aluminum particles that can be entrained in the vertical flow of the gas as it vents out of the silencer. The burning particles must persist until they are carried to the edges of the natural gas cloud where mixing with air has created a flammable mixture. The burning behavior of single aluminum particles has been summarized in the literature. [4] Given the flow characteristics of the silencer, the minimum particle size of a burning aluminum particle must be 140 microns, based on Exponent calculations. These considerations bound the particle size range that the particle impact hypothesis must satisfy to be feasible.

Furthermore, the bowl of the silencer was found to contain ten to fifteen gallons of water after the incident, likely a result of fire suppression activity. However, the observation of water in the silencer demonstrates its ability to accumulate water inside the unit. Given the quantity of rain in the days prior to the event, at the time of the incident, any debris was likely submerged in the rain water, which had accumulated in the bowl of the silencer. It was considered unlikely that sufficient debris was present to account for the plume observed during the discharge event. The most likely explanation for the plume was water from the bowl that was entrained in the gas stream. The pooled water and entrained water droplets would have served to quench any hot particles liberated by discharge. Thus, the hot particle ignition scenario was deemed unlikely.

4.5 Ignition source: electrostatic ignition mechanisms

Static electricity develops when there is an imbalance of charges on a particle. Electrostatic charging of particles is a phenomenon that generates electrical charges or static electricity by the contact and separation of materials, [10, Section 1.1] typically between different atomic materials (i.e. metal vs. plastic, carbon vs. copper), that allows an electrical charge to build-up on a particle. Electrostatic charging can result from induction, friction between two materials (i.e. rubbing a piece of cloth on a glass rod), or separation of two materials after physical contact. [10, Section 1.4] This charged particle then carries the non-neutral charge until it can find a path to discharge the excessive charge and reach neutrality (i.e. equilibrium) again.

There appear to be two methods of electrostatic charging that could apply to this incident: water droplets and solid particles.

4.5.1 Water Droplets

Liquid droplets formed by water vapor condensation can obtain an electrostatic charge [11, Section 14.5] as the water droplets move throughout the pipe and silencer by colliding with each other, with the inner steel pipe, the filter media in the silencer, and with the gas particles themselves. However, the gas stream had been dehumidified such that insufficient moisture was present to form condensate during the blowdown.

Another possibility is that the water in the bowl of the silencer was the source of water droplets. This hypothesis requires that the high pressure gas flow exits the pipe into the silencer, that some of the gas flow would disturb the pool of water, and that this would in turn create water droplets that are carried out of the silencer. The formation of aerosolized droplets of water from high velocity wind shear is a known phenomenon, particularly relevant in ocean/sea spray formation. [12] Droplets can also be formed from a liquid interface due to direct impingement by a gas jet, [13] [14] mechanical perturbation, [15] and bubble formation and collapse. [12] The creation of water droplets from the pool of water in the silencer would likely create splashing and misting which could generate a charge on the water droplets. [11, Section 5.2.1]

The recorded video of the incident shows a cloud of material exiting the top of the silencer just before the fire is ignited. This cloud may have been a cloud of water droplets directly above the unit before ignition. If an aerosolized liquid such as water is entrained in the gas flow, a charge can be generated and an electrostatic ignition hazard is possible. [16, pp. 293-294]

4.5.2 Solid Particles

As particles are transported through the pipe and out of the system through the silencer, each individual particle may develop an electrostatic charge. [11, Section 14.5.1] The developed charge can originate from the collisions between different particles such as the natural gas particles colliding with the inner metal pipe, with each individual particle, with other foreign particles in the system, or with the filter media used in the silencer. In addition, considerations, such as particle diameter and size distribution, impurities, water content and separation speed, complicate the process of determining the amount of charging. [17, p. 16/121] The cloud recorded exiting the top of the silencer may be a cloud of foreign material directly above the unit before ignition. Additionally, upwards of ten blowdowns may be needed before no visible dust cloud exits a pipe during a blowdown. [2, Chapter 5] It has been reported to Exponent that only a few blowdowns of the incident pipe had been conducted. While the initial pressures of previous blowdowns were not available for all blowdown events, the pressures were likely lower than the initial pressure on the day of the incident. For example, it was reported to Exponent that on one occasion the source pressure of the blowdown was 25 psig and on another occasion it was 500 psig. This is in contrast to the incident blowdown pressure of 1,097 psig.

If a stream of natural gas is completely clean (i.e., no moisture) and free of foreign particles, it is unlikely to develop a large enough charge to result in a discharge. [11, Sections 5.2 & 14.5.1] However, if natural gas is contaminated with foreign particles, as shown in the material analysis, the foreign particles may develop an electrostatic charge. Volume resistivity of a foreign material or powder also plays a role in determining the potential electrical charge on the material. For example, if the material has a resistivity equal to or above 10^8 ohm-m, charging is possible; most organic powders meet this requirement. [17, p. 16/121] The material appeared to be organic and, as such, would be expected to generate a charge.

4.6 Feasibility of ignition by electrostatic discharge

Flammable gas ignition by an electrostatic discharge can occur under the following conditions: [10, Section 2.4]

- *Combustible material, gas, vapor or dust, must be present in the flammable range of fuel-to-air ratios for ignition to be possible;*
- *A static electric charge must have built up on a non-conducting object, or a conducting object that is insulated from earth, and this charge must have sufficient [potential]to [discharge] to a [neighboring object with a potential difference];*
- *The [discharge] must have sufficient energy to ignite the surrounding flammable mixture.*

In this specific incident, the fire occurred when the natural gas exited the silencer and the gas mixed with the outside air. The mixture would have had to have been within its flammable range (approximately 5%-15%) for ignition to occur (i.e. bullet point #1). [11, Table B.1] Since the fire occurred, it is known that this circumstance was met. Solid debris was discovered in the system that could have generated an electrostatic charge due to movement in the system and resulting collisions with other particles (i.e. bullet point #2). Additionally, water droplets could have formed due to the agitation of any pooled water in the bowl of the silencer and this process would likely have generated a charge on the droplets (i.e. bullet point #2).

The minimum ignition energy for an electrostatic discharge to ignite the natural gas flammable environment requires 0.21 mJ of energy at 8.5%. [11, Table B.1] There are several electrostatic discharge methods that were considered. They include a spark discharge, corona discharge, and brush discharge. Each discharge mechanism was evaluated based on the facts of this case and the potential energy in a given discharge method (i.e. bullet point #3).

- A spark discharge is an electrostatic discharge that typically occurs between two conductive objects that are isolated from each other. [17, p. 16/97] This type of discharge has the ability to have an energy greater than 1000 mJ depending on the materials present. [11, Figure 5.5.1] A spark discharge is a competent ignition source for an environment containing natural gas within its flammable limits. Depending on the exact resistivity of the foreign material, a spark discharge may or may not be possible. [11, Sections 15.4.3.2 & 15.4.3.3] A spark discharge could be ruled out as a discharge mechanism if the resistivity of the foreign material entrained in the gas stream is above 10^{10} ohm-m. [11, Section 15.4.3 & 15.4.3.3]

- A corona discharge is an electrostatic discharge that does not release as much energy as a spark discharge. A corona discharge is defined as "a low energy electrical discharge that results from localized electrical breakdown of gases near sharp conductive edges, needle points, and wires." [11, Section 5.5.2] The reported maximum energy in a corona discharge is on the order of 0.1 mJ. [11, Figure 5.5.1] The limited energy in this type of discharge decreases the likelihood that it is a competent ignition source [11, Section 14.5.2]; however, it cannot be ruled out due to the minimum ignition energy needed to ignite an environment containing natural gas within its flammable limits.
- A brush discharge is an electrostatic discharge that is similar to a corona discharge. However, it has a reported maximum energy that is greater than 1 mJ. [11, Figure 5.5.1] The brush discharge may occur from a single grounded electrode and extend into a charged space. The charged space may be a cloud of spray or dust. [17, pp. 16/97-98] This type of discharge is likely to have the energy needed to ignite an environment containing natural gas within its flammable limits. For this reason, it is a likely competent ignition source.

While the recording of the incident provided to Exponent does not show the exact time of ignition, the recording does show that the fire at a point after ignition is close to the metal structure of the silencer (the first observed fire in the video footage appears to be at the exit of the silencer and toward the outside edge). The ignition point is likely to have been near the metal structure and this is generally confirmed by the video footage. This is also consistent with the hypothesis that a static electricity discharge was the ignition source for the fire that is the subject of this report.

In summary, a high velocity discharge of clean gases (where neither liquid droplets nor solid particles are present) rarely acquires sufficient charge which results in ignition. However, when gases contain solid particles or liquid droplets, or when these particles are formed during discharge, flammable vapors can be ignited by sufficient charges. [10, Section 4.3.1]

5 Conclusions and Lessons Learned

Through the course of this investigation, Exponent reached the following three conclusions related to the ignition event:

1. Ignition by particle impact is unlikely in this specific scenario due to the role of water.
2. Ignition by electrostatic discharge in clean natural gas is unlikely.
3. The presence of foreign material (either solid particles or liquid droplets) in the gas stream poses an electrostatic hazard. This ignition scenario is the probable ignition source for the subject fire.
4. More likely than not, the ignition occurred due to electrostatic discharge from aerosolized water droplets that originated in the "bowl" of the silencer.

Foreign material entrained in a flammable gas stream may result in electrostatic discharge and, ultimately, ignition of the gas. However, it is exceedingly difficult to determine a strict lower limit for the quantity of contaminant necessary to cause a fire or explosion. The prevailing guidance is to avoid intentional discharges of natural gas at high pressure, when possible. In the case study discussed above, a blowdown was performed as a safety measure to de-inventory at-risk equipment. For systems where safety-critical blowdowns may occur, other strategies—such as design considerations, preventive maintenance, and facility siting—may be necessary to adequately control the hazard.

6 References

[1] NFPA 56: Standard for Fire and Explosion Prevention During Cleaning and Purging of Flammable Gas Piping Systems, National Fire Protection Association, Quincy, MA, 2017.

[2] D. Sakar, Thermal Power Plant: Pre-Operational Activities. Elsevier, 2017.

[3] U.S. Chemical Safety Board Kleen Energy Report, Available at http://www.csb.gov/assets/1/19/FINAL_Urgent_Recommendation[1].pdf (accessed on September 6, 2017).

[4] R.A. Ogle. Dust Explosion Dynamics, Elsevier, New York, NY, 2017.

[5] B. Lewis and G. von Elbe. Combustion, Flames and Explosions of Gases. 3rd Ed., Academic Press, New York, NY, 1987.

[6] F. Powell. "Ignition of gases and vapors," Industrial and Engineering Chemistry, 61:29-37, 1969.

[7] W. Bartknecht. "Ignition capabilities of hot surfaces and mechanically generated sparks in flammable gas and dust/air mixtures," Plant/Operations Progress, 7:114-121, 1988.

[8] N. Gibson, F.C. Lloyd, and G.R. Perry. "Fire hazards in chemical plant from friction sparks involving the thermite reaction," IChemE Symposium Series, 25:26-35, 1968.

[9] R. Eckhoff. Explosion Hazards in the Process Industries. Gulf Publishing Company, Houston, TX, 2005.

[10] Guidelines for the Control of Static Electricity in Industry, Occupational Safety and Health Service, Department of Labor, New Zealand, 1990.

[11] NFPA 77: Recommended Practice on Static Electricity, National Fire Protection Association, Quincy, MA, 2014.

[12] S. J. Norris, I. M. Brooks, B. I. Moat, M. J. Yelland, G. de Leeuw, R. W. Pascal, and B. Brooks. "Near-surface measurements of sea spray aerosol production over whitecaps in the open ocean." Ocean Sci., 9:133-145, 2013.

[13] A. He and A. Belmonte. "Deformation of a liquid surface due to an impinging gas jet: A conformal mapping approach." Physics of Fluids, 22(4):042103-042103-7, 2010.

[14] C.W. Berendsen, J.C. Zeegers, and A.A. Darhuber. "Deformation and dewetting of thin liquid films induced by moving gas jets." Journal of Colloid and Interface Science, 407:505-515, 2013.

[15] A.J. James, B. Vukasinovic, M.K. Smith, and A. Glezer. "Vibration-induced drop atomization and bursting." J. Fluid Mech, 476:1-28, 2003.

[16] T. Kletz. What Went Wrong: Case Histories of Process Plant Disasters. 4th Ed., Gulf Professional Publishing, Houston, TX, 1998.

[17] F.P. Lees. Loss Prevention in the Process Industries. 2nd Ed., Butterworth-Heinemann, Oxford, 1996.

Expect the Unexpected:
Hydrogen Sulfide Exposure during Troubleshooting

Katherine Pearson
BP Americas
501 Westlake Park Blvd
Houston, TX 77079

John Champion
The Dow Chemical Company
4460 Highway 225
Deer Park Dow Center
Deer Park, TX 77536

Prepared for Presentation at
American Institute of Chemical Engineers
2018 Spring Meeting and 14th Global Congress on Process Safety
Orlando, Florida
April 22 – 25, 2018

Expect the Unexpected:
Hydrogen Sulfide Exposure during Troubleshooting

Katherine Pearson
BP Americas

John Champion
The Dow Chemical Company

Keywords: Hydrogen Sulfide, Exposure, Troubleshooting

Abstract

During routine troubleshooting work involving multi-stage vacuum jets, an unexpected release of Hydrogen Sulfide (H_2S) at the indoor job site led to exposure of two people, one of whom later died. The worker who subsequently died was attending to his coworker, whom he believed was suffering from a heart attack. Several additional individuals were exposed when they rushed to the room without protective equipment; fortunately, all made a full recovery.

This paper discusses the circumstances that led to this tragic incident as well as the management system and cultural issues that contributed to the event and to the subsequent emergency response. Key learnings included:

- Troubleshooting and maintenance activities require robust pre-job planning and hazard recognition processes.
- Hazardous by-products can be generated in a facility, even from non-hazardous chemicals, and can result in exposures unless well understood. This facility was not subject to the OSHA 1910.119 PSM regulation, and the raw materials and products were non-hazardous.
- A strong Incident Command structure can reduce risk to personnel during emergency response.

Note that this incident did not occur at a facility owned by BP Americas, or by The Dow Chemical Company at the time of the incident. However, both authors were employed by the company that owned this facility at the time of the incident and Ms. Pearson was the lead investigator.

1. Authors' Note

This incident did not occur at a facility owned by BP Americas, or by The Dow Chemical Company at the time of the incident. However, both authors were employed by the company that owned this facility at the time of the incident and Ms. Pearson was the lead investigator.

This event was summarized primarily from the memories of people involved in the incident investigation and a set of summary slides created after the event. The slides were shared within the company that owned this facility and with several industry safety groups. In the years since this incident occurred, Dow acquired the facility, then divested it, and the process where the incident occurred was shut down. The original investigation report, which included the fact-hypothesis matrix and investigation tree, were no longer available at the time this paper was written.

Most case histories papers focus solely on the facts and the management system failures of an incident. Memories fade over time, but those that were directly involved in this incident and the aftermath will never forget the impact it had on them. We have included a section on the emotional impact because we believe that making the story personal most effectively drives change and improvement.

2. The Incident

February 15, 2006 started out as just another day at work at the plant. One of the reactors was down and James and Bobby[1] had just started troubleshooting a steam ejector system. It was a job they had done before. After opening a flange on the ejector system, Bobby collapsed. Bobby had recently returned to work after heart surgery, and James assumed Bobby was having a heart attack. James called for assistance on the radio saying, "Bobby's had a heart attack."

A large group quickly rushed to the room to respond. The responders were attempting to use a defibrillator on Bobby. When James collapsed, then another responder collapsed seconds later, the response changed to an evacuation. The responders' focus was on getting the others out, not on protecting themselves. No one put on breathing air. Luckily, the area they were in had a door leading directly to an open roof giving the responders a short and clear evacuation route.

Everyone in the incident area, a total of 13 people, was exposed to some level of H_2S. Most of the exposed were released from the hospital quickly, except for James and Bobby. Tragically, James passed away three days later. Fortunately, Bobby was released from the hospital and recovered from his exposure.

3. Investigation

After operations ceased and the area was deemed safe, the investigation began with a survey of the area and interviews with people involved. An experienced investigator was brought in from outside the facility.

The investigation team was large and included many of the people who had responded to the event as well as engineering and other support staff. Several people were brought in from outside the facility to provide fresh eyes perspective. OSHA came onsite to conduct its own investigation. Company legal

[1] Names have been changed.

counsel was included in the company investigation team. Protocols for formal evidence collection and chain of custody were established.

3.1 Incident Sequence

The mechanics were working on the third floor of a building on some vacuum jets above a process reactor. Figure 1 shows the vacuum jet system and how it connects to the process.

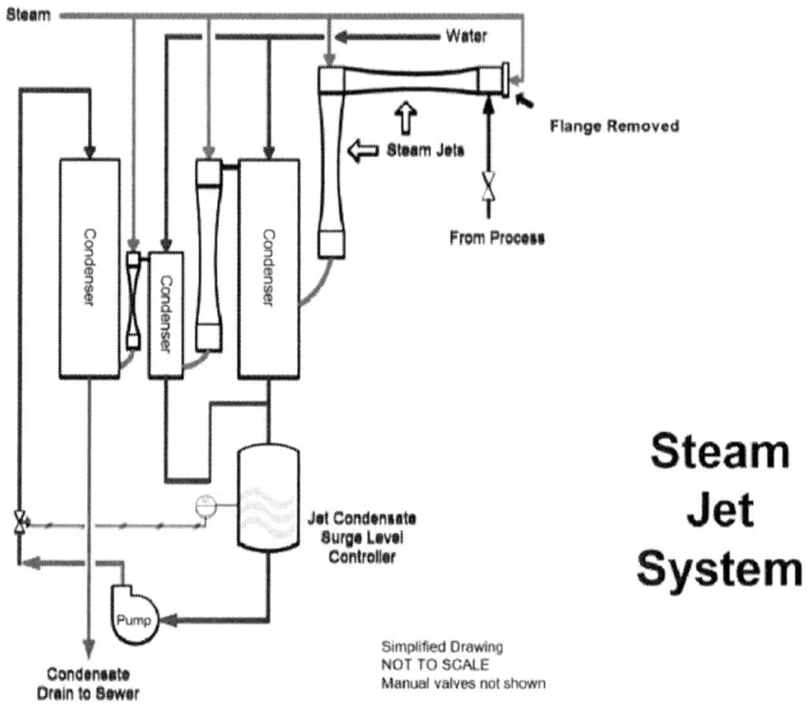

Figure 1. Steam Jet System

The vacuum system needed occasional maintenance. When proper vacuum couldn't be reached, air leaks into the system were suspected. First, the mechanics tightened flanges. If that didn't resolve the problem, the throat of the ejectors was checked by using valve connections to rod out the steam orifice nozzles on the jets. The orifice clearance was very small, and there had been previous times where a small chunk of metal or debris got hung up in the throat. When the rod would not go through the 1st stage jet, the mechanics determined that there was a blockage at this point and they needed to open the 1st stage jet.

To clear this system, the mechanics had isolated the process connection (on the right in Figure 1) and lined up the steam to the jets for a period. The mechanics did not obtain a permit because they were "troubleshooting."

Bobby collapsed after opening the flange. James stayed with him after calling in the emergency. Bobby was in a corner of the room in a congested area. James and the responder who collapsed were both near Bobby's head. The responder backed away from the area before collapsing saying "there is something wrong in that corner." Others were also close to the fallen, but did not feel any ill effects.

At the time the other victims collapsed, the responders were attempting to use a defibrillator on Bobby. The responders thought the defibrillator was broken because it wouldn't shock. The defibrillator was working correctly, being equipped with sensors to detect a heartbeat, being equipped with sensors to detect a heartbeat that prevented the unit from operating.

When the responders and victims were interviewed, only one person recalled smelling anything. Olfactory fatigue can occur with H_2S at concentrations greater than 100 ppm [1].

Only three of the responders reported hearing an H_2S alarm sounding about 15 ft. below them through the grating floor. One person thought the alarm was a vacuum pump alarm; the sound of the vacuum pump alarm was similar. Another tried to shut off the alarm because it was so annoying. The alarms were tested and found to be loud enough for people above the grating to hear.

The responders reported they were focused on the medical emergency. The possibility of another cause was something that didn't enter their minds until other people collapsed.

In hind-sight, it may seem obvious that the H_2S came from the opened pipe. However, it was not obvious to the employees involved. Emotions were high throughout the incident investigation; the employees were connected both through work and through the community. The plant had about 110 employees, which means about 10% of the employees were directly involved with the event. By the time the incident investigation started, there were already many theories as to the source of the H_2S: the jets weren't isolated from the reactor, the gas came up through the building from the sewer, it came from offsite and was not H_2S, etc.

3.2 Fact-Hypothesis Matrix

A fact-hypothesis matrix [3] was used to organize the data in a logical fashion. A simplified, small example of the matrix is shown in Table 1. (Some facts have been combined where, on the original matrix, each fact was a separate entry.) The goal was to help manage emotion and instill some logic and data-driven decision-making into team discussions. The team reviewed each hypothesis and asked what data needed to be collected to rule out each theory. All data collected was added to the matrix and evaluated whether it refuted or supported each hypothesis.

Most of the theories were quickly ruled out. The vacuum jets were confirmed to be isolated from the adjacent reactor by multiple valves. There were no indications that any type of gas release had occurred in other plant locations or offsite.

It was known that one of the raw materials could release H_2S if the material was subject to low pH conditions such as could exist in the sewer system. There were no indications that a spill to the sewer of any type had occurred. The sewer pH was steady. There were no unexplained volume changes in other process or raw material tanks.

Table 1. Fact-Hypothesis Matrix Example

	Jets were found isolated from reactor. Corroborated by interviews.	No indications of offsite release of any toxic gas. Plant detectors outside did not alarm. No ill effects on any persons outside building.	No other building H2S detectors alarmed. Detectors confirmed working.	No indication of releases to the sewer to cause H2S. Sewer pH and conductivity stable.	Condensate drain line from the jets not isolated.	Condensate drain line to sewer vents found plugged
Gas came from reactor feeding Jets.	Does not support.	N/A	N/A	N/A	N/A	N/A
Reaction in the sewer that came up through the building trenches	N/A	N/A	Does not support.	Does not support.	N/A	N/A
H2S gas generated offsite	N/A	Does not support.	Does not support.	N/A	N/A	N/A
Other gas generated offsite	N/A	Does not support.	N/A	N/A	N/A	N/A
Gas came from the jet condensate drain line.	N/A	N/A	N/A	N/A	Supports	Supports

Figure 2 represents the conditions during the work. Only the condensate drain line was found to not be isolated from the vacuum jet system. The jet condensate surge tank head space and liquid were both sampled and no H_2S or related compounds were found.

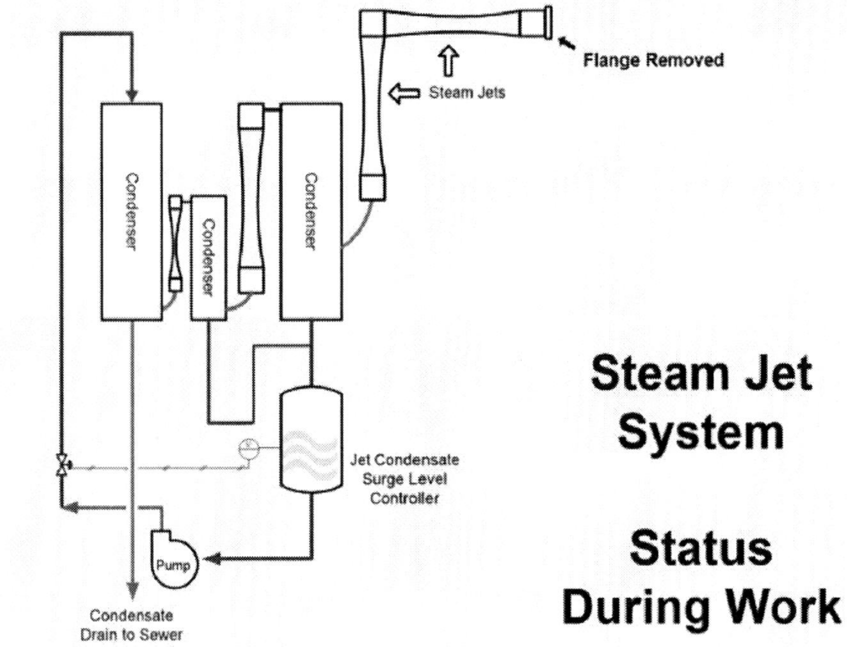

Figure 2. Status During Work

During the discussion, one of the mechanics remembered the condensate drain line joined another drain line from an unrelated process in a separate building. Others were unaware of this connection. Figure 3 shows this interconnection. The steam jet system where the incident occurred is on the right.

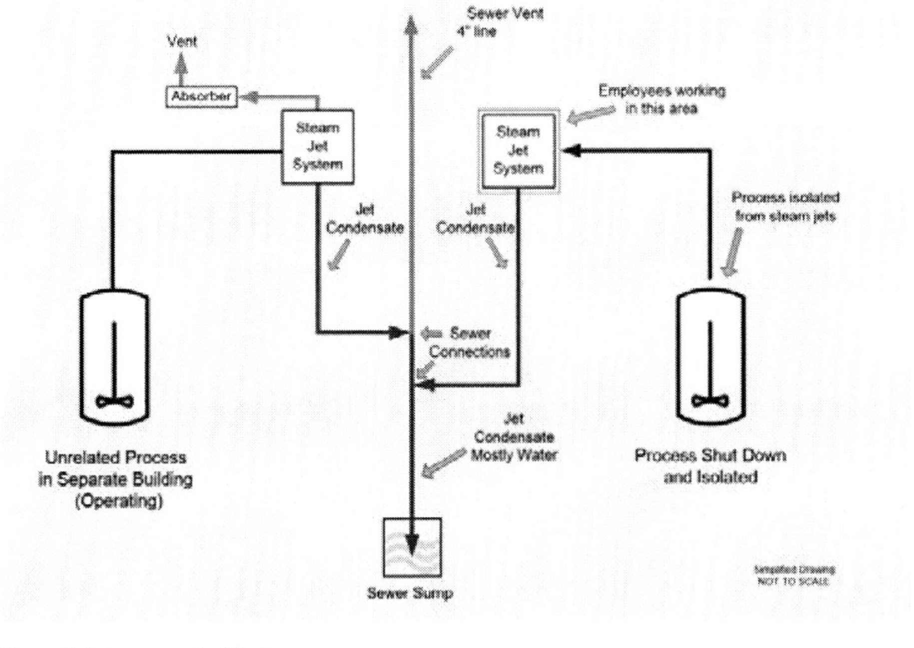

Figure 3. Interconnected System

The sewer line connection was not shown on any drawings. The team postulated that some H_2S from the other reactor system (on the left in Figure 3) had somehow made its way up the vacuum jet drain line. The reactor steam jet system had a 1" vent and the drain lines had a 4" piping vent (shown in Figure 3 in pink) to vent any gas generated in the process. The team felt they had ruled out all other potential sources and suspected the vents might have plugged.

The investigation team wanted to gather additional data to test the theory. The other reactor had been shut down after the incident, and the team wanted to start it back up and sample for H_2S at the open jet. After much discussion and review of the data, the Occupational Safety and Health Administration (OSHA) representatives, who retained control of the incident scene, agreed to the experiment. With the building secured, and with involved employees equipped with breathing air, the other reactor was restarted. H_2S was detected near the open jet. The path of the gas is shown in Figure 4.

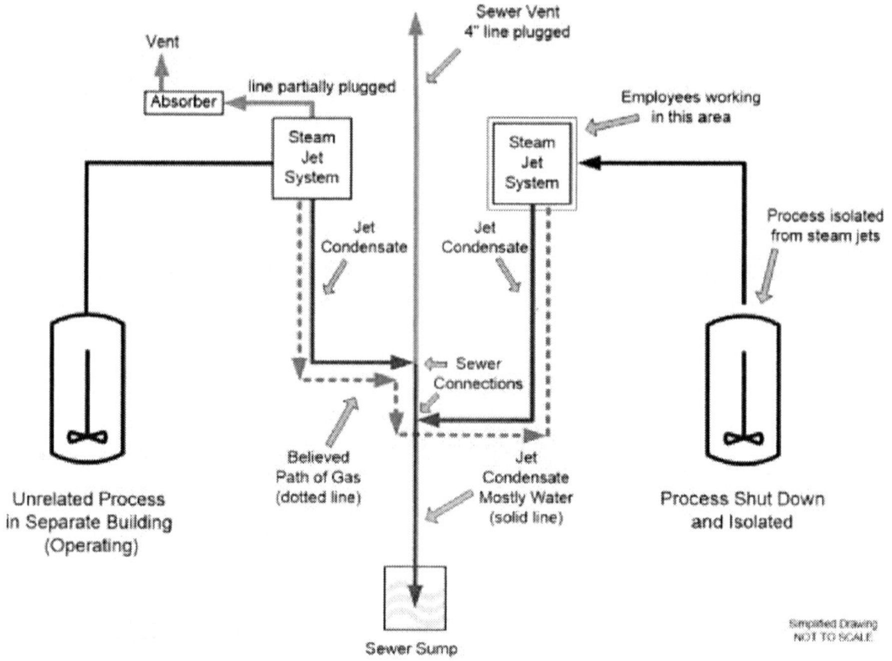

Figure 4. Flow Path for H_2S

Based on the fact-hypothesis matrix and the subsequent testing, the investigation team concluded that the H_2S formed in the unrelated reactor system (shown on the left in figure 4). Normally, the small volume of gases that the process periodically generated would be vented from the last jet condenser and absorbed in a connected scrubber. However, the 1" line to the scrubber was found about half-plugged, which would have restricted vapor flow.

Gases released from the jet condensate would normally be vented out the sewer vent above the roof. However, the 4" sewer vent was also found to be plugged with a polymer-like material, preventing normal escape of the vapor.

3.3 Root Cause Investigation

Once the team determined how the incident occurred, the investigation focus shifted to why the incident occurred. The team employed a logic tree method [2] called Multiple Cause Systems Oriented Incident Investigation [3][4] to identify causal and contributing factors and document the investigation. The logic tree is not recreated here.

Some of the key "why" discussions were:

- Why were there so many people in the incident area? Eleven people responded to the work area, although the actual response team needed for the emergency was far less. This placed more people in a hazardous exposure zone. Some of the people who responded were supposed to go to the field incident command post, including the area manager, the maintenance manager, and the foreman. Unfortunately, incident command was set up right beside the victim, instead of farther away, because the responders assumed it was a medical emergency.
- Why didn't the mechanics get a permit? The mechanic said they didn't get a permit because they were "troubleshooting." This was not consistent with what the plant leadership thought they had established as an expectation. The plant had put considerable effort into implementation of a safe work permit system starting about four years prior to the incident following acquisition of the plant. However, the investigation team was not sure the event would have been avoided if the mechanics had gotten a permit; would the need for drain line isolation have been identified by the permit writer/approver?
- Why was the jet condensate line not recognized as a line that needed to be isolated? The condensate line was not equipped with an isolation valve, and no one had identified it as a lockout/tagout (LOTO) requirement on prior jobs. This job had been done in the past with no incident, indicating a lack of hazard recognition for potential releases from this line.
- Why did no one react to the H_2S alarm? Besides being focused on a medical emergency, there were other factors that may have contributed to the lack of response to the H_2S alarms. The plant had many "false alarms" on the H_2S detectors and, as a result, workers didn't trust the detectors. With the knowledge that the reactor condensate sewer vent was plugged, the team realized some of those "false alarms" may not have been false. In addition, the fixed detector that alarmed only did so locally and audibly. The sound of the H_2S alarm was very similar to the vacuum pump alarm, and the mechanics were working on the vacuum system. The mechanics likely did not interpret the sound of the alarm as a potential H_2S release in the area.

4. Emotional Impact

The impact of this event on James' family, the survivors, the plant personnel, and the investigation team was considerable. The investigation team was working while the two mechanics remained in the hospital. Many people on the investigation team had worked closely with the mechanics. Employees were leaving work and going to the hospital to visit the families. When James was taken off life support, the investigation team had to take a break while everyone digested the news of James' death and regained sufficient composure to focus on the investigation again.

The investigation team worked long hours for about three weeks to complete the investigation. There were a lot of stakeholders who wanted answers including local and corporate legal counsel, high-level corporate personnel, and OSHA. Disagreements among the team seemed amplified; even agreements could feel contentious. The team was mentally and emotionally exhausted after the first week. Managing team fatigue became key to progressing the investigation.

The entire site shut down for James' funeral, about a week into the investigation. Two corporate representatives stayed to answer potential phone calls, but virtually everyone else went to the funeral. The investigation team was emotionally invested in what had happened, and the funeral made the team even more determined to make changes to prevent future similar events.

5. Lessons Learned

Learnings from this event are applicable to many facilities. Key learnings included an understanding of reactive chemistry, equipment connectivity, safe work practices, gas detection, and emergency response.

5.1 Reactive Chemistry

This facility manufactured low-hazard products using low-hazard raw materials and was not covered by the OSHA 1910.119 Process Safety Management (PSM) rule. Even though not required by regulation, a thorough Process Hazard Analysis (PHA) had been recently conducted on the process equipment.

The PHA identified that H_2S could be generated in the reactor under abnormal conditions such as low pH or high temperatures. However, it was not known that H_2S could be generated in significant amounts under normal operating conditions. In the months after the incident, experiments conducted showed that a normal batch went through about a 15-minute period of low pH about an hour into the batch and generated significant amounts of H_2S during that time.

Key learning: Understand the process and reactive chemical hazards, including all phases of batch reactions.

5.2 Equipment Connectivity

The PHA did not cover the interconnectivity of the condensate drain lines. In fact, it would have been very difficult to do so since the interconnectivity was not shown on any drawings. Recognition of the hazard of the condensate drain line interconnectivity would likely have driven a more stringent isolation and equipment opening plan.

Key learning: Know how the equipment is connected, including through utilities and sewer systems and accurately reflect connections on the Piping & Instrument Drawings (P&IDs).

The piping for this system was revised prior to restarting operation as shown in Figure 5. A block valve was added to each condensate drain line along with a bleed for clearing. The two sewers do eventually connect, but the connection is downstream of these sewer sumps and outside of the building. After the piping changes were made, the facility experienced significantly fewer H_2S alarms.

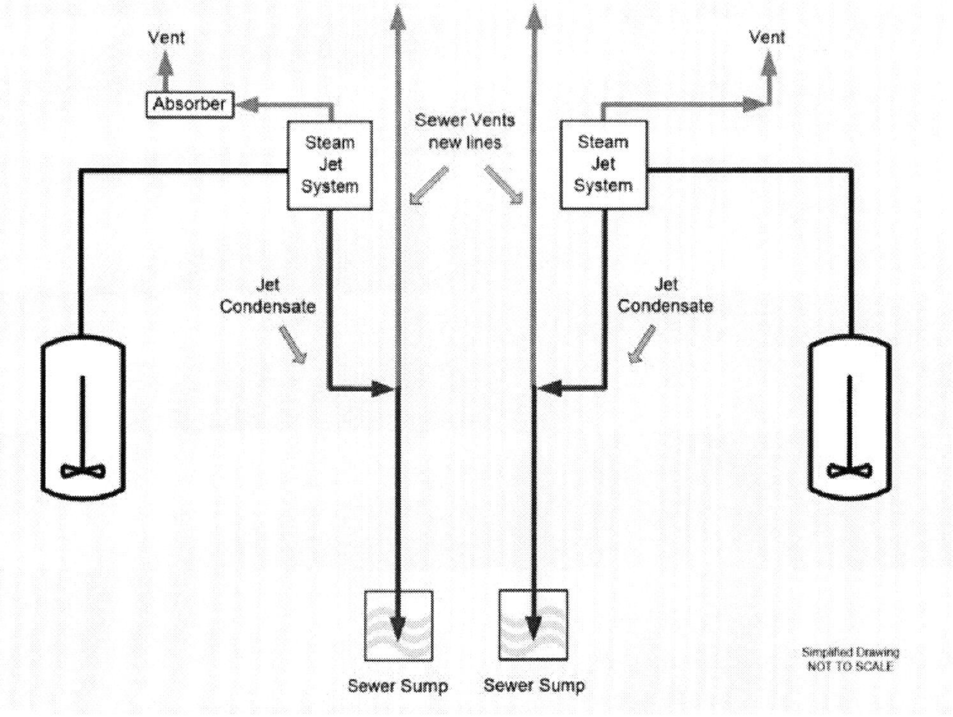

Figure 5. Revised System after the Incident

5.3 Safe Work Practices

At its core, this incident is a LOTO incident. The incident tragically demonstrated why equipment should be isolated from every potential source of energy every single time. With a strong LOTO culture and discipline, the system would have been isolated from the sewer, which would have prevented the event.

In combination with a strong LOTO system, excellent Equipment and Pipe Opening, and Safe Work Permit systems and cultures can prevent hazards, both known and unforeseen, from impacting people. To prevent incidents like this, there needs to be a culture where these work processes are always used when opening systems, even when "troubleshooting."

When equipment and piping is opened, the Personal Protective Equipment (PPE) and monitoring equipment should reflect all potentially hazardous materials, including possible contaminants that could be released.

A plant safety stand-down was held to communicate the learnings of the incident. Safe work permit, LOTO, and equipment and pipe opening procedures were reviewed with all employees.

Key Learning: Apply LOTO, Safe Work Permitting, and Equipment and Line Opening work practices consistently and completely.

5.4 Gas Detection

Chemical facilities where highly toxic materials are handled or potentially generated should utilize a combination of fixed detectors and personal gas monitors. The primary benefit of fixed detectors is to alert personnel to gas releases before people enter an area where a release has taken place. Personal monitors can reveal a direct and immediate threat to a person or work crew and can cover more area than is practical with fixed detectors.

This facility immediately added the use of personal H_2S detectors for the processes involved. A personal monitor would have warned James and Bobby right away that there was H_2S present. Personal monitors can also alert when conditions change during work.

Fixed detectors should alarm in a continuously manned control area that is remote from the process. This allows alarms to be communicated to plant personnel and, if needed, to emergency response personnel. In the field, the audible alarms should be unique-sounding and clearly heard in all potentially impacted locations. Audible alarms can be enhanced with a visual indication, such as a strobe light.

The facility did make several changes to the fixed detector system. The facility set up an H_2S alarm panel in a continuously occupied control room outside of the process area. The local alarms were changed so that if one detector alarmed, all the detector audible alarms sounded. Flashing lights were also added in key locations. With the revised system, an H_2S alarm was obvious throughout the building.

The root causes of toxic or flammable gas alarms need be investigated and addressed. Repeated "false" alarms can undermine the effectiveness of the monitoring system. After repeated false alarms, personnel tend to disregard the alarms and may ignore a real alarm.

Key learning: Utilize a combination of fixed detectors and personal gas monitors where toxic gas can be present and investigate all alarms.

5.5 Emergency Response/Drills

Bobby had a medical history that led his co-workers to assume a medical emergency was occurring. In any facility where there is potential for toxic materials or an oxygen-deficient atmosphere to be present, precautions should be taken by responders whenever a person goes down. The emergency responders should approach cautiously with the right PPE and gas detection, under the assumption that a chemical emergency is taking place. The medical emergency can be treated when the area is proven safe.

Excellent emergency response systems are critical for any chemical manufacturing or handling facility. Many facilities in the United States utilize the National Incident Management Systems (NIMS) practices established by the US Federal Emergency Management Agency or something very similar. [5] Under NIMS, an Incident Command is established outside of the "hot zone" and a carefully coordinated response is executed when entering the area where the emergency is taking place. Better utilization of NIMS practices would have limited the number of people that were at risk in this event.

In many incidents, a completely remote Emergency Operations Center (EOC) is also established to support Incident Command. Technical and leadership support staff should assemble in a remote location with communication access to provide support to the response team. The EOC might have recognized more quickly that this was not a medical emergency.

Finally, to be proficient in emergency response, regular drills are needed. Tabletop drills can provide an opportunity to talk through potential scenarios and get people thinking of the right way to respond. Nothing is better than live drills where the responders are forced to don protective response equipment and go through the process of responding to a particular scenario in the field. Where there are gas detectors in use, one of the scenarios that should be drilled on routinely is a response to a gas detector alarm.

Key learning: Plan and train for emergency situations, to develop proficiency both in the first responders and the support systems such as the EOC.

6. Conclusion

Prior to this incident, the facility was going through a significant culture and work process change because of the acquisition of the plant. The basic safety management systems had been implemented, but this incident shows there were gaps. The existing plant culture and long-standing work practices can make changes to new work processes quite difficult.

To avoid an incident like this, these fundamental safety management systems are critical:

- Understand the process and reactive chemical hazards, including all phases of batch reactions.
- Know how the equipment is connected, including through utilities and sewer systems and accurately reflected on the P&IDs.
- Apply LOTO, Safe Work Permitting, and Equipment and Line Opening work practices consistently and completely.
- Utilize a combination of fixed detectors and personal gas monitors where toxic gas can be present and investigate all alarms.
- Plan and train for emergency situations, to develop proficiency both in the first responders and the support systems such as the EOC.

These systems, as well as the other systems that make up a comprehensive process safety management program, need to be in place and continuously audited and improved. Believing that existing systems are "good enough" leads to complacency. Maintaining a sense of vulnerability to process hazards is critical to avoid complacency, drive continuous improvement, and ultimately eliminate all process safety incidents.

7. References

[1] New Jersey Department of Health Hazardous Substance Fact Sheet on Hydrogen Sulfide, Available at http://nj.gov/health/eoh/rtkweb/documents/fs/1017.pdf Accessed on February 16, 2018.

[2] CCPS. *Guidelines for Investigating Chemical Process Incidents, 2nd Edition.* Center for Chemical Process Safety, American Institute of Chemical Engineers, New York, NY, 2003.

[3] A. M. Dowell, III, S. E. Anderson, "A Methodology That Ensures Getting The Most From Accident Investigations", Presentation to Texas Chemical Council Safety Seminar, June 6, 1994.

[4] A. M. Dowell, III, "Incident Investigation: Process to Identify Root Causes of Mechanical Failures", *Process Safety Progress*, **Vol. 25**, No. 1, pp 8-15, 2006.

[5] U.S. Federal Emergency Management Agency. *National Incident Management System, 3rd Edition*. Available at https://www.fema.gov/media-library-data/1508151197225-ced8c60378c3936adb92c1a3ee6f6564/FINAL_NIMS_2017.pdf Accessed on February 16, 2018.

Importance of Properly Designing Dust Explosion Protection Systems: Case Study – 2014 Georgia Pacific Corrigan Facility Fire and Explosion

Scott G. Davis
Gexcon US
4833 Rugby Avenue Suite 100
Bethesda, Maryland 20814
Scott.G.Davis@gexcon.com

John L. Pagliaro
Gexcon US
4833 Rugby Avenue Suite 100
Bethesda, Maryland 20814
John.Pagliaro@gexcon.com

Prepared for Presentation at
American Institute of Chemical Engineers
2018 Spring Meeting and 14th Global Congress on Process Safety
Orlando, Florida
April 22 – 25, 2018

Importance of Properly Designing Dust Explosion Protection Systems: Case Study – 2014 Georgia Pacific Corrigan Facility Fire and Explosion

Scott G. Davis
Gexcon US
4833 Rugby Avenue Suite 100
Bethesda, Maryland 20814
Scott.G.Davis@gexcon.com

John L. Pagliaro
Gexcon US

Keywords: dust hazard analysis (DHA), dust explosions, dust explosion protective systems (suppression, isolation, deflagration venting); isolation dampers; incident investigation

Abstract

When designing handling and conveying equipment for combustible dusts, it is crucial to properly implement protective measures (e.g., deflagration venting, suppression, and isolation) capable of mitigating the potential consequences in the event the dust is ignited. Generally speaking, dust processing, storage, and collection equipment, such as cyclones, mills, baghouses, dryers, and silos are typically connected via pipes, pneumatic conveyors, and dust extraction or aspiration lines. If a fire or dust explosion occurs in one vessel, or the interconnecting piping, the flame front or pressure can quickly propagate through these conduits to other parts of the plant and potentially escalate to even worse consequences than the original event. More specifically, pressure piling or flame jet ignition in connected vessels can result in a series of more catastrophic secondary events. To effectively prevent or mitigate these phenomena, deflagration suppression or venting is provided for the source vessel, and explosion isolation is used to prevent the explosions or fire from propagating to neighboring vessels. Improper design of isolation devices can be costly, as illustrated by this case study on the fire and explosion that occurred in 2014 at a plywood manufacturing facility in Corrigan, Texas.

On April 26, 2014, a fire that originated at a plywood sander eventually propagated through a pneumatic dust conveying system resulting in an explosion in a baghouse (dust collector), fatally injuring two employees, and seriously injuring others who were responding to the incident. Sparks from the sander entered the extraction pipe, which was protected by an active suppression/isolation system (spark-sprinkler-abort gate) upstream of the baghouse. The sparks ignited dust piles that accumulated in the extraction pipe (prior to the first spark detection zone) due to an insufficient flow velocity through the system. When the fire in the extraction pipe was discovered, the extraction blower downstream the baghouse (negative pressure system) was turned off, which allowed smoldering and burning to continue within the pipe. When the blower was subsequently turned back on, a flame front developed that propagated into the baghouse due to an improperly designed isolation system. This incident occurred because of: (1) an insufficient flow velocity, which allowed dust to accumulate in the extraction pipe; (2) the improper design of a passive damper as an isolation device; (3) and operational errors associated with the blower being turned off and then back on prior to extinguishing the burning material in the pipe. This paper analyzes the root causes of the incident as well as key lessons learned related to: system design; performing a dust hazard analysis; required air stream flow rates; proper back blast designs; impeding deflagration vents; and exclusion zones in the path of a vented deflagration in the baghouse.

1 Introduction

Current nationally recognized safety standards for processes involving combustible dust require that owners/operators of a facility: (1) determine whether potentially combustible dusts in their facility are indeed combustible or explosible, and, if so, characterize their properties as required to support a Dust Hazard Analysis (DHA); and (2) where dust materials that have been determined to be combustible or explosible are handled, shall be responsible to ensure a DHA is completed for existing processes that are within a 3-year period from the effective date (September 25, 2015) of NFPA 652 - *The Standard on the Fundamentals of Combustible Dust* (2016 Edition), or that are undergoing material modification [1]. DHAs are systematic reviews to identify and evaluate the potential fire, flash fire, or explosion hazards associated with the presence of one or more combustible particulate solids in a process or facility. The main purpose of the DHA is to evaluate the fire, deflagration, and explosion hazards and provide recommendations to manage those hazards with safeguards and mitigation measures.

When designing handling and conveying equipment for combustible dusts, it is crucial to properly implement both preventive (e.g., housekeeping, eliminate ignition sources, ensure adequate air-gas velocities to prevent accumulation, hazardous area classification) and protective measures (e.g., deflagration venting, suppression, isolation, and exclusion zones) capable of mitigating the potential consequences in the event the dust is ignited. Generally speaking, dust processing, storage, and collection equipment, such as cyclones, mills, baghouses, dryers, and silos are typically connected via pipes, pneumatic conveyors, and dust extraction or aspiration lines. If a fire or dust explosion occurs in one vessel, or the interconnecting piping, the flame front or pressure can quickly propagate through these conduits to other parts of the plant and potentially escalate to even worse consequences than the original event. More specifically, pressure piling, flame jet ignition, or turbulent flame acceleration in connected vessels can result in a series of more catastrophic secondary events. To effectively prevent and mitigate these phenomena, deflagration suppression or venting is provided for the source vessel, and explosion isolation is used to prevent the explosions or fire from propagating to neighboring vessels. The importance of designing mitigation measures that protect both the individual vessels as well as the interconnected system will be shown next.

1.1 Source Vessel – Deflagration Safe Guards

There are a number of ways to manage and mitigate deflagration and explosion hazards associated with combustible dust for a given process or piece of equipment. These include operating the process at conditions that cannot support combustion (i.e., where the dust concentration is below the combustible limit or in an inert atmosphere), avoiding the accumulation of material in conveying equipment, or avoiding potential ignition sources in the system. One method of avoiding ignition sources is to "detect" sparks or embers within the systems and initiate a water mist spray system to extinguish the potential ignition source. While these measures may prevent an explosion from initiating for most conditions, there may be times during normal or abnormal operation when an explosive atmosphere may exist in an enclosed vessel, which could be subsequently ignited resulting in a deflagration.

A dust-air deflagration that is left unmitigated can generate significant pressures within a closed vessel, where the theoretical maximum pressure for most dusts, as determined using standard closed vessel experiments (e.g., 20 liter sphere, 1 m^3 sphere), can reach up to 8-9 barg (118 -131 psig) [2]. These unmitigated deflagration pressures can catastrophically fail handling and process equipment, resulting in external fireballs, blast hazards, and missile hazards (projectiles) that can impact personnel or property. There are two principal means of mitigating the hazardous pressures resulting from the deflagration or explosion within a piece of equipment or vessel: (1) explosion suppression (chemical or inerting like water); and (2) deflagration venting. The principal behind both methods is essentially the same – reduce the pressure from the deflagration event to levels that are within the structural limits of the equipment, hence avoiding damage to the equipment and loss of containment.

Unlike spark suppression, which prevents the ignition from occurring by extinguishing the ignition source, deflagration suppression systems are designed to detect the early phase of the deflagration (flame) development and activate a high rate discharge system that extinguishes the flame and protects the integrity of the equipment. Deflagration venting provides a burst panel or door that is designed to open at pressures well below the maximum deflagration pressure (~8 barg or 118 psig for wood dust) and relieve the pressure generated in the vessel via the vent opening, resulting in significantly lower pressures within the vessel. For example, consider the vessel in Figure 1 that is designed to withstand pressures up to 0.55 barg or 8 psig. This means that the reduced explosion pressure (maximum pressure during a vented explosion) P_{red} is equal to 0.55 barg (8 psig). Hence, the vessel is designed with an explosion vent that opens in this case at a pressure P_{stat} = 0.2 barg (3 psig) with sufficient vent area to reduce the theoretical maximum explosion pressure from 8 barg to 0.55 barg (or 118 psig to 8 psig) as shown in Figure 1.

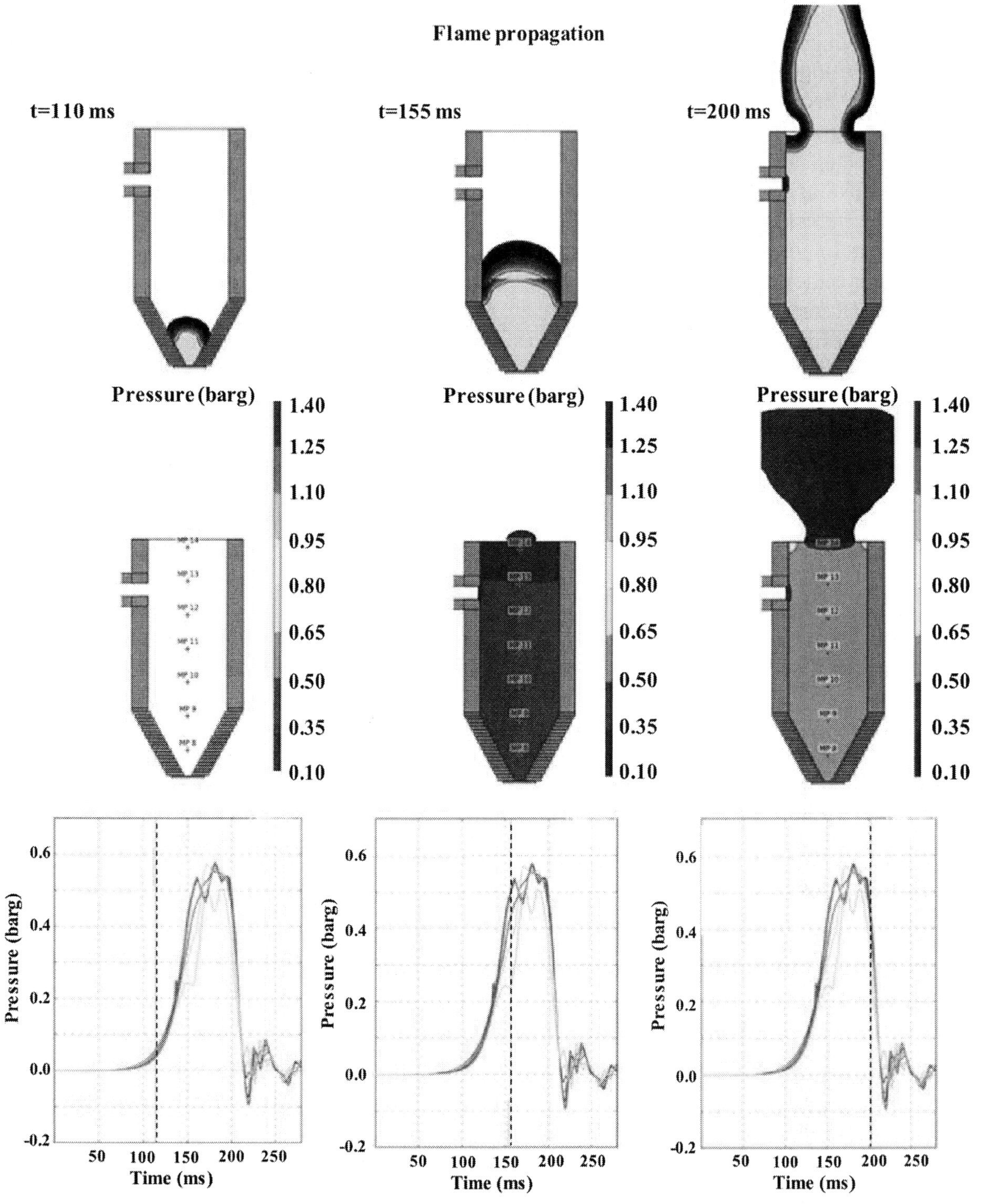

Figure 1: Deflagration development in a "vented" vessel, where the maximum explosion pressure P_{max} = 8 barg (118 psig), the vent opening pressure P_{stat} = 0.3 barg (4.4 psig), and the reduced explosion pressure is P_{red} = 0.55 barg (8 psig). Top, middle, and bottom images show flame propagation, 3D pressure map, and pressure vs. time at various heights inside the vessel.

1.2 Process System – Interconnected Vessel Safeguards

While protecting the source vessel is extremely important, of equal or greater importance is protecting interconnected vessels. If a fire or dust explosion occurs in one vessel, or the interconnecting piping, the flame front or pressure can quickly propagate through these conduits to other parts of the plant and potentially escalate to even worse consequences than the original event. For example, consider two vessels that are individually protected with deflagration venting for a reduced pressure of 0.55 barg (or 8 psig), meaning each vessel has a deflagration vent capable of reducing the explosion pressure to 0.55 barg (8 psig) for an ignition in the respective vessel. However, if they are interconnected and the explosion initiates in the leftmost vessel (see Figure 2), it pre-pressurizes the connected vessel and causes turbulence in the interconnected piping, resulting in a flame that accelerates into the rightmost vessel and over-pressurizes the vessel to greater than 3 barg (44 psig), despite the rightmost vessel having a deflagration vent properly sized for an explosion originating in the vessel.

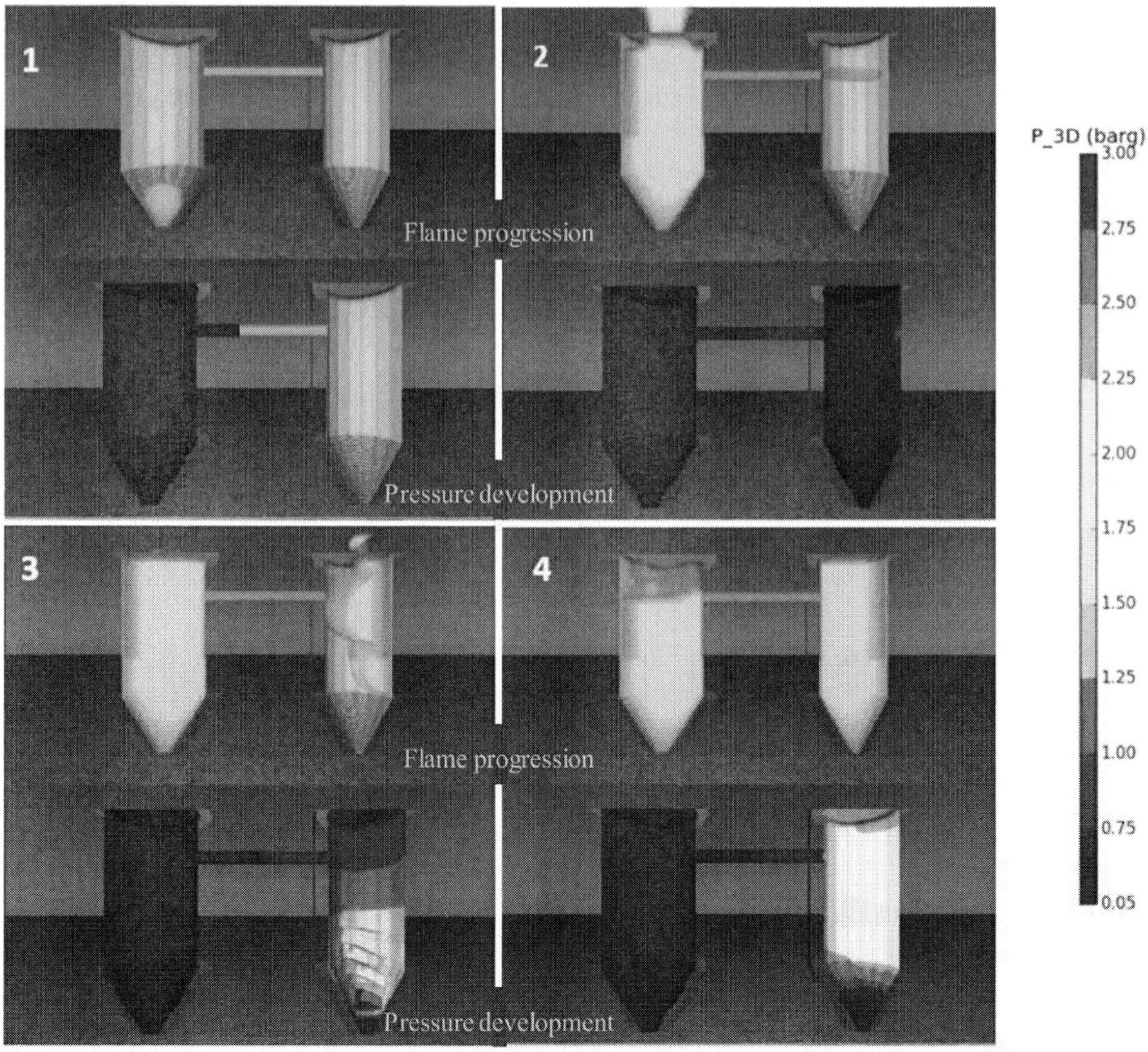

Figure 2: Deflagration development in two interconnected vessels, both individually protected with deflagration venting for P_{red} = 0.55 barg (8 psig).

Given the potential for such severe escalation, explosion isolation is required on process systems where a deflagration can originate in one section and propagate into neighboring areas. Isolation systems can vary from passive systems, (i.e., self-closing valves, rotary valves, passive explosion diverters) to active systems where the deflagration must be detected for system activation (i.e., guillotine or slam shut valves, extinguishing barriers).

Lastly, isolation systems can also be designed to prevent ignition sources from entering a neighboring vessel. For example, in systems where spark or ember generation can be relatively high, the detection of such sparks or embers can activate an abort gate that opens to divert the potential ignition sources away from the connected vessel.

2 Background

This paper will present an incident that occurred on April 26, 2014 at the Georgia Pacific Corrigan facility, where a fire that originated at a plywood sander eventually propagated through a pneumatic dust conveying system and resulted in an explosion in a baghouse (dust collector), fatally injuring two employees, and seriously injuring others who were responding to the incident. The paper will discuss the proximate causes of the incident as well as key root causes that let to this tragic event.

The Georgia Pacific Corrigan facility manufactured plywood. In the facility, plywood sheets were assembled and cut to size, then sent to the finishing department where the top and bottom surfaces were sanded to the required specification. This was accomplished using a larger belt sander that was automatically fed with plywood sheets.

2.1 The finishing sander dust collection system

The belt sander in the finishing department had its own dedicated dust collection system. The system removed wood dust via a pneumatic conveying system that operated under negative pressure. The system comprised conveying pipes, a baghouse, and a blower located downstream of the baghouse. The blower created a vacuum that extracted the dust from the sander through the conveying pipes and into the baghouse.

Vertical pipes extracted dust from multiple locations around the sander (left image in Figure 3). These smaller diameter pipes merged into a larger 36" diameter extraction pipe (referred to herein as the blowpipe), which carried dust across the facility to the baghouse (see right image of Figure 3). Lower than design flow rates were measured in the dust conveying duct work, which often led to material accumulation in the ducts.

Figure 3: Sander and inlet pipes to dust collection system (left) and overhead blowpipe (right).

The flow through the baghouse is illustrated in Figure 4. The dust laden airstream entered the baghouse at the base of the cylindrical section. The baghouse then separated the dust from the flow and the clean air exited near the top of the baghouse, passed through the blower, and exhausted out of a vertical stack.

Figure 4: Sander baghouse with arrows showing the flow of air (green) and dust (orange).

2.2 Dust conveying safety systems

Automatically fed belt sanders occasionally produce sparks. Therefore, the dust collection system was equipped with a spark detection system designed to prevent sparks from entering the baghouse. There were two zones intended to detect and suppress sparks using a water mist system and one additional zone intended to detect and redirect sparks using an abort gate if sparks persisted in the flow beyond the suppression zones. Figure 5 shows a two-dimensional diagram of the system. Note that the first spark detection/suppression zone was located in the blowpipe approximately 225 ft from the sander.

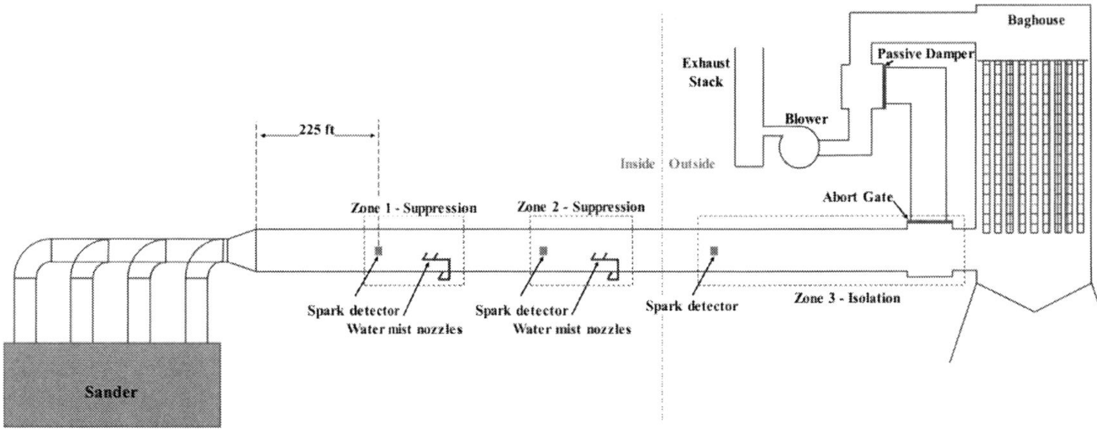

Figure 5: Dust conveying system with spark suppression and isolation components highlighted.

Figure 6 illustrates the flow through the abort gate during normal operation (left images) and the change in flow when a spark is detected beyond the suppression zones (right images) for both the actual configuration and 2D schematic of the system. During normal operation, the gate is positioned horizontally and the flow enters the baghouse. When a spark is detected, the spring-loaded abort gas closes, bypassing flow around the baghouse (see Figure 6).

There was a "back-blast" damper, or explosion isolation damper, installed in the ducting to isolate the clean air outlet of the baghouse from the aborted flow. In general, a back-blast damper is a passive isolation flap valve used on pneumatic systems to prevent flow counter to normal flow. A back-blast damper is designed to keep the valve open when the flow is in the direction normal to the process flow, and when an explosion initiates downstream of the valve and begins to propagate in the opposite direction, the valve closes automatically due to the pressure increase and prevents the flame from propagating past the valve (see example in Figure 7).

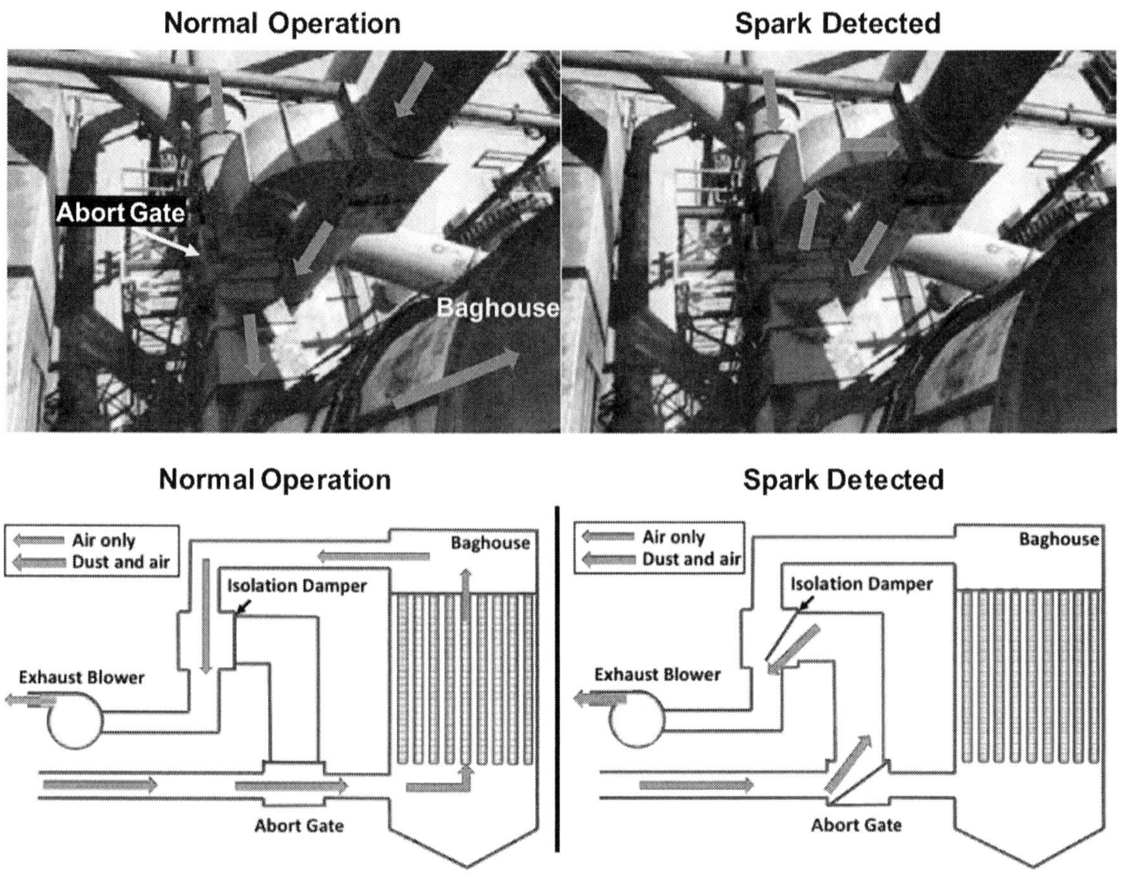

Figure 6: Flow through the abort gate during normal operation (left) and when a spark is detected (right). Arrows show the flow of air (green) and dust (orange)

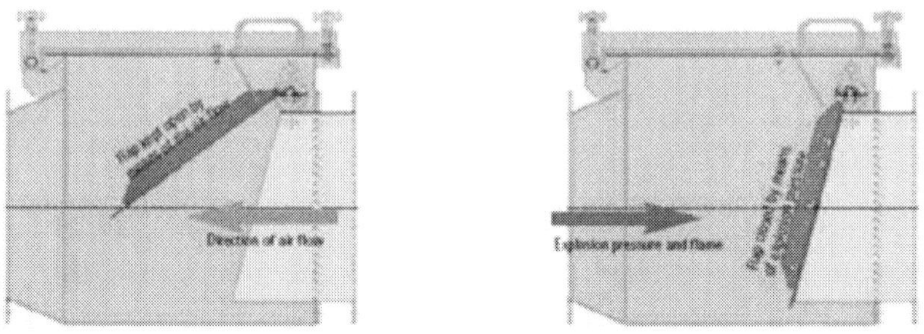

Figure 7: A back-blast damper or passive isolation flap valve normally open when flow (left image) and closed when a downstream explosion reverses the flow (right image)

A "back-blast damper", or isolation damper, was included to prevent potential ignition sources, fire, and/or flame fronts exiting the activated abort gate from entering the baghouse through the clean air outlet. Figure 8 shows where the damper was in the ducting (left image) and provides a close-up image of it (right image).

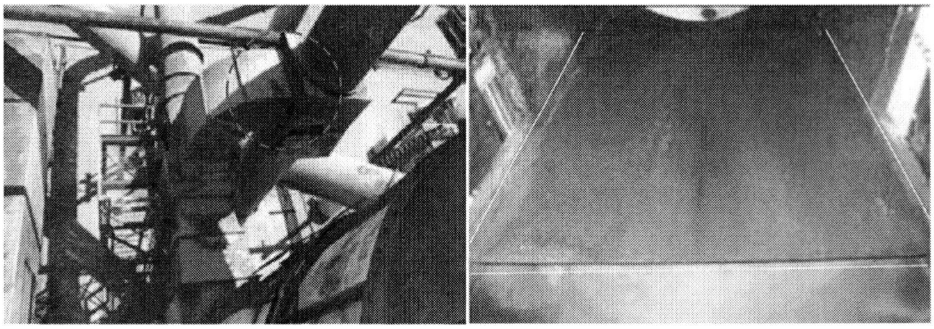

Figure 8: Isolation damper location (left) and close-up image (right).

Figure 9 illustrates the intended abort gate and isolation damper positions during normal operation (left image), and when the abort gate is activated and the flow bypasses the baghouse (right image). Arrows in the figure indicate the flow path of both the dust/air streams and clean air. While the intent of the isolation damper was to prevent aborted ignition sources and deflagrations in the conveying system from entering the clean side outlet of the baghouse, when used in a negative pressure system, a damper does not actually create a seal or effectively isolate the baghouse (more details in a following section). Therefore, if a deflagration or fire began in the aborted flow or just upstream the abort gate, it could pass the damper and enter the clean air side of the baghouse (which will be discussed further below), and thus the system had a critical design error.

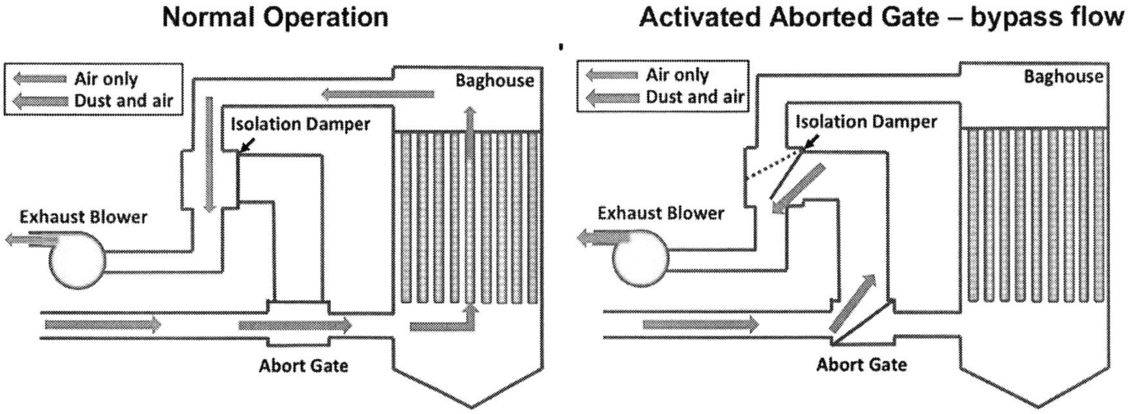

Figure 9: Left image shows the abort gate and damper position during normal operation and right image shows the abort gate position and the intended (dotted line) and actual (solid line) isolation damper position when the flow is aborted.

Figure 10 shows the location of the control panel for the blower, which was directly underneath the baghouse.

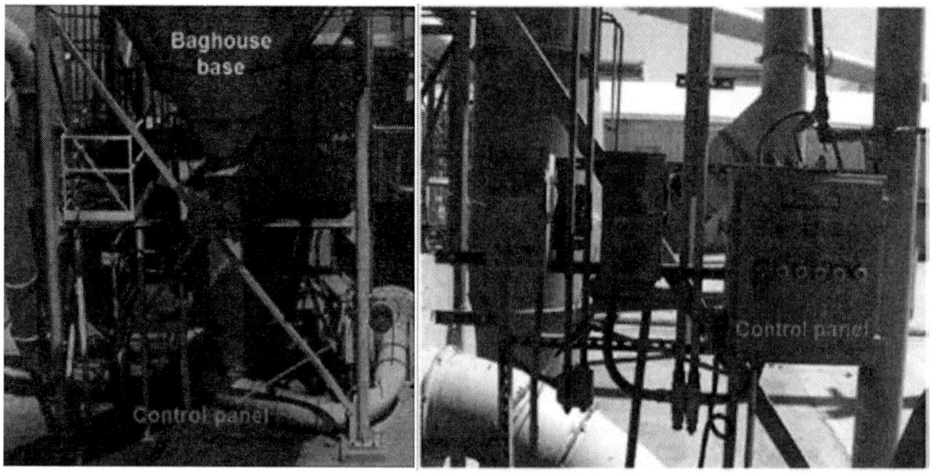

Figure 10: Location of the control panel in relation to the baghouse.

2.3 Baghouse Deflagration Venting

Deflagration vents were installed on the sander baghouse and portions of the ductwork to the baghouse to minimize structural damage in the event of an internal deflagration. The sander baghouse was equipped with 8 vents that were evenly spaced and located at approximately mid-height (see Figure 11). As seen in Figure 11, the vents were also attached to the baghouse with chains so that they would not become projectiles. Vertical filter bags covering steel cartridges were used as the media separator in the baghouse (see right image in Figure 11).

Figure 11: Deflagration vents located on the baghouse (left), close-up of a vent (center), and bag cartridges blown outside (right).

3 The Incident

On April 26, 2014, two plywood sheets were simultaneously pulled into the finishing sander leading to a stoppage. The two employees operating the sander noticed that the it had stopped, so they turned off the power and removed the two sheets. One of the operators then opened an access door to the sander and noticed a fire. The operators fought the fire with hand-held extinguishers and eventually extinguished it. All three zones in the ducting detected sparks and reached maximum spark counts (999 count) near the time of the fire. The two sprinkler zones activated and the abort gate opened.

At some point after the initial fire was discovered at the sander, the main blower motor for the dust conveying system was shut off by an employee, resulting in the loss of airflow through the system. Around the same time, employees noticed that a significant portion of the overhead blowpipe was smoking and glowing red, so they began applying water to the blowpipe using firehoses.

Meanwhile, several other employees responded to smoke coming from the blower exhaust stack located near the baghouse. These employees connected firehoses and attempted to spray water into the smoking exhaust stack. Two employees got on an aerial lift to try to gain a better vantage point for spraying water down the stack.

A supervisor then instructed the men to get down from the aerial lift so that he could turn the main blower back on. The controls for the main blower were located under the baghouse. When the supervisor turned the main blower back on, there was a small boom and flames shot out of the exhaust stack. Seconds later, there was a bigger boom, the deflagration vents on the baghouse yielded, and a large fireball propagated out of the baghouse engulfing the immediate area and those within it. The resulting deflagration fatally injured two employees and seriously injured several others responding to the incident.

3.1 Cause and Origin

3.1.1 Ignition source for the duct fire and explosion

Sparks were generated when two sheets of plywood were pulled into the sander at the same time. The sander is designed to finish boards to a specified thickness, and when the two boards were pulled in together, the greater-than-usual friction between the sander belt and the plywood sheets created sparks.

3.1.2 Cause of the Explosion in the Baghouse

Sparks were sucked up into the blowpipe when the two boards were simultaneously pulled into the sander. When sparks were sucked up into the blow pipe, stationary dust accumulations in the pipe caught fire. This was evident from: (1) witness statements; (2) the pipe becoming hot enough to ignite the paint on the outside surface; (3) the spark sensors continually outputting a spark count of 999 even though the sander was not in operation; and (4) the burnt accumulations of dust found in certain sections of the blowpipe during post-incident inspections. The abort opened (as confirmed by the as-found position during post-incident inspections) and there was no evidence of a fire in the baghouse (i.e., the baghouse deluge was not initiated). Dust accumulated in the blowpipe because the airstream flowrate through the system was too low due to changes to the equipment during the original installation.

The main blower motor for the system was then turned off by an employee and the airflow through the system stopped. This exacerbated the fire in the blowpipe and allowed it to grow. Due to unclear emergency procedures, the blower was turned back on, and the flow through the blowpipe lifted burning dust nests, embers, and accumulated dust piles and carried them downstream toward the baghouse. These burning accumulations were not extinguished by the spark suppression system as it was designed to: (1) only extinguish sparks; and (2) use a minimum amount of water to not disrupt production every time a spark was detected.

The burning nests and embers initiated a suspended dust cloud deflagration in the dust conveying system which propagated out the exhaust stack and back upstream into the baghouse through the clean air outlet duct (see Figure 12 for illustration). The initial deflagration that entered the baghouse penetrated through the filter bags into the dirty side of the baghouse. It also agitated dust in the bags and created a suspended dust cloud inside the baghouse that ignited and caused a secondary deflagration that was accompanied by the loud boom that yielded the vent panels. When the panels yielded, a large fireball extended downward engulfing the employees that were attempting to control the fire.

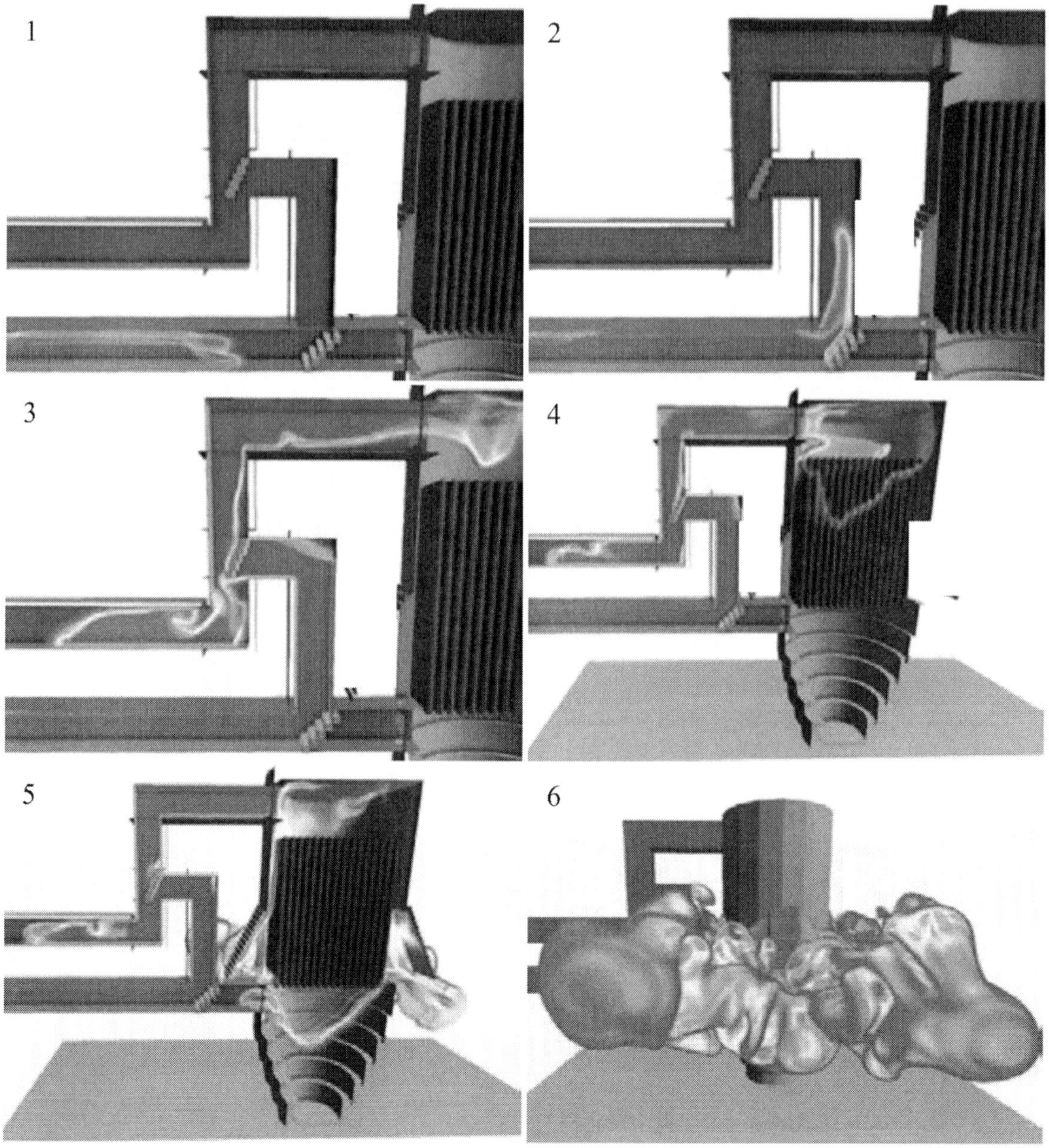

Figure 12: Illustration of the deflagration occurring in the conveying duct and propagating up through the clean air outlet duct into the baghouse

3.1.3 Contributing factors

There were numerous proximate causes of the incident or which led to an escalation of the initiating event. For example, numerous aspects of the dust collection system design were inadequately designed and failed to comply with NFPA Safety Standards, more specifically:

Accumulation of Material

Proper sizing of air-moving devices and ducts to properly capture, control, and convey material and to keep the duct interior clean and free of residual material. Insufficient airstream flowrate in the conveying system allowed dust to accumulate in conveying pipes. A design change to the sheave (i.e., the connection between the electric motor and blower fan) during system start-up resulted in a flow velocity of 3,768 ft/min, well below the original system design of 5,000 ft/min and the minimum recommended velocity of 4,000 ft/min per NFPA 91- *Standard for Exhaust Systems for Air Conveying of Vapors, Gases, Mists, and Noncombustible Particulate Solids* [3]. Those installing the system and the company never verified the flow velocity through the system after making the change to the sheave.

Lack of Access Ports

Lack of access ports prevented the employees from conducting routine inspections that would have revealed accumulated dust in the conveying pipes. NFPA 91 requires that access doors are provided in horizontal ducts adjacent to elbows, junctions, and vertical ducts

Improper Isolation of the dust collector (baghouse)

As found in the company records, the original design requirements included, "shall supply by-pass duct from the abort valve to the blower unit which isolates the filter unit in case of an emergency". NFPA 664 - *Standard for the Prevention of Fires and Explosions in Wood Processing and Woodworking Facilities* [4] further requires isolation for conveying systems with fire and deflagration hazards to prevent propagation of fire and deflagration both upstream and downstream into occupied areas or other critical process equipment. As discussed above, the inadequate back-blast damper design allowed sparks/burning dust/flames to enter the baghouse and initiate a dust explosion in the baghouse.

As recommended in NFPA 664 [4], the negative pressure system should have had a second abort gate or equivalent isolation device to prevent deflagrations from traveling back up the clean air outlet duct of the baghouse (see Figure 13). For a negative pressure system, NFPA 664 [4] does not recommend using a single abort gate upstream of the dust collection as this type of system will not clear the ductwork of sparks and embers once the abort gate activates. This is due to leakage through the baghouse that reduces flow in the conveying duct to levels not able to completely purge the sparks and embers from the ductwork.

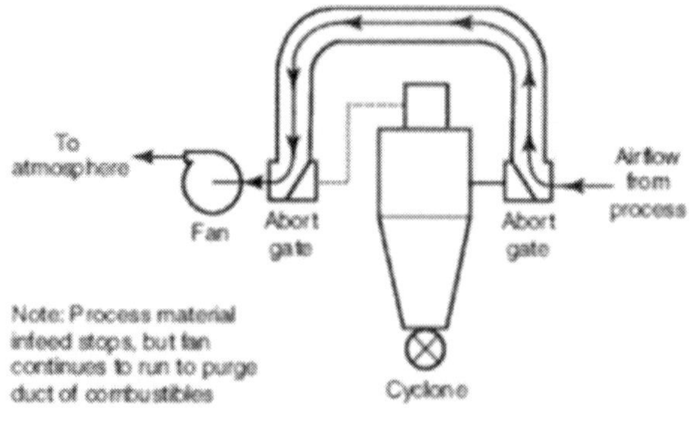

Figure 13: Example of an air-material separator with a negative pressure system and two abort gates.

Deflagration venting –exclusion zones

The facility failed to identify exclusion zones in the path of a vented deflagration in the baghouse, and in fact located the system controls underneath the deflagration vents of the baghouse. NFPA 68 requires that facilities consider the size of the fireball that can extend from the dust deflagration vents when choosing the location of vents and vent ducts (see Figure 12). The maximum fireball dimensions also need to be calculated to establish exclusion zones. In addition, appropriate signs need to be provided to warn against deflagration vent location.

Inadequate emergency procedures

The facility owner had contradictory procedures when a fire was observed in the sander, where one procedure indicated to turn off the blower and the other indicated to keep it on. NFPA 652 requires the owner/operator to establish clear written procedures for operating its facility and equipment during emergency events. The explosion in the baghouse was initiated by stopping the blower and turning it back on. More specifically, stopping the dust extraction fan to the baghouse and stopping the exhaust of the sparks and fire to the atmosphere, exacerbated the fire condition in the blowpipe. When the exhaust fan was turned back on, air was reintroduced into the blowpipe and lifted burning dusts nests and clouds. The burning material eventually transitioned to a deflagration just upstream or past the abort gate. There were no engineering controls to prevent the blower from being turned off and restarted in the event of a fire in the blowpipe. FM Design Standard 7-73 [5] states that for important pneumatic dust handling systems with multi-zone spark detection, conveying fans or blowers should be left running in all cases.

Deflagration venting – impeded deflagration vent

The cartridges for the bags in the baghouse extended below the base of the deflagration vents. The sander baghouse design drawings indicate that 10' long bags and cartridges are to be installed. However, measurements made during a post-incident inspection show that 12' long bags and cartridges were improperly installed. The longer bags reduced the effectiveness of the vent openings and this issue is cautioned against in NFPA 68 - *Guide for Venting of Deflagrations* [2]. This arrangement impeded the venting flow when the deflagration inside the baghouse occurred and could have resulted in a failure of the dust collector with the potential for additional consequences.

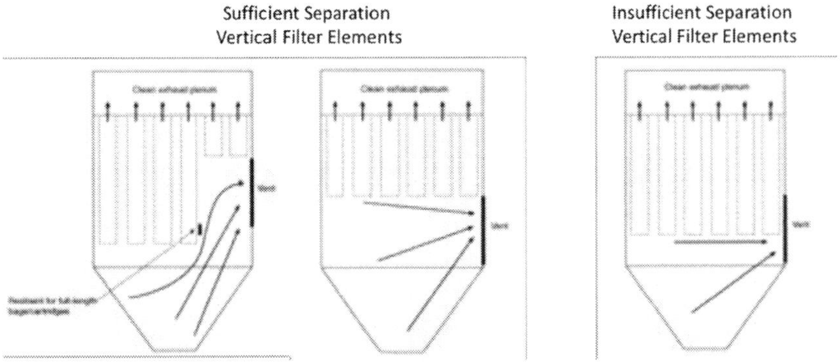

Figure 14: Sufficient (left and center image) and insufficient (right image) for vertical filter elements in a baghouse.

4 Root Cause – Lessons Learned

NFPA 654 - *Standard for the Prevention of Fire and Dust Explosions from the Manufacturing, Processing, and Handling of Combustible Particulate Solids* (2006 and 2012 versions) [6] requires process hazard analyses to be reviewed and updated every five years. There is no evidence that the owner performed, reviewed or updated the hazard analysis of the dust conveying systems at the plant after they acquired it in 2007. If they had performed a proper process hazard analysis, they could have identified the critical design deficiencies in the sander dust collection system. More specifically, a thorough hazard analysis and associated evaluation of the existing controls could have identified deficiencies in the system design, such as:

- required air stream flow rates
- proper isolation designs
- exclusion zones in the anticipated path of a vented deflagration
- impeding deflagration vents

Had these hazards been properly identified in a dust hazard analysis, the explosion in the baghouse could have been avoided because:

a. Dust accumulation in the blowpipe would have been detected and the low flow rate issue may have been resolved.

b. The baghouse would have been isolated from the ductwork. Even if dust still accumulated in the blowpipe and a deflagration occurred inside it, it would not have propagated into the baghouse, resulting in a larger secondary explosion that vented via the deflagration vents.

c. Exclusion zones around the deflagration vents would have prevented personnel and controls from being present in the direct path of an ensuing fireball from a vented deflagration.

In addition, NFPA standards require that the owner/operator shall establish written procedures for operating its facility and equipment to prevent or mitigate fires, deflagrations, and explosions from combustible particulate solids. Either the blower would have been kept on to exhaust the potential ignition sources or left off and the fire remain localized in the blow pipe.

5 Conclusion

This incident exemplifies the importance of conducting DHAs in order to not only identify, but prevent or mitigate fires, deflagrations, and explosions from combustible particulate solids.

6 References

[1] NFPA 652: Standard on the Fundamentals of Combustible Dust, 2016 Edition. In *NFPA National Fire Codes Online.* Retrieved from http://codesonline.nfpa.org

[2] NFPA 68: Guide for Venting of Deflagrations, 2013 Edition. In *NFPA National Fire Codes Online.* Retrieved from http://codesonline.nfpa.org

[3] NFPA 91: Standard for Exhaust Systems for Air Conveying of Vapors, Gases, Mists, and Noncombustible Particulate Solids, 2015 Edition. In *NFPA National Fire Codes Online.* Retrieved from http://codesonline.nfpa.org

[4] NFPA 664: Standard for the Prevention of Fires and Explosions in Wood Processing and Woodworking Facilities, 2012 Edition. In *NFPA National Fire Codes Online.* Retrieved from http://codesonline.nfpa.org

[5] FM 7-73: Dust Collectors and Collection Systems, 2000 Edition. Retrieved from http://www.fmapprovals.com/

[6] NFPA 654: Standard for the Prevention of Fire and Dust Explosions from the Manufacturing, Processing, and Handling of Combustible Particulate Solids, 2013 Edition. In *NFPA National Fire Codes Online.* Retrieved from http://codesonline.nfpa.org

Deflagration in Empty Tank being Rinsed with Water

Laurence G. Britton
Consulting Scientist
Charleston WV

Benjamin McDavid
Senior Process Safety Engineer
Dublin OH
bcmcdavid@ashland.com

Prepared for Presentation at
American Institute of Chemical Engineers
2018 Spring Meeting and 14th Global Congress on Process Safety
Orlando, Florida
April 22 – 25, 2018

Deflagration in Empty Tank being Rinsed with Water

Benjamin McDavid
Ashland Inc.,
Presenter Email address bcmcdavid@ashland.com

Laurence G. Britton
Process Safety Consultant

Keywords: Deflagration, TBHP, Peroxide, Bleach, Oxygen enrichment, Static electricity, Water slug

Abstract

A deflagration occurred inside a 200-gallon atmospheric vessel at a manufacturing facility during the course of apparently normal operations. The vessel had previously contained an aqueous solution of tert-butyl hydroperoxide (t-BHP) which had been transferred out about 10 minutes earlier. The incident occurred about one minute after starting a cold water rinse. Although the hinged vessel lid was forced open and damaged there were no injuries or chemical exposures since no personnel were in the area at the time. Venting of gases through the opened lid might have limited damage to the vessel. The incident was caused by an overcharge of sodium hypochlorite (bleach) solution to the deionized water system and subsequent fast, gassy reaction of the bleach with t-BHP. During the incident investigation the only ignition scenario not ruled out was a spark from an electrically charged slug of water falling from the protruding water inlet nozzle through a charged mist that had formed during the gassy reaction. Despite the small size of the vessel the "water slug" scenario may have been made possible by the very small ignition energy of the gas-phase reaction products, comprising acetone and other flammable vapors mixed with about 90% oxygen.

Incident

The vessel mixes deionized water and tert-butyl hydroperoxide (t-BHP) then pumps the dilute solution to another vessel. The t-BHP mixture was present in the vessel for about 1 hour before the deflagration. The mixture was pumped out and nine minutes later a rinse of deionized water was started. About one minute into the rinse step a deflagration was heard by plant staff. Flammable gas detectors did not indicate flammable gas in any area. The control system was still operating normally with no alarms or signs of a process upset. Plant staff reviewed all the ongoing processes to help determine the source of the upset. The process was then put into hold. When the conditions in the plant were found to be safe, operators walked through the unit for a visual check of the process. The vessel was found to have a damaged hinged lid, but no other signs of damage, smoke or soot.

Deionized water system

The deionized water system is designed to treat plant water with 12.5% sodium hypochlorite (bleach) solution to limit biological growth. This utility water supply is shared by various processes at the plant. The normal concentration of bleach is 30 ppm. On the following Monday, a tank that supplies bleach to the process water was found to be empty. Simultaneous failures of both free chlorine analyzers plus a control system programming error caused the dosing pump to continuously add bleach solution. The concentration of bleach was found to be approximately 3,800 ppm instead of the desired 30 ppm.

Timeline of events

Thursday, the day before the event the control system program for bleach addition is put into hold due to program update. Also at 11:00 AM both chlorine analyzers malfunction leading to an excess of bleach distributed to the system. The deflagration occurs during the first batch operation following the analyzer failures.

300 gallons of 12.5 percent bleach solution is charged to the 9000-gallon deionized water tank. The work week for the Maintenance group responsible for operation of the deionized water system ends the day before the event.

On Friday there is a shift change for the Operations group at 7:00 AM on the day of the event. At 7:59 AM, the vessel is rinsed with 100 pounds of deionized water and drained to ensure the tank does not have any residual material. At 8:02 AM deionized water charge is started. Four minutes later 23 pounds of t-BHP is added. At this point the reaction between the bleach and the t-BHP begins. The deionized water charge is completed at 8:55 AM.

The mixture is transferred out and is completed at 9:14 AM. The control system starts waiting for deionized water to be available. At 9:23 AM a rinse step begins a charge amount of 460 pounds at a rate of 30 pounds per minute.

The rinse step was in progress for about 1 minute when a loud noise was heard throughout the plant. The operator evaluates all the processes and cannot determine the source of the noise. The reactor system is put into Hold to cease all processes. Field investigation finds the damaged lid at 9:40 AM. At 10:05, the bleach solution system is found to be in Hold and turned off.

The following Monday, when the Maintenance group starts the work week, the bleach solution tank is found to be empty during routine checks. The chlorine concentration is found to be very high and reported to supervision.

Figure 1 shows a sketch of the mixing vessel. The agitator is mounted near the side and the shaft is on an angle. The agitator shaft and deionized water nozzles are on opposite sides of the fixed half of the hinged lid. The outlet is on the side at the bottom of the vessel and goes to a pump. The vessel has a diameter of 762 mm with a straight side of 1677 mm. The working volume is 200 gallons.

This vessel has been in operation for more than 20 years. The gas phase concentration of flammable material was very small and there were no apparent ignition sources inside the vessel. Several Process Hazard Analyses (PHAs) had been performed through the years and no fire or explosion hazards had been identified for this part of the process. The primary hazard previously identified was the decomposition of t-BHP. The teams identified excessive chlorination of the deionized water as a potential quality problem, but did not believe that the concentration could reach high enough levels to cause a reaction with the t-BHP.

Figure 1: Sketch of Vessel showing Water Inlet and Off-Axis Stirrer

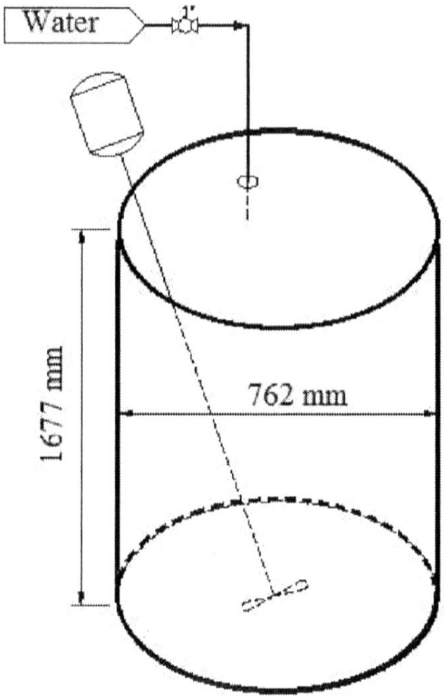

Investigation of chemistry

Hydrogen peroxide reacts with bleach generating water, sodium chloride and oxygen. t-BHP also reacts with bleach, but the products from the reactions are not well known. Verhoeff noted that potential decomposition products include t-butyl alcohol, acetone, methane, methanol, formaldehyde, carbon monoxide, oxygen, and water [1]. The reaction was simulated in the laboratory to obtain a better understanding of the chemistry. An Automatic Pressure Tracking Adiabatic Calorimeter (APTAC) and a Reaction Calorimeter RC1 MP6 were initially used for the experiments.

The first step was to replicate the conditions that happened in the plant. The APTAC test cell was initially evacuated to ~5 psia and the materials were at ambient temperature. The test cell began pressurizing immediately upon injection of the t-BHP, reaching a maximum within 15 minutes. The temperature also climbed 4.5°C. The pressure rise was rapid, but not fast enough to explain the event. The pressure rise is shown in Figure 2. The bleach concentration was varied to find the minimum concentration to initiate the reaction at a measurable rate. It was found that at ambient conditions the reaction rate becomes significant when the free chlorine exceeds 800 ppm.

A sample of the vapor space was withdrawn into an evacuated Parr bomb. The gas was successfully ignited with a fuse wire indicating a flammable atmosphere. The calculated heat of combustion was 150 kJ/mol. Further work was planned in the RC1 MP6 to generate a larger sample size in order to better analyze the vapor composition. Shortly after the reactants were added to the RC1, an ignition occurred inside the vessel.

Separate laboratory experiments in an open beaker showed that the reaction between the bleach-containing deionized water from the supply system and t-BHP is immediate, vigorous and very gassy with significant bubble formation. However no persistent froth is formed.

Figure 2: APTAC T&P Histories

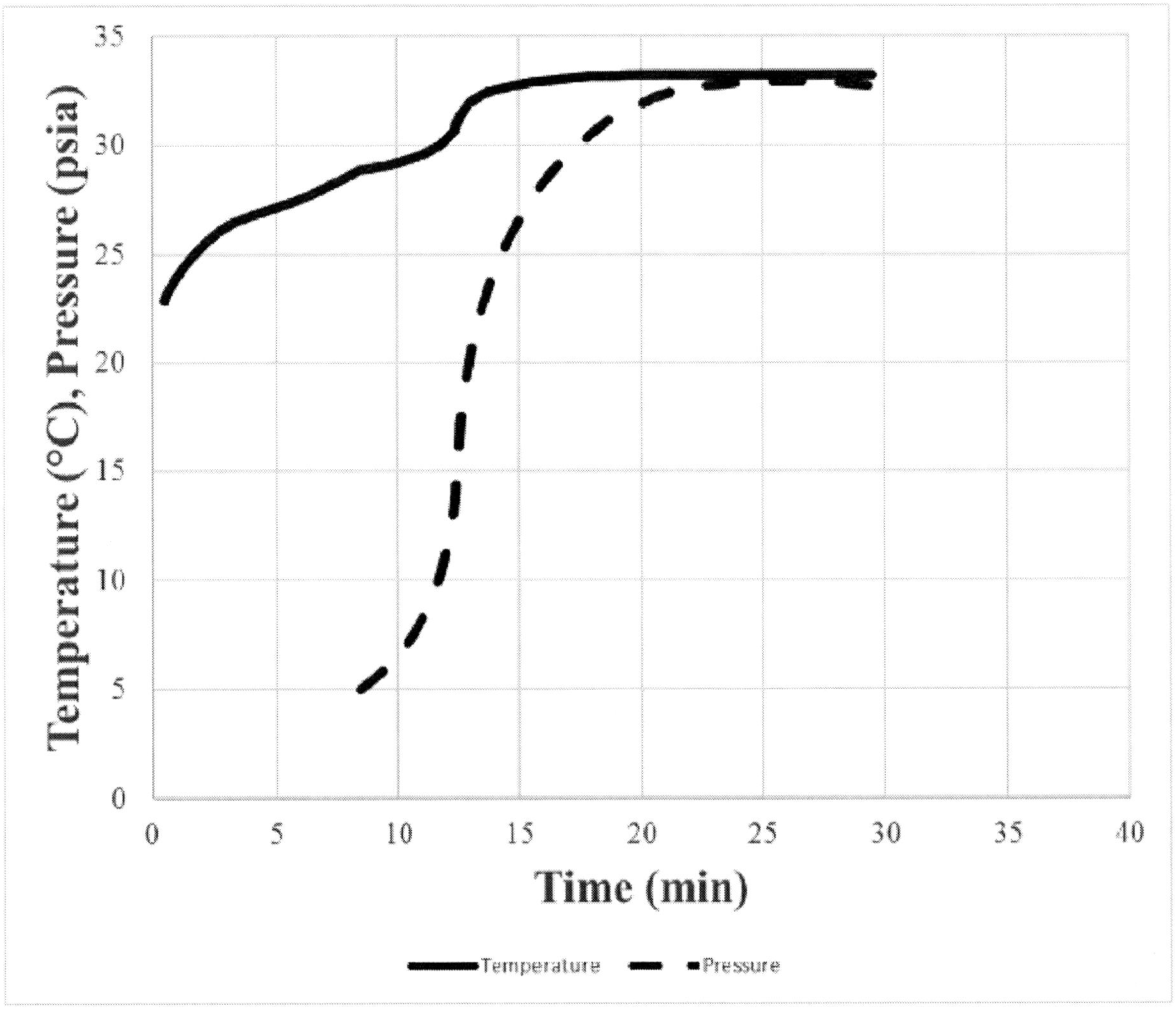

Analysis of products in the gas phase

The experimental work was continued in a different laboratory that utilized a Vent Sizing Package 2 (VSP2) adiabatic calorimeter and had increased analytical capabilities. The results from the VSP2 experiments were consistent with previous work showing an increase in pressure due to non-condensable gas generation. The headspace gas was analyzed by a gas chromatography-mass spectrometer (GC-MS). The major components detected were acetone, t-butanol, residual t-BHP and oxygen. Minor components detected were propylene, methanol and carbon dioxide. Trace amounts of isobutyraldehyde, isobutylene, formaldehyde, carbon dioxide, methyl hydroperoxide, dimethyl peroxide and *t*-butyl isopropyl ether were also found [2].

For an approximate concentration estimate it was assumed that the vapor contained only acetone, which is the most volatile of the major products. Using the measured heat of combustion from the Parr bomb (150 kJ/mol) and knowing the high heat of combustion of acetone (1730 kJ/mol), the equivalent acetone concentration was about 8.7%. This concentration is within the flammable range for acetone, whose lower limit is 2.5% and upper limit is about 13.0%.

In addition to GC-MS, the gas phase of the reaction was analyzed using an MSA ALTAIR 5X Multigas Detector. The gas was diluted with nitrogen to reduce the concentration to the measurable range. The analysis indicated that 94% of the vapor was oxygen. The gas mixture that deflagrated apparently comprised at least 90% oxygen with up to about 9% acetone plus other reaction products.

Analysis of products in the liquid phase

The liquid phase was also analyzed by the GC-MS. As in the gas phase, the major components detected were acetone, t-butanol and residual t-BHP. Quantification was performed by GC with flame ionization detection (GC-FID). In a control experiment with t-BHP alone, it was demonstrated that t-BHP survived the injector port (230 °C), decomposing <1% thermally to acetone and t-butanol.

Apart from the 98.12 percent water, the liquid from the experiment contained 0.09 weight percent acetone, 0.89 weight percent t-butanol and 0.87 weight percent t-BHP, accounting for 96 percent of t-BHP introduced [2].

Determination of Minimum Ignition Energy (MIE)

A test to determine the minimum ignition energy (MIE) of a gas or vapors in air at atmospheric pressure was carried out in accordance with ASTM E582 "Standard Test Method for Minimum Ignition Energy and Quenching Distance in Gaseous Mixtures". The test was conducted by mixing the chemicals involved in the incident and transferring the generated vapors directly to the testing apparatus [3].

A capacitive discharge spark was used as the ignition source. The spark gap was optimized and an optimized amount of capacitance was added to the circuit. The capacitors were then charged with a high voltage generator until the breakdown voltage was reached and the capacitor discharged in the form of a spark between the electrodes. The minimum energy to ignite the vapor was determined to be 0.015 mJ. The high oxygen concentration reduced the MIE to less than one-tenth the MIE value of acetone in air, which is probably about 0.2 mJ although its stoichiometric value of 1.15 mJ often appears in MIE tabulations.

The 0.015 mJ MIE for the product gases is not however the lowest MIE (LMIE) for the mixture involved in the mixer deflagration. This is because the fuel:oxygen ratio was not systematically varied and optimized; the mixture tested might have been far from optimum. Neither was the fuel composition varied for the range of possible fuel components generated by the reaction. This topic is resumed later.

Ignition sources

(a) External Ignition. Potential ignition sources outside the vessel include the agitator motor plus miscellaneous electrical and mechanical equipment. The agitator motor is closest to the vessel. Using the vapor pressures of the major reaction products to estimate their concentrations and then applying Le Chatelier's rule, the lower flammable limit (LFL) for the mixture was estimated to be about 3.1 mol%. Using a commercial gas dispersion program the investigation team estimated that at distances more than 15 cm from tank roof openings the vapor concentration would fall below the LFL. There was an air circulation fan in the area which could have further decreased the flammable vapor concentration.

(b) Unstable Materials. One possible explosion source could have been chlorine dioxide, which becomes unstable at vapor concentrations above 30 percent. The concentration of free chorine in the liquid phase was considered too low to allow chlorine dioxide gas to be released at levels approaching 30 percent, even allowing for generation of foam during the earlier gassy reaction.

(c) Static sparks or friction. The ball valve at the bottom of the tank could have been the source of a static charge or friction, but at the time of the incident about 1 inch of water was present in the bottom of the vessel and so the valve was effectively isolated. The ball valves at the top of the tank were not actuated during the incident.

The laboratory work determined that a flammable atmosphere was present in the vessel, but no ignition source could be identified. The vessel is bonded and grounded. The area is electrically classified (although not for flammable gases in enriched oxygen atmospheres) and there are no electrical devices located close enough to the vessel to be a source of ignition. There is an agitator in the vessel, but it was not operating when the ignition occurred. The approach taken relied on Sherlock Holmes' observation to Dr. Watson that ".........when you have eliminated the impossible, whatever remains, *however improbable*, must be the truth...." The only identified source of energy inside the vessel at the time of the deflagration was the falling rinse water.

Water Slug Sparking Mechanism

After three supertankers blew up in December 1969 during water washing of center tanks a large amount of work was carried out to investigate the electrostatic properties of water mists. The most likely ignition mechanism appeared to be the formation of a water slug from the washing jet nozzle located near the center of a tank. Owing to the surrounding charged water mist the grounded nozzle represented a region of high potential gradient so the water slug would carry away a net charge induced by the local electric field. When the slug impacted the grounded tank wall, representing a region of low potential gradient, a spark might occur, hence igniting the residual crude oil vapors in the tank. During cold washing the steady-state mist concentration in large tanks is about 1 g/m^3 and a typical mist charge density is 30 nC/m^3. Center tanks on supertankers have volumes of many thousands of cubic meters so the space potential caused by even modest charge densities can attain several tens of kilovolts [5].

That such a mechanism could occur in a 200 gallon tank initially seemed quite implausible since the upset condition (high bleach concentration) leading to the explosion was apparently a one-off event; the ignition source was therefore expected to be continuous or frequent during normal operation. This is quite apart from the high mist charge densities that might be required for a sufficiently energetic spark in such a small volume.

A simple model was developed comprising (1) water slug capacitance estimates (2) space potential estimates using solutions to Poisson's equation given in [5] for both infinite cylinder and infinite coaxial cylinders (3) solution for mist charge densities to satisfy the spark ignition energy requirement. The mist charge densities were then compared with reported values from the literature to determine the viability of the mechanism [6]. The "infinite cylinder" results were later checked against accurate series solutions obtained for a finite cylinder having the same dimensions as the mixing tank and the error in mist charge density was found to be small compared with the uncertainty in slug capacitance. Details of the model are beyond the scope of this paper so only a brief summary is presented.

The range of slug capacitances "C" was estimated to be $4 \leq C \geq 10$ pF, taking into account the likely dimensions of water slugs plus their proximity to the grounded tank wall. The worst case ignition energy of the gas mixture in the mixing vessel was assumed to be 2 microjoules (0.002 mJ), based on the measured LMIE of optimized mixtures of propane and oxygen. Using the spark energy equation $E = 5*10^{-10} * CV^2$ (where E is spark energy in mJ, C is capacitance in pF and V is voltage in volts) the corresponding voltage range required for charged, isolated water slugs is 632-1000 volts.

Neglecting vessel end effects and the presence of the stirrer this range of space potentials (632-1000V) might be expected at the radius of the water inlet pipe for mist charge densities between 330 and 520 nC/m^3. In the presence of an axial stirrer shaft the requisite charge densities would be increased by the factor 2.70. In actuality the stirrer shaft was not axial and was close to the tank wall at the level of the water inlet pipe. Furthermore it was significantly removed from the inlet pipe. Consequently its effect on space potential in the vicinity of the water inlet pipe might have been small. Analytical solution is not possible for asymmetrical geometry but this was not a major problem since the objective was merely to determine if the requisite charge densities could reasonably be attained. Assuming a simple 50% reduction of space potential by the stirrer shaft the requisite charge density range was increased to 850 ± 190 nC/m^3. However, without stirrer shaft effects and with a large water slug the requisite charge density might have been as small as 330 nC/m^3.

Literature review showed that for cold water jet washing of large tanks, the steady-state mist charge density generally increases as tank volume decreases [6]. Furthermore, charge densities as high as 1700 nC/m^3 have been reported for small scale water mists [4] while solvent mists in a 1.2 m diameter tank were measured at 169 nC/m^3 [7]. In the present case the tank diameter was 0.76 m and the droplets might have comprised both aqueous and organic components; hence charge densities of several hundred nC/m^3 might have been possible given a realistic mist generation mechanism.

It was hypothesized that a relatively stable charged mist might have been present in the tank before the water rinse was started. Vigorous gassy reactions (such as between t-BHP and bleach) form a large proportion of droplets similar in size to those found in sea fogs and smokescreens (typically 0.3 to 3 microns), whose coagulation rates are slow at the mass concentrations of interest [8]. The explosion occurred within about a minute of starting the water rinse; at this time the residual air in the water feed line could have caused the inlet nozzle to sputter and generate a succession of water slugs.

The mechanism outlined above requires incendiary sparks at breakdown voltages below 1000V. The minimum quenching distance for propane-oxygen is 0.2 mm which, from the well-known Paschen curve, requires a breakdown voltage of at least 1400V. However standard quenching distance tests utilize flanged electrodes whereas the smallest MIE values are found with unflanged, pointed electrodes [5]. Should the quenching distance be decreased by a factor of 3 or 4 from those measured using glass flanged electrodes, its value would approach 0.05 mm; the corresponding sparking voltage of about 700V would be towards the low end of the range required by the model. It was concluded that although the water slug scenario is unlikely it could not be ruled out.

The general analysis of the water slug mechanism [6] might have relevance for other "closed vessel" explosions involving gassy reactions and elevated oxygen concentrations. A case in point is Allied-Signal's wastewater tank explosion discussed a quarter century ago at the 1993 Loss Prevention Symposium [9]. The investigation included the scenario "corona discharge generated by cascading liquids" but this was considered unlikely. A companion paper considered corona discharges above frothy surfaces [10].

Plant changes

A number of changes were made as a result of the incident:

- The bleach solution feed pump to the deionized water system was changed to a smaller pump that is inherently limited to the required flow rate.
- A level indicator was installed on the bleach solution tank. An interlock was installed to stop the pump if there is a high consumption deviation alarm.
- Operation department now checks the bleach solution tank level when the maintenance group not on site.
- The colorimetric free chlorine meters were replaced with amperometric free chlorine meters.
- The order of addition was changed to add t-BHP last prior to charging to minimize the time that t-BHP is present in the vessel
- The area was reclassified as Class I Division II and all hardware was upgraded. Although this is not a direct result of the investigation, the change was made as a response to the potential flammable atmosphere.

- All feed nozzles at the top of the vessel were reconfigured to be directed at the vessel walls to prevent splash filling.
- The vapor space was connected directly to the ventilation fan to remove all vapors. An interlock was installed to require fan operation while t-BHP is present in the vessel.

Conclusions

An upset in the bleach metering system for a deionized water supply caused an unplanned reaction in a t-BHP mixing tank. This reaction created a condition where an emptied vessel contained flammable vapors highly enriched in oxygen. The atmosphere in the tank had a very low ignition energy and found an ignition source inside the vessel. The only ignition source identified was a spark from an electrically charged, falling slug of water.

The Root Cause of the incident was failure to appreciate the hazard of adding too much bleach to the deionized water supply tank. The consequence was lack of adequate safeguards in the bleach metering system. The primary causal factor was simultaneous failure of both chlorine analyzers at the end of a four-day Maintenance work week, allowing the failures to go undetected for at least three days.

Acknowledgements

The authors would like to thank Andy Appleman, Christina Aultman, Dr. Keith Chenault, Dr. Enio Kumpinsky, Tom Olechnowitz, Matthew Pfeifer and Kevin Wadsworth for assistance with this work.

References

[1] Verhoeff, J., “Explosion Hazards of Tertiary Butyl Hydroperoxide (TBHP)”, IChemE Symposium Series No. 68, 1981.

[2] Chenault, H. Keith (2017), “Products Generated by Reaction of *tert*-Butyl Hydroperoxide with Sodium Hypochlorite,” Ashland Internal Memorandum.

[3] Johnson, Tom and Osterberg,, Paul. (2017), “Minimum Ignition Energy Testing of Tert-Butyl Hydroperoxide Reaction Gases” Fauske and Associates, Report no. FAI-2017-0864

[4] Vos, B., “Electrostatic Charge Generation during the Washing of Tanks IV: Mechanism Studies”, Paper 15, Proc 3rd. Conf. on Static Electrification, Institute of Physics, London 1971).

[5] Britton, L.G., “Avoiding Static Ignition Hazards in Chemical Operations”, AIChE-CCPS Concept Book (1999).

[6] Britton, L.G., “Ashland Tank Explosion Scenario: Ignition via Electrically Charged Water Slug”, Jacobs Technical Report to Ashland, Project Number SGAS0190, July 13th (2017).

[7] Williams, T.J., and Jones, "Modelling the Electrostatic Ignition Hazards associated with the Cleaning of Tanks containing Flammable Atmospheres", J. Loss Prev. Process Ind. 14 (2001) 129-133.
[8] Britton, L.G., and Harrison, B.K., "Minimum Explosible Concentrations of Mist and Dust Clouds", Process Safety Progress, Vol. 37, No. 1, 2018, pp. 4-17.
[9] Wehrum, W., "Case Study of a Hydrogen Peroxide Related Deflagration in a Wastewater Treatment Tank", Process Safety Progress, Vol. 12, No. 4, 1993, pp. 199-202.
[10] Pratt, T.H., "Electrostatic Ignitions in Enriched Oxygen Atmospheres: A Case Study", ibid., pp. 203-205.

AUTHOR INDEX

AUTHOR INDEX

AUTHOR INDEX

AIChE
120 Wall Street, FL 23
New York, NY 10005-4020

ISBN 978-1-5108-6668-3

9 781510 866683

Astronomy at High Angular Resolution 2011

The Central Kiloparsec in Galactic Nuclei

Journal of Physics: Conference Series Volume 372

Bad Honnef, Germany
28 August - 2 September 2011

ISBN: 978-1-62276-806-6
ISSN: 1742-6588

IOP Institute of Physics